Alphaherpesviruses

Molecular Virology

Edited by

Sandra K. Weller, PhD

Board of Trustees Distinguished Professor and
Chair of Molecular, Microbial and Structural Biology
University of Connecticut Health Center
Farmington, CT
USA

Caister Academic Press

Caister Academic Press
Norfolk, UK

www.caister.com

British Library Cataloguing-in-Publication Data
A catalogue record for this book is available from the British Library

ISBN: 978-1-904455-76-9

Cover image adapted from Figure 10.7

Printed and bound in Great Britain

Contents

Contributors

Allison Abendroth
Department of Infectious Diseases and Immunology
University of Sydney
Sydney, NSW
Australia

allison.abendroth@sydney.edu.au

Ann M. Arvin
Departments of Pediatrics and Microbiology and Immunology
Stanford University School of Medicine
Stanford, CA
USA

AArvin@Stanford.edu

Joel D. Baines
Cornell University College of Veterinary Medicine
Veterinary Medical Center
Cornell University
Ithaca, NY
USA

jdb11@cornell.edu

John A. Blaho
Department of Microbiology
Mount Sinai School of Medicine
New York, NY
USA

john.blaho@mssm.edu

David C. Bloom
Department of Molecular Genetics and Microbiology
University of Florida College of Medicine
Gainesville, FL
USA

dbloom@mgm.ufl.edu

Gary H. Cohen
Department of Microbiology
School of Dental Medicine
University of Pennsylvania
Philadelphia, PA
USA

ghc@dental.upenn.edu

Jeffrey I. Cohen
Laboratory of Infectious Diseases
National Institutes of Health
Bethesda, MD
USA

jcohen@niaid.nih.gov

James F. Conway
Department of Structural Biology
University of Pittsburgh School of Medicine
Pittsburgh, PA
USA

jxc100@pitt.edu

Christopher R. Cotter
Department of Microbiology
Mount Sinai School of Medicine
New York, NY
USA

christopher.cotter@mssm.edu

Bryan R. Cullen
Department of Molecular Genetics and Microbiology and Center for Virology
Duke University Medical Center
Durham, NC
USA

bryan.cullen@duke.edu

Neal A. DeLuca
Department of Molecular Genetics and Biochemistry
University of Pittsburgh School of Medicine
Pittsburgh, PA
USA

ndeluca@pitt.edu

Timothy E. Dudek
Ragon Institute of Massachusetts General Hospital
MIT and Harvard
Charlestown, MA
USA

tdudek@partners.org

Roselyn J. Eisenberg
Department of Pathobiology
School of Veterinary Medicine
University of Pennsylvania
Philadelphia, PA
USA

roselyn@dental.upenn.edu

Roger D. Everett
MRC-University of Glasgow Centre for Virus Research
Glasgow
UK

roger.everett@glasgow.ac.uk

Philipe A.M. Gobeil
Department of Microbiology and Immunology
Dartmouth Medical School
Lebanon, NH
USA

philipe.a.m.gobeil@dartmouth.edu

Charles Grose
University of Iowa
Children's Hospital
Iowa City, IA
USA

charles-grose@uiowa.edu

Ekaterina E. Heldwein
Department of Molecular Biology and Microbiology
Tufts University School of Medicine
Boston, MA
USA

katya.heldwein@tufts.edu

Fred L. Homa
Department of Microbiology and Molecular Genetics
University of Pittsburgh School of Medicine
Pittsburgh, PA
USA

flhoma@pitt.edu

Keith R. Jerome
Vaccine and Infectious Disease Division
Fred Hutchinson Cancer Research Center
Department of Laboratory Medicine
University of Washington
Seattle, WA
USA

kjerome@fhcrc.org

Odd Andre Karlsen
University of Bergen
Bergen
Norway

odd.karlsen@mbi.uib.no

Paul R Kinchington
Departments of Ophthalmology and Molecular Microbiology and Genetics
University of Pittsburgh
Pittsburgh, PA

kinch@pitt.edu

David M. Knipe
Department of Microbiology and Molecular Genetics
Harvard Medical School
Boston, MA
USA

david_knipe@hms.harvard.edu

Claude Krummenacher
Department of Biochemistry
School of Dental Medicine
University of Pennsylvania
Philadelphia, PA
USA

ckrummen@dental.upenn.edu

Dacia L. Kwiatkowski
Department of Molecular Genetics and Microbiology
University of Florida College of Medicine
Gainesville, FL
USA

dak2015@med.cornell.edu

Christos Kyratsous
College of Physicians and Surgeons
Columbia University
New York, NY
USA

christos.kyratsous@gmail.com

David A. Leib
Department of Microbiology and Immunology
Dartmouth Medical School
Lebanon, NH
USA

david.a.leib@dartmouth.edu

Christine M. Livingston
University of Connecticut Health Center
Farmington, CT
USA

cmliving@stanford.edu

Eduardo A. Montalvo
Visiting Scientist
University of Iowa
Iowa City, IA
USA

montalve@csr.nih.gov

Ian Mohr
Department of Microbiology and NYU Cancer Institute
New York University School of Medicine
New York, NY
USA

ian.mohr@med.nyu.edu

Karen L. Mossman
Department of Pathology and Molecular Medicine
Michael DeGroote Centre for Learning and Discovery
McMaster University
Hamilton, ON
Canada

mossk@mcmaster.ca

Samuel Rabkin
Department of Neurosurgery
Massachusetts General Hospital and Harvard Medical School
Boston, MA
USA

rabkin@helix.mgh.harvard.edu

Stephen A. Rice
Department of Microbiology
University of Minnesota Medical School
Minneapolis, MN
USA

ricex019@umn.edu

Kari L. Roberts
Cornell University College of Veterinary Medicine
Veterinary Medical Center
Cornell University
Ithaca, NY
USA

klr54@cornell.edu

Rozanne M. Sandri-Goldin
Department of Microbiology and Molecular Genetics
School of Medicine
University of California, Irvine
Irvine, CA
USA

rmsandri@uci.edu

Saul Silverstein
College of Physicians and Surgeons
Columbia University
New York, NY
USA

sjs6@columbia.edu

Paul T. Sobol
Department of Pathology and Molecular Medicine
Michael DeGroote Centre for Learning and Discovery
McMaster University
Hamilton, ON
Canada

sobolpt@mcmaster.ca

Susan Vleck
Stanford University
Stanford, CA
USA

sevleck@stanford.edu

Jennifer L. Umbach
Department of Molecular Genetics & Microbiology and Center for Virology
Duke University Medical Center
Durham, NC
USA

JL56@duke.edu

Stacey A. Ward
Department of Molecular, Microbial and Structural Biology
University of Connecticut Health Center
Farmington, CT
USA

ward_s@kids.wustl.edu

Matthew D. Weitzman
The Salk Institute for Biological Studies
La Jolla, CA
USA

weitzman@salk.edu

Sandra K. Weller
Department of Molecular, Microbial and Structural Biology
University of Connecticut Health Center
Farmington, CT
USA

weller@nso2.uchc.edu

Leigh Zerboni
Department of Pediatrics
Stanford University School of Medicine
Stanford, CA
USA

zerboni@stanford.edu

Dedication

Dr Priscilla Ann Schaffer – 29 December 1941 to 18 November 2009

Dr Priscilla Schaffer was internationally recognized for her prolific scientific contributions focusing primarily on herpes simplex virus, its life cycle, its ability to establish lifelong latent infections and its relation to neurological diseases. From her early work as a postdoctoral fellow and assistant professor of Virology at the Baylor College of Medicine, her pioneering work with temperature sensitive mutants of HSV paved the way for a more thorough understanding of the genome organization of herpes simplex virus. She held many prestigious titles since leaving Baylor in 1976, including Professor of Microbiology and Molecular Genetics at Harvard Medical School, Chief of the Laboratory of Tumor Virus Genetics at the Dana-Farber Cancer, Chair of the Department of Microbiology at University of Pennsylvania School of Medicine, Chief of the Laboratory of Molecular Virology at the Beth Israel Deaconess Medical Center and finally Research Professor of Molecular and Cellular Biology at the Cancer Center at the University of Arizona in Tucson.

Her scientific work focused on the molecular and biological aspects of herpes simplex virus (HSV) gene expression, DNA replication and latency. Using principles of genetics, specifically complementation and recombination, her work led the way for the identification of viral gene products with roles at all stages of the viral life cycle. Her early work with the HSV thymidine kinase and DNA polymerase contributed prominently to our understanding of drug resistance and the development of antiviral drug therapies still used to combat herpes virus infections. Her seminal work in viral gene expression, DNA replication, glycoprotein synthesis, establishment of latency and pathogenesis moved the study of HSV biology forward in so many ways, including opening the way for countless others to also make important contributions to this field. She was a brilliant and dedicated investigator with over 150 peer-reviewed papers, countless reviews and grants and prominent positions throughout her life. However her devotion and support of young scientists will also be remembered as part of her scientific legacy. Those of you who knew Priscilla either personally or by reputation realize the great loss to the scientific community of a beloved and devoted mentor and supporter of young scientists.

Varicella Zoster Virus Transcriptional Regulation and the Roles of VZV IE Proteins

1

Jeffrey I. Cohen

Abstract

Varicella-zoster virus (VZV) encodes three immediate-early proteins, IE4, IE62, and IE63; however, only IE62 has TAATGARAT-like sequences in its promoter which are present in the promoters of each of the herpes simplex virus immediate-early proteins. The TAATGARAT-like elements on the IE62 promoter bind to VZV ORF10 protein, Oct1, and HCF-1. In addition, histone methyltransferases are recruited to the IE62 promoter to modify chromatin to a transcriptionally active form. VZV IE62, the major VZV transactivator binds to VZV IE4 and IE63, and Med25, part of the mediator complex which upregulates gene expression. VZV IE62, IE4, and IE63 are present in the viral tegument where they may help to regulate transcription early in infection. IE63 binds to several cellular proteins including ASF1 and RNA polymerase II.

Two hypotheses have been proposed for regulation of VZV gene expression during latency. First, relocalization of HCF-1 from the cytoplasm to the nucleus of sensory ganglia in response to stimuli associated with reactivation may help to augment transcription of IE62 to reactivate VZV from latency. Second, promoters of latent genes are maintained in a euchromatic state allowing their transcription, while promoters of genes not associated with latency are in a heterochromatic state resulting in repression of transcription.

Introduction

Primary infection with varicella-zoster virus (VZV) causes varicella or chickenpox. The virus spreads throughout the body in lymphocytes and establishes a latent infection in cranial nerve and dorsal root ganglia where it can reactivate to cause zoster or shingles. Live attenuated VZV vaccines are licensed to prevent varicella and to reduce the incidence and severity of shingles. Antiviral medication, including acyclovir, valacyclovir, and famciclovir which are phosphorylated by the viral thymidine kinase and inhibit the viral DNA polymerase are effective in treating varicella and zoster.

VZV and herpes simple virus (HSV) are the two human members of the alphaherpesvirus subfamily. While most VZV genes are highly conserved with those of HSV, the genes in these two viruses often have different promoter regions and the pattern of gene expression during latency is very different for VZV and HSV. Some of these differences are likely responsible for the very different diseases they cause. Varicella causes a disseminated rash, while primary HSV infection results in more localized disease. VZV reactivates once if at all in most persons, whereas HSV reactivates very frequently. These biological differences may be reflected in differences in gene regulation and expression. This chapter will review the current status of VZV transcriptional regulation and the roles of the viral immediate-early (IE) proteins.

VZV genome and virion

The 125 kbp VZV genome consists of a unique long (UL) region with adjacent terminal long (TRL) and internal long (IRL) repeats, and a unique short (US) region with adjacent internal short (IRS) and terminal short (TRS) repeats

(Davison and Scott, 1986). The UL region is usually fixed, while the US region can invert in either of two orientations; therefore, there are two isomers of the viral genome.

The genome encodes at least 70 unique genes, and all but six (ORFs 1, 2, 13, 32, 57, and S/L) have homologues in HSV. VZV ORFs 1 to 61 and ORFs 65 to 68 are located in UL and US, respectively, while VZV ORFs 62 to 64 and their identical copies – ORFs 69 to 71 – are located in IRS and TRS, respectively. VZV ORF S/L (also known as ORF0) is located at both ends of the genome and is formed when the termini of the genome ligate to form an episome (Kemble *et al.*, 2000). ORFs 42 and 45 are spliced segments of a single gene, while ORFS/L is spliced in its 3′ non-coding region. In addition to the IE genes that encode IE4, IE62, and IE63, VZV contains putative early genes that encode proteins important for viral DNA replication including the viral thymidine kinase (ORF36) which phosphorylates acyclovir, the viral DNA polymerase (ORF28) which is inhibited by acyclovir, the single-stranded DNA binding protein (ORF29), the origin binding protein (ORF51), and the viral protein kinases (ORF47 and ORF66). The putative late genes encode structural VZV proteins.

The VZV virion consists of a DNA core surrounded by the viral nucleocapsid which includes the major nucleocapsid protein (ORF40) and ORF21. This in turn is surrounded by the virion tegument (which includes the products of ORF4, ORF9, ORF10, ORF47, ORF62, and ORF63). The outer layer of the virion consists of a lipid envelope with nine glycoproteins (gB, gC, gE, gH, gI, gK, gL, gM, gN).

VZV replication

VZV enters cells either by fusion of the viral envelope to the plasma membrane or by endocytosis and entry into the cytoplasm (reviewed in Cohen *et al.*, 2007b). The VZV nucleocapsid and tegument proteins (including the transactivating proteins IE4, IE62 and ORF10 protein) are transported to the nucleus where transcription begins. The linear viral genome in virions has an unpaired nucleotide at each 3′ end, and the ends base-pair to form a circular genome in infected cells. VZV IE gene transcription occurs in the nucleus, IE proteins are translated in the cytoplasm, and the proteins are transported back to the nucleus to activate transcription of viral early genes. Early genes are transcribed in the nucleus, translated in the cytoplasm, and the early viral proteins traffic to the nucleus to initiate viral DNA replication. VZV encodes homologues of each of the 7 HSV genes needed for viral origin-dependent DNA replication (VZV ORF6, 16, 28, 29, 51, 52, and 55); however, these have not been formally shown to be required for VZV replication. Late genes are then transcribed in the nucleus, translated in the cytoplasm, and capsid proteins are transported to the nucleus. Viral DNA is replicated in the nucleus, encapsidated and transported out of the nucleus. The nucleocapsids acquire an envelope in the cytoplasm and virions are released from the plasma membrane.

VZV transcripts

The inability to produce high titres of cell-free virus has limited the ability to classify VZV genes as immediate-early (IE), early, or late. Three VZV genes, ORF4 (Defechereux *et al.*, 1993), ORF62 (Forghani *et al.*, 1990), and ORF63 (Debrus *et al.*, 1995) have been classified as IE genes. IE genes are expressed at the time of infection in the absence of new virus protein synthesis. Genes are experimentally defined as IE based on expression of their mRNA in the presence of the protein synthesis inhibitor cycloheximide followed by actinomycin D. Early gene transcription requires IE protein expression. Optimal levels of late gene transcription require both IE protein expression and viral DNA replication. Due to the lack of high titred cell-free virus, early and late transcripts of VZV are putatively assigned to kinetic classes based on their homologues in HSV. Recent studies indicate that transcripts for gC are markedly delayed relative to other glycoproteins (Storlie *et al.*, 2008). VZV genes are transcribed by host cell RNA polymerase II.

The transcripts for several VZV genes have been mapped, including ORF4 (Kinchington *et al.*, 1994), ORF14 (gC, Ling *et al.*, 1991), ORF21 (Cohrs *et al.*, 1998; Xia and Straus, 1999), ORF28 (Meier and Straus, 1993), ORF29 (Meier and Straus, 1993), ORF36 (Davison and Scott, 1986), ORF61 (Stevenson *et al.*, 1992), ORF62

(McKee *et al.*, 1990), ORF63 (Kinchington *et al.*, 1994; Kost *et al.*, 1995), ORF67 (Kato *et al.*, 1989; Ling *et al.*, 1992), and ORF68 (Kato *et al.*, 1989; Rahaus and Wolff, 2000b). Many VZV genes are arranged in groups of up to four genes with a single common polyadenylation site. Many of the polyadenylation sites, as well as TATA elements, differ from consensus sequences (Davison and Scott, 1986). Some VZV proteins are encoded from several co-terminal transcripts. For example VZV glycoprotein E (gE) is encoded by two transcripts (Kato *et al.*, 1989), while VZV gI is encoded by three transcripts (Ling *et al.*, 1992); however, one transcript is thought to be the predominant transcript for protein expression. VZV gE and gI have been shown to be transcribed both monocistronically and bistronically, although most transcription is monocistronic (Rauhaus and Wolff, 2000). Different strains of virus can express different levels of transcripts. The VZV Oka vaccine strain expresses 20-fold lower levels of ORF14 (gC) transcripts than does VZV strain Scott (Ling *et al.*, 1991).

VZV promoters

While each of the HSV IE genes has an upstream TAATGARAT element, this element is lacking from the VZV IE4, ORF61, and IE63 promoters (Kinchington *et al.*, 1994); however IE62 has TAATGARAT-like elements (Moriuchi *et al.*, 1995a). Most VZV promoters have a low basal level of activity, but the level is markedly increased in the presence of VZV IE62.

VZV IE promoters

The VZV IE62 promoter sequence was initially mapped to nucleotides −410 to +57 relative to the mRNA start site (McKee *et al.*, 1990). The promoter contains three TAATGARAT-like motifs; one motif (ATGTAAATGAAAT), has an overlapping octamer binding sequence while the other two (TAAACTAAT, AATAGTAAT) lack overlapping octamer binding sequences (Fig. 1.1). These TAATGARAT-like motifs are important for transactivation of IE62 by the VZV ORF10 protein (Moriuchi *et al.*, 1995a). Additional motifs including cyclic AMP-responsive elements (CRE) (TGACGACG, TGACGACA) and a GA-rich motif (CGGAAA) are also required for full transactivation of the IE62 promoter by ORF10 protein. ORF10 protein forms a complex with the cellular octamer binding protein 1 (Oct1) and the transcriptional coactivator host cell factor 1 (HCF-1) and TAATGARAT-like elements on the IE62 promoter (Fig. 1.2). This complex of proteins is referred to as the enhanceosome (Kristie *et al.*, 2010). By analogy with HSV, VZV IE62, Sp1, and GABP are thought to bind to a domain near the amino terminus of HCF-1, and along with ORF10 protein and Oct-1 form the enhancer core complex. ORF10 protein binds to the two TAATGARAT-like elements that lack overlapping octamer-binding motifs, but not the TAATGARAT-like element containing an overlapping octamer-binding motif. Oct2 also binds to a TAATGARAT-like element lacking an octamer-binding motif on the IE62 promoter and inhibits Oct 1 binding to the promoter and ORF10-induced transactivation of the IE62 promoter. HCF-1 is recruited to the IE62 promoter and is essential for transactivation of the IE62 promoter by ORF10 protein or by IE62 (Narayanan *et al.*, 2005). HCF-1 has also been shown to be critical for VZV replication; in HCF-depleted cells VZV plaque size is reduced (Narayanan *et al.*, 2005).

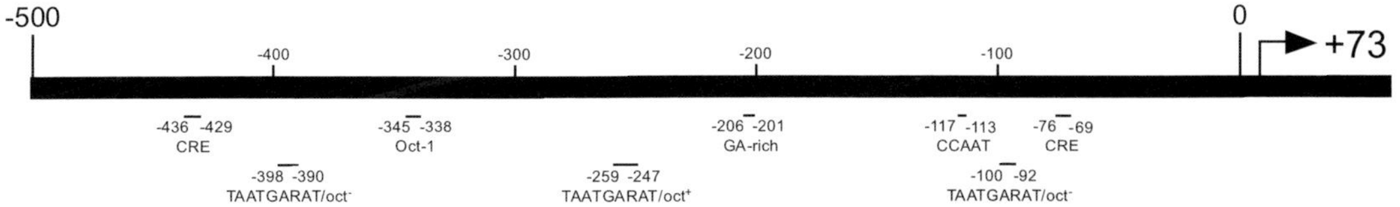

Figure 1.1 Structure of the VZV IE62 promoter. Numbers shown are relative to the start site of transcription (arrow). CRE (cyclic AMP response elements), Oct-1, GA-rich (or CGGAAR), CCAAT, and TAATGARAT sites are shown.

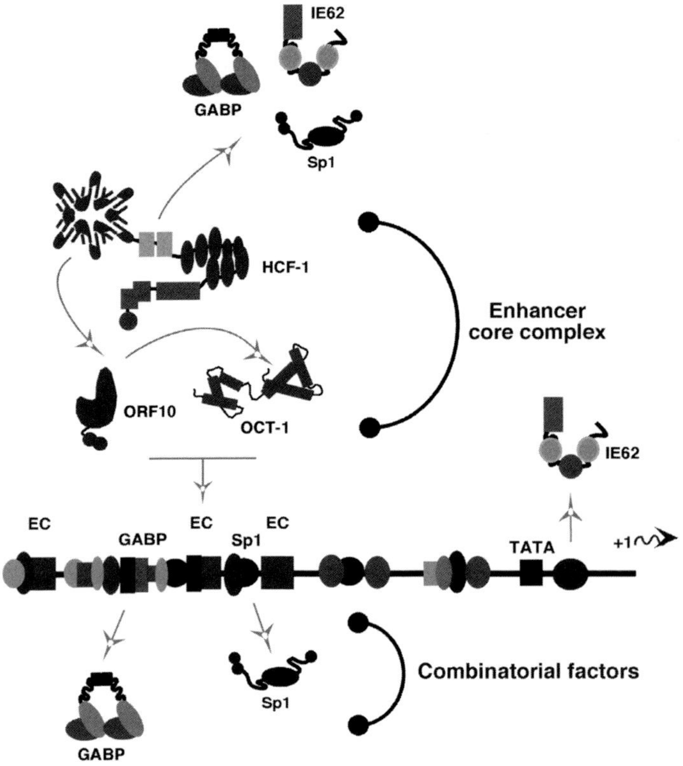

Figure 1.2 Model for activation of the VZV IE62 promoter by ORF10 protein, HCF-1, Oct-1, Sp1, and GABP. EC, enhancer core containing TAATGARAT sequence, GABP GA-binding protein. (adapted from Kristie TM, Liang L, and Vogel JL. Control of alphaherpesvirus IE gene expression by HCF-1 coupled chromatin modification activities. Biochim. Biophys. Acta *1799*, 257–265, 2010).

Recruitment of basal transcription factors (Sp1, TATA box-binding protein [TBP], transcription factor [TF] IIB, TFIIH) and RNA polymerase II to the IE62 promoter is enhanced in the presence of IE62 (Narayanan *et al.*, 2007). Transcription of the IE62 promoter is regulated by modification of chromatin. In the absence of IE62, the IE62 promoter is repressed by dimethyl-H3K9. In contrast, in the presence of IE62, the promoter is activated with reduced dimethyl-H3K9 and increased trimethyl-H3K4 and recruitment of the H3K4 histone methyltransferases Set1 and MLL1 to the promoter. HCF-1 is required for efficient recruitment of Set1 and MLL1 to the IE62 promoter and for H3K4 trimethylation. These findings indicate that IE62 recruits HCF-1 and histone methyltransferases Set1 and MLL1 to the IE62 promoter to modify chromatin to a transcriptionally active form.

The VZV IE62 promoter also contains a CCAAT motif that binds to the NF-Y transcription factor. Mutation of the CCAAT motif enhances IE62 promoter activity; therefore the CCAAT motif is a negative regulatory element on the promoter (Moriuchi *et al.*, 1995b). The VZV IE62 promoter also contains an SP1 site; SP1 binds to IE62 and deletion of the SP1 site prevents IE62 transcription (Narayanan *et al.*, 2007). The IE62 promoter contains a GA-binding protein site and a motif that binds to a member of the ATF/AP-1 transcription factor family (McKee and Preston, 1991).

Brn-3a and Brn-3b activate the VZV IE62 promoter (Brownlees *et al.*, 1999). Brn-3a and Brn-3b are members of the POU family of transcription factors that are expressed in sensory ganglia and bind to TAATGARAT elements. The TAATGARAT-like elements in the IE62 promoter are required for the response of the IE62 promoter to Brn-3b. The Oct 2.1, 2.4, and 2.5 isoforms of Oct 2, which are expressed in neuronal cells, inhibit the transactivation of the IE62 promoter (Patel *et al.*, 1998). These isoforms inhibit basal activity of IE62 in neuronal cells, but not in other cells.

VZV IE62 and IE63 are expressed from a bidirectional promoter. This promoter contains putative binding sites for NF-KB, heat shock factor 1 (HSF-1), nuclear factor 1 (NF-1), CRE, and Oct-1. Using a recombinant VZV containing a cassette with the IE62/63 bidirectional promoter driving reporter genes, Jones *et al.* (2006) showed that while mutation of the NF-KB binding site had no effect on IE62 or IE63 transcription, mutation of the NF-1 and HSF-1 binding sites increased IE62 transcription in skin and decreased IE62 transcription in fibroblasts. Mutation of the Oct-1 binding site enhanced IE62 transcription in skin, while mutation of one of the CRE binding sites inhibited IE62 transcription in fibroblasts. Mutation of one of the CRE binding sites and the NF-1 site enhanced IE63 transcription in fibroblasts, while mutation of the Oct-1 binding site increased IE63 transcription in skin.

The USF transcription factor binds to the VZV IE4 promoter (Michael *et al.*, 1998). USF is a member of the basic-helix–loop–helix-leucine zipper family of proteins and binds a consensus sequence- CACGTG. The IE4 promoter contains a CATGTG sequence which differs from the USF binding site at only one nucleotide. The IE4 promoter is activated by the USF transcription factor and IE62.

VZV putative early promoters (VZVORF28/29)

ORF28 (which encodes the viral DNA polymerase) and ORF29 (which encodes the major DNA binding protein), like IE62 and IE63, are expressed from a bidirectional promoter (Meier *et al.*, 1993). This promoter contains a palindrome, AATCACGTGATT, which has a USF consensus sequence (CACGTG) at its centre. The ORF28/29 promoter binds to USF and is activated by the protein in concert with VZV IE62 (Meier *et al.*, 1994; Rahaus *et al.*, 2003). USF was shown to directly interact with IE62. HCF-1 is important for IE62 mediated activation of both the ORF28 and ORF29 promoters; depletion of HCF-1 reduces transactivation of these promoters by IE62 (Narayanan *et al.*, 2007). One TATA box regulates expression of the ORF28 gene, while two TATA boxes regulate ORF29; both genes share a binding site for the Sp1 transcription factor which regulates their activity (Yang *et al.*, 2004). While both genes share a bidirectional promoter, ORF29 is expressed during both latency and lytic infection while ORF28 is expressed only during lytic infection. The size of the ORF29 transcript differs during lytic and latent infection (Meier *et al.*, 1993).

VZV putative late promoters (glycoproteins)

Sp1 binds to two Sp1 binding sites (GC elements) on the VZV gE promoter, and TBP binds to the TATA box on the VZV gE promoter (Rahaus and Wolff, 1999; 2000a). Mutation of these binding sites reduced the basal activity of the gE promoter. These two Sp1 binding sites together are required for transactivation of the promoter by IE62 (Berarducci *et al.*, 2007). When recombinant viruses were constructed with mutations in the Sp1 sites, mutating one of the sites impaired viral replication, while mutation of the other site abolished replication.

The VZV gI promoter contains a 19 base pair sequence termed an activating upstream sequence (AUS) and a 40 base pair sequence termed a 29 responsive element (29RE) (He *et al.*, 2001). Transcription factors Sp1 and USF bind to AUS and the ORF29 protein interacts with 29RE to enhance transactivation of the gI promoter by IE62. Mutation of the Sp1 site in the gI promoter prevents IE62 transactivation of the promoter (Peng *et al.*, 2003). Recombinant viruses with mutations in the Sp1 or USF elements of the gI promoter replicate to wild-type levels *in vitro*, although the viruses are impaired for replication *in vivo* (Ito *et al.*, 2003). Viruses with mutations in both the SP1 and USF elements replicate to wild-type levels, but have a small plaque phenotype *in vitro*. HCF-1 is important for IE62 mediated activation of the gI promoter; depletion of HCF-1 reduces transactivation of this promoter by IE62 (Narayanan *et al.*, 2007).

Other promoters

ORF21 is transactivated by IE62 and a 40 bp region, 96 to 56 nucleotides upstream of the 5′ start site of the gene 21 transcript is required for transactivating activity (Cohrs *et al.*, 1998).

The VZV ORF61 promoter contains three Sp1 binding sites, each of which interact with Sp1 (Wang *et al.*, 2009). Disruption of each of the three individual Sp1 sites in recombinant VZVs, reduced ORF61 protein expression by 26%, 22%, or 80%. Alteration of two of the Sp1 sites reduced ORF61 protein expression by 95%.

The ORF10 promoter contains a USF binding site that interacts with USF; mutation of the USF site markedly reduces transactivation of the promoter by IE62 (Che *et al.*, 2007). VZV with a mutation in the USF binding site of the ORF10 promoter failed to express ORF10 protein. While the ORF10 promoter contains two putative Oct-1 binding sites, deletion of these sites did not effect IE62-mediated transactivation of the promoter.

USF binds to the ORF8/9 and ORF22 promoters (Rahaus *et al.*, 2003).

A search of the VZV genome for Sp1 and USF binding sites identified 9 VZV gene promoters with USF1 sites, 10 with Sp1 sites, and 8 with both USF1 and Sp1 sites (Ruyechan *et al.*, 2003). Other transcription factors also have a role in VZV transcriptional regulation. Knockdown of ATF-2 or jun levels in cells reduced the activity of the ORF4, ORF21, ORF29, and ORF68 promoters (Rahaus and Wolff, 2003).

Role of transcription factors in VZV DNA replication

In addition to their role in regulating transcription of viral genes, the Sp transcription factors have been shown to suppress VZV DNA replication (Khalil *et al.*, 2008). Sp1 and Sp3 bind to a GC box in the downstream region of the VZV origin of DNA replication (oriS); mutation of this GC box prevented binding of the Sp proteins and enhanced origin-dependent viral DNA replication. The YY1 transcription factor binds to a different site in the downstream region of VZV oriS; mutation of this site had no effect on origin-dependent viral DNA replication.

VZV immediate early (IE) proteins

VZV IE62

VZV IE62 encodes a phosphoprotein located in the virion tegument (Kinchington *et al.*, 1992). IE62 is required for virus infectivity (Sato *et al.*, 2003a). The ORF 66 protein kinase is required for localization of IE62 in the virion (Kinchington *et al.*, 2001). VZV IE62 localizes predominantly to the nuclei of infected cells and transactivates VZV IE (including IE62), early, and late promoters and enhances the infectivity of virion DNA in transient expression assays (Inchauspe *et al.*, 1989; Moriuchi *et al.*, 1994c; Perera *et al.*, 1992a). IE62 can activate or repress expression of its own promoter in transient expression assays depending on the cell type (Disney *et al.*, 1990; Perera *et al.*, 1992b).

IE62 interacts with several viral and cellular proteins (Fig. 1.3). IE62 binds to VZV ORF9 protein and the two proteins colocalize in the cytoplasm when transfected with the ORF66 protein kinase (Cilloniz *et al.*, 2007) IE62 binds to VZV IE4 and this interaction, which occurs only with a dephosphorylated form of IE62 (Spengler *et al.*, 2000), results in nuclear localization of IE4 (Defechereux *et al.*, 1996). IE62 also binds to IE63 (Lynch *et al.*, 2002).

IE62 binds to TBP and TFIIB and results in different levels of transcriptional activation depending on the sequence of the TATA element within a promoter (Perera, 2000), which may occur either directly or indirectly by recognizing conformational variations in DNA-bound TBP, TBP-transcription factor IIA/B, or TBP-TATA-associated factor complexes. Amino acids 238 to 258 of IE62 are required for binding to the USF1 transcription factor (Yang *et al.*, 2006). IE62 binds to Sp1 which likely stabilizes the interaction of HCF-1 with the IE62 promoter. Sp1 is recruited to the VZV glycoprotein I promoter to activate its transcription (Peng *et al.*, 2003).

The transcriptional activation domain of IE62 is located at its amino terminus (Cohen *et al.*, 1993; Perera *et al.*, 1993) and has been mapped

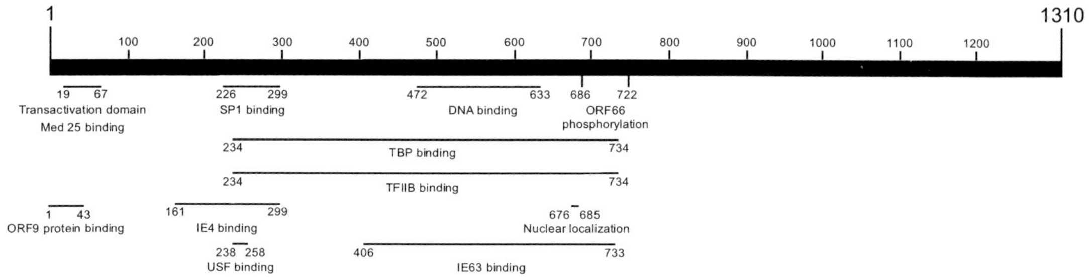

Figure 1.3 Structure of VZV IE62. Numbers refer to amino acid positions in the protein. Binding sites for Med25, SP1, DNA, TBP, TFII B, ORF9 protein, IE4, USF, and IE63 are shown, as well as the transactivation domain, site of phosphorylation by ORF66 protein kinase, and nuclear localization signal.

to amino acids 19 to 67 (Yamamoto *et al.*, 2009). The transcriptional activation domain of IE62 interacts with the human mediator complex by binding to the Med25 subunit (Yang *et al.*, 2008; Yamamoto *et al.*, 2009). Mediator complexes consist of about 30 subunits and act as a bridge between RNA polymerase II and transcriptional activators to assemble pre-initiation complexes and up-regulate gene expression. Reduction of Med25 levels, by siRNA or ectopic expression of Med25, inhibit transcriptional activation by the IE62 activation domain. Amino acids 895 to 1097 of IE62 also have a role in the transactivating activity of the protein, as evidenced by the effect of deleting these amino acids on reporter genes in transient transfection assays (Baudoux *et al.*, 1995). The amino and carboxy regions of IE62 have domains that are rich in serine and acidic amino acids, analogous to the SEAC domains in the PC4 cellular transcriptional regulatory protein (Kretschmar *et al.*, 1994); at present the function of these domains in VZV is unknown.

The DNA binding domain of IE62 is located at amino acids 417 to 646 (Tyler *et al.*, 1993) and can form a homodimer (Tyler *et al.*, 1994b). The DNA binding domain contains the WLQN sequence which is conserved with its HSV ICP4 homologue. Mutation of a lysine adjacent to this sequence reduces DNA binding activity and abrogates IE62 transactivating activity (Tyler *et al.*, 1994a). IE62 has a nuclear localization signal at amino acids 676 to 685 which is rich in arginine and leucine residues (Kinchington and Turse, 1998).

IE62 from the Oka vaccine strain of VZV had weaker transcriptional activity for VZV IE and putative early and late promoters than that of the parental virus in transient transfection assays in one study (Gomi *et al.*, 2000), but no difference in activation of putative early promoters in another study (Cohrs *et al.*, 2006). The varying results were thought to be due to differences in the length of the promoters in the reporter constructs with varying amounts of upstream regulatory sites. Analysis of VZV transcripts in cells infected with vaccine versus parental Oka virus showed lower levels of transcripts for ORF 62, 65, and 66 in cells infected with the vaccine virus (Cohrs *et al.*, 2006).

VZV IE4

IE4 is located in the viral tegument (Kinchington *et al.*, 1995). IE4 localizes to the nucleus in the presence of IE62; however, in the absence of IE62 most of the protein is in the cytoplasm (Defechereux *et al.*, 1996). IE4 transactivates several VZV IE, putative early, and putative late genes, but not IE63 or ORF28 protein in transient expression assays (Defechereux *et al.*, 1993; Moriuchi *et al.*, 1994b). Transactivation of VZV promoters has been reported to occur at both the transcriptional and post-transcriptional levels (Defechereux *et al.*, 1997; Perera *et al.*, 1994). IE4 enhances the ability of IE62 to transactivate VZV genes. The amino terminus of IE4 is required for transactivation, while several carboxy terminal cysteines, which contain a putative zinc domain finger, are dispensable (Moruichi *et al.*, 1995d,

Perera *et al.*, 1994). IE4 forms a dimer which is required for its transactivation (Baudoux *et al.*, 2000). IE4 interacts with the NF-KB p50 and p65 subunits, TBP, TFIIB (deMaisieres *et al.*, 1998), and IE62 (Spengler *et al.*, 2000). IE4 is essential for VZV replication (Cohen *et al.*, 2005b; Sato *et al.*, 2003b; Zhang *et al.*, 2007).

VZV IE63

IE63 is a phosphoprotein that is present in the virion tegument (Kinchington *et al.*, 1995). The protein localizes to the nucleus of infected cells and is phosphorylated by CDK1 and CDK5 (Habran *et al.*, 2005), casein kinases 1 and 2 (Bontems *et al.*, 2002), and ORF47 protein (Kenyon *et al.*, 2001). The most heavily phosphorylated region of the protein lies between amino acids 142 to 210, especially amino acids 165 to 186 (Fig. 1.4). Phosphorylation of IE63, particularly serine 224, is important for its localization when transfected into cells (Bontems *et al.*, 2002; Habran *et al.*, 2005); however, construction of a VZV mutant with substitution of alanines for multiple serine and threonine residues at phosphorylation sites (including mutation of serine 224) did not affect the localization of IE63 in the context of virus infection (Cohen *et al.*, 2005a). IE63 contains two nuclear localization signals at amino acids 226 to 229 and 260 to 263 (Baiker *et al.*, 2004).

Early studies indicated that VZV IE63 repressed IE62 and activated VZV thymidine kinase promoters (Jackers *et al.*, 1993), or that IE63 had a minimal effect on modulating VZV promoters (Kost *et al.*, 1995). These studies used a retroviral promoter, SV40 promoter, or the IE63 promoter to express IE63. Recent studies using the more potent CMV promoter indicate that IE63 represses VZV IE4 and putative early (DNA polymerase, thymidine kinase) promoters in transient transfection assays (Bontems *et al.*, 2002; DiValentin *et al.*, 2005). Examination of the role of IE63 in the context of VZV infection showed that a deletion mutant of IE63 results in increased expression of only IE62 (Hoover *et al.*, 2006). In contrast, IE63 enhances the activity of IE62 on the VZV gI promoter (Lynch *et al.*, 2002). The *trans*-repressing activity of IE63 in transient expression assays has been correlated with its phosphorylation status (Bontems *et al.*, 2002; Habran *et al.*, 2005). IE63 amino acids 151 to 213 are required for its *trans*-repressing activity in epithelial cells, but a different domain (amino acids 214 to 278) is necessary for *trans*-repression in neural cells (diValentin *et al.*, 2005).

IE63 represses expression of IL-6 and IL-8 in transient expression assays (diValentin *et al.*, 2005); however, expression of IE63 in a stable cell line failed to show repression of these genes (Habran *et al.*, 2007). IE63 reduces cellular gene transcription (IL-6, IL-8) induced in the presence of TNF-α (Habran *et al.*, 2007). IE63 upregulates the EF-1α promoter in transient expression assays (Zuranski *et al.*, 2005). Two analyses of cellular genes regulated by IE63, using stable cell lines or adenovirus expressing IE63, showed that the protein regulates a very limited set of genes and that the effects are dependent on the cell type used.

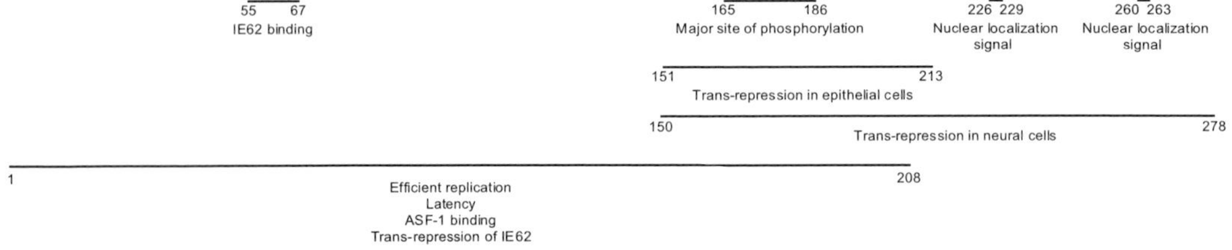

Figure 1.4 Structure of VZV IE63. Numbers refer to amino acid positions in the protein. IE62 and ASF-1 binding sites, nuclear localization signals, major site of phosphorylation, and domains important for trans-repression, latency, and efficient replication are shown.

IE63 down-regulates genes encoding heat shock proteins and TNF associated factor (TRAF)1, and upregulates genes encoding histones, monocyte chemotactic protein (MCP)–1, and HLA-DRB1 in melanoma cells, but not in HeLa cells (Habran *et al.*, 2007). Similarly, IE63 down-regulates heat shock proteins and up-regulates histones in human fibroblasts, but not in neuroblastoma cells (Hoover *et al.*, 2006).

VZV lacking one copy of IE63 increased levels of apoptosis in primary human neurons in cell culture compared with wild-type virus (Hood *et al.*, 2006). However, no differences in apoptosis were noted in rodent ganglia after infection with wild-type versus an IE63 null virus (Cohen *et al.*, 2004a).

IE63 binds to anti-silencing function (ASF)–1 and enhances the interaction of ASF-1 with histones H3.1 and H3.3 which may modulate the level of histones in VZV-infected cells (Ambagala *et al.*, 2009). IE63 interacts with RNA polymerase II, TFIIE, and TFIIH (diValentin *et al.*, 2005; Lynch *et al.*, 2002) and with IE62 (Baiker *et al.*, 2004).

IE63 inhibits phosphorylation of elongation initiation factor (eIF)-2α and inhibits the activity of interferon-α on virus replication (Ambagala *et al.*, 2007). IE63 is not required for VZV replication, but a deletion mutant is very impaired for replication in cell culture (Cohen *et al.*, 2004a). The carboxy terminal 70 amino acids are dispensable for efficient replication and for establishment of latency. Several amino acids that are phosphorylated are important for efficient replication (Baiker *et al.*, 2004; Cohen *et al.*, 2005a).

Other VZV activators of transcription

VZV ORF61 protein

ORF61 encodes a phosphoprotein that is not located in the virion tegument (Kinchington *et al.*, 1995). While ORF61 is not required for VZV infectivity, deletion of the gene reduces expression of VZV gE and impairs virus replication in cell culture (Cohen and Nguyen, 1998). ORF61 protein localizes to the nucleus of infected cells and transactivates VZV IE, early, and late promoters and enhances the infectivity of virion DNA in transient expression assays (Moriuchi *et al.*, 1993b). The amino terminus of ORF61 protein, which is required for transactivation, contains a RING finger domain that binds zinc (Everett *et al.*, 1993; Moriuchi *et al.*, 1994a). Deletion of the carboxy terminus of ORF61 protein results in a protein that inhibits the infectivity of virion DNA and acts as a dominant negative mutant in transient transfection assays (Moriuchi *et al.*, 1994a). ORF61 protein can enhance or repress the ability of IE62 and IE4 to up-regulate other viral genes depending on the cell type (Moriuchi *et al.*, 1993b; Perera *et al.*, 1992).

VZV ORF10 protein

ORF10 protein is a tegument protein (Kinchington *et al.*, 1992) and transactivates IE62 in transient expression assays (Moriuchi *et al.*, 1993a) and enhances the infectivity of virion DNA (Moriuchi *et al.*, 1994c). The transactivation domain of ORF10 protein is located near its amino terminus in a cluster of hydrophobic amino acids surrounded by acidic amino acids (Moriuchi *et al.*, 1995c). ORF10 is not required for virus replication (Cohen *et al.*, 1994b).

VZV ORF29 protein

ORF29 protein binds to single-stranded DNA (Kinchington *et al.*, 1988) and to the gI promoter (Boucaud *et al.*, 1998). While ORF29 protein does not modulate VZV gene expression by itself in transient expression assays, ORF29 protein regulates transactivation by IE62 which is dependent on the cell line tested (Boucaud *et al.*, 1998).

VZV regulators of transcription

VZV ORF47 and 66 protein kinases

ORF47 encodes a phosphoprotein present in the tegument of virions (Stevenson *et al.*, 1994). ORF47 protein is a serine-threonine protein kinase that phosphorylates several VZV proteins including ORF32 protein (Reddy *et al.*, 1998), ORF47 protein (Ng and Grose, 1992), IE62 (Ng *et al.*, 1994), IE63 (Kenyon *et al.*, 2001), VZV gE (Kenyon *et al.*, 2002) and VZV gI (Sato *et al.*, 2003c). VZV ORF47 regulates localization of IE62 in the cell; in the absence of the kinase domain of ORF47, IE62 remains almost exclusively in the nucleus and there is little protein in the cytoplasm (Besser *et al.*, 2003). ORF47 protein and IE62 coimmunoprecipitate. ORF47 is dispensable for replication *in vitro* (Heineman and Cohen, 1995); the kinase domain of the protein is required for normal virion assembly (Besser *et al.*, 2004).

ORF66 protein, a phosphoprotein (Stevenson *et al.*, 1994), is a serine-threonine protein kinase that phosphorylates IE62 at serine 686 and serine 722 (Eisfeld *et al.*, 2006). Expression of VZV ORF66 protein in concert with IE62 results in cytoplasmic localization of the latter protein (Kinchington *et al.*, 2001); ORF66 also is required for localization of IE62 in virions (Kinchington *et al.*, 2001). ORF66 is dispensable for replication *in vitro* (Heineman *et al.*, 1996).

Transcriptional regulation during latency

Several hypotheses have been proposed to explain VZV gene transcriptional regulation during latent infection. HCF-1 is normally present in the cytoplasm of sensory neurons, but localizes to the nucleus in these neurons in response to stimuli that result in reactivation of HSV (Kristie *et al.*, 1999). HCF-1 is required for ORF10-mediated transactivation of the VZV IE62 promoter (Narayanan *et al.*, 2005) and HCF-1 plays a role in chromatin remodelling of VZV promoters during lytic infection (Narayanan *et al.*, 2007). Therefore, relocalization of HCF-1 to the nucleus in response to stimuli associated with reactivation may help to augment IE62 gene transcription to reactivate VZV from latency.

Gene regulation during latency may be controlled by modification of chromatin (Gary *et al.*, 2006). Examination of latently infected human ganglia showed that promoters of latent genes (ORF62 and ORF63) are maintained in a euchromatic state (with acetylation of histone H3 at lysine 9) allowing transcription during latency, while promoters not associated with latency (ORF14 and ORF36) are in a heterochromatic state (without acetylation of histone H3 at lysine 9) resulting in repression of transcription during latency. In contrast, analysis of cells lytically infected with VZV showed that both the latent and the lytic promoters are maintained in a euchromatic state. Therefore chromatain modifications may be responsible for regulation of viral gene expression during latency.

Viral microRNAs are 21 to 24 nucleotides long and share sequence homology with the viral genome and bind to mRNA and inhibit protein expression. Viral microRNAs have been detected in many human herpesviruses, including HSV during latency and inhibit expression of immediate-early viral genes (Umbach *et al.*, 2008). While deep sequencing of transcripts from human trigeminal ganglia latently infected with HSV-1 and VZV identified several HSV microRNAs, no VZV microRNAs were detected (Umbach *et al.*, 2009). Unlike HSV, VZV rarely reactivates, and it is possible that VZV does not encode microRNAs.

VZV genes expressed during latency

Transcripts from VZV ORF4 (Kennedy *et al.*, 2000), ORF21 (Kennedy *et al.*, 2000), ORF29 (Kennedy *et al.*, 2000; Meier *et al.*, 1993), ORF62 (Kennedy *et al.*, 2000; Meier *et al.*, 1993), ORF63 (Kennedy *et al.*, 2000), and ORF66 (Cohrs *et al.*, 2003) have been detected in human ganglia during latency. ORF63 is the most abundant transcript expressed during latency. IE4 (Lungu *et al.*, 1998), ORF21 protein (Grinfeld and Kennedy, 2004; Lungu *et al.*, 1998), ORF 29 protein (Grinfeld and Kennedy, 2004; Lungu *et al.*, 1998), IE62 (Grinfeld and Kennedy, 2004; Lungu *et al.*, 1998), IE63 (Grinfeld and Kennedy, 2004; Lungu *et al.*, 1998; Mahalingham *et al.*, 1996), and ORF66 protein (Cohrs *et al.*, 2003) have been detected in human ganglia. A recent study casts doubt on the presence of IE63 in human ganglia (Zerboni *et al.*, 2010). While ORF21, ORF29, IE62, IE63 are located predominantly in the nucleus during virus replication, these proteins were found in the cytoplasm of neurons during latency (Grinfeld and Kennedy, 2004; Lungu *et al.*, 1998; Mahalingam *et al.*, 1996). Therefore sequestration of VZV IE62 and IE63, which normally regulate gene expression in the nucleus, to the cytoplasm during latency may prevent additional gene expression during latent infection. IE4 and ORF66 protein are located in both the cytoplasm and the nucleus during replication; however, both proteins localized to the cytoplasm of neurons during latency

(Cohrs *et al.*, 2003; Grinfeld and Kennedy, 2004; Lungu *et al.*, 1998). In contrast, IE4, ORF21, ORF29, IE62, and IE63 were located in the both the nucleus and the cytoplasm during reactivation in human ganglia (Lungu *et al.*, 1998).

Using a rodent model, VZV deleted for ORF4, ORF29, or ORF63 (Cohen *et al.*, 2004a; 2005b; 2007a) could enter ganglia, but latency was impaired. In contrast VZV unable to express ORF66 was not impaired for latency (Sato *et al.*, 2003c).

Conclusions and future trends

While much has been learned about VZV transcriptional regulation, the inability to produce high-titred cell-free virus has limited the ability to classify VZV genes in the three kinetic classes. Production of high titres of cell virus would enhance our knowledge of VZV transcription. While transient transfections using VZV genes driven by a CMV promoter and reporter constructs containing VZV gene promoters has helped to identify VZV transactivator proteins, more work is needed studying the effect of VZV promoter mutations in the context of the whole viral genome.

Additional studies are also needed to clarify the mechanisms of latency and reactivation of VZV. VZV gene transcription during latency is markedly different than HSV in which only a single gene, the HSV LAT is expressed. The roles of the six gene products expressed during VZV latency needs to be further explored. Due to the inability of a small animal model to recapitulate varicella and zoster, the VZV latency field has had to rely on autopsy tissue from humans. The limited material available and concerns about postmortem reactivation have slowed progress in this field. Simian varicella virus is the closest homologue to human VZV. The study of simian varicella virus latency and reactivation in African green or rhesus monkeys should be a useful model system to generate hypotheses about VZV latency. While microRNAs have been reported in human ganglia latently infected with HSV, these have not yet been described for VZV. Therefore, additional studies are needed to determine if VZV encodes microRNAs.

The mechanism of attenuation of the Oka vaccine strain is not known, and many of the nucleotide differences are located outside of the open reading frames; some of these differences may affect gene expression. Differences in transcriptional activity of IE62 from the Oka VZV vaccine strain and wild-type virus have yielded conflicting results. Since VZV IE62 is the viral protein that has the most amino acid differences between the vaccine and parental strain, it will be important to determine if these differences are important for transcription in the context of the viral genome and in animal models, and whether they contribute to attenuation of the vaccine virus.

Additional knowledge of how VZV IE genes are regulated and how IE proteins interact with cellular proteins might provide additional targets for the treatment of varicella. Currently all of the licensed antivirals for varicella and zoster act by inhibiting the viral DNA polymerase, and new drugs with different mechanisms of action would be helpful for treatment of virus resistant to antiviral agents. While a zoster vaccine has recently been licensed, it is only about 50% effective in preventing zoster. Further study of the regulation of transcription of VZV latency genes might yield new approaches to abrogating latency or reducing reactivation and thereby preventing or reducing zoster.

References

Ambagala, A.P., Bosma, T., Ali, M.A., Poustovoitov, M., Chen, J.J., Gershon, M.D., Adams, P.D., and Cohen, J.I. (2009). Varicella-zoster virus immediate-early 63 protein interacts with human antisilencing function 1 protein and alters its ability to bind histones H3.1 and H3.3. J. Virol. *83*, 200–209.

Ambagala, A.P., and Cohen, J.I. (2007). Varicella-Zoster virus IE63, a major viral latency protein, is required to inhibit the alpha interferon-induced antiviral response. J. Virol. *81*, 7844–7851.

Baiker, A., Bagowski, C., Ito, H., Sommer, M., Zerboni, L., Fabel, K., Hay, J., Ruyechan, W., and Arvin, A.M. (2004). The immediate-early 63 protein of varicella-zoster virus: analysis of functional domains required for replication *in vitro* and for T-cell and skin tropism in the SCIDhu model *in vivo*. J. Virol. 78, 1181–1194.

Baudoux, L., Defechereux, P., Rentier, B., and Piette, J. (2000). Gene activation by Varicella-zoster virus IE4 protein requires. its dimerization and involves both the arginine-rich sequence, the central part, and the carboxyl-terminal cysteine-rich region. J. Biol. Chem. *275*, 32822–32831.

Baudoux, L., Defechereux, P., Schoonbroodt, S., Merville, M.P., Rentier, B., and Piette, J. (1995). Mutational analysis of varicella-zoster virus major immediate-early protein IE62. Nucleic Acids Res. *23*, 1341–1349.

Berarducci, B., Sommer, M., Zerboni, L., Rajamani, J., and Arvin, A.M. (2007). Cellular and viral factors regulate the varicella-zoster virus gE promoter during viral replication. J. Virol. *81*, 10258–10267.

Besser, J., Ikoma, M., Fabel, K., Sommer, M.H., Zerboni, L., Grose, C., Arvin, A.M.. (2004). Differential requirement for cell fusion and virion formation in the pathogenesis of varicella-zoster virus infection in skin and T cells. J. Virol. *78*, 13293–13305.

Besser, J., Sommer, M.H., Zerboni, L., Bagowski, C.P., Ito, H., Moffat, J., Ku, C.C., Arvin, A.M.. (2003). Differentiation of varicella-zoster virus ORF47 protein kinase and IE62 protein binding domains and their contributions to replication in human skin xenografts in the SCID-hu mouse. J. Virol. *77*, 5964–5974.

Boucaud, D., Yoshitake, H., Hay, J., and Ruyechan, W. (1998). The varicella-zoster virus (VZV) open-reading frame 29 protein acts as a modulator of a late VZV gene promoter. J. Infect. Dis. *178 Suppl 1*, S34–38.

Bontems, S., Di Valentin, E., Baudoux, L., Rentier, B., Sadzot-Delvaux, C., and Piette, J. (2002). Phosphorylation of varicella-zoster virus IE63 protein by casein kinases influences its cellular localization and gene regulation activity. J. Biol. Chem. *277*, 21050–21060.

Brownlees, J., Gough, G., Thomas, S., Watts, P., Cohen, J., Coffin, R., and Latchman, D.S. (1999). Distinct responses of the herpes simplex virus and varicella zoster virus immediate early promoters to the cellular transcription factors Brn-3a and Brn-3b. Int. J. Biochem. Cell Biol. *31*, 451–461.

Che, X., Berarducci, B., Sommer, M., Ruyechan, W.T., and Arvin, A.M. (2007). The ubiquitous cellular transcriptional factor USF targets the varicella-zoster virus open reading frame 10 promoter and determines virulence in human skin xenografts in SCIDhu mice *in vivo*. J. Virol. *81*, 3229–3239.

Cilloniz, C., Jackson, W., Grose, C., Czechowski, D., Hay, J., and Ruyechan, W.T. (2007). The varicella-zoster virus (VZV) ORF9 protein interacts with the IE62 major VZV transactivator. J. Virol. *81*, 761–774.

Cohen, J.I., Cox, E., Pesnicak, L., Srinivas, S., and Krogmann, T. (2004a). The varicella-zoster virus ORF63 latency-associated protein is critical for establishment of latency. J. Virol. *78*, 11833–11840.

Cohen, J.I., Heffel, D., and Seidel, K. (1993). The transcriptional activation domain of varicella-zoster virus open reading frame 62 protein is not conserved with its herpes simplex virus homolog. J. Virol. *67*, 4246–4251.

Cohen, J.I., Krogmann, T., Bontems, S., Sadzot, C., and Pesnicak, L. (2005a). Regions of the varicella-zoster virus ORF63 latency-associated protein important for efficient replication *in vitro* are also critical for efficient establishment of latency. J. Virol. *79*, 5069–5077.

Cohen, J.I., Krogmann, T., Pesnicak, L., and Ali, M.A. (2007a). Absence or overexpression of the varicella-zoster virus (VZV) ORF29 latency-associated protein impairs late gene expression and reduces latency in a rodent model. J. Virol. *81*, 1586–1591.

Cohen, J.I., Krogmann, T., Ross, J.P., Pesnicak, L.P., and Prikhod'ko, E.A. (2005b). The varicella-zoster virus ORF4 latency associated protein is important for establishment of latency. J. Virol. *79*, 6969–6975.

Cohen, J.I., and Nguyen, H. (1998). Varicella-zoster virus ORF61 deletion mutants replicate in cell culture, but a mutant with stop codons in ORF61 reverts to wild-type virus. Virology *246*, 306–316.

Cohen, J.I., and Seidel, K.E. (1994b). Varicella-zoster virus (VZV) open reading frame 10 protein, the homolog of the essential herpes simplex virus protein VP16, is dispensable for VZV replication *in vitro*. J. Virol. *68*, 7850–7858.

Cohen, J.I., Straus, S.E., and Arvin, A.M. (2007b). Varicella-zoster virus: Replication, pathogenesis, and management. In: Fields Virology, Knipe DM, Howley PM, *et al.*, eds. (Philadelphia, Lipincott-Williams & Wilkins) pp. 2773–2818.

Cohrs, R.J., Barbour, M., and Gilden, D.H. (1998). Varicella-zoster virus gene 21: transcriptional start site and promoter region. J. Virol. *72*, 42–47.

Cohrs, R.J., Gilden, D.H., Gomi, Y., Yamanishi, K., Cohen, J.I.. (2006). Comparison of virus transcription during lytic infection of the Oka parental and vaccine strains of Varicella-Zoster virus. J. Virol. *80*, 2076–2082.

Cohrs, R.J., Gilden, D.H., Kinchington, P.R., Grinfeld, E., and Kennedy, P.G. (2003). Varicella-zoster virus gene 66 transcription and translation in latently infected human Ganglia. J. Virol. *77*, 6660–6665.

Davison, A., J., and Scott, J.E. (1986). The complete DNA sequence of varicella-zoster virus. J. Gen. Virol. *67*, 1759–1816.

Debrus, S., Sadzot-Delvaux, C., Nikkels, A.F., Piette, J., and Rentier, B. (1995). Varicella-zoster virus gene 63 encodes an immediate-early protein that is abundantly expressed during latency. J. Virol. *69*, 3240–3245.

Defechereux, P., Debrus, S., Baudoux, L., Rentier, B., and Piette, J. (1997). Varicella-zoster virus open reading frame 4 encodes an immediate-early protein with posttranscriptional regulatory properties. J. Virol. *71*, 7073–7079.

Defechereux, P., Debrus, S., Baudoux, L., Schoonbroodt, S., Merville, M.P., Rentier, B., and Piette, J. (1996). Intracellular distribution of the ORF4 gene product of varicella-zoster virus is influenced by the IE62 protein. J. Gen. Virol. *77*, 1505–1513.

Defechereux, P., Melen, L., Baudoux, L., Merville-Louis, M.P., Rentier, B., and Piette, J. (1993). Characterization of the regulatory functions of varicella-zoster virus open reading frame 4 gene product. J. Virol. *67*, 4379–4385.

de Maisieres, P.D., Baudoux-Tebache, L., Merville, M.P., Rentier, B., Bours, V., and Piette, J. (1998). Activation of the human immunodeficiency virus long terminal repeat by varicella-zoster virus IE4 protein requires. nuclear factor-kappaB and involves both the amino-terminal and the carboxyl-terminal cysteine-rich region. J. Biol. Chem. *273*, 13636–13644.

Di Valentin, E., Bontems, S., Habran, L., Jolois, O., Markine-Goriaynoff, N., Vanderplasschen, A., Sadzot-Delvaux, C., and Piette, J. (2005). Varicella-zoster virus IE63 protein represses the basal transcription machinery by disorganizing the pre-initiation complex. Biol. Chem. *386*, 255–267.

Disney, G.H., McKee, T.A., Preston, C.M., Everett, R.D.(1990). The product of varicella-zoster virus gene 62 autoregulates its own promoter. J. Gen. Virol. *71*, 2999–3003.

Eisfeld, A.J., Turse, S.E., Jackson, S.A., Lerner, E.C., and Kinchington, P.R. (2006). Phosphorylation of the varicella-zoster virus (VZV) major transcriptional regulatory protein IE62 by the VZV open reading frame 66 protein kinase. J. Virol. *80*, 1710–1723.

Everett, R.D., Barlow, P., Milner, A., Luisi, B., Orr, A., Hope, G., and Lyon, D. (1993). A novel arrangement of zinc-binding residues and secondary structure in the C3HC4 motif of an alpha herpes virus protein family. J. Mol. Biol. *234*, 1038–1047.

Forghani, B., Mahalingam, R., Vafai, A., Hurst, J.W., and Dupuis, K.W. (1990). Monoclonal antibody to immediate early protein encoded by varicella-zoster virus gene 62. Virus Res. *16*, 195–210.

Gary, L., Gilden, D.H., and Cohrs, R.J. (2006). Epigenetic regulation of varicella-zoster virus open reading frames 62 and 63 in latently infected human trigeminal ganglia. J. Virol. *80*, 4921–4926.

Gomi, Y., Imagawa, T., Takahashi, M., and Yamanishi, K. (2000). Oka varicella vaccine is distinguishable from its parental virus in DNA sequence of open reading frame 62 and its transactivation activity. J. Med. Virol. *61*, 497–503.

Grinfeld, E., and Kennedy, P.G. (2004). Translation of varicella-zoster virus genes during human ganglionic latency. Virus Genes *29*, 317–319.

Habran, L., Bontems, S., Di Valentin, E., Sadzot-Delvaux, C., and Piette, J. (2005). Varicella-zoster virus IE63 protein phosphorylation by roscovitine-sensitive cyclin-dependent kinases modulates its cellular localization and activity. J. Biol. Chem. *280*, 29135–29143.

Habran, L., El Mjiyad, N., Di Valentin, E., Sadzot-Delvaux, C., Bontems, S., and Piette, J. (2007). The varicella-zoster virus immediate-early 63 protein affects chromatin-controlled gene transcription in a cell-type dependent manner. BMC Mol. Biol. *8*, 99.

He, H., Boucaud, D., Hay, J., and Ruyechan, W.T. (2001). Cis and trans elements regulating expression of the varicella zoster virus gI gene. Arch. Virol. Suppl.17, 57–70.

Heineman, T.C., and Cohen, J.I. (1995). The varicella-zoster virus (VZV) open reading frame 47 (ORF47) protein kinase is dispensable for viral replication and is not required for phosphorylation of ORF63 protein, the VZV homolog of herpes simplex virus ICP22. J. Virol. *69*, 7367–7370.

Heineman, T.C., Seidel, K., and Cohen, J.I. (1996), The varicella-zoster virus ORF66 protein induces kinase activity and is dispensable for viral replication. J. Virol. *70*, 7312–7317.

Hood, C., Cunningham, A.L., Slobedman, B., Arvin, A.M., Sommer, M.H., Kinchington, P.R., and Abendroth, A. (2006). Varicella-zoster virus ORF63 inhibits apoptosis of primary human neurons. J. Virol. *80*, 1025–1031.

Hoover, S.E., Cohrs, R.J., Rangel, Z.G., Gilden, D.H., Munson, P., and Cohen, J.I. (2006). Downregulation of varicella-zoster virus (VZV) immediate-early ORF62 transcription by VZV ORF63 correlates with virus replication *in vitro* and with latency. J. Virol. *80*, 3459–3468.

Inchauspe, G., and Ostrove, J.M. (1989). Differential regulation by varicella-zoster virus (VZV) and herpes simplex virus type-1 trans-activating genes. Virology *173*, 710–714.

Ito, H., Sommer, M.H., Zerboni, L., He, H., Boucaud, D., Hay, J., Ruyechan, W., and Arvin, A.M. (2003). Promoter sequences of varicella-zoster virus glycoprotein I targeted by cellular transactivating factors Sp1 and USF determine virulence in skin and T cells in SCIDhu mice *in vivo*. J. Virol. 77, 489–498.

Jackers, P., Defechereux, P., Baudoux, L., Lambert, C., Massaer, M., Merville-Louis, M.P., Rentier, B., and Piette, J. (1992). Characterization of regulatory functions of the varicella-zoster virus gene 63-encoded protein. J. Virol. *66*, 3899–3903.

Jones, J.O., Sommer, M., Stamatis, S., and Arvin, A.M. (2006). Mutational analysis of the varicella-zoster virus ORF62/63 intergenic region. J. Virol. *80*, 3116–3121.

Kato, T., Kitamura, K., Hayakawa, Y., Takahashi, M., Kojima, A., Sato, S., and Yamanishi, K. (1989). Transcription mapping of glycoprotein I (gpI) and gpIV of varicella-zoster virus and immunological analysis of the gpI produced in cells infected with the recombinant vaccinia virus. Microbiol. Immunol. *33*, 299–312.

Kennedy, P.G., Grinfeld, E., and Bell, J.E. (2000). Varicella-zoster virus gene expression in latently infected and explanted human ganglia. J. Virol. *74*, 11893–11898.

Kenyon, T.K., Cohen, J.I., and Grose, C. (2002). Phosphorylation by the varicella-zoster virus ORF47 protein serine kinase determines whether. endocytosed viral gE traffics to the trans-Golgi network or recycles to the cell membrane. J. Virol. *76*, 10980–10993.

Kenyon, T.K., Lynch, J., Hay, J., Ruyechan, W., and Grose, C. (2001). Varicella-zoster virus ORF47 protein serine kinase: characterization of a cloned, biologically active phosphotransferase and two viral substrates, ORF62 and ORF63. J. Virol. *75*, 8854–8858.

Khalil, M.I., Hay, J., and Ruyechan, W.T. (2008). Cellular transcription factors Sp1 and Sp3 suppress varicella-zoster virus origin-dependent DNA replication. J. Virol. *82*, 11723–11733.

Kinchington, P.R., Bookey, D., and Turse, S.E. (1995). The transcriptional regulatory proteins encoded by varicella-zoster virus open reading frames (ORFs) 4 and 63, but not ORF 61, are associated with purified virus particles. J. Virol. *69*, 4274–4282.

Kinchington, P.R., Fite, K., Seman, A., and Turse, S.E. (2001). Virion association of IE62, the varicella-zoster virus (VZV) major transcriptional regulatory protein, requires. expression of thee VZV open reading frame 66 protein kinase. J. Virol. 75, 9106–9113.

Kinchington, P.R., Fite, K., and Turse, S.E. (2000). Nuclear accumulation of IE62, the varicella-zoster virus (VZV) major transcriptional regulatory protein, is inhibited by phosphorylation mediated by the VZV open reading frame 66 protein kinase. J. Virol. *74*, 2265–2277.

Kinchington, P.R., Hougland, J.K., Arvin, A.M., Ruyechan, W.T., and Hay, J. (1992). The varicella-zoster virus immediate-early protein IE62 is a major component of virus particles. J. Virol. *66*, 359–366.

Kinchington, P.R., Inchauspe, G., Subak-Sharpe, J.H., Robey, F., Hay, J., and Ruyechan, W.T. (1988). Identification and characterization of a varicella-zoster virus DNA-binding protein by using antisera directed against a predicted synthetic oligopeptide. J. Virol. *62*, 802–809.

Kinchington, P.R., and Turse, S.E. (1998). Regulated nuclear localization of the varicella-zoster virus major regulatory protein, IE62. J. Infect. Dis. 178 Suppl *1*, S16–21.

Kinchington, P.R., Vergnes, J.P., Defechereux, P., Piette, J., and Turse, S.E. (1994). Transcriptional mapping of the varicella-zoster virus regulatory genes encoding open reading frames 4 and 63. J. Virol. *68*, 3570–3581.

Kost, R.G., Kupinsky, H., and Straus, S.E. (1995). Varicella-zoster virus gene 63: transcript mapping and regulatory activity. Virology *209*, 218–224.

Kretzschmar, M., Kaiser, K., Lottspeich, F., and Meisterernst, M. (1994). A novel mediator of class II gene transcription with homology to viral immediate-early transcriptional regulators. Cell 78, 525–534.

Kristie, T.M., Liang, L., and Vogel, J.L. (2010). Control of alpha-herpesvirus IE gene expression by HCF-1 coupled chromatin modification activities. Biochim. Biophys. Acta *1799*, 257–265.

Kristie, T.M., Vogel, J.L., and Sears, A.E. (1999). Nuclear localization of the C1 factor (host cell factor) in sensory neurons correlates with reactivation of herpes simplex virus from latency. Proc. Natl. Acad. Sci. U.S.A. *96*, 1229–1233.

Ling, P., Kinchington, P.R., Ruyechan, W.T., and Hay, J. (1991). A detailed analysis of transcripts mapping to varicella zoster virus gene 14 (glycoprotein V). Virology *184*, 625–635.

Ling, P., Kinchington, P.R., Sadeghi-Zadeh, M., Ruyechan, W.T., and Hay, J. (1992). Transcription from varicella-zoster virus gene 67 (glycoprotein IV). J. Virol. *66*, 3690–3698.

Lungu, O., Panagiotidis, C.A., Annunziato, P.W., Gershon, A.A., and Silverstein, S.J. (1998). Aberrant intracellular localization of Varicella-Zoster virus regulatory proteins during latency. Proc. Natl. Acad. Sci. U.S.A. *95*, 7080–7085.

Lynch, J.M., Kenyon, T.K., Grose, C., Hay, J., and Ruyechan, W.T. (2002). Physical and functional interaction between the varicella zoster virus IE63 and IE62 proteins. Virology *302*, 71–82.

Mahalingam, R., Wellish, M., Cohrs, R., Debrus, S., Piette, J., Rentier, B., and Gilden, D.H. (1996). Expression of protein encoded by varicella-zoster virus open reading frame 63 in latently infected human ganglionic neurons. Proc. Natl. Acad. Sci. U.S.A. *93*, 2122–2124.

McKee, T.A.., Disney, G.H., Everett, R.D., and Preston, C.M. (1990). Control of expression of the varicella-zoster virus major immediate early gene. J. Gen. Virol. *71*, 897–906.

McKee, T.A., and Preston, C.M. (1991). Identification of two protein binding sites within the varicella-zoster virus major immediate early gene promoter. Virus Res. *20*, 59–69.

Meier, J.L., and Straus, S.E. (1993). Varicella-zoster virus DNA polymerase and major DNA-binding protein genes have overlapping divergent promoters. J. Virol. *67*, 7573–7581.

Meier, J.L., Holman, R.P., Croen, K.D., Smialek, J.E., and Straus, S.E. (1993). Varicella-zoster virus transcription in human trigeminal ganglia. Virology *193*, 193–200.

Meier, J.L., Luo, X., Sawadogo, M., and Straus, S.E. (1994). The cellular transcription factor USF cooperates with varicella-zoster virus immediate-early protein 62 to symmetrically activate a bidirectional viral promoter. Mol. Cell Biol. *14*, 6896–6906.

Michael, E.J., Kuck, K.M., and Kinchington, P.R. (1998). Anatomy of the varicella-zoster virus open-reading frame 4 promoter. J. Infect. Dis. *178* (Suppl 1), S27–33.

Moriuchi, H., Moriuchi, M., and Cohen, J.I. (1995a). Proteins and cis-acting elements associated with transactivation of the varicella-zoster virus (VZV) immediate-early gene 62 promoter by VZV open reading frame 10 protein. J. Virol. *69*, 4693–4701.

Moriuchi, H., Moriuchi, M., and Cohen, J.I. (1994a). The RING finger domain of the varicella-zoster virus open reading frame 61 protein is required for its transregulatory functions. Virology *205*, 238–426.

Moriuchi, H., Moriuchi, M., and Cohen, J.I. (1995b). The varicella-zoster virus immediate-early 62 promoter contains a negative regulatory element that binds transcriptional factor NF-Y. Virology *214*, 256–258.

Moriuchi, H., Moriuchi, M., Pichyangkura, R., Triezenberg, S.J., Straus, S.E., and Cohen, J.I. (1995c). Hydrophobic cluster analysis predicts an amino-terminal domain of varicella-zoster virus open reading frame 10 required for transcriptional activation. Proc. Natl. Acad. Sci. U.S.A. *92*, 9333–9337.

Moriuchi, H., Moriuchi, M., Smith, H.A., and Cohen, J.I. (1994b). Varicella-zoster virus open reading frame 4 protein is functionally distinct from and does not complement its herpes simplex virus type 1 homolog, ICP27. J. Virol. *68*, 1987–1992.

Moriuchi, H., Moriuchi, M., Straus, S.E., and Cohen, J.I. (1993a). Varicella-zoster virus open reading frame 10 protein, the herpes simplex virus VP16 homolog, transactivates herpesvirus immediate-early gene promoters. J. Virol. *67*, 2739–2746.

Moriuchi, H., Moriuchi, M., Straus, S.E., and Cohen, J.I. (1993b). Varicella-zoster virus (VZV) open reading frame 61 protein transactivates VZV gene promoters and enhances the infectivity of VZV DNA. J. Virol. *67*, 4290–4295.

Moriuchi, M., Moriuchi, H., Debrus, S., Piette, J., and Cohen, J.I. (1995d). The acidic amino-terminal region of varicella-zoster virus open reading frame 4 protein is required for transactivation and can functionally replace the corresponding region of herpes simplex virus ICP27. Virology *208*, 376–382.

Moriuchi, M., Moriuchi, H., Straus, S.E., and Cohen, J.I. (1994c). Varicella-zoster virus (VZV) virion-associated transactivator open reading frame 62 protein enhances the infectivity of VZV DNA. Virology *200*, 297–300.

Narayanan, A., Nogueira, M.L., Ruyechan, W.T., and Kristie, T.M. (2005). Combinatorial transcription of herpes simplex virus and varicella zoster virus immediate early genes is strictly determined by the cellular coactivator HCF-1. J. Biol. Chem. *280*, 1369–1375.

Narayanan, A., Ruyechan, W.T., and Kristie, T.M. (2007). The coactivator host cell factor-1 mediates Set1 and MLL1 H3K4 trimethylation at herpesvirus immediate early promoters for initiation of infection. Proc. Natl. Acad. Sci. U.S.A. *104*, 10835–10840.

Ng, T.I., and Grose, C. (1992). Serine protein kinase associated with varicella-zoster virus ORF 47. Virology *191*, 9–18.

Ng, T.I., Keenan, L., Kinchington, P.R., and Grose, C. (1994). Phosphorylation of varicella-zoster virus open reading frame (ORF) 62 regulatory product by viral ORF 47-associated protein kinase. J. Virol. *68*, 1350–1359.

Patel, Y., Gough, G., Coffin, R.S., Thomas, S., Cohen, J.I., and Latchman, D.S. (1998). Cell type specific repression of the varicella zoster virus immediate early gene 62 promoter by the cellular Oct-2 transcription factor. Biochim. Biophys Acta *1397*, 268–274.

Peng, H., He, H., Hay, J., and Ruyechan, W.T. (2003). Interaction between the varicella zoster virus IE62 major transactivator and cellular transcription factor Sp1. J. Biol. Chem. *278*, 38068–38075.

Perera, L.P. (2000). The TATA motif specifies the differential activation of minimal promoters by varicella zoster virus immediate-early regulatory protein IE62. J. Biol. Chem. *275*, 487–496.

Perera, L.P., Kaushal, S., Kinchington, P.R., Mosca, J.D., Hayward, G.S., and Straus, S.E. (1994). Varicella-zoster virus open reading frame 4 encodes a transcriptional activator that is functionally distinct from that of herpes simplex virus homology ICP27. J. Virol. *68*, 2468–2477.

Perera, L.P., Mosca, J.D., Ruyechan, W.T., and Hay, J. (1992a). Regulation of varicella-zoster virus gene expression in human T lymphocytes. J. Virol. *66*, 5298–5304.

Perera, L.P., Mosca, J.D., Ruyechan, W.T., Hayward, G.S., Straus, S.E., and Hay, J. (1993). A major transactivator of varicella-zoster virus, the immediate-early protein IE62, contains a potent N-terminal activation domain. J. Virol. *67*, 4474–4483.

Perera, L.P., Mosca, J.D., Sadeghi-Zadeh, M., Ruyechan, W.T., and Hay, J. (1992b). The varicella-zoster virus immediate early protein, IE62, can positively regulate its cognate promoter. Virology *191*, 346–354.

Rahaus, M., Desloges, N., Yang, M., Ruyechan, W.T., and Wolff, M.H. (2003a). Transcription factor USF, expressed during the entire phase of varicella-zoster virus infection, interacts physically with the major viral transactivator IE62 and plays a significant role in virus replication. J. Gen. Virol. *84*, 2957–2967.

Rahaus, M., and Wolff, M.H. (1999). Influence of different cellular transcription factors on the regulation of varicella-zoster virus glycoproteins E (gE) and I (gI) UTR's activity. Virus Res. *62*, 77–88.

Rahaus, M., Wolff, M.H.. (2003). Reciprocal effects of varicella-zoster virus (VZV) and AP1: activation of jun, fos and ATF-2 after VZV infection and their importance for the regulation of viral genes. Virus Res. *92*, 9–21.

Rahaus, M., and Wolff, M.H. (2000a). Transcription factor Sp1 is involved in the regulation of varicella-zoster virus glycoprotein E. Virus Res. *69*, 69–81.

Rahaus, M., and Wolff, M.H. (2000b). Transcriptional pattern of varicella-zoster virus glycoproteins E and I. Intervirology *43*, 180–184.

Reddy, S.M., Cox, E., Iofin, I., Soong, W., and Cohen, J.I. (1998). Varicella-zoster virus (VZV) ORF32 encodes a phosphoprotein that is posttranslationally modified by the VZV ORF47 protein kinase. J. Virol. *72*, 8083–8088.

Ruyechan, W.T., Peng, H., Yang, M., Hay, J.. (2003). Cellular factors and IE62 activation of VZV promoters. Med, J. . Virol. 70 (Suppl 1), S90–94.

Sato, B., Ito, H., Hinchliffe, S., Sommer, M.H., Zerboni, L., and Arvin, A.M. (2003a). Mutational analysis of open reading frames 62 and 71, encoding the varicella-zoster virus immediate-early transactivating protein, IE62, and effects on replication *in vitro* and in skin xenografts in the SCID-hu mouse *in vivo*. J. Virol. *77*, 5607–5620.

Sato, B., Sommer, M., Ito, H., and Arvin, A.M. (2003b). Requirement of varicella-zoster virus immediate-early 4 protein for viral replication. J. Virol. *77*, 12369–12372.

Sato, H., Pesnicak, L., and Cohen, J.I. (2003c). Varicella-zoster virus ORF47 protein kinase, which is required for replication in human T cells, and ORF66 protein kinase, which is expressed during latency, are dispensable for establishment of latency. J. Virol. *77*, 11180–11185.

Spengler, M.L., Ruyechan, W.T., Hay, J.. (2000). Physical interaction between two varicella zoster virus gene regulatory proteins, IE4 and IE62. Virology *272*, 375–381.

Stevenson, D., Colman, K.L., and Davison, A.J. (1992). Characterization of the varicella-zoster virus gene 61 protein. J. Gen. Virol. *73*, 521–530.

Stevenson, D., Colman, K.L., and Davison, A.J. (1994). Characterization of the putative protein kinases specified by varicella-zoster virus genes 47 and 66. J. Gen. Virol. *75*, 317–326.

Storlie, J., Jackson, W., Hutchinson, J., and Grose, C. (2006). Delayed biosynthesis of varicella-zoster virus glycoprotein C: upregulation by hexamethylene bisacetamide and retinoic acid treatment of infected cells. J. Virol. *80*, 9544–9556.

Tyler, J.K., Allen, K.E., and Everett, R.D. (1994a). Mutation of a single lysine residue severely impairs the DNA recognition and regulatory functions of the VZV gene 62 transactivator protein. Nucleic Acids Res. *22*, 270–278.

Tyler, J.K., and Everett, R.D. (1993). The DNA binding domain of the varicella-zoster virus gene 62 protein interacts with multiple sequences which are similar to the binding site of the related protein of herpes simplex virus type 1. Nucleic Acids Res. *21*, 513–522.

Tyler, J.K., and Everett, R.D. (1994b). The DNA binding domains of the varicella-zoster virus gene 62 and herpes simplex virus type 1 ICP4 transactivator proteins heterodimerize and bind to DNA. Nucleic Acids Res. *22*, 711–721.

Umbach, J.L., Kramer, M.F., Jurak, I., Karnowski, H.W., Coen, D.M., Cullen, B.R.. (2008). MicroRNAs expressed by herpes simplex virus 1 during latent infection regulate viral mRNAs. Nature. *454*, 780–783.

Umbach, J.L., Nagel, M.A., Cohrs, R.J., Gilden, D.H., and Cullen, B.R. (2009). Analysis of human alphaherpesvirus microRNA expression in latently infected human trigeminal ganglia. J. Virol. *80*, 10677–10683.

Wang, L., Sommer, M., Rajamani, J., and Arvin, A.M. (2009). Regulation of the ORF61 promoter and ORF61 functions in varicella-zoster virus replication and pathogenesis. J. Virol. *83*, 7560–7572.

Xia, D., and Straus, S.E. (1999). Transcript mapping and transregulatory behavior of varicella-zoster virus gene 21, a latency-associated gene. Virology *258*, 304–313.

Yamamoto, S., Eletsky, A., Szyperski, T., Hay, J., and Ruyechan, W.T. (2009). Analysis of the varicella-zoster virus IE62 N-terminal acidic transactivating domain and its interaction with the human mediator complex. J. Virol. *83*, 6300–6305.

Yang, M., Hay, J., and Ruyechan, W.T. (2004). The DNA element controlling expression of the varicella-zoster virus open reading frame 28 and 29 genes consists of two divergent unidirectional promoters which have a common USF site. J. Virol. *78*, 10939–10952.

Yang, M., Hay, .J., and Ruyechan, W.T. (2008). Varicella-zoster virus IE62 protein utilizes the human mediator complex in promoter activation. J. Virol. *82*, 12154–12163.

Yang, M., Peng, H., Hay, J., and Ruyechan, W.T. (2006). Promoter activation by the varicella-zoster virus major transactivator IE62 and the cellular transcription factor USF. J. Virol. *80*, 7339–7353.

Zerboni, L., Sobel, R.A., Romachandron, V., Rajamani, J., Royechan, W., Abendroth, A., and Arvin, A. (2010). Expression of varicella-zoster virus immediate-early regulatory protein IE63 in neurons of latenty infected human sensory ganglia. J. Virol. *84*, 3421–3430.

Zhang, Z., Rowe, J., Wang, W., Sommer, M., Arvin, A., Moffat, J., and Zhu, H. (2007). Genetic analysis of varicella-zoster virus ORF0 to ORF4 by use of a novel luciferase bacterial artificial chromosome system. J. Virol. *81*, 9024–9033.

Zuranski, T., Nawar, H., Czechowski, D., Lynch, J.M., Arvin, A., Hay, J., and Ruyechan, W.T. (2005). Cell-type-dependent activation of the cellular EF-1alpha promoter by the varicella-zoster virus IE63 protein. Virology *338*, 35–42.

Functions and Mechanism of Action of the Herpes Simplex Virus Regulatory Protein, ICP4

2

Neal A. DeLuca

Abstract
ICP4 is expressed from the HSV genome very early in infection. It is a large structurally complex nuclear phosphoprotein that is essential for viral growth largely due to its requirement for the transcriptional activation of most HSV early and late genes. It also acts a repressor of transcription under certain circumstances. The HSV genome is transcribed by RNA polII, and ICP4 interacts with components of the RNA polII transcription machinery to carry out is functions in transcription. The interactions that are important for its functions can be genetically defined implicating a modular composition of the ICP4 protein. ICP4 also plays a specific role in virus growth in sympathetic neurons implicating a specific function in pathogenesis. This is a review about what is known about ICP4 from many genetic, biological and biochemical studies, from many laboratories.

Introduction

There are about 80 genes encoded for by the approximately 152 kb linear double stranded DNA genome of herpes simplex virus type 1 (McGeoch *et al.*, 1986, 1988). All of the known genes to date are transcribed by cellular RNA polII (Alwine *et al.*, 1974). Therefore, like all nuclear replicating DNA viruses, the expression of the viral genome is entirely dependant on the cellular transcription machinery. However, there are numerous viral functions that modify the cellular machinery affecting polII transcription, or that act as transcription factors themselves. In productive infection the resulting interplay between the cellular transcription machinery and viral functions results in a regulated cascade of viral gene expression. Viral genes can be roughly characterized into three kinetic classes defined by their peak rates of protein and mRNA synthesis, and how they are expressed in the presence of inhibitors of macromolecular synthesis. These are the: immediate early (IE) or α, early (E) or β, and late (L) or γ genes (Honess and Roizman, 1974, 1975).

IE genes are defined as those that are expressed in the absence of de novo viral protein synthesis. When inhibitors of translation, such as cycloheximide are added to cells prior to and during infection IE mRNA accumulates. Washing out the cycloheximide and subsequently adding actinomycin D, an inhibitor of polII transcription, results in the synthesis of 5 IE proteins (Honess and Roizman, 1974). These are infected cell polypeptides (ICPs): 4, 0, 22, 27 and 47. In untreated cells, the peak synthesis of IE proteins is between 2 and 4 h post infection. If actinomycin D is not added after reversal of the translation block, then E and L genes are expressed. Furthermore if an amino acid analogue, such as canavanine, is added at the outset of infection only IE genes are expressed (Honess and Roizman, 1975). These observations demonstrate that one or more IE protein(s) is necessary for the expression of E and L genes. Further kinetic and inhibitor studies established that E genes are maximally expressed between and 4 and 6 h post infection, and that the expression of L genes requires DNA synthesis long after 6 h post infection. With the exception of ICP47, the remaining IE proteins are nuclear phosphoproteins (Pereira *et al.*, 1977). One of the IE proteins, ICP4, was found to be crucial for the progression from the transcription of IE genes to the transcription of the remainder of the viral genome.

Genetics of ICP4

Early genetic studies mostly with temperature sensitive (ts) mutants revealed a considerable amount of information about ICP4. Mutants in the complementation group 1–2 (Schaffer *et al.*, 1973) were found to overexpress a 175 hDa nuclear protein and very little of the major proteins expressed during wt virus infection at the non-permissive temperature (Courtney and Benyesh-Melnick, 1974). This 175 kd protein was thought to be the IE protein, ICP4 (Honess and Roizman, 1974), which was inferred to map to the short repeated regions of the genome (Morse *et al.*, 1978; Preston *et al.*, 1978). Subsequent marker rescue experiments firmly established that ts mutants defining the 1–2 complementation group map within in the gene for ICP4, which is present in the short repeat region of the genome and therefore is diploid (Fig. 2.1) (Dixon and Schaffer, 1980; Preston, 1981; Stow *et al.*, 1978; Stow and Wilkie, 1978).

Phenotypic analysis of ts mutants in ICP4 revealed that it is involved in the induced expression of HSV E and L proteins, and is also involved in the shutoff of IE protein (including itself) expression (Courtney and Benyesh-Melnick, 1974; Dixon and Schaffer, 1980; Preston, 1979a). The basis for this regulatory effect was found to be at the level of accumulation of mRNA, possibly transcription (Preston, 1979b; Watson and Clements, 1978, 1980). It was later shown that ICP4 possibly exerts its regulatory effect at the level of transcription initiation as determined by measurements of transcription rates of viral genes in nuclear run-on experiments (DeLuca and Schaffer, 1988; Godowski and Knipe, 1986). In addition, while most of the ts mutants examined have the phenotype where IE proteins are overexpressed and E and L proteins are uniformly very poorly expressed at the non-permissive temperature, there are some mutants possessing the phenotype where E proteins are quite abundantly expressed and L proteins remain under-expressed (DeLuca *et al.*, 1984). This result suggests that ICP4 may activate E and L gene transcription by different mechanisms, requiring different domains or regions of the ICP4 molecule. It was also shown that these mutants map in the C-terminal region of ICP4, an observation that would later prove to be consistent with studies of deletion and nonsense mutation and biochemical analysis.

In summary early studies, mostly with ts mutants, revealed the following: ICP4, (i) is expressed as an IE protein, (ii) is 175 kDa on SDS-PAGE gels and localizes to the nucleus of infected cells, (iii) is heavily phosphorylated in a complex pattern (Pereira *et al.*, 1977; Preston, 1979a; Wilcox *et al.*, 1980), (iv) is crucial for productive expression of viral E and L proteins, (v) regulates E and L genes differently, (vi) is involved in the shutoff of IE gene expression, and (vii) functions at the level of synthesis of viral mRNA.

The application of recombinant (Post *et al.*, 1980) and sequencing (McGeoch *et al.*, 1986; Wagner *et al.*, 1981) approaches to the study of HSV has lead to a more detailed understanding of the molecular genetics of the virus. The former

Figure 2.1 The location of the ICP4 genes on the HSV genome. The 152 kb genome is shown along with the following features; (i) the unique long and short segments (U_S and U_L), (ii) the directly repeated terminal a sequence, which is inverted (a′) at the joint of the long and short segments, (iii) the inverted repeat segments bracketing U_L (b and b′), and (iv) the inverted repeat segments bracketing Us (c and c′). The message from the ICP4 genes is also shown. It is present in the c sequences and therefore is diploid.

allowed for the targeted manipulation of the HSV genome. This facilitated structure function studies on individual gene products. The later revealed gene structure, primary protein structure, and evolutionary relationships. McGeoch *et al.* (1986) compared the sequence of HSV-1 strain 17 to that of the VZV ICP4 analogue, IE62 (McGeoch *et al.*, 1986). Despite the complete lack of nucleotide sequence conservation between the two viruses, several regions of amino acid similarity were noted, implying that these were important regions for protein function. Fig. 2.2 is a representation of this type of analysis aligning and comparing the deduced amino acid sequence for HSV-1 strain 17 ICP4 (McGeoch *et al.*, 1986) and the ICP4 analogues from 6 other alphaherpesvirus, HSV-2 (McGeoch *et al.*, 1991), VZV (Davison and Scott, 1985), PRV (Ou *et al.*, 2002), EHV (Grundy *et al.*, 1989), BHV, and MDV (Afonso *et al.*, 2001). The primary sequence of the analogues ranges from 1298 amino acids for HSV-1 strain 17 to 1488 amino acids for EHV. Unless indicated, the residue numbers sited from here on will be relative to the HSV-1 sequence. The regions of amino acid conservation are considerably less widespread than those observed when HSV-1 and VZV were compared. There are two broad regions of similarity, one between 300 and 500 and another between 800 and 1200. In between these regions there is very little similarity. In fact there are large gaps in the alignment between similar regions, suggesting that the different parts of the molecule are not structurally linked in a precise way. The alignment in Fig. 2.2 suggests that regions outside of the two blocks of conservation are not important for function. We know from genetic studies that this is not generally the case. It could be that the ICP4 analogues possess regions that are species or cell type specific, so that these regions would only be conserved in subsets of these viruses. The regions of broad conservation seen in Fig. 2.2 would represent core functions of interactions of ICP4. For example, the region between 300 and 500 is the DNA binding/dimerization region (see below).

In order to examine the regions that are specifically important for HSV and its related viruses to function, the amino acids sequences of HSV-1, HSV-2 and B virus ICP4 were compared (Fig. 2.3). In this case the HSV-1 sequence is from strain KOS. It is 95% homologous to strain 17, but is 4 residues smaller (1294 vs. 1298). The nucleotide sequence will be published elsewhere. As would be expected there are the two conserved regions indicated in Fig. 2.2, however there are additional regions of conservation when only these three viruses are compared. There is an additional region of similarity extending from residue 1200 to the end of the molecule. There is a curious region of considerable similarity between amino acid 550 and 700. On the broader alignment in Fig. 2.2, this region was highly divergent, with large gaps in the alignment between the analogues. In both alignments, the amino terminal 300 amino acids are the most divergent. However, in the alignment of the primate simplex viruses (Fig. 2.3), discrete blocks of conservation are revealed. These blocks are present around amino acids 20, 90, and 200. There is also a serine, acidic amino acid block between amino acids 175 and 215 that is conserved. Interestingly, this region bares similarity to a cellular transcriptional coactivator called p15 or PC4 (Meisterernst *et al.*, 1991).

The conserved regions of ICP4 represent regions of the molecule that specify activities or interactions that are important for its function(s). The remainder of this discussion about the structure, function and genetics of ICP4 will be with respect to the conserved regions. To facilitate the discussion, the alignments of Figs. 2.2 and 2.3 are generally summarized in Fig. 2.4. The black rectangles represent primary sequence that is highly conserved in the broader alphaherpesvirus alignment (Fig. 2.2). The dark and light grey rectangles represent sequences that are highly and loosely conserved in the alignment of the three primate simplex viruses, respectively (Fig. 2.3).

Recombinant techniques have also allowed for more detailed genetic analysis. There are two general approaches that have been used to study viral regulatory proteins: (i) transient transfection assays to examine activity *in trans*, and (ii) the construction and phenotypic analysis of mutant viruses. ICP4 has been extensively studied by both approaches. ICP4 expressed from a transfected expression plasmid, activates expression of a reporter construct driven by a variety of

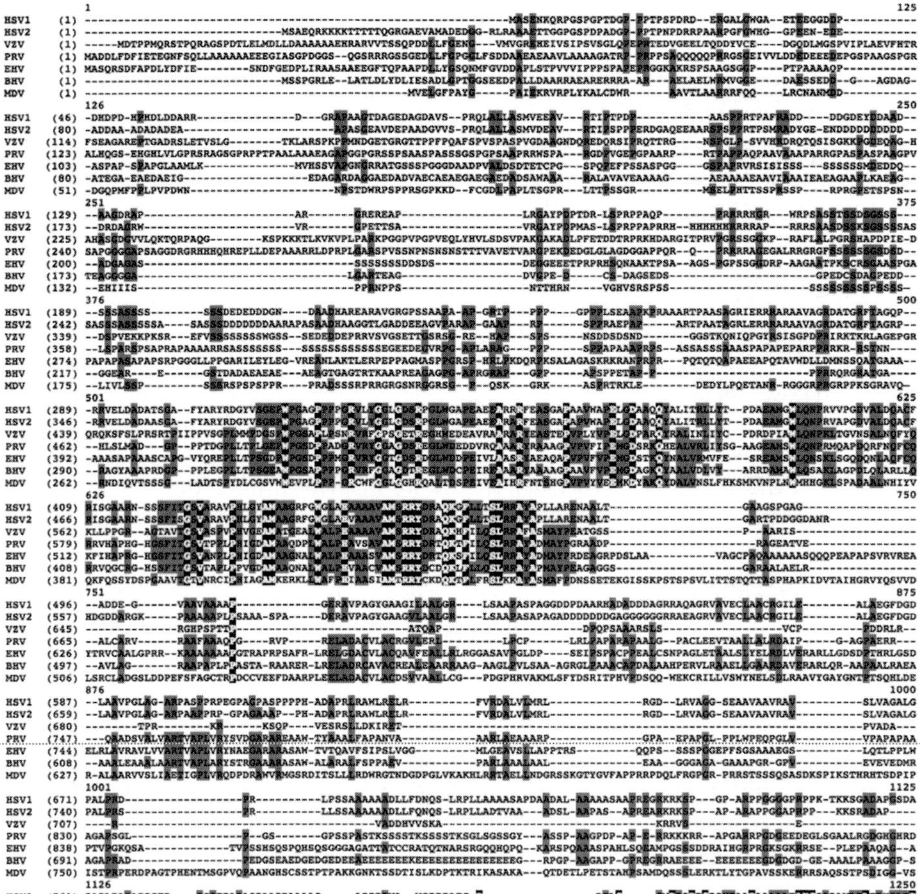

Figure 2.2 Comparison of the primary protein structures of the ICP4 homologues from alphaherpesviruses. Alignment of the indicated sequences was performed with Vector NTI. Residue numbers are shown relative to the sequence of each virus. Gaps in the sequences are indicated by dashes. Residues that are conserved in all the viruses are highlighted in black. Residues conserved in the majority of the viruses are highlighted in grey.

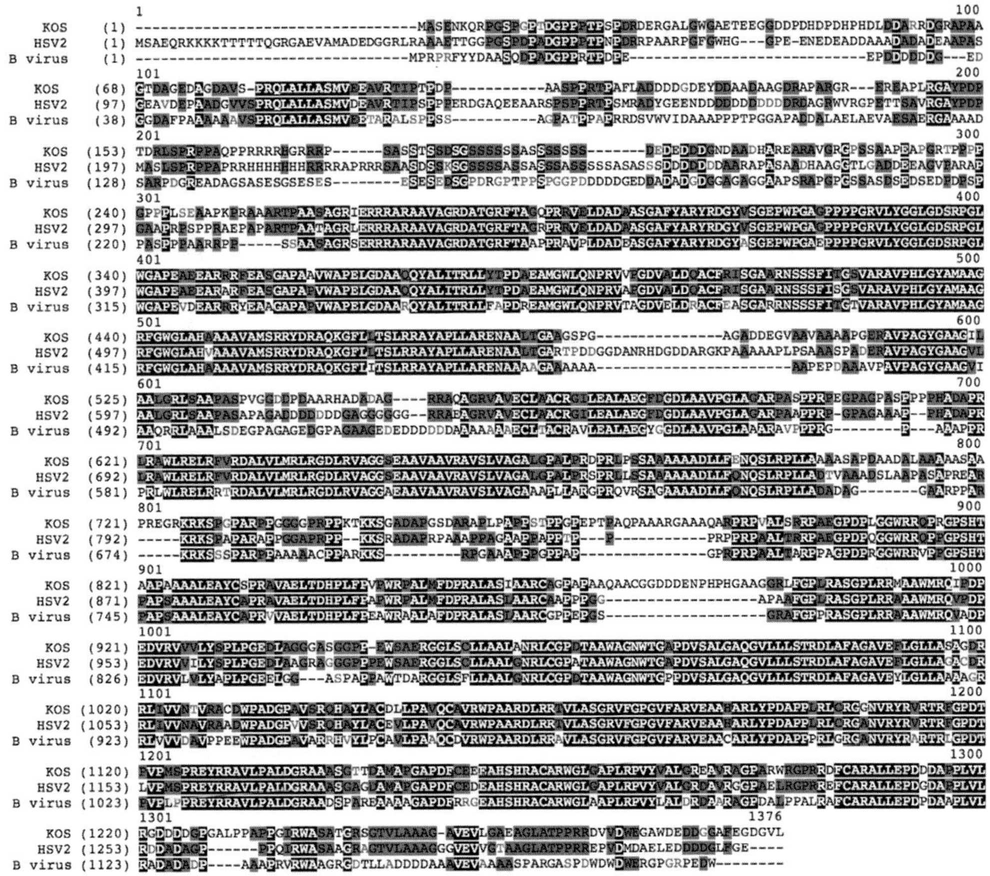

Figure 2.3 Comparison of the primary protein structures of the ICP4 homologues from HSV-1 (strain KOS), HSV-2, and B virus. Alignment was performed and is displayed as in Fig. 2.2.

promoters, and repress expression from its own promoter (DeLuca and Schaffer, 1985; Everett, 1984b; Gelman and Silverstein, 1985; O'Hare and Hayward, 1985a,b). Several studies have also used similar transfection assays to analyse mutant version of ICP4 (DeLuca and Schaffer, 1985, 1987; Paterson and Everett, 1988a; Shepard *et al.*, 1989). Since these genetic studies performed by the transient transfection approach are consistent with those conducted in the context of the virus, this review will describe results obtained with experiments conducted in the course of virus infection.

The targeted manipulation of essential genes requires an approach to provide the corresponding function *in trans*. Cell lines that are permissive hosts for HSV and are genetically transformed with a wild-type copy of the targeted gene provided the means to construct and propagate most engineered mutants in essential genes. ICP4 was the first essential gene of HSV to be studied in this way (DeLuca *et al.*, 1985). Many studies (Bruce and Wilcox, 2002; Compel and DeLuca, 2003; DeLuca and Schaffer, 1987, 1988; Paterson and Everett, 1988a,b; Shepard *et al.*, 1989; Xiao *et al.*, 1997) have shown that ICP4 has a modular like composition similar to other transcription activators (Keegan *et al.*, 1986; Ptashne, 1988). It has a region that directs the protein to DNA, or a DNA binding domain, and regions that are also needed in addition to the DNA binding domain for transcriptional activation, or repression. Fig. 2.4 shows how these domains are situated relative to each other on the primary sequence of ICP4.

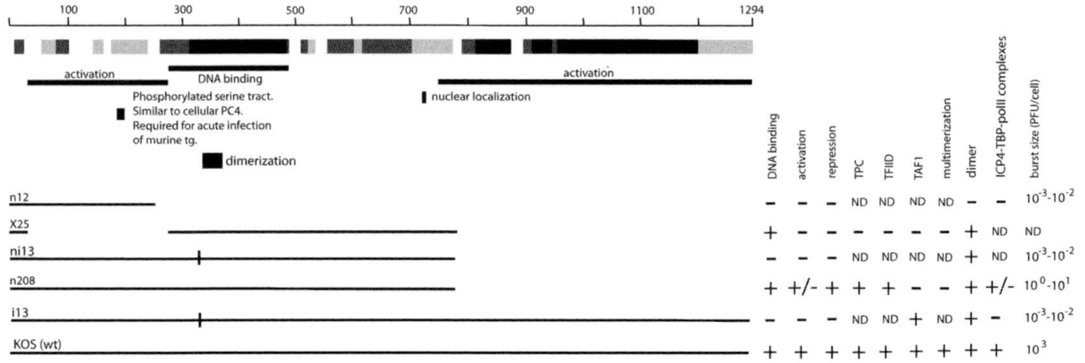

	DNA binding	activation	repression	TPC	TFIID	TAF1	multimerization	dimer	ICP4-TBP-polII complexes	burst size (PFU/cell)
n12	–	–	–	ND	ND	ND	ND	–	–	10^{-3}-10^{-2}
X25	+	–	–	–	–	–	–	+	ND	ND
ni13	–	–	–	ND	ND	ND	ND	+	ND	10^{-3}-10^{-2}
n208	+	+/-	+	+	+	–	–	+	+/-	10^{0}-10^{1}
i13	–	–	–	ND	ND	+	ND	+	–	10^{-3}-10^{-2}
KOS (wt)	+	+	+	+	+	+	+	+	+	10^{3}

Figure 2.4 Domains of ICP4. The key for sequence conservation (black, light and dark grey) as well as the domains are described in the text. Several ICP4 proteins are shown with respect to their ability to bind to DNA, activate transcription, repress transcription, form a tripartite complex with TBP/TFIIB on DNA (TPC), interact with TFIID and TAF1 in solution, multimerize on DNA, form dimers, and form transcription preinitiation complexes *in vitro* and *in vivo* (ICP4–TBP-polII complex), and the burst size of the virus expressing the indicated protein (PFU/cell).

A region described as a sequence responsible for ICP4 dimerization is also indicated (Gallinari *et al.*, 1994).

While many mutants have been constructed and analysed Fig. 2.4, diagrams 5 mutants that that roughly define the known activities of ICP4. With the exception of X25, all are expressed from both ICP4 loci on viral genomes. X25 is transdominant mutant (Shepard *et al.*, 1990), making the construction of a mutant virus difficult. It is expressed from a transformed cell line. It exists as a dimer and binds to DNA. However, it does not activate E and L genes and does not repress IE genes. Therefore it lacks domains that are involved in interactions necessary for transcription regulation in addition to the DNA binding domain.

The virus n12 expresses the amino-terminal 251 amino acids (DeLuca and Schaffer, 1987, 1988). It is lacking the large regions of conservation with the other alphaherpesviruses, does not bind to DNA and does not activate or repress transcription. The yield of n12 on non-complementing cells is 10^{-5}–10^{-6} that of wt virus. It is questionable weather this represents true yield or simply a small fraction of the inoculum.

The virus n208 expresses the amino-terminal 774 amino acids of the ICP4 protein (DeLuca and Schaffer, 1987, 1988). It lacks a large region of considerable conservation with the other alphaherpesviruses. Interestingly, the yield of this virus on non-complementing cells is only reduced 10^{2}- to 10^{3}-fold relative to wt virus. n208 retains the ability to repress transcription of the ICP4 gene (DeLuca and Schaffer, 1988; Gu *et al.*, 1995) and efficiently activates E genes, but is somewhat defective in the activation of L genes. The restriction on late gene expression is not absolute since there is some virus production. However, the phenotype of n208 suggests that there are interactions specified by the C-terminal 520 amino acids that are crucial for late gene expression (see below). ts mutants mapping within this 520 amino acid interval exist that are considerably less permissive for gene expression at the non-permissive temperature than n208 (DeLuca *et al.*, 1984; Dixon and Schaffer, 1980; Preston, 1981), suggesting that such mutations have a more global effect on structure, affecting the activity of the transactivation region in the amino terminus of the molecule. However, ts mutants and engineered mutants in this region exist (Compel and DeLuca, 2003; DeLuca *et al.*, 1984) that have a phenotype similar to n208, suggesting that there are activities specified by this region that function independently or in addition to those of the amino terminus to activate transcription.

A number of studies have examined the consequences of more subtle changes (amino acid) in the block of conservation between amino acids 300 and 500 (Davison *et al.*, 1984; Paterson and

Everett, 1988b; Paterson *et al.*, 1990; Shepard *et al.*, 1989). These changes most often reduce the ability of ICP4 to bind to DNA and activate or repress transcription. Fig. 2.4 shows one such mutant, i13, which is a two amino acid insertion at amino acid 338 (Shepard *et al.*, 1989). The insertion completely abolishes the activity of the n208 and full length molecules. The insertion does not affect the ability of the molecule to dimerize (Shepard and DeLuca, 1991). Mutations in the region between amino acids 410 and 490, where the greatest extent of conservation exists (Fig. 2.2), have the greatest effects on DNA binding (Davison *et al.*, 1984; Shepard *et al.*, 1989). From all these studies, it is generally accepted that the ability of ICP4 to bind to DNA is required to activate viral E and L genes and repress transcription of its own promoter. The specific nature of this requirement is not as well defined and is discussed below.

The modular structure of ICP4 is further substantiated by complementation studies. For example X25, which can bind to DNA but not regulate transcription, and i13, which cannot bind to DNA or regulate transcription, can dimerize, bind to DNA and activate early genes (Shepard and DeLuca, 1991). As during infection with n208, late genes are not activated. Earlier examples of intragenic complementation of ICP4 mutants by heterodimer formation have also been demonstrated (Shepard and DeLuca, 1989). In addition to complementation by heterodimer formation, transdominant inhibition of wt ICP4 by the X25 molecule has also been demonstrated (Shepard *et al.*, 1990). wt HSV forms about 1% of the number of plaques on X25-expressing Vero cells than on non-transformed Vero cells. The wild-type monomer of wt ICP4 is sequestered into heterodimers with X25. The monomeric C-terminal region in the heterodimer was shown to be altered in conformation relative to the wt homodimer. This is the region lacking in n208. Consistent with this observation, late genes are more poorly expressed in wt-virus infected X25 cells. When the X25 gene was inserted into transgenic mice, the resulting mice where more resistant to infection by HSV-1 (Smith and DeLuca, 1992).

In summary, genetic studies have revealed that ICP4 (i) is modular in composition, (ii) has DNA binding and dimerization domains that are necessary but not sufficient for activation and repression of transcription, (iii) has regions that contribute to activation both N-terminal and C-terminal to the DNA binding domain, (iv) the C-terminal region is more important for the activation of late genes, and (v) functions as a dimer since activities of the ICP4 can be assembled by the dimerization of specific mutants resulting in intragenic complementation.

Biochemical properties of ICP4

Sequence and genetic analyses of ICP4 suggest that ICP4 is modular in composition. There are blocks of conservation with other alphaherpesviruses, and activation and DNA binding functions are easily separated. Therefore, these regions or domains can be described in terms of their activities and the molecules that ICP4 interacts with. This section will review what is known about the properties of ICP4 in general and also in terms of the regions of ICP4 that specify specific interactions.

ICP4 is a nuclear phosphoprotein (Courtney and Benyesh-Melnick, 1974; Pereira *et al.*, 1977). It has an apparent molecular weight of 175 kDa on SDS-PAGE. However, its primary translation product is approximately 133 kDa. The larger apparent size on SDS-PAGE is most likely due to the extensive and complicated patterns of phosphorylation (Wilcox *et al.*, 1980) and poly(ADP-ribosyl)ation (Preston and Notarianni, 1983). The extensive phosphorylation of ICP4 was substantiated by subsequent 2-dimentional gel and phosphotryptic peptide analysis (Ackermann *et al.*, 1984; Xia *et al.*, 1996a,b). The serine tract located around amino acid 160 (Figs. 2.3 and 2.4) is extensively phosphorylated. This region bares similarity to the cellular coactivator PC4 (Meisterernst *et al.*, 1991), is important for transactivation in the absence of the C-terminal 520 amino acids (Shepard *et al.*, 1989), and has been shown to be important for function in the trigeminal ganglia of mice (Bates and DeLuca, 1998; Xia *et al.*, 1996b). There is much that still needs to be investigated regarding the functions of the post-translational modifications of ICP4.

ICP4 was first shown to binding to DNA cellulose (Purifoy and Powell, 1976), eluting from single and double stranded DNA with approximately 0.3 M NaCl. This is in contrast to ICP8, which elutes from single and double stranded DNA with approximately 1.0 M and 0.3 M NaCl respectively. This binding is due to non-specific DNA protein interactions. ICP4 also has a sequence specific DNA-binding activity. It was first shown to bind to the sequence ATCGTCNNNNYCGRC (N=any base, Y=pyrimidine, R=purine) (Faber and Wilcox, 1986). This site was first described to be present near the mRNA start site of the ICP4 message itself (Faber and Wilcox, 1988; Muller, 1987), raising possibilities for mechanisms of repression of the ICP4 promoter (discussed below). The affinity of ICP4 for this site is approximately 1 nM (Kattar-Cooley and Wilcox, 1989). However, there appeared to be other sites in the promoter regulatory domains of other HSV genes to which ICP4 bound that did not obviously match the above consensus and appeared to be of lower affinity (Imbalzano *et al.*, 1990; Kristie and Roizman, 1986a,b). In one case these sites were shown not to affect the expression of the gene they were near (Imbalzano *et al.*, 1990). Later, it was shown that all these sites fit a looser consensus, RTCGTCNNYNYSG, where R is purine, Y is pyrimidine, S is C or G, and N is any base (DiDonato *et al.*, 1991). The sequence specific DNA binding activity is inherent in a bacterial produced dimer consisting of the residues between approximately 300 and 500 (Everett *et al.*, 1991; Wu and Wilcox, 1990). This activity is conserved in all the alphaherpesviruses. The essential involvement in DNA binding in gene activation and repression is discussed below.

ICP4 was first shown to be a dimer with hydrodynamic properties consistent with a very elongated molecule (Metzler and Wilcox, 1985). It probably exists in cells as an obligatory dimer. The elongated structure of ICP4 is largely determined by the amino terminal 300 amino acids (Shepard *et al.*, 1990), since small changes in this region have a greater effect on its Stokes Radius than deletion of the entire C-terminal region of conservation (n208) (Fig. 2.4). (unpublished observations). The elongated structure of ICP4, may provide the basis for ICP4 to function at a distance on the DNA from a given promoter.

While the DNA binding activity is important for ICP4 to function in activation and repression, it is not sufficient for either activity. There are regions of the molecule that are required due to their ability to interact with other molecules involved in transcription. Nuclear extracts from cells can be prepared containing the sufficient polII transcription factors to reconstitute transcription *in vitro* (Dignam *et al.*, 1983). The addition of partially purified ICP4 to such reactions resulted in increased transcription of specific promoter templates (Beard *et al.*, 1986; Pizer *et al.*, 1986). Individual polII transcription factors have been produced in bacteria, or further purified from nuclear extracts and combined with purified ICP4 and the appropriate mutant ICP4 molecules to reconstitute both ICP4-activated and repressed transcription *in vitro* (Gu and DeLuca, 1994; Gu *et al.*, 1993, 1995; Smith *et al.*, 1993). These experiments resulted in the minimum set of cellular factors that ICP4 will function with to activate and repress transcription. Using this system, epitope tagged versions of wt and n208 ICP4 were used to affinity purify cellular transcription factors from the minimized set of factors (Carrozza and DeLuca, 1996). The only component of the general transcription that interacted with ICP4 in this study was TFIID. The component of TFIID most likely responsible for the interaction was TAF1. Furthermore, n208 did not interact with TAF1 or TFIID, whereas i13 interacted with both despite its inability to activate transcription. There, ICP4 interacts with TFIID through its TAF1 component as a function of the C-terminal 520 amino acids (Fig. 2.4), and this may be part of the basis for the full activity of wt ICP4. More recently, using a virus expressing a TAP tagged version of ICP4, we have observed that ICP4 interacts with TFIID and the mediator complex in the context of viral infection (Lester and DeLuca, 2009, unpublished observations). It will be interesting to see what regions of ICP4 specify these interactions.

The above interactions can occur in solution, not requiring DNA. Two additional interactions that possibly share interacting interfaces occur between ICP4 and the cellular transcription machinery. Both interactions require the DNA binding domain, DNA, and the amino-terminal

274 amino acids of ICP4. Neither requires the domain described above that interacts with TFIID through TAF1. The first is a more specific interaction. ICP4 quantitatively supershifts a TBP/TFIIB complex on the TATA box in a gel shift experiment provided a specific ICP4 binding site is located in a specific position with respect to the TATA box (Gu *et al.*, 1995; Kuddus *et al.*, 1995; Smith *et al.*, 1993). This tripartite complex (TPC, Fig. 2.4) is required for a mechanism of repression and is described below. The second interaction is similar to the first with respect to requirements except it does not require a specific ICP4 binding site. Again, the DNA binding domain of ICP4 is required. When HSV promoter templates are immobilized on magnetic beads and incubated with cellular transcription factors, TFIID, and subsequently polII will bind to the template as a function of the TATA box to form a preinitiation complex. Purified, ICP4 proteins possessing an intact DNA binding domain and the amino-terminal 274 amino acids will facilitate the recruitment of TFIID and subsequently polII to the promoter as a function of the TATA box (Grondin and DeLuca, 2000). This activity is consistent with the role of ICP4 in activation (discussed below).

Another biochemical property that may possibly affect the ability of ICP4 to regulate transcription is the ability to multimerize on DNA. One ICP4 dimer can form a discrete complex with DNA in a mobility shift experiment. It was noticed that multiple protein DNA complexes were formed with the DNA probe so long as the ICP4 contained the C-terminal 520 amino acids. Mutants that do not contain this region only formed one protein DNA complex (Kuddus and DeLuca, 2007). It was determined that the additional protein DNA complexes formed when the C-terminal region was present represented multimerized dimers on the DNA. Importantly, the ability to multimerize on the DNA increased the affinity of ICP4 for the DNA. The increased affinity of ICP4 for DNA afforded by the C-terminal region may allow for the occupancy of late promoters after the onset of DNA replication. The molecule n208 does not form the replication compartments (Quinlan *et al.*, 1984) characteristic of wt ICP4 (DeLuca and Schaffer, 1988). Wt ICP4 also binds to incoming genomes forming discrete foci at ND10 structures referred to as viral nucleoprotein complexes (Everett *et al.*, 2003). The ability of n208 ICP4 to form nuclear foci is less efficient than wild type, and the formation of replication compartment precursors called prereplicative sites is impaired (Livingston *et al.*, 2008). As previously stated n208 does activate early genes, but its inability to concentrate on the DNA may limit its ability to activate transcription.

In summary, the known biochemical activities of ICP4 include (Fig 2.4) (i) the ability to bind specifically to DNA as a function of a region between amino acids 300 and 500, (ii) binding to TFIID through the TAF1 subunit as a function of the 520 C-terminal amino acids, (iii) the ability to form a tripartite complex on DNA requiring the DNA binding domain and the N-terminal 274 amino acids, (iv) the facilitation of preinitiation complex formation on promoters requiring DNA binding and the 274 N-terminal amino acids, and (v) the ability to interact with TFIID and mediator in the context of viral infection.

Mechanism of ICP4 function

PolII transcription

As previously stated, HSV genes are transcribed by the RNA polymerase II transcription machinery of the cell. All HSV genes contain a TATA box located about 25 base pairs upstream of the mRNA start site. The proteins involved in the transcription process can be grouped into three different classes of factors; the general transcription factors (GTFs), gene specific regulators, and coactivators. Since ICP4 functions with these factors, the following is a brief review of the polII transcription machinery.

The basal factors are the minimal factors that will direct the accurately initiated transcription. These factors have been defined by biochemical fractionation and reconstitution of transcription in *in vitro* reactions. They are TFIIA, B, D, E, F, H, and RNA polII (Reinberg and Roeder, 1987; Sawadogo and Roeder, 1985a). With the exception of TFIIB, all of these factors are comprised of multiple proteins. Of these factors, TFIID was thought to provide for the initial promoter recognition (Sawadogo and Roeder, 1985b).

It contains the TATA-binding protein (TBP), which recognizes the TATA box and by itself can substitute for TFIID to promote basal transcription (Buratowski, 1995). TFIID also contains approximately 14 other proteins, called TBP-associated factors or TAFs (Martinez, 2002) (Zhou *et al.*, 1992). TAFs are generally thought to interact with gene specific regulators in order to repress or activate basal transcription (Buratowski, 1995). They also provide the means to recognize down-stream elements such as the initiator or DPE element elements present near to or downstream of the start sites of some eukaryotic genes and HSV late genes (Burke and Kadonaga, 1997). The binding of TFIID to the core promoter element is thought to serve as a nucleation site for the remainder of the GTFs and polII. In addition, one of the TAFs, TAF1, has several enzymatic activities, including ubiquitin conjugating, kinase, and HAT activities (Naar *et al.*, 2001). There are TBP containing complexes that differ in their TAF compositions, TBP-free TAF complexes (TFTC), and acetyltransferase complexes (see below) that contain some of the TAFs found in TFIID and some additional TAFs. It is thought that different core promoter recognition complexes are present in different cell types, vary as a function of differentiation, and respond to environmental stimuli to activate different promoter elements (Muller *et al.*, 2007). HSV E and L promoters possess different core promoter structure. One intriguing hypothesis is that ICP4 interacts with different core promoter recognition complexes to differentially activate E and L genes.

Sp1 and VP16 are examples of the second group of transcription factors mentioned above. Members of this group of transcription factors can act as both activators and repressors of transcription. In the case of Sp1, it binds to specific upstream sites in promoters and contacts the GTFs through TAF130 or TAF4 (Hoey *et al.*, 1993; Tanese *et al.*, 1996) to activate Sp1 binding site containing promoters.

The third group of factors are the coactivators. The most universal of these is referred to as Mediator complexes. First identified in yeast through genetic and biochemical analysis, mediator complexes were found to be reversibly associated with large subunit of RNA polII (Thompson *et al.*, 1993), and essential for mediating high levels of transcriptional activation by specific activators in *in vitro* reconstituted experiments (Kim *et al.*, 1994). These complexes are composed of approximately 20 proteins serving in both coactivator and corepressor roles (Hampsey, 1998; Myers and Kornberg, 2000). Subsequently various human versions of the yeast Mediator were discovered and purified (hMediator, CRSP, PC2, SMCC/TRAP, etc) through interactions with different activators (reviewed in Malik and Roeder, 2005; Rachez and Freedman, 2001). These complexes ranged in size due to variations in subunit composition suggesting that either different subcomplexes exist or that differences in purification strategies yielded alternative complexes (Conaway *et al.*, 2005). Mediator complexes in both yeast and metazoans are thought to play a coactivator function similar to TAFs in TFIID in that they act as a bridge between activators and repressors bound at promoter elements to modulate interactions with RNA polII and affect its function. Recently adenovirus E1A was found to interact with a mediator component and that this interaction facilitated activation (Boyer *et al.*, 1999). Interestingly, HSV immediate early and early promoters contain binding sites for some of the activators that have been shown to interact with mediator components such as the cellular activator Sp1 found in both IE and E promoters and the viral activator VP16 found only in IE promoters. Sp1 requires the CRSP co-activator complex in addition to TFIID for efficient activation (Ryu and Tjian, 1999; Ryu *et al.*, 1999).

Nucleosome remodelling complexes and HAT coactivator complexes are mainly involved in genes that are regulated or affected by nucleosome structure. Lytic replication of HSV is not thought to involve nucleosomal structure (Leinbach and Summers, 1980). However, recently it was found that the presence of the VP16 activation domain is responsible for the presence of acetylated histones on IE promoters (Herrera and Triezenberg, 2004). Therefore, while it seems unlikely that there is substantial classic nucleosomal structure on replicating HSV genomes, it remains possible that some histone involvement may be significant.

Repression of transcription by ICP4

The phenotypic analysis of temperature-sensitive (ts) mutants in the complementation group that corresponded to the gene for ICP4 resulted in the conclusion that ICP4 was 'autoregulatory'. At the non-permissive temperature, not only was the expression of viral early and late genes dramatically reduced, IE genes were overexpressed. Temperature shift-up experiments with ICP4 ts mutants resulted in the resumption of ICP4 protein synthesis (Courtney and Benyesh-Melnick, 1974; Dixon and Schaffer, 1980; Preston, 1979a; Watson and Clements, 1978). These trends were also observed with observations of mRNA abundance. Later, it was reported that the transcription rates of IE genes were elevated in ICP4 ts mutant infected-cells (Godowski and Knipe, 1986). These observations support the general conclusion that ICP4 represses the transcription of IE genes by a general mechanism.

ICP4 represses its own transcription by binding to a relatively strong ICP4 binding site near the start site of ICP4 transcription. There is a binding site for ICP4, the core of which is the sequence ATCGTC, spanning the start site of ICP4 transcription (Faber and Wilcox, 1988; Muller, 1987). In transient transfection assays, viral infection, and in reconstituted *in vitro* transcription reactions, ICP4 represses its own transcription in a binding site dependant manner (DeLuca and Schaffer, 1985; Gu *et al.*, 1993; Michael and Roizman, 1993; O'Hare and Hayward, 1985b; Rivera-Gonzalez *et al.*, 1994; Roberts *et al.*, 1988). In the context of viral infection, it was demonstrated that repression of the ICP4 gene by ICP4 is conditioned by the distance and stereoaxial alignment of the ICP4 DNA binding site relative to the TATA box (Leopardi *et al.*, 1995). This was also shown to be the case in *in vitro* transcription reactions (Gu *et al.*, 1995; Kuddus *et al.*, 1995). These later studies suggest that the mechanism of repression is more complicated than ICP4 simply blocking or impeding RNA polII.

Biochemical studies revealed that the mechanism by which ICP4 represses transcription involves the formation of a tripartite complex (TPC) on repressed promoters involving TBP or TFIID, TFIIB and ICP4 (Smith *et al.*, 1993). This is consistent with several findings:

1. Both repression and TPC formation depend upon the correct relative direction of the ICP4 binding site and the TATA box (Gu *et al.*, 1995), since ICP4 and TBP bind in the major and minor grove of B DNA, both repression and TPC formation are optimized when the sites are n+1/2 turns of the helix away (Kuddus *et al.*, 1995).
2. Both TPC formation and repression diminish with increasing distance between the ICP4 binding site and the TATA box (Kuddus *et al.*, 1995).
3. Mutants of ICP4 that can bind DNA but not form the TPC cannot repress transcription (Kuddus *et al.*, 1995; Smith *et al.*, 1993). ICP4 does not block transcription, but does not allow other activators to activate transcription (Gu *et al.*, 1995; Gu *et al.*, 1993). This is diagrammed in Fig. 2.5. The ICP4 promoter is activated by cellular transcription factors that bind to the region upstream of the ICP4 TATA box as well as by the virion transactivator VP16 (Campbell *et al.*, 1984; Cordingley *et al.*, 1983; Mackem and Roizman, 1982). ICP4 does not repress transcription mediated by TBP or TFIID alone, or transcription mediated solely by the TATA box, demonstrating that it does not block basal transcription (Gu *et al.*, 1993). Therefore, the formation of the TPC precludes the function of activators. Interestingly, the DNAse I footprint of TFIID, TFIIB, and ICP4 bound DNA resembles that of TBP, TFIIB, and ICP4 bound DNA (Smith *et al.*, 1993). The DNAse I footprints of TFIID/TFIIB and TBP/TFIIB DNA are different with that of the former extending downstream of the TATA box. In other words the formation of the TPC results in a footprint for TFIID that more resembles that of TBP, the later of which cannot participate in activator mediated transcription. This supports the hypothesis that a mechanism by which ICP4 represses its own transcription involves the formation of the TPC, which in turn alters the conformation of TFIID such that

it cannot communicate with activators or co-activators, or that it masks such interactions (Fig. 2.5). The result of which is that transcription is repressed to the level one would observe for basal transcription.

The mechanism described above is probably one way by which ICP4 represses its own transcription. The details of that mechanism are very specific, requiring precisely located ICP4 binding sites with respect to the TATA box. None of the other IE genes have an ICP4 binding site located as it is in the ICP4 promoter. However there are two other HSV promoters that do contain such sites; one is the latency associated transcript (LAT) and the other is for orfP (Batchelor and O'Hare, 1990) or L/STs (Bohenzky *et al.*, 1993; Yeh and Schaffer, 1993). LAT is the most abundant RNA found in latently infected ganglia (Stevens *et al.*, 1987). It is a stable intron that is spliced from a larger primary transcript, the majority of which is not stable (Farrell *et al.*, 1991; Wagner *et al.*, 1988b). The primary transcript is derived from a TATA box-containing promoter that has an ICP4 binding positioned similarly to the ICP4 promoter (Batchelor and O'Hare, 1990; Dobson *et al.*, 1989). Unlike ICP4 it is expressed late in productive virus infection (Wagner *et al.*, 1988a). Mutation of the ICP4 binding site changes the kinetics of expression from late to early suggesting that DNA replication can overcome the repression by ICP4 (Rivera-Gonzalez *et al.*, 1994). The L/STs may be transcripts for the orfP gene (Lagunoff and Roizman, 1995). This promoter is abundantly expressed in the absence of ICP4 (Lagunoff and Roizman, 1995) and is very effectively repressed by ICP4, which forms a tripartite complex with TBP/TFIID on the orfP promoter (Gu *et al.*, 1995). Therefore, both LAT and the orfP or L/STs promoters are relatively

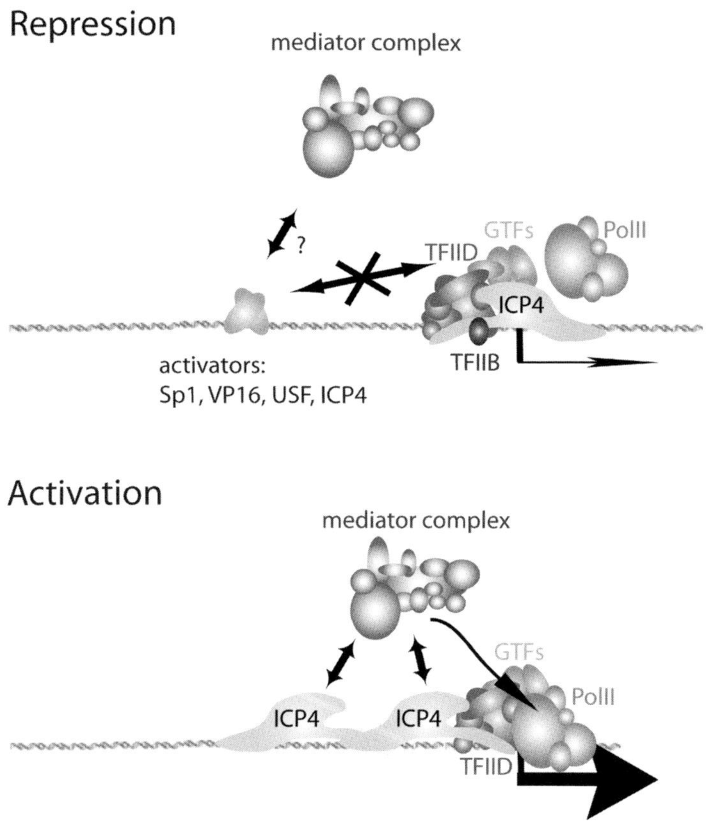

Figure 2.5 Model for ICP4 mediated repression and activation of polII transcription.

active in the absence of ICP4, suggesting a role for these genes *in situ*ations where ICP4 is not generally present, such as during latency. The roles of the LAT and orfP genes are under considerable study, but to date are not uniformly agreed on. More recently, micro RNAs have been found in productive and latent infection that map to the region of the genome, which would be consistent with their synthesis from the LAT and/or OrfP promoters. These miRNAs target ICP0, gamma 34.5, and most interestingly ICP4 (Umbach *et al.*, 2008). The biological significance of these observations remains to be elucidated. However, it is intriguing to note that ICP4 and specific miRNAs could regulate each other, possibly to control the balance between the latent and productive state of gene expression.

The transcripts corresponding to the five IE genes all accumulate to greater abundance in the absence of ICP4 activity, due to an increase in their transcription rates (DeLuca and Schaffer, 1988; Godowski and Knipe, 1986). ICP4 binding sites of varying affinity can be found at many places in the HSV genome (DiDonato *et al.*, 1991). The mechanism by which ICP4 represses its own, LAT and OrfP transcription requires a properly positioned ICP4 binding site as described above. While the remaining IE genes all contain ICP4 binding sites of varying affinity, the position of these sites is not consistent with the mechanism of repression described above. An interesting case is that of the ICP0 promoter. A relatively strong ICP4 binding site is present within the ICP0 promoter and ICP4 will repress this promoter as a function of this site in transient assays (Resnick *et al.*, 1989). However, the presence of this site does not affect the transcription of the ICP0 gene in the context of viral infection (Everett and Orr, 1991). In transient contransfection assays the relative levels of ICP4 to transcription template are different from that in virus infection and this may explain the difference in outcomes. With the tripartite complex mechanism described above, the association of ICP4 with the template is not simply a function of the ICP4–DNA interactions, but also a function of protein–protein interactions, and the added interactions of these proteins with DNA (TFIID). One might expect this mechanism to function at lower ICP4 to genome ratios than a mechanism that solely involved ICP4–DNA interactions. It remains possible that ICP4 may repress transcription as a direct function of DNA binding by a mechanism that does not involve a properly positioned ICP4 binding site or TPC as described above. We have also shown that ICP4 can multimerize on DNA as a function of ICP4 dimer–dimer interactions requiring a domain in the C-terminal one third of the protein. It is conceivable that under certain circumstances of infection the concentration of ICP4 may be such that multimerization of ICP4 at a particular genomic locus may repress transcription at that locus.

There also appears to be a more general mechanism by which the transcription of IE genes is attenuated as a function of ICP4 activity. The main difference between IE and E genes is the presence of the TAATGARAT elements in the IE promoters, which render them responsive to activation by VP16 in the incoming virion. Both IE and E promoters contain TATA boxes and upstream binding sites for cellular activators, such as Sp1. In the absence of the TAATGARAT elements, IE promoters (without ICP4 binding sites at the transcription start site) are qualitatively expressed in a similar manner to E promoters in that they are induced by ICP4 and subsequently shut off. Therefore, the main mechanism by which IE genes are shut off may be the same as that for E genes. ICP4 is required but may or may not be directly involved. ICP4 induces the transcription of many early and late viral genes. The activities of these gene products may affect the activities of components of the cellular transcription machinery. The abundance, modification state, location and hence activities of cellular factors will certainly be affected during the course of infection. The abundance of TFIIA, a general transcription factor associated with activated transcription may be reduced (Zabierowski and DeLuca, 2004). Sp1 is phosphorylated and its activity altered during infection (Kim and DeLuca, 2002). It is interesting to note that the most HSV promoters active late after infection only contain a TATA box and an initiator element, a combination that ICP4 effectively activates (Gu and DeLuca, 1994). Therefore, it may be that as infection progresses the activity of the cellular transcription machinery may less effectively support transcription that is aided by activators such as Sp1 and VP16 found in both E and/or IE promoters, respectively, thereby attenuating transcription from these promoters. The effects of infection on the activities of specific cellular transcription complexes/factors requires further study.

Activation of transcription by ICP4

During virus infection ICP4 activates the transcription of most early and late genes. While all HSV promoters contain a TATA box, early promoters also contain binding sites for cellular activators such as Sp1 upstream of the TATA box (Coen *et al.*, 1986). Late promoters are those mostly transcribed after viral DNA replication and often contain an initiator element (INR) located at the start site of transcription. The following section summarizes the biological and biochemical studies that relate to the mechanism by which ICP4 activates polII transcription.

One of the more puzzling aspects regarding the mechanism of ICP4 activation is the requirement for DNA binding. The DNA binding property of ICP4 as measured by the ability to bind to specific sites referred to above, correlates with the ability to activate transcription. Mutations that inactivate specific DNA binding also render the molecule unable to activate transcription (Paterson and Everett, 1988b; Shepard *et al.*, 1989). While ICP4 binding sites can contribute to activation by ICP4 *in vitro* (Tedder *et al.*, 1989), the same sites are not required for activated transcription in the context of viral infection (Smiley *et al.*, 1992). *In vitro* transcription studies on the gC promoter showed that ICP4 can activate transcription in the absence of specific binding sites (Gu and DeLuca, 1994). Further studies were unable to demonstrate activation- or viral-specific sequences within the gD and tk promoters (Coen *et al.*, 1986; Everett, 1984a). More recent chromatin immunoprecipitation (ChIP) studies conducted on virus infected cells demonstrate that ICP4 binds, as a function of its DNA binding domain, near the promoters of genes it activates. In so doing, it also recruits TFIID and polII to these promoters (Sampath and Deluca, 2008). It is important to note that there are not specific ICP4 binding sites present in the vicinity of the sequences to which ICP4 was found to locate. As mentioned above, ICP4 does have the ability to multimerize on DNA and this property increases its affinity for DNA (Kuddus and DeLuca, 2007). In addition, ICP4 was found to bind to the genome very early in infection forming a type of viral nucleoprotein (Everett *et al.*, 2003). Given these observations, it is possible that ICP4 does not need to bind to the higher affinity sites people have typically studied by gel shift analysis, but rather its binding to DNA in the process of activation is less discriminate, or possibly non-specific.

Biochemical studies on ICP4 have revealed a number of insights into the mechanism by which ICP4 activates early and late genes. ICP4-mediated regulation of transcription has been reconstituted *in vitro* with chromatographically separated and purified human and/or bacterially produced transcription factors (Carrozza and DeLuca, 1996; Gu and DeLuca, 1994; Gu *et al.*, 1993). In the process of chromatographically simplifying the transcription system that still allowed ICP4 to activate transcription of the gC promoter, it became clear that ICP4 could activate transcription with a relatively simple set of transcription factors (Carrozza and DeLuca, 1996). This is likely due to the size and complexity of ICP4. ICP4 contains a sequence that is similar to a cellular coactivator, PC4 (Meisterernst *et al.*, 1991). It interacts with TFIID trough the TAF1 subunit as a function of the C-terminal 520 amino acids (Fig. 2.4). It also interacts with TBP/TFIIB (Smith *et al.*, 1993) and promotes the formation of transcription preinitiation complexes (Grondin and DeLuca, 2000) independent of the C-terminal 520 amino acids but as a function of the amino terminal transactivation region (Fig. 2.4). Its affinity for DNA is increased by the ability to multimerize on DNA, another function of the C-terminal 520 amino acids (Kuddus and DeLuca, 2007). Lastly, we have recently found by the analysis of TAP purified ICP4 that ICP4 not only interacts with TFIID, but also with the mediator complex in virus infected cells (Lester and DeLuca, 2009, unpublished). The ability of ICP4 to interact with multiple cellular factors that are crucial for polII transcription, and its ability to bind and multimerize on DNA support a general model by which ICP4 can activate polII promoters (Fig. 2.5). Fig. 2.5 shows ICP4 binding upstream of the start site, but it is also possible that it can function downstream of the start site as well. It may also be that ICP4 concentrates on the viral genome in the vicinity of promoters it activates. During infection ICP4 was found near the start site of tk transcription and the integrity of the TATA box increased the abundance of ICP4 near the promoter (Sampath and Deluca, 2008). This suggests that ICP4 may bind cooperatively to the promoter with TFIID rather than generally binding all over the genome.

The ability of ICP4 to efficiently activate the gC promoter in the absence of TFIIA is largely due to the presence of the initiator element at the start site of transcription (Cook *et al.*, 1995; Gu and DeLuca, 1994; Kim *et al.*, 2002). Inr elements are present at the start site of many late gene promoters. Mutation of the Inr can be compensated for by addition of TFIIA. This is the situation with the tk promoter (Zabierowski and DeLuca, 2004; Zabierowski and Deluca, 2008). This suggests that the mechanism by which ICP4 activates early and late genes may be slightly different. The observation that an ICP4 molecule (n208) lacking the C-terminal 520 amino acids can activate early genes, but poorly activates late genes, further supporting this idea. TAF1 is required for recognition of the Inr by TFIID. Removal of the region of ICP4 that interacts with TAF1 (as in n208, Fig. 2.4) may preserve the ability to activate the early promoter, but decrease the ability of ICP4 to activate late genes through the Inr. In addition, while wt ICP4 recruited TFIID and polII to the tk (E) and gC(L) promoters, n208 only recruited TFIID and polII to the tk promoter during viral infection further supports this idea (Sampath and Deluca, 2008). This observation may not only be a function of the inability of the n208 ICP4 to interact with TAF1, but also due to its inability to multimerize on DNA.

In summary, ICP4 is both an activator and repressor of polII transcription. It specifically represses transcription by (Fig. 2.5, top) (i) forming a tripartite complex with TFIID and TFIIB on a promoter containing an ICP4 binding site specifically positioned relative to the TATA box, (ii) not allowing other activators to communicate with the core transcription complex thereby repressing transcription to the level observed for basal transcription, and (iii) directly or indirectly repressing other genes by an alternative mechanism(s). ICP4 activates transcription by (Fig. 2.5, bottom) (i) interacting with DNA, possibly non-specifically, (ii) multimerizing on DNA, thereby increasing its affinity for DNA, (iii) interacting with TFIID to promote transcription initiation complex formation, and (iv) interacting with mediator. These interactions explain why ICP4 is a general activator of polII transcription. Lastly, considerable evidence exist that ICP4 activates E and L promoters by subtly different mechanisms.

Activity of ICP4 *in vivo*

ICP4 is generally considered as a regulator of productive cycle transcription, and therefore it is not often studied in the context of infection of animals, or in a latency model. Indeed, impairment of its activator function would certainly attenuate the virus *in vivo*. This may be for trivial reasons; if it doesn't grow well in cells because later genes are under expressed, then it certainly won't grow well *in vivo*. However, ICP4 is large, possesses multiple activities, and is extensively phosphorylated in a complicated way. It may possess functions that specifically function *in vivo*, or in regulating the balance between lytic and latent infection. The repression function may be one such activity given that ICP4 represses both LAT and the possibly the transcripts that ultimately specify miRNAs to ICP0, gamma 34.5 and ICP4 (Umbach *et al.*, 2008). The expression of LAT or L/S/Ts (OrfP) might affect the levels of latent genomes (Randall *et al.*, 2000; Sawtell and Thompson, 1992) by affecting the abundance of productive cycle activators. Any accumulation of ICP4 would repress the expression of these transcripts, diminishing the attenuating activities of these transcripts, thus favouring virus growth. It is interesting to note that the expression of ICP4 from adenovirus in latently infected ganglia resulted in reactivation (Halford *et al.*, 2001). This could be from the de novo activation of early genes on the viral genome. However, the genome is likely to be repressed by chromatin (Deshmane and Fraser, 1989), and the provision of ICP4 may not be sufficient to activate early genes repressed in this way. It is possible that the regions around the inverted repeats are partially active from a transcription standpoint (LAT). This activity may extend to ICP4 and ICP0, resulting in low levels of these proteins. The levels of these proteins would be reduced due to the possible action of LAT and miRNAs. The resulting levels may not be sufficient for transcription activities. However, if enough ICP4 accumulated or if it were exogenously provided *in trans* (from the adenovirus), it would attenuate the expression of LAT or L/STs, resulting in the increased abundance of ICP0 and ICP4. ICP0 has been shown to relieve repression of genes repressed by epigenetic mechanisms (Cliffe and Knipe, 2008; Ferenczy and DeLuca, 2009), rendering more susceptible to activation by the viral transactivators, including ICP4 (Terry-Allison *et al.*, 2007).

There is also the possibility that there is an activation function of ICP4 that is more important for the virus life cycle in the neurons of sensory ganglia. The transcriptional environment of the sympathetic neuron most likely differs from that in the susceptible cells at the periphery where the virus initially replicates, or in tissue culture. Yet the transcription of viral IE, E and L genes is necessary for the life cycle of the virus in neurons, as are the activation function(s) of ICP4. ICP4 may have evolved to activate and repress viral genes in different cellular environments. Therefore, it is possible that there are functions or domains specifying activities that are crucial for function in neurons, which are dispensable, or redundant in non-neuronal cells. One region of ICP4 has been demonstrated to possibly function in this manner. There is a region rich in serine and charged residues between 170 and 210 (Figs. 2.3 and 2.4) that is dispensable for growth in culture (Paterson and Everett, 1990) or at the periphery of infection in the mouse eye model, but is absolutely required for growth in the trigeminal ganglia (Bates and DeLuca, 1998; Xia *et al.*, 1996b). The mutant virus grew on the cornea, and was transported back to the ganglia, but was defective in early and late gene expression and did not grow. Instead, it directly established latency. The region that is deleted in the mutant viruses (the serine tract) is conserved in the ICP4 analogues of alphaherpesviruses (Fig 2.3). This is the region that has similarity to a cellular coactivator, PC4. It is tempting to speculate that this region of ICP4 provides coactivator function that is dispensable in non-neuronal cells, but required in neuronal cells.

Alternatively, ICP4 may participate in multiple interactions with the cellular transcription machinery that together comprise its activator functions, and some of these interactions may be functionally redundant in non-neuronal cells. This would imply that there are possibly other regions of ICP4 that are also crucial for function in neuronal cells. A possible clue comes from the observation that there are other domains within the amino terminal domains of ICP4 that are conserved but dispensable in non-neuronal tissue culture cells, but are very defective for activation in the absence of the C-terminal activation region. It will be interesting to examine the phenotype of such mutants in the mouse model or in cultured primary neuronal cells.

Conclusions

ICP4 is essential for the expression of most of the HSV genome, and therefore is absolutely essential for viral growth. It activates early and late gene transcription possibly by different mechanisms. ICP4 also represses the transcription of several genes, including itself. The multiple mechanisms by which ICP4 effects transcription involve interactions with DNA and different components of the general polII transcription machinery. ICP4 is a large and structurally complex molecule possessing discrete domains that confer the interactions important for its function. Given its size and the observations that many HSV gene products have multiple functions, it is highly likely that ICP4 interacts with many other cellular and viral gene products to carry out its functions. It is also likely that it is involved in other processes aside from transcriptional regulation.

Studies thus far have focused on interactions with transcription factors. Most of the biochemical studies have been performed with transcription factors from uninfected cells. It is well known that HSV infection affects the abundance, activities and modification state of many cellular proteins. It is also known that different core promoter recognition complexes (i.e. TFIID) exist in cells to allow for the recognition of different promoters, which become active as a response to environmental stimuli or during different stages in differentiation (Muller *et al.*, 2007). It is possible that HSV changes the general polII transcription machinery as infection proceeds, necessitating different interactions between ICP4 and the core promoter recognition complexes present to allow for the observed temporal cascade of viral gene expression.

The repression function of ICP4 is less likely to have a significant effect on acute viral replication, particularly in cell culture. Its ability to repress LAT, itself, and the L/STs suggest a possible role in some aspect of its *in vivo* life cycle, or during latency. The observation that there are conserved domains of ICP4 that are specifically important for growth in the trigeminal ganglia of mice suggests an *in vivo* specific role for ICP4. The nature or this requirement is unknown. This activity may affect the repression function of ICP4 such that these transcripts are dysregulated and the balance between viral activators and cellular repressors is affected. It may also be an activation function that is redundant in most cell types, but crucial for growth in neurons. ICP4 was also shown to be an inner tegument protein of the virion (Yao and Courtney, 1989). Perhaps in neurons the ICP4 associated with the nucleocapsid is more important for establishing infection following transport to the nucleus of the sensory neuron.

In summary, genetic and biochemical studies have taught us a great deal about the essential functions of ICP4 in the regulation of HSV gene expression. However, there remains much to learn about this complicated regulatory protein from a biochemical standpoint and from the standpoint of its role in pathogenesis.

References

Ackermann, M., Braun, D.K., Pereira, L., and Roizman, B. (1984). Characterization of herpes simplex virus 1 alpha proteins 0, 4, and 27 with monoclonal antibodies. J. Virol. 52, 108–118.

Afonso, C.L., Tulman, E.R., Lu, Z., Zsak, L., Rock, D.L., and Kutish, G.F. (2001). The genome of turkey herpesvirus. J. Virol. 75, 971–978.

Alwine, J.C., Steinhart, W.L., and Hill, C.W. (1974). Transcription of herpes simplex type 1 DNA in nuclei isolated from infected HEp-2 and KB cells. Virology *60*, 302–307.

Batchelor, A.H., and O'Hare, P. (1990). Regulation and cell-type-specific activity of a promoter located upstream of the latency-associated transcript of herpes simplex virus type 1. J. Virol. *64*, 3269–3279.

Bates, P.A., and DeLuca, N.A. (1998). The polyserine tract of herpes simplex virus ICP4 is required for normal viral gene expression and growth in murine trigeminal ganglia. J. Virol. 72, 7115–7124.

Beard, P., Faber, S., Wilcox, K.W., and Pizer, L.I. (1986). Herpes simplex virus immediate early infected-cell polypeptide 4 binds to DNA and promotes transcription. Proc. Natl. Acad. Sci. U.S.A. *83*, 4016–4020.

Bohenzky, R.A., Papavassiliou, A.G., Gelman, I.H., and Silverstein, S. (1993). Identification of a promoter mapping within the reiterated sequences that flank the herpes simplex virus type 1 UL region. J. Virol. *67*, 632–642.

Boyer, T.G., Martin, M.E., Lees, E., Ricciardi, R.P., and Berk, A.J. (1999). Mammalian Srb/Mediator complex is targeted by adenovirus E1A protein. Nature *399*, 276–279.

Bruce, J.W., and Wilcox, K.W. (2002). Identification of a motif in the C terminus of herpes simplex virus regulatory protein ICP4 that contributes to activation of transcription. J. Virol. *76*, 195–207.

Buratowski, S. (1995). Mechanisms of gene activation. Science *270*, 1773–1774.

Burke, T.W., and Kadonaga, J.T. (1997). The downstream core promoter element, DPE, is conserved from Drosophila to humans and is recognized by TAFII60 of Drosophila. Genes Dev. *11*, 3020–3031.

Campbell, M.E., Palfreyman, J.W., and Preston, C.M. (1984). Identification of herpes simplex virus DNA sequences which encode a trans-acting polypeptide responsible for stimulation of immediate early transcription. J. Mol. Biol. *180*, 1–19.

Carrozza, M.J., and DeLuca, N.A. (1996). Interaction of the viral activator protein ICP4 with TFIID through TAF250. Mol. Cell Biol. *16*, 3085–3093.

Cliffe, A.R., and Knipe, D.M. (2008). Herpes simplex virus ICP0 promotes both histone removal and acetylation on viral DNA during lytic infection. J. Virol. *82*, 12030–12038.

Coen, D.M., Weinheimer, S.P., and McKnight, S.L. (1986). A genetic approach to promoter recognition during trans induction of viral gene expression. Science *234*, 53–59.

Compel, P., and DeLuca, N.A. (2003). Temperature-dependent conformational changes in herpes simplex virus ICP4 that affect transcription activation. J. Virol. *77*, 3257–3268.

Conaway, R.C., Sato, S., Tomomori-Sato, C., Yao, T., and Conaway, J.W. (2005). The mammalian Mediator complex and its role in transcriptional regulation. Trends Biochem. Sci. *30*, 250.

Cook, W.J., Lin, S.M., DeLuca, N.A., and Coen, D.M. (1995). Initiator elements and regulated expression of the herpes simplex virus thymidine kinase gene. J. Virol. *69*, 7291–7294.

Cordingley, M.G., Campbell, M.E., and Preston, C.M. (1983). Functional analysis of a herpes simplex virus type 1 promoter: identification of far-upstream regulatory sequences. Nucleic Acids Res. *11*, 2347–2365.

Courtney, R.J., and Benyesh-Melnick, M. (1974). Isolation and characterization of a large molecular-weight polypeptide of herpes simplex virus type 1. Virology *62*, 539–551.

Davison, A.J., and Scott, J.E. (1985). DNA sequence of the major inverted repeat in the varicella-zoster virus genome. J. Gen. Virol. *66*, 207–220.

Davison, M.J., Preston, V.G., and McGeoch, D.J. (1984). Determination of the sequence alteration in the DNA of the herpes simplex virus type 1 temperature-sensitive mutant ts K. J. Gen. Virol. *65*, 859–863.

DeLuca, N.A., Courtney, M.A., and Schaffer, P.A. (1984). Temperature-sensitive mutants in herpes simplex virus type 1 ICP4 permissive for early gene expression. J. Virol. *52*, 767–776.

DeLuca, N.A., McCarthy, A.M., and Schaffer, P.A. (1985). Isolation and characterization of deletion mutants of herpes simplex virus type 1 in the gene encoding immediate-early regulatory protein ICP4. J. Virol. *56*, 558–570.

DeLuca, N.A., and Schaffer, P.A. (1985). Activation of immediate-early, early, and late promoters by temperature-sensitive and wild-type forms of herpes simplex virus type 1 protein ICP4. Mol. Cell Biol. *5*, 1997–1208.

DeLuca, N.A., and Schaffer, P.A. (1987). Activities of herpes simplex virus type 1 (HSV-1) ICP4 genes specifying nonsense peptides. Nucleic Acids Res. *15*, 4491–4511.

DeLuca, N.A., and Schaffer, P.A. (1988). Physical and functional domains of the herpes simplex virus transcriptional regulatory protein ICP4. J. Virol. *62*, 732–743.

Deshmane, S.L., and Fraser, N.W. (1989). During latency, herpes simplex virus type 1 DNA is associated with nucleosomes in a chromatin structure. J. Virol. *63*, 943–947.

DiDonato, J.A., Spitzner, J.R., and Muller, M.T. (1991). A predictive model for DNA recognition by the herpes simplex virus protein ICP4. J. Mol. Biol. *219*, 451–470.

Dignam, J.D., Lebovitz, R.M., and Roeder, R.G. (1983). Accurate transcription initiation by RNA polymerase II in a soluble extract from isolated mammalian nuclei. Nucleic Acids Res. *11*, 1475–1489.

Dixon, R.A., and Schaffer, P.A. (1980). Fine-structure mapping and functional analysis of temperature-sensitive mutants in the gene encoding the herpes simplex virus type 1 immediate early protein VP175. J. Virol. *36*, 189–203.

Dobson, A.T., Sederati, F., Devi-Rao, G., Flanagan, W.M., Farrell, M.J., Stevens, J.G., Wagner, E.K., and Feldman, L.T. (1989). Identification of the latency-associated transcript promoter by expression of rabbit beta-globin mRNA in mouse sensory nerve ganglia latently infected with a recombinant herpes simplex virus. J. Virol. *63*, 3844–3851.

Everett, R.D. (1984a). A detailed analysis of an HSV-1 early promoter: sequences involved in trans-activation by viral immediate-early gene products are not early-gene specific. Nucleic Acids Res. *12*, 3037–3056.

Everett, R.D. (1984b). Trans activation of transcription by herpes virus products: requirement for two HSV-1 immediate-early polypeptides for maximum activity. EMBO J. *3*, 3135–3141.

Everett, R.D., Elliott, M., Hope, G., and Orr, A. (1991). Purification of the DNA binding domain of herpes simplex virus type 1 immediate-early protein Vmw175 as a homodimer and extensive mutagenesis of its DNA recognition site. Nucleic Acids Res. *19*, 4901–4908.

Everett, R.D., and Orr, A. (1991). The Vmw175 binding site in the IE-1 promoter has no apparent role in the expression of Vmw110 during herpes simplex virus type 1 infection. Virology *180*, 509–517.

Everett, R.D., Sourvinos, G., and Orr, A. (2003). Recruitment of herpes simplex virus type 1 transcriptional regulatory protein ICP4 into foci juxtaposed to ND10 in live, infected cells. J. Virol. *77*, 3680–3689.

Faber, S.W., and Wilcox, K.W. (1986). Association of the herpes simplex virus regulatory protein ICP4 with specific nucleotide sequences in DNA. Nucleic Acids Res. *14*, 6067–6083.

Faber, S.W., and Wilcox, K.W. (1988). Association of herpes simplex virus regulatory protein ICP4 with sequences spanning the ICP4 gene transcription initiation site. Nucleic Acids Res. *16*, 555–570.

Farrell, M.J., Dobson, A.T., and Feldman, L.T. (1991). Herpes simplex virus latency-associated transcript is a stable intron. Proc. Natl. Acad. Sci. U.S.A. *88*, 790–794.

Ferenczy, M.W., and DeLuca, N.A. (2009). Epigenetic modulation of gene expression from quiescent herpes simplex virus genomes. J. Virol. *83*, 8514–8524.

Gallinari, P., Wiebauer, K., Nardi, M.C., and Jiricny, J. (1994). Localization of a 34-amino-acid segment implicated in dimerization of the herpes simplex virus type 1 ICP4 polypeptide by a dimerization trap. J. Virol. *68*, 3809–3820.

Gelman, I.H., and Silverstein, S. (1985). Identification of immediate early genes from herpes simplex virus that transactivate the virus thymidine kinase gene. Proc. Natl. Acad. Sci. U.S.A. *82*, 5265–5269.

Godowski, P.J., and Knipe, D.M. (1986). Transcriptional control of herpesvirus gene expression: gene functions required for positive and negative regulation. Proc. Natl. Acad. Sci. U.S.A. *83*, 256–260.

Grondin, B., and DeLuca, N. (2000). Herpes simplex virus type 1 ICP4 promotes transcription preinitiation complex formation by enhancing the binding of TFIID to DNA. J. Virol. *74*, 11504–11510.

Grundy, F.J., Baumann, R.P., and O'Callaghan, D.J. (1989). DNA sequence and comparative analyses of the equine herpesvirus type 1 immediate early gene. Virology *172*, 223–236.

Gu, B., and DeLuca, N. (1994). Requirements for activation of the herpes simplex virus glycoprotein C promoter *in vitro* by the viral regulatory protein ICP4. J. Virol. *68*, 7953–7965.

Gu, B., Kuddus, R., and DeLuca, N.A. (1995). Repression of activator-mediated transcription by herpes simplex virus ICP4 via a mechanism involving interactions with the basal transcription factors TATA-binding protein and TFIIB. Mol. Cell Biol. *15*, 3618–3626.

Gu, B., Rivera-Gonzalez, R., Smith, C.A., and DeLuca, N.A. (1993). Herpes simplex virus infected cell polypeptide 4 preferentially represses Sp1-activated over basal transcription from its own promoter. Proc. Natl. Acad. Sci. U.S.A. *90*, 9528–9532.

Halford, W.P., Kemp, C.D., Isler, J.A., Davido, D.J., and Schaffer, P.A. (2001). ICP0, ICP4, or VP16 expressed from adenovirus vectors induces reactivation of latent herpes simplex virus type 1 in primary cultures. of latently infected trigeminal ganglion cells. J. Virol. *75*, 6143–6153.

Hampsey, M. (1998). Molecular genetics of the RNA polymerase II general transcriptional machinery. Microbiol. Mol. Biol. Rev. *62*, 465–503.

Herrera, F.J., and Triezenberg, S.J. (2004). VP16-dependent association of chromatin-modifying coactivators and underrepresentation of histones at immediate-early gene promoters during herpes simplex virus infection. J. Virol. *78*, 9689–9696.

Hoey, T., Weinzierl, R.O., Gill, G., Chen, J.L., Dynlacht, B.D., and Tjian, R. (1993). Molecular cloning and functional analysis of Drosophila TAF110 reveal properties expected of coactivators. Cell *72*, 247–260.

Honess, R.W., and Roizman, B. (1974). Regulation of herpesvirus macromolecular synthesis. I. Cascade regulation of the synthesis of three groups of viral proteins. J. Virol. *14*, 8–19.

Honess, R.W., and Roizman, B. (1975). Regulation of herpesvirus macromolecular synthesis: sequential transition of polypeptide synthesis requires. functional viral polypeptides. Proc. Natl. Acad. Sci. U.S.A. *72*, 1276–1280.

Imbalzano, A.N., Shepard, A.A., and DeLuca, N.A. (1990). Functional relevance of specific interactions between herpes simplex virus type 1 ICP4 and sequences from the promoter-regulatory domain of the viral thymidine kinase gene. J. Virol. *64*, 2620–2631.

Kattar-Cooley, P., and Wilcox, K.W. (1989). Characterization of the DNA-binding properties of herpes simplex virus regulatory protein ICP4. J. Virol. *63*, 696–704.

Keegan, L., Gill, G., and Ptashne, M. (1986). Separation of DNA binding from the transcription-activating function of a eukaryotic regulatory protein. Science *231*, 699–704.

Kim, D.B., and DeLuca, N.A. (2002). Phosphorylation of transcription factor Sp1 during herpes simplex virus type 1 infection. J. Virol. *76*, 6473–6479.

Kim, D.B., Zabierowski, S., and DeLuca, N.A. (2002). The initiator element in a herpes simplex virus type 1 late-gene promoter enhances activation by ICP4, resulting in abundant late-gene expression. J. Virol. *76*, 1548–1558.

Kim, Y.J., Bjorklund, S., Li, Y., Sayre, M.H., and Kornberg, R.D. (1994). A multiprotein mediator of transcriptional activation and its interaction with the C-terminal repeat domain of RNA polymerase II. Cell *77*, 599–608.

Kristie, T.M., and Roizman, B. (1986a). Alpha 4, the major regulatory protein of herpes simplex virus type 1, is stably and specifically associated with promoter-regulatory domains of alpha genes and of selected other. viral genes. Proc. Natl. Acad. Sci. U.S.A. *83*, 3218–3222.

Kristie, T.M., and Roizman, B. (1986b). DNA-binding site of major regulatory protein alpha 4 specifically associated with promoter-regulatory domains of alpha genes of herpes simplex virus type 1. Proc. Natl. Acad. Sci. U.S.A. *83*, 4700–4704.

Kuddus, R., Gu, B., and DeLuca, N.A. (1995). Relationship between TATA-binding protein and herpes simplex virus type 1 ICP4 DNA-binding sites in complex formation and repression of transcription. J. Virol. *69*, 5568–5575.

Kuddus, R.H., and DeLuca, N.A. (2007). DNA-dependent oligomerization of herpes simplex virus type 1 regulatory protein ICP4. J. Virol. *81*, 9230–9237.

Lagunoff, M., and Roizman, B. (1995). The regulation of synthesis and properties of the protein product of open reading frame P of the herpes simplex virus 1 genome. J. Virol. *69*, 3615–3623.

Leinbach, S.S., and Summers, W.C. (1980). The structure of herpes simplex virus type 1 DNA as probed by micrococcal nuclease digestion. J. Gen. Virol. *51*, 45–59.

Leopardi, R., Michael, N., and Roizman, B. (1995). Repression of the herpes simplex virus 1 alpha 4 gene by its gene product (ICP4) within the context of the viral genome is conditioned by the distance and stereoaxial alignment of the ICP4 DNA binding site relative to the TATA box. J. Virol. *69*, 3042–3048.

Livingston, C.M., DeLuca, N.A., Wilkinson, D.E., and Weller, S.K. (2008). Oligomerization of ICP4 and rearrangement of heat shock proteins may be important for herpes simplex virus type 1 prereplicative site formation. J. Virol. *82*, 6324–6336.

Mackem, S., and Roizman, B. (1982). Structural features. of the herpes simplex virus alpha gene 4, 0, and 27 promoter-regulatory sequences which confer alpha regulation on chimeric thymidine kinase genes. J. Virol. *44*, 939–949.

Malik, S., and Roeder, R.G. (2005). Dynamic regulation of pol II transcription by the mammalian Mediator complex. Trends Biochem. Sci. *30*, 256.

Martinez, E. (2002). Multi-protein complexes in eukaryotic gene transcription. Plant Mol. Biol. *50*, 925–947.

McGeoch, D.J., Cunningham, C., McIntyre, G., and Dolan, A. (1991). Comparative sequence analysis of the long repeat regions and adjoining parts of the long unique regions in the genomes of herpes simplex viruses types 1 and 2. J. Gen. Virol. *72*, 3057–3075.

McGeoch, D.J., Dalrymple, M.A., Davison, A.J., Dolan, A., Frame, M.C., McNab, D., Perry, L.J., Scott, J.E., and Taylor, P. (1988). The complete DNA sequence of the long unique region in the genome of herpes simplex virus type 1. J. Gen. Virol. *69*, 1531–1574.

McGeoch, D.J., Dolan, A., Donald, S., and Brauer, D.H. (1986). Complete DNA sequence of the short repeat region in the genome of herpes simplex virus type 1. Nucleic Acids Res. *14*, 1727–1745.

Meisterernst, M., Roy, A.L., Lieu, H.M., and Roeder, R.G. (1991). Activation of class II gene transcription by regulatory factors is potentiated by a novel activity. Cell *66*, 981–993.

Metzler, D.W., and Wilcox, K.W. (1985). Isolation of herpes simplex virus regulatory protein ICP4 as a homodimeric complex. J. Virol. *55*, 329–337.

Michael, N., and Roizman, B. (1993). Repression of the herpes simplex virus 1 alpha 4 gene by its gene product occurs within the context of the viral genome and is associated with all three identified cognate sites. Proc. Natl. Acad. Sci. U.S.A. *90*, 2286–2290.

Morse, L.S., Pereira, L., Roizman, B., and Schaffer, P.A. (1978). Anatomy of herpes simplex virus (HSV) DNA. X. Mapping of viral genes by analysis of polypeptides and functions specified by HSV-1 X HSV-2 recombinants. J. Virol. *26*, 389–410.

Muller, F., Demeny, M.A., and Tora, L. (2007). New problems in RNA polymerase II transcription initiation: matching the diversity of core promoters with a variety of promoter recognition factors. J. Biol. Chem. *282*, 14685–14689.

Muller, M.T. (1987). Binding of the herpes simplex virus immediate-early gene product ICP4 to its own transcription start site. J. Virol. *61*, 858–865.

Myers, L.C., and Kornberg, R.D. (2000). Mediator of transcriptional regulation. Annu. Rev. Biochem. *69*, 729–749.

Naar, A.M., Lemon, B.D., and Tjian, R. (2001). Transcriptional coactivator complexes. Annu. Rev. Biochem. *70*, 475–501.

O'Hare, P., and Hayward, G.S. (1985a). Evidence for a direct role for both the 175, 000- and 110, 000-molecular-weight immediate-early proteins of herpes simplex virus in the transactivation of delayed-early promoters. J. Virol. *53*, 751–760.

O'Hare, P., and Hayward, G.S. (1985b). Three trans-acting regulatory proteins of herpes simplex virus modulate immediate-early gene expression in a pathway involving positive and negative feedback regulation. J. Virol. *56*, 723–733.

Ou, C.J., Wong, M.L., and Chang, T.J. (2002). A TEF-1-element is required for activation of the promoter of pseudorabies virus glycoprotein X gene by IE180. Virus Genes *25*, 241–253.

Paterson, T., and Everett, R.D. (1988a). Mutational dissection of the HSV-1 immediate-early protein Vmw175 involved in transcriptional transactivation and repression. Virology *166*, 186–196.

Paterson, T., and Everett, R.D. (1988b). The regions of the herpes simplex virus type 1 immediate early protein Vmw175 required for site specific DNA binding closely correspond to those involved in transcriptional regulation. Nucleic Acids Res. *16*, 11005–11025.

Paterson, T., and Everett, R.D. (1990). A prominent serine-rich region in Vmw175, the major transcriptional regulator protein of herpes simplex virus type 1, is not essential for virus growth in tissue culture. J. Gen. Virol. *71*, 1775–1783.

Paterson, T., Preston, V.G., and Everett, R.D. (1990). A mutant of herpes simplex virus type 1 immediate early polypeptide Vmw175 binds to the cap site of its own promoter *in vitro* but fails to autoregulate *in vivo*. J. Gen. Virol. *71*, 851–861.

Pereira, L., Wolff, M.H., Fenwick, M., and Roizman, B. (1977). Regulation of herpesvirus macromolecular synthesis. V. Properties of alpha polypeptides made in HSV-1 and HSV-2 infected cells. Virology *77*, 733–749.

Pizer, L.I., Tedder, D.G., Betz, J.L., Wilcox, K.W., and Beard, P. (1986). Regulation of transcription *in vitro* from herpes simplex virus genes. J. Virol. *60*, 950–959.

Post, L.E., Conley, A.J., Mocarski, E.S., and Roizman, B. (1980). Cloning of reiterated and nonreiterated herpes simplex virus 1 sequences as BamHI fragments. Proc. Natl. Acad. Sci. U.S.A. *77*, 4201–4205.

Preston, C.M. (1979a). Abnormal properties of an immediate early polypeptide in cells infected with the herpes simplex virus type 1 mutant tsK. J. Virol. *32*, 357–369.

Preston, C.M. (1979b). Control of herpes simplex virus type 1 mRNA synthesis in cells infected with wild-type virus or the temperature-sensitive mutant tsK. J. Virol. *29*, 275–284.

Preston, C.M., and Notarianni, E.L. (1983). Poly(ADP-ribosyl)ation of a herpes simplex virus immediate early polypeptide. Virology *131*, 492–501.

Preston, V.G. (1981). Fine-structure mapping of herpes simplex virus type 1 temperature-sensitive mutations within the short repeat region of the genome. J. Virol. *39*, 150–161.

Preston, V.G., Davison, A.J., Marsden, H.S., Timbury, M.C., Subak-Sharpe, J.H., and Wilkie, N.M. (1978). Recombinants between herpes simplex virus types 1 and 2: analyses of genome structures. and expression of immediate early polypeptides. J. Virol. *28*, 499–517.

Ptashne, M. (1988). How eukaryotic transcriptional activators work. Nature *335*, 683–689.

Purifoy, D.J., and Powell, K.L. (1976). DNA-binding proteins induced by herpes simplex virus type 2 in HEp-2 cells. J. Virol. *19*, 717–731.

Quinlan, M.P., Chen, L.B., and Knipe, D.M. (1984). The intranuclear location of a herpes simplex virus DNA-binding protein is determined by the status of viral DNA replication. Cell *36*, 857–868.

Rachez, C., and Freedman, L.P. (2001). Mediator complexes and transcription. Curr. Opin. Cell Biol. *13*, 274–280.

Randall, G., Lagunoff, M., and Roizman, B. (2000). Herpes simplex virus 1 open reading frames O and P are not necessary for establishment of latent infection in mice. J. Virol. *74*, 9019–9027.

Reinberg, D., and Roeder, R.G. (1987). Factors involved in specific transcription by mammalian RNA polymerase II. Purification and functional analysis of initiation factors IIB and IIE. J. Biol. Chem. *262*, 3310–3321.

Resnick, J., Boyd, B.A., and Haffey, M.L. (1989). DNA binding by the herpes simplex virus type 1 ICP4 protein is necessary for efficient down regulation of the ICP0 promoter. J. Virol. *63*, 2497–2503.

Rivera-Gonzalez, R., Imbalzano, A.N., Gu, B., and Deluca, N.A. (1994). The role of ICP4 repressor activity in temporal expression of the IE-3 and latency-associated transcript promoters during HSV-1 infection. Virology *202*, 550–564.

Roberts, M.S., Boundy, A., O'Hare, P., Pizzorno, M.C., Ciufo, D.M., and Hayward, G.S. (1988). Direct correlation between a negative autoregulatory response element at the cap site of the herpes simplex virus type 1 IE175 (alpha 4) promoter and a specific binding site for the IE175 (ICP4) protein. J. Virol. *62*, 4307–4320.

Ryu, S., and Tjian, R. (1999). Purification of transcription cofactor complex CRSP. Proc. Natl. Acad. Sci. U.S.A. *96*, 7137–7142.

Ryu, S., Zhou, S., Ladurner, A.G., and Tjian, R. (1999). The transcriptional cofactor complex CRSP is required for activity of the enhancer-binding protein Sp1. Nature *397*, 446–450.

Sampath, P., and Deluca, N.A. (2008). Binding of ICP4, TATA-binding protein, and RNA polymerase II to herpes simplex virus type 1 immediate-early, early, and late promoters in virus-infected cells. J. Virol. *82*, 2339–2349.

Sawadogo, M., and Roeder, R.G. (1985a). Factors involved in specific transcription by human RNA polymerase II: analysis by a rapid and quantitative *in vitro* assay. Proc. Natl. Acad. Sci. U.S.A. *82*, 4394–4398.

Sawadogo, M., and Roeder, R.G. (1985b). Interaction of a gene-specific transcription factor with the adenovirus major late promoter upstream of the TATA box region. Cell *43*, 165–175.

Sawtell, N.M., and Thompson, R.L. (1992). Herpes simplex virus type 1 latency-associated transcription unit promotes anatomical site-dependent establishment and reactivation from latency. J. Virol. *66*, 2157–2169.

Schaffer, P.A., Aron, G.M., Biswal, N., and Benyesh-Melnick, M. (1973). Temperature-sensitive mutants of herpes simplex virus type 1: isolation, complementation and partial characterization. Virology *52*, 57–71.

Shepard, A.A., and DeLuca, N.A. (1989). Intragenic complementation among partial peptides of herpes simplex virus regulatory protein ICP4. J. Virol. *63*, 1203–1211.

Shepard, A.A., and DeLuca, N.A. (1991). Activities of heterodimers composed of DNA-binding- and trans-activation-deficient subunits of the herpes simplex virus regulatory protein ICP4. J. Virol. *65*, 299–307.

Shepard, A.A., Imbalzano, A.N., and DeLuca, N.A. (1989). Separation of primary structural components conferring autoregulation, transactivation, and DNA-binding properties to the herpes simplex virus transcriptional regulatory protein ICP4. J. Virol. *63*, 3714–3728.

Shepard, A.A., Tolentino, P., and DeLuca, N.A. (1990). trans-dominant inhibition of herpes simplex virus transcriptional regulatory protein ICP4 by heterodimer formation. J. Virol. *64*, 3916–3926.

Smiley, J.R., Johnson, D.C., Pizer, L.I., and Everett, R.D. (1992). The ICP4 binding sites in the herpes simplex virus type 1 glycoprotein D (gD) promoter are not essential for efficient gD transcription during virus infection. J. Virol. *66*, 623–631.

Smith, C.A., Bates, P., Rivera-Gonzalez, R., Gu, B., and DeLuca, N.A. (1993). ICP4, the major transcriptional regulatory protein of herpes simplex virus type 1, forms a tripartite complex with TATA-binding protein and TFIIB. J. Virol. *67*, 4676–4687.

Smith, C.A., and DeLuca, N.A. (1992). Transdominant inhibition of herpes simplex virus growth in transgenic mice. Virology *191*, 581–588.

Stevens, J.G., Wagner, E.K., Devi-Rao, G.B., Cook, M.L., and Feldman, L.T. (1987). RNA complementary to a herpesvirus alpha gene mRNA is prominent in latently infected neurons. Science *235*, 1056–1059.

Stow, N.D., Subak-Sharpe, J.H., and Wilkie, N.M. (1978). Physical mapping of herpes simplex virus type 1 mutations by marker rescue. J. Virol. *28*, 182–192.

Stow, N.D., and Wilkie, N.M. (1978). Physical mapping of temperature-sensitive mutations of herpes simplex virus type 1 by intertypic marker rescue. Virology *90*, 1–11.

Tanese, N., Saluja, D., Vassallo, M.F., Chen, J.L., and Admon, A. (1996). Molecular cloning and analysis of two subunits of the human TFIID complex: hTAFII130 and hTAFII100. Proc. Natl. Acad. Sci. U.S.A. *93*, 13611–13616.

Tedder, D.G., Everett, R.D., Wilcox, K.W., Beard, P., and Pizer, L.I. (1989). ICP4-binding sites in the promoter and coding regions of the herpes simplex virus gD gene contribute to activation of *in vitro* transcription by ICP4. J. Virol. *63*, 2510–2520.

Terry-Allison, T., Smith, C.A., and DeLuca, N.A. (2007). Relaxed repression of herpes simplex virus type 1 genomes in Murine trigeminal neurons. J. Virol. *81*, 12394–12405.

Thompson, C.M., Koleske, A.J., Chao, D.M., and Young, R.A. (1993). A multisubunit complex associated with the RNA polymerase II CTD and TATA-binding protein in yeast. Cell *73*, 1361–1375.

Umbach, J.L., Kramer, M.F., Jurak, I., Karnowski, H.W., Coen, D.M., and Cullen, B.R. (2008). MicroRNAs expressed by herpes simplex virus 1 during latent infection regulate viral mRNAs. Nature *454*, 780–783.

Wagner, E.K., Devi-Rao, G., Feldman, L.T., Dobson, A.T., Zhang, Y.F., Flanagan, W.M., and Stevens, J.G. (1988a). Physical characterization of the herpes simplex virus latency-associated transcript in neurons. J. Virol. *62*, 1194–1202.

Wagner, E.K., Flanagan, W.M., Devi-Rao, G., Zhang, Y.F., Hill, J.M., Anderson, K.P., and Stevens, J.G. (1988b). The herpes simplex virus latency-associated transcript is spliced during the latent phase of infection. J. Virol. *62*, 4577–4585.

Wagner, M.J., Sharp, J.A., and Summers, W.C. (1981). Nucleotide sequence of the thymidine kinase gene of herpes simplex virus type 1. Proc. Natl. Acad. Sci. U.S.A. *78*, 1441–1445.

Watson, R.J., and Clements, J.B. (1978). Characterization of transcription-deficient temperature-sensitive mutants of herpes simplex virus type 1. Virology *91*, 364–379.

Watson, R.J., and Clements, J.B. (1980). A herpes simplex virus type 1 function continuously required for early and late virus RNA synthesis. Nature *285*, 329–330.

Wilcox, K.W., Kohn, A., Sklyanskaya, E., and Roizman, B. (1980). Herpes simplex virus phosphoproteins. I. Phosphate cycles on and off some viral polypeptides and can alter their affinity for DNA. J. Virol. *33*, 167–182.

Wu, C.L., and Wilcox, K.W. (1990). Codons 262 to 490 from the herpes simplex virus ICP4 gene are sufficient to encode a sequence-specific DNA binding protein. Nucleic Acids Res. *18*, 531–538.

Xia, K., DeLuca, N.A., and Knipe, D.M. (1996a). Analysis of phosphorylation sites of herpes simplex virus type 1 ICP4. J. Virol. *70*, 1061–1071.

Xia, K., Knipe, D.M., and DeLuca, N.A. (1996b). Role of protein kinase A and the serine-rich region of herpes simplex virus type 1 ICP4 in viral replication. J. Virol. *70*, 1050–1060.

Xiao, W., Pizer, L.I., and Wilcox, K.W. (1997). Identification of a promoter-specific transactivation domain in the herpes simplex virus regulatory protein ICP4. J. Virol. *71*, 1757–1765.

Yao, F., and Courtney, R.J. (1989). A major transcriptional regulatory protein (ICP4) of herpes simplex virus type 1 is associated with purified virions. J. Virol. *63*, 3338–3344.

Yeh, L., and Schaffer, P.A. (1993). A novel class of transcripts expressed with late kinetics in the absence of ICP4 spans the junction between the long and short segments of the herpes simplex virus type 1 genome. J. Virol. *67*, 7373–7382.

Zabierowski, S., and DeLuca, N.A. (2004). Differential cellular requirements for activation of herpes simplex virus type 1 early (tk) and late (gC) promoters by ICP4. J. Virol. *78*, 6162–6170.

Zabierowski, S.E., and Deluca, N.A. (2008). Stabilized binding of TBP to the TATA box of herpes simplex virus type 1 early (tk) and late (gC) promoters by TFIIA and ICP4. J. Virol. *82*, 3546–3554.

Zhou, Q., Lieberman, P.M., Boyer, T.G., and Berk, A.J. (1992). Holo-TFIID supports transcriptional stimulation by diverse activators and from a TATA-less promoter. Genes Dev. *6*, 1964–1974.

The Functions and Activities of HSV-1 ICP27, a Multifunctional Regulator of Gene Expression

3

Rozanne M. Sandri-Goldin

Abstract

Herpes simplex virus 1 (HSV-1) protein ICP27 is a multifunctional regulator that is essential for HSV-1 infection. ICP27 performs a number of different functions during infection that include inhibiting cellular pre-mRNA splicing, stimulating viral early and late gene transcription by recruiting cellular RNA polymerase II to viral replication sites, binding and exporting viral RNA to the cytoplasm and stimulating translation of some HSV-1 transcripts by binding translation initiation factors. ICP27 also recruits Hsc70 to nuclear foci (VICE domains) that are enriched in chaperones and components of the proteasome, and which are believed to be involved in nuclear protein quality control. ICP27 interacts with a number of proteins and it binds RNA. Post-translational modifications have been demonstrated to regulate ICP27's interactions with several proteins. NMR analysis of the N-terminus showed that it is highly flexible, which may be necessary for switching between different protein interactions. Further, ICP27 undergoes a head-to-tail intramolecular association that may also regulate its interactions, especially with proteins that require that both the N- and C-termini of ICP27 be intact for interaction. This review will cover the different activities of ICP27 and what we know about how these activities are regulated.

Introduction

ICP27 is an immediate early protein of 512 amino acids that is required for productive infection. ICP27 regulates both cellular and viral gene expression through transcriptional and post-transcriptional activities (Sandri-Goldin, 2008). At early times after infection, beginning at about 3 hours, ICP27 mediates the inhibition of host cell splicing, which contributes to the shut off of host protein synthesis because cellular pre-mRNAs are not properly processed (Hardwicke and Sandri-Goldin, 1994; Hardy and Sandri-Goldin, 1994; Bryant *et al.*, 2001). ICP27 interacts with a number of splicing factors (Sandri-Goldin and Hibbard, 1996; Bryant *et al.*, 2001) including with members of a family of highly conserved splicing factors, termed SR proteins because they contain serine-arginine dipeptide repeats (Sciabica *et al.*, 2003). SR proteins play essential roles in spliceosomal assembly (Graveley, 2000). The activity of SR proteins is regulated by phosphorylation and dephosphorylation, such that SR proteins must be properly phosphorylated to perform their roles in spliceosome assembly (Misteli *et al.*, 1998; Prasad *et al.*, 1999). ICP27 recruits a predominantly cytoplasmic SR protein specific kinase termed SRPK1 (Zhong *et al.*, 2009) to the nucleus and this results in the aberrant phosphorylation of SR proteins (Sciabica *et al.*, 2003). As a consequence, spliceosome assembly becomes stalled at an early stage before the first step in the splicing reaction (Linberg and Kreivi, 2002; Sciabica *et al.*, 2003).

Also at early times after infection, beginning around 4 hours, ICP27 interacts with cellular RNA polymerase II (RNAP II), and specifically with the C-terminal domain (CTD) of RNAP II (Zhou and Knipe, 2001; Dai-Ju *et al.*, 2006). ICP27 recruits RNAP II to viral transcription-replication sites (Dai-Ju *et al.*, 2006), and this recruitment is necessary for elevated viral transcription. ICP27 interacts with RNAP II CTD through both its N-terminal leucine-rich region and its C-terminal zing-finger domain (Dai-Ju *et al.*, 2006). ICP27 mutants that cannot interact with RNAP II fail to recruit RNAP II to viral replication sites and early and late gene transcription is severely decreased in infections with these mutants (Dai-Ju *et al.*, 2006; Li *et al.*, 2008). Thus, it is through the recruitment of RNAP II that ICP27 stimulates early and late gene transcription.

ICP27 also participates in another aspect of HSV-1 transcription. During highly active viral transcription, a degradation of RNAP II occurs and specifically the phospho-serine-2 form of the CTD, which is the phosphorylated form found in the elongation complex (Dai-Ju *et al.*, 2006). The degradation of RNAP II occurs through ubiquitination and proteasomal degradation and can be prevented with proteasome inhibitors (Dai-Ju *et al.*, 2006). We have proposed that this occurs during robust viral transcription as a result of stalling of RNAP II elongation complexes in regions of high density on the viral genome. In addition to ICP27 recruiting RNAP II to viral replication centres, ICP27 also recruits Hsc70 to nuclear foci that lie at the periphery of the replication compartments (Li *et al.*, 2008). These foci have been dubbed VICE domains for virus-induced chaperone-enriched because they contain Hsc70, other chaperone proteins and components of the proteasome (Burch and Weller, 2004). These VICE domains appear to function as nuclear protein quality centres during HSV-1 infection (Livingston *et al.*, 2009). We have provided evidence that they play a role in the degradation of stalled RNAP II complexes because expression of a dominant negative Hsc70 mutant prevented proteasomal degradation of RNAP II during HSV-1 infection (Li *et al.*, 2008). Thus, ICP27 also plays an indirect role in nuclear protein quality control.

Later in infection, beginning at about 5 to 6 hours, ICP27 begins to shuttle between the nucleus and cytoplasm in its role an RNA export protein (Sandri-Goldin, 1998; Johnson *et al.*, 2009). ICP27 interacts with and recruits the cellular TREX complex RNA export adaptor protein Aly/REF to viral replication compartments (Koffa *et al.*, 2001; Chen *et al.*, 2002; Chen *et al.*, 2005). ICP27 also interacts with another TREX complex protein, UAP56 (Johnson, Swesey and Sandri-Goldin, unpublished results). ICP27 binds viral RNA through its RGG box RNA binding motif (Mears and Rice, 1996; Sandri-Goldin, 1998; Corbin-Lickfett *et al.*, 2009) and escorts the RNA to the cellular mRNA export receptor TAP/NXF1, to which ICP27 binds directly through both the N- and C-termini (Chen *et al.*, 2002; Chen *et al.*, 2005). The TREX complex-ICP27-RNP is then escorted by TAP/NXF1 through the nuclear pore complex to the cytoplasm for translation of the mRNA. The interaction of ICP27 with TAP/NXF1 is essential for the export of ICP27 to the cytoplasm and for viral mRNA export. Expression of a TAP/NXF1 dominant negative mutant caused retention of ICP27 in the nucleus (Chen *et al.*, 2002; Chen *et al.*, 2005). In infections with ICP27 N-terminal and C-terminal mutants that are unable to interact with TAP/NXF1, ICP27 is retained in the nucleus and viral transcripts are also retained in the nucleus (Johnson and Sandri-Goldin, 2009). Knockdown of TAP/NXF1 by siRNA also results in the retention of ICP27 and viral RNA in the nucleus (Johnson *et al.*, 2009). Conversely, during infection with an ICP27 mutant that has the RGG box RNA binding domain deleted, ICP27 is exported to the cytoplasm because this deletion does not affect its interaction with TAP/NXF1, but viral RNA accumulates in the nucleus. Thus, ICP27 must be able to bind to TAP/NXF1 and to be able to bind RNA for efficient viral mRNA export to occur.

ICP27 has also been shown to stimulate the translation of some viral mRNAs (Ellison *et al.*, 2005; Larralde *et al.*, 2006; Fontaine-Rodriquez and Knipe, 2008). ICP27 has been found to associate with translation initiation factors PABP, eIF3 and eIF4G (Fontaine-Rodriquez *et al.*, 2004; Dobrikova *et al.*, 2009). Although the mechanism by which the translational stimulation occurs has not been elucidated, it is possible that ICP27 bound to its RNA cargo upon entering the cytoplasm attracts translation initiation factors to its bound transcript. It is not known how or when ICP27 disassociates from the mRNA and this remains an important question.

A schematic of ICP27 is shown in Fig. 3.1 with the domains involved in its interactions with several interacting proteins and with RNA illustrated. A model depicting the activities of ICP27 is depicted in Fig. 3.2.

ICP27 post-translational modifications

A central question in understanding the many roles of ICP27 is how its activities and interactions are regulated. Post-translational modifications can affect protein–protein interactions. ICP27 is phosphorylated (Zhi and Sandri-Goldin, 1999; Corbin-Lickfett *et al.*, 2010; Rojas *et al.*, 2010) and is methylated on arginine residues (Mears and Rice, 1996; Souki *et al.*, 2009). We have analysed the importance of these modifications to ICP27's functional interactions.

ICP27 is phosphorylated by CK2 and protein kinase A (PKA)

ICP27 is phosphorylated during infection and phosphopeptide mapping studies revealed that it is phosphorylated on serine residues at positions 16, 18 and 114 in the N-terminal portion of the protein (Zhi and Sandri-Goldin, 1999) (Fig. 3.3). Serines at positions 16 and 18 are phosphorylated by CK2 and serine 114 is phosphorylated by PKA (Zhi and Sandri-Goldin, 1999). The serines at positions 16 and 18 are adjacent to the leucine-rich region in the N-terminus that is required for interaction with TAP/NXF1 and for ICP27 export to the cytoplasm. The serine at position 114 is within the nuclear localization signal (Mears *et al.*, 1995). To determine the role of phosphorylation in modulating the activities of ICP27, we created viral mutants bearing substitutions at each these sites (Rojas *et al.*, 2010). The serine residues were substituted with alanine to mimic the unphosphorylated protein in one set of mutants and with glutamic acid to mimic the negative charge of the phosphate group in the other set of mutants. Single, double and triple mutants were constructed and analysed (Corbin-Lickfett *et al.*, 2010; Rojas *et al.*, 2010). In one step viral curves, all of the phosphorylation site mutants replicated at levels at least two logs lower than wild type HSV-1 KOS, regardless of whether the substitution was alanine or glutamic acid and regardless of whether it was a single, double or triple mutant. DNA replication, early and late mRNA and protein levels were greatly reduced for all of the mutants. Viral replication compartment formation was aberrant with the formation of ring-like structures that never progressed into full-blown replication compartments. Further, in infections with the phosphorylation site mutants, ICP27 did not colocalize with RNAP II nor did it recruit RNAP II to viral transcription/replication sites. Hsc70 was not recruited to nuclear VICE domains, but remained diffusely distributed throughout the cytoplasm (Rojas *et al.*, 2010). In

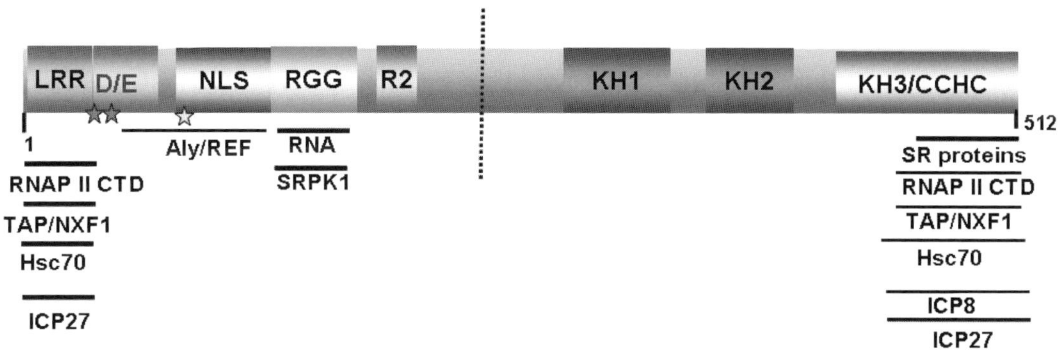

Figure 3.1 Schematic representation of ICP27. The 512 amino acid ICP27 protein is shown with the leucine-rich region (LRR), acidic region (D/E), the nuclear localization signal (NLS), the RGG box motif, a second arginine rich region (R2), three predicted KH domains and the zinc-finger region (CCHC) depicted. Interacting proteins and the regions of ICP27 with which each interacts are also shown. The three asterisks mark the phosphorylation sites.

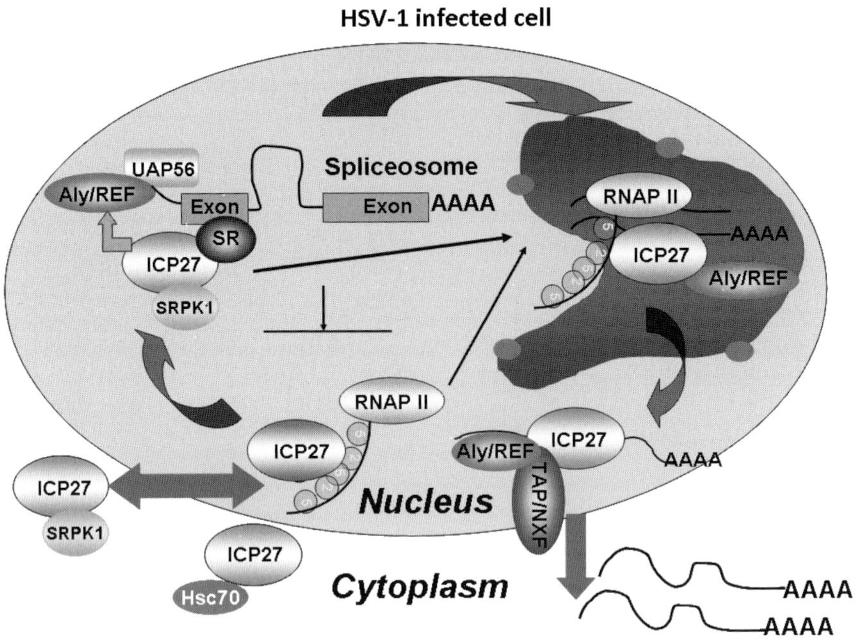

Figure 3.2 Model of the activities of ICP27. The interactions and activities of ICP27 are depicted. Cytoplasmic ICP27 is shown recruiting SRPK1 and Hsc70 to the nucleus. ICP27 is also shown interacting with the CTD of RNAP II. ICP27 recruits RNAP II and Aly/REF to viral replication compartments, which have Hsc70–containing nuclear foci or VICE domains at the periphery. ICP27 bound to RNA interacts directly with TAP/NXF1 to export RNA to the cytoplasm.

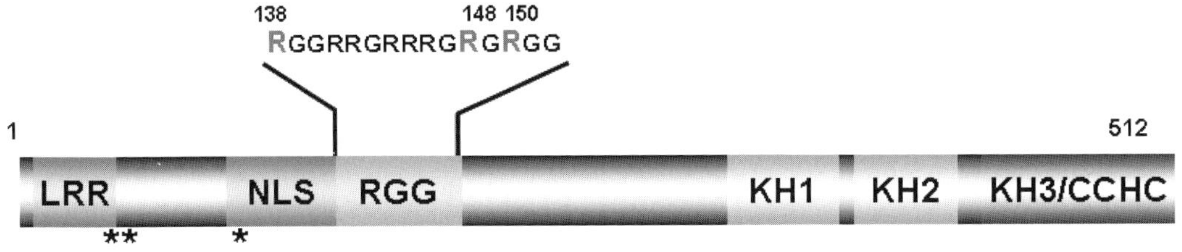

Figure 3.3 ICP27 phosphorylation sites and arginine methylation sites. The three phosphorylation sites of ICP27 are depicted by asterisks and the arginine residues within the RGG box that are methylated are shown.

infections with the phosphorylation site mutants ICP27 also did not shuttle but remained in the nucleus even at 16 hours after infection. Further, ICP27 did not interact with TAP/NXF1 (Corbin-Lickfett *et al.*, 2010), which accounts for why ICP27 was confined to the nucleus.

Phosphorylation induces local structural changes that are reversible upon dephosphorylation, thus conferring flexibility on a protein so that it can undergo different interactions. Because the phosphorylation site mutants would be locked into whatever structural change might have occurred with the substitution of alanine or glutamic acid, we looked at the solution structure of the N-terminal 160 amino acids of ICP27 for the triple alanine substitution mutant and for the triple glutamic acid substitution compared to wild type ICP27 by NMR (Corbin-Lickfett *et al.*, 2010). The N-terminal region of wild type ICP27 was found to be highly flexible, although not in a random coil. The structure was compact, yet flexible in a structure that is described as molten globular. NMR analysis of the ICP27 alanine and glutamic acid substitution mutants showed that there was a higher degree of disorder. These changes increased the flexibility of ICP27 such that association with several proteins, including RNAP II and TAP/NXF1 was impaired (Corbin-Lickfett *et al.*, 2010). Thus, reversible phosphorylation appears to be important for ICP27 to be able to associate and then dissociate from binding partners so that it can perform its many roles as a multifunctional regulator.

ICP27 is methylated on arginine residues

Methylation of arginine residues by cellular protein arginine methyltransferases (PMRT) has been shown to affect protein–protein interactions and to regulate protein export (Bedford and Clarke, 2009). Arginine methylation by PRMT1, the most abundant PRMT, occurs on glycine-arginine-rich motifs or so called GAR motifs. ICP27 encodes an RGG box motif that is required for its RNA binding and for its interaction with SRPK1 (Mears and Rice, 1996; Sandri-Goldin, 1998; Sciabica *et al.*, 2003; Corbin-Lickfett *et al.*, 2009), and ICP27 was shown to be methylated *in vivo* (Mears and Rice, 1996). To determine the sites of arginine methylation, we performed mass spectrometry on ICP27 that was immunopurified from HSV-1 KOS-infected cells (Souki *et al.*, 2009). Arginines at position 138, 148 and 150 within the RGG box were found to be the major sites of arginine methylation (Fig. 3.3). Viral mutants were constructed in which each of these arginines was substituted with lysine, which has the same positive charge but which is not a substrate for PRMTs. Single, double and triple substitutions were made to analyse the role of arginine methylation during infection. In one step growth curves, the single substitution mutants replicated at levels about 5 fold lower than wild type HSV-1 KOS and the double and triple mutants replicated about 10 fold lower (Souki *et al.*, 2009). Global transcription was decreased as measured by microarray analysis for the ICP27 arginine-to-lysine viral mutant infections and DNA replication was also reduced. Replication compartment formation was delayed, but unlike the situation with the phosphorylation site mutants that formed aberrant ring-like structures, full-blown replication compartments were seen but not until 12 hours after infection (Souki *et al.*, 2009). In accord with previous reports that showed that arginine methylation regulates protein export (Shen *et al.*, 1998; Yu *et al.*, 2004), ICP27 was exported to the cytoplasm at earlier times during infection with the arginine substitution mutants. This effect could be mimicked by using the methylation inhibitor adenosine dialdehyde (AdOx), which caused a hypomethylation of ICP27 in KOS infected cells (Souki *et al.*, 2009).

In addition to regulating ICP27 export, arginine methylation also affects ICP27 protein interactions (Souki and Sandri-Goldin, 2009). ICP27 interacts with SRPK1 thorough the RGG box motif and co-immunoprecipitation of SRPK1 with ICP27 was reduced for the arginine double and triple substitution mutants and for wild type ICP27 treated with AdOx during infection (Souki and Sandri-Goldin, 2009). Further, ICP27 recruits SRPK1 from the cytoplasm to the nucleus in its role as a mediator of splicing inhibition (Sciabica *et al.*, 2003). In cells infected with the arginine substitution mutants and in KOS-infected cells treated with AdOx, SRPK1 remained in the cytoplasm, thus indicating that the functional interaction of ICP27 with SRPK1 is regulated by arginine methylation. In addition, the TREX RNA export protein Aly/REF interacts with ICP27 through a region that includes the RGG box (Chen *et al.*, 2002). Aly/REF co-immunoprecipitation with ICP27 was reduced in infections with the arginine substitution mutants and Aly/REF did not colocalize with ICP27 and was not recruited to viral replication compartments as it is in KOS-Infected cells (Souki and Sandri-Goldin, 2009). Thus, arginine methylation regulates ICP27 export and its functional interactions with SRPK1 and Aly/REF.

RNA binding by ICP27

ICP27 binds RNA *in vitro* and *in vivo* (Mears and Rice, 1996; Ingram *et al.*, 1996; Sandri-Goldin, 1998; Sokolowski *et al.*, 2003; Corbin-Lickfett *et al.*, 2009). ICP27 encodes an RGG box RNA binding motif in the N-terminal half of the protein and three predicted KH domains in the C-terminal half of the protein (Fig. 3.1) (Soliman and Silverstein, 2000). Studies so far have shown that ICP27 RNA binding occurs through the RGG box motif. The RGG box of Fragile X Mental Retardation Protein (FMRP) has been shown to specifically recognize G-quartet or G-quadruplex RNA structures (Ramos *et al.*, 2003). G-quartets are complex tertiary structures that can readily form in DNA or RNA containing three or more consecutive guanines. Four guanosine bases can interact via Hoogsteen base–base interactions to create a planar ring, which can stack upon each other vertically with additional G-quartet planes to form a higher order structure (Siddiqui-Jain *et al.*, 2002; Cocco *et al.*, 2003). The Epstein–Barr Virus protein EBNA-1 has also been shown to bind G-quartets through an RGG box (Norseen *et al.*, 2009).

To determine the RNA binding specificity of ICP27, two approaches were used. A yeast three-hybrid screen was performed using overlapping fragments from the HSV-1 glycoprotein C (gC) gene. The gC locus was chosen because *in vivo* UV cross-linking studies showed that ICP27 bound gC mRNA (Sandri-Goldin, 1998). A positive clone was selected that maps within the gC coding sequence and which would be present in the RNA. Overlapping 30mers were designed from this 330 nucleotide sequence to narrow down the binding motif. Electrophoretic mobility shift assays (EMSA) were performed using a bacterially expressed N-terminal polypeptide of ICP27 from amino 1 to 160, which includes the RGG box (Corbin-Lickfett *et al.*, 2009). In the second approach, SELEX was performed for 10 rounds and positive sequences were also tested by EMSAs. Strong binders, weak binders and non-binders were identified among these sequences and the gC sequences. The most surprising finding was that a G-quartet substrate showed no shift in the EMSAs. There was no readily identifiable consensus sequence among the strong binders but the sequences to which ICP27 bound were GC-rich. To determine if ICP27 bound some other structure, NMR analysis was performed on RNAs to which ICP27 bound, as well as to RNAs to which it did not bind or bound weakly. This analysis revealed why ICP27 did not bind to the highly structured G-quartet RNA. The RNAs to which ICP27 bound were flexible and relatively unstructured and ICP27 did not bind to structured RNAs (Corbin-Lickfett *et al.*, 2009). That the RGG box confers the specificity of ICP27 binding was shown by performing EMSAs with the arginine to lysine triple substitution mutant compared to wild type ICP27 and compared to the phosphorylation site triple substitution mutants. The arginine to lysine substitution mutant failed to bind to RNAs that wild type ICP27 and the phosphorylation site mutants bound (Corbin-Lickfett, Souki and Sandri-Goldin, unpublished results). Further, *in situ* hybridization studies looking at poly(A+) RNA in cells infected for 8 hours with HSV-1 KOS compared to infections with the double and triple arginine to lysine substitution mutants showed that poly(A+) RNA accumulated in the nucleus in the arginine substitution mutant infections but was efficiently exported to the cytoplasm in KOS infections (Corbin-Lickfett, Souki and Sandri-Goldin, unpublished results). We conclude that ICP27 binds RNA through the RGG motif and that ICP27 specifically binds GC-rich sequences that are relatively unstructured.

We propose a model to account for how ICP27 preferentially binds viral RNA over cellular RNA, which is shown in Fig. 3.4. ICP27 binds RNAs that are GC-rich and the HSV-1 genome is 68% GC-rich, whereas the average GC content of the cellular genome is 44%. Thus, ICP27 would preferentially bind viral mRNA over cellular mRNA. There are GC-rich cellular RNAs, namely ribosomal RNAs. But these are highly structured and ICP27 binds relatively unstructured RNAs, so again it would prefer to bind to viral RNAs. It appears that HSV-1 has evolved a very clever strategy to export its RNAs in encoding ICP27, which not only prefers to bind viral RNAs over cellular mRNAs but which also hijacks cellular RNA export proteins, and binds directly to TAP/NXF1 the cellular mRNA export receptor to facilitate viral RNA export.

ICP27 undergoes a head-to-tail intramolecular association

A number of cellular proteins have been shown to undergo interactions of their N-and C-termini in a reversible manner that regulates their activities or protein interactions. These include

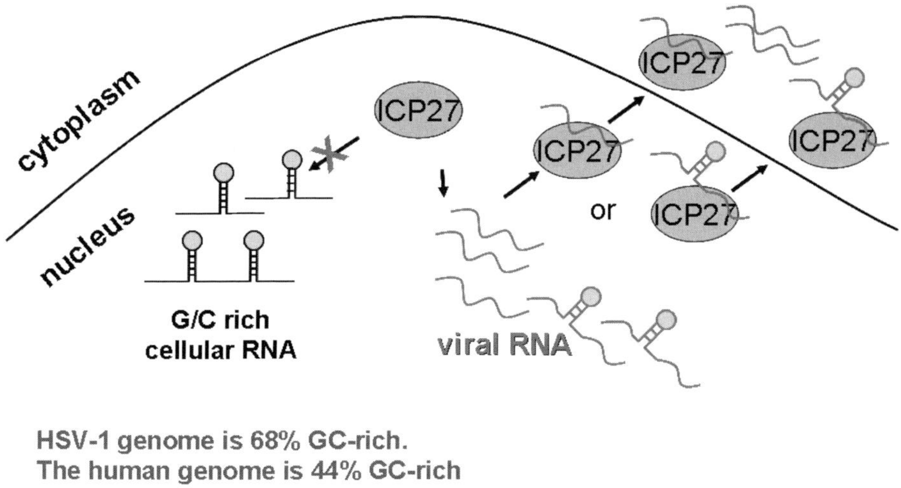

Figure 3.4 Model for viral RNA binding by ICP27. ICP27 preferentially binds to GC-rich RNA and the HSV-1 genome is 68% GC-rich whereas the cellular genome is 44% GC. ICP27 does not bind highly structured RNA but prefers flexible, relatively unstructured RNA and therefore would not bind to GC-rich cellular ribosomal RNA. Thus, ICP27 preferentially binds viral RNA through the RGG box.

zyxin, an adhesion protein that regulates actin assembly (Moody *et al.*, 2009); myosin VIIA, an unconventional myosin that has ATPase activity (Umeki *et al.*, 2009); a scaffold protein PDZK1 (LaLonde and Bretscher, 2009); talin, a large, integrin-associated cytoskeletal protein (Roberts and Critchley, 2009); the membrane associated Na^+/H^+ exchanger regulatory factor NHERF1/EBP50 (Morales *et al.*, 2007); kinesin, a motor protein (Dietrich *et al.*, 2008; Wong *et al.*, 2009), and Raf kinase, which is involved in regulating the MAPK pathway (Chong and Guan, 2003). Head-to-tail interaction generally results in a negative regulation of the protein's interactions with other proteins or a regulation of its activity. An HSV-1 protein, UL9 has also been shown to undergo head-to-tail interactions but as a dimer not as a monomer (Chattopadhyay and Weller, 2007). Because head-to-tail intramolecular interaction would alter protein conformation, such interactions may serve as another mechanism for regulating protein interactions.

Several proteins with which ICP27 interacts require that both the N-terminal leucine-rich region and the C-terminal zinc-finger domain be intact. These include RNAP II, TAP/NXF1 and Hsc70. We tested the possibility that ICP27 might undergo a head-to-tail interaction using bimolecular fluorescence complementation (BiFC) (Kerppola, 2006; Kerppola, 2009). In this procedure, the yellow fluorescent protein YFP is cut in half and the N-terminal half is fused to the N-terminus of a protein of interest and the C-terminal half is fused to the C-terminal half of the same protein or of a different protein thought to interact with the protein of interest. If the two proteins or two domains of the same protein come into direct contact, the two halves of the YFP molecule will renature through covalent bonding and fluorescence will occur. As shown in the scheme in Fig. 3.5, we fused the N-terminal half of YFP onto the N-terminus of ICP27 and the C-terminal half of YFP onto the C-terminus of ICP27 to determine if ICP27 can undergo an intramolecular interaction. We also fused the N-terminal half of YFP onto the N-terminus of one molecule of ICP27 and the C-terminal half onto the C-terminus of another molecule of ICP27 to determine if ICP27 could undergo a head-to-tail intermolecular interaction, thus forming a dimer. We have previously shown that ICP27 can form dimers (Zhi *et al.*, 1999). BiFC analysis showed that indeed ICP27 can undergo a head-to-tail intramolecular interaction but not a head-to-tail intermolecular interaction (Hernandez and Sandri-Goldin, 2010).

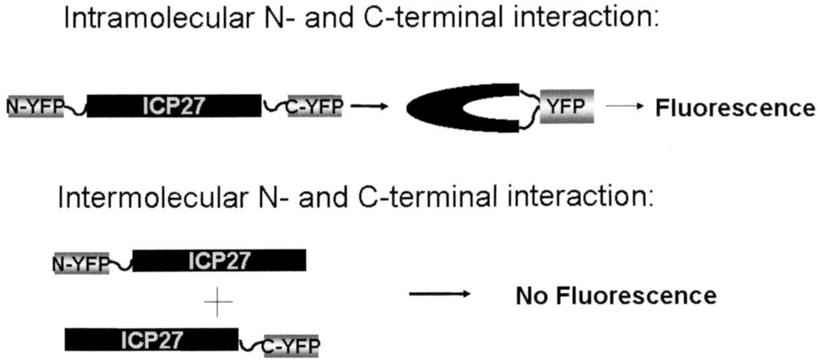

Figure 3.5 ICP27 undergoes a head-to tail intramolecular interaction. The constructs that were made for BiFC analysis are depicted. The N-YFP-ICP27-C-YFP protein fluoresced in the BiFC analysis whereas N-YFP-ICP27+ICP27-CYP did not fluoresce, indicating that intramolecular but not intermolecular head-to-tail associations occur.

Substitution of leucines at positions 8, 11 and 15 in the N-terminal leucine-rich region with alanine abolished BiFC as did substitution of cysteines with serine at 483 and 488 in the zinc-finger domain, indicating that these changes caused structural perturbations that did allow the head and tail of ICP27 to associate. These results were confirmed using another approach, fluorescence resonance energy transfer (FRET) after photo-bleaching (Karpova and McNally, 2006). In this approach, CFP was fused to the N-terminus of ICP27 and YFP was fused to the C-terminus. As controls, CFP was fused to the N-terminus of one molecular of ICP27 and YFP was fused to the C-terminus of another and the constructs were cotransfected into cells. In this procedure, the acceptor protein, YFP, is photobleached at a specific location within the cell using the 514 nm laser line. The change in donor fluorescence is quantified by comparing pre-bleach and post-bleach images. FRET is detected if donor fluorescence increases significantly more compared to what is detected in negative controls. Donor CFP fluorescence increased significantly upon bleaching (post-bleach) in the CFP-ICP27-YFP samples but FRET was not detected in the samples that were co-transfected with CFP-ICP27 and ICP27-YFP (Hernandez and Sandri-Goldin, 2010). This result confirms that ICP27 undergoes an intramolecular head-to-tail interaction but not an intermolecular interaction.

A viral recombinant was constructed with N-YFP-ICP27-C-YFP replacing the wild type ICP27 gene (Hernandez and Sandri-Goldin, 2010). The resulting virus is severely compromised for viral replication and must be propagated on an ICP27 complementing cell line. The N-YFP-ICP27-C-YFP fusion protein is abundantly expressed at later times during infection and BiFC is observed, however, because the renaturation of YFP occurs through covalent bonding, the N-YFP-ICP27-C-YFP fusion protein becomes locked in the head-to-tail configuration once BiFC occurs. This likely impedes ICP27 from undergoing a number of its protein interactions. And in fact, we found that N-YFP-ICP27-C-YFP does not colocalize with RNAP II or TAP/NXF1 (Hernandez and Sandri-Goldin, 2010). These findings indicate that ICP27 undergoes a head-to-tail intramolecular interaction that may regulate its interactions with proteins such as RNAP II, TAP/NXF1 and Hsc70, all of which require the N- and C-termini of ICP27 for interaction.

Summary and perspectives

The many interactions that ICP27 undergoes require it to be highly flexible. Post-translation modifications, namely phosphorylation and arginine methylation regulate some of ICP27's functional interactions as does its head-to-tail association. ICP27 is a multifunctional protein that has evolved to serve HSV-1 well during infection. It stacks the deck against cellular gene expression and promotes viral gene expression at every level from transcription of viral RNA through translation. ICP27 is a truly remarkable protein. Although we have uncovered the mechanisms for how it performs a number of functions and have begun to discern how its activities may be regulated, there is much that we still don't know about how these various activities are orchestrated at the right tempo throughout infection, and we will be concentrating on unravelling more of the protein's mysteries in the future.

Acknowledgements

I thank the members of my research group for their hard work on ICP27. The studies described here that were performed in my laboratory were funded by NIH grants AI21215 and AI61397.

References

Bedford, M.T., and Clarke, S.G. (2009). Protein arginine methylation in mammals: Who, what and why. Mol. Cell *33*, 1–13.

Bryant, H.E., Wadd, S., Lamond, A.I., Silverstein, S.J., and Clements, J.B. (2001). Herpes simplex virus IE63 (ICP27) protein interacts with spliceosome-associated protein 145 and inhibits splicing prior to the first catalytic step. J. Virol. *75*, 4376–4385.

Burch, A.D., and Weller, S.K. (2004). Nuclear sequestration of cellular chaperone and proteasomal machinery during herpes simplex virus type 1 infection. J. Virol. *78*, 7175–7185.

Chattopadhyay, S., and Weller, S.K. (2007). Direct interaction between the N- and C-terminal portions of the herpes simplex virus type 1 origin binding protein UL9 implies the formation of a head to tail dimer. J. Virol. *81*, 13659–13667.

Chen, I.B., Li, L., Silva, L., and Sandri-Goldin, R.M. (2005). ICP27 recruits Aly/REF but not TAP/NXF1 to herpes simplex virus type 1 transcription sites although TAP/NXF1 is required for ICP27 export. J. Virol. *79*, 3949–3961.

Chen, I.B., Sciabica, K.S., and Sandri-Goldin, R.M. (2002). ICP27 interacts with the export factor Aly/REF to direct herpes simplex virus 1 intronless RNAs to the TAP export pathway. J. Virol. *76*, 12877–12889.

Chong, H., and Guan, K. (2003). Regulation of Raf through phosphorylation and N terminus-C terminus interaction. J. Biol. Chem. *218*, 36269–36276.

Cocco, M.J., Hanakahi, L.A., Huber, M.D., and Maizels, N. (2003). Specific interactions of distamycin with G-quadruplex DNA. Nucleic Acids Res. *31*, 2944–2951.

Corbin-Lickfett, K., Chen, I.B., Cocco, M.J., and Sandri-Goldin, R.M. (2009). The HSV-1 ICP27 RGG box specifically binds flexible, GC-rich sequences but not G-Quartet structures. Nucleic Acids Res. *37*, 7290–7301.

Corbin-Lickfett, K., Rojas, S., Li, L., Cocco, M.J., and Sandri-Goldin, R.M. (2010). ICP27 Phosphorylation Site Mutants Display Altered Functional Interactions with Cellular Export Factors Aly/REF and TAP/NXF1 but Are Able to Bind Herpes Simplex Virus 1 RNA. J. Virol. *84*, 2212–2222.

Dai-Ju, J.Q., Li, L., Johnson, L.A., and Sandri-Goldin, R.M. (2006). ICP27 interacts with the C-terminal domain of RNA polymerase II and facilitates its recruitment to herpes simplex virus-1 transcription sites, where it undergoes proteasomal degradation during infection. J. Virol. *80*, 3567–3581.

Dietrich, K.A., Sindelar, C.V., Brewer, P.D., Downing, K.H., Cremo, C.R., and Rice, S.A. (2008). The kinesin-1 motor protein is regulated by a direct interaction of its head and tail. Proc. Natl. Acad. Sci. U.S.A. *105*, 8938–8943.

Dobrikova, E., Shveygert, M., Walters, R., and Gromeier, M. (2009). Herpes simplex virus proteins ICP27 and UL47 associate with polyadenylate binding protein and control its sub-cellular distribution. J. Virol. doi:10.1128/JVI.01740-09.

Ellison, K.S., Maranchuk, R.A., Mottet, K.L., and Smiley, J.R. (2005). Control of VP16 translation by the herpes simplex virus type 1 immediate-early protein ICP27. J. Virol. *79*, 4120–4131.

Fontaine-Rodriquez, E.C., and Knipe, D.M. (2008). Herpes simplex virus ICP27 increases translation of a subset of viral late mRNAs. J. Virol. doi 10.1128/JVI, 2395–2407.

Fontaine-Rodriquez, E.C., Taylor, T.J., Olesky, M., and Knipe, D.M. (2004). Proteomics of herpes simplex virus infected cell protein 27: association with translation initiation factors. Virology *330*, 487–492.

Graveley, B.R. (2000). Sorting out the complexity of SR protein functions. RNA *6*, 1197–1211.

Hardwicke, M.A., and Sandri-Goldin, R.M. (1994). The herpes simplex virus regulatory protein ICP27 can cause a decrease in cellular mRNA levels during infection. J. Virol. *68*, 4797–4810.

Hardy, W.R., and Sandri-Goldin, R.M. (1994). Herpes simplex virus inhibits host cell splicing, and regulatory protein ICP27 is required for this effect. J. Virol. *68*, 7790–7799.

Hernandez, F.P., and Sandri-Goldin, R.M. (2010). Herpes simplex virus 1 regulatory protein ICP27 undergoes a head-to-tail intramolecular interaction. J. Virol. 84, doi:10.1128/JVI.02319-09.

Ingram, A., Phelan, A., Dunlop, J., and Clements, J.B. (1996). Immediate early protein IE63 of herpes simplex virus type-1 binds RNA directly. J. Gen. Virol. *77*, 1847–1851.

Johnson, L.A., Li, L., and Sandri-Goldin, R.M. (2009). The cellular RNA export receptor TAP/NXF1 is required for ICP27- mediated export of herpes simplex virus 1 RNA, whereas, the TREX-complex adaptor protein Aly/REF appears to be dispensable. J. Virol. *83*, 6335–6346.

Johnson, L.A., and Sandri-Goldin, R.M. (2009). Efficient nuclear export of herpes simplex virus 1 transcripts requires. both RNA binding by ICP27 and ICP27 interaction with TAP/NXF1. J. Virol. *83*, 1184–1192.

Karpova, T., and McNally, J.G. (2006). Detecting protein–protein interactions with CFP-YFP FRET by acceptor photobleaching. Curr. . Protoc. Cytom. *12*, 12–17.

Kerppola, T.K. (2006). Visualization of molecular interactions by fluorescence complementation. Nat. Rev. Mol. Cell Biol. *7*, 449–456.

Kerppola, T.K. (2009). Visualization of molecular interactions using bimolecular fluorescence complementation analysis: Characteristics of protein fragment complementation. Chem. Soc. Rev. *38*, 2876–2886.

Koffa, M.D., Clements, J.B., Izaurralde, E., Wadd, S., Wilson, S.A., Mattaj, I.W., and Kuersten, S. (2001). Herpes simplex virus ICP27 protein provides viral mRNAs with access to the cellular mRNA export pathway. EMBO J. *20*, 5769–5778.

LaLonde, D.P., and Bretscher, A. (2009). The scaffold protein PDZK1 undergoes a head to tail intramolecular association that negatively regulates its interaction with EBP50. Biochem. *48*, 2261–2271.

Larralde, O., Smith, R.W., Wilkie, G.S., Gray, N.K., and Clements, J.B. (2006). Direct stimulation of translation by the multifunctional herpesvirus ICP27 protein. J. Virol. *80*, 1588–1591.

Li, L., Johnson, L.A., Dai-Ju, J.Q., and Sandri-Goldin, R.M. (2008). Hsc70 focus formation at the periphery of HSV-1 transcription sites requires. ICP27. PLoS ONE 3 doi:10.1371/journal.pone.0001491, e1491.

Linberg, A., and Kreivi, J.P. (2002). Splicing inhibition at the level of spliceosome assembly in the presence of herpes simplex virus protein ICP27. Virology *294*, 189–198.

Livingston, C.M., Ifrim, M.F., Cowan, A.E., and Weller, S.K. (2009). Virus-induced chaperone-enriched (VICE) domains function as nuclear protein quality controlk centers during HSV-1 infection. PLoS Pathog. *5*, e1000619.

Mears, W.E., Lam, V., and Rice, S.A. (1995). Identification of nuclear and nucleolar localization signals in the herpes simplex virus regulatory protein ICP27. J. Virol. *69*, 935–947.

Mears, W.E., and Rice, S.A. (1996). The R.G.G box motif of the herpes simplex virus I.C.P27 protein mediates an R.N.A.-binding activity and determines *in vivo* methylation. J. Virol. *70*, 7445–7474.

Misteli, T., Caceres, J.F., Clement, J.Q., Krainer, A.R., Wilkinson, M.F., and Spector, D.L. (1998). Serine phosphorylation of SR proteins is required for their recruitment to sites of transcription *in vivo*. J. Cell Biol. *143*, 297–307.

Moody, J.D., Grange, J., Ascione, M.P.A., Boothe, D., Bushnell, E., and Hansen, M.D.H. (2009). A zyxin head to tail interaction regulates zyxin–VASP complex formation. Biochim. Biophys. Res. Commun. *378*, 625–628.

Morales, F.C., Takahashi, Y., Momin, S., Adams, H., Chen, X., and Georgescu, M. (2007). NHERF1/EBP50 head to tail intramolecular interaction masks association with PDZ domain ligands. Mol. Cell Biol. *27*, 2527–2537.

Norseen, J., Johnson, F.B., and Lieberman, P.M. (2009). Role for G-quadruplex RNA binding by Epstein–Barr virus nuclear antigen 1 in DNA replication and metaphase chromosome attachment. J. Virol. *83*, 10336–10346.

Prasad, J., Colwill, K., Pawson, T., and Manley, J.L. (1999). The protein kinase Clk/Sty directly modulates SR protein activity: Both hyper- and hypophosphorylation inhibit splicing. Mol. Cell Biol. *19*, 6991–7000.

Ramos, A., Hollingworth, D., and Pastore, A. (2003). G-quartet-dependent recognition between the FMRP RGG box and RNA. RNA *9*, 1198–1207.

Roberts, G.C.K., and Critchley, D.R. (2009). Structural and biophysical properties of the integrin-associated cytoskeletal protein talin. Biophys Rev. *1*, 61–69.

Rojas, S., Corbin-Lickfett, K., Escudero-Paunetto, L., and Sandri-Goldin, R.M. (2010). ICP27 phosphorylation site mutants are defective in herpes simplex virus 1 replication and gene expression. J. Virol. *84*, 2200–2211.

Sandri-Goldin, R.M. (1998). ICP27 mediates herpes simplex virus RNA export by shuttling through a leucine-rich nuclear export signal and binding viral intronless RNAs through an RGG motif. Genes Dev. *12*, 868–879.

Sandri-Goldin, R.M. (2008). The many roles of the regulatory protein ICP27 during herpes simplex virus infection. Front. Biosci. . 13, 5241–5256.

Sandri-Goldin, R.M., and Hibbard, M.K. (1996). The herpes simplex virus type 1 regulatory protein ICP27 coimmunoprecipitates with anti-Sm antiserum and an the C-terminus appears to be required for this interaction. J. Virol. *70*, 108–118.

Sciabica, K.S., Dai, Q.J., and Sandri-Goldin, R.M. (2003). ICP27 interacts with SRPK1 to mediate HSV-1 inhibition of pre-mRNA splicing by altering SR protein phosphorylation. EMBO J. *22*, 1608–1619.

Shen, E.C., Henry, M.F., Weiss, V.H., Valentini, S.R., Silver, P.A., and Lee, M.S. (1998). Arginine methylation facilitates the nuclear export of hnRNP proteins. Genes Dev. *12*, 679–691.

Siddiqui-Jain, A., Grand, C.L., Bearss, D.J., and Hurley, L.H. (2002). Direct evidence for a G-quadruplex in a promoter region and its targeting with a small molecule to repress c-myc transcription. Proc. Natl. Acad. Sci. U.S.A. *99*, 11593–11598.

Sokolowski, M., Scott, J.E., Heaney, R.P., Patel, A.H., and Clements, J.B. (2003). Identification of herpes simplex virus RNAs that interact specifically with regulatory protein ICP27 *in vivo*. J. Biol. Chem. *278*, 33540–33549.

Soliman, T.M., and Silverstein, S.J. (2000). Identification of an export control sequence and a requirement for the KH domains in ICP27 from herpes simplex virus type 1. J. Virol. *74*, 7600–7609.

Souki, S.K., Gershon, P.D., and Sandri-Goldin, R.M. (2009). Arginine methylation of the ICP27 RGG box regulates ICP27 export and is required for efficient herpes simplex virus 1 replication. J. Virol. *83*, 5309–5320.

Souki, S.K., and Sandri-Goldin, R.M. (2009). Arginine methylation of the ICP27 RGG box regulates the functional interactions of ICP27 with SRPK1 and Aly/REF during herpes simplex virus 1 infection. J. Virol. *83*, 8970–8975.

Umeki, N., Jung, H.S., Watanabe, S., Sakai, T., Li, X., Ikebe, R., Craig, R., and Ikebe, M. (2009). The tail binds to the head-neck domain, inhibiting ATPase activity of myosin VIIA. Proc. Natl. Acad. Sci. U.S.A. *106*, 8483–8488.

Wong, Y.L., Dietrich, K.A., Naber, N., Cooke, R., and Rice, S.A. (2009). The kinesin-1 tail conformationally restricts the nucleotide pocket. Biophys J. *96*, 2799–2807.

Yu, M.C., Bachand, F., McBride, A.E., Komili, S., Casolari, J.M., and Silver, P.A. (2004). Arginine methyltransferase affects interactions and recruitment of mRNA processing and export factors. Genes Dev. *18*, 2024–2035.

Zhi, Y., and Sandri-Goldin, R.M. (1999). Analysis of the phosphorylation sites of the herpes simplex virus type 1 regulatory protein ICP27. J. Virol. *73*, 3246–3257.

Zhi, Y., Sciabica, K.S., and Sandri-Goldin, R.M. (1999). Self interaction of the herpes simplex virus type 1 regulatory protein ICP27. Virology *257*, 341–351.

Zhong, X., Ding, J., Adams, J.A., Ghosh, G., and Fu, X.D. (2009). Regulation of SR protein phosphorylation and alternative splicing by modulating kinetic interactions of SRPK1 with molecular chaperones. Genes Dev. *23*, 482–495.

Zhou, C., and Knipe, D.M. (2001). Association of herpes simplex virus 1 ICP8 and ICP27 with cellular RNA polymerase II holoenzyme. J. Virol. *76*, 5893–5904.

The Role of ICP0 in Counteracting Intrinsic Cellular Resistance to Virus Infection

4

Roger D. Everett

Abstract

In recent years it has become apparent that, in addition to the acquired and innate defences against virus infection, there is also a third aspect to antiviral defences that operates at the intracellular level. This concept is known as intrinsic resistance, intrinsic antiviral defence or intrinsic immunity. Its key features include constitutively expressed cellular proteins that restrict viral gene expression, and viral regulatory proteins that counteract the actions of the cellular inhibitors. This chapter reviews the cellular proteins and pathways that are thought to be involved in intrinsic resistance to HSV-1 infection, and the mechanisms by which these are inactivated by ICP0, an important viral regulatory protein. The phenotype of ICP0 null mutant HSV-1 is described to give a background to the phenomenon, then the principal properties of ICP0 itself are summarized. The effects of ICP0 on components of cellular nuclear structures known as ND10 or PML nuclear bodies are reviewed, then the possible roles of these proteins in intrinsic resistance are discussed. The relationships between ICP0, intrinsic resistance and the regulation of viral chromatin structure are considered, and finally the parallels between ICP0 and related proteins expressed by other alphaherpesviruses are described. Intrinsic resistance and the manner in which viruses overcome it are important aspects of the biology of virus infection, but we have much to learn before we achieve a complete understanding of the viral and cellular proteins that are involved.

Introduction

ICP0 is one of the most extensively studied HSV-1 proteins because of its important roles in fundamental aspects of HSV-1 biology. This chapter in part updates a review of the roles of ICP0 in HSV-1 infection that was included in a previous edition of this book (Everett, 2006b), but there are also several other reviews available that summarize general aspects of ICP0 or its roles in specific functions (Efstathiou and Preston, 2005; Everett, 2000a,b, 2001, 2006a; Hagglund and Roizman, 2004; Preston, 2000; Roizman and Taddeo, 2006). This chapter is not intended to present a comprehensive review of all aspects of the protein, especially since other articles in this volume will cover connections between ICP0 and the interferon response (Chapter 13), the DNA damage response (Chapter 14), latency (Chapter 17), and the stress response (Chapter 15). Instead this chapter will concentrate on the role of ICP0 in counteracting the so-called intrinsic cellular resistance to virus infection (also known as intrinsic cellular antiviral defence or intrinsic immunity). To put this in context it is first necessary to describe the basic properties of ICP0-null HSV-1 mutants and the light these have shed on the roles of ICP0 in HSV-1 biology.

Overview of the role of ICP0 in HSV-1 biology

ICP0 is a 775-amino acid protein encoded by Immediate-Early (IE) gene 1 that is required for efficient initiation of lytic HSV-1 infection and productive reactivation from latency (see reviews cited above). A notable feature of ICP0 is its RING finger domain, a cysteine-rich motif near the N-terminus of the polypeptide that is stabilized by coordination with two zinc atoms in a protein fold characteristic of a large family of viral and cellular proteins that have E3 ubiquitin ligase activity (Joazeiro and Weissman, 2000). HSV-1 mutants that express an ICP0 protein lacking the RING finger are as defective as a complete null mutant in all core biological assays, and therefore the RING domain is essential for ICP0 function (Chen and Silverstein, 1992; Everett, 1989; Everett *et al.*, 2009b; Harris *et al.*, 1989; Lium and Silverstein, 1997; Thompson and Sawtell, 2006). Members of the alphaherpesvirus sub-family encode RING finger regulatory proteins related to ICP0, and therefore ICP0 is the prototype member of a family of herpesvirus proteins that are thought to confer related biological functions.

The phenotype of ICP0 null mutant HSV-1 during lytic infection

Ever since ICP0 null mutant HSV-1 mutants were first described in the 1980s (Sacks and Schaffer, 1987; Stow and Stow, 1986), the functions of the protein and the phenotypes of the mutant viruses have stimulated an intriguing, complex and at times controversial literature.

ICP0-null mutant viruses are able to replicate in cultured cells, but their ability to initiate productive infection is decreased compared to wild type (wt) virus. The extent of this defect varies widely between cell types, being absent in osteosarcoma U2OS cells (Yao and Schaffer, 1995) and of the order of 1000-fold in human diploid fibroblasts (HFs) (Everett *et al.*, 2004a and references therein). Many common laboratory cell types restrict replication of ICP0-null mutant HSV-1 to intermediate extents (Everett, 1989; Everett *et al.*, 2004a; Stow and Stow, 1986). Therefore it is important to consider cell type when interpreting published studies.

The simplest way to envisage the defect of an ICP0 null mutant is in terms of probability. Mutant and wt virus particles are equally likely to initiate plaque formation in U2OS cells, while in HFs application of 1000 plaque forming units (PFU) based on U2OS cell titres would result, on average, in only a single mutant virus plaque (Everett *et al.*, 2004a). Thus the plaque forming efficiency of an ICP0 mutant in a given cell type, compared to that in U2OS cells, is perhaps the most reliable measure of the relative activities of a set of ICP0 mutants. It follows that when assessing the defect of an ICP0 mutant virus it is important to use input multiplicities according to titres in U2OS cells to ensure that equal amounts of potentially infectious virus are used. If inputs are based on titres in Vero cells, for example, much more mutant than wt virus will be used in real terms, resulting in misleading outcomes.

Another key feature of the ICP0-null mutant phenotype is its multiplicity dependence. This means that as increased amounts of a mutant are used the defect in replication becomes less severe. This multiplicity dependence is again affected by cell type, but crucially it is non-linear. Thus an input of ICP0 null mutant virus of around 10 to 20 PFU per cell (based on titres in U2OS cells) is sufficient to infect HFs as efficiently as a wt infection at 1 PFU per cell (Everett *et al.*, 2004a). In this situation, all the cells in the monolayer will be infected with the mutant virus, and eventual virus yields in terms of PFU determined in U2OS cells will be little different from those of a wt infection. Therefore the rare cells in a population of HFs that become productively infected after a low multiplicity ICP0 null mutant infection will produce large numbers of progeny particles which will infect surrounding cells at high multiplicity. All such cells will readily engage in lytic infection, and thus even the mutant virus will eventually spread through the HF cell sheet.

The above deceptively simple explanation of the phenotype of ICP0 null mutant HSV-1 has important practical implications. In particular, it is extremely difficult to compare wt and mutant virus infections in restrictive cell types. At low multiplicity most cells in the mutant virus infected population will not enter lytic infection, and therefore any phenotype in the wt sample may not be a direct effect of ICP0 itself, but instead an indirect effect of efficient expression of other viral proteins and/or the progression of infection. At higher multiplicities, especially in many common transformed laboratory cell types, the threshold of input multiplicity may have been crossed and the mutant might replicate as efficiently as the wt. Therefore experiments should be designed according to whether an effect of ICP0 itself is being assessed (adjust multiplicities so that all cells are infected and are expressing other viral proteins at equivalent efficiencies), or whether the extent of the ICP0 null mutant defect is being estimated (use low multiplicities based on U2OS cell titres).

To summarize, there is a cell type dependent restriction in the replication of ICP0-null mutant HSV-1 which reduces the probability that an infected cell will progress to productive infection. This restriction can be overcome at increased multiplicity, implying that it is conferred by host factors that can be titrated out. One way of interpreting these observations is to propose that there is a competition between the virus and the cell, with cell factors that restrict the initiation of productive infection and viral factors (principally ICP0) that compete with or overcome the host cell restriction. These findings have contributed to the development of the concept of intrinsic cellular resistance that is defined below.

ICP0 and latent/quiescent HSV-1 infection

The role of ICP0 during HSV-1 latency is a complex issue (Coleman *et al.*, 2008; Efstathiou and Preston, 2005; Knipe and Cliffe, 2008; Preston, 2000). It is not intended to discuss this topic in any detail here, but it is pertinent to summarize the current understanding of the role ICP0 plays in this critical biological property of the virus. It follows from the discussion above that conducting well controlled experiments using ICP0-null mutant HSV-1 in animal models is not straightforward due to potential difficulties of normalizing the viral input with the wt virus control, and the likelihood of differential replication of wt and mutant viruses in the periphery prior to the establishment of latency. These factors would be expected to result in differences in the genome copy number of wt and mutant viruses in latently infected neurones. However, a careful study taking these factors into account confirmed that ICP0-null mutant HSV-1 can establish latency as efficiently as the wt virus (Halford and Schaffer, 2000). The situation is more complex during reactivation. If reactivation is measured in terms of detection of viral gene expression in single latently infected neurones, it appears that ICP0 is not required (Thompson and Sawtell, 2006). If, however, reactivation is measured by production of infectious virus, the expression of ICP0 is very important (Halford and Schaffer, 2001). It is likely that the initial events of reactivation leading to viral gene expression are not ICP0 dependent, but that ICP0 is required to allow progression from this early stage of reactivation to a productive infection, particularly as replication spreads to non-neuronal cells.

The situation is clearer in models of quiescent infection in cultured HFs or similar human cells. As noted above, restrictive cells infected at low multiplicity with ICP0-null mutant HSV-1 have a very low probability (of the order of 1 in a 1000) of engaging in productive infection. Although viral gene expression is not detectable in 80–90% of HFs receiving a potentially infectious ICP0-null mutant particle, IE gene expression is not completely repressed and about 100-fold more cells express ICP4 in detectable quantities than the number that initiate plaque formation (Everett *et al.*, 2004a). Of these ICP4 positive HFs, around 10% express readily detectable levels of ICP8 at 24 h after infection, yet 90% of these ICP8 positive cells do not progress to initiate a plaque (Everett *et al.*, 2004a). Therefore there is non-uniformity in the repression of the viral genome in the absence of ICP0 expression, not only in terms of the cells that allow viral gene expression but also the viral gene products that are expressed in those cells (Everett *et al.*, 2004a). Despite this limited viral gene expression in some cells, around 99.9% of cells in total do not engage in lytic infection and instead the genomes enter a quiescent state in which no viral protein expression is detectable.

ICP0 null mutant viruses that have additional lesions in other important viral regulatory proteins are repressed extremely efficiently. Even reporter genes under the control of a strong promoter become silenced by 24 h after infection with these viral mutants (Mossman and Smiley, 1999; Preston and Nicholl, 1997; Samaniego *et al.*, 1998; reviewed in Preston, 2000). These mutants have allowed the development of models of quiescent infection in which cell cultures can be infected at relatively high multiplicity such that the majority of cells contain repressed, quiescent viral genomes. Importantly, the quiescent viral genomes can be de-repressed by subsequent expression of ICP0 provided by a variety of different vector systems (Everett *et al.*, 2009; Harris *et al.*, 1989; Samaniego *et al.*, 1998; Zhu *et al.*, 1990). At one stage it appeared that ICP0 might be the only factor that could accomplish this de-repression, but it has since emerged that cell type and experimental method are important issues. De-repression can be achieved in rodent cells, and to a limited extent even in HFs, by VP16, ICP4 or other treatments (Halford *et al.*, 2001; Miller *et al.*, 2006; Preston, 2007), and in rodent neurones and (to a limited extent) Vero cells by the histone deacetylase inhibitor TSA (Arthur *et al.*, 2001; Danaher *et al.*, 2005; Terry-Allison *et al.*, 2007). However, de-repression by ICP0 is extremely efficient in all systems, which demonstrates that ICP0 can reverse the mechanism by which viral genome repression is established and maintained. How this is achieved remains a topic of considerable interest and debate.

Acquired, innate and intrinsic antiviral defences

The phenotypes of ICP0-null mutant HSV-1, and those of equivalent mutants of human cytomegalovirus (HCMV) in its pp71 (UL82) and IE1 (pp72) regulatory proteins, have led to the hypothesis of intracellular intrinsic cellular resistance to virus infection that functions at the single cell level (Saffert and Kalejta, 2008; Tavalai and Stamminger, 2008). Organisms and cells have a variety of means of combating virus infections, including the acquired immunity conferred by the humoral and cellular arms of the immune system, and innate immunity in which virus infection leads to the production of interferons and other cytokines that stimulate expression of proteins that have antiviral properties. The hypothesis of intrinsic antiviral defence, or intrinsic immunity, is quite distinct in that it proposes the existence of cellular proteins that are expressed constitutively and which interfere with virus infection in individual infected cells. Examples include inhibition of HIV uncoating by TRIM5a and the effects of APOBEC proteins on HIV genomes (reviewed in Huthoff and Towers, 2008; Sebastian and Luban, 2007; Sokolskaja and Luban, 2006; Takeuchi and Matano, 2008), and the inhibition of herpesvirus gene expression at the early stages of infection discussed here. Intrinsic resistance is a recent concept, and much remains to be learned about the cellular proteins involved and their detailed mechanisms of action. It is useful at this stage to discuss features of intrinsic resistance that shed some light on its nature.

Firstly, because the cellular proteins involved in intrinsic resistance are expressed constitutively it is likely that they have roles in the uninfected cell. Indeed, all of the proteins discussed below in connection with intrinsic resistance to herpesvirus infections have significant functions in regulating cellular gene expression or other processes. Secondly, the intrinsic resistance hypothesis invokes viral proteins that counteract the actions of the cellular proteins. It follows that viruses that are defective in these functions are more susceptible to intrinsic resistance, while cells that do not express important intrinsic resistance proteins serve as better hosts for viruses that are unable to express their counteracting proteins. Thirdly, in the case of herpesvirus infections in which latency is an essential part of their life cycle, it is possible that the virus utilizes the intrinsic resistance machinery in order to allow repression of viral gene expression and thus facilitate the establishment of quiescence and latency. If so, it follows that the virus must have evolved its countermeasures in concert with the host, and therefore species specificity in the effectiveness of intrinsic resistance is likely. Finally, there is a clear distinction between viral regulatory proteins that counteract intrinsic resistance and others that stimulate viral gene expression more directly (such as the HSV-1 proteins VP16, ICP4, and ICP27), in that the former act on the inhibitory cellular proteins, while the latter act positively on viral genomes or transcripts, usually in concert with cellular proteins that facilitate gene expression.

The characteristics of intrinsic resistance proposed above fit well with the phenotype of ICP0-null mutant HSV-1 and the properties of

ICP0. The following sections will describe some of the experimental evidence that is relevant to this model.

The functional domains and biochemical properties of ICP0

ICP0 functional domains

ICP0 has been subjected to extensive analyses that have mapped several functional domains, motifs and interaction regions. The most notable of these is the RING finger in the N-terminal third of the protein which confers E3 ubiquitin ligase activity (Boutell *et al.*, 2002). The RING finger is completely essential for ICP0's core biological functions in stimulating lytic replication and de-repression of quiescence (Everett *et al.*, 2004a; Everett *et al.*, 2009; Harris *et al.*, 1989; Zhu *et al.*, 1990). ICP0 RING finger mediated E3 ubiquitin ligase activity results in the proteasome-dependent degradation of a number of cellular proteins, including some of the components of nuclear structures known as ND10, PML nuclear bodies or PML NBs (Everett *et al.*, 1998; Parkinson and Everett, 2000), centromeres (Everett *et al.*, 1999; Lomonte and Morency, 2007; Lomonte *et al.*, 2001), and some proteins involved in the DNA damage response (Lilley *et al.*, 2010; Parkinson *et al.*, 1999). As discussed in more detail below, a number of these proteins have been implicated in intrinsic antiviral defence.

Phosphopeptide mapping has identified 3 major regions of phosphorylation of ICP0, named Phos1, Phos2 and Phos3 (Davido *et al.*, 2005). Serine to alanine mutations within Phos1 (which lies close to the RING finger) cause a modest defect in plaque formation efficiency, while mutations in Phos2 and Phos3 affect certain ICP0 functions in transfection based experiments but have not as yet been found to cause a phenotype in cultured cell infections (Boutell *et al.*, 2008).

ICP0 is located in the nucleus at early times of infection by virtue of a nuclear localization signal (nls) at around residue 500 (Ciufo *et al.*, 1994; Meredith *et al.*, 1995), and following that is a serine-rich region that has some sequence similarity with the cellular co-repressor protein CoREST (Gu *et al.*, 2005). The nls is essential for ICP0 activity in lytic infection and de-repression/reactivation assays (Everett *et al.*, 2009). The C-terminal third of the protein is a complex region that includes a motif that binds strongly to the ubiquitin specific protease USP7 (Everett *et al.*, 1997), sequences required for self interaction (multimerization) (Meredith *et al.*, 1995; Mullen *et al.*, 1995), localization to ND10 (Everett and Maul, 1994), and interaction with the cellular co-repressor CoREST (Gu and Roizman, 2007). Mutations in the USP7 interaction region reduce ICP0 stability and cause an intermediate defect in ICP0 activity (Canning *et al.*, 2004; Everett *et al.*, 2009). Mutations in regions required for efficient ND10 localization and multimerization also cause intermediate defects. For example, the plaque formation efficiency of a mutant lacking the entire C-terminal third of the protein is reduced to around 10% of that of the wt in HFs. This is a substantial decrease, but it should borne in mind that this mutant retains around 100-fold greater replicative ability than null or RING finger mutants (Everett *et al.*, 2009).

The E3 ubiquitin ligase activity of ICP0

The ubiquitin-proteasome pathway is the major mechanism by which protein stability is regulated and as such it underlies at some stage the operation of almost all cellular processes. In summary, the pathway involves an E1 ubiquitin activating enzyme that forms a thiol-ester linkage with the C-terminus of ubiquitin, followed by transfer of the ubiquitin to another thiol-ester linkage to the active site cysteine of a member of the E2 ubiquitin conjugating enzymes. The E3 ubiquitin ligase regulates and provides the substrate specificity of a reaction to conjugate the ubiquitin moiety via an isopeptide linkage to lysine side chains in the target protein. This cycle can be reiterated to produce polyubiquitin chains, and proteins thus marked are targeted by the proteasome for degradation. E3 ubiquitin ligases fall into a number of families, of which the RING finger class has a large number of cellular and viral members (Joazeiro and Weissman, 2000). All E3 ubiquitin ligases function in concert with one or more of the E2 ubiquitin conjugation enzymes, and in the case of ICP0 its partners are those of the UbcH5 and UbcH6 families (Boutell *et al.*, 2002; Gu and Roizman, 2003). As noted above, mutations in the RING finger that eliminate the ubiquitin E3 ligase activity cripple the protein in terms of its core biological activities, and therefore any hypothesis put forward to account for the mechanism of action of ICP0 must include a rationale for the necessity for its RING finger.

Early work suggested that ICP0 includes a second E3 ubiquitin ligase domain in the C-terminal third of the protein that stimulates autoubiquitination of the E2 ubiquitin conjugating enzyme cdc34 (Hagglund and Roizman, 2002). However, this observation was not replicated (Boutell *et al.*, 2002) and has not since been followed up in any detail.

The targets of ICP0 activity

The ubiquitin ligase activity of ICP0 can be studied *in vitro*, and a number of proteins have been identified as being susceptible to ubiquitination by ICP0 in combination with UbcH5a in such assays. These include ICP0 itself (Canning *et al.*, 2004), USP7 (Boutell *et al.*, 2005), RNF8 (Lilley *et al.*, 2010) and p53 (Boutell and Everett, 2003). For the first three of these there is strong evidence that such ubiquitination also occurs during HSV-1 infection. Like many RING finger proteins, ICP0 is unstable as a result of autoubiquitination. In normal circumstances it is protected from this self-destructive activity through its interaction with the ubiquitin specific protease USP7, which removes ubiquitin adducts on the ICP0 polypeptide (Canning *et al.*, 2004). However, perhaps because it interacts so strongly with ICP0, USP7 is also partially susceptible to ICP0-mediated ubiquitination and proteasome mediated degradation, implying reciprocal activities of the two proteins (Boutell *et al.*, 2005). RNF8, a cellular RING finger ubiquitin ligase that monoubiquitinates histone H2A during the DNA damage response, is readily ubiquitinated by ICP0 *in vitro* and is degraded during HSV-1 infection in an ICP0 RING dependent manner (Lilley *et al.*, 2010). RNF168, another cellular E3 ubiquitin ligase that is related to RNF8 and which is also involved in the DNA damage response, is similarly degraded during wt but not ICP0 null mutant infection (Lilley *et al.*, 2010). Although p53 can be ubiquitinated by ICP0 *in vitro* (Boutell and Everett, 2003), global levels of p53 are stabilized during HSV-1 infection via phosphorylation events (Boutell and Everett, 2004) and through induction of hdm2 degradation in an ICP0-independent manner (Boutell *et al.*, 2005). These stabilizing influences appear to be dominant over any potential ICP0-mediated ubiquitination of p53 in infected cells.

As well as the substrates noted above, ICP0 also induces the degradation of several other cellular proteins during HSV-1 infection. These include PML, the core component of ND10 (a process that results in complete disruption of these structures) (Chelbi-Alix and de The, 1999; Everett *et al.*, 1998), the SUMO-modified forms of Sp100 (another major ND10 component) (Chelbi-Alix and de The, 1999; Muller and Dejean, 1999; Parkinson and Everett, 2000), centromere proteins CENP-A, -B, and -C (Everett *et al.*, 1999; Lomonte and Morency, 2007; Lomonte *et al.*, 2001) (resulting in centromere destruction and mitotic failure), and the catalytic subunit of DNA-PK (Lees-Miller *et al.*, 1996; Parkinson *et al.*, 1999). Another protein recently found to be degraded in response to ICP0 expression is ZBTB38 (Rauch, N. and Everett, R.D., unpublished observations), a member of the ZBTB family of heterochromatin binding proteins that are involved in transcriptional repression (Filion *et al.*, 2006). Undoubtedly more substrates, direct or indirect, of the E3 ubiquitin ligase activity of ICP0 will be revealed by further studies. This greatly complicates interpretation of the biological significance of the degradation of any individual defined substrate. The following sections will discuss links between these activities of ICP0 and its role in combating the intrinsic cellular antiviral defence.

The role of ND10 proteins in intrinsic defence

ND10 – a brief description

ND10 (also known as PML Nuclear Bodies or PML NBs) are small distinct nuclear substructures that are defined by the presence of the promyelocytic leukaemia protein PML. These structures and their component proteins have been implicated in several pathways including the regulation of gene expression, apoptosis, chromatin metabolism, the DNA damage response, oncogenesis, the stress response, and cellular responses to virus infection (for reviews, see Bernardi and Pandolfi, 2003, Dellaire and Bazett-Jones, 2004; Eskiw and Bazett-Jones, 2002; Everett, 2001; Everett, 2006a; Maul, 1998; Negorev and Maul, 2001; Reineke and Kao, 2009; Saffert and Kalejta, 2008; Tavalai and Stamminger, 2008; Zhong *et al.*, 1999, 2000b). Over 50 different proteins have been detected in ND10 but not all are constitutively present and some have only been observed to co-localize with PML when highly expressed or in specific circumstances (Negorev and Maul, 2001).

There is a substantial literature linking ND10 with the cellular response to infection by a wide variety of viruses. That concerning RNA viruses, small DNA viruses such as the papovaviruses, and adenoviruses has been reviewed elsewhere and will not be considered further here (Everett, 2001; Everett and Chelbi-Alix, 2007; Maul, 1998). The hypothesis that ND10 might be involved in intrinsic defences against DNA virus infection was stimulated by two straightforward early observations: firstly, that the input genomes and early replication compartments of several DNA viruses were found to be associated with ND10 (Ishov and Maul, 1996; Maul, 1998; Maul *et al.*, 1996); secondly, that many DNA viruses, notably HSV-1 and HCMV, encoded regulatory proteins (ICP0 of HSV-1 and IE72 of HCMV) that first co-localized with then disrupted ND10 structures (reviewed in Everett, 2001; Everett and Chelbi-Alix, 2007). However, on the basis of this evidence it was equally possible that viruses engaged ND10 not because their effects are inhibitory, but rather because components of ND10 are beneficial for virus replication. It is only recently that new approaches have strengthened the former hypothesis. The following sections will consider the evidence that a number of ND10 components have roles in intrinsic defence.

The promyelocytic leukaemia protein PML

PML is the key component of ND10 that is required for their assembly (Ishov *et al.*, 1999; Zhong *et al.*, 2000a). PML is the prototype member of the TRIM (tripartite motif) family of proteins that include an N-terminal RING finger (not thought to confer E3 ubiquitin ligase activity in the case of PML), two further zinc stabilized folds known as B-boxes, and a coiled-coil motif involved in protein–protein interactions (Jensen *et al.*, 2001). Alternative splicing downstream of the TRIM coding sequences gives rise to 6 major PML isoforms that range in size from 560 to 882 amino acid residues (Jensen *et al.*, 2001). A characteristic of PML is that it is modified by SUMO conjugation, principally at lysine residues 160 and 490 within the region conserved in all isoforms (Kamitani *et al.*, 1998). SUMO modification at other lysine residues may be isoform dependent, since isoform VI has been reported to be modified also at lysine 65 (Kamitani *et al.*, 1998), while recent evidence suggests that this residue is not modified in isoform I, but instead at lysine 616 in sequences common to isoforms I and IV (Cuchet *et al.*, 2010). Other residues are likely to be subject to SUMO conjugation in certain circumstances, as the complex modification patterns of isoforms III to VI cannot be explained on the basis of the known modification sites, while PML IV lacking all known SUMO modification sites still exhibits modified bands of mobility consistent with SUMO conjugation (Cuchet *et al.*, 2010). Both SUMO conjugation and a SUMO interaction motif (SIM) in the core PML sequences common to isoforms I to V are required for proper ND10 assembly (Shen *et al.*, 2006; Zhong *et al.*, 2000a).

ICP0 mediated disruption of ND10 occurs because it induces the proteasome dependent degradation of PML, and there is evidence that the SUMO modified forms of the protein may be preferentially susceptible (Boutell *et al.*, 2003; Chelbi-Alix and de The, 1999; Everett *et al.*, 1998; Gu and Roizman, 2003; Muller and Dejean, 1999). In this respect, it is intriguing that IE1 mediated disruption of ND10 during HCMV infection also correlates with the loss of the SUMO modified forms of PML (Lee *et al.*, 2004). The RING finger of ICP0 is essential for induction of PML degradation and ND10 disruption, and therefore this process correlates well with the functions of ICP0 in stimulating lytic infection and efficient reactivation from latency.

However, an important consideration is that, as noted above, several other cellular proteins are also susceptible to ICP0-mediated RING finger dependent degradation, so despite the prominence of its effects on PML, it cannot be concluded on the basis of this evidence alone that PML is a biologically relevant target.

The link between PML and intrinsic antiviral defence was made when it was found that depletion of PML using RNAi technology improved the replication efficiency of ICP0 null mutant HSV-1 while having no effect on that of the wt virus (Everett *et al.*, 2008a; Everett *et al.*, 2006; Everett *et al.*, 2008b). These findings, subsequently confirmed independently (Kyratsous and Silverstein, 2009; Kyratsous *et al.*, 2009), provide good evidence that PML contributes to an intrinsic antiviral defence, as defined in an early section of this chapter. Also consistent with this conclusion is the correlation between improved replication of ICP0 null mutant HSV-1 and deficient PML depletion in a number of cancer cell lines, and greatly improved replication of the mutant virus in cells expressing a dominant negative PML–RARα fusion protein (Sobol *et al.*, 2009). Nonetheless, it is important to stress that RNAi mediated depletion of PML has but a relatively modest effect on improving ICP0 null mutant HSV-1 plaque formation, of the order of 5 to 10-fold, whereas if PML were the only factor a 1000-fold improvement would be expected. Nonetheless, yields of ICP0 null mutant virus 24 h after a low multiplicity infection can be improved by as much as two orders of magnitude in PML depleted cells (Everett *et al.*, 2008b). Parallel studies demonstrated that PML depletion also increases plaque formation by IE1 defective HCMV (Tavalai *et al.*, 2006, 2008), suggesting that similar mechanisms may be involved in these two herpesvirus infections.

How PML might act in an intrinsic antiviral defence remains to be understood. PML itself can repress transcription in some assays (Zhong *et al.*, 2000b), but PML also interacts with many other proteins with roles in the regulation of gene expression and in chromatin pathways (Bernardi and Pandolfi, 2007; Ishov *et al.*, 1999; Zhong *et al.*, 1999). Since PML is required for the formation of ND10, depletion of the protein could in principle affect the behaviour of many other ND10 components, some of which also have roles in transcriptional repression and intrinsic antiviral defence. Therefore depletion of PML might improve ICP0 null mutant replication through direct or indirect effects, or both. It is pertinent to note that PML is an interferon responsive gene (Chelbi-Alix *et al.*, 1995), potentially providing a link between innate and intrinsic antiviral defences. It has been suggested that PML is required for the increased sensitivity to interferon of ICP0-null mutant compared to the wt virus (Chee *et al.*, 2003).

Sp100

Sp100 is another interferon inducible major ND10 component that is heavily SUMO modified and expressed from a family of alternatively spliced transcripts (Guldner *et al.*, 1999; Sternsdorf *et al.*, 1997). It has strong links to chromatin modification pathways, as all isoforms bind to the heterochromatin binding protein HP1 (Seeler *et al.*, 1998), and the longer isoforms (Sp100-B, -C and -HMG) include domains characteristic of chromatin associated proteins (Guldner *et al.*, 1999). These longer Sp100 isoforms, but not the most highly expressed isoform, Sp100-A (which is effectively a truncated version of the others), have been linked with repression of gene expression (Negorev *et al.*, 2006; Wilcox *et al.*, 2005). High level expression of the longer isoforms of Sp100 reduces HSV-1 IE gene expression (Negorev *et al.*, 2006; Negorev *et al.*, 2009), while depletion of all isoforms using RNAi improves plaque formation efficiency and gene expression by ICP0 null mutant HSV-1 (Everett *et al.*, 2008a).

ICP0 expression results in loss of the SUMO modified forms of Sp100-A (Muller and Dejean, 1999; Parkinson *et al.*, 1999). Therefore Sp100 appears, like PML, to be a target of ICP0 RING finger activity. However, this effect could be indirect because RNAi mediated depletion of PML also independently causes loss of SUMO modified Sp100-A (Everett *et al.*, 2006); the apparent effect of ICP0 on Sp100 might primarily be caused by loss of PML in cells expressing ICP0. Taking these studies together, the evidence suggests that Sp100 also plays a role in intrinsic antiviral defence.

ATRX and hDaxx

ATRX and hDaxx are two prominent constitutive components of ND10, although they are also located throughout the nucleoplasm. Unlike PML and Sp100 they do not appear to be abundantly SUMO-modified, but hDaxx interacts with SUMO-modified PML (Ishov *et al.*, 1999), probably through a SIM near its C-terminus (Santiago *et al.*, 2009). ATRX and hDaxx interact strongly with each other and they can be co-purified in a complex that includes a number of proteins involved in chromatin modification (Hollenbach *et al.*, 2002; Xue *et al.*, 2003). In addition, localization of ATRX to ND10 is dependent on hDaxx (Ishov *et al.*, 2004; Lukashchuk *et al.*, 2008). Several studies report that hDaxx is involved in transcriptional repression of selected cellular genes in model assays (see for example Chang *et al.*, 2005; Ecsedy *et al.*, 2003; Hollenbach *et al.*, 1999; Michaelson and Leder, 2003; Puto and Reed, 2008), and it has been established to be an important component of intrinsic defence against HCMV (Cantrell and Bresnahan, 2006; Preston and Nicholl, 2006; Saffert and Kalejta, 2006; 2007; Woodhall *et al.*, 2006). In this instance, hDaxx was demonstrated to inhibit HCMV IE gene expression but in normal circumstances this activity is eliminated by the actions of the viral tegument protein pp71. An HCMV mutant lacking this protein expresses its IE genes very inefficiently, but RNAi mediated depletion of hDaxx increases gene expression by pp71 deficient virus significantly.

Several pieces of evidence point to a role of viral chromatin structure in this mode of regulation: firstly, hDaxx interacts with histone deacetylase (HDAC) enzymes (Hollenbach *et al.*, 2002); secondly, pp71 viral mutant gene expression can be improved by treatment of cells with the HDAC inhibitor TSA (Saffert and Kalejta, 2006); and thirdly, hDaxx is required for the establishment of a viral chromatin structure with repressive histone marks (Woodhall *et al.*, 2006).

A role for SWI/SNF related chromatin modifier protein ATRX in regulating HCMV IE gene expression was also identified by an RNAi approach (Lukashchuk *et al.*, 2008). Although pp71 interacts with hDaxx (Hofmann *et al.*, 2002), the earliest event that can be observed on expression of pp71 is removal of ATRX from ND10 without any apparent effect on hDaxx (Lukashchuk *et al.*, 2008). In that depletion of ATRX improves substantially pp71 mutant HCMV IE gene expression and subsequent plaque formation (Lukashchuk *et al.*, 2008), it is possible that this aspect of intrinsic defence involves the hDaxx/ATRX complex.

Recent work on the roles of these two proteins in HSV-1 infection has revealed a similar story. Depletion of either hDaxx or ATRX improves plaque forming efficiency by ICP0 null mutant but not wt HSV-1, and in the case of hDaxx depletion there is a marked increase in the efficiency of Early HSV-1 gene expression (Lukashchuk *et al.*, 2010). Therefore, as with PML and Sp100, the implication is that these proteins take part in the cell mediated repression of HSV-1 gene expression that occurs in the absence of ICP0. Using technologies to deplete endogenous hDaxx then to reconstitute expression at close to endogenous levels, it was found that sequences within hDaxx that are required for its interaction with ATRX are necessary for the repression of ICP0 null mutant gene expression (Lukashchuk *et al.*, 2010). Therefore it appears that this aspect of repression requires the interaction between the two proteins.

Unlike in the cases of PML and Sp100, there is as yet no evidence for a direct biochemical or destabilizing effect of ICP0 on either hDaxx or ATRX but, as described below, ICP0 does affect the spatial behaviour of the two proteins during HSV-1 infection.

Recruitment of ND10 proteins to the sites of parental HSV-1 genomes

The ND10 proteins described above have all been found to contribute to cell mediated repression of HSV-1 gene expression in the absence of ICP0, in that replication of ICP0 null mutant virus is more efficient in the absence of each individual one. Despite these clear and reproducible findings, it is not yet understood how this repression occurs and how such apparently diverse proteins can accomplish this function. There is however one aspect in the behaviour of these proteins that is shared: during the very early stages of infection, all are recruited to sites that are closely associated with parental HSV-1 genomes and early replication compartments. In wt infections this is very short lived and difficult to observe because it is rapidly inhibited by the actions of ICP0. In ICP0 null mutant infections, the recruitment of these proteins into novel ND10-like structures associated with the viral genomes is readily observable (Everett and Murray, 2005; Everett *et al.*, 2003, 2004b). Furthermore, this is an extremely rapid cellular response to entry of the viral genome into the nucleus, since it does not require *de novo* viral gene expression (Everett and Murray, 2005) and it can be detected within 30 minutes of adding virus to cells (Everett *et al.*, 2007).

The fact that the intrinsic cellular defence renders viral gene expression undetectable in the majority of cells infected at low multiplicity with an ICP0 null mutant virus implies that it begins to act at the earliest stages of infection. Therefore the cellular response that leads to the recruitment of ND10 component proteins to viral genomes has the hallmarks one might expect of an intrinsic resistance mechanism. It may be significant that examination of a panel of ICP0 mutants has revealed a tight link between their relative abilities to complement ICP0 null mutant replication and to counteract this recruitment process (Everett *et al.*, 2009). Nonetheless, it must be borne in mind that other proteins, not necessarily only those known to be associated with ND10, may also be recruited in a similar manner through some common mechanism. Therefore if the recruitment process is indeed a hallmark of intrinsic defence, the involvement of other as yet undefined proteins cannot be excluded.

Given the potential importance of the recruitment phenomenon, it is of interest to understand the mechanisms that enable it. Although PML is required for nucleation of normal ND10 in uninfected cells (Ishov *et al.*, 1999; Zhong *et al.*, 2000a), it is not required for the recruitment to viral genome foci of the other ND10 proteins described above (Everett *et al.*, 2007; Everett *et al.*, 2008a; Everett *et al.*, 2006). Similarly, recruitment of PML, hDaxx and ATRX is not dependent on Sp100 (Everett *et al.*, 2008a and unpublished data). Recruitment of PML and Sp100 is not dependent on hDaxx, but recruitment of ATRX is (Lukashchuk *et al.*, 2010). Therefore, while recruitment of some proteins may be dependent on others, there must be some shared property of PML, Sp100 and hDaxx that enables their independent recruitment to viral genome foci. Interestingly, it has been observed that as yet uncharacterised SUMO modified proteins are also recruited to sites associated with HSV-1 genomes, which implies that SUMO related pathways may be involved in the recruitment process (Boutell *et al.*, 2010). Recent work has found that the SUMO interaction motifs of PML, Sp100, and hDaxx are required for recruitment of these proteins to HSV-1 genome associated sites (Cuchet *et al.*, 2010). Taken together, these observations suggest that the recruitment process and assembly of novel ND10-like foci in close association with incoming HSV-1 genomes might involve bridging interactions between SUMO modified and SIM containing proteins, as proposed for the assembly of normal ND10 (Shen *et al.*, 2006). Given that many other proteins contain SIMs and/or are modified by SUMO conjugation, it is possible that the known list of proteins that are recruited to sites closely associated with incoming HSV-1 genomes will be extended considerably. It will be a lengthy task to determine whether any individual recruited protein is involved in regulating HSV-1 gene expression or is recruited merely because it simply shares a characteristic in common with other recruited proteins.

The involvement of SIMs in the recruitment process implies that the SUMO conjugation pathway may play a role in the mechanism of intrinsic resistance. This is of interest given that a number of transcriptional repressors can be SUMO modified and indeed must be so modified in order to enable their repressive properties (Hay, 2005). Furthermore, ICP0 causes the degradation or deconjugation of the SUMO conjugated forms of many as yet uncharacterised proteins in a RING finger dependent manner (Boutell *et al.*, 2010; Everett *et al.*, 1998). One can speculate that there may be a unifying theme of cellular protein recruitment to sites of HSV-1 genomes, intrinsic resistance, and the actions of ICP0: recruitment involves SUMO modified and/or SIM containing proteins, the relevant repression proteins require SUMO conjugation or SIM mediated interactions for their activity, and ICP0 eliminates the SUMO conjugates thereby inhibiting both recruitment and repression. This hypothesis requires further investigation to test its validity.

Chromatin structure, histone modification, ICP0 and intrinsic defence

The role of chromatin structure in the ICP0 phenotype

Whatever the mechanism of intrinsic resistance, it is clear that the end result is long term repression of viral gene expression, and it is reasonable to assume that this represents a stage in the establishment of a latency or quiescence. Numerous studies have led to the conclusion that quiescent or latent HSV-1 genomes are in a state resembling facultative heterochromatin with histone marks characteristic of a repressed chromatin structure (reviewed in Knipe and Cliffe, 2008). Therefore the phenomenon of intrinsic resistance could encompass the pathways that package viral genomes into an initial chromatin structure then modify the chromatin with histone marks and non-histone chromatin proteins in order to maintain a repressed state. It is also clear that ICP0 can reverse this state, since it readily reactivates or de-represses gene expression from quiescent, repressed genomes. The regulation of viral chromatin is discussed elsewhere in this volume in Chapters 2 and 17, so the topic will not be reviewed here in any detail. However, the possible connections between viral chromatin and the activities of ICP0 have attracted much research activity. The intention here is to discuss these issues in relation to intrinsic resistance.

Histone deacetylases

One model that is commonly put forward to explain intrinsic resistance involves rapid chromatinization of the viral genome after its entry into the nucleus, followed by establishment of a repressed chromatin state through the activities of histone deacetylase (HDAC) enzymes. This model posits that ICP0 functions by inhibiting chromatin assembly and/or HDACs, thereby maintaining an active chromatin structure. The evidence in favour of this model includes: (i) chromatin precipitation (ChIP) assays have detected an increased histone load on ICP0 null mutant compared to wt HSV-1 genomes after infection of cultured cells, associated with increased levels of repressive and decreased levels of activating histone marks (such as acetylation of histone H3) (Cliffe and Knipe, 2008); (ii) detection of interactions between ICP0 and a number of HDACs (Lomonte *et al.*, 2004); (iii) improved replication of ICP0 null mutant HSV-1 in certain cell lines after treatment with HDAC inhibitors (Poon *et al.*, 2006; Terry-Allison *et al.*, 2007); and (iv) induction of reactivation of quiescent or latent HSV-1 in rodent neurones and some cultured cell types by treatment with HDAC inhibitors (see for example Danaher *et al.*, 2005; Terry-Allison *et al.*, 2007).

Compelling though this evidence may be, there are some significant caveats. For example, the ability of HDAC inhibitors to increase ICP0 null mutant gene expression or induce de-repression in certain cell types does not necessarily mean that ICP0 has the same function, particularly in view of the fact that HDAC inhibitors such as sodium butyrate have long been used as a means of non-specifically augmenting gene expression. The evidence that runs counter to the hypothesis that ICP0 functions through inhibition of HDACs includes: (i) whereas the HDAC inhibitor TSA causes a substantial increase in total cellular acetylated histone H4, ICP0 has no effect and therefore does not have a global inhibitory effect on HDAC activity (Lomonte *et al.*, 2004); (ii) in cell types such as HFs and HepaRG hepatocytes, in which the ICP0 defective phenotype is very pronounced and therefore intrinsic resistance highly active, treatment with TSA has no effect on ICP0 null mutant replication (Everett *et al.*, 2008a); (iii) quiescent HSV-1 in non-neuronal cells is not induced to reactivate by TSA (Terry-Allison *et al.*, 2007).

This counter evidence formally discounts in its simplest form the hypothesis that intrinsic resistance functions through the activities of HDACs and is counteracted by an effect of ICP0 on these proteins. Nonetheless, the role of chromatin structure in the regulation of many aspects of HSV-1 infection is self evident, especially the establishment and maintenance of latency and during reactivation. Thus HDACs are likely to play a role in aspects of HSV-1 biology, but inhibition of their activity does not render ICP0 redundant in human fibroblasts and other restrictive human cell types.

The CoREST–REST-HDAC complex

A variation on the HDAC hypothesis arose from the observation that a serine and alanine rich sequence of ICP0 between residues 560 and 600 exhibits about 50% identity with a sequence in the cellular protein CoREST, a corepressor involved in the repression of neuronal specific genes in non-neuronal cells (Gu *et al.*, 2005). CoREST forms a complex with REST and HDACs 1 and 2, and it was hypothesized that ICP0 might, by virtue of its sequence similarity motif with CoREST, disrupt this complex and thereby relieve repression of HSV-1 gene expression. In support of this hypothesis, evidence was presented in favour of an interaction between ICP0 and CoREST (subsequently mapped to a distinct region of ICP0 further towards its carboxy terminus; Gu and Roizman, 2007). Furthermore, the CoREST/REST/HDAC complex is disrupted during HSV-1 infection, with CoREST being phosphorylated and both CoREST and HDAC1 becoming translocated to the cytoplasm (Gu *et al.*, 2005). High level expression of a truncated form of CoREST also causes disruption of the CoREST/REST/HDAC complex and partially complemented an ICP0 null mutant virus (Gu and Roizman, 2007), and a mutation in the region of ICP0 reported to interact with CoREST delayed viral replication kinetics (Gu and Roizman, 2009).

It is an attractive hypothesis that the CoREST/REST/HDAC1 complex represses HSV-1 gene expression and thus imparts intrinsic resistance, since it invokes known repressors and the end result would be predicted to be repressive modification of viral chromatin structure. However, as a note of caution it should be pointed out that the observed disruption of the complex occurs only at mid to late times of infection (Gu *et al.*, 2005), long after the principal effects of ICP0 have been accomplished, and the mutation in the proposed CoREST binding region has a much less marked effect on viral replication efficiency than a RING finger mutation (Gu and Roizman, 2009). Furthermore, a mutant form of ICP0 that lacks entirely the CoREST binding region increases ICP0 null mutant replication by up to two orders of magnitude and decreases stimulation of reactivation/de-repression of quiescent HSV-1 by only around 2-fold (Everett *et al.*, 2009). Therefore the proposed interaction between ICP0 and CoREST is dispensable for significant ICP0 activity in its core functions.

DNA damage response proteins

An emerging and intriguing field is the possible link between intrinsic resistance and the cellular DNA damage response. This latter topic is reviewed in detail by Sandy Weller and Matt Weitzman elsewhere in this volume, but it is worthwhile to summarize some key points here. It has been established that HSV-1 infection activates the DNA damage response, perhaps because the cell recognizes the HSV-1 genome, a substantial quantity of unchromatinized DNA with free ends, as a concentration of damaged DNA (Lilley *et al.*, 2005). Consistent with this hypothesis, the histone variant H2AX becomes phosphorylated during HSV-1 infection to produce γH2AX, and this modified protein accumulates within the cellular chromatin surrounding the viral genomes (Wilkinson and Weller, 2006).

In the course of the DNA damage response in an uninfected cell, γH2AX formation is also stimulated and the protein accumulates at the sites of DNA damage (for example irradiation induced DNA damage foci, IRIF) as a consequence of a cascade of activities and the sequential recruitment of a large number of DNA damage response proteins. An essential event in IRIF formation is the recruitment of RNF8 and RNF168, two cellular RING finger ubiquitin ligases that catalyse the monoubiquitination of histone H2A. This stabilizes the recruitment of DNA repair factors that lead to restructuring of chromatin around the break and subsequent repair of the damaged DNA. It is intriguing that PML and a number of other ND10 proteins are also recruited to IRIF (reviewed in Everett, 2006a), and it has recently been found that SUMO conjugates are also present at sites of damaged DNA (Morris *et al.*, 2009). Taken together, these observations open the possibility of significant parallels between IRIF formation and the assembly of the novel ND10-like foci that are associated with HSV-1 genomes.

Given these parallels, it is extremely interesting that, in addition to its ability to disrupt ND10 and centromeres, ICP0 also disrupts irradiation induced DNA damage foci (IRIF) (Lilley *et al.*, 2010). Just as ICP0 targets specific proteins within ND10 and centromeres, ICP0 mediated disruption of IRIF occurs because it induces the degradation of proteins that are important for their formation, namely RNF8 and RNF168. In the case of RNF8 this was shown to be due to a direct RNF8–ICP0 interaction and consequent ubiquitination of the cellular protein (Lilley *et al.*, 2010). Since IRIF and HSV-1 induced ND10-like foci have components in common (PML and SUMO conjugates, for example), and since ICP0 efficiently destabilizes both structures, it is possible that the recruitment of ND10 proteins to sites associated with viral genomes may be related in some way to a cellular DNA damage response. In a previous section the hypothesis was put forward that the assembly of novel ND10-like complexes at sites associated with viral genomes is connected with intrinsic resistance. If this is correct, and there are also connections between the mechanisms of formation of these foci and the DNA damage response, an ensuing speculation is that the latter may also be involved in intrinsic resistance. This model is supported by improved replication of ICP0 null mutant HSV-1 in RNF8 depleted cells (Lilley *et al.*, 2010). These exciting and promising findings open novel avenues to explore in order to understand intrinsic resistance to HSV-1 infection.

Comparative properties of other members of the ICP0 family

A characteristic of members of the alphaherpesvirus subfamily is the expression of a RING finger protein related to ICP0, of which the most extensively studied are those expressed by bovine herpesvirus 1 (BHV-1), equine herpes virus 1 (EHV-1), pseudorabies virus (PRV) and varicella zoster virus (VZV). The bovine and equine virus proteins have long been referred to as BICP0 and EICP0, so it would make sense to refer to the others as PICP0 and VICP0 despite other names (particularly EP0 for the PRV and orf61 for the VZV proteins) being in common usage. The following paragraphs describe the similarities between these four proteins and ICP0.

All of these proteins activate gene expression in reporter assays (Bowles *et al.*, 1997; Cabirac *et al.*, 1990; Fraefel *et al.*, 1994; Ho *et al.*, 1999; Kim *et al.*, 1999; Moriuchi *et al.*, 1993; Ono *et al.*, 1998; Watanabe *et al.*, 1995) and, where tested, can at least in part complement an ICP0-null HSV-1 mutant (Everett *et al.*, 1995; Kyratsous *et al.*, 2009; Moriuchi *et al.*, 1992, 1995). The gene expression activation functions of the proteins are RING finger dependent (Bowles *et al.*, 2000; Inman *et al.*, 2001; Moriuchi *et al.*, 1994; Saira *et al.*, 2008; Watanabe *et al.*, 1996; Zhang *et al.*, 2005). Virus mutants for each of the above proteins have been isolated and their phenotypes found to be analogous to that of ICP0 null mutant HSV-1 in terms of reduced efficiency of replication (Boldogkoi *et al.*, 2000; Brukman and Enquist, 2006b; Cheung *et al.*, 1994; Cohen and Nguyen, 1998; Geiser *et al.*, 2005; Koppel *et al.*, 1996; Wang *et al.*, 2009; Yao *et al.*, 2003).

There is evidence that BICP0 can interact with components of the interferon response pathway and inhibit interferon induced gene expression (Henderson *et al.*, 2005; Saira *et al.*, 2007, 2009), while lack of PICP0 renders PRV more sensitive to interferon pre-treatment (Brukman and Enquist, 2006a; Brukman and Enquist, 2006b). These activities have parallels with the effects of interferon treatment on ICP0 null mutant HSV-1 replication (see Chapter 13 and references therein). However, it should be borne in mind that inactivation of key components of the interferon response pathway does not overcome the ICP0 null mutant defect in cell culture (Everett *et al.*, 2008b) and that when expressed at levels sufficient to complement completely an ICP0 null mutant, ICP0 does not affect interferon stimulated gene expression (although it does enable the mutant virus to replicate more efficiently in the face of an interferon induced state) (Everett and Orr, 2009). Likewise, expression of BICP0, EICP0 and PICP0 in amounts that complement

ICP0 null mutant HSV-1 does not interfere with interferon stimulated gene expression or affect the abundance of STAT-1, STAT-2 or IRF-3 (Everett *et al.*, 2010).

Again by analogy with the studies on ICP0 and histone modification cited above, BICP0 interacts with HDAC1 (Zhang and Jones, 2001) and the histone acetyltransferase p300 (Zhang *et al.*, 2006) in co-transfected cells expressing high levels of exogenous proteins. On the same theme, sodium butyrate (an HDAC inhibitor and general activator of gene expression) increases gene expression during infection with VICP0 deficient VZV (Wang *et al.*, 2009). These observations contribute to the general debate of whether or not the primary function of the ICP0 family of proteins is to regulate chromatin structure through affects on histone modification pathways.

In terms of biochemical commonalties with ICP0, it is known that the RING domain of BICP0 has E3 ubiquitin ligase activity *in vitro* (Diao *et al.*, 2005), and this is true for all of the above named proteins (Everett *et al.*, 2010). All four of the proteins discussed in this section are capable, like ICP0, of inducing the formation of co–localizing conjugated ubiquitin in transfected cells, which implies that they act as E3 ubiquitin ligases *in vivo* (Parkinson and Everett, 2001). All members of the group localize to and/or affect ND10, although in detail their effects were distinguishable (Kyratsous and Silverstein, 2009; Parkinson and Everett, 2000; Reichelt *et al.*, 2009).

A major question that arises from the above studies is whether these other members of the ICP0 family also interfere with intrinsic resistance to virus infection, and if so, whether they utilize the same mechanisms as ICP0. An important consideration is that while all are related by the presence of a RING finger, they have limited homology elsewhere, especially if all five proteins are compared with each other. If the specificity of RING finger mediated E3 ubiquitin ligase activity is determined by separate interaction motifs, this group of proteins might exhibit variations in their targets specificities. The available evidence, albeit limited, indicates that this is indeed the case. Although EICP0 expressed in the context of recombinant HSV-1 reduces the levels of the SUMO modified forms of PML and Sp100, none of the four ICP0 related proteins affect PML in a model co-transfection assay (Parkinson and Everett, 2000). Whereas VICP0 induces the loss of certain isoforms of Sp100, it has no obvious affect on PML (Kyratsous *et al.*, 2009). Nonetheless, depletion of PML increases wt VZV plaque formation, and depletion of either PML or hDaxx increases VZV yields at 24 h after infection (Kyratsous and Silverstein, 2009). These data suggest that PML and hDaxx play some role in restricting VZV replication but as yet there is no conclusive evidence that VICP0 is involved in counteracting these activities.

Recent work on the ICP0 related proteins using an inducible cell line system based on that used for a study of ICP0 (Everett *et al.*, 2009) has allowed a comparison of several properties of this group of proteins, with the exception of VICP0 which was expressed at levels insufficient for reliable analysis. As might be expected from the studies cited above, significant differences were found in the abilities of BICP0, EICP0, PICP0 and ICP0 to degrade PML and SUMO modified forms of Sp100. However, all these proteins inhibit very efficiently the recruitment of ND10 component proteins to the sites of incoming HSV-1 genomes (Everett *et al.*, 2010). The hypothesis that this rapid cellular response to entry of viral genomes into the nucleus is an important component of intrinsic resistance has been discussed above. The correlation between complementation of ICP0 null mutant HSV-1, induction of reactivation of quiescent HSV-1 and inhibition of the recruitment of ND10 component proteins to sites associated with viral genomes presents a promising avenue for future research.

Conclusions and perspectives

The concept of cellular intrinsic resistance to herpesvirus infections has grown out of an understanding of the phenotypes of HSV-1 and HCMV mutants defective in ICP0, IE1 and pp71. While the model of repression of viral gene expression through the actions of pre-existing cellular proteins is clear, we have much to learn about the cellular proteins that are involved and how they accomplish their actions. The evidence indicates that several ND10 component proteins are involved in this process. However, in the case of HSV-1, depletion of individual ND10 proteins results in improvements in the replication of the ICP0 null mutant virus that are modest compared to the maximum possible. There are a number of potential explanations for these observations. Firstly, these proteins might act in parallel, such that depletion of multiple proteins would be required to inactivate intrinsic resistance. There is evidence in favour of this model in the case of PML and Sp100 (Everett *et al.*, 2008a). Secondly, they may act in complex, as apparently in the case of hDaxx and ATRX (Lukashchuk *et al.*, 2008; Lukashchuk *et al.*, 2010). A third possibility is that they act on sequential steps of the viral transcriptional programme, as suggested for hDaxx and PML in the case of HCMV (Tavalai *et al.*, 2008; Tavalai and Stamminger, 2008). Fourthly, it can be by no means excluded that a cellular protein that is the core component of intrinsic resistance remains to be discovered.

We also lack a clear understanding of how the actions of these proteins lead to transcriptional repression of the viral genome. While in the final analysis chromatin structure must play an important role, particularly during the various stages of latency and quiescence, there are significant pieces of evidence that are incompatible with a simple model based on histone acetylation and deacetylation events. There are also uncertainties about the detailed mechanisms by which viral regulatory proteins accomplish their roles in overcoming intrinsic resistance. Given the diversity between important proteins such as ICP0, IE1 and pp71 (and even between different members of the ICP0 family), it seems that the precise mechanisms by which they accomplish their related functions must differ – there must be alternative routes that lead to the same destination. Unravelling the viral and cellular complexities of intrinsic resistance and how it is overcome will be a difficult challenge, but one that is worthwhile since it is likely to underlie fundamental aspects of virus biology.

Acknowledgements

The work in the author's laboratory is funded by the Medical Research Council. The author is grateful for the constructive criticisms of Chris Boutell, Caroline Lilley and Matt Weitzman.

References

Arthur, J.L., Scarpini, C.G., Connor, V., Lachmann, R.H., Tolkovsky, A.M., and Efstathiou, S. (2001). Herpes simplex virus type 1 promoter activity during latency establishment, maintenance, and reactivation in primary dorsal root neurons *in vitro*. J. Virol. *75*, 3885–3895.

Bernardi, R., and Pandolfi, P.P. (2003). Role of PML and the PML-nuclear body in the control of programmed. cell death. Oncogene *22*, 9048–9057.

Bernardi, R., and Pandolfi, P.P. (2007). Structure, dynamics and functions of promyelocytic leukaemia nuclear bodies. Nat. Rev. Microbiol. 8, 1006–1016.

Boldogkoi, Z., Braun, A., and Fodor, I. (2000). Replication and virulence of early protein 0 and long latency transcript deficient mutants of the Aujeszky's disease (pseudorabies) virus. Microbes Infect. *2*, 1321–1328.

Boutell, C., Canning, M., Orr, A., and Everett, R.D. (2005). Reciprocal activities between herpes simplex virus type 1 regulatory protein ICP0, a ubiquitin E3 ligase, and ubiquitin-specific protease USP7. J. Virol. *79*, 12342–12354.

Boutell, C., Cuchet, D., Vanni, E., Orr, A., and Everett, R.D. (2010). Herpes simplex virus type 1 regulatory protein ICP0 exhibits SUMO-targeted ubiquitin ligase activity. Submitted for publication.

Boutell, C., Everett, R., Hilliard, J., Schaffer, P., Orr, A., and Davido, D. (2010). Herpes simplex virus type 1 ICP0 phosphorylation mutants impair the E3 ubiquitin ligase activity of ICP0 in a cell type-dependent manner. J. Virol. *82*, 10647–10656.

Boutell, C., and Everett, R.D. (2003). The herpes simplex virus type 1 (HSV-1) regulatory protein ICP0 interacts with and Ubiquitinates p53. J. Biol. Chem. *278*, 36596–36602.

Boutell, C., and Everett, R.D. (2004). Herpes simplex virus type 1 infection induces the stabilization of p53 in a USP7- and ATM-independent manner. J. Virol. *78*, 8068–8077.

Boutell, C., Orr, A., and Everett, R.D. (2003). PML residue lysine 160 is required for the degradation of PML induced by herpes simplex virus type 1 regulatory protein ICP0. J. Virol. *77*, 8686–8694.

Boutell, C., Sadis, S., and Everett, R.D. (2002). Herpes simplex virus type 1 immediate-early protein ICP0 and its isolated RING finger domain act as ubiquitin E3 ligases *in vitro*. J. Virol. *76*, 841–850.

Bowles, D.E., Holden, V.R., Zhao, Y., and O'Callaghan, D.J. (1997). The ICP0 protein of equine herpesvirus 1 is an early protein that independently transactivates expression of all classes of viral promoters. J. Virol. *71*, 4904–4914.

Bowles, D.E., Kim, S.K., and O'Callaghan, D.J. (2000). Characterization of the trans-activation properties of equine herpesvirus 1 EICP0 protein. J. Virol. *74*, 1200–1208.

Brukman, A., and Enquist, L.W. (2006a). Pseudorabies virus EP0 protein counteracts an interferon-induced antiviral state in a species-specific manner. J. Virol. *80*, 10871–10873.

Brukman, A., and Enquist, L.W. (2006b). Suppression of the interferon-mediated innate immune response by pseudorabies virus. J. Virol. *80*, 6345–6356.

Cabirac, G.F., Mahalingam, R., Wellish, M., and Gilden, D.H. (1990). Trans-activation of viral tk promoters by proteins encoded by varicella zoster virus open reading frames 61 and 62. Virus Res. *15*, 57–68.

Canning, M., Boutell, C., Parkinson, J., and Everett, R.D. (2004). A RING Finger Ubiquitin Ligase Is Protected from Autocatalyzed Ubiquitination and Degradation by Binding to Ubiquitin-specific Protease USP7. J. Biol. Chem. *279*, 38160–38168.

Cantrell, S.R., and Bresnahan, W.A. (2006). Human cytomegalovirus (HCMV) UL82 gene product (pp71) relieves hDaxx-mediated repression of HCMV replication. J. Virol. *80*, 6188–6191.

Chang, C.C., Lin, D.Y., Fang, H.I., Chen, R.H., and Shih, H.M. (2005). Daxx mediates the small ubiquitin-like modifier-dependent transcriptional repression of Smad4. J. Biol. Chem. *280*, 10164–10173.

Chee, A.V., Lopez, P., Pandolfi, P.P., and Roizman, B. (2003). Promyelocytic leukemia protein mediates interferon-based anti-herpes simplex virus 1 effects. J. Virol. *77*, 7101–7105.

Chelbi-Alix, M.K., and de The, H. (1999). Herpes virus induced proteasome-dependent degradation of the nuclear bodies-associated PML and Sp100 proteins. Oncogene *18*, 935–941.

Chelbi-Alix, M.K., Pelicano, L., Quignon, F., Koken, M.H.M., Venturini, L., Stadler, M., Pavlovic, J., Degos, L., and de Thé, H. (1995). Induction of the PML protein by interferons in normal and APL cells. Leukemia *9*, 2027–2033.

Chen, J., and Silverstein, S. (1992). Herpes simplex viruses with mutations in the gene encoding ICP0 are defective in gene expression. J. Virol. *66*, 2916–2927.

Cheung, A.K., Fang, J., and Wesley, R.D. (1994). Characterization of a pseudorabies virus that is defective in the early protein 0 and latency genes. Am. J. Vet. Res. *55*, 1710–1716.

Ciufo, D.M., Mullen, M.A., and Hayward, G.S. (1994). Identification of a dimerization domain in the C-terminal segment of the IE110 transactivator protein from herpes simplex virus. J. Virol. *68*, 3267–3282.

Cliffe, A.R., and Knipe, D.M. (2008). Herpes Simplex Virus ICP0 Promotes Both Histone Removal and Acetylation on Viral DNA During Lytic Infection. J. Virol. *82*, 12030–12038.

Cohen, J.I., and Nguyen, H. (1998). Varicella-zoster virus ORF61 deletion mutants replicate in cell culture, but a mutant with stop codons in ORF61 reverts to wild-type virus. Virology *246*, 306–316.

Coleman, H.M., Connor, V., Cheng, Z.S., Grey, F., Preston, C.M., and Efstathiou, S. (2008). Histone modifications associated with herpes simplex virus type 1 genomes during quiescence and following ICP0-mediated de-repression. J. Gen. Virol. *89*, 68–77.

Cuchet, D., Lukashchuk, V., Sykes, A., Murray, J., Orr, A., and Everett, R.D. (2010). SUMO interaction techniques are required for the recruitment of ND10 components to MSV-1 induced foci. Submitted for publication.

Danaher, R.J., Jacob, R.J., Steiner, M.R., Allen, W.R., Hill, J.M., and Miller, C.S. (2005). Histone deacetylase inhibitors induce reactivation of herpes simplex virus type 1 in a latency-associated transcript-independent manner in neuronal cells. J. NeuroVirol. *11*, 306–317.

Davido, D.J., von Zagorski, W.F., Lane, W.S., and Schaffer, P.A. (2005). Phosphorylation site mutations affect herpes simplex virus type 1 ICP0 function. J. Virol. *79*, 1232–1243.

Dellaire, G., and Bazett-Jones, D.P. (2004). PML nuclear bodies: dynamic sensors of DNA damage and cellular stress. Bioessays *26*, 963–977.

Diao, L., Zhang, B., Fan, J., Gao, X., Sun, S., Yang, K., Xin, D., Jin, N., Geng, Y., and Wang, C. (2005). Herpes virus proteins ICP0 and BICP0 can activate NF-kappaB by catalyzing IkappaBalpha ubiquitination. Cell Signal *17*, 217–229.

Ecsedy, J.A., Michaelson, J.S., and Leder, P. (2003). Homeodomain-interacting protein kinase 1 modulates Daxx localization, phosphorylation, and transcriptional activity. Mol. Cell Biol. *23*, 950–960.

Efstathiou, S., and Preston, C.M. (2005). Towards an understanding of the molecular basis of herpes simplex virus latency. Virus Res. *111*, 108–119.

Eskiw, C.H., and Bazett-Jones, D.P. (2002). The promyelocytic leukemia nuclear body: sites of activity? Biochem. Cell Biol. *80*, 301–310.

Everett, R., Orr, A., and Elliott, M. (1995). The equine herpesvirus 1 gene 63 RING finger protein partially complements Vmw110, its herpes simplex virus type 1 counterpart. J. Gen. Virol. *76*, 2369–2374.

Everett, R.D. (1989). Construction and characterization of herpes simplex virus type 1 mutants with defined lesions in immediate early gene 1. J. Gen. Virol. *70*, 1185–1202.

Everett, R.D. (2000a). ICP0, a regulator of herpes simplex virus during lytic and latent infection. Bioessays *22*, 761–770.

Everett, R.D. (2000b). ICP0, a regulator of herpes simplex virus during lytic and latent infection. Bioessays *22*, 761–770.

Everett, R.D. (2001). DNA viruses and viral proteins that interact with PML nuclear bodies. Oncogene *20*, 7266–7273.

Everett, R.D. (2006a). Interactions between DNA viruses, ND10 and the DNA damage response. Cell. Microbiol. *8*, 365–374.

Everett, R.D. (2006b). The roles of ICP0 during HSV-1 infection. In Alphaherpesviruses. Molecular and Cellular Biology, Sandri-Goldin, R.M., ed. (Wymondham, Caister Academic Press), pp. 39–64.

Everett, R.D., Boutell, C., and Orr, A. (2004a). Phenotype of a herpes simplex virus type 1 mutant that fails to express immediate-early regulatory protein ICP0. J. Virol. *78*, 1763–1774.

Everett, R.D., and Chelbi-Alix, M.K. (2007). PML and PML nuclear bodies: Implications in antiviral defence. Biochimie *89*, 819–830.

Everett, R.D., Earnshaw, W.C., Findlay, J., and Lomonte, P. (1999). Specific destruction of kinetochore protein CENP-C and disruption of cell division by herpes simplex virus immediate-early protein Vmw110. EMBO J. *18*, 1526–1538.

Everett, R.D., Freemont, P., Saitoh, H., Dasso, M., Orr, A., Kathoria, M., and Parkinson, J. (1998). The disruption of ND10 during herpes simplex virus infection correlates with the Vmw110- and proteasome-dependent loss of several PML isoforms. J. Virol. *72*, 6581–6591.

Everett, R.D., and Maul, G.G. (1994). HSV-1 IE protein Vmw110 causes redistribution of PML. EMBO J. *13*, 5062–5069.

Everett, R.D., McNair, C., Grant, L., Boutell, C., and Orr, A. (2010). Comparison of the biological and biochemical activities of several members of the herpes simplex virus ICP0 family of proteins. J. Virol. *84*, 3476–3487.

Everett, R.D., Meredith, M., Orr, A., Cross, A., Kathoria, M., and Parkinson, J. (1997). A novel ubiquitin-specific protease is dynamically associated with the PML nuclear domain and binds to a herpesvirus regulatory protein. EMBO J. *16*, 1519–1530.

Everett, R.D., and Murray, J. (2005). ND10 Components Relocate to Sites Associated with Herpes Simplex Virus Type 1 Nucleoprotein Complexes during Virus Infection. J. Virol. *79*, 5078–5089.

Everett, R.D., Murray, J., Orr, A., and Preston, C.M. (2007). Herpes simplex virus type 1 genomes are associated with ND10 nuclear substructures in quiescently infected human fibroblasts. J. Virol. *81*, 10991–11004.

Everett, R.D., and Orr, A. (2009). Herpes simplex virus type 1 regulatory protein ICP0 aids infection in cells with a pre-induced interferon response but does not impede interferon induced gene expression. J. Virol. *83*, 4978–4983.

Everett, R.D., Parada, C., Gripon, P., Sirma, H., and Orr, A. (2008a). Replication of ICP0-null mutant herpes simplex virus type 1 is restricted by both PML and Sp100. J. Virol. *82*, 2661–2672.

Everett, R.D., Parsy, M.L., and Orr, A. (2009). Analysis of the functions of herpes simplex virus type 1 regulatory protein ICP0 that are critical for lytic infection and derepression of quiescent viral genomes. J. Virol. *83*, 4963–4977.

Everett, R.D., Rechter, S., Papior, P., Tavalai, N., Stamminger, T., and Orr, A. (2006). PML contributes to a cellular mechanism of repression of herpes simplex virus type 1 infection that is inactivated by ICP0. J. Virol. *80*, 7995–8005.

Everett, R.D., Sourvinos, G., Leiper, C., Clements, J.B., and Orr, A. (2004b). Formation of nuclear foci of the herpes simplex virus type 1 regulatory protein ICP4 at early times of infection: localization, dynamics, recruitment of ICP27, and evidence for the de novo induction of ND10-like complexes. J. Virol. *78*, 1903–1917.

Everett, R.D., Sourvinos, G., and Orr, A. (2003). Recruitment of herpes simplex virus type 1 transcriptional regulatory protein ICP4 into foci juxtaposed to ND10 in live, infected cells. J. Virol. *77*, 3680–3689.

Everett, R.D., Young, D.F., Randall, R.E., and Orr, A. (2008b). STAT-1- and IRF-3-dependent pathways are not essential for repression of ICP0-null mutant herpes simplex virus type 1 in human fibroblasts. J. Virol. *82*, 8871–8881.

Filion, G.J., Zhenilo, S., Salozhin, S., Yamada, D., Prokhortchouk, E., and Defossez, P.A. (2006). A family of human zinc finger proteins that bind methylated DNA and repress transcription. Mol. Cell Biol. *26*, 169–181.

Fraefel, C., Zeng, J., Choffat, Y., Engels, M., Schwyzer, M., and Ackermann, M. (1994). Identification and zinc dependence of the bovine herpesvirus 1 transactivator protein BICP0. J. Virol. *68*, 3154–3162.

Geiser, V., Zhang, Y., and Jones, C. (2005). Analysis of a bovine herpesvirus 1 recombinant virus that does not express the bICP0 protein. J. Gen. Virol. *86*, 1987–1996.

Gu, H., Liang, Y., Mandel, G., and Roizman, B. (2005). Components of the REST/CoREST/histone deacetylase repressor complex are disrupted, modified, and translocated in HSV-1-infected cells. Proc. Natl. Acad. Sci. U.S.A. *102*, 7571–7576.

Gu, H., and Roizman, B. (2003). The degradation of promyelocytic leukemia and Sp100 proteins by herpes simplex virus 1 is mediated by the ubiquitin-conjugating enzyme UbcH5a. Proc. Natl. Acad. Sci. U.S.A. *100*, 8963–8968.

Gu, H., and Roizman, B. (2007). Herpes simplex virus-infected cell protein 0 blocks the silencing of viral DNA by dissociating histone deacetylases from the CoREST–REST complex. Proc. Natl. Acad. Sci. U.S.A. *104*, 17134–17139.

Gu, H., and Roizman, B. (2009). The two functions of herpes simplex virus 1 ICP0, inhibition of silencing by the CoREST/REST/HDAC complex and degradation of PML, are executed in tandem. J. Virol. *83*, 181–187.

Guldner, H.H., Szostecki, C., Schroder, P., Matschl, U., Jensen, K., Luders, C., Will, H., and Sternsdorf, T. (1999). Splice variants of the nuclear dot-associated Sp100 protein contain homologies to HMG-1 and a human nuclear phosphoprotein-box motif. J. Cell Sci. *112*, 733–747.

Hagglund, R., and Roizman, B. (2002). Characterization of the novel E3 ubiquitin ligase encoded in exon 3 of herpes simplex virus-1-infected cell protein 0. Proc. Natl. Acad. Sci. U.S.A. *99*, 7889–7894.

Hagglund, R., and Roizman, B. (2004). Role of ICP0 in the Strategy of Conquest of the Host Cell by Herpes Simplex Virus 1. J. Virol. *78*, 2169–2178.

Halford, W.P., Kemp, C.D., Isler, J.A., Davido, D.J., and Schaffer, P.A. (2001). ICP0, ICP4, or VP16 expressed from adenovirus vectors induces reactivation of latent herpes simplex virus type 1 in primary cultures. of latently infected trigeminal ganglion cells. J. Virol. *75*, 6143–6153.

Halford, W.P., and Schaffer, P.A. (2000). Optimized viral dose and transient immunosuppression enable herpes simplex virus ICP0-null mutants To establish wild-type levels of latency *in vivo*. J. Virol. *74*, 5957–5967.

Halford, W.P., and Schaffer, P.A. (2001). ICP0 is required for efficient reactivation of herpes simplex virus type 1 from neuronal latency. J. Virol. *75*, 3240–3249.

Harris, R.A., Everett, R.D., Zhu, X.X., Silverstein, S., and Preston, C.M. (1989). Herpes simplex virus type 1 immediate-early protein Vmw110 reactivates latent herpes simplex virus type 2 in an *in vitro* latency system. J. Virol. *63*, 3513–3515.

Hay, R.T. (2005). SUMO: a history of modification. Mol. Cell *18*, 1–12.

Henderson, G., Zhang, Y., and Jones, C. (2005). The Bovine herpesvirus 1 gene encoding infected cell protein 0 (bICP0) can inhibit interferon-dependent transcription in the absence of other. viral genes. J. Gen. Virol. *86*, 2697–2702.

Ho, T.Y., Wu, S.L., Chang, T.J., Hsiang, C.H., Chang, S.H., and Hsiang, C.Y. (1999). Pseudorabies virus early protein 0 trans-activates the TATA-associated promoter by stimulating the transcription initiation. Virus Res. *61*, 77–86.

Hofmann, H., Sindre, H., and Stamminger, T. (2002). Functional interaction between the pp71 protein of human cytomegalovirus and the PML-interacting protein human Daxx. J. Virol. *76*, 5769–5783.

Hollenbach, A.D., McPherson, C.J., Mientjes, E.J., Iyengar, R., and Grosveld, G. (2002). Daxx and histone deacetylase II associate with chromatin through an interaction with core histones and the chromatin-associated protein Dek. J. Cell Sci. *115*, 3319–3330.

Hollenbach, A.D., Sublett, J.E., McPherson, C.J., and Grosveld, G. (1999). The Pax3-FKHR oncoprotein is unresponsive to the Pax3-associated repressor hDaxx. EMBO J. *18*, 3702–3711.

Huthoff, H., and Towers, G.J. (2008). Restriction of retroviral replication by APOBEC3G/F and TRIM5alpha. Trends Microbiol. *16*, 612–619.

Inman, M., Zhang, Y., Geiser, V., and Jones, C. (2001). The zinc ring finger in the bICP0 protein encoded by bovine herpesvirus-1 mediates toxicity and activates productive infection. J. Gen. Virol. *82*, 483–492.

Ishov, A.M., and Maul, G.G. (1996). The periphery of nuclear domain 10 (ND10) as site of DNA virus deposition. J. Cell Biol. *134*, 815–826.

Ishov, A.M., Sotnikov, A.G., Negorev, D., Vladimirova, O.V., Neff, N., Kamitani, T., Yeh, E.T., Strauss, J.F., and Maul, G.G. (1999). PML is critical for ND10 formation and recruits the PML-interacting protein daxx to this nuclear structure when modified by SUMO-1. J. Cell Biol. *147*, 221–234.

Ishov, A.M., Vladimirova, O.V., and Maul, G.G. (2004). Heterochromatin and ND10 are cell-cycle regulated and phosphorylation-dependent alternate nuclear sites of the transcription repressor Daxx and SWI/SNF protein ATRX. J. Cell Sci. *117*, 3807–3820.

Jensen, K., Shiels, C., and Freemont, P.S. (2001). PML protein isoforms and the RBCC/TRIM motif. Oncogene *20*, 7223–7233.

Joazeiro, C.A., and Weissman, A.M. (2000). RING finger proteins: mediators of ubiquitin ligase activity. Cell *102*, 549–552.

Kamitani, T., Kito, K., Nguyen, H.P., Wada, H., Fukuda-Kamitani, T., and Yeh, E.T. (1998). Identification of three major sentrinization sites in PML. J. Biol. Chem. *273*, 26675–26682.

Kim, S.K., Bowles, D.E., and O'Callaghan D., J. (1999). The gamma2 late glycoprotein K promoter of equine herpesvirus 1 is differentially regulated by the IE and EICP0 proteins. Virology *256*, 173–179.

Knipe, D.M., and Cliffe, A. (2008). Chromatin control of herpes simplex virus lytic and latent infection. Nat. Rev. Microbiol. *6*, 211–221.

Koppel, R., Fraefel, C., Vogt, B., Bello, L.J., Lawrence, W.C., and Schwyzer, M. (1996). Recombinant bovine herpesvirus-1 (BHV-1) lacking transactivator protein BICPO entails lack of glycoprotein C and severely reduced infectivity. Biol. Chem. *377*, 787–795.

Kyratsous, C.A., and Silverstein, S.J. (2009). Components of nuclear domain 10 bodies regulate varicella-zoster virus replication. J. Virol. *83*, 4262–4274.

Kyratsous, C.A., Walters, M.S., Panagiotidis, C.A., and Silverstein, S.J. (2009). Complementation of a herpes simplex virus ICP0 null mutant by varicella zoster Virus ORF61p. J. Virol. *83*, 10637–10643.

Lee, H.R., Kim, D.J., Lee, J.M., Choi, C.Y., Ahn, B.Y., Hayward, G.S., and Ahn, J.H. (2004). Ability of the human cytomegalovirus IE1 protein to modulate sumoylation of PML correlates with its functional activities in transcriptional regulation and infectivity in cultured fibroblast cells. J. Virol. *78*, 6527–6542.

Lees-Miller, S.P., Long, M.C., Kilvert, M.A., Lam, V., Rice, S.A., and Spencer, C.A. (1996). Attenuation of DNA-dependent protein kinase activity and its catalytic subunit by the herpes simplex virus type 1 transactivator ICP0. J. Virol. *70*, 7471–7477.

Lilley, C.E., Carson, C.T., Muotri, A.R., Gage, F.H., and Weitzman, M.D. (2005). DNA repair proteins affect the lifecycle of herpes simplex virus 1. Proc. Natl. Acad. Sci. U.S.A. *102*, 5844–5849.

Lilley, C.E., Chaurushiya, M.S., Boutell, C., Landry, S., Suh, J., Panier, S., Everett, R.D., Stewart, G.S., Durocher, D., and Weitzman, M.D. (2010). A viral E3 ligase targets RNF8 and RNF168 to control histone ubiquitination and the cellular response to DNA damage. EMBO J. *29*, 943–955.

Lium, E.K., and Silverstein, S. (1997). Mutational analysis of the herpes simplex virus type 1 ICP0 C3HC4 zinc ring finger reveals a requirement for ICP0 in the expression of the essential alpha27 gene. J. Virol. *71*, 8602–8614.

Lomonte, P., and Morency, E. (2007). Centromeric protein CENP-B proteasomal degradation induced by the viral protein ICP0. FEBS Lett. *581*, 658–662.

Lomonte, P., Sullivan, K.F., and Everett, R.D. (2001). Degradation of nucleosome-associated centromeric histone H3-like protein CENP-A induced by herpes simplex virus type 1 protein ICP0. J. Biol. Chem. *276*, 5829–5835.

Lomonte, P., Thomas, J., Texier, P., Caron, C., Khochbin, S., and Epstein, A.L. (2004). Functional interaction between class II histone deacetylases and ICP0 of herpes simplex virus type 1. J. Virol. *78*, 6744–6757.

Lukashchuk, V., McFarlane, S., Everett, R.D., and Preston, C.M. (2008). Human cytomegalovirus protein pp71 displaces the chromatin-associated factor ATRX from nuclear domain 10 at early stages of infection. J. Virol. *82*, 12543–12554.

Lukashchuk, V., Orr, A., and Everett, R.D. (2010). Regulation of HSV-1 infection by ND10 components ATRX and hDaxx. J. Virol. *84*, 4026–4040.

Maul, G.G. (1998). Nuclear domain 10, the site of DNA virus transcription and replication. Bioessays *20*, 660–667.

Maul, G.G., Ishov, A.M., and Everett, R.D. (1996). Nuclear domain 10 as preexisting potential replication start sites of herpes simplex virus type-1. Virology *217*, 67–75.

Meredith, M., Orr, A., Elliott, M., and Everett, R. (1995). Separation of sequence requirements for HSV-1 Vmw110 multimerisation and interaction with a 135-kDa cellular protein. Virology *209*, 174–187.

Michaelson, J.S., and Leder, P. (2003). RNAi reveals anti-apoptotic and transcriptionally repressive activities of DAXX. J. Cell Sci. *116*, 345–352.

Miller, C.S., Danaher, R.J., and Jacob, R.J. (2006). ICP0 is not required for efficient stress-induced reactivation of herpes simplex virus type 1 from cultured quiescently infected neuronal cells. J. Virol. *80*, 3360–3368.

Moriuchi, H., Moriuchi, M., and Cohen, J.I. (1994). The RING finger domain of the varicella-zoster virus open reading frame 61 protein is required for its transregulatory functions. Virology *205*, 238–246.

Moriuchi, H., Moriuchi, M., Dean, H., Cheung, A.K., and Cohen, J.I. (1995). Pseudorabies virus EPO is functionally homologous to varicella-zoster virus ORF61 protein and herpes simplex virus type 1 ICPO. Virology *209*, 281–283.

Moriuchi, H., Moriuchi, M., Smith, H.A., Straus, S.E., and Cohen, J.I. (1992). Varicella-zoster virus open reading frame 61 protein is functionally homologous to herpes simplex virus type 1 ICP0. J. Virol. *66*, 7303–7308.

Moriuchi, H., Moriuchi, M., Straus, S.E., and Cohen, J.I. (1993). Varicella-zoster virus (VZV) open reading frame 61 protein transactivates VZV gene promoters and enhances the infectivity of VZV DNA. J. Virol. *67*, 4290–4295.

Morris, J.R., Boutell, C., Keppler, M., Densham, R., Weekes, D., Alamshah, A., Butler, Y., Galanty, Y., Pangon, L., Kiuchi, T., *et al.* (2009). The SUMO modification pathway is involved in the BRCA1 response to genotoxic stress. Nature *462*, 886–890.

Mossman, K.L., and Smiley, J.R. (1999). Truncation of the C-terminal acidic activation domain of herpes simplex virus VP16 renders expression of the immediate-early genes almost entirely dependent on ICP0. J. Virol. *73*, 9726–9733.

Mullen, M.A., Gerstberger, S., Ciufo, D.M., Mosca, J.D., and Hayward, G.S. (1995). Evaluation of colocalization interactions between the IE110, IE175, and IE63 transactivator proteins of herpes simplex virus within subcellular punctate structures. J. Virol. *69*, 476–491.

Muller, S., and Dejean, A. (1999). Viral immediate-early proteins abrogate the modification by SUMO-1 of PML and Sp100 proteins, correlating with nuclear body disruption. J. Virol. *73*, 5137–5143.

Negorev, D., and Maul, G.G. (2001). Cellular proteins localized at and interacting within ND10/PML nuclear bodies/PODs suggest functions of a nuclear depot. Oncogene *20*, 7234–7242.

Negorev, D.G., Vladimirova, O.V., Ivanov, A., Rauscher, F., 3rd, and Maul, G.G. (2006). Differential role of Sp100 isoforms in interferon-mediated repression of herpes simplex virus type 1 immediate-early protein expression. J. Virol. *80*, 8019–8029.

Negorev, D.G., Vladimirova, O.V., and Maul, G.G. (2009). Differential functions of interferon-upregulated Sp100 isoforms: herpes simplex virus type 1 promoter-based immediate-early gene suppression and PML protection from ICP0-mediated degradation. J. Virol. *83*, 5168–5180.

Ono, E., Watanabe, S., Nikami, H., Tasaki, T., and Kida, H. (1998). Pseudorabies virus (PRV) early protein 0 activates PRV gene transcription in combination with the immediate-early protein IE180 and enhances the infectivity of PRV genomic DNA. Vet. Microbiol. *63*, 99–107.

Parkinson, J., and Everett, R.D. (2000). Alphaherpesvirus proteins related to herpes simplex virus type 1 ICP0 affect cellular structures. and proteins. J. Virol. *74*, 10006–10017.

Parkinson, J., and Everett, R.D. (2001). Alphaherpesvirus proteins related to herpes simplex virus type 1 ICP0 induce the formation of colocalizing, conjugated ubiquitin. J. Virol. *75*, 5357–5362.

Parkinson, J., Lees-Miller, S.P., and Everett, R.D. (1999). Herpes simplex virus type 1 immediate-early protein vmw110 induces the proteasome-dependent degradation of the catalytic subunit of DNA-dependent protein kinase. J. Virol. *73*, 650–657.

Poon, A.P., Gu, H., and Roizman, B. (2006). ICP0 and the US3 protein kinase of herpes simplex virus 1 independently block histone deacetylation to enable gene expression. Proc. Natl. Acad. Sci. U.S.A. *103*, 9993–9998.

Preston, C.M. (2000). Repression of viral transcription during herpes simplex virus latency. J. Gen. Virol. *81*, 1–19.

Preston, C.M. (2007). Reactivation of expression from quiescent herpes simplex virus type 1 genomes in the absence of immediate-early protein ICP0. J. Virol. *81*, 11781–11789.

Preston, C.M., and Nicholl, M.J. (1997). Repression of gene expression upon infection of cells with herpes simplex virus type 1 mutants impaired for immediate-early protein synthesis. J. Virol. *71*, 7807–7813.

Preston, C.M., and Nicholl, M.J. (2006). Role of the cellular protein hDaxx in human cytomegalovirus immediate-early gene expression. J. Gen. Virol. *87*, 1113–1121.

Puto, L.A., and Reed, J.C. (2008). Daxx represses RelB target promoters via DNA methyltransferase recruitment and DNA hypermethylation. Genes Dev. *22*, 998–1010.

Reichelt, M., Brady, J., and Arvin, A.M. (2009). The replication cycle of varicella-zoster virus: analysis of the kinetics of viral protein expression, genome synthesis, and virion assembly at the single-cell level. J. Virol. *83*, 3904–3918.

Reineke, E.L., and Kao, H.Y. (2009). Targeting promyelocytic leukemia protein: a means to regulating PML nuclear bodies. Int. J. Biol. Sci. *5*, 366–376.

Roizman, B., and Taddeo, B. (2006). HSV-1 and the host cell: A story of global conspiracies, plots and counterplots. In Alphaherpesviruses. Molecular and Cellular Biology, Sandri-Goldin, R.M., ed. (Wymondham, Caister Academic Press), pp. 261–282.

Sacks, W.R., and Schaffer, P.A. (1987). Deletion mutants in the gene encoding the herpes simplex virus type 1 immediate-early protein ICP0 exhibit impaired growth in cell culture. J. Virol. *61*, 829–839.

Saffert, R.T., and Kalejta, R.F. (2006). Inactivating a cellular intrinsic immune defense mediated by Daxx is the mechanism through which the human cytomegalovirus pp71 protein stimulates viral immediate-early gene expression. J. Virol. *80*, 3863–3871.

Saffert, R.T., and Kalejta, R.F. (2007). Human cytomegalovirus gene expression is silenced by Daxx-mediated intrinsic immune defense in model latent infections established *in vitro*. J. Virol. *81*, 9109–9120.

Saffert, R.T., and Kalejta, R.F. (2008). Promyelocytic leukemia-nuclear body proteins: herpesvirus enemies, accomplices, or both? Future Virology *3*, 265–277.

Saira, K., Chowdhury, S., Gaudreault, N., da Silva, L., Henderson, G., Doster, A., and Jones, C. (2008). The zinc RING finger of bovine herpesvirus 1-encoded bICP0 protein is crucial for viral replication and virulence. J. Virol. *82*, 12060–12068.

Saira, K., Zhou, Y., and Jones, C. (2007). The infected cell protein 0 encoded by bovine herpesvirus 1 (bICP0) induces degradation of interferon response factor 3 and, consequently, inhibits beta interferon promoter activity. J. Virol. *81*, 3077–3086.

Saira, K., Zhou, Y., and Jones, C. (2009). The infected cell protein 0 encoded by bovine herpesvirus 1 (bICP0) associates with interferon regulatory factor 7 and consequently inhibits beta interferon promoter activity. J. Virol. *83*, 3977–3981.

Samaniego, L.A., Neiderhiser, L., and DeLuca, N.A. (1998). Persistence and expression of the herpes simplex virus genome in the absence of immediate-early proteins. J. Virol. *72*, 3307–3320.

Santiago, A., Godsey, A.C., Hossain, J., Zhao, L.Y., and Liao, D. (2009). Identification of two independent SUMO-interacting motifs in Daxx: evolutionary conservation from Drosophila to humans and their biochemical functions. Cell Cycle *8*, 76–87.

Sebastian, S., and Luban, J. (2007). The Retroviral Restriction Factor TRIM5alpha. Curr. Infect. Dis. Rep. *9*, 167–173.

Seeler, J.S., Marchio, A., Sitterlin, D., Transy, C., and Dejean, A. (1998). Interaction of SP100 with HP1 proteins: a link between the promyelocytic leukemia-associated nuclear bodies and the chromatin compartment. Proc. Natl. Acad. Sci. U.S.A. *95*, 7316–7321.

Shen, T.H., Lin, H.K., Scaglioni, P.P., Yung, T.M., and Pandolfi, P.P. (2006). The mechanisms of PML-nuclear body formation. Mol. Cell *24*, 331–339.

Sobol, P.T., Hummel, J.L., Rodrigues, R.M., and Mossman, K.L. (2009). PML has a predictive role in tumor cell permissiveness to interferon-sensitive oncolytic viruses. Gene Ther.

Sokolskaja, E., and Luban, J. (2006). Cyclophilin, TRIM5, and innate immunity to HIV-1. Curr. Opin. Microbiol. *9*, 404–408.

Sternsdorf, T., Jensen, K., and Will, H. (1997). Evidence for covalent modification of the nuclear dot-associated proteins PML and Sp100 by PIC1/SUMO-1. J. Cell Biol. *139*, 1621–1634.

Stow, N.D., and Stow, E.C. (1986). Isolation and characterization of a herpes simplex virus type 1 mutant containing a deletion within the gene encoding the immediate early polypeptide Vmw110. J. Gen. Virol. *67*, 2571–2585.

Takeuchi, H., and Matano, T. (2008). Host factors involved in resistance to retroviral infection. Microbiol. Immunol. *52*, 318–325.

Tavalai, N., Papior, P., Rechter, S., Leis, M., and Stamminger, T. (2006). Evidence for a role of the cellular ND10 protein PML in mediating intrinsic immunity against human cytomegalovirus infections. J. Virol. *80*, 8006–8018.

Tavalai, N., Papior, P., Rechter, S., and Stamminger, T. (2008). Nuclear domain 10 components promyelocytic leukemia protein and hDaxx independently contribute to an intrinsic antiviral defense against human cytomegalovirus infection. J. Virol. *82*, 126–137.

Tavalai, N., and Stamminger, T. (2008). New insights into the role of the subnuclear structure ND10 for viral infection. Biochim. Biophys. Acta *1783*, 2207–2221.

Terry-Allison, T., Smith, C.A., and DeLuca, N.A. (2007). Relaxed repression of herpes simplex virus type 1 genomes in Murine trigeminal neurons. J. Virol. *81*, 12394–12405.

Thompson, R.L., and Sawtell, N.M. (2006). Evidence that the herpes simplex virus type 1 ICP0 protein does not initiate reactivation from latency *in vivo*. J. Virol. *80*, 10919–10930.

Wang, L., Sommer, M., Rajamani, J., and Arvin, A.M. (2009). Regulation of the ORF61 promoter and ORF61 functions in varicella-zoster virus replication and pathogenesis. J. Virol. *83*, 7560–7572.

Watanabe, S., Ono, E., Shimizu, Y., and Kida, H. (1995). Pseudorabies virus early protein 0 transactivates the viral gene promoters. J. Gen. Virol. *76*, 2881–2885.

Watanabe, S., Ono, E., Shimizu, Y., and Kida, H. (1996). Mapping of transregulatory domains of pseudorabies virus early protein 0 and identification of its dominant-negative mutant. Arch. Virol. *141*, 1001–1009.

Wilcox, K.W., Sheriff, S., Isaac, A., and Taylor, J.L. (2005). SP100B is a repressor of gene expression. J. Cell Biochem. *95*, 352–365.

Wilkinson, D.E., and Weller, S.K. (2006). Herpes simplex virus type I disrupts the ATR-dependent DNA-damage response during lytic infection. J. Cell Sci. *119*, 2695–2703.

Woodhall, D.L., Groves, I.J., Reeves, M.B., Wilkinson, G., and Sinclair, J.H. (2006). Human Daxx-mediated repression of human cytomegalovirus gene expression correlates with a repressive chromatin structure around the major immediate early promoter. J. Biol. Chem. *281*, 37652–37660.

Xue, Y., Gibbons, R., Yan, Z., Yang, D., McDowell, T.L., Sechi, S., Qin, J., Zhou, S., Higgs, D., and Wang, W. (2003). The ATRX syndrome protein forms a chromatin remodeling complex with Daxx and localizes in promyelocytic leukemia nuclear bodies. Proc. Natl. Acad. Sci. U.S.A. *100*, 10635–10640.

Yao, F., and Schaffer, P.A. (1995). An activity specified by the osteosarcoma line U2OS can substitute functionally for ICP0, a major regulatory protein of herpes simplex virus type 1. J. Virol. *69*, 6249–6258.

Yao, H., Osterrieder, N., and O'Callaghan, D.J. (2003). Generation and characterization of an EICP0 null mutant of equine herpesvirus 1. Virus Res. *98*, 163–172.

Zhang, Y., Jiang, Y., Geiser, V., Zhou, J., and Jones, C. (2006). Bovine herpesvirus 1 immediate-early protein (bICP0) interacts with the histone acetyltransferase p300, which stimulates productive infection and gC promoter activity. J. Gen. Virol. *87*, 1843–1851.

Zhang, Y., and Jones, C. (2001). The bovine herpesvirus 1 immediate-early protein (bICP0) associates with histone deacetylase 1 to activate transcription. J. Virol. *75*, 9571–9578.

Zhang, Y., Zhou, J., and Jones, C. (2005). Identification of functional domains within the bICP0 protein encoded by bovine herpesvirus 1. J. Gen. Virol. *86*, 879–886.

Zhong, S., Hu, P., Ye, T.Z., Stan, R., Ellis, N.A., and Pandolfi, P.P. (1999). A role for PML and the nuclear body in genomic stability. Oncogene *18*, 7941–7947.

Zhong, S., Muller, S., Ronchetti, S., Freemont, P.S., Dejean, A., and Pandolfi, P.P. (2000a). Role of SUMO-1-modified PML in nuclear body formation. Blood *95*, 2748–2752.

Zhong, S., Salomoni, P., and Pandolfi, P.P. (2000b). The transcriptional role of PML and the nuclear body. Nat. Cell Biol. *2*, E85–90.

Zhu, X.X., Chen, J.X., Young, C.S., and Silverstein, S. (1990). Reactivation of latent herpes simplex virus by adenovirus recombinants encoding mutant IE-0 gene products. J. Virol. *64*, 4489–4498.

Multiple Roles of Immediate-early Protein ICP22 in HSV-1 Replication

5

Stephen A. Rice

Abstract

ICP22 is the least characterized of the five herpes simplex virus type 1 (HSV-1) immediate-early (IE) proteins. However, accumulating evidence indicates that it carries out a number of interesting regulatory activities inside the infected cell. These include the enhancement of viral gene expression, the modification of RNA polymerase II (RNAP II), and the reorganization of host cell molecular chaperones into nuclear inclusion bodies. Recent studies of engineered HSV-1 mutants indicate that certain of ICP22's activities are genetically separable from each other. Thus, similar to several other of the IE proteins, ICP22 appears to be a multifunctional, multi-domain polypeptide. This chapter summarizes the current state of knowledge concerning ICP22 and its varied regulatory roles during the productive HSV-1 infection.

Introduction

Upon initially infecting a susceptible cell, HSV-1 expresses five IE genes. The encoded polypeptides are crucial regulators that help shape the cellular environment into one that is conducive for viral replication. For example, the IE protein ICP0, through its E3 ubiquitin ligase activity, targets the host PML protein for proteolytic degradation, in this way likely counteracting a host defence mechanism (Everett and Chelbi-Alix, 2007). Another IE protein, ICP47, interferes with adaptive immunity by preventing major histocompatibility class I antigen presentation (Hill *et al.*, 1995). In addition to disrupting host defences, the viral IE proteins are crucial for inducing the expression of the remaining 75 or so viral delayed-early (DE) and late (L) genes. In fact, four of the five IE proteins (ICP0, ICP4, ICP22, and ICP27) are nuclear proteins that have demonstrated roles in viral gene expression.

Among the four IE proteins that regulate viral gene expression, ICP22 has received by far the least attention in the published literature. This likely stems from the fact that early studies showed ICP22 to be non-essential for HSV-1 growth in Vero cells (Post and Roizman, 1981), one of the most common cell lines used to study viral replication. Moreover, other early studies failed to demonstrate that the cloned ICP22 gene (unlike the ICP0, 4 and 27 genes) could mediate regulatory effects on co-expressed genes in plasmid transfection assays (Gelman and Silverstein, 1985; O'Hare and Hayward, 1985). However, as discussed herein, multiple lines of evidence accumulated over the years demonstrate that ICP22 is indeed an important HSV-1 regulatory protein. It is needed for viral growth in many cell lines (Sears *et al.*, 1985; Poffenberger *et al.*, 1993; Orlando *et al.*, 2006a) and is required for efficient viral replication and latency in mice and guinea pig models of infection (Sears *et al.*, 1985; Poffenberger *et al.*, 1994; Orlando *et al.*, 2006b). Moreover, as detailed below, ICP22 has intriguing and seemingly novel effects on several host cell factors, including RNAP II and molecular chaperones.

Expression and properties of ICP22

ICP22 is encoded by a spliced transcript originating from the viral US1 gene (Fig. 5.1A). Although the coding region for ICP22 is in the unique U_S component of the HSV-1 genome, the promoter and 5′ portion of the transcribed region (including the intron) reside in the adjoining repeated sequence, known as R_S. Because of this arrangement, a second copy of the promoter drives expression of a distinct gene, US12, which encodes the IE protein ICP47. The ICP22/47 promoter is quite unusual in that it contains a viral origin of DNA replication, OriS, firmly embedded in the midst of the transcriptional regulatory sequences, situated between the TATA box and the TAATGARAT motifs that confer IE kinetics to the gene (Preston *et al.*, 1984). However, OriS does not appear to play a major role in ICP22/47 promoter regulation (Summers and Leib, 2002). On the other hand, recent data do indicate that the ICP22/47 promoter has some interesting regulatory properties, in that it is induced by heat-shock treatment (Kushnir *et al.*, 2010). This is intriguing given recent evidence linking ICP22 to the function of host cell chaperones (Bastian *et al.*, 2010), several of which are themselves induced by heat-shock.

The ICP22 open reading frame (ORF) encodes a polypeptide composed of 420 amino acids (Fig. 5.1B). Homologues of ICP22 are encoded by all known members of the *Alphaherpesvirinae*, with the conserved region corresponding to residues 161–292 of ICP22 (Schwyzer *et al.*, 1994). A region C-terminal to this, corresponding to ICP22 residues 295–420, is conserved throughout the genus *Simplexvirus*, one of the four genera of the *Alphaherpesvirinae*, but it is not conserved across genera (Bastian and Rice, 2009). Most of the known simplexviruses are primate viruses associated with urogenital and neuronal infections (van Regenmortel *et al.*, 2000).

ICP22 localizes to the nucleus soon after its synthesis (Leopardi *et al.*, 1997). Consistent

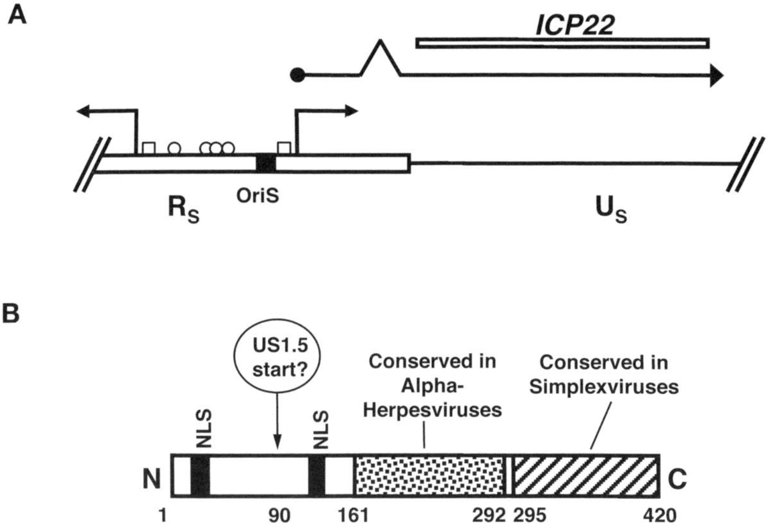

Figure 5.1 Schematic diagram of the ICP22 gene and polypeptide. (A) Structure of the US1 gene. R_S and U_S components or the HSV-1 genome are indicated. The leftward and rightward arrows indicate the transcription start sites of the ICP4 and ICP22 transcripts, respectively. TATA boxes and TAATGARAT motifs are shown as small open boxes and circles, respectively. Above, the spliced ICP22 transcript is shown, along with the location of the ICP22 ORF (narrow bar). (B) ICP22 polypeptide. Amino acid residues are shown at the bottom. The two mapped NLS motifs are indicated, as are ICP22 sequences conserved throughout the alphaherpesviruses or in the *Simplexvirus* genus only. The arrow indicates the likely translational initiation site of the U_S1.5 protein.

with this, two nuclear localization signals (NLSs) have been mapped in the N-terminal part of the protein (Fig. 5.1B) (Stelz *et al.*, 2002). However, a recent study of engineered viral ICP22 mutants suggests that there is at least one additional NLS in the C-terminal portion (Bastian and Rice, 2009). In terms of its intranuclear localization, much of ICP22 localizes diffusely, but some of it localizes to small nuclear bodies (Leopardi *et al.*, 1997; Jahedi *et al.*, 1999) that are distinct from PML foci (Salsman *et al.*, 2008). Recently, these structures were shown to correspond to previously identified entities known as VICE (for 'virus-induced, chaperone-enriched') domains (Bastian *et al.*, 2010). In fact, it was demonstrated that ICP22 is essential for VICE domain formation. The relationship between ICP22 and VICE domains is discussed in more detail later in this chapter.

Translation of ICP22 mRNA in an *in vitro* translation system yields a protein with an apparent molecular weight of ~68 kDa (Watson *et al.*, 1979). However, during infection, ICP22 becomes extensively post-translationally modified, primarily by phosphorylation (Wilcox *et al.*, 1980; Purves and Roizman, 1992). Thus, when ICP22 is isolated from infected cells, it migrates on SDS-PAGE as several species that range in apparent molecular weight from 65–90 kDa. Both of HSV-1's known protein kinases, UL13 and US3, have been implicated in phosphorylating ICP22 on serine and/or threonine residues (Purves and Roizman, 1992; Asai *et al.*, 2007). ICP22 has also been reported to be phosphorylated on one or more tyrosine residues (O'Toole *et al.*, 2003), but the kinase or kinases responsible have not been identified. There is also evidence that ICP22 is nucleotidylated (Blaho *et al.*, 1993).

To date, no enzymatic activities or biochemical functions have been ascribed to ICP22. Thus, it is likely that many of its functions are mediated through interactions with other proteins. However, few such interactions have been characterized. ICP22 has been shown to bind to two cell-cycle-regulated cellular proteins, p60 and p78 (Bruni and Roizman, 1998; Bruni *et al.*, 1999). The p60 protein is also known as Pict-1 (Okahara *et al.*, 2004), while p78 is known as MSP58 (Okumura *et al.*, 2005). Interestingly, both of these cellular proteins interact with PTEN, a protein and lipid phosphatase that functions as a tumour suppressor (Keniry and Parsons, 2008). The relationship of the PTEN pathway to ICP22 function remains an unexplored area. There is also evidence that ICP22 interacts with cyclin-dependent kinase cdk9 (Durand *et al.*, 2005), a protein involved in transcriptional regulation. This interaction and its potential significance are discussed in more detail later.

The US1 gene harbours a second gene, US1.5, that encodes an N-terminally truncated form of ICP22 (Carter and Roizman, 1996). Initially, it was proposed that the US1.5 protein is encoded by a distinct transcript derived from a promoter downstream of ICP22/47 promoter, and that translation initiates on this transcript at an AUG codon that corresponds to codon 147 of the ICP22 ORF (Carter and Roizman, 1996). Later, the same group amended the proposed initiation site to codon 171 (Poon *et al.*, 2000). However, more recent work by Bowman and Schaffer provides strong evidence that translation of US1.5 begins at codon 90 of ICP22 (Bowman and Schaffer, 2009). Moreover, despite efforts from this latter group, no evidence of a second transcript that could encode the US1.5 protein was obtained. It was therefore suggested that the US1.5 protein arises from alternative translational initiation on the ICP22 transcript, possibly involving a ribosomal scanning or shunting mechanism. Like ICP22, the US1.5 protein is expressed as a number of post-translationally modified species, and the production of these is dependent on UL13 (Ogle and Roizman, 1999). For simplicity, the US1.5 protein will not be discussed further in this chapter, but it should be remembered that this protein may contribute to some of ICP22's functions. As well, the US1.5 protein could have functions not shared with ICP22.

Enhancement of viral gene expression by ICP22

Although HSV-1 mutants replicate efficiently in some cell lines such as Vero cells, in other cells they show significant defects in viral growth (Sears *et al.*, 1985; Poffenberger *et al.*, 1993;

Orlando *et al.*, 2006a). For example, rodent and primary human lines tend to be restrictive for ICP22 mutant growth, whereas growth-transformed human lines tend to be permissive. However, the molecular basis of this cell type-dependence is unknown. It has been suggested that permissive cell lines express a factor that substitutes functionally for ICP22 (Sears *et al.*, 1985). Alternatively, restrictive cell lines might contain a factor that imposes a block to replication that must be overcome by ICP22. These two possibilities could be potentially distinguished by asking whether ICP22 mutants replicate efficiently in heterokaryons formed from fusion of permissive and restrictive cells. This approach has been used successfully to investigate the basis of the cell type-dependent replication of HSV-1 ICP0 and VP16 mutants (Hancock *et al.*, 2006).

Early studies revealed why ICP22 mutants do not replicate efficiently in restrictive cells: they have a defect in viral L gene expression (Sears *et al.*, 1985; Poffenberger *et al.*, 1993). In contrast, viral DNA replication occurs efficiently in such infections (Sears *et al.*, 1985; Rice *et al.*, 1995; Orlando *et al.*, 2006a). The defect in L gene expression in ICP22 mutant-infected cells appears to be due to a reduction in the levels of several L gene mRNAs, although the ICP0 mRNA is also reduced in abundance at late times after infection (Purves *et al.*, 1993). Nuclear run-on assays suggest that the reduced accumulation of L gene mRNAs is due to a defect in their transcription (Rice *et al.*, 1995). Curiously, not all L genes require ICP22 for efficient expression in restrictive cells. For example, the UL48 gene does not. L genes that show a strong requirement for ICP22 include US11, UL38, and UL41. The molecular basis of the differential requirements of L genes for ICP22 is not understood.

ICP22's ability to enhance L gene expression in restrictive cells requires the UL13 protein kinase (Purves *et al.*, 1993; Long *et al.*, 1999). Thus, HSV-1 UL13 mutants exhibit a cell type defect in viral growth and L gene expression that is nearly identical to that of ICP22 mutants. Given that UL13 has been implicated in phosphorylating ICP22, a likely model is that ICP22 must be phosphorylated by UL13 in order to carry out its L gene enhancement function.

Alteration of cell-cycle components by ICP22

A substantial amount of work, much of it from Bernard Roizman's laboratory, indicates that ICP22 alters certain components of the host cell-cycle machinery during infection. Moreover, at least some of these effects appear linked to L gene enhancement by ICP22. One striking effect of ICP22 is on the activity and regulation of the cyclin-dependent kinase cdc2 (cdk1). In cycling human cells, cdc2 interacts with cyclins A and B to regulate the G2/M transition of the cell-cycle. However, in HSV-1 infected cells, where the cell-cycle is known to be dysregulated (Ehmann *et al.*, 2000; Song *et al.*, 2000), cyclins A and B levels fall precipitously, while paradoxically, cdc2 activity increases despite the absence of a known cyclin partner (Advani *et al.*, 2000a). These changes require both ICP22 and UL13. The finding that cdc2 is active despite lacking its normal cyclin partners suggests that it may acquire a new co-factor in infected cells. Consistent with this, Advani *et al.* showed that cdc2 binds to an HSV-1 protein, the DNA polymerase processivity factor UL42, which possesses a cyclin box-like motif (Advani *et al.*, 2000b). The ICP22-dependent activation of cdc2 kinase may also involve a host cell phosphatase, cdc25C, which normally regulates cdc2 by removing inhibitory phosphates. Smith-Donald and Roizman showed that cdc25C can interact with both ICP22 and cdc2, suggesting that HSV-1 uses cdc25C to help maintain the activity of the modified cdc2 (Smith-Donald and Roizman, 2008). In summary, these studies indicate that ICP22, in a UL13-dependent fashion, physically and functionally alters the cdc2 kinase during HSV-1 infection.

What could be the function of the HSV-1-altered cdc2 kinase? Using a dominant-negative cdc2, Advani *et al.* showed that cdc2 activity is needed during infection for efficient expression of US11, one of the L genes known to be dependent upon ICP22 (Advani *et al.*, 2000b). Thus, ICP22's ability to activate and alter the cdc2 kinase could be central to its role in enhancing viral L gene expression. Could the altered kinase be having its effects by phosphorylating novel substrates? Consistent with this, it was found that cdc2/UL42 complex binds to and phosphorylates the host cell topoisomerase IIα (Advani *et al.*, 2003). It was suggested that activated topoisomerase IIα might enhance viral L gene expression by untangling replicated viral genomes and enabling them to serve more efficiently as templates for L gene transcription (Advani *et al.*, 2003). It is also possible that the putative stimulatory effect of topoisomerase IIα on gene expression is more direct, as this enzyme has been shown to interact with RNA and mRNA processing factors (Lee *et al.*, 2004; Park *et al.*, 2008).

Work from Priscilla Schaffer's laboratory has uncovered another interesting link between ICP22 and the cell-cycle (Orlando *et al.*, 2006a). Although Vero cells are normally permissive for ICP22 mutant replication, it was found that they become restrictive when they are synchronized in S phase and then infected. Moreover, under these conditions, the expression of S-phase cyclins is altered in a striking ICP22-dependent fashion. The authors concluded that ICP22 deregulates the cell cycle upon infection of S-phase cells by modulating the expression of key cell-cycle regulatory proteins.

Modification of RNAP II phosphorylation by ICP22

Similarly to other nuclear-replicating DNA viruses, HSV-1 uses the host RNAP II to transcribe its genes. Interestingly, HSV-1 infection causes a dramatic change to the phosphorylation of RNAP II (Rice *et al.*, 1994), and ICP22 is required for this effect (Rice *et al.*, 1995; Durand *et al.*, 2005). Moreover, ICP22 can alter RNAP II phosphorylation even when expressed in uninfected cells (Fraser and Rice, 2007). Before describing the effects of ICP22 on RNAP II, a brief review of RNAP II transcription and its regulation by phosphorylation is in order.

RNAP II, of course, is the large multi-subunit enzyme that synthesizes all eukaryotic cell mRNAs. Decades of work continue to yield ever more sophisticated views of RNAP II transcription and its regulation (see Fuda *et al.*, 2009, for a recent review). One key aspect of this regulation involves the phosphorylation of RNAP II on its carboxy-terminal domain (CTD), an unusual repeating structure found at the C-terminus of the enzyme's large subunit (LS) (for recent reviews see Brookes and Pombo, 2009; Buratowski, 2009). The CTD consists of multiple repeats (52 in humans) of the heptapeptide consensus sequence YSPTSPS. It is conserved throughout eukaryotes and is essential for cell viability. CTD phosphorylation is mediated by multiple cellular CTD kinases and is reversed by a corresponding set of cellular CTD phosphatases. Serines-2 (Ser-2) and -5 (Ser-5) are the predominant residues phosphorylated, although recent work shows that Serine-7 can also targeted.

What is the role of the CTD in transcription? In short, the CTD serves as a binding platform for cellular proteins involved in transcription regulation, mRNA processing, and chromatin modification (reviewed in Brookes and Pombo, 2009; Buratowski, 2009). Since the phosphorylation state of the CTD changes continually throughout the transcription cycle, the CTD acts to coordinate the recruitment of necessary cellular factors to the transcript as it matures. This is illustrated in the diagram shown in Fig. 5.2A, which shows how RNAP II phosphorylation changes during the cycle of transcription on a typical eukaryotic gene. First, a hypophosphorylated RNAP II complex is recruited to the promoter via contacts between the unphosphorylated CTD and transcriptional initiation factors. Coincident with or shortly after initiation, the cdk7 subunit of the general transcription factor TFIIH phosphorylates the CTD on Ser-5, disrupting the association of RNAP II with initiation factors and allowing it to clear the promoter region. Ser-5 phosphorylation also aids in the recruitment of capping factors needed for the modification of the pre-mRNA's 5′ end. As transcription proceeds, RNAP II often becomes stalled, due to inherent signals in the gene and/or the action of negative transcriptional elongation factors. Resumption of transcription is associated with phosphorylation of the CTD on Ser-2 by the cdk9 subunit of the transcriptional elongation factor P-TEFb. Ser-2 modification also helps to recruit polyadenylation and possibly splicing factors to the RNAP II complex. The sequential nature of CTD phosphorylation explains why the ratio of Ser-2/Ser-5-phosphorylation on RNAP II increases as the enzyme moves from the 5′ to the 3′ end of the gene (Komarnitsky *et al.*, 2000).

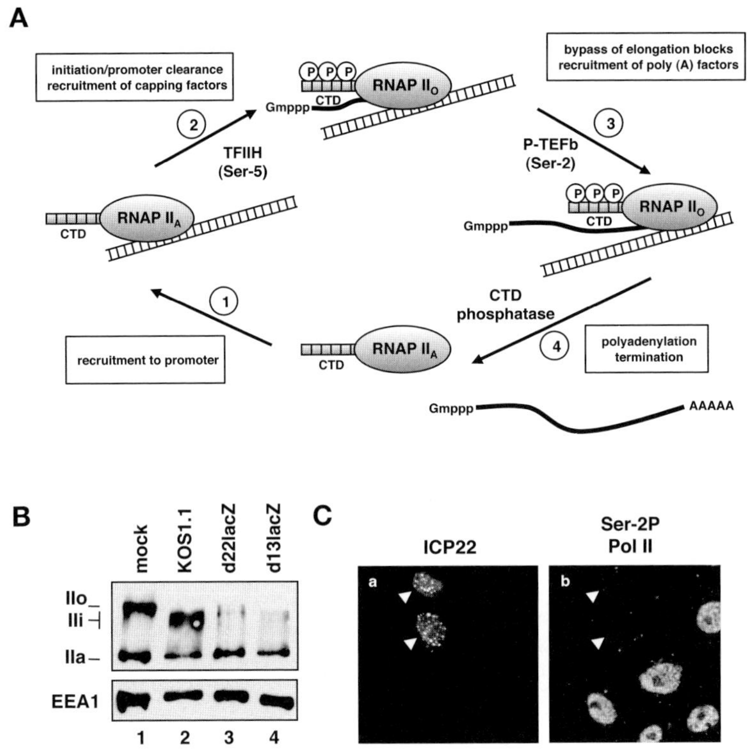

Figure 5.2 Alteration of RNAP II phosphorylation by ICP22. (A) Regulation of transcription by RNAP II CTD phosphorylation. See text for details. (B) Altered phosphorylation of RNAP II in HSV-1 infected cells. Vero cells were mock-infected or infected with HSV-1 strain KOS1.1, the ICP22 null mutant d22lacZ, or the UL13 mutant d13lacZ. At 6 hpi, total proteins were collected and analysed by immunoblotting using monoclonal antibody ARNA3, which binds to the body of the RNAP II LS. EEA1 is a cellular protein that serves as a loading control. (C) ICP22 induces loss of Ser-2 phosphorylated RNAP II. Vero cells were transfected with an ICP22 expression plasmid. Two days later, the cells were fixed and stained with H5, a monoclonal antibody that reacts with the Ser-2-phosphorylated CTD. The images show the same field of cells stained for ICP22 (panel a) and Ser-2-phosphorylated RNAP II (panel b). Note that cells expressing ICP22 (arrowheads) are negative for Ser-2-phosporylated RNAP II.

Upon cleavage/polyadenylation of the primary transcript, termination of transcription occurs, and CTD phosphatases must dephosphorylate RNAP II prior to a new round of transcription. CTD phosphatases are also likely active during transcriptional elongation.

Because of the above CTD modifications, RNAP II in growing mammalian cells exists in two major forms that differ in the extent of CTD phosphorylation. RNAP II$_A$ is the hypophosphorylated form, whereas RNAP II$_O$ is the hyperphosphorylated form, having both Ser-2 and Ser-5 modifications. Phosphorylation on RNAP II$_O$ is extensive; it has been estimated that most of the 52 CTD repeats contain at least one phosphate moiety. Because of this, the LS species from RNAP II$_A$ and RNAP II$_O$ (known as IIa and IIo, respectively) migrate as distinct bands on SDS-PAGE and can be readily distinguished by immunoblotting.

As mentioned above, HSV-1 infection dramatically alters RNAP II phosphorylation (Rice *et al.*, 1994). This can be seen in the experiment show in Fig. 5.2B, in which Vero cells were infected with HSV-1 and the RNAP II forms present at 6 h post-infection (hpi) were examined by immunoblotting. In the mock infection (lane 1), there are comparable levels of IIa and IIo, indicating that both RNAP II$_A$ and RNAP II$_O$ are abundant in uninfected cells. After wild-type (WT) HSV-1 infection (lane 2), however, IIo is replaced by a new set of faster migrating species that have electrophoretic mobilities intermediate between that of IIa and IIo. These forms were collectively designated IIi (for intermediately migrating), and the form of RNAP II having them was designated RNAP II_I. IIi is a phosphorylation variant of the LS, since it can be converted to IIa by *in vitro* treatment with calf intestinal phosphatase. Based on its intermediate electrophoretic mobility, it is likely that IIi is partially phosphorylated compared to IIo. Induction of RNAP II_I by HSV-1 is observed in numerous cell lines including Vero, HeLa, and human embryonic lung (HEL) cells (Rice *et al.*, 1994; Rice *et al.*, 1995), and is also seen in cells infected with the closely related HSV-2 (Rice and Fraser, 2006).

Does RNAP II_I resemble the normal hyperphosphorylated RNAP II_O form in having both Ser-5 and Ser-2 modifications? Fraser and Rice used phosphospecific anti-CTD antibodies to address this question (Fraser and Rice, 2005). Surprisingly, RNAP II_I was found to be severely depleted in Ser-2 phosphorylation compared to RNAP II_O. Quantitative immunoblotting indicated that the level of Ser-2 phosphorylation is reduced by ~40-fold at 8 hpi. Thus, it appears that HSV-1 infection generates a highly unusual form of RNAP II that lacks one of its standard post-translational modifications. Incompletely phosphorylated forms of RNAP II have been observed previously in eukaryotic cells, e.g. during mammalian or amphibian development (Bellier *et al.*, 1997; Palancade *et al.*, 2001), but these forms are suspected to be transcriptionally inactive. In contrast, photoaffinity cross-linking experiments demonstrate that RNAP II_I is transcriptionally engaged (Spencer *et al.*, 1997). Together, these results raise the possibility that HSV-1 transcription involves a different CTD phosphorylation cycle than that shown in Fig. 5.2A. Alternatively, it has been suggested that Ser-2-phosphorylated RNAP II is generated during HSV-1 transcription, but that this form is rapidly degraded (Dai-Ju *et al.*, 2006).

Studies with HSV-1 mutants have shown that both ICP22 and UL13 are required for the induction of RNAP II_I (Rice *et al.*, 1995; Long *et al.*, 1999; Durand *et al.*, 2005). The requirement for these proteins does not correlate with the permissivity of the infections, i.e. ICP22 and UL13 are required irrespective of whether the infections are restrictive or permissive. The importance of these two proteins in HSV-1-mediated RNAP II alterations can be seen in lanes 3 and 4 of Fig. 5.2b, wherein the RNAP II forms present in ICP22 and UL13 mutant-infected Vero cells were examined. In these cases, there is a dramatic reduction in the level of RNAP II_I compared to the WT infection (lane 2). A key point to remember is that Vero cells are permissive for these mutants. Therefore, it can be concluded that RNAP II_I is not essential for a productive HSV-1 infection. However, it remains possible that RNAP II_I is required in cells that are restrictive for ICP22 and UL13 mutants.

The experiment shown in Fig. 5.2b illustrates another important point about HSV-1's effects on RNAP II. Note that even in the absence of ICP22 or UL13 (lanes 3 and 4), the IIo form is dramatically decreased in abundance compared to mock-infected cells (lane 2). Similar results are seen in an ICP22/UL13 double mutant infection (K. Fraser and S. Rice, unpublished). Thus, there is clearly an ICP22/UL13-independent mechanism that leads to loss of hyperphosphorylated RNAP II. Sandri-Goldin and colleagues provided a potential explanation for this by showing that viral L gene expression results in a significant decrease in Ser-2-phosphorylated RNAP II (Dai-Ju *et al.*, 2006). Interestingly, this effect appears to involve the proteasome-dependent degradation of the enzyme. It was proposed that transcription of the HSV-1 genome on opposing strands leads to the collision and stalling of RNAP II complexes, which must then be removed by proteolysis.

How does ICP22 alter RNAP II? Somewhat surprisingly, a recent study demonstrates that ICP22 has two distinct effects on RNAP II, and that these differ with respect to their UL13-dependence (Fraser and Rice, 2007). The first effect, which is UL13-independent, is to trigger the rapid loss of Ser-2 phosphorylated RNAP II. As discussed above, viral L gene expression seems to do the same, although presumably by a different mechanism. The ability of ICP22 to induce depletion of RNAP II forms having Ser-2 phosphorylation can be demonstrated even in uninfected cells. This is illustrated in the experiment shown in Fig. 5.2C, in which Vero cells were transfected with an ICP22 expression plasmid. Two days later the cells were stained with an antibody specific for Ser-2-phosphorylated CTD. Although most mammalian cells express abundant amounts of this antigen in their nuclei (Bregman *et al.*, 1995), those expressing ICP22 (arrowheads) are practically devoid of this antigen (compare panels a and b). The ability of ICP22 to trigger the loss of Ser-2-phosphorylated RNAP II can also be demonstrated in infected cells, particularly when metabolic inhibitors are used to enhance the IE phase of gene expression (Fraser and Rice, 2007).

The second effect of ICP22, which is UL13-dependent, is to induce the accumulation of RNAP II_I (Rice *et al.*, 1995; Long *et al.*, 1999). The mechanism by which it does this is unclear. However, it is quite interesting that an RNAP II_I-like form accumulates in ICP22 and UL13 mutant infections when proteasome inhibitors are added to the overlay media (K. Fraser and S. Rice, unpublished). This suggests that RNAP II_I is produced during ICP22/UL13 mutant infections, but that it is rapidly degraded by the proteasome. It is known that the proteasome is a key component of a cellular rescue pathway that removes RNAP II complexes that have stalled on genes for various reasons (Svejstrup, 2007). Given this, it is possible that the ICP22-induced loss of Ser-2 phosphorylation causes RNAP II to stall (perhaps on cellular genes) and/or to be perceived as abnormal. This might initiate a host response that has the potential to degrade this form of RNAP II. If so, then the second, UL13-dependent function of ICP22 may be to inhibit this response, thus allowing RNAP II_I accumulation. Further experiments will be required to test this model.

As can be appreciated from the above discussion, there are numerous unanswered questions about ICP22's effects on RNAP II. To close this section, two of these will be considered. The first is based on the finding that RNAP II_I lacks Ser-2 CTD phosphorylation. Does this indicate that HSV-1 transcription does not require P-TEFb, the major factor responsible for adding this modification to RNAP II? This would be somewhat surprising as P-TEFb is essential for most host cell transcription (Chao and Price, 2001) Recall that Ser-2 phosphorylation by P-TEFb promotes transcriptional elongation and helps to recruit polyadenylation and possibly splicing factors to the transcript. In regards to elongation, there are several reasons to think that HSV-1 genes may be less sensitive to elongation blocks than cellular genes. They are significantly shorter, and generally intronless. In addition, they are packaged into a non-standard form of chromatin (Lacasse and Schang, 2010) that may not be subject to the same sorts of transcriptional elongation controls that govern cellular genes. In regards to polyadenylation, there is evidence that the IE protein ICP27, which itself interacts with RNAP II (Zhou and Knipe, 2002; Dai-Ju *et al.*, 2006), promotes polyadenylation (McGregor *et al.*, 1996). This could explain how HSV-1 mRNAs become polyadenylated despite the lack of Ser-2 phosphorylation on RNAP II. For these reasons, it does seem at least feasible that transcription of HSV-1 genes does not require P-TEFb. However, recent work from Durand and Roizman challenges this notion (Durand and Roizman, 2008). These investigators found that cdk9, the kinase component of P-TEFb, is required for efficient HSV-1 L gene expression. Moreover, ICP22 was shown to physically interact with cdk9, and to stimulate its ability to phosphorylate the CTD *in vitro* (Durand *et al.*, 2005). One possible scenario that is consistent with much of the data is that *in vivo*, ICP22's interaction with cdk9 inhibits its activity, thus explaining ICP22's ability to trigger the loss of Ser-2 phosphorylation. However, this model does not explain why cdk9 is needed for viral L gene expression. Obviously, more work is needed to resolve these issues, but certainly the ICP22–cdk9 interaction is an intriguing clue that should be pursued vigorously.

The second question is what replication benefit, if any, does HSV-1 derive from the ICP22-dependent modification of RNAP II? One possibility is that it is important for the virus to generate RNAP II_I, as this form of the enzyme may be specialized for HSV-1 genome transcription. However, to date there is little evidence to support this hypothesis. It is certainly hard to deny that ICP22 and UL13 mutants replicate quite well in permissive cells despite their inability to induce RNAP II_I. An alternative possibility is that the important biological effect of ICP22 is not to induce RNAP II_I, but rather to trigger the loss of Ser-2-phosphorylated RNAP II. This may serve to inhibit host genome transcription. This effect, even if only transient, could blunt the host's innate immune response to infection, and might therefore provide a replication advantage to HSV-1. Supporting this model, multiple studies indicate that ICP22 can inhibit the expression of several reporter genes in plasmid co-transfection assays (Prod'hon *et al.*, 1996; Cun *et al.*, 2006; Bowman *et al.*, 2009).

ICP22 is required for induction of VICE domains

A recent study by Bastian *et al.* has identified a new function for ICP22: it alters components of the host cell chaperone machinery (Bastian *et al.*, 2010). This finding arose from an investigation of the small nuclear bodies to which ICP22 localizes to during infection. It was discovered that these bodies are the same structures as those previously identified as VICE domains (Burch and Weller, 2004; Burch and Weller, 2005). VICE domains are virus-induced nuclear foci that contain cellular chaperones (Hsc70, Hsp70, Hsp90, Hsp40), proteosomal components, and ubiquitinated proteins. They appear to play an important role in protein quality-control in the infected cell nucleus (Livingston *et al.*, 2009). Bastian *et al.* also showed that ICP22 is absolutely essential for the formation of VICE domains. This was found to be true in cells that are permissive for ICP22 mutants, as well as in those that are not. Moreover, unlike many of the other functions of ICP22, its ability to induce VICE domains is independent of UL13.

The requirement for ICP22 in VICE domain formation can be seen in the experiment shown in Fig. 5.3A. In this study, Vero cells were mock-infected or infected with WT HSV-1 or an ICP22

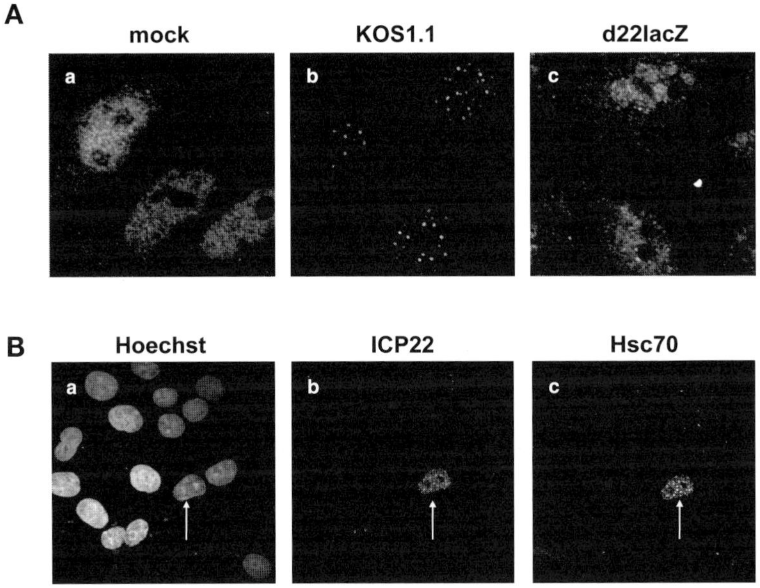

Figure 5.3 ICP22 induces VICE domains in infected cells and reorganizes Hsc70 in transfected cells. (A) ICP22 is required for VICE domain formation. Vero cells were mock-infected or infected with WT HSV-1 strain KOS1.1 or ICP22 deletion mutant d22lacZ. At 6 hpi, the cells were fixed and stained for Hsc70. Note that WT HSV-1 but not ICP22 mutant infection induces reorganization of Hsc70 into punctate VICE domains. (B) ICP22 reorganizes Hsc70 in uninfected cells. Vero cells were transfected with an ICP22 expression construct. One day after transfection, the cells were fixed and stained for DNA (using the Hoechst reagent), ICP22, or Hsc70 (panels a-c). The images show a single microscopic field. The arrows point to a cell expressing high levels of ICP22. Note that the Hsc70 in this cell has been reorganized into nuclear inclusion bodies.

deletion mutant. At 6 hpi, the cells were stained for Hsc70, a major component of VICE domains. In mock-infected cells (Fig. 5.3A), Hsc70 is found widely dispersed throughout the cell, in both the nucleus and cytoplasm. After WT HSV-1 infection (FIG 5.3B), Hsc70 becomes concentrated in several nuclear VICE domains, which typically are found on the periphery of viral replication compartments (Burch and Weller, 2004). However, in cells infected with an ICP22 deletion mutant (panel c), Hsc70 localization resembles that of mock-infected cells. Since ICP22 null mutants grow productively in Vero cells, these results indicate that VICE domain formation is not an essential feature of HSV-1 productive infection. However, it is possible that VICE domain formation is required in cells that are restrictive for the growth of ICP22 mutants.

It is not known how ICP22 induces VICE domains. However, the studies of Bastian *et al.* suggest that it plays a direct role, as it was found that ICP22 can induce Hsc70-containing nuclear inclusions even in the absence of viral infection. This can be seen in the experiment shown in Fig. 5.3B, in which Vero cells were transfected with an ICP22 expression plasmid. One day later the cells were fixed and stained for ICP22 and Hsc70. It can be seen that in cells expressing high levels of ICP22 (panel b, white arrow), Hsc70 becomes concentrated in nuclear inclusion bodies (panel c) that resemble VICE domains. It is not yet known whether other host chaperones also localize to these inclusions.

The functional significance of ICP22's effects on Hsc70/VICE domains remains to be determined. One possibility is that, during infection, ICP22 relocalizes Hsc70 from the cytoplasm to the nucleus, possibly to assist in the massive reorganization of nuclear material that accompanies HSV-1 replication. It is worth noting that another viral protein, the human papillomavirus late capsid protein 2, also induces the nuclear accumulation of Hsc70 in discrete foci, in this case to PML bodies (Florin *et al.*, 2004). It is thought to do so via a direct physical interaction. Does ICP22 physically interact with Hsc70? The answer is not yet known. Bastian *et al.* showed that mutant forms of ICP22 can bind to Hsc70, but they were unable to detect an interaction between the WT protein and Hsc70 (Bastian *et al.*, 2010). However, this negative result does not preclude that such an interaction occurs, and thus this question should probably be pursued further.

Multiple functions and functional domains of ICP22

Recent studies of a set of engineered HSV-1 ICP22 mutants (Fig. 5.4) provide a rough outline of the functional domain structure of ICP22 (Bastian and Rice, 2009; Bastian *et al.*, 2010). This work focused on four activities of ICP22: (i) its ability to promote viral growth and L gene expression in a restrictive cell line, HEL cells; (ii) its capacity to induce RNAP II_I; (iii) its ability to trigger loss of Ser-2-phosphorylated RNAP II; and (iv) its ability to induce VICE domains. It should be noted that the first two of these functions are dependent upon UL13, while the latter two are not. Analysis of the ICP22 mutants led to several conclusions. First, both of ICP22's effects on RNAP II map to the C-terminal half of the protein. Although more mapping needs to be done, the fact that the two effects of ICP22 on RNAP II were not genetically separated suggests that they may be mediated by the same functional domain. The activity of this domain may be modulated by UL13-dependent phosphorylation. Second, the ability of ICP22 to induce VICE domains requires an N-terminal portion of ICP22 that is dispensable for ICP22's effects on RNAP II. This suggests that ICP22 has at least two separate functional domains, one that affects host chaperones and one that affects RNAP II. Third, both N- and C-terminal portions of ICP22 are needed for its L gene enhancement function. The fact that this function does not track genetically with ICP22's effects on RNAP II demonstrates that L gene enhancement is not simply due to RNAP II modifications. It should be noted that a genetic study from Roizman's lab concluded that the N-terminal portion of ICP22 is not required for viral growth and L gene expression in restrictive cells (Ogle and Roizman, 1999), and thus is at odds with the results of Bastian and Rice. The reason for this difference is not clear at present.

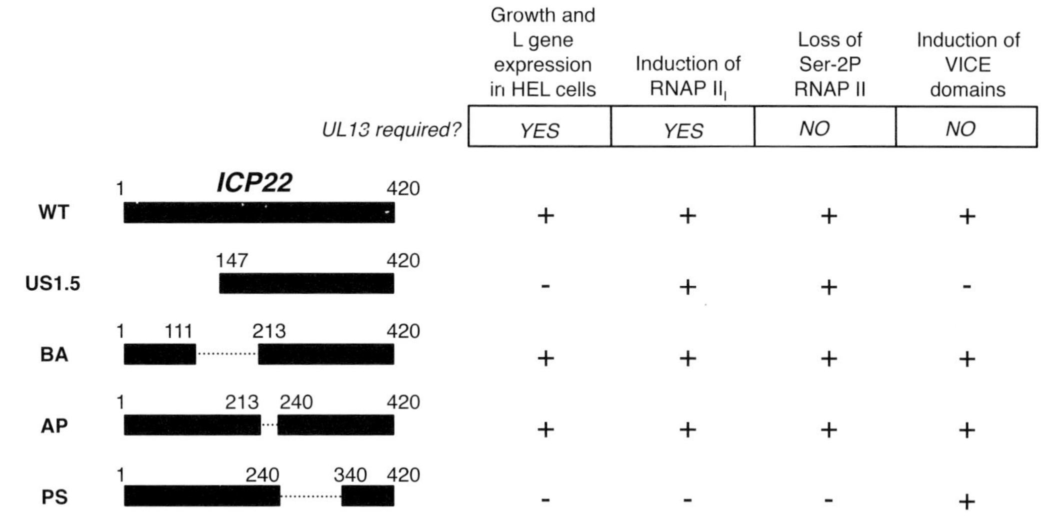

Figure 5.4 Genetic evidence for multiple functional domains in ICP22. Recombinant viruses expressing WT or mutant forms of ICP22 were analysed for four functions of ICP22: (i) ability to promote viral growth and L gene expression in HEL cells; (ii) capacity to induce RNAP II$_I$; (iii) ability to trigger loss of Ser-2-phosphorylated RNAP II; and iv) ability to induce VICE domains. Note that UL13 is required for only the first two of these functions. The US1.5 mutant was engineered assuming that translation of the US1.5 protein begins at codon 147 of ICP22. However, more recent information suggests that the US1.5 protein encoded by HSV-1 is initiated at codon 90.

Remaining questions, future directions

Although much has been learned about ICP22 over the years, many puzzles still remain. Foremost among these is the question of how this protein enhances viral L gene expression. This is the one function of ICP22 that is indisputably critical for viral replication. Regarding this function, there are a number of more specific questions that need to be answered. What is the basis of the cell-type dependence of ICP22's effects on viral L genes? Why do only a subset of L genes require ICP22 for expression? To what extent do ICP22's effects on the cell-cycle machinery, RNAP II, and molecular chaperones play a role in L gene enhancement? Likewise, there are many questions relating to ICP22's dramatic effects on RNAP II, which lead to a novel form of this enzyme. Is the cycle of transcription inherently different on HSV-1 genes, as suggested by the lack of Ser-2 phosphorylation on the CTD? What is the mechanism or mechanisms by which ICP22 mediates its two seemingly distinct effects on RNAP II? Is the modification of RNAP II by ICP22 important for enhancing viral transcription, or does it play a role in repressing host genome transcription? The recent findings linking ICP22 to VICE domains open up a whole new set of questions. How does ICP22 induce the reorganization of Hsc70 in both uninfected and infected cells? Does it physically interact with Hsc70? Do ICP22's effects on VICE domains enable the virus to better manage protein quality-control events in the infected cell nucleus, or do these effects have some other role in virus biology?

In terms of future directions for ICP22 research, three general areas seem especially promising. The first is viral genetics, i.e. studying the phenotypes of engineered viral ICP22 mutants. To date, this approach has been underutilized. However, now that there are a number of ICP22 activities that can be readily assayed, mutational analysis should be straightforward and yield a much more complete understanding of ICP22's functions and domain structure. Second, since ICP22 presumably carries out many of its activities by interacting with other proteins, it will be important to more completely characterize its protein binding-partners, both cellular and viral. Third, as some of ICP22's activities may not be readily evident during tissue-culture infections, ICP22 should be studied in the context of small animal models of infection. This approach will illuminate ICP22's effects on *in vivo* replication, pathogenesis, and latency.

A final point worth discussing is the functional relationship between ICP22 and the homologues encoded by other members of the alphaherpesviruses. Although it is clear that several of these homologues have interesting regulatory activities (e.g. see Hood *et al.*, 2006; Ambagala and Cohen, 2007; Breitenbach *et al.*, 2009), it is not known whether all family members share a common function. Relevant to this, recent experiments indicate that the varicella-zoster virus homologue of ICP22, ORF63, is dissimilar to ICP22 in that it does not alter RNAP II phosphorylation (Fraser and Rice, 2007) or Hsc70 localization (Bastian and Rice, unpublished) in transfected cells. These results suggest that members of the ICP22 family of proteins have evolved virus- or genus-specific functions. However, it is still quite possible that they share one or more common activities, perhaps encoded in the central conserved region of ICP22 (Fig. 5.1B). The continued molecular analysis of ICP22 will provide a highly useful framework for functional comparisons with other family members.

Acknowledgements

The studies on ICP22 from the author's laboratory were supported by Public Health Service grants from the National Institute of Allergy and Infectious Disease.

References

Advani, S.J., Brandimarti, R., Weichselbaum, R.R., and Roizman, B. (2000a). The disappearance of cyclins A and B and the increase in activity of the G(2)/M-phase cellular kinase cdc2 in herpes simplex virus 1-infected cells require expression of the alpha22/U(S)1.5 and U(L)13 viral genes. J. Virol. *74*, 8–15.

Advani, S.J., Weichselbaum, R.R., and Roizman, B. (2000b). The role of cdc2 in the expression of herpes simplex virus genes. Proc. Natl. Acad. Sci. USA, 97, 10996–11001.

Advani, S.J., Weichselbaum, R.R., and Roizman, B. (2003). Herpes simplex virus 1 activates cdc2 to recruit topoisomerase IIalpha for post-DNA synthesis expression of late genes. Proc. Natl. Acad. Sci. U.S.A. *100*, 4825–4830.

Ambagala, A.P., and Cohen, J.I. (2007). Varicella-Zoster virus IE63, a major viral latency protein, is required to inhibit the alpha interferon-induced antiviral response. J. Virol. *81*, 7844–7851.

Asai, R., Ohno, T., Kato, A., and Kawaguchi, Y. (2007). Identification of proteins directly phosphorylated by UL13 protein kinase from herpes simplex virus 1. Microbes Infect. *9*, 1434–1438.

Bastian, T.W., Livingston, C.M., Weller, S.K., and Rice, S.A. (2010). Herpes simplex virus type 1 immediate-early protein ICP22 is required for VICE domain formation during productive viral infection. J. Virol. *84*, 2384–2394.

Bastian, T.W., and Rice, S.A. (2009). Identification of sequences in herpes simplex virus type 1 ICP22 that influence RNA polymerase II modification and viral late gene expression. J. Virol. *83*, 128–139.

Bellier, S., Chastant, S., Adenot, P., Vincent, M., Renard, J.P., and Bensaude, O. (1997). Nuclear translocation and carboxyl-terminal domain phosphorylation of RNA polymerase II delineate the two phases of zygotic gene activation in mammalian embryos. EMBO J. *16*, 6250–6262.

Blaho, J.A., Mitchell, C., and Roizman, B. (1993). Guanylylation and adenylylation of the alpha regulatory proteins of herpes simplex virus require a viral beta or gamma function. J. Virol. *67*, 3891–3900.

Bowman, J.J., Orlando, J.S., Davido, D.J., Kushnir, A.S., and Schaffer, P.A. (2009). Transient expression of herpes simplex virus type 1 ICP22 represses viral promoter activity and complements the replication of an ICP22 null virus. J. Virol. *83*, 8733–8743.

Bowman, J.J., and Schaffer, P.A. (2009). Origin of expression of the herpes simplex virus type 1 protein U(S)1.5. J. Virol. *83*, 9183–9194.

Bregman, D.B., Du, L., van der Zee, S., and Warren, S.L. (1995). Transcription-dependent redistribution of the large subunit of RNA polymerase II to discrete nuclear domains. J. Cell. Biol. *129*, 287–298.

Breitenbach, J.E., Ebner, P.D., and O'Callaghan, D.J. (2009). The IR4 auxiliary regulatory protein expands the *in vitro* host range of equine herpesvirus 1 and is essential for pathogenesis in the murine model. Virology *383*, 188–194.

Brookes, E., and Pombo, A. (2009). Modifications of RNA polymerase II are pivotal in regulating gene expression states. EMBO Rep. *10*, 1213–1219.

Bruni, R., Fineschi, B., Ogle, W.O., and Roizman, B. (1999). A novel cellular protein, p60, interacting with both herpes simplex virus 1 regulatory proteins ICP22 and ICP0 is modified in a cell-type-specific manner and Is recruited to the nucleus after infection. J. Virol. *73*, 3810–3817.

Bruni, R., and Roizman, B. (1998). Herpes simplex virus 1 regulatory protein ICP22 interacts with a new cell cycle-regulated factor and accumulates in a cell cycle-dependent fashion in infected cells. J. Virol. *72*, 8525–8531.

Buratowski, S. (2009). Progression through the RNA polymerase II CTD cycle. Mol. Cell *36*, 541–546.

Burch, A.D., and Weller, S.K. (2004). Nuclear sequestration of cellular chaperone and proteasomal machinery during herpes simplex virus type 1 infection. J. Virol. *78*, 7175–7185.

Burch, A.D., and Weller, S.K. (2005). Herpes simplex virus type 1 DNA polymerase requires. the mammalian chaperone hsp90 for proper localization to the nucleus. J. Virol. *79*, 10740–10749.

Carter, K.L., and Roizman, B. (1996). The promoter and transcriptional unit of a novel herpes simplex virus 1 alpha gene are contained in, and encode a protein in frame with, the open reading frame of the alpha 22 gene. J. Virol. *70*, 172–178.

Chao, S.H., and Price, D.H. (2001). Flavopiridol inactivates P-TEFb and blocks most RNA polymerase II transcription *in vivo*. J. Biol. Chem. *276*, 31793–31799.

Cun, W., Hong, M., Liu, L.D., Dong, C.H., Luo, J., and Li, Q.H. (2006). Structural and functional characterization of herpes simplex virus 1 immediate-early protein infected-cell protein 22. J. Biochem. *140*, 67–73.

Dai-Ju, J.Q., Li, L., Johnson, L.A., and Sandri-Goldin, R.M. (2006). ICP27 interacts with the C-terminal domain of RNA polymerase II and facilitates its recruitment to herpes simplex virus 1 transcription sites, where it undergoes proteasomal degradation during infection. J. Virol. *80*, 3567–3581.

Durand, L.O., Advani, S.J., Poon, A.P., and Roizman, B. (2005). The carboxyl-terminal domain of RNA polymerase II is phosphorylated by a complex containing cdk9 and infected-cell protein 22 of herpes simplex virus 1. J. Virol. *79*, 6757–6762.

Durand, L.O., and Roizman, B. (2008). Role of cdk9 in the optimization of expression of the genes regulated by ICP22 of herpes simplex virus 1. J. Virol. *82*, 10591–10599.

Ehmann, G.L., McLean, T.I., and Bachenheimer, S.L. (2000). Herpes simplex virus type 1 infection imposes a G(1)/S block in asynchronously growing cells and prevents G(1) entry in quiescent cells. Virology *267*, 335–349.

Everett, R.D., and Chelbi-Alix, M.K. (2007). PML and PML nuclear bodies: implications in antiviral defence. Biochimie *89*, 819–830.

Florin, L., Becker, K.A., Sapp, C., Lambert, C., Sirma, H., Muller, M., Streeck, R.E., and Sapp, M. (2004). Nuclear translocation of papillomavirus minor capsid protein L2 requires. Hsc70. J. Virol. *78*, 5546–5553.

Fraser, K.A., and Rice, S.A. (2005). Herpes simplex virus type 1 infection leads to loss of serine-2 phosphorylation on the carboxyl-terminal domain of RNA polymerase II. J. Virol. *79*, 11323–11334.

Fraser, K.A., and Rice, S.A. (2007). Herpes simplex virus immediate-early protein ICP22 triggers loss of serine 2-phosphorylated RNA polymerase II. J. Virol. *81*, 5091–5101.

Fuda, N.J., Ardehali, M.B., and Lis, J.T. (2009). Defining mechanisms that regulate RNA polymerase II transcription *in vivo*. Nature *461*, 186–192.

Gelman, I.H., and Silverstein, S. (1985). Identification of immediate early genes from herpes simplex virus that transactivate the virus thymidine kinase gene. Proc. Natl. Acad. Sci. USA, 82, 5265–5269.

Hancock, M.H., Corcoran, J.A., and Smiley, J.R. (2006). Herpes simplex virus regulatory proteins VP16 and ICP0 counteract an innate intranuclear barrier to viral gene expression. Virology *352*, 237–252.

Hill, A., Jugovic, P., York, I., Russ, G., Bennink, J., Yewdell, J., Ploegh, H., and Johnson, D. (1995). Herpes simplex virus turns off the TAP to evade host immunity. Nature *375*, 411–415.

Hood, C., Cunningham, A.L., Slobedman, B., Arvin, A.M., Sommer, M.H., Kinchington, P.R., and Abendroth, A. (2006). Varicella-zoster virus ORF63 inhibits apoptosis of primary human neurons. J. Virol. *80*, 1025–1031.

Jahedi, S., Markovitz, N.S., Filatov, F., and Roizman, B. (1999). Colocalization of the herpes simplex virus 1 UL4 protein with infected cell protein 22 in small, dense nuclear structures. formed. prior to onset of DNA synthesis. J. Virol. *73*, 5132–5138.

Keniry, M., and Parsons, R. (2008). The role of PTEN signaling perturbations in cancer and in targeted therapy. Oncogene *27*, 5477–5485.

Komarnitsky, P., Cho, E.J., and Buratowski, S. (2000). Different phosphorylated forms of RNA polymerase II and associated mRNA processing factors during transcription. Genes Dev. *14*, 2452–2460.

Kushnir, A.S., Davido, D.J., and Schaffer, P.A. (2010). Role of nuclear factor Y in stress-induced activation of the herpes simplex virus type 1 ICP0 promoter. J. Virol. *84*, 188–200.

Lacasse, J.J., and Schang, L.M. (2010). During lytic infections, herpes simplex virus type 1 DNA is in complexes with the properties of unstable nucleosomes. J. Virol. *84*, 1920–1933.

Lee, C.G., Hague, L.K., Li, H., and Donnelly, R. (2004). Identification of toposome, a novel multisubunit complex containing topoisomerase IIalpha. Cell Cycle *3*, 638–647.

Leopardi, R., Ward, P.L., Ogle, W.O., and Roizman, B. (1997). Association of herpes simplex virus regulatory protein ICP22 with transcriptional complexes containing EAP, ICP4, RNA polymerase II, and viral DNA requires. posttranslational modification by the U(L)13 protein kinase. J. Virol. *71*, 1133–1139.

Livingston, C.M., Ifrim, M.F., Cowan, A.E., and Weller, S.K. (2009). Virus-Induced Chaperone-Enriched (VICE) domains function as nuclear protein quality control centers during HSV-1 infection. PLoS Pathog. *5*, e1000619.

Long, M.C., Leong, V., Schaffer, P.A., Spencer, C.A., and Rice, S.A. (1999). ICP22 and the UL13 protein kinase are both required for herpes simplex virus-induced modification of the large subunit of RNA polymerase II. J. Virol. *73*, 5593–5604.

McGregor, F., Phelan, A., Dunlop, J., and Clements, J.B. (1996). Regulation of herpes simplex virus poly (A) site usage and the action of immediate-early protein IE63 in the early-late switch. J. Virol. *70*, 1931–1940.

O'Hare, P., and Hayward, G.S. (1985). Evidence for a direct role for both the 175, 000- and 110, 000-molecular-weight immediate-early proteins of herpes simplex virus in the transactivation of delayed-early promoters. J. Virol. *53*, 751–760.

O'Toole, J.M., Aubert, M., Kotsakis, A., and Blaho, J.A. (2003). Mutation of the protein tyrosine kinase consensus site in the herpes simplex virus 1 alpha22 gene alters ICP22 posttranslational modification. Virology *305*, 153–167.

Ogle, W.O., and Roizman, B. (1999). Functional anatomy of herpes simplex virus 1 overlapping genes encoding infected-cell protein 22 and US1.5 protein. J. Virol. *73*, 4305–4315.

Okahara, F., Ikawa, H., Kanaho, Y., and Maehama, T. (2004). Regulation of PTEN phosphorylation and stability by a tumor suppressor candidate protein. J. Biol. Chem. *279*, 45300–45303.

Okumura, K., Zhao, M., Depinho, R.A., Furnari, F.B., and Cavenee, W.K. (2005). Cellular transformation by the MSP58 oncogene is inhibited by its physical interaction with the PTEN tumor suppressor. Proc. Natl. Acad. Sci. USA, 102, 2703–2706.

Orlando, J.S., Astor, T.L., Rundle, S.A., and Schaffer, P.A. (2006a). The products of the herpes simplex virus type 1 immediate-early US1/US1.5 genes downregulate levels of S-phase-specific cyclins and facilitate virus replication in S-phase Vero cells. J. Virol. *80*, 4005–4016.

Orlando, J.S., Balliet, J.W., Kushnir, A.S., Astor, T.L., Kosz-Vnenchak, M., Rice, S.A., Knipe, D.M., and Schaffer, P.A. (2006b). ICP22 Is required for wild-type composition and infectivity of herpes simplex virus type 1 virions. J. Virol. *80*, 9381–9390.

Palancade, B., Bellier, S., Almouzni, G., and Bensaude, O. (2001). Incomplete RNA polymerase II phosphorylation in Xenopus laevis early embryos. J. Cell. Sci. *114*, 2483–2489.

Park, S.W., Parrott, A.M., Fritz, D.T., Park, Y., Mathews, M.B., and Lee, C.G. (2008). Regulation of the catalytic function of topoisomerase II alpha through association with RNA. Nucleic Acids Res. *36*, 6080–6090.

Poffenberger, K.L., Idowu, A.D., Fraser-Smith, E.B., Raichlen, P.E., and Herman, R.C. (1994). A herpes simplex virus type 1 ICP22 deletion mutant is altered for virulence and latency *in vivo*. Arch. Virol. *139*, 111–119.

Poffenberger, K.L., Raichlen, P.E., and Herman, R.C. (1993). *In vitro* characterization of a herpes simplex virus type 1 ICP22 deletion mutant. Virus Genes *7*, 171–186.

Poon, A.P., Ogle, W.O., and Roizman, B. (2000). Posttranslational processing of infected cell protein 22 mediated by viral protein kinases is sensitive to amino acid substitutions at distant sites and can be cell-type specific. J. Virol. *74*, 11210–11214.

Post, L.E., and Roizman, B. (1981). A generalized technique for deletion of specific genes in large genomes: alpha gene 22 of herpes simplex virus 1 is not essential for growth. Cell *25*, 227–232.

Preston, C.M., Cordingley, M.G., and Stow, N.D. (1984). Analysis of DNA sequences which regulate the transcription of a herpes simplex virus immediate early gene. J. Virol. *50*, 708–716.

Prod'hon, C., Machuca, I., Berthomme, H., Epstein, A., and Jacquemont, B. (1996). Characterization of regulatory functions of the HSV-1 immediate-early protein ICP22. Virology *226*, 393–402.

Purves, F.C., Ogle, W.O., and Roizman, B. (1993). Processing of the herpes simplex virus regulatory protein alpha 22 mediated by the UL13 protein kinase determines the accumulation of a subset of alpha and gamma mRNAs and proteins in infected cells. Proc. Natl. Acad. Sci. USA, 90, 6701–6705.

Purves, F.C., and Roizman, B. (1992). The UL13 gene of herpes simplex virus 1 encodes the functions for posttranslational processing associated with phosphorylation of the regulatory protein alpha 22. Proc. Natl. Acad. Sci. USA, 89, 7310–7314.

Rice, S.A., and Fraser, K.A. (2006). The modification of cellular RNA polymerase II during HSV-1 infection, In Alphaherpesviruses, Molecular and Cellular Biology, Sandri-Goldin, R.M., ed. (Norfolk, U.K.: Caister Academic Press), pp. 21–37.

Rice, S.A., Long, M.C., Lam, V., Schaffer, P.A., and Spencer, C.A. (1995). Herpes simplex virus immediate-early protein ICP22 is required for viral modification of host RNA polymerase II and establishment of the normal viral transcription program. J. Virol. *69*, 5550–5559.

Rice, S.A., Long, M.C., Lam, V., and Spencer, C.A. (1994). RNA polymerase II is aberrantly phosphorylated and localized to viral replication compartments following herpes simplex virus infection. J. Virol. *68*, 988–1001.

Salsman, J., Zimmerman, N., Chen, T., Domagala, M., and Frappier, L. (2008). Genome-wide screen of three herpesviruses for protein subcellular localization and alteration of PML nuclear bodies. PLoS Pathog. *4*, e1000100.

Schwyzer, M., Wirth, U.V., Vogt, B., and Fraefel, C. (1994). BICP22 of bovine herpesvirus 1 is encoded by a spliced 1.7 kb RNA which exhibits immediate early and late transcription kinetics. J. Gen. Virol. *75*, 1703–1711.

Sears, A.E., Halliburton, I.W., Meignier, B., Silver, S., and Roizman, B. (1985). Herpes simplex virus 1 mutant deleted in the alpha 22 gene: growth and gene expression in permissive and restrictive cells and establishment of latency in mice. J. Virol. *55*, 338–346.

Smith-Donald, B.A., and Roizman, B. (2008). The interaction of herpes simplex virus 1 regulatory protein ICP22 with the cdc25C phosphatase is enabled *in vitro* by viral protein kinases US3 and UL13. J. Virol. *82*, 4533–4543.

Song, B., Liu, J.J., Yeh, K..C., and Knipe, D.M. (2000). Herpes simplex virus infection blocks events in the G1 phase of the cell cycle. Virology *267*, 326–334.

Spencer, C.A., Dahmus, M.E., and Rice, S.A. (1997). Repression of host RNA polymerase II transcription by herpes simplex virus type 1. J. Virol. *71*, 2031–2040.

Stelz, G., Rucker, E., Rosorius, O., Meyer, G., Stauber, R.H., Spatz, M., Eibl, M.M., and Hauber, J. (2002). Identification of two nuclear import signals in the alpha-gene product ICP22 of herpes simplex virus 1. Virology *295*, 360–370.

Summers, B.C., and Leib, D.A. (2002). Herpes simplex virus type 1 origins of DNA replication play no role in the regulation of flanking promoters. J. Virol. *76*, 7020–7029.

Svejstrup, J.Q. (2007). Contending with transcriptional arrest during RNAPII transcript elongation. Trends Biochem. Sci. *32*, 165–171.

van Regenmortel, M.H.V., Fauquet, C.M., Bishop, D.H.L., Carstens, E.B., Estes, M.K., Lemon, S.M., Maniloff, J., Mayo, M.A., McGeoch, D.J., Pringle, C.R., and Wickner, R.B., eds. (2000). Virus taxonomy: seventh report of the International Committee on Taxonomy of Viruses, 1st ed. (San Diego, Calif.: Academic Press).

Watson, R.J., Preston, C.M., and Clements, J.B. (1979). Separation and characterization of herpes simplex virus type 1 immediate-early mRNAs. J. Virol. *31*, 42–52.

Wilcox, K.W., Kohn, A., Sklyanskaya, E., and Roizman, B. (1980). Herpes simplex virus phosphoproteins. I. Phosphate cycles on and off some viral polypeptides and can alter their affinity for DNA. J. Virol. *33*, 167–182.

Zhou, C., and Knipe, D.M. (2002). Association of herpes simplex virus type 1 ICP8 and ICP27 proteins with cellular RNA polymerase II holoenzyme. J. Virol. *76*, 5893–5904.

HSV-1 DNA Replication

6

Stacey A. Ward and Sandra K. Weller

Abstract

The *cis*- and *trans*-acting elements required for DNA synthesis of Herpes Simplex Virus (HSV) have been identified, and genetic and biochemical analyses have provided important insights into how they work together to replicate the large double-stranded viral genome. Furthermore, viral enzymes involved in DNA replication have provided a rich store of useful targets for antiviral therapy against herpesviruses. Despite these advances, many questions remain unresolved concerning the overall mechanism of genome replication. For instance, it has long been recognized that the products of viral DNA replication are head-to-tail concatemers; however, it is not clear how these concatemers are generated. In this chapter we will summarize the known functions of viral replication proteins and explore the possibility that these viral proteins may function in combination with cellular proteins to produce concatemers suitable for packaging into preformed viral capsids.

Introduction

The duplication of genetic material is one of the most fundamental and important processes that organisms must perform. The molecular mechanisms involved in herpesvirus DNA replication and its regulation are of interest for several reasons. Herpesviruses provide important models for the study of eukaryotic DNA replication. Many of the replication proteins encoded by herpesviruses represent functional analogues of the eukaryotic DNA replication machinery. Furthermore, viral proteins required for viral DNA replication provide attractive targets for antiviral chemotherapy. In this chapter we will discuss recent progress on the *cis*- and *trans*-acting factors shown to be essential for viral DNA synthesis as well as models for how recombination-dependent replication may contribute to the overall mechanism of viral replication. The virus encodes seven proteins necessary for origin-dependent DNA synthesis – UL9 (origin-binding protein), ICP8 (ssDNA binding protein), UL30/UL42 (polymerase complex), and UL5/UL8/UL52 (helicase/primase) (Challberg, 1986; Wu *et al.*, 1988). We will focus on DNA replication in the extensively studied herpes simplex virus 1 (HSV-1), although DNA replication by the other herpesviruses resembles that of HSV in many ways.

HSV genome

The dsDNA genome of HSV is large (162 kbp) and structurally complex, containing two unique regions (U_L and U_S) that are flanked by the inverted repeat sequences '*b*' and '*c*', respectively. The *a* repeat is present at both termini and an inverted copy is present at the junction between U_L and U_S (Fig. 6.1). The unique long and short regions of HSV have been shown to invert relative to each other, producing four possible genomic combinations (Hayward *et al.*, 1975; Sheldrick and Berthelot, 1975).

Origins of replication

DNA replication generally begins at specific sites on viral genomes known as origins of replication,

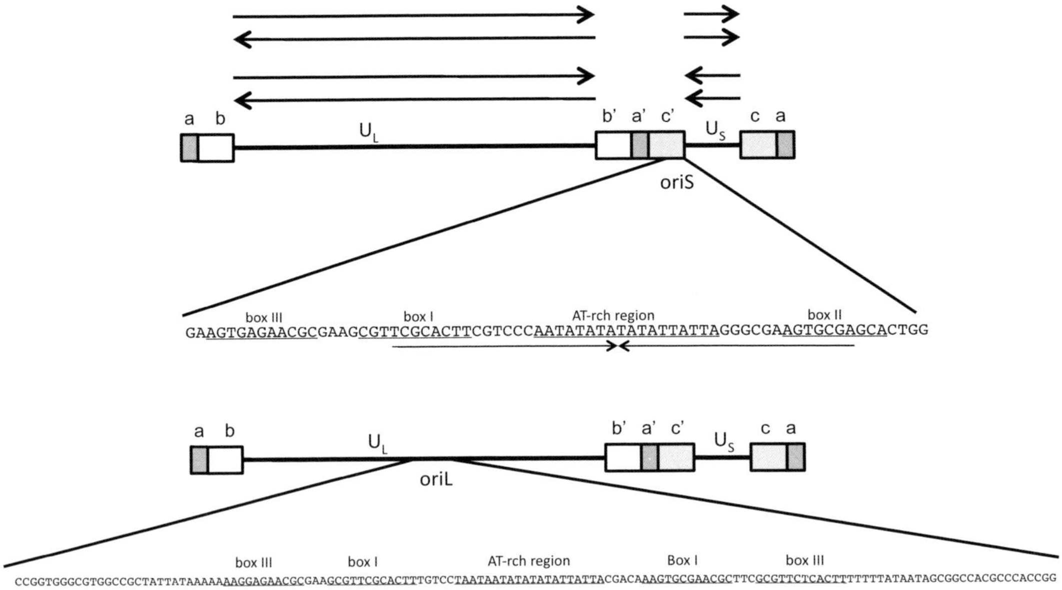

Figure 6.1 Viral genome and replication origins. The HSV-1 genome contains two unique regions (U_L and U_S) flanked by inverted repeat regions. Two origins of replication have been identified (oriS and oriL). Two copies of OriS are present found within the *c* repeat region. OriL is located in the U_L region. OriS contains an imperfect palindrome (arrows) and three UL9-binding sites (boxes I, II and III). OriL contains a perfect 144bp palindrome and four UL9-binding sites (two copes if box I and two copies of box III).

and many viruses encode specific origin binding proteins responsible for initiation of viral DNA synthesis by inducing local unwinding or melting of the duplex to provide access to template strands. Viral origins often consist of a core region and auxiliary sequences that are important for optimal viral replication. The core regions contain binding sites for viral initiator proteins and A/T-rich sequences that facilitate melting during initial unwinding reactions. The auxiliary sequences enhance the efficiency of viral DNA replication and interestingly, often contain binding sites for cellular or viral transcription factors. The importance of auxiliary regions in origins was originally discovered in polyoma viruses where the presence of the transcriptional enhancer stimulates DNA replication up to 1000-fold (DePamphilis, 1988).

Three origins of replication have been identified within the HSV genome: one within U_L (oriL) and two copies of an origin within the *c* repeats flanking U_S (oriS) (Stow, 1985; Stow and Davison, 1986; Weller *et al.*, 1985). The presence of three origins of replication in the HSV genome is intriguing. HSV-1 mutants lacking one or two origins are competent for lytic replication in cell culture systems (Igarashi *et al.*, 1993; Polvino-Bodnar *et al.*, 1987); however, mutants containing only one origin frequently undergo recombination resulting in the duplication of one or more origins (Igarashi *et al.*, 1993), suggesting that it may be beneficial to have multiple origins.

It has been suggested that the two origins may function differently during lytic and latent infection. Despite considerable similarity in sequence between oriS and oriL, some differences exist with respect to overall palindromic length as well as the number of UL9 binding sites. One additional difference between oriL and oriS sequences is the presence of a *g*lucocorticoid *r*esponse *e*lement (GRE) in oriL (Hardwicke and Schaffer, 1997). In transfection based assays, replication of oriS-containing plasmids has been reported to be more efficient than that of oriL-containing plasmids (Hardwicke and Schaffer, 1995). Jiang *et al.* reported that deletion of oriL resulted in higher-fidelity viral DNA replication (Jiang *et al.*, 2006); however, the biological significance of this finding is not clear. In an attempt to determine whether oriL and oriS perform distinct functions, Balliet and Schaffer studied the *in vivo* behaviour of mutant viruses containing point mutations in site I of oriL or oriS that are defective for initiation in a plasmid based assay. Titres of the oriL mutant virus in tear film, trigeminal ganglia and hindbrain were reduced while the oriS mutation behaved like wild-type virus (Balliet and Schaffer, 2006). These observations imply that oriL plays a role in *in vivo* replication or pathogenesis.

Both oriS and oriL contain A/T rich sequences flanked by binding sites for the origin recognition protein, UL9 (Fig. 6.1). OriS consists of a 45-bp palindrome containing two recognition sites for UL9 (box I and box II) on either side of the A/T rich region (Hazuda *et al.*, 1991; Lockshon and Galloway, 1988; Martin *et al.*, 1991; Weir and Stow, 1990). A third weaker binding site (box III) is located to the left of box I outside the palindromic region (Fig. 6.1a). OriL consists of a longer palindrome, 144bp in length, that contains four recognition sites for UL9 (Fig. 6.1b) (Hardwicke and Schaffer, 1995; Weller *et al.*, 1985). Both origins are situated between transcriptional initiation sites: oriS between two immediate-early genes (ICP4 and ICP22/47) and oriL in the unique long region between two early genes, the major DNA binding protein (UL29 or ICP8) and the DNA polymerase (UL30). Deletion of regions surrounding the HSV core palindromic origins results in decreased replication of an ori-containing plasmid, suggesting that DNA sequences flanking the origins play a role in enhancing replication (Hardwicke and Schaffer, 1995; Wong and Schaffer, 1991). The presence of cellular transcription factors at viral origins may facilitate recruitment of other replication proteins such as polymerases. Reports that cellular proteins OF-1 and Sp1 can bind to HSV oriS are consistent with this model (Dabrowski *et al.*, 1994; Murata *et al.*, 2004; Nguyen-Huynh and Schaffer, 1998). Alternatively, it is possible that viral and/or cellular proteins change the conformation or activity of the initiator protein UL9 at the origin affecting its global architecture or nucleosome structure, allowing access by other DNA binding proteins.

The arrangement of repeated UL9 binding sites and the palindromic nature of both origins led to the suggestion that structural changes in the origin may be important for the initiation of viral DNA synthesis (Martin *et al.*, 1991; Weller *et al.*, 1985). As will be discussed in more detail below, experimental support for the formation of a hairpin structure within the viral origin has been reported (Aslani *et al.*, 2001; Aslani *et al.*, 2000; Makhov *et al.*, 1996a; Olsson *et al.*, 2009).

Trans-acting viral replication proteins

The analysis of temperature-sensitive mutant viruses defective for HSV DNA replication was pioneered by Priscilla Schaffer and others and led to the identification of seven complementation groups, each representing a viral gene product important for viral DNA synthesis (Conley *et al.*, 1981; Marchetti *et al.*, 1988; Parris *et al.*, 1988; Purifoy *et al.*, 1977; Schaffer *et al.*, 1973; Weller *et al.*, 1983a,b). The importance of the seven viral genes was confirmed by an origin-based transfection assay that allowed the identification of viral genes capable of supporting origin-dependent replication (Wu *et al.*, 1988). These seven gene products include the single-stranded binding protein ICP8 (UL29), a two-subunit polymerase complex (UL30/UL42), a three subunit helicase/primase complex (UL5/UL8/UL52) and the origin recognition protein, UL9 (McGeoch *et al.*, 1988; Wu *et al.*, 1988). The first six proteins are core replication proteins shared among all human and animal herpesviruses while the origin recognition protein is unique to HSV and other alphaherpesviruses (Fig. 6.2). It is likely that the overall strategy of lytic viral DNA replication is generally conserved in all herpesviruses; however, the mechanism and regulation of initiation of viral DNA synthesis may differ somewhat. The origins of replication in beta- and gammaherpesviruses are more complex, and no clear UL9 homologues have been identified.

UL9

Genetic analysis indicates that the origin binding protein UL9 is essential for viral DNA replication (Carmichael *et al.*, 1988; Malik *et al.*, 1992; Weir *et al.*, 1989). Interestingly, however, the analysis of temperature-sensitive (*ts*) mutants indicates that

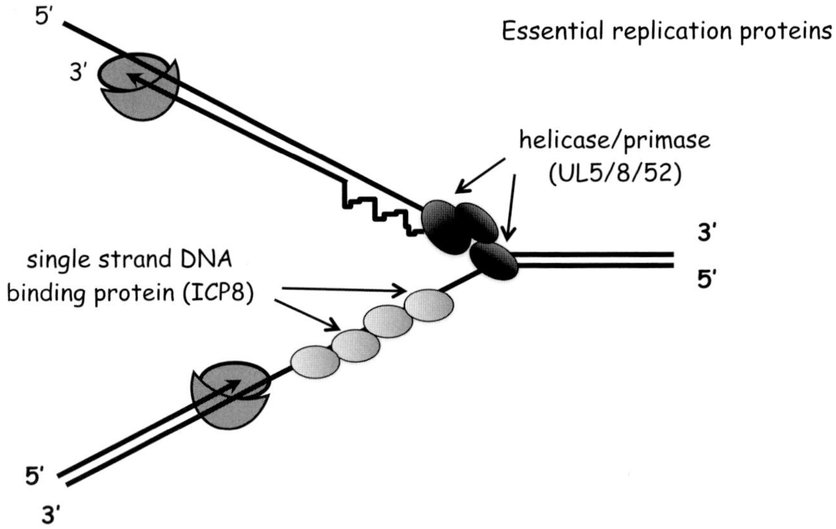

Figure 6.2 Model replication fork. An HSV-1 replication fork would be expected to contain the helicase–primase complex (UL5/UL52/UL8) at the fork: UL5 would be expected to unwind duplex DNA ahead of the fork and U52 would be expected to lay down RNA primers which could then be extended by the two-subunit DNA polymerase (UL30/UL42). The HSV-1 Pol would also be expected to carry out leading strand synthesis. ICP8 (UL29, SSB) would be expected to bind to ssDNA generated during HSV DNA synthesis.

although UL9 is indispensable early in HSV-1 infection, it does not appear to be required late in infection, once DNA synthesis has initiated (Blumel and Matz, 1995; Schildgen *et al.*, 2005).

UL9 forms a dimer in solution and exhibits several biochemical activities including nucleoside triphosphatase, DNA helicase on partially double-stranded substrates, non-specific ssDNA binding, and cooperative binding at an origin of replication (Bruckner *et al.*, 1991; Dodson and Lehman, 1993; Elias *et al.*, 1990; Elias *et al.*, 1992; Fierer and Challberg, 1992; Hazuda *et al.*, 1992; Marintcheva and Weller, 2001a). Seven conserved helicase motifs characteristic of the SF2 family of helicases reside in the N-terminal domain of UL9 (between residues 1–534), and a C-terminally truncated UL9 protein is capable of helicase activity (Abbotts and Stow, 1995; Martinez *et al.*, 1992). The conserved helicase motifs are essential both *in vivo* and for the *in vitro* ATPase and helicase activities (Malik and Weller, 1996; Marintcheva and Weller, 2001b, 2003a; Marintcheva and Weller, 2003b; Martinez *et al.*, 1992).

Specific and non-specific DNA binding activities of UL9

UL9 appears to contact DNA in two different ways: non-specific binding to ssDNA, a feature shared by superfamily I and II helicases, and sequence-specific binding within the origins of replication. The non-specific ssDNA binding region resides within the N-terminus of UL9 (Murata and Dodson, 1999). Mutational analysis of the helicase motifs within the N-terminal portion of UL9 demonstrated that resides 105–124 (motif Ia) were required for ssDNA binding and ssDNA stimulated ATPase activity (Marintcheva and Weller, 2003b). Motif Ia mutations R112A and R113A/F115A retained intrinsic ATPase activity but exhibited reduced levels of ssDNA binding and ssDNA-stimulated ATPase activity (Marintcheva and Weller, 2003b).

The domain responsible for specific origin binding was first mapped to the C-terminus of UL9 (residues 564–832) (Stow *et al.*, 1998; Weir *et al.*, 1989). Recently, a conserved sequence element in UL9, RVKNL (amino acids 756–760), was shown to be required for origin specific DNA binding (Olsson *et al.*, 2009). Interestingly, residues in other regions of the UL9 protein are capable of modulating origin specific DNA binding activity of UL9 (Chattopadhyay and Weller, 2006; Olsson *et al.*, 2009) (see below).

Mechanism of UL9 activation at the origin

By analogy with another viral origin binding protein, SV40 large T antigen, UL9 is expected to be able to unwind duplex DNA in an ATP dependent fashion. The specific unwinding at the SV40 origin by T antigen was critical in the establishment of the *in vitro* replication assay that led to the identification of the replication proteins required for SV40 DNA replication (Li and Kelly, 1984). Although UL9 binding to an origin has been shown to result in distortion of the origin structure as measured by increased sensitivity of bound DNA to potassium permanganate or DNase I (Koff *et al.*, 1991), it has not been possible to reconstitute origin-dependent replication *in vitro*. Recent studies, however, have now begun to shed some light on how UL9 binds to and disrupts origin sequences. One important observation is that the single-strand DNA binding protein ICP8 interacts specifically with UL9 (Boehmer *et al.*, 1994; Boehmer and Lehman, 1993) and stimulates its helicase activity (Boehmer, 1998; Fierer and Challberg, 1992). A 13-amino acid sequence (WPXXXGAXXFXXL) in the extreme C-terminus of UL9 has been shown to be responsible for the UL9–ICP8 interaction (Olsson *et al.*, 2009).

Another important set of observations relates to the oligomeric status of UL9. UL9 forms dimers in solution that can bind cooperatively at an origin of replication (Elias *et al.*, 1992). The C-terminus of UL9 can bind origin DNA with a similar footprint to that of full-length UL9 but cannot induce the A/T-rich distorted structure seen with full-length UL9 (Stabell and Olivo, 1993) suggesting that sequences important for cooperativity reside in the N- terminus. EM studies of UL9 bound to oriS suggest the existence of a pair of dimers assembled on oriS (Makhov *et al.*, 1996b). It has been proposed that one subunit of the dimer contacts a pentanucleotide repeat of the UL9 recognition site in the major groove of one strand, while the other subunit contacts the partially overlapping pentanucleotide site on the other strand (Aslani *et al.*, 2002; Macao *et al.*, 2004). Cooperative binding of a second dimer requires not only the appropriate spacing of the two binding sites flanking the A/T rich region but also sequences within the N-terminus of the UL9 protein (Bruckner *et al.*, 1991; Elias *et al.*, 1990; Elias *et al.*, 1992; Fierer, 1995; Fierer and Challberg, 1992; Hazuda *et al.*, 1992). Based on these observations, it had been assumed that the N-terminal 534 residues contain a region necessary for efficient dimerization and that UL9 dimers adopt a head-to-head configuration. Our laboratory reported a direct interaction between the N- and C-terminal portions of UL9 suggesting that UL9 dimers may be oriented in a head-to-tail arrangement in which the N-terminus is in contact with the C-terminus (Chattopadhyay and Weller, 2007).

Finally, it is now becoming apparent that structural changes in the origin region contribute to the initiation process. The palindromic nature of HSV origins and their potential ability to form hairpin structures was originally predicted based on early sequence analysis (Martin *et al.*, 1991; Weller *et al.*, 1985). Interestingly, Elias and colleagues have shown that an 80-bp double-stranded oligonucleotide containing minimal oriS can be converted to a novel conformation termed oriS* in a reaction that requires ATP hydrolysis, UL9 and ICP8 (Aslani *et al.*, 2000, 2002, 2000; Macao *et al.*, 2004). OriS* contains a hairpin formed by complementary base pairing between box I and box III in oriS and, once formed, can be stably bound by UL9. The ability to form hairpin structures has been conserved in the origins of alphaherpesviruses, and mutations which disrupt hairpin formation decrease the ability of origin containing plasmids to replicate in transfected cells. Second site mutations that restore the ability to form a hairpin structure also restore the ability of mutated oriS to support DNA replication (Aslani *et al.*, 2001).

These observations support a two-step mechanism for origin activation. The initial cooperative binding of UL9 leads to distortion of the A/T rich regions that can be measured by sensitivity to P1 nuclease (He and Lehman, 2001). In a second step, UL9 (in conjunction with ICP8) in the presence of ATP induces the formation of oriS* (Aslani *et al.*, 2002). A model proposed by Elias and colleagues suggests that conformational changes occur in UL9 during this process to shift the binding specificity from duplex origin binding to non-specific ssDNA (Macao *et al.*, 2004; Olsson *et al.*, 2009). Since origin binding is specified by the C-terminus and the ssDNA binding domain requires motif Ia within the N-terminus (Marintcheva and Weller, 2003b), this reaction would be expected to require a sigificant conformational change in UL9 on the origin. The exposure of single-stranded DNA may also cause a conformational change in ICP8 that releases it from UL9 and positions it onto single-stranded DNA (Gustafsson *et al.*, 1995). ICP8 may then act to prevent reannealing of complementary strands. The proposed conformational change of UL9 at the origin is reminiscent of the initiation of origin unwinding by T antigen, the origin binding protein of SV40 which shifts from dimer to hexamer during the initiation reaction (Enemark and Joshua-Tor, 2006; Meinke *et al.*, 2007). Although superfamily II helicases do not form hexamers, both helicases transition from site-specific dimer binding to non-specific ssDNA binding by a higher order structure presumably by adopting an altered conformation. Thus, a dimer of SV40 T antigen binds the origin, leading to a distortion of AT rich elements (Bramhill and Kornberg, 1988) followed by conformational changes in T antigen, facilitating the formation of a hexamer and double hexamer configuration required for active unwinding by the helicase activity (Mastrangelo *et al.*, 1989; Meinke *et al.*, 2007). Similar conformational changes have been reported for the papilloma virus E1 protein (Enemark and Joshua-Tor, 2006).

The current models for origin unwinding by ICP8 and UL9 are based on a series of *in vitro* experiments using partially duplex DNA or supercoiled origin-containing plasmid DNA. Thus, ICP8 and UL9 can unwind artificial oriS-containing duplex DNA if it contains some single-stranded DNA, either a single-strand DNA tail or an unpaired AT-rich spacer region (He and Lehman, 2001). Koff *et al.* originally showed that UL9 can distort an oriS-containing plasmid as long as the plasmid is supercoiled (Koff *et al.*, 1991), and more recent work from the Griffith laboratory indicates that prewinding complexes containing ICP8 and UL9 can be visualized on negatively supercoiled oriS plasmids (Makhov *et al.*, 2003). Negative supertwisting may facilitate the extrusion of the oriS* hairpin. It is not clear how well these *in vitro* experiments mimic the conditions that arise in an infected cell. Although the topological configuration of the HSV genome in an infected cell prior to DNA synthesis is not known, it seems unlikely that it would be negatively supercoiled. Other viral or cellular factors may be required for origin activation on the viral genome in the context of viral infection (Makhov *et al.*, 2003).

Regulation of UL9 function

The initiation of DNA synthesis is a very tightly regulated step in most organisms and complex mechanisms have evolved to ensure that cellular origins fire once and only once per cell cycle. During replication of viruses, it is not necessary to impose such strict regulation on the firing of viral origins; however, initiation events are generally subject to regulation. As described above, the analysis of temperature-sensitive mutants indicates that UL9 is only required during the first five hours of lytic infection (Blumel and Matz, 1995; Schildgen *et al.*, 2005), suggesting that UL9 is not required continuously for DNA replication. Moreover, overexpression of wild-type UL9 has been shown to be inhibitory to HSV-1 infection. In cell lines containing high copy numbers of the UL9 gene, HSV exhibits a decreased efficiency of plaque formation (Malik and Weller, 1996). Furthermore, transfection of cells with full-length UL9 or the C-terminal DNA binding domain results in transdominant inhibition of plaque formation (Arbuckle and Stow, 1993; Chattopadhyay and Weller, 2006; Marintcheva and Weller, 2003a; Perry *et al.*, 1993). Interestingly, UL9 levels do not diminish appreciably during infection [(Elias *et al.*, 1986; Isler and Schaffer, 2001) and S. Ward and S. Weller, unpublished data]. We have thus proposed that one or more of the activities of UL9 may be modulated during infection.

The inhibitory properties of UL9 can be mediated at least in part by its ability to bind to the origins of replication, since inhibition can be partially relieved by mutations in the C-terminus which abrogate DNA binding (Arbuckle and Stow, 1993; Chattopadhyay and Weller, 2006; Malik and Weller, 1996; Perry *et al.*, 1993). We have suggested that once DNA replication has initiated, it may be necessary to clear excess UL9 from the origin. We have shown that the N-terminal region of UL9 can regulate the DNA binding ability of the C-terminus (Chattopadhyay and Weller, 2006). A series of constructs that contain the C-terminal DNA binding domain and varying lengths of the N terminus were shown to exhibit surprising differences in their ability to bind to the origin. Versions whose N-terminus is at residue 304 or 394 exhibited diminished ability to bind to the origin, and these mutants were able to relieve inhibition. On the other hand, mutants whose N-terminal residue was amino acid 494 or 535 were able to bind the origin efficiently and were inhibitory to plaque formation. Interestingly, at least three different forms of UL9 have been observed during infection – full-length protein (94kDa, amino acids 1–851), OBPC-1 (53 kDa), and OBPC-2 (35 kDa, amino acids 571–851) (Link and Schaffer, 2007; Link *et al.*, 2007). OBPC-1 is reportedly formed by cathepsin–B cleavage of full-length UL9 protein resulting in an N-terminally truncated fragment (Link *et al.*, 2007), but efforts to determine the exact cleavage site within UL9 have been unsuccessful. It is intriguing to speculate that the presence of naturally occurring truncated forms of UL9 may act to influence DNA binding or other activities of UL9 and perhaps abrogate its potentially inhibitory properties. Alternatively, UL9 may be regulated by other types of post-translational modifications such as phosphorylation (Isler and Schaffer, 2001) or ubiquitination (Eom *et al.*, 2004; Eom and Lehman, 2003).

It is also possible that UL9 activity is regulated by binding to viral and cellular proteins. As detailed above, ICP8 can bind to and stimulate the helicase activity of UL9 *in vitro* (Boehmer and Lehman, 1993). In addition, cellular proteins have also been reported to interact with UL9, including the large subunit of DNA polymerase alpha and human neural F-Box 42-kDa protein (NFB42 or FBX2) (Eom *et al.*, 2004; Lee *et al.*, 1995). Furthermore, multiple *in vitro* activities of UL9 (oriS affinity, origin-dependent ATPase activity, and distortion of the origin A/T-rich region) have been reported to be stimulated by the addition of cellular heat shock proteins (Eom and Lehman, 2002; Tanguy Le Gac and Boehmer, 2002). Recently, sequence analysis has indicated that UL9 contains conserved motifs predicted to interact with cellular proteins MDM2 and LC8 (Ward and Weller, unpublished data). Experiments are in progress to assess whether these putative cellular interaction partners of UL9 are important for the initiation or regulation of viral DNA replication in infected cells.

ICP8

ICP8 (encoded by the UL29 gene), also known as the HSV single-strand DNA binding protein, was one of the first viral proteins recognized as being essential for HSV DNA synthesis. A series of DNA replication-negative temperature-sensitive mutants defining complementation group 1–1 (Schaffer *et al.*, 1973) were later mapped to the UL29 gene (Conley *et al.*, 1981; Weller *et al.*, 1983b). ICP8 is a zinc metalloprotein which binds non-specifically to ssDNA (Gupte *et al.*, 1991) in a cooperative manner; furthermore, ICP8 exhibits helix-destabilizing activities that are consistent with a role in unwinding duplex DNA during DNA synthesis (Dudas and Ruyechan, 1998; Mapelli *et al.*, 2000). Purified ICP8 forms dimers and higher order filaments *in vitro* and when bound to ssDNA holds the DNA in an extended conformation (Makhov *et al.*, 2009; Mapelli *et al.*, 2000, 2005; Ruyechan, 1983).

Structural information has revealed that ICP8 contains two separate domains, an N- (aa 9–1038) and C- (aa 1049–1129) terminal domain, connected by a short linker region of loose electron density (Mapelli *et al.*, 2005). The N-terminal domain consists of head, neck and shoulder regions, and a number of aromatic and positively charged residues from the front side of the neck (including Tyr543, Asn551, Arg772, Lys774, Arg776, Tyr988, Phe998, and Asn1002) are exposed on the surface and may contact DNA either by base stacking or electrostatic interactions. Modelling experiments suggest that the region between amino acids 530 and 1028 is responsible for contacting ssDNA. Zinc is apparently coordinated by three cysteines (Cys499, Cys502, Cys510) and a histidine (His512), but the zinc binding domain is not predicted to be part of the DNA binding domain (Mapelli *et al.*, 2005). Residues in the C-terminal region are predicted to be involved in protein–protein interactions, including Glu1129, which is exposed and available for interaction as well as a conserved F(N/D)F motif (amino acids 1142–1144). A similar F(N/D)F motif was shown in the adenovirus DNA binding protein to be involved in protein–protein interactions (Mapelli *et al.*, 2005). Further genetic analysis will be required to test these predictions.

ICP8 plays multiple roles in HSV infection

In addition to its role in conjunction with UL9 in activating DNA synthesis at the origins of replication, ICP8 plays many other roles in the viral life cycle. Helix destabilizing properties and the ability to stimulate HSV polymerase and helicase/primase activities are consistent with a role during the elongation phase of DNA synthesis (Boehmer, 1998; Hamatake *et al.*, 1997). A role in the coordination of leading and lagging strand synthesis at the replication fork has also been suggested (Falkenberg *et al.*, 2000). ICP8 also plays a critical role in remodelling nuclear architecture in HSV infected cells and is essential for the formation of prereplicative sites and replication compartments (de Bruyn Kops and Knipe, 1994; Quinlan *et al.*, 1984). ICP8 has also been reported to regulate viral gene expression by repressing transcription from parental genomes and stimulating late gene expression from progeny genomes (Gao and Knipe, 1991; Godowski and Knipe, 1986; McNamee *et al.*, 2000). The recently reported interaction between ICP8 and the immediate-early protein ICP27 may play a role in the stimulation of late gene transcription (Olesky *et al.*, 2005; Zhou and Knipe, 2002).

UL30/42

HSV DNA polymerase is comprised of two subunits: a catalytic subunit encoded by the UL30 gene and a processivity subunit encoded by the UL42 gene (Gottlieb *et al.*, 1990). The analysis of temperature-sensitive mutants revealed that both UL30 (Pol) and UL42 are essential for viral replication (Marchetti *et al.*, 1988; Parris *et al.*, 1988; Purifoy *et al.*, 1977). UL30 contains several motifs conserved among members of the α-like DNA polymerase family (Liu *et al.*, 2006). Mutations in these conserved motifs affect binding of dNTPs and/or their incorporation (Hwang *et al.*, 1999; Tian *et al.*, 2009). UL30 also contains 3′–5′ exonuclease activity, and mutations that reduce nuclease activity reduce the fidelity of replication (Hwang *et al.*, 1997). UL30 has been reported to exhibit apurinic/apyrimidinic lyase activities, which are thought to be involved base-excision repair mechanisms (Bogani and Boehmer, 2008). It will be of interest to determine whether this activity plays a in role recombination-dependent replication (described below).

The accessory protein UL42 increases the affinity of the polymerase for primer/template DNA and stimulates long chain DNA synthesis (Gottlieb and Challberg, 1994; Gottlieb *et al.*, 1990). Unlike 'sliding clamp' processivity factors such as PCNA, which form multimeric rings in solution and require clamp loaders for loading onto DNA, UL42 binds to DNA by itself as a monomer with relatively high affinity (Randell and Coen, 2004; Weisshart *et al.*, 1999 and references therein). Despite high affinity binding for DNA, UL42 can diffuse linearly ('slide') on DNA (Randell and Coen, 2001). Interestingly, the crystal structure of UL42 reveals similarities in the overall structural fold between UL42 and sliding clamp processivity factors such as PCNA, despite biochemical and mechanistic differences (Bridges *et al.*, 2001; Zuccola *et al.*, 2000). The crystal structure of UL42 also reveals that one surface of UL42 interacts with UL30 while the other surface has positively charged residues and is now thought to bind DNA. In fact, mutations in which the conserved arginines on this putative DNA binding surface are changed to alanines decreases DNA binding and processive DNA synthesis *in vitro* (Bridges *et al.*, 2001; Randell *et al.*, 2005; Zuccola *et al.*, 2000). It has been proposed that UL42 utilizes these electrostatic interactions to slide on DNA (Randell and Coen, 2001, 2004; Randell *et al.*, 2005; Zuccola *et al.*, 2000).

These data are consistent with a model in which UL42 promotes processivity by tethering UL30 to DNA and sliding with UL30 as it synthesizes long DNA chains (Jiang *et al.*, 2007b; Randell *et al.*, 2005). Interestingly, UL42 may be mechanistically similar to the T7 DNA pol processivity factor, thioredoxin in *E. coli* (Ghosh *et al.*, 2008). Mutations which reduce DNA binding of UL42 are associated with defects in viral DNA replication and replication fidelity suggesting that UL42 can also influence replication fidelity in infected cells (Jiang *et al.*, 2007a).

UL5/8/52

The HSV-1 helicase/primase is a heterotrimeric complex consisting of the products of the UL5, UL52 and UL8 genes (Crute *et al.*, 1989). All three are essential for viral DNA synthesis in cell culture (Carmichael and Weller, 1989; Goldstein and Weller, 1988a; Lee and Knipe, 1983; Weller *et al.*, 1987; Zhu and Weller, 1992b). The amino acid sequence of UL5 reveals seven super-family 1 conserved helicase motifs, and mutants in these motifs are compromised for viral growth and for helicase and ATPase activity *in vitro* (Zhu and Weller, 1992a). UL52 contains a signature DID motif conserved in many primases which is required for its primase activity (Dracheva *et al.*, 1995; Klinedinst and Challberg, 1994). The C-terminus of UL52 contains a putative zinc-binding motif reminiscent of zinc binding regions in prokaryotic and eukaryotic primases (Carrington-Lawrence and Weller, 2003; Chen *et al.*, 2005; Chen *et al.*, 2007).

The UL5 and UL52 subunits form a subcomplex that exhibits helicase, NTPase and primase activities (Calder and Stow, 1990; Dodson and Lehman, 1991; Sherman *et al.*, 1992). The presence of conserved signature motifs suggest that the helicase and primase activities of the complex likely reside in the UL5 and UL52 subunits, respectively; however, several lines of evidence suggest a complex interdependence between the two subunits (Biswas and Weller, 1999; Biswas and Weller, 2001; Chen *et al.*, 2005; Graves-Woodward *et al.*, 1997). For instance, mutations in the putative zinc finger at the C-terminus of UL52 influence not only primase but also ATPase, helicase, and DNA-binding activities of the UL5/UL52 subcomplex (Biswas and Weller, 1999; Biswas and Weller, 2001; Cavanaugh *et al.*, 2009a; Chen *et al.*, 2005). These results suggest that UL52 binding to DNA via the zinc finger may be necessary for loading UL5 onto DNA. Alternatively, it is possible that UL5 and UL52 share a DNA-binding site created by interaction between the two subunits. The roles of each subunit in DNA binding and primer synthesis was recently probed in a series of kinetic and photo-crosslinking studies with the UL5, UL52 and UL8 subunits expressed in various pair–wise combinations (Cavanaugh *et al.*, 2009a). This study indicated that UL52 likely contains the entire primase active site but also contributes to helicase activity. Furthermore it appears that UL5 contacts DNA and contributes amino acids needed for the initiation of primer synthesis (Cavanaugh *et al.*, 2009a), thereby confirming a complex interdependence between the UL5 and UL52 subunits.

The UL8 subunit does not exhibit any catalytic activities on its own but has been reported to facilitate nuclear uptake of the complex and to modulate the enzymatic functions of the UL5 and UL52 subcomplex (Barnard *et al.*, 1997; Sherman *et al.*, 1992; Tenney *et al.*, 1994). More recently UL8 was shown to increase the rate of primer synthesis by the UL5–UL52 subcomplex by increasing the rate of primer initiation (Cavanaugh *et al.*, 2009a). UL8 may also act to recruit other viral replication proteins to the replication fork based on reported interactions with UL9, UL30, and ICP8 (Falkenberg *et al.*, 1997; Marsden *et al.*, 1996; McLean *et al.*, 1994; Tanguy Le Gac *et al.*, 1996; Tenney *et al.*, 1994).

Non-essential viral gene products

Besides the seven essential replication proteins encoded by HSV-1, several viral proteins appear to play auxiliary roles which are only essential for virus growth under certain circumstances. These include enzymes involved in nucleotide biosynthesis and DNA metabolism such as ribonucleotide reductase, thymidine kinase, deoxyuridine triphosphatase and uracil-DNA-glycosylase. The virally encoded ribonucleotide reductase complex, UL39 (ICP6) and UL40, can produce dNTPs used in DNA synthesis and is essential for replication of HSV in serum-starved cells in culture and for the establishment of latent infection animal models (Brandt *et al.*, 1991; Cameron *et al.*, 1988; Goldstein and Weller, 1988b; Idowu *et al.*, 1992; Jacobson *et al.*, 1989). As discussed in Chapter 7, the HSV RR1 subunit (ICP6) may also act as a chaperone, perhaps to prevent the induction of apoptosis and/or to promote the assembly of the translational machinery (Chabaud *et al.*, 2003; Langelier *et al.*, 2002; Perkins *et al.*, 2002a,b; Walsh and Mohr, 2006). The virally encoded thymidine kinase (TK) protein functions to phosphorylate thymidine and other nucleosides and is not required for viral growth in culture or in the clinical setting. Interestingly, TK also phosphorylates antiviral compounds such as acyclovir, and mutations within the TK gene which confer resistance to acyclovir do not hinder the replication kinetics of HSV-1 (Sergerie and Boivin, 2006). In fact, TK mutant viruses have been found in latently infected neurons of immunocompromised patients, indicating that these viruses can become latent (Wang *et al.*, 2007). The deoxyuridine triphosphatase (dUTPase) and uracil DNA-glycosylase may be important in preventing misincorporation of uracil residues into the viral genome. Although HSV-1 mutants lacking dUTPase and uracil N-glycosylase can replicate in cell culture, they exhibit more severe defects in non-dividing cells and *in vivo* (Pyles and Thompson, 1994a,b).

Overall model for HSV DNA replication

The overall mechanism of HSV DNA replication is poorly understood. The products of DNA replication are larger-than-unit length concatemers consisting of tandem repeats of the viral genome. Although it was previously thought that linear virion DNA circularized in infected cells and that concatemers were generated by rolling circle replication (reviewed in Boehmer and Lehman, 1997), this model has never been directly confirmed experimentally.

As described above, the initial step in viral DNA replication likely involves the recognition of one or more of the viral origins of replication by UL9. UL9 is required early but is dispensable at later times (Blumel and Matz, 1995), leading to a two stage model of DNA replication: a UL9-dependent mode followed by a mechanism that is not dependent on UL9. UL9, in conjunction with ICP8, is believed to cause the initial distortion or destabilization at a viral origin allowing the recruitment of the rest of the replication fork proteins, possibly via protein–protein interactions. For instance, UL9 has been reported to interact with ICP8, UL8, and UL42 (Boehmer *et al.*, 1994; McLean *et al.*, 1994; Trego and Parris, 2003). ICP8 has been reported to stimulate the helicase and ATPase activities of UL9 (Arana *et al.*, 2001; Aslani *et al.*, 2002; Makhov *et al.*, 2003), and UL42 has also been reported to enhance the 3′ to 5′ helicase activity of UL9 (Trego and Parris, 2003).

Once the origin is activated, it has been proposed that the helicase/primase complex is recruited and acts to unwind the duplex DNA (helicase function) and synthesize short RNA primers to initiate DNA replication (primase function) (Fig. 6.2). UL5 is required continuously during infection as evidenced by a block in DNA synthesis using temperature-sensitive mutants in the UL5 gene (Zhu and Weller, 1988). In addition to its catalytic activities, the helicase/primase subunits themselves may be required to form a scaffold which can recruit other replication proteins (Carrington-Lawrence and Weller, 2003; Livingston *et al.*, 2008). The last of the herpes replication proteins recruited to the fork appears to be the two-subunit polymerase, UL30 and UL42. Recruitment of UL30 requires the presence of an active primase, indicating that conformational changes that occur in an active helicase/primase complex and/or the RNA primer itself are important to bring the polymerase to viral DNA (Carrington-Lawrence and Weller, 2003). Once the polymerase complex is recruited to the replication fork, it is believed to catalyse leading and lagging strand DNA synthesis as well as 3′–5′ exonuclease proofreading (Hwang *et al.*, 1997). Once DNA synthesis has initiated at one or more of the origins of replication, it appears that the mode of viral DNA synthesis shifts to a UL9-independent mechanism which could proceed by rolling circle replication (reviewed in

Boehmer and Lehman, 1997) and/or a recombination-dependent replication step (reviewed in Wilkinson and Weller, 2003). Rolling-circle replication has been reported on plasmid DNA in baculovirus-infected insect cells expressing six HSV replication proteins – ICP8, UL30/UL42, and UL5/UL8/UL52 (Skaliter and Lehman, 1994). Interestingly, the addition of UL9 and an oriS-containing plasmid clone in this assay exerts an inhibitory effect on DNA synthesis, providing additional evidence for inhibitory properties of UL9 and support for the notion of an UL9-independent mode of DNA synthesis. The rolling circle mechanism has also been reconstituted *in vitro* on a preformed replication fork using the six core HSV replication proteins (HSV polymerase, single-strand DNA binding protein and the helicase/primase) (Falkenberg *et al.*, 2000; Graves-Woodward *et al.*, 1997; Klinedinst and Challberg, 1994). Rolling circle replication may generate the head-to tail concatemers observed in HSV-infected cells (reviewed in Boehmer and Lehman, 1997); however, this model has never been confirmed directly in infected cells. Furthermore, simple rolling circle replication does not explain the observation that genomic inversions occur as soon as viral DNA synthesis can be detected (Lamberti and Weller, 1996; Severini *et al.*, 1996; Zhang *et al.*, 1994). By analogy with the better-studied DNA bacteriophages, is possible that HSV uses a recombination-dependent mechanism to generate head-to-tail concatemers (reviewed in Wilkinson and Weller, 2003).

DNA replication and recombination are linked

Several lines of evidence suggest that HSV DNA replication and recombination are tightly linked (Dutch *et al.*, 1992, 1995; Honess *et al.*, 1980; Smiley *et al.*, 1990; Weber *et al.*, 1988). Recombination is a frequent event between two co-infecting HSV genomes (Brown *et al.*, 1973, 1992; Honess *et al.*, 1980; Schaffer *et al.*, 1974). Recombination also occurs within repeated regions of the genome resulting in the inversion of the L and S components relative to one another (Fig. 6.1a) (Hayward *et al.*, 1975; Sheldrick and Berthelot, 1975). High frequency inversion events can also be observed in cells transfected with plasmids containing the bacterial transposable element Tn5 in the presence of the seven essential viral replication genes (Weber *et al.*, 1988). Thus, inversion during infection appears to be the result of generalized recombination mediated in part by viral replication proteins. Interestingly, the six core HSV-encoded replication factors can induce concatemer formation on SV40 genomes if SV40 large T-antigen is also provided (Blumel *et al.*, 2000). Since SV40 replication normally results in the appearance of two circular daughter molecules, it is noteworthy that the presence of HSV replication proteins can alter the mode of replication to generate concatemeric DNA.

Additional evidence linking replication and recombination comes from physical studies of replicating DNA molecules: viral DNA replication results in the accumulation of head-to-tail concatemers that are highly branched, with X- and Y-junctions (Jacob and Roizman, 1977; Severini *et al.*, 1996; Shlomai *et al.*, 1976). Furthermore, pulsed-field gel electrophoresis (PFGE) suggests that HSV DNA replication products adopt a complex, perhaps branched, structure which cannot enter the gel, even after digestion with a restriction enzyme which recognizes a single site within the HSV genome (Martinez *et al.*, 1996; Severini *et al.*, 1996; Zhang *et al.*, 1994).

Interestingly, high rates of homologous recombination in alphaherpesviruses have also been demonstrated *in vivo* in animal infection models (Kintner *et al.*, 1995; Lingen *et al.*, 1997) as well as in human populations (Bowden *et al.*, 2004; Meurens *et al.*, 2004; Muylkens *et al.*, 2009; Norberg *et al.*, 2007). High rates of recombination may participate in the evolution of diversity in herpesviruses. Furthermore, these studies indicate that recombination plays an important role in herpesvirus epidemiology.

Recombination-dependent replication in bacteriophage

The linkage between replication and recombination in DNA viruses was first described for the DNA bacteriophages lambda (Skalka, 1974) and T4 (Mosig, 1994, 1998; reviewed in Stahl *et al.*, 1997). During T4 DNA replication, origin-dependent replication occurs during the early stages of infection and then switches to a recombination-dependent mode of infection (Mosig, 1994, 1998). During this later stage, DNA synthesis is primed from the ends of DNA strands that have invaded homologous segments of other chromosomes. T4 encodes its own RecA homologue which can carry out strand invasion and a helicase which can promote branch migration (Mosig, 1994, 1998). Lambda DNA replication is believed to initiate by theta replication on a circular chromosome; however, concatemeric DNA is generated either by rolling circle DNA replication or by recombination. The observation that lambda recombines at high frequency even in *rec*A mutants led to the identification of a virally encoded specialized recombination system, the lambda Red system encoded by three adjacent genes, gam, bet and exo (Brooks and Clark, 1967; Court *et al.*, 2003; Stahl *et al.*, 1997).

The beta and exo functions together are able to perform a single-strand annealing reaction; however, it is possible that in the presence of cellular RecA, the lambda Red system participates in strand invasion reactions as well (Stahl *et al.*, 1997). Thus both T4 and lambda have evolved recombination-dependent replication strategies essential for the formation of concatemeric DNA which is in turn essential for the production of viral progeny. We and others have suggested that HSV also utilizes a recombination-dependent replication mechanism to replicate its DNA (Wilkinson and Weller, 2003).

Several models have been proposed for the initiation of recombination events during herpesviral infection. The double-stranded ends of DNA are known to be recombinogenic. One mechanism for the production of such ends may involve a cellular enzyme, EndoG, capable of producing double-stranded breaks in viral DNA (Huang *et al.*, 2002, 2006). It is also possible that recombination could be initiated by the ends generated during cleavage and packaging reactions. In addition, the HSV genome within virions is known to contain nicks and gaps (reviewed in Wilkinson and Weller, 2003). When a replication fork passes through a region containing a nick, double-strand breaks are predicted to occur (Kowalczykowski, 2000), and double-strand breaks generated in this way would be expected to stimulate recombination.

The roles of viral and cellular repair proteins

As described above, genomic inversions can be induced in the presence of the seven replication proteins (Weber *et al.*, 1988); however, the possible roles of other viral and cellular proteins are unclear. Unlike T4, HSV does not encode its own ATP-dependent RecA homologue; however, it does encode a two-subunit homologue to the lambda red beta/exo proteins. This virally encoded recombinase consists of the alkaline nuclease (UL12) and the major single-strand DNA binding protein (ICP8) (Reuven *et al.*, 2003; Reuven *et al.*, 2004). UL12 exhibits 5′–3′ exonuclease activity and interacts specifically with ICP8. It has been known for some time that ICP8 can promote the annealing of complementary ssDNA and therefore can function as a single-strand DNA annealing protein (Dutch and Lehman, 1993; Makhov and Griffith, 2006). ICP8 can also mediate strand exchange of short double-stranded substrates with circular ssDNA (Bortner *et al.*, 1993; Nimonkar and Boehmer, 2002). In conjunction with UL12, ICP8 can induce strand transfer of much longer substrates (Reuven *et al.*, 2003, 2004; Reuven and Weller, 2005). This two-subunit complex may constitute a simple recombinase analogous to the red alpha/beta recombinase system encoded by bacteriophage lambda.

It is also possible that cellular proteins contribute to recombination in infected cells. Several repair proteins have been detected in prereplicative sites as well as replication compartments, including Mre11, Rad50, Nbs1, RPA, MSH2, BRCA1, BLM, WRN, DNA ligase I, RAD51 and Ku86 (Gregory and Bachenheimer, 2008; Lilley *et al.*, 2005; Shirata *et al.*, 2005; Taylor and Knipe, 2004; Wilcock and Lane, 1991; Wilkinson and Weller, 2004). Mutant and RNAi knockdown cells lines deficient in ATM, Mre11, Nbs1, ligase IV exhibit defects in viral growth suggesting that one or more of these proteins may play an important role in viral replication (Lilley *et al.*, 2005; Muylaert and Elias, 2007; Shirata *et al.*, 2005;

Yao and Elias, 2001; Yao *et al.*, 1997). It is also possible that by analogy with the lambda system, the UL12/ICP8 complex may act in conjunction with components of the cellular repair/recombination machinery to promote homologous recombination in infected cells. Specific interactions between cellular recombination proteins and ICP8 (Taylor and Knipe, 2004) and UL12 (Balasubramanian and Weller, unpublished data) have been observed. Additional experiments will be necessary to determine the precise roles of viral and cellular proteins in recombination and whether recombination-dependent replication plays an essential role in genome duplication.

Formation of replication compartments

Herpes simplex virus infection results in the dramatic reorganization of the infected cell nucleus involving relocalization of cellular proteins and the ordered assembly of replication compartments, large globular domains within the nucleus of infected cells in which gene expression, DNA replication and cleavage and packaging are thought to occur (de Bruyn Kops and Knipe, 1994; Lamberti and Weller, 1996; Quinlan *et al.*, 1984). Following entry of viral genomes into the nucleus, nucleoprotein complexes containing viral immediate early proteins ICP4 and ICP27 can be detected followed by the recruitment of ND10 proteins such as PML adjacent to the ICP4 nucleoprotein complexes on viral genomes (Everett and Murray, 2005; Everett *et al.*, 2004; Livingston *et al.*, 2008; Sourvinos and Everett, 2002). ND10 are dynamic nuclear bodies containing cellular proteins such as PML, and the recruitment of ND10 proteins to incoming viral genomes is now thought to be repressive to viral gene expression (see Chapters 4 and 13). The disruption of ND10-like foci is mediated by the HSV immediate early protein ICP0 which induces the degradation of PML and other ND10 proteins (Chelbi-Alix and de The, 1999; Everett *et al.*, 1998).

The formation of replication compartments occurs in at least five stages and involves at least three distinguishable intermediate types of subassemblies (stage II microfoci, stage IIIa and IIIb prereplicative sites) (Burkham *et al.*, 2001; Carrington-Lawrence and Weller, 2003; Livingston *et al.*, 2008; Lukonis *et al.*, 1997; Lukonis and Weller, 1996) (Fig. 6.3). Cells in stage I contain ND10-like structures and no visible ICP8 staining. Progression to stage II requires the disruption of ND10-like structures and the oligomerization of the immediate early protein ICP4 on viral genomes (Livingston *et al.*, 2008). Stage II cells contain microfoci of ICP8, UL9 and the helicase/primase complex that form adjacent to ICP4 nucleoprotein complexes (Livingston *et al.*, 2008). If the helicase/primase complex is active, stage IIIa foci can form which contain the

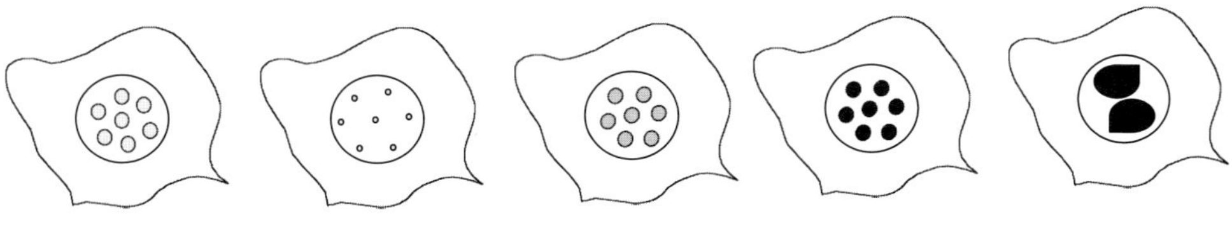

Figure 6.3 Progression of replication compartment formation. Stage I cells are indistinguishable from uninfected cells, and contain intact ND10 foci. Stage II cells exhibit dispersed ND10 proteins and ICP8 microfoci containing ICP8, UL9 and the UL5/UL8/UL52 complex. Stage IIIa foci also contain ICP8, UL9 and the UL5/UL8/UL52 complex. Stage IIIb foci are identical to stage IIa foci except they also contain the UL30/UL42 complex and some isoforms of PML. Stage IV replication compartments represent viral replication centres in which DNA replication and cleavage/packaging reactions occur. Figure adapted from Livingston *et al.* (2008).

same five viral replication proteins (ICP8, UL5, UL8, UL52, UL9) but are somewhat larger than the stage II microfoci. Stage IIIb foci contain all seven replication proteins and some isoforms of PML (Burkham *et al.*, 2001); the viral polymerase is recruited to these foci only in the presence of active primase (Burkham *et al.*, 1998; Carrington-Lawrence and Weller, 2003).

Stage IV cells contain mature replication compartments in which viral DNA replication takes place. Cellular proteins such as RPA, Rad51, Nbs1 and some isoforms of PML are also recruited to prereplicative sites and replication compartments (Burkham *et al.*, 2001; Lilley *et al.*, 2005; Shirata *et al.*, 2005; Wilkinson and Weller, 2004). Replication compartments are formed by the coalescence of pre-replicative foci first into small replication compartments which eventually evolve into larger replication compartments which take up most of the nucleus (Taylor *et al.*, 2003).

Antiviral therapeutics targeting viral DNA replication

Viral proteins required for viral DNA replication provide very attractive targets for antiviral chemotherapy. Most antiherpesvirus drugs have been aimed towards inhibiting the viral polymerase, and the most successful has been acyclovir and its derivatives; however, other replication proteins also provide intriguing and potentially novel targets. It is now recognized that acyclovir-resistant strains of HSV occur in a clinical setting, especially in immunocompromised patients co-infected with HIV or treated with immunosuppressants in a post-organ transplantation setting (Englund *et al.*, 1990). In addition, high doses of acyclovir use to treat patients with herpes encephalitis have been associated with side effects such as nephrotoxicity (Izzedine *et al.*, 2005). These findings underscore the need to develop additional antiviral strategies.

DNA polymerase inhibitors

Agents such as the nucleoside analogue acyclovir (9-(2-hydroxyethoxymethyl)guanine) (Elion *et al.*, 1977) and its derivatives, which target the viral DNA polymerase, represent one of the most successful antiviral drug programmes in history (reviewed in Coen and Schaffer, 2003). Acyclovir is selectively activated by the viral thymidine kinase to a monophosphate form and subsequently converted by cellular kinases to the triphosphate form, which inhibits the polymerase. Acyclovir triphosphate is both a competitive inhibitor and a chain-terminating substrate of the polymerase (Chatis and Crumpacker, 1992). A series of derivatives of acyclovir have been developed with improved solubility, oral bioavailability or an improved ability to be phosphorylated by TK including penciclovir, valacyclovir, famciclovir and a homodimeric form of acyclovir (Brandi *et al.*, 2009; Coen and Schaffer, 2003; Earnshaw *et al.*, 1992). A recent paper identifying the features of purine dNTPs that can be recognized by the HSV polymerase may also prove useful in the future development of nucleoside analogues (Cavanaugh *et al.*, 2009b).

A second class of anti-HSV polymerase inhibitor is exemplified by phosphonoformic acid (Foscarnet, AstraZeneca) which mimics the pyrophosphate product of polymerization. Foscarnet has been approved for use in patients infected with acyclovir-resistant strains, but it is more toxic than acyclovir and must be given intravenously. A new class of non-nucleoside drugs that target herpesvirus DNA polymerases were recently described which not only inhibit the HSV-1 and -2 polymerases, but also have been shown to inhibit HCMV, VZV, and HHV-8 (Oien *et al.*, 2002; Schnute *et al.*, 2005; Thomsen *et al.*, 2003). Studies of resistant mutant viruses combined with the crystal structure of HSV-1 Pol suggest that these act by binding near the dNTP binding site (Liu *et al.*, 2006; Thomsen *et al.*, 2003). It is also theoretically possible to target the polymerase with inhibitors that target the interactions between the UL30 and UL42 subunits (Loregian and Coen, 2006; Pilger *et al.*, 2004), although these have shown limited clinical promise.

Helicase/primase inhibitors (HPI)

Other replication proteins have also been targeted for drug development. In particular, three related classes of compounds – 2-amino-thiazole, thiazolylphenyl, and thiazolylsulfonamide derivatives – have been described that target the helicase/

primase complex (reviewed in Coen and Schaffer, 2003). Several compounds, including BAY 57-1293 and BILS 22 BS, demonstrate good efficacy against HSV-1 in animal models and clinical trials. BAY 57-1293 can be administered orally, exhibits low toxicity and better efficacy than that of valcyclovir (Valtrex, GlaxoSmithKline) (Biswas *et al.*, 2007; Kaufman *et al.*, 2008). This compound also reduced reoccurrences of both herpes keratitis in rabbits and HSV-1 genital lesions in guinea pigs when administered after primary infection, indicating potential benefits in limiting reactivation from latency (Baumeister *et al.*, 2007; Kaufman *et al.*, 2008). Drug-resistant mutants can be recovered both in lab stocks and clinical isolates of HSV-1, and the location of resistance-conferring mutations map to single amino acid mutations within the UL5 gene and/or the UL52 gene (Betz *et al.*, 2002; Biswas *et al.*, 2008; Kleymann *et al.*, 2002; Liuzzi *et al.*, 2004) raising the possibility that these inhibitors may bind to the helicase–primase complex at the interface shared by the two subunits.

Short hairpin oligonucleotides

The discovery of siRNA and shRNA pathways in mammalian cells has opened a potential field of anti-viral nucleotides. Inhibitory RNA can be engineered to target viral genes essential for replication and thus block early points in the viral life cycle with minimal impact on the host cell. These oligonucleotides have been used to decrease viral DNA load in cell culture, with replication targets of ICP8, UL30 (polymerase), and UL5 (helicase) (Falkenhagen *et al.*, 2009; Zhang *et al.*, 2008). This strategy may be complicated, however, by indications that HSV-1 and other viruses may shut down portions of the RNAi gene silencing pathway (Wu *et al.*, 2009). Appropriate delivery vehicles as well as an ability to inactivate the ability of the virus to counteract RNAi will be needed in order to take advantage of RNAi antiviral strategies.

Cellular targets

The concept of inhibiting viral infection by the inhibition of required cellular proteins has several potential advantages (Fauci and Challberg, 2005). Antiviral agents that target cellular proteins could be active against multiple viruses, since it is becoming clear that cellular proteins are required for replication of several different viruses. Another potential advantage is that drug-resistance would not develop with drugs that target cellular proteins.

Virtually any cellular protein or process which is essential for viral replication could theoretically be targeted in this scenario, and several potential targets have been discussed in this chapter. Pharmacological cyclin-dependent kinase inhibitors (PCIs) have been shown to exhibit antiviral activity against HSV-1, HSV-2, HCMV, and VZV (Schang *et al.*, 2002). Drugs such as roscovitine were originally developed as anti-cancer drugs and they show promise for this purpose as well. Roscovitine appears to inhibit transcription of all kinetic classes of HSV proteins; inhibition is observed as early as 2 hours post infection (Schang *et al.*, 2006). PCIs are currently in Phase II clinical trials for the treatment of various cancers, and it will be of considerable interest to determine whether treated patients exhibit decreased incidences of herpes infections (Malumbres and Barbacid, 2009; Schang *et al.*, 2006).

Summary

In summary, despite the identification and analysis of the *cis*- and *trans*-acting elements required for viral DNA synthesis in infected cells, several important questions remain regarding the mechanism and regulation of viral DNA replication. To date, an *in vitro* replication system which recapitulates origin and UL9-dependent DNA replication has not been reported, suggesting that we have not as yet identified all required cellular proteins or structural elements. Furthermore, the mechanism by which head-to-tail concatemers are formed remains elusive. Despite these holes in our knowledge, we have learned a great deal about the activities of *cis*- and *trans*-acting viral proteins, identified potential cellular interaction partners, and during the next several years the answers to many of these unresolved questions are expected to be forthcoming. A better understanding of herpesvirus DNA replication will also lead to the development of novel antiviral therapies and more effective treatments.

References

Abbotts, A.P., and Stow, N.D. (1995). The origin-binding domain of the herpes simplex virus type 1 UL9 protein is not required for DNA helicase activity. J. Gen. Virol. *76*, 3125–3130.

Arana, M.E., Haq, B., Tanguy Le Gac, N., and Boehmer, P.E. (2001). Modulation of the Herpes Simplex Virus Type-1 UL9 DNA Helicase by Its Cognate Single-strand DNA-binding Protein, ICP8. J. Biol. Chem. *276*, 6840–6845.

Arbuckle, M.I., and Stow, N.D. (1993). A mutational analysis of the DNA-binding domain of the herpes simplex virus type 1 UL9 protein. J. Gen. Virol. *74*, 1349–1355.

Aslani, A., Macao, B., Simonsson, S., and Elias, P. (2001). Complementary intrastrand base pairing during initiation of Herpes simplex virus type 1 DNA replication. Proc. Natl. Acad. Sci. U.S.A. *98*, 7194–7199.

Aslani, A., Olsson, M., and Elias, P. (2002). ATP-dependent unwinding of a minimal origin of DNA replication by the origin-binding protein and the single-strand DNA-binding protein ICP8 from herpes simplex virus type I. J. Biol. Chem. *277*, 41204–41212.

Aslani, A., Simonsson, S., and Elias, P. (2000). A novel conformation of the herpes simplex virus origin of DNA replication recognized by the origin binding protein. J. Biol. Chem. *275*, 5880–5887.

Balliet, J.W., and Schaffer, P.A. (2006). Point mutations in herpes simplex virus type 1 oriL, but not in oriS, reduce pathogenesis during acute infection of mice and impair reactivation from latency. J. Virol. *80*, 440–450.

Barnard, E.C., Brown, G., and Stow, N.D. (1997). Deletion mutants of the herpes simplex virus type 1 UL8 protein: effect on DNA synthesis and ability to interact with and influence the intracellular localization of the UL5 and UL52 proteins. Virology *237*, 97–106.

Baumeister, J., Fischer, R., Eckenberg, P., Henninger, K., Ruebsamen-Waigmann, H., and Kleymann, G. (2007). Superior efficacy of helicase-primase inhibitor BAY 57-1293 for herpes infection and latency in the guinea pig model of human genital herpes disease. Antivir. Chem. Chemother. *18*, 35–48.

Betz, U.A., Fischer, R., Kleymann, G., Hendrix, M., and Rubsamen-Waigmann, H. (2002). Potent *in vivo* antiviral activity of the herpes simplex virus primase-helicase inhibitor BAY 57-1293. Antimicrob. Agents Chemother. *46*, 1766–1772.

Biswas, N., and Weller, S.K. (1999). A mutation in the C-terminal putative $Zn2^{+}$ finger motif of UL52 severely affects the biochemical activities of the HSV-1 helicase-primase subcomplex. J. Biol. Chem. *274*, 8068–8076.

Biswas, N., and Weller, S.K. (2001). The UL5 and UL52 Subunits of the Herpes Simplex Virus Type 1 Helicase-Primase Subcomplex Exhibit a Complex Interdependence for DNA Binding. J. Biol. Chem. *276*, 17610–17619.

Biswas, S., Jennens, L., and Field, H.J. (2007). The helicase primase inhibitor, BAY 57-1293 shows potent therapeutic antiviral activity superior to famciclovir in BALB/c mice infected with herpes simplex virus type 1. Antiviral Res. *75*, 30–35.

Biswas, S., Kleymann, G., Swift, M., Tiley, L.S., Lyall, J., Aguirre-Hernandez, J., and Field, H.J. (2008). A single drug-resistance mutation in HSV-1 UL52 primase points to a difference between two helicase-primase inhibitors in their mode of interaction with the antiviral target. J. Antimicrob. Chemother. *61*, 1044–1047.

Blumel, J., Graper, S., and Matz, B. (2000). Structure of simian virus 40 DNA replicated by herpes simplex virus type 1. Virology *276*, 445–454.

Blumel, J., and Matz, B. (1995). Thermosensitive UL9 gene function is required for early stages of herpes simplex virus type 1 DNA synthesis. J. Gen. Virol. *76*, 3119–3124.

Boehmer, P., Craigie, M., Stow, N., and Lehman, I. (1994). Association of origin binding protein and single strand DNA-binding protein, ICP8, during herpes simplex virus type 1 DNA replication *in vivo*. J. Biol. Chem. *269*, 29329–29334.

Boehmer, P., and Lehman, I. (1993). Physical interaction between the herpes simplex virus 1 origin-binding protein and single-stranded DNA-binding protein ICP8. Proc. Natl. Acad. Sci. *90*, 8444–8448.

Boehmer, P.E. (1998). The herpes simplex virus type-1 single-strand DNA-binding protein, ICP8, increases the processivity of the UL9 protein DNA helicase. J. Biol. Chem. *273*, 2676–2683.

Boehmer, P.E., and Lehman, I.R. (1997). Herpes simplex virus DNA replication. Annu. Rev. Biochem. *66*, 347–384.

Bogani, F., and Boehmer, P. (2008). The replicative DNA polymerase of herpes simplex virus 1 exhibits apurinic/apyrimidinic and 5′-deoxyribose phosphate lyase activities. Proc. Natl. Acad. Sci. U.S.A. *105*, 11709–11714.

Bortner, C., Hernandez, T.R., Lehman, I.R., and Griffith, J. (1993). Herpes simplex virus 1 single-strand DNA-binding protein (ICP8) will promote homologous pairing and strand transfer. J. Mol. Biol. *231*, 241–250.

Bowden, R., Sakaoka, H., Donnelly, P., and Ward, R. (2004). High recombination rate in herpes simplex virus type 1 natural populations suggests significant co-infection. Infect. Genet. Evol. *4*, 115–123.

Bramhill, D., and Kornberg, A. (1988). A model for initiation at origins of DNA replication. Cell *54*, 915–918.

Brandi, G., Rossi, L., Schiavano, G.F., Millo, E., and Magnani, M. (2009). A new homodimer of aciclovir as a prodrug with increased solubility and antiviral activity. Int. J. Antimicrob. Agents *34*, 177–180.

Brandt, C.R., Kintner, R.L., Pumfery, A.M., Visalli, R.J., and Grau, D.R. (1991). The herpes simplex virus ribonucleotide reductase is required for ocular virulence. J. Gen. Virol. *72*, 2043–2049.

Bridges, K.G., Chow, C.S., and Coen, D.M. (2001). Identification of crucial hydrogen-bonding residues for the interaction of herpes simplex virus DNA polymerase subunits via peptide display, mutational, and calorimetric approaches. J. Virol. *75*, 4990–4998.

Brooks, K., and Clark, A.J. (1967). Behavior of lambda bacteriophage in a recombination deficienct strain of *Escherichia coli*. J. Virol. *1*, 283–293.

Brown, S.M., Ritchie, D.A., and Subak-Sharpe, J.H. (1973). Genetic studies with herpes simplex virus type 1. The isolation of temperature-sensitive mutants, their arrangement into complementation groups and recombination analysis leading to a linkage map. J. Gen. Virol. *18*, 329–346.

Brown, S.M., Subak-Sharpe, J.H., Harland, J., and MacLean, A.R. (1992). Analysis of intrastrain recombination in herpes simplex virus type 1 strain 17 and herpes simplex virus type 2 strain HG52 using restriction endonuclease sites as unselected markers and temperature-sensitive lesions as selected markers. J. Gen. Virol. *73*, 293–301.

Bruckner, R., Crute, J., Dodson, M., and Lehman, I. (1991). The herpes simplex virus 1 origin binding protein: a DNA helicase. J. Biol. Chem. *266*, 2669–2674.

Burkham, J., Coen, D.M., Hwang, C.B., and Weller, S.K. (2001). Interactions of herpes simplex virus type 1 with ND10 and recruitment of PML to replication compartments. J. Virol. *75*, 2353–2367.

Calder, J.M., and Stow, N.D. (1990). Herpes simplex virus helicase-primase: the UL8 protein is not required for DNA-dependent ATPase and DNA helicase activities. Nucleic Acids Res. *18*, 3573–3578.

Cameron, J.M., McDougall, I., Marsden, H.S., Preston, V.G., Ryan, D.M., and Subak-Sharpe, J.H. (1988). Ribonucleotide reductase encoded by herpes simplex virus is a determinant of the pathogenicity of the virus in mice and a valid antiviral target. J. Gen. Virol. *69*, 2607–2612.

Carmichael, E.P., Kosovsky, M.J., and Weller, S.K. (1988). Isolation and characterization of herpes simplex virus type 1 host range mutants defective in viral DNA synthesis. J. Virol. *62*, 91–99.

Carmichael, E.P., and Weller, S.K. (1989). Herpes simplex virus type 1 DNA synthesis requires. the product of the UL8 gene: isolation and characterization of an ICP6::lacZ insertion mutation. J. Virol. *63*, 591–599.

Carrington-Lawrence, S.D., and Weller, S.K. (2003). Recruitment of polymerase to herpes simplex virus type 1 replication foci in cells expressing mutant primase (UL52) proteins. J. Virol. *77*, 4237–4247.

Cavanaugh, N.A., Ramirez-Aguilar, K.A., Urban, M., and Kuchta, R.D. (2009a). Herpes Simplex Virus-1 Helicase-Primase: Roles of Each Subunit in DNA Binding and Phosphodiester Bond Formation. Biochemistry.

Cavanaugh, N.A., Urban, M., Beckman, J., Spratt, T.E., and Kuchta, R.D. (2009b). Identifying the features. of purine dNTPs that allow accurate and efficient DNA replication by herpes simplex virus I DNA polymerase. Biochemistry *48*, 3554–3564.

Chabaud, S., Lambert, H., Sasseville, A.M., Lavoie, H., Guilbault, C., Massie, B., Landry, J., and Langelier, Y. (2003). The R1 subunit of herpes simplex virus ribonucleotide reductase has chaperone-like activity similar to Hsp27. FEBS Lett. *545*, 213–218.

Challberg, M.D. (1986). A Method for Identifying the Viral Genes Required for Herpesvirus DNA Replication. Proc. Natl. Acad. Sci. U.S.A. *83*, 9094–9098.

Chatis, P.A., and Crumpacker, C.S. (1992). Resistance of herpesviruses to antiviral drugs. Antimicrob. Agents Chemother. *36*, 1589–1595.

Chattopadhyay, S., and Weller, S.K. (2006). DNA Binding activity of the herpes simplex virus type 1 origin binding protein, UL9, can be modulated by sequences in the N terminus: correlation between transdominance and DNA binding. J. Virol. *80*, 4491–4500.

Chattopadhyay, S., and Weller, S.K. (2007). Direct Interaction between the N- and C-terminal portions of the herpes simplex virus type 1 origin binding protein UL9 implies the formation of a head-to-tail dimer. J. Virol. *81*, 13659–13667.

Chelbi-Alix, M.K., and de The, H. (1999). Herpes virus induced proteasome-dependent degradation of the nuclear bodies-associated PML and Sp100 proteins. Oncogene *18*, 935–941.

Chen, Y., Carrington-Lawrence, S.D., Bai, P., and Weller, S.K. (2005). Mutations in the putative zinc-binding motif of UL52 demonstrate a complex interdependence between the UL5 and UL52 subunits of the human herpes simplex virus type 1 helicase/primase complex. J. Virol. *79*, 9088–9096.

Chen, Y., Livingston, C.M., Carrington-Lawrence, S.D., Bai, P., and Weller, S.K. (2007). A mutation in the human herpes simplex virus type 1 UL52 zinc finger motif results in defective primase activity but can recruit viral polymerase and support viral replication efficiently. J. Virol. *81*, 8742–8751.

Coen, D.M., and Schaffer, P.A. (2003). Antiherpesvirus drugs: a promising spectrum of new drugs and drug targets. Nat. Rev. Drug Discov *2*, 278–288.

Conley, A.J., Knipe, D.M., Jones, P.C., and Roizman, B. (1981). Molecular genetics of herpes simplex virus. VII. Characterization of a temperature-sensitive mutant produced by *in vitro* mutagenesis and defective in DNA synthesis and accumulation of gamma polypeptides. J. Virol. *37*, 191–206.

Court, D.L., Sawitzke, J.A., and Thomason, L.C. (2003). Genetic Engineering Using Homologous Recombination. Annu. Rev. Genet. *36*, 361–388.

Crute, J.J., Tsurumi, T., Zhu, L.A., Weller, S.K., Olivo, P.D., Challberg, M.D., Mocarski, E.S., and Lehman, I.R. (1989). Herpes simplex virus 1 helicase-primase: a complex of three herpes-encoded gene products. Proc. Natl. Acad. Sci. U.S.A. *86*, 2186–2189.

Dabrowski, C.E., Carmillo, P.J., and Schaffer, P.A. (1994). Cellular protein interactions with herpes simplex virus type 1 oriS. Mol. Cell Biol. *14*, 2545–2555.

de Bruyn Kops, A., and Knipe, D.M. (1994). Preexisting nuclear architecture defines the intranuclear location of herpesvirus DNA replication structures. J. Virol. *68*, 3512–3526.

DePamphilis, M.L. (1988). Transcriptional elements as components of eukaryotic origins of DNA replication. Cell *52*, 635–638.

Dodson, M.S., and Lehman, I.R. (1991). Association of DNA helicase and primase activities with a subassembly of the herpes simplex virus 1 helicase-primase composed of the UL5 and UL52 gene products. Proc. Natl. Acad. Sci. U.S.A. *88*, 1105–1109.

Dodson, M.S., and Lehman, I.R. (1993). The herpes simplex virus type I origin binding protein. DNA-dependent nucleoside triphosphatase activity. J. Biol. Chem. *268*, 1213–1219.

Dracheva, S., Koonin, E.V., and Crute, J.J. (1995). Identification of the primase active site of the herpes simplex virus type 1 helicase-primase. J. Biol. Chem. *270*, 14148–14153.

Dudas, K.C., and Ruyechan, W.T. (1998). Identification of a region of the herpes simplex virus single-stranded DNA-binding protein involved in cooperative binding. J. Virol. *72*, 257–265.

Dutch, R.E., Bianchi, V., and Lehman, I.R. (1995). Herpes simplex virus type 1 DNA replication is specifically required for high-frequency homologous recombination between repeated sequences. J. Virol. *69*, 3084–3089.

Dutch, R.E., Bruckner, R.C., Mocarski, E.S., and Lehman, I.R. (1992). Herpes simplex virus type 1 recombination: role of DNA replication and viral a sequences. J. Virol. *66*, 277–285.

Dutch, R.E., and Lehman, I.R. (1993). Renaturation of complementary DNA strands by herpes simplex virus type 1 ICP8. J. Virol. *67*, 6945–6949.

Earnshaw, D.L., Bacon, T.H., Darlison, S.J., Edmonds, K., Perkins, R.M., and Vere Hodge, R.A. (1992). Mode of antiviral action of penciclovir in MRC-5 cells infected with herpes simplex virus type 1 (HSV-1), HSV-2, and varicella-zoster virus. Antimicrob. Agents Chemother. *36*, 2747–2757.

Elias, P., Gustafsson, C.M., and Hammarsten, O. (1990). The origin binding protein of herpes simplex virus 1 binds cooperatively to the viral origin of replication oris. J. Biol. Chem. *265*, 17167–17173.

Elias, P., Gustafsson, C.M., Hammarsten, O., and Stow, N.D. (1992). Structural elements required for the cooperative binding of the herpes simplex virus origin binding protein to oriS reside in the N-terminal part of the protein. J. Biol. Chem. *267*, 17424–17429.

Elias, P., O'Donnell, M.E., Mocarski, E.S., and Lehman, I.R. (1986). A DNA binding protein specific for an origin of replication of herpes simplex virus type 1. Proc. Natl. Acad. Sci. U.S.A. *83*, 6322–6326.

Elion, G.B., Furman, P.A., Fyfe, J.A., de Miranda, P., Beauchamp, L., and Schaeffer, H.J. (1977). Selectivity of action of an antiherpetic agent, 9-(2-hydroxyethoxymethyl) guanine. Proc. Natl. Acad. Sci. U.S.A. *74*, 5716–5720.

Enemark, E.J., and Joshua-Tor, L. (2006). Mechanism of DNA translocation in a replicative hexameric helicase. Nature *442*, 270–275.

Englund, J.A., Zimmerman, M.E., Swierkosz, E.M., Goodman, J.L., Scholl, D.R., and Balfour, H.H., Jr. (1990). Herpes simplex virus resistant to acyclovir. A study in a tertiary care center. Ann. Intern. Med. *112*, 416–422.

Eom, C.-Y., Heo, W.D., Craske, M.L., Meyer, T., and Lehman, I.R. (2004). The neural F-box protein NFB42 mediates the nuclear export of the herpes simplex virus type 1 replication initiator protein (UL9 protein) after viral infection. Proc. Natl. Acad. Sci. *101*, 4036–4040.

Eom, C.-Y., and Lehman, I.R. (2002). The human DnaJ protein, hTid-1, enhances binding of a multimer of the herpes simplex virus type 1 UL9 protein to oris, an origin of viral DNA replication. Proc. Natl. Acad. Sci. *99*, 1894–1898.

Eom, C.-Y., and Lehman, I.R. (2003). Replication-initiator protein (UL9) of the herpes simplex virus 1 binds NFB42 and is degraded via the ubiquitin-proteasome pathway. Proc. Natl. Acad. Sci. *100*, 9803–9807.

Everett, R.D., Freemont, P., Saitoh, H., Dasso, M., Orr, A., Kathoria, M., and Parkinson, J. (1998). The disruption of ND10 during herpes simplex virus infection correlates with the Vmw110- and proteasome-dependent loss of several PML isoforms. J. Virol. *72*, 6581–6591.

Everett, R.D., and Murray, J. (2005). ND10 components relocate to sites associated with herpes simplex virus type 1 nucleoprotein complexes during virus infection. J. Virol. *79*, 5078–5089.

Everett, R.D., Sourvinos, G., Leiper, C., Clements, J.B., and Orr, A. (2004). Formation of nuclear foci of the herpes simplex virus type 1 regulatory protein ICP4 at early times of infection: localization, dynamics, recruitment of ICP27, and evidence for the de novo induction of ND10-Like complexes. J. Virol. *78*, 1903–1917.

Falkenberg, M., Bushnell, D.A., Elias, P., and Lehman, I.R. (1997). The UL8 subunit of the heterotrimeric herpes simplex virus type 1 helicase-primase is required for the unwinding of single strand DNA-binding protein (ICP8)-coated DNA substrates. J. Biol. Chem. *272*, 22766–22770.

Falkenberg, M., Lehman, I.R., and Elias, P. (2000). Leading and lagging strand DNA synthesis *in vitro* by a reconstituted herpes simplex virus type 1 replisome. Proc. Natl. Acad. Sci. U.S.A. *97*, 3896–3900.

Falkenhagen, A., Heinrich, J., and Moelling, K. (2009). Short hairpin-loop-structured oligodeoxynucleotides reduce HSV-1 replication. Virology J. *6*, 43.

Fauci, A.S., and Challberg, M.D. (2005). Host-based antipoxvirus therapeutic strategies: turning the tables. J. Clin. Invest. *115*, 231–233.

Fierer, D.S. (1995). The Stoichiometry of Binding of the Herpes Simplex Virus Type 1 Origin Binding Protein, UL9, to Ori(S). J. Biol. Chem. *270*, 7330–7334.

Fierer, D.S., and Challberg, M.D. (1992). Purification and characterization of UL9, the herpes simplex virus type 1 origin-binding protein. J. Virol. *66*, 3986–3995.

Frenkel, N., and Roizman, B. (1972). Separation of the herpesvirus deoxyribonucleic acid duplex into unique fragments and intact strand on sedimentation in alkaline gradients. J. Virol. *10*, 565–572.

Gao, M., and Knipe, D.M. (1991). Potential role for herpes simplex virus ICP8 DNA replication protein in stimulation of late gene expression. J. Virol. *65*, 2666–2675.

Ghosh, S., Hamdan, S.M., Cook, T.E., and Richardson, C.C. (2008). Interactions of Escherichia coli thioredoxin, the processivity factor, with bacteriophage T7 DNA polymerase and helicase. J. Biol. Chem. *283*, 32077–32084.

Godowski, P.J., and Knipe, D.M. (1986). Transcriptional control of herpesvirus gene expression: gene functions required for positive and negative regulation. Proc. Natl. Acad. Sci. U.S.A. *83*, 256–260.

Goldstein, D.J., and Weller, S.K. (1988a). An ICP6::lacZ insertional mutagen is used to demonstrate that the UL52 gene of herpes simplex virus type 1 is required for virus growth and DNA synthesis. J. Virol. *62*, 2970–2977.

Goldstein, D.J., and Weller, S.K. (1988b). Herpes simplex virus type 1-induced ribonucleotide reductase activity is dispensable for virus growth and DNA synthesis: isolation and characterization of an ICP6 lacZ insertion mutant. J. Virol. *62*, 196–205.

Gottlieb, J., and Challberg, M.D. (1994). Interaction of herpes simplex virus type 1 DNA polymerase and the UL42 accessory protein with a model primer template. J. Virol. *68*, 4937–4945.

Gottlieb, J., Marcy, A.I., Coen, D.M., and Challberg, M.D. (1990). The herpes simplex virus type 1 UL42 gene product: a subunit of DNA polymerase that functions to increase processivity. J. Virol. *64*, 5976–5987.

Graves-Woodward, K.L., Gottlieb, J., Challberg, M.D., and Weller, S.K. (1997). Biochemical analyses of mutations in the HSV-1 helicase-primase that alter ATP hydrolysis, DNA unwinding, and coupling between hydrolysis and unwinding. J. Biol. Chem. *272*, 4623–4630.

Gregory, D.A., and Bachenheimer, S.L. (2008). Characterization of mre11 loss following HSV-1 infection. Virology *373*, 124–136.

Gupte, S.S., Olson, J.W., and Ruyechan, W.T. (1991). The major herpes simplex virus type-1 DNA-binding protein is a zinc metalloprotein. J. Biol. Chem. *266*, 11413–11416.

Gustafsson, C.M., Falkenberg, M., Simonsson, S., Valadi, H., and Elias, P. (1995). The DNA ligands influence the interactions between the herpes simplex virus 1 origin binding protein and the single strand DNA-binding protein, ICP-8. J. Biol. Chem. *270*, 19028–19034.

Hamatake, R.K., Bifano, M., Hurlburt, W.W., and Tenney, D.J. (1997). A functional interaction of ICP8, the herpes simplex virus single-stranded DNA-binding protein, and the helicase–primase complex that is dependent on the presence of the UL8 subunit. J. Gen. Virol. *78*, 857–865.

Hardwicke, M.A., and Schaffer, P.A. (1995). Cloning and characterization of herpes simplex virus type 1 oriL: comparison of replication and protein–DNA complex formation by oriL and oriS. J. Virol. *69*, 1377–1388.

Hardwicke, M.A., and Schaffer, P.A. (1997). Differential effects of nerve growth factor and dexamethasone on herpes simplex virus type 1 oriL- and oriS-dependent DNA replication in PC12 cells. J. Virol. *71*, 3580–3587.

Hayward, G.S., Jacob, R.J., Wadsworth, S.C., and Roizman, B. (1975). Anatomy of herpes simplex virus DNA: evidence for four populations of molecules that differ in the relative orientations of their long and short components. Proc. Natl. Acad. Sci. U.S.A. *72*, 4243–4247.

Hazuda, D.J., Perry, H.C., and McClements, W.L. (1992). Cooperative interactions between replication origin-bound molecules of herpes simplex virus origin-binding protein are mediated via the amino terminus of the protein. J. Biol. Chem. *267*, 14309–14315.

Hazuda, D.J., Perry, H.C., Naylor, A.M., and McClements, W.L. (1991). Characterization of the herpes simplex virus origin binding protein interaction with OriS. J. Biol. Chem. *266*, 24621–24626.

He, X., and Lehman, I.R. (2001). An initial ATP-independent step in the unwinding of a herpes simplex virus type I origin of replication by a complex of the viral origin-binding protein and single-strand DNA-binding protein. Proc. Natl. Acad. Sci. U.S.A. *98*, 3024–3028.

Honess, R.W., Buchan, A., Halliburton, I.W., and Watson, D.H. (1980). Recombination and linkage between structural and regulatory genes of herpes simplex virus type 1: study of the functional organization of the genome. J. Virol. *34*, 716–742.

Huang, K.J., Ku, C.C., and Lehman, I.R. (2006). Endonuclease G: a role for the enzyme in recombination and cellular proliferation. Proc. Natl. Acad. Sci. U.S.A. *103*, 8995–9000.

Huang, K.J., Zemelman, B.V., and Lehman, I.R. (2002). Endonuclease G, a candidate human enzyme for the initiation of genomic inversion in herpes simplex type 1 virus. J. Biol. Chem. *277*, 21071–21079.

Hwang, Y.T., Liu, B.Y., Coen, D.M., and Hwang, C.B. (1997). Effects of mutations in the Exo III motif of the herpes simplex virus DNA polymerase gene on enzyme activities, viral replication, and replication fidelity. J. Virol. *71*, 7791–7798.

Hwang, Y.T., Liu, B.Y., Hong, C.Y., Shillitoe, E.J., and Hwang, C.B. (1999). Effects of exonuclease activity and nucleotide selectivity of the herpes simplex virus DNA polymerase on the fidelity of DNA replication *in vivo*. J. Virol. 73, 5326–5332.

Idowu, A.D., Fraser-Smith, E.B., Poffenberger, K.L., and Herman, R.C. (1992). Deletion of the herpes simplex virus type 1 ribonucleotide reductase gene alters virulence and latency *in vivo*. Antiviral Res. *17*, 145–156.

Igarashi, K., Fawl, R., Roller, R.J., and Roizman, B. (1993). Construction and properties of a recombinant herpes simplex virus 1 lacking both S-component origins of DNA synthesis. J. Virol. *67*, 2123–2132.

Isler, J.A., and Schaffer, P.A. (2001). Phosphorylation of the Herpes Simplex Virus Type 1 Origin Binding Protein. J. Virol. *75*, 628–637.

Izzedine, H., Launay-Vacher, V., and Deray, G. (2005). Antiviral drug-induced nephrotoxicity. Am J. Kidney Dis. *45*, 804–817.

Jacob, R.J., and Roizman, B. (1977). Anatomy of herpes simplex virus DNA VIII. Properties of the replicating DNA. J. Virol. *23*, 394–411.

Jacobson, J.G., Leib, D.A., Goldstein, D.J., Bogard, C.L., Schaffer, P.A., Weller, S.K., and Coen, D.M. (1989). A herpes simplex virus ribonucleotide reductase deletion mutant is defective for productive acute and reactivatable latent infections of mice and for replication in mouse cells. Virology *173*, 276–283.

Jiang, C., Hwang, Y.T., and Hwang, C.B. (2006). Herpes simplex virus type 1 recombinants without the oriL sequence replicate DNA with increased fidelity. Virology *347*, 277–285.

Jiang, C., Hwang, Y.T., Randell, J.C., Coen, D.M., and Hwang, C.B. (2007a). Mutations that decrease DNA binding of the processivity factor of the herpes simplex virus DNA polymerase reduce viral yield, alter the kinetics of viral DNA replication, and decrease the fidelity of DNA replication. J. Virol. *81*, 3495–3502.

Jiang, C., Hwang, Y.T., Wang, G., Randell, J.C., Coen, D.M., and Hwang, C.B. (2007b). Herpes simplex virus mutants with multiple substitutions affecting DNA binding of UL42 are impaired for viral replication and DNA synthesis. J. Virol. *81*, 12077–12079.

Kaufman, H.E., Varnell, E.D., Gebhardt, B.M., Thompson, H.W., Atwal, E., Rubsamen-Waigmann, H., and Kleymann, G. (2008). Efficacy of a helicase-primase inhibitor in animal models of ocular herpes simplex virus type 1 infection. J. Ocul. Pharmacol. Ther. *24*, 34–42.

Kintner, R.L., Allan, R.W., and Brandt, C.R. (1995). Recombinants are isolated at high frequency following *in vivo* mixed ocular infection with two avirulent herpes simplex virus type 1 strains. Arch. Virol. *140*, 231–244.

Kleymann, G., Fischer, R., Betz, U.A., Hendrix, M., Bender, W., Schneider, U., Handke, G., Eckenberg, P., Hewlett, G., Pevzner, V., *et al.* (2002). New helicase-primase inhibitors as drug candidates for the treatment of herpes simplex disease. Nat. Med. *8*, 392–398.

Klinedinst, D.K., and Challberg, M.D. (1994). Helicase–primase complex of herpes simplex virus type 1: a mutation in the UL52 subunit abolishes primase activity. J. Virol. *68*, 3693–3701.

Koff, A., Schwedes, J.F., and Tegtmeyer, P. (1991). Herpes simplex virus origin-binding protein (UL9) loops and distorts the viral replication origin. J. Virol. *65*, 3284–3292.

Kowalczykowski, S.C. (2000). Initiation of genetic recombination and recombination-dependent replication. Trends Biochem. Sci. *25*, 156–165.

Lamberti, C., and Weller, S.K. (1996). The herpes simplex virus type 1 UL6 protein is essential for cleavage and packaging but not for genomic inversion. Virology *226*, 403–407.

Langelier, Y., Bergeron, S., Chabaud, S., Lippens, J., Guilbault, C., Sasseville, A.M., Denis, S., Mosser, D.D., and Massie, B. (2002). The R1 subunit of herpes simplex virus ribonucleotide reductase protects cells against apoptosis at, or upstream of, caspase-8 activation. J. Gen. Virol. *83*, 2779–2789.

Lee, C.K., and Knipe, D.M. (1983). Thermolabile *in vivo* DNA-binding activity associated with a protein encoded by mutants of herpes simplex virus type 1. J. Virol. *46*, 909–919.

Lee, S.S., Dong, Q., Wang, T.S., and Lehman, I.R. (1995). Interaction of herpes simplex virus 1 origin-binding protein with DNA polymerase alpha. Proc. Natl. Acad. Sci. U.S.A. *92*, 7882–7886.

Li, J.J., and Kelly, T.J. (1984). Simian Virus 40 DNA Replication *in vitro*. Proc. Natl. Acad. Sci. U.S.A. *81*, 6973–6977.

Lilley, C.E., Carson, C.T., Muotri, A.R., Gage, F.H., and Weitzman, M.D. (2005). DNA repair proteins affect the lifecycle of herpes simplex virus 1. Proc. Natl. Acad. Sci. U.S.A. *102*, 5844–5849.

Lingen, M., Hengerer, F., and Falke, D. (1997). Mixed vaginal infections of Balb/c mice with low virulent herpes simplex type 1 strains result in restoration of virulence properties: vaginitis/vulvitis and neuroinvasiveness. Med. Microbiol. Immunol. *185*, 217–222.

Link, M.A., and Schaffer, P.A. (2007). Herpes simplex virus type 1 C-terminal variants of the origin binding protein (OBP), OBPC-1 and OBPC-2, cooperatively regulate viral DNA levels *in vitro*, and OBPC-2 affects mortality in mice. J. Virol. *81*, 10699–10711.

Link, M.A., Silva, L.A., and Schaffer, P.A. (2007). Cathepsin B mediates cleavage of herpes simplex virus type 1 origin binding protein (OBP) to yield OBPC-1, and cleavage is dependent upon viral DNA replication. J. Virol. *81*, 9175–9182.

Liu, S., Knafels, J.D., Chang, J.S., Waszak, G.A., Baldwin, E.T., Deibel, M.R., Jr., Thomsen, D.R., Homa, F.L., Wells, P.A., Tory, M.C., *et al.* (2006). Crystal structure of the herpes simplex virus 1 DNA polymerase. J. Biol. Chem. *281*, 18193–18200.

Liuzzi, M., Kibler, P., Bousquet, C., Harji, F., Bolger, G., Garneau, M., Lapeyre, N., McCollum, R.S., Faucher, A.M., Simoneau, B., *et al.* (2004). Isolation and characterization of herpes simplex virus type 1 resistant to aminothiazolylphenyl-based inhibitors of the viral helicase-primase. Antiviral Res. *64*, 161–170.

Livingston, C.M., DeLuca, N.A., Wilkinson, D.E., and Weller, S.K. (2008). Oligomerization of ICP4 and rearrangement of heat shock proteins may be important for herpes simplex virus type 1 prereplicative site formation. J. Virol. *82*, 6324–6336.

Lockshon, D., and Galloway, D.A. (1988). Sequence and structural requirements of a herpes simplex viral DNA replication origin. Mol. Cell Biol. *8*, 4018–4027.

Loregian, A., and Coen, D.M. (2006). Selective anti-cytomegalovirus compounds discovered by screening for inhibitors of subunit interactions of the viral polymerase. Chem. Biol. 13, 191–200.

Lukonis, C.J., Burkham, J., and Weller, S.K. (1997). Herpes simplex virus type 1 prereplicative sites are a heterogeneous population: only a subset are likely to be precursors to replication compartments. J. Virol. *71*, 4771–4781.

Lukonis, C.J., and Weller, S.K. (1996). Characterization of nuclear structures. in cells infected with herpes simplex virus type 1 in the absence of viral DNA replication. J. Virol. *70*, 1751–1758.

Macao, B., Olsson, M., and Elias, P. (2004). Functional properties of the herpes simplex virus type I origin-binding protein are controlled by precise interactions with the activated form of the origin of DNA replication. J. Biol. Chem. *279*, 29211–29217.

Makhov, A.M., Boehmer, P.E., Lehman, I.R., and Griffith, J.D. (1996a). The herpes simplex virus type 1 origin-binding protein carries out origin specific DNA unwinding and forms unwound stem–loop structures. EMBO J. *15*, 1742–1750.

Makhov, A.M., Boehmer, P.E., Lehman, I.R., and Griffith, J.D. (1996b). Visualization of the unwinding of long DNA chains by the herpes simplex virus type 1 UL9 protein and ICP8. J. Mol. Biol. *258*, 789–799.
Makhov, A.M., and Griffith, J.D. (2006). Visualization of the annealing of complementary single-stranded DNA catalyzed by the herpes simplex virus type 1 ICP8 SSB/recombinase. J. Mol. Biol. *355*, 911–922.
Makhov, A.M., Lee, S.S., Lehman, I.R., and Griffith, J.D. (2003). Origin-specific unwinding of herpes simplex virus 1 DNA by the viral UL9 and ICP8 proteins: visualization of a specific preunwinding complex. Proc. Natl. Acad. Sci. U.S.A. *100*, 898–903.
Makhov, A.M., Sen, A., Yu, X., Simon, M.N., Griffith, J.D., and Egelman, E.H. (2009). The bipolar filaments formed. by herpes simplex virus type 1 SSB/recombination protein (ICP8) suggest a mechanism for DNA annealing. J. Mol. Biol. *386*, 273–279.
Malik, A., and Weller, S. (1996). Use of transdominant mutants of the origin-binding protein (UL9) of herpes simplex virus type 1 to define functional domains. J. Virol. *70*, 7859–7866.
Malik, A.K., Martinez, R., Muncy, L., Carmichael, E.P., and Weller, S.K. (1992). Genetic analysis of the herpes simplex virus type 1 UL9 gene: Isolation of a lacZ insertion mutant and expression in eukaryotic cells. Virology *190*, 702–715.
Malumbres, M., and Barbacid, M. (2009). Cell cycle, CDKs and cancer: a changing paradigm. Nat. Rev. Cancer *9*, 153–166.
Mapelli, M., Muhleisen, M., Persico, G., van der Zandt, H., and Tucker, P.A. (2000). The 60-residue C-terminal region of the single-stranded DNA binding protein of herpes simplex virus type 1 is required for cooperative DNA binding. J. Virol. *74*, 8812–8822.
Mapelli, M., Panjikar, S., and Tucker, P.A. (2005). The crystal structure of the herpes simplex virus 1 ssDNA-binding protein suggests the structural basis for flexible, cooperative single-stranded DNA binding. J. Biol. Chem. *280*, 2990–2997.
Marchetti, M.E., Smith, C.A., and Schaffer, P.A. (1988). A temperature-sensitive mutation in a herpes simplex virus type 1 gene required for viral DNA synthesis maps to coordinates 0.609 through 0.614 in UL. J. Virol. *62*, 715–721.
Marintcheva, B., and Weller, S.K. (2001a). A tale of two HSV-1 helicases: roles of phage and animal virus helicases in DNA replication and recombination. Prog. Nucleic Acid Res. Mol. Biol. *70*, 77–118.
Marintcheva, B., and Weller, S.K. (2001b). Residues within the conserved helicase motifs of UL9, the origin-binding protein of herpes simplex virus-1, are essential for helicase activity but not for dimerization or origin binding activity. 10.1074/jbc.M007743200. J. Biol. Chem. *276*, 6605–6615.
Marintcheva, B., and Weller, S.K. (2003a). Existence of transdominant and potentiating mutants of UL9, the herpes simplex virus type 1 origin-binding protein, suggests that levels of UL9 protein may be regulated during infection. J. Virol. *77*, 9639–9651.
Marintcheva, B., and Weller, S.K. (2003b). Helicase motif Ia is involved in single-strand DNA-binding and helicase activities of the herpes simplex virus type 1 origin-binding protein, UL9. J. Virol. *77*, 2477–2488.
Marsden, H.S., Cross, A.M., Francis, G.J., Patel, A.H., MacEachran, K., Murphy, M., McVey, G., Haydon, D., Abbotts, A., and Stow, N.D. (1996). The herpes simplex virus type 1 UL8 protein influences the intracellular localization of the UL52 but not the ICP8 or POL replication proteins in virus-infected cells. J. Gen. Virol. *77*, 2241–2249.
Martin, D.W., Deb, S.P., Klauer, J.S., and Deb, S. (1991). Analysis of the herpes simplex virus type 1 OriS sequence: mapping of functional domains. J. Virol. *65*, 4359–4369.
Martinez, R., Sarisky, R.T., Weber, P.C., and Weller, S.K. (1996). Herpes simplex virus type 1 alkaline nuclease is required for efficient processing of viral DNA replication intermediates. J. Virol. *70*, 2075–2085.
Martinez, R., Shao, L., and Weller, S.K. (1992). The conserved helicase motifs of the herpes simplex virus type 1 origin-binding protein UL9 are important for function. J. Virol. *66*, 6735–6746.
Mastrangelo, I.A., Hough, P.V., Wall, J.S., Dodson, M., Dean, F.B., and Hurwitz, J. (1989). ATP-dependent assembly of double hexamers of SV40 T antigen at the viral origin of DNA replication. Nature *338*, 658–662.
McGeoch, D.J., Dalrymple, M.A., Dolan, A., McNab, D., Perry, L.J., Taylor, P., and Challberg, M.D. (1988). Structures. of herpes simplex virus type 1 genes required for replication of virus DNA. J. Virol. *62*, 444–453.
McLean, G.W., Abbotts, A.P., Parry, M.E., Marsden, H.S., and Stow, N.D. (1994). The herpes simplex virus type 1 origin-binding protein interacts specifically with the viral UL8 protein. J. Gen. Virol. *75*, 2699–2706.
McNamee, E.E., Taylor, T.J., and Knipe, D.M. (2000). A dominant-negative herpesvirus protein inhibits intranuclear targeting of viral proteins: effects on DNA replication and late gene expression. J. Virol. *74*, 10122–10131.
Meinke, G., Phelan, P., Moine, S., Bochkareva, E., Bochkarev, A., Bullock, P.A., and Bohm, A. (2007). The crystal structure of the SV40 T-antigen origin binding domain in complex with DNA. PLoS Biol. *5*, e23.
Meurens, F., Keil, G.M., Muylkens, B., Gogev, S., Schynts, F., Negro, S., Wiggers, L., and Thiry, E. (2004). Interspecific recombination between two ruminant alphaherpesviruses, bovine herpesviruses 1 and 5. J. Virol. *78*, 9828–9836.
Mosig, G. (1994). Homologous Recombination. In Bacteriophage T4, Karam, J.D., ed. (Washington D.C., ASM Press), pp. 54–82.
Mosig, G. (1998). Recombination and recombination-dependent DNA replication in bacteriophage T4. Annu. Rev. Genet. *32*, 379–413.
Murata, L.B., and Dodson, M.S. (1999). The herpes simplex virus type 1 origin-binding protein. sequence-specific activation of adenosine triphosphatase activity

by a double-stranded DNA containing box I. J. Biol. Chem. *274*, 37079–37086.
Murata, L.B., Dodson, M.S., and Hall, J.D. (2004). A human cellular protein activity (OF-1), which binds herpes simplex virus type 1 origin, contains the Ku70/Ku80 heterodimer. J. Virol. *78*, 7839–7842.
Muylaert, I., and Elias, P. (2007). Knockdown of DNA Ligase IV/XRCC4 by RNA Interference Inhibits Herpes Simplex Virus Type I DNA Replication. J. Biol. Chem. *282*, 10865–10872.
Muylkens, B., Farnir, F., Meurens, F., Schynts, F., Vanderplasschen, A., Georges, M., and Thiry, E. (2009). Coinfection with two closely related alphaherpesviruses results in a highly diversified recombination mosaic displaying negative genetic interference. J. Virol. *83*, 3127–3137.
Nguyen-Huynh, A.T., and Schaffer, P.A. (1998). Cellular transcription factors enhance herpes simplex virus type 1 oriS-dependent DNA replication. J. Virol. *72*, 3635–3645.
Nimonkar, A.V., and Boehmer, P.E. (2002). *In vitro* strand exchange promoted by the herpes simplex virus type-1 single strand DNA-binding protein (ICP8) and DNA helicase-primase. J. Biol. Chem. *277*, 15182–15189.
Norberg, P., Kasubi, M.J., Haarr, L., Bergstrom, T., and Liljeqvist, J.A. (2007). Divergence and recombination of clinical herpes simplex virus type 2 isolates. J. Virol. *81*, 13158–13167.
Oien, N.L., Brideau, R.J., Hopkins, T.A., Wieber, J.L., Knechtel, M.L., Shelly, J.A., Anstadt, R.A., Wells, P.A., Poorman, R.A., Huang, A., *et al.* (2002). Broad-spectrum antiherpes activities of 4-hydroxyquinoline carboxamides, a novel class of herpesvirus polymerase inhibitors. Antimicrob. Agents Chemother. *46*, 724–730.
Olesky, M., McNamee, E.E., Zhou, C., Taylor, T.J., and Knipe, D.M. (2005). Evidence for a direct interaction between HSV-1 ICP27 and ICP8 proteins. Virology *331*, 94–105.
Olsson, M., Tang, K.W., Persson, C., Wilhelmsson, L.M., Billeter, M., and Elias, P. (2009). Stepwise evolution of the herpes simplex virus origin binding protein and origin of replication. J. Biol. Chem. *284*, 16246–16255.
Parris, D.S., Cross, A., Haarr, L., Orr, A., Frame, M.C., Murphy, M., McGeoch, D.J., and Marsden, H.S. (1988). Identification of the gene encoding the 65-kilodalton DNA-binding protein of herpes simplex virus type 1. J. Virol. *62*, 818–825.
Perkins, D., Pereira, E.F., Gober, M., Yarowsky, P.J., and Aurelian, L. (2002a). The herpes simplex virus type 2 R1 protein kinase (ICP10 PK) blocks apoptosis in hippocampal neurons, involving activation of the MEK/MAPK survival pathway. J. Virol. *76*, 1435–1449.
Perkins, D., Yu, Y., Bambrick, L.L., Yarowsky, P.J., and Aurelian, L. (2002b). Expression of herpes simplex virus type 2 protein ICP10 PK rescues neurons from apoptosis due to serum deprivation or genetic defects. Exp. Neurol. 174, 118–122.
Perry, H.C., Hazuda, D.J., and McClements, W.L. (1993). The DNA binding domain of herpes simplex virus type 1 origin binding protein is a transdominant inhibitor of virus replication. Virology *193*, 73–79.
Pilger, B.D., Cui, C., and Coen, D.M. (2004). Identification of a small molecule that inhibits herpes simplex virus DNA Polymerase subunit interactions and viral replication. Chem. Biol. 11, 647–654.
Polvino-Bodnar, M., Orberg, P.K., and Schaffer, P.A. (1987). Herpes simplex virus type 1 oriL is not required for virus replication or for the establishment and reactivation of latent infection in mice. J. Virol. *61*, 3528–3535.
Purifoy, D.J., Lewis, R.B., and Powell, K.L. (1977). Identification of the herpes simplex virus DNA polymerase gene. Nature *269*, 621–623.
Pyles, R.B., and Thompson, R.L. (1994a). Evidence that the herpes simplex virus type 1 uracil DNA glycosylase is required for efficient viral replication and latency in the murine nervous system. J. Virol. *68*, 4963–4972.
Pyles, R.B., and Thompson, R.L. (1994b). Mutations in accessory DNA replicating functions alter the relative mutation frequency of herpes simplex virus type 1 strains in cultured murine cells. J. Virol. *68*, 4514–4524.
Quinlan, M.P., Chen, L.B., and Knipe, D.M. (1984). The intranuclear location of a herpes simplex virus DNA-binding protein is determined by the status of viral DNA replication. Cell *36*, 857–868.
Randell, J.C., and Coen, D.M. (2001). Linear diffusion on DNA despite high-affinity binding by a DNA polymerase processivity factor. Mol. Cell *8*, 911–920.
Randell, J.C., and Coen, D.M. (2004). The herpes simplex virus processivity factor, UL42, binds DNA as a monomer. J. Mol. Biol. 335, 409–413.
Randell, J.C., Komazin, G., Jiang, C., Hwang, C.B., and Coen, D.M. (2005). Effects of substitutions of arginine residues on the basic surface of herpes simplex virus UL42 support a role for DNA binding in processive DNA synthesis. J. Virol. *79*, 12025–12034.
Reuven, N.B., Staire, A.E., Myers, R.S., and Weller, S.K. (2003). The herpes simplex virus type 1 alkaline nuclease and single-stranded DNA binding protein mediate strand exchange *in vitro*. J. Virol. *77*, 7425–7433.
Reuven, N.B., and Weller, S.K. (2005). Herpes simplex virus type 1 single-strand DNA binding protein ICP8 enhances the nuclease activity of the UL12 alkaline nuclease by increasing its processivity. J. Virol. *79*, 9356–9358.
Reuven, N.B., Willcox, S., Griffith, J.D., and Weller, S.K. (2004). Catalysis of strand exchange by the HSV-1 UL12 and ICP8 proteins: potent ICP8 recombinase activity is revealed upon resection of dsDNA substrate by nuclease. J. Mol. Biol. *342*, 57–71.
Ruyechan, W.T. (1983). The major herpes simplex virus DNA-binding protein holds single-stranded DNA in an extended configuration. J. Virol. *46*, 661–666.
Schaffer, P.A., Aron, G.M., Biswal, N., and Benyesh-Melnick, M. (1973). Temperature-sensitive mutants of herpes simplex virus type 1: isolation, complementation and partial characterization. Virology *52*, 57–71.
Schaffer, P.A., Tevethia, M.J., and Benyesh-Melnick, M. (1974). Recombination between

temperature-sensitive mutants of herpes simplex virus type 1. Virology 58, 219–228.

Schang, L.M., Bantly, A., Knockaert, M., Shaheen, F., Meijer, L., Malim, M.H., Gray, N.S., and Schaffer, P.A. (2002). Pharmacological cyclin-dependent kinase inhibitors inhibit replication of wild-type and drug-resistant strains of herpes simplex virus and human immunodeficiency virus type 1 by targeting cellular, not viral, proteins. J. Virol. *76*, 7874–7882.

Schang, L.M., St Vincent, M.R., and Lacasse, J.J. (2006). Five years of progress on cyclin-dependent kinases and other. cellular proteins as potential targets for antiviral drugs. Antivir. Chem. Chemother. *17*, 293–320.

Schildgen, O., Graper, S., Blumel, J., and Matz, B. (2005). Genome replication and progeny virion production of herpes simplex virus type 1 mutants with temperature-sensitive lesions in the origin-binding protein. J. Virol. *79*, 7273–7278.

Schnute, M.E., Cudahy, M.M., Brideau, R.J., Homa, F.L., Hopkins, T.A., Knechtel, M.L., Oien, N.L., Pitts, T.W., Poorman, R.A., Wathen, M.W., *et al.* (2005). 4-Oxo-4, 7-dihydrothieno[2, 3-b]pyridines as non-nucleoside inhibitors of human cytomegalovirus and related herpesvirus polymerases. J. Med. Chem. *48*, 5794–5804.

Sergerie, Y., and Boivin, G. (2006). Thymidine kinase mutations conferring acyclovir resistance in herpes simplex type 1 recombinant viruses. Antimicrob. Agents Chemother. *50*, 3889–3892.

Severini, A., Scraba, D.G., and Tyrrell, D.L. (1996). Branched structures. in the intracellular DNA of herpes simplex virus type 1. J. Virol. *70*, 3169–3175.

Sheldrick, P., and Berthelot, N. (1975). Inverted repetitions in the chromosome of herpes simplex virus. Cold Spring Harb. Symp. Quant. Biol. *39 Pt 2*, 667–678.

Sherman, G., Gottlieb, J., and Challberg, M.D. (1992). The UL8 subunit of the herpes simplex virus helicase–primase complex is required for efficient primer utilization. J. Virol. *66*, 4884–4892.

Shirata, N., Kudoh, A., Daikoku, T., Tatsumi, Y., Fujita, M., Kiyono, T., Sugaya, Y., Isomura, H., Ishizaki, K., and Tsurumi, T. (2005). Activation of ataxia telangiectasia-mutated DNA damage checkpoint signal transduction elicited by herpes simplex virus infection. J. Biol. Chem. *280*, 30336–30341.

Shlomai, J., Strauss, B., Asher, Y., Friedmann, A., and Becker, Y. (1976). Analysis of herpes simplex virus DNA synthesized in infected nuclei by chromatography on benzoylated naphthoylated DEAE cellulose columns. J. Gen. Virol. *32*, 189–204.

Skaliter, R., and Lehman, I.R. (1994). Rolling circle DNA replication *in vitro* by a complex of herpes simplex virus type 1-encoded enzymes. Proc. Natl. Acad. Sci. U.S.A. *91*, 10665–10669.

Skalka, A. (1974). A replicator's view of recombination (and repair). In Mechanisms of Recombination, Grell, R.F., ed. (Plenum Press), pp. 421–432.

Smiley, J.R., Duncan, J., and Howes, M. (1990). Sequence requirements for DNA rearrangements induced by the terminal repeat of herpes simplex virus type 1 KOS DNA. J. Virol. *64*, 5036–5050.

Sourvinos, G., and Everett, R.D. (2002). Visualization of parental HSV-1 genomes and replication compartments in association with ND10 in live infected cells. EMBO J. *21*, 4989–4997.

Stabell, E.C., and Olivo, P.D. (1993). A truncated herpes simplex virus origin binding protein which contains the carboxyl terminal origin binding domain binds to the origin of replication but does not alter its conformation. Nucleic Acids Res. *21*, 5203–5211.

Stahl, M.M., Thomason, L., Poteete, A.R., Tarkowski, T., Kuzminov, A., and Stahl, F.W. (1997). Annealing vs. invasion in phage lambda recombination. Genetics *147*, 961–977.

Stow, N.D. (1985). Mutagenesis of a herpes simplex virus origin of DNA replication and its effect on viral interference. J. Gen. Virol. *66*, 31–42.

Stow, N.D., and Davison, A.J. (1986). Identification of a varicella-zoster virus origin of DNA replication and its activation by herpes simplex virus type 1 gene products. J. Gen. Virol. *67*, 1613–1623.

Tanguy Le Gac, N., and Boehmer, P.E. (2002). Activation of the Herpes Simplex Virus Type-1 Origin-binding Protein (UL9) by Heat Shock Proteins. J. Biol. Chem. *277*, 5660–5666.

Tanguy Le Gac, N., Villani, G., Hoffmann, J.S., and Boehmer, P.E. (1996). The UL8 subunit of the herpes simplex virus type-1 DNA helicase-primase optimizes utilization of DNA templates covered by the homologous single-strand DNA-binding protein ICP8. J. Biol. Chem. *271*, 21645–21651.

Taylor, T.J., and Knipe, D.M. (2004). Proteomics of herpes simplex virus replication compartments: association of cellular DNA replication, repair, recombination, and chromatin remodeling proteins with ICP8. J. Virol. *78*, 5856–5866.

Taylor, T.J., McNamee, E.E., Day, C., and Knipe, D.M. (2003). Herpes simplex virus replication compartments can form by coalescence of smaller compartments. Virology *309*, 232–247.

Tenney, D.J., Hurlburt, W.W., Micheletti, P.A., Bifano, M., and Hamatake, R.K. (1994). The UL8 component of the herpes simplex virus helicase–primase complex stimulates primer synthesis by a subassembly of the UL5 and UL52 components. J. Biol. Chem. *269*, 5030–5035.

Thomsen, D.R., Oien, N.L., Hopkins, T.A., Knechtel, M.L., Brideau, R.J., Wathen, M.W., and Homa, F.L. (2003). Amino acid changes within conserved region III of the herpes simplex virus and human cytomegalovirus DNA polymerases confer resistance to 4-oxo-dihydroquinolines, a novel class of herpesvirus antiviral agents. J. Virol. *77*, 1868–1876.

Tian, W., Hwang, Y.T., Lu, Q., and Hwang, C.B. (2009). Finger domain mutation affects enzyme activity, DNA replication efficiency, and fidelity of an exonuclease-deficient DNA polymerase of herpes simplex virus type 1. J. Virol. *83*, 7194–7201.

Trego, K.S., and Parris, D.S. (2003). Functional interaction between the herpes simplex virus type 1 polymerase processivity factor and origin-binding proteins: enhancement of UL9 helicase activity. J. Virol. *77*, 12646–12659.

Walsh, D., and Mohr, I. (2006). Assembly of an active translation initiation factor complex by a viral protein. Genes Dev. *20*, 461–472.

Wang, K., Mahalingam, G., Hoover, S.E., Mont, E.K., Holland, S.M., Cohen, J.I., and Straus, S.E. (2007). Diverse herpes simplex virus type 1 thymidine kinase mutants in individual human neurons and ganglia. J. Virol. *81*, 6817–6826.

Weber, P.C., Challberg, M.D., Nelson, N.J., Levine, M., and Glorioso, J.C. (1988). Inversion events in the HSV-1 genome are directly mediated by the viral DNA replication machinery and lack sequence specificity. Cell *54*, 369–381.

Weir, H.M., Calder, J.M., and Stow, N.D. (1989). Binding of the herpes simplex virus type 1 UL9 gene product to an origin of viral DNA replication. Nucleic Acids Res. *17*, 1409–1425.

Weir, H.M., and Stow, N.D. (1990). Two binding sites for the herpes simplex virus type 1 UL9 protein are required for efficient activity of the oriS replication origin. J. Gen. Virol. *71*, 1379–1385.

Weisshart, K., Chow, C.S., and Coen, D.M. (1999). Herpes simplex virus processivity factor UL42 imparts increased DNA-binding specificity to the viral DNA polymerase and decreased dissociation from primer-template without reducing the elongation rate. J. Virol. *73*, 55–66.

Weller, S.K., Aschman, D.P., Sacks, W.R., Coen, D.M., and Schaffer, P.A. (1983a). Genetic analysis of temperature-sensitive mutants of HSV-1: the combined use of complementation and physical mapping for cistron assignment. Virology *130*, 290–305.

Weller, S.K., Carmichael, E.P., Aschman, D.P., Goldstein, D.I., and Schaffer, P.A. (1987). Genetic and phenotypic characterization of mutants in four essential genes that map to the left half of HSV-1 UL DNA. Virology *161*, 198–210.

Weller, S.K., Lee, K.J., Sabourin, D.J., and Schaffer, P.A. (1983b). Genetic analysis of temperature-sensitive mutants which define the gene for the major herpes simplex virus type 1 DNA-binding protein. J. Virol. *45*, 354–366.

Weller, S.K., Spadaro, A., Schaffer, J.E., Murray, A.W., Maxam, A.M., and Schaffer, P.A. (1985). Cloning, sequencing, and functional analysis of oriL, a herpes simplex virus type 1 origin of DNA synthesis. Mol. Cell Biol. *5*, 930–942.

Wilcock, D., and Lane, D.P. (1991). Localization of p53, retinoblastoma and host replication proteins at sites of viral replication in herpes-infected cells. *349*, 429–431.

Wilkinson, D., and Weller, S. (2003). The Role of DNA recombination in herpes simplex virus DNA replication. IUBMB Life *55*, 451–458.

Wilkinson, D.E., and Weller, S.K. (2004). Recruitment of cellular recombination and repair proteins to sites of herpes simplex virus type 1 DNA replication is dependent on the composition of viral proteins within prereplicative sites and correlates with the induction of the DNA damage response. J. Virol. *78*, 4783–4796.

Wong, S.W., and Schaffer, P.A. (1991). Elements in the transcriptional regulatory region flanking herpes simplex virus type 1 oriS stimulate origin function. J. Virol. *65*, 2601–2611.

Wu, C.A., Nelson, N.J., McGeoch, D.J., and Challberg, M.D. (1988). Identification of herpes simplex virus type 1 genes required for origin-dependent DNA synthesis. J. Virol. *62*, 435–443.

Wu, Z., Zhu, Y., Bisaro, D.M., and Parris, D.S. (2009). Herpes simplex virus type 1 suppresses RNA-induced gene silencing in mammalian cells. J. Virol. *83*, 6652–6663.

Yao, X.D., and Elias, P. (2001). Recombination during early herpes simplex virus type 1 infection is mediated by cellular proteins. J. Biol. Chem. *276*, 2905–2913.

Yao, X.D., Matecic, M., and Elias, P. (1997). Direct repeats of the herpes simplex virus a sequence promote non-conservative homologous recombination that is not dependent on XPF/ERCC4. J. Virol. *71*, 6842–6849.

Zhang, X., Efstathiou, S., and Simmons, A. (1994). Identification of novel herpes simplex virus replicative intermediates by field inversion gel electrophoresis: implications for viral DNA amplification strategies. Virology *202*, 530–539.

Zhang, Y.Q., Lai, W., Li, H., and Li, G. (2008). Inhibition of herpes simplex virus type 1 by small interfering RNA. Clin. and Exp. Dermatol. *33*, 56–61.

Zhou, C., and Knipe, D.M. (2002). Association of herpes simplex virus type 1 ICP8 and ICP27 proteins with cellular RNA polymerase II holoenzyme. J. Virol. *76*, 5893–5904.

Zhu, L., and Weller, S.K. (1988). UL5, a protein required for HSV DNA synthesis: genetic analysis, overexpression in Escherichia coli, and generation of polyclonal antibodies. Virology *166*, 366–378.

Zhu, L.A., and Weller, S.K. (1992a). The six conserved helicase motifs of the UL5 gene product, a component of the herpes simplex virus type 1 helicase-primase, are essential for its function. J. Virol. *66*, 469–479.

Zhu, L.A., and Weller, S.K. (1992b). The UL5 gene of herpes simplex virus type 1: isolation of a lacZ insertion mutant and association of the UL5 gene product with other. members of the helicase–primase complex. J. Virol. *66*, 458–468.

Zuccola, H.J., Filman, D.J., Coen, D.M., and Hogle, J.M. (2000). The crystal structure of an unusual processivity factor, herpes simplex virus UL42, bound to the C terminus of its cognate polymerase. Mol. Cell *5*, 267–278.

Translational Control in Herpes Simplex Virus-infected Cells

7

Ian Mohr

Abstract

Like all viruses, alphaherpesviruses are completely reliant upon the protein synthesis machinery resident in their host cells. In particular, viral mRNAs must effectively compete with cellular mRNAs to engage ribosomes. To ensure high-level production of the polypeptides required for their lytic replication, multiple independent gene products expressed by the model alphaherpesvirus HSV-1 effectively seize control of critical host cell translational control pathways. Surprisingly, while host protein synthesis is profoundly suppressed by global changes in mRNA metabolism, the assembly of a multi-subunit, cap-binding translation initiation factor complex required to recruit 40S subunits to mRNA is directly stimulated. This involves both inactivation of a cellular translational repressor by viral functions, and direct interaction between specific viral proteins and select cellular translation initiation factors. In addition to their dependence on cellular components required for mRNA translation, virus-encoded functions must preserve its activity by neutralizing potent host responses capable of incapacitating the translation machinery, one of which senses stress within the endoplasmic reticulum lumen and another of which functions as a host innate defence component by sensing double-stranded RNA, a molecular signature of viral infection. This chapter discusses in detail the many virus–host interactions that are presently known to control translation in cells productively infected with HSV-1 and highlights recent developments in this area.

Introduction

Nowhere is the dependence of viruses on functions resident in their hosts more evident than their absolute reliance on the cellular protein synthesis machinery. Indeed, whereas nucleic acid polymerases, protein and nucleoside kinases, and transcription factors may all be included amongst the constellation of known virus-encoded functions, the ability of all viruses, including the prototypical alphaherpesvirus HSV-1, to produce the protein products required for their replication requires recruiting host ribosomes to viral mRNAs and ensuring that host innate defences capable of inactivating the translation machinery are effectively neutralized. Failure to accomplish either of these critical goals would compromise their ability to complete their productive replication program and curtail spread to new hosts. The fundamental nature of this problem is amply illustrated by the collection of viral functions dedicated to capturing, controlling and maintaining the activity of the cellular translation apparatus. Not only does this include strategies to directly manipulate cellular translation factors, but potentially involves altering host mRNA metabolism and nearly always necessitates intervening in complex cellular signalling pathways at key junctures to effectively wrestle control of decisive regulatory mechanisms away from their hosts.

Control of translation initiation by a multi-subunit, cap-binding complex

Unlike their prokaryotic counterparts, eukaryotic ribosomes are unable to directly recognize the vast majority of their mRNA targets and position themselves on their 5′ termini. Instead, the 7-methyl GTP cap (m^7GTP) structure found on the mRNA 5′-terminus nucleates assembly of eIF4F, a multi-subunit translation initiation factor, which in turn recruits 40S ribosome subunits to the mRNA 5′ end (Fig. 7.1A and B). Given its role in marking the 5′ end of the mRNA and loading 40S subunits, it is no wonder eIF4F is a prime target for regulating translation initiation in response to a diverse array of stimuli including viral infection.

The translation initiation factor eIF4F (reviewed in Sonenberg & Hinnebusch, 2009) is a cellular polypeptide complex composed of a m^7GTP cap-binding subunit (eIF4E), a large scaffolding component (eIF4G), and a prototypical DEAD box-containing RNA helicase (eIF4A; Fig. 7.1A). Two of three eIF4A isoforms (eIF4AI and II) encoded by discrete genes appear to function equivalently in translation initiation (Yoder Hill *et al.*, 1993; Li *et al.*, 1999), whereas eIF4AIII is required for nonsense mediated mRNA decay (Shibuya *et al.*, 2004). Ancillary associated factors eIF4B or eIF4H stimulate the helicase activity of eIF4A, functioning as helicase processivity factors (reviewed in Rogers *et al.*, 2002). Depending upon their abundance and phosphorylation state, a family of small eIF4E-binding proteins (4E-BPs) can sequester eIF4E and limit binding of eIF4E to eIF4G (reviewed in Sonenberg and Hinnebusch, 2009; Ma and Blenis, 2009). In their hypophosphorylated form, 4E-BP's bind tightly to eIF4E and can suppress cap-dependent translation. Hyperphosphorylation of the 4E-BP's by the kinase mTORC1 in response to a variety of environmental cues results in the release of eIF4E, allowing the cap-binding protein to associate with eIF4G and assemble an active eIF4F complex (Haghihat *et al.*, 1995; Mader *et al.*, 1995; Marcotrigiano *et al.*, 1999; Fig. 7.2). In addition, the cap-binding protein eIF4E is phosphorylated by the eIF4G-associated kinase Mnk1 (Fukunaga and Hunter, 1997; Waskiewicz *et al.*, 1997, 1999; Pyrronet *et al.*, 1999; Scheper *et al.*, 2001). While not essential for translation initiation, eIF4E phosphorylation appears to play a regulatory role, although the details of how this is achieved and which mRNAs are most susceptible are poorly understood (reviewed in Scheper & Proud, 2002; Buxade *et al.*, 2008). Finally, the cellular polyadenylate binding protein (PABP) bound to poly-A sequences found at mRNA 3′ termini, while itself not an eIF4F-core component, physically interacts with eIF4G in the eIF4F initiation complex (Imataja *et al.*, 1998; Fig. 7.1B). This association between the 5′ and 3′ ends of the mRNA generates a circular or 'closed loop' topology that is thought to be important for translation (Wells *et al.*, 1998; Kahvejian *et al.*, 2001; Amrani *et al.*, 2008). Such interactions between proteins bound at mRNA 5′ and 3′ ends stimulate translation and are thought to act as a checkpoint, ensuring that the mRNA

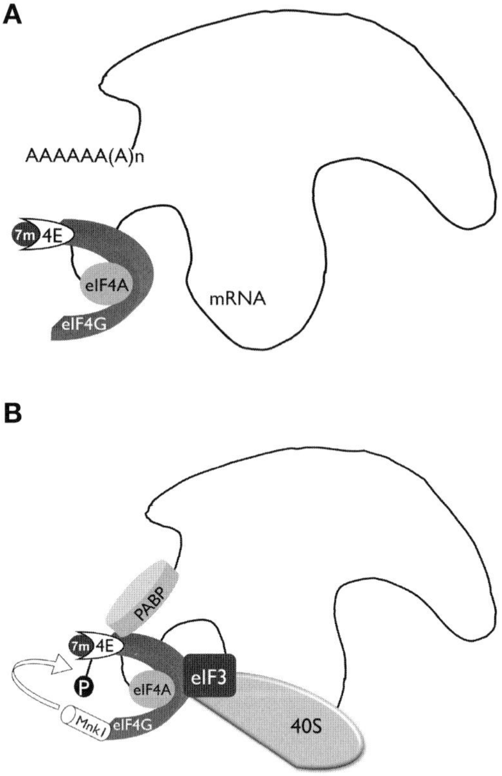

Figure 7.1 The eIF4F complex and associated factors bound to the mRNA. (A) Upon binding the 7-methyl guanine cap (7m) at the 5′ end of the mRNA, eIF4E (4E) recruits eIF4G and eIF4A to assemble the eIF4F complex. (B) The 40S ribosome is recruited through its association with eIF3. eIF4G also interacts with the polyadenylate binding protein (PABP) bound to the 3′ end of the mRNA and the eIF4E kinase Mnk1, which phosphorylates eIF4E (depicted as a P attached to 4E).

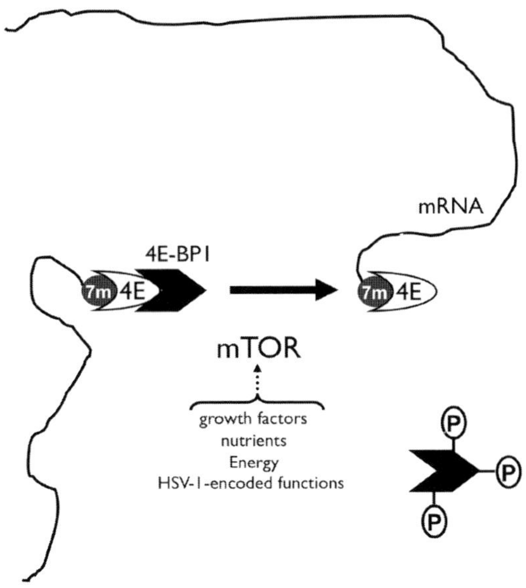

Figure 7.2 Repression of cap-dependent mRNA translation by 4E-BP1 phosphorylation. The cellular cap-binding protein (eIF4E) is depicted bound to the 7-methyl GTP cap (7m) at the mRNA 5′ end. eIF4E, in turn, is bound to the translational repressor 4E-BP1 (eIF4E-binding protein 1) and is unable to assemble into an eIF4F complex with the other translation initiation factors, eIF4G and eIF4A. Activation of the kinase mTOR in response to a variety of cues such as HSV-1 infection, growth factor signalling, alterations to the nutrient pool, or changes in cellular energy reserves results in phosphorylation of the translational repressor protein 4E-BP1 and the release of eIF4E from 4E-BP1.

is both capped and polyadenylated prior to translation initiation. Finally, PABP can be found associated with two cellular polypeptides, Paip1, an eIF4G-related protein that stimulates translation (Craig *et al.*, 1998), or Paip2, a translational repressor that prevents PABP from interacting with eIF4G, Paip1 and the mRNA polyA tail (Khaleghpour *et al.*, 2001; Karim *et al.*, 2006).

All of these components are tethered in place at the 5′ end of the mRNA through an interaction with eIF4G (Fig. 7.1b). In addition to providing a scaffold on which the eIF4F components eIF4E and eIF4A are assembled, eIF4G also binds to the eIF4E kinase Mnk1 and PABP (Pyronnet *et al.*, 1999; Imataja *et al.*, 1998). Aside from our perceived role for eIF4G as a static molecular scaffold, little is known regarding its function. This is further highlighted by the existence of at least five eIF4GI isoforms, generated by translation initiation at alternative ATG codons, coupled with differential phosphorylation, all of which are of unknown functional significance (Byrd *et al.*, 2002; Raught *et al.*, 2000). Furthermore, mammalian cells contain a homologue of eIF4GI, eIF4GII, which appears to be selectively recruited to capped mRNA upon differentiation (Caron *et al.*, 2004). Finally, recent studies demonstrating that the N-terminus of eIF4G undergoes a conformational change upon engaging eIF4E suggest that the role of eIF4G may indeed go beyond that of an inert scaffold upon which other initiation factors assemble (Gross *et al.*, 2003).

Regulating eIF4F in HSV-1 infected cells

In many instances, translation of host capped, polyadenylated mRNAs is impaired in virus-infected cells. Often, this results from inactivation or modification of one or more eIF4F components by a virus-encoded gene product (reviewed in Mohr *et al.*, 2007). Notably, picornaviruses, whose mRNA is not capped and initiates translation via alternative means, encode proteinases that cleave eIF4G and PABP (reviewed in Lloyd, 2006). Alternatively, suppression of host protein synthesis is reportedly associated with dephosphorylation of the cap-binding protein eIF4E in cells infected with either adenovirus, VSV, or influenza virus, all of which produce capped mRNAs (Feigenblum & Schneider, 1993; Cuesta *et al.*, 2000; Connor & Lyles, 2002). Significantly, host mRNA translation is also strongly suppressed in HSV-1-infected cells and HSV-1 mRNAs are capped and polyadenylated like their cellular counterparts (Bartowski & Roizman, 1976). However, in stark contrast with these viral systems, HSV-1-infection effectively stimulates the cellular cap-dependent translation machinery.

While the steady-state levels of eIF4F core components eIF4E, eIF4G, and eIF4A remain unchanged in HSV-1 infected cells, phosphorylation of 4E-BP1 relieves translational repression by liberating eIF4E, making it available to associate with eIF4G. Phosphorylation of 4E-BP1 in cells infected with HSV-1 occurs early in its productive replication cycle and is sensitive to rapamycin, an inhibitor of the cellular kinase mTORC1. Moreover, following its phosphorylation, steady state levels of 4E-BP1 were observed to precipitously decline in infected cells. Proteasome inhibitors or rapamycin effectively prevented the decrease in 4E-BP1 abundance. Thus, following its phosphorylation by mTOR, the translational repressor 4E-BP1 is degraded by the proteasome in HSV-1 infected cells (Walsh and Mohr, 2004). In addition to inactivating the translational repressor 4E-BP1, rapid phosphorylation of the cap-binding protein eIF4E was observed following infection of primary human cells or a variety of established cell lines with HSV-1. eIF4E phosphorylation occurred early in the viral lifecycle, was sustained throughout the productive growth cycle, and was sensitive to small molecule inhibitors of p38 and Mnk, the latter being the cellular eIF4E kinase. More importantly, inhibiting the eIF4E kinase Mnk impaired viral protein synthesis and reduced viral replication by 100-fold in primary human cells (Walsh and Mohr, 2004).

As eIF4E and 4E-BP1 phosphorylation have been reported to stimulate translation in other biological settings, this raised the possibility that the changes to eIF4F in HSV-1 infected cells might be required to stimulate the cellular translation machinery. Indeed, in growth arrested primary human cells infected with HSV-1, eIF4E is released from the 4E-BP1 repressor and the amount of eIF4G bound to eIF4E increases dramatically (Walsh and Mohr, 2004). This suggests that HSV-1 can promote the assembly of eIF4F complexes in quiescent resting cells. The ability to stimulate eIF4F complex assembly is likely important for productive viral growth in quiescent cells, as the capacity of resting cells to support high levels of protein synthesis are reduced relative to their exponentially growing counterparts (Stanners and Becker, 1971; Thomas and Gordon, 1979; Thomas *et al.*, 1981; Thomas and Thomas, 1986). Strikingly, although the amount of eIF4G bound to eIF4E increases substantially upon infection of quiescent cells with HSV-1, a concomitant increase in the amount of PABP bound to eIF4G was not observed (Walsh *et al.*, 2005). Interaction of PABP with HSV-1 encoded proteins could potentially contribute to this. PABP has been found associated with the viral ICP27 protein, although this interaction is markedly diminished upon treatment with ribonuclease (Fontaine-Rodriguez *et al.*, 2004). Whereas ICP27 reportedly stimulates mRNA translation when tethered to a reporter mRNA as a fusion protein containing a MS2 site-specific RNA binding domain (Larrralde *et al.*, 2006), other studies in infected cells suggest only translation of VP16 and ICP5 mRNAs appears responsive to ICP27 (Ellison *et al.*, 2005; Fontaine-Rodriguez *et al.*, 2008). Furthermore, while PABP normally shuttles between the nucleus and cytoplasm in uninfected cells, PABP accumulates in infected cell nuclei through the action of a multiprotein complex containing ICP27 and the UL47 gene product (Dobrikova *et al.*, 2010).

Although it is not a core component of the eIF4F complex, PABP bound to the poly-A tail at the 3′ end of the mRNA is known to associate with eIF4G bound at the 5′ end of the mRNA; moreover, this association between eIF4G and PABP stimulates translation. While the significance of failing to recruit PABP into eIF4F complexes and nuclear PABP accumulation following HSV-1 infection of quiescent fibroblasts is presently not clear, similar observations have been reported when KSHV latently infected cells are induced to reactivate (Arias *et al.*, 2009). However, upon infection of quiescent fibroblasts with the betaherpesvirus HCMV, both eIF4G and PABP are recruited into eIF4E-containing complexes (Walsh *et al.*, 2005). As host polypeptide synthesis is impaired in cells productively infected with HSV-1 or KSHV, but *not* HCMV, it is tempting to speculate that the differential association of PABP with eIF4F complexes and its nuclear accumulation might somehow play a role in host shut-off. Further work will hopefully illuminate the significance of this finding and either challenge or support the aforementioned hypothesis.

All of the changes to the eIF4F complex that occur in HSV-1 infected cells require viral gene expression (Walsh and Mohr, 2004). However, phosphorylation of eIF4E as well as 4E-BP1 was observed in infected cells treated with PAA, an inhibitor of viral DNA replication that blocks the expression of 'true late' or γ_2 genes. Therefore, this class of viral genes is not required for phosphorylation of eIF4E together with 4E-BP1. Consistent with this finding, viruses with mutations in the immediate early ICP0 gene were unable to promote phosphorylation of eIF4E along with 4E-BP1 or stimulate the assembly of eIF4F complexes in quiescent cells (Walsh and Mohr, 2004). Although genetic analysis initially established that the modifications to eIF4F observed in cells productively infected with HSV-1 require the multi-functional regulatory protein ICP0, ICP0 itself is not directly involved. Instead, eIF4E phosphorylation and eIF4F assembly directly require ICP6 (Walsh and Mohr, 2006), and ICP6 expression is largely ICP0-dependent (DeLuca *et al.*, 1985; Goldstein and Weller, 1988). Protein synthesis is slowed in growth-arrested, primary cells infected with an ICP6-deficient mutant. While inactivation of the 4E-BP1 repressor proceeds normally, binding of eIF4E to eIF4G is substantially reduced, demonstrating that release of eIF4E from the 4E-BP1 repressor is insufficient to promote eIF4F assembly in HSV-1-infected cells. Instead, the ICP6 gene product binds to eIF4G (Fig. 7.3) and directly promotes the association of eIF4E with the amino terminus of eIF4G (Walsh and Mohr, 2006). This is the first example of an initiation factor-associated protein that actively promotes complex assembly, and it defines a new controllable step in translation.

Failure to accumulate phosphorylated eIF4E in cells infected with an ICP6 mutant likely also results from impaired eIF4F assembly (Fig. 7.4). Since the eIF4E kinase Mnk1 is bound to eIF4G, the observed reduction in phosphorylated eIF4E could potentially be accounted for by reduced binding of the eIF4G-bound kinase (Mnk1) to the substrate eIF4E. Thus, by promoting eIF4F assembly, ICP6 also stimulates eIF4E phosphorylation by ensuring that the kinase Mnk1 is correctly positioned relative to its substrate, eIF4E. The ability to stimulate eIF4F complex assembly is likely to be important for productive viral growth in quiescent cells that have diminished ability to support high levels of protein synthesis (Stanners and Becker, 1971). Indeed, asynchronous primary cells contain an activity that stimulates protein synthesis in the absence of ICP6 (Walsh and Mohr, 2006). Finally, this establishes a new activity for ICP6, the large subunit of the virus-encoded ribonucleotide reductase. While reductase activity requires a holoenzyme comprised of a large and small subunit (UL40), it has been puzzling why the large subunit, which is devoid of ribonucleotide reductase activity, accumulates prior to the small subunit. Early expression of ICP6 appears to be required to stimulate eIF4F assembly and ICP6 on its own is capable of promoting eIF4F assembly and stimulating protein synthesis in quiescent cells.

While ICP6 binds to eIF4G to stimulate eIF4F assembly, other viral functions are required to inactivate the 4E-BP1 translational repressor. Genetic and pharmacological experiments suggests that 4E-BP1 phosphorylation is not

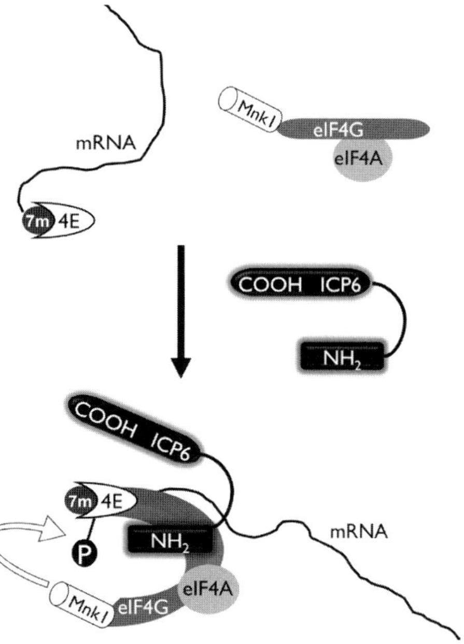

Figure 7.3 ICP6 promotes eIF4F assembly in HSV-1-infected cells. While 4E-BP1 phosphorylation inactivates 4E-BP1 and dissolves the eIF4E–4E-BP1 protein complex, the release of eIF4E from the repressor 4E-BP1 is not sufficient for eIF4F assembly. In growth-arrested cells infected with HSV-1, the assembly of eIF4E together with eIF4G and eIF4A into an eIF4F complex requires the assistance of ICP6. Through its N-terminal domain, ICP6 recognizes eIF4G and stimulates binding of eIF4G to eIF4E, thus promoting assembly of an eIF4F complex in HSV-1-infected cells.

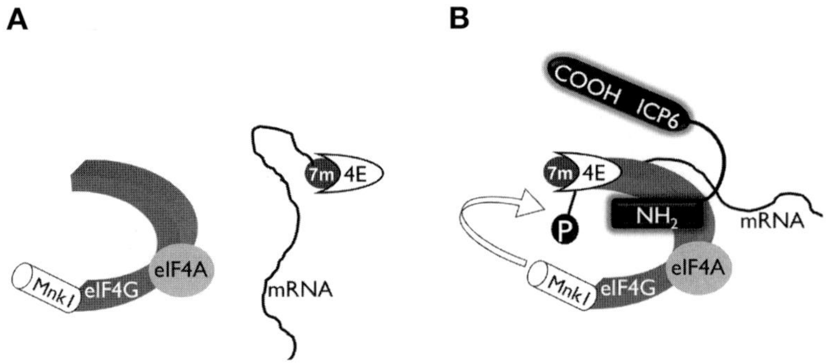

Figure 7.4 Stimulation of eIF4E phosphorylation by ICP6. The molecular architecture of the eIF4E complex allows for the correct apposition of the cellular kinase Mnk with its substrate, eIF4E. (A) Without ICP6, eIF4F assembly is impaired, as the cap-binding protein eIF4E is not bound to eIF4G. Moreover, phosphorylated eIF4E does not accumulate since eIF4E is not anchored near the eIF4G-associated eIF4E kinase Mnk. (B) By stimulating eIF4F assembly, ICP6 ensures that the kinase Mnk and its substrate, eIF4E, are both correctly positioned on the eIF4G platform. Thus, eIF4F assembly also promotes eIF4E phosphorylation.

ICP6-dependent, occurs after ICP4 expression yet prior to the onset of DNA synthesis (Walsh and Mohr, 2006). Given the prominent role ser/thr phosphorylation plays in regulating 4E-BP1 activity, it is tempting to speculate that either one or both HSV-1-encoded viral ser/thr kinases (Us3, UL13) effectively intervene and deregulate mTORC1 signalling in infected cells.

While some of the HSV-1-encoded players responsible for regulating eIF4F have been identified, much remains to be learned from studying their detailed mechanism of action. Such mechanistic insights into how viral factors promote eIF4F complex assembly may not only have relevance for understanding the productive growth cycle of alphaherpesviruses, but will likely contribute to our understanding of how translation is regulated during the lifecycle of related beta- and gammaherpesviruses. Although the members of the herpesvirus family are certainly diverse, it appears that the basic function of stimulating the cellular translation machinery has been conserved across sub-family lines. Indeed, eIF4F complex assembly along with eIF4E and 4E-BP1 phosphorylation are observed in HCMV infected cells; in addition, inhibiting the eIF4E kinase Mnk reduces HCMV replication by 300-fold (Kudchodkar *et al.*, 2004; Walsh *et al.*, 2005). Similarly, inhibiting Mnk impairs KSHV lytic reactivation (Arias *et al.*, 2009). Although herpesvirus sub-family members all stimulate the cellular translation machinery, the mechanisms utilized by the alpha, beta and gamma viruses to achieve this goal appear to be quite different (Walsh *et al.*, 2005; Moorman *et al.*, 2008). Finally, while promoting assembly of eIF4F complexes along with phosphorylation of eIF4E and 4E-BP1 is important for productive viral replication, herpesvirus biology raises additional considerations, as these viruses have adapted to colonize a variety of terminally differentiated cells (reviewed in Roizman *et al.*, 2007; Mocarski *et al.*, 2007; Keiff and Rickinson, 2007; Cohen *et al.*, 2007; Yamanishi *et al.*, 2007; Ganem, 2007). In many cases, cells latently infected with a herpesvirus produce either small or undetectable levels of viral polypeptides (reviewed in Roizman *et al.*, 2007; Mocarski *et al.*, 2007). It is therefore conceivable that the switch to a productive mode of growth requiring intensive translation of viral mRNAs necessitates a mechanism to mobilize the translational machinery in quiescent cells.

Shutting down host protein synthesis by targeting mRNA metabolism

Host protein synthesis is dramatically curtailed in HSV-1-infected cells. However, this is not achieved by inactivating host translation initiation factors, in marked contrast with some viruses (i.e. poliovirus) whose mRNAs do not contain a 5′ cap and initiate translation through alternative means (reviewed in Mohr *et al.*, 2007). As reviewed in the preceding section, HSV-1 infection actually stimulates the host's translational machinery. So what then contributes to the preferential translation of viral mRNAs in HSV-1 infected cells? Indeed, during its productive growth cycle, HSV-1 exerts a powerful effect on host macromolecular metabolism. Using multiple tactics requiring the action of several viral gene products, HSV-1 is able to suppress the synthesis of most host polypeptides, as the progressive movement of mRNAs from large polyribosomes to smaller ones is observed (Laurent *et al.*, 1998). This process begins almost immediately after the viral tegument proteins are deposited into the host cell cytoplasm. Upon liberation from the tegument compartment in the virion, vhs, a FEN-1 related nuclease, enhances mRNA turnover (Everly *et al.*, 2002). *In vitro*, vhs possesses endonuclease activity and promotes degradation of mRNA substrates in the 5′ to 3′ direction (Elgadi *et al.* 1999; Karr and Read, 1999; Perez-Parada *et al.*, 2004). Vhs is specifically targeted to mRNA through an interaction with cellular translation initiation factors eIF4AII (Feng *et al.*, 2005) together with eIF4H and eIF4B (Feng *et al.*, 2001, 2005; Doepker at. al., 2004). Moreover, vhs has recently been found tethered to at least a subset of eIF4F cap-binding complexes (Feng *et al.*, 2005). The requirement for eIF4H is important for vhs function, as siRNA depletion of eIF4H impedes vhs-mediated mRNA degradation in infected cells (Sarma *et al.*, 2008).

Despite clearly accelerating turnover of mRNAs in HSV-1-infected cells, vhs is surprisingly *not* selective for cellular mRNAs as opposed to viral mRNAs. Indeed, both cellular and viral mRNAs are substrates for vhs *in vitro* and *in vivo* (Kwong and Frenkel, 1987). Although not essential for growth in cultured cells (Read and Frenkel, 1983), host shut-off is delayed and the transition between different kinetic classes of viral mRNAs, such as IE (or α) to DE (or β), is not as discrete in cells infected with a vhs mutant virus (Kwong and Frenkel, 1987; Oroskar and Read, 1987). The observed effects on viral and cellular mRNA stability support the contention that vhs accelerates the turnover of *both* viral and host mRNAs. Indeed, vhs cleaves mRNA in polyribosomes and plays a role in polysome disassembly observed early on in infected cells (Taddeo *et al.*, 2009). Finally, vhs is also important for pathogenesis, as vhs mutants are attenuated in animal models (Strelow and Leib, 1995; Smith *et al.*, 2002). It has been suggested that enhanced turnover of host mRNAs important for innate responses, such as those encoding interferons, underlies the reduced virulence of these mutants (Leib *et al.*, 1999; Murphy *et al.*, 2003; Duerst *et al.*, 2004). Significantly, a single amino acid substitution resulting in an RNAse-deficient HSV-2 vhs protein is sufficient to attenuate the virus (Korom *et al.*, 2008). Furthermore, vhs antagonizes establishment of the antiviral state and vhs-deficient viruses display enhanced pathogenesis in mice unable to respond to type I interferons (Pasieka *et al.*, 2008a,b).

If vhs does not display any overt preference for cellular mRNAs, how then is host translation selectively silenced? While the mechanism underlying host shutoff is not completely understood, it clearly involves multiple viral components. Unlike vhs, which is carried in the virion, ICP27 is the product of an IE gene. In addition to its role promoting export of viral mRNA out of the nucleus late in the productive growth cycle (Soliman *et al.*, 1997), the essential, multifunctional ICP27 protein also inhibits mRNA splicing at early times (Hardy and Sandri-Golding, 1994). As most cellular mRNAs are spliced, while only 4 of the 84 known HSV-1 genes contain introns, ICP27 has a profound effect on the proper maturation of host mRNAs and in turn their subsequent translation. (Sciabica *et al.*, 2003; reviewed in Sandri-Goldin, 2008). In addition to inhibiting host mRNA processing, transcription of host genes is reduced to levels less than 50% of those measured in uninfected cells within the first 4 hours of infection (Preston and Newton, 1976). This may, perhaps in part, reflect the recruitment of the cellular transcription machinery to viral promoters along with modification of the cellular RNA pol II holoenzyme by viral IE gene products (Spencer *et al.*, 1997; Zhou and Knipe, 2002). Thus, while vhs enhances the turnover of all mRNAs indiscriminately, the inhibition of splicing together with the global reduction in cellular gene transcription ensures that the supply of host mRNA in the cytosol is not replenished. This stands in sharp contrast to the high level of transcripts being produced from viral genes. Of course, it also remains possible that additional uncharacterized virus-encoded functions will be described that contribute to the powerful suppression of host mRNA translation in HSV-1-infected cells.

Unlike Adenovirus, whose late mRNAs all contain a common tri-partite leader sequence found at their 5′ end that promotes their translation while cellular mRNA translation is suppressed (Xi *et al.*, 2004), shared sequences at 5′ or 3′ ends of HSV-1 mRNAs have not been identified in existing untranslated regions. It is possible however, given the GC rich nature of the HSV-1 genome, that the 5′ end of viral mRNAs are capable of adopting a related conformation and this structure might aid in recruiting ribosomes. Alternatively, the efficient translation of HSV-1 mRNAs, which occurs irrespective of the presence or absence of vhs-mediated mRNA turnover, may simply be accounted for by the sheer abundance of viral transcripts against a pool of cellular mRNAs that are not being continually replenished. It is important to remember, however, that while the inhibition of host mRNA translation is potent, it is not absolute as select cellular mRNAs are translated in HSV-1 infected cells. Further investigation will determine if these mRNAs are either important components of the host response or somehow contribute to viral replication (Esclatine *et al.*, 2004).

eIF2α phosphorylation and translational control

While impairing host polypeptide synthesis and activating eIF4F in resting cells play an important role in enabling viral protein production, these events in and of themselves are not sufficient to ensure the sustained protein production required to complete the viral productive growth cycle. In particular, an innate host response designed to incapacitate the critical translation initiation factor eIF2 must be effectively thwarted.

The capacity of eukaryotic cells to respond to numerous forms of stress, ranging from thermal intolerance or nutrient deprivation to viral infection, involves, in part, their ability to regulate the activity of eIF2. Upon phosphorylation on S51 of its α subunit by one of four known eIF2α kinases, eIF2 is inactivated and translation is inhibited. Composed of three subunits, eIF2 is a G – protein which functions to bring the initiator methionine tRNA to the 40S ribosomal subunit (reviewed in Sonenberg & Hinnebusch, 2009; Fig. 7.5A). The association of eIF2 with the initiator tRNA and GTP results in formation of what is known as the ternary complex, which subsequently engages eIF3 and eIF1A bound to a 40S ribosome giving rise to a 43S complex (Fig. 7.5A). Hydrolysis of GTP bound to eIF2 in a reaction that requires the GTPase-activating factor eIF5 results from conformational changes associated with ternary complex binding to the 40S subunit (Algire *et al.*, 2005). After the 43S complex is loaded onto the 5′ end of the mRNA through an interaction between eIF3 and eIF4G, a component of the eIF4F cap-binding complex, the 40S ribosome along with eIF2, eIF3 and the RNA helicase eIF4A proceed to locate the AUG initiation codon through an ATP-dependent unidirectional translocation process referred to as scanning. Recognition of the AUG codon by the scanning ribosome complex is coupled to discharge of inorganic phosphate (P_i) from eIF2 followed by both 60S subunit joining and the release of inactive eIF2 bound to GDP (Algire *et al.*, 2005; Fig. 7.5A). Translation elongation directed by the nascent 80S ribosome proceeds, while eIF2•GDP is recycled into its active GTP-bound form through the actions of eIF2B, a guanine nucleotide exchange factor (Fig. 7.5A). Phosphorylation of eIF2 on its α subunit, however, stabilizes the interaction between eIF2B and eIF2 and prevents the exchange of GDP for GTP (Fig. 7.5B). Moreover, as eIF2B is present in limiting amounts, small changes in eIF2α phosphorylation can exert large effects on overall translation rates by sequestering available supplies of eIF2B.

Inactivation of translation initiation factors by host defences is antagonized by viral functions

Not surprisingly, host innate defences frequently target cellular translation initiation factors hoping to prevent viral protein synthesis, thus prematurely terminating the viral life cycle and

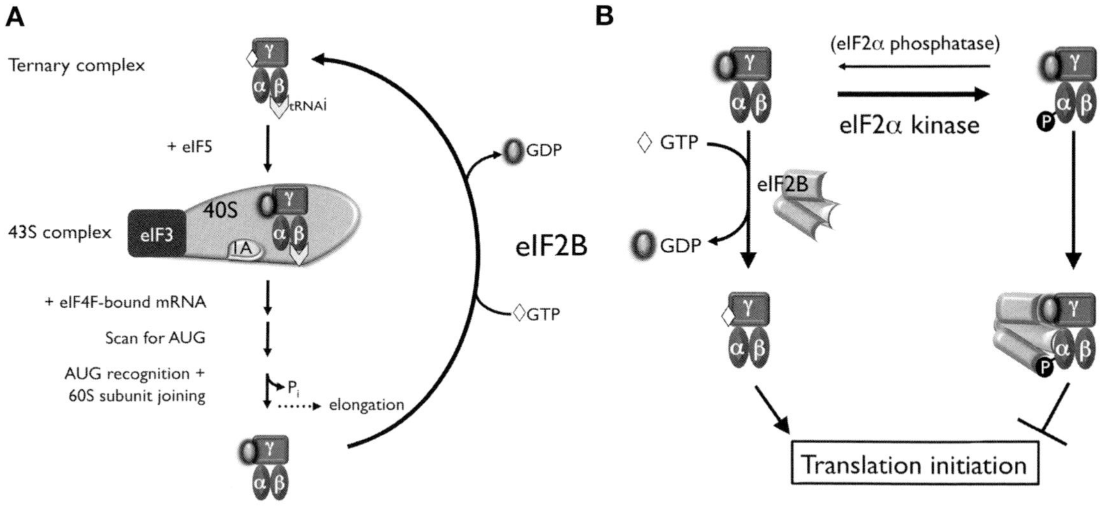

Figure 7.5 Regulation of translation by phosphorylation of eIF2, a critical translation initiation factor. (A) Composed of α,β, and γ subunits, eukaryotic translation initiation factor 2 (eIF2) forms a ternary complex with GTP and the initiator tRNA (tRNAi). The ternary complex (eIF2•GTP•tRNAi) subsequently associates with a 40S ribosome subunit complexed with eIF3 and eIF1A (labelled 1A in the diagram) to form the 43S complex, and the GTP-bound to eIF2 is hydrolysed in a reaction that requires the GTPase activating factor eIF5 and results in eIF2 bound to GDP+Pi. An interaction between eIF3 and eIF4F allows the 43S complex to recognize the 5′ end of the mRNA. Once the AUG codon in the mRNA has been identified by a unidirectional translocation process termed scanning, Pi is rapidly released from eIF2, 60S subunit joining subsequently results in the release of the eIF2•GDP complex, and translation elongation commences. The guanine nucleotide exchange factor eIF2B is required to exchange the GDP bound to eIF2 and replace it with GTP, thus recycling the active form of eIF2. (B) After phosphorylation of eIF2 on its alpha subunit by PKR, an eIF2α kinase, eIF2B remains tightly bound and cannot exchange the GDP bound to eIF2 for GTP. This failure to recycle eIF2 to its active, GTP bound form inhibits the initiation of translation.

impeding the progress of the infection (reviewed in Schneider and Mohr, 2003). Copious quantities of dsRNA or RNA with extensive secondary or tertiary structure produced in virus-infected cells are thought to be responsible for alerting the cell to the viral intruder (reviewed in Mohr *et al.*, 2007). Some viral genomes are composed of dsRNA, while others such as single strand RNA viruses form dsRNA as replicative intermediates. However, dsDNA viruses, including HSV-1, in their efforts to accommodate large blocks of genetic information in limited spaces, contain overlapping open reading frames on opposite DNA strands that produce mRNA transcripts capable of forming dsRNA (Jacquemont and Roizman, 1975). Abundant viral dsRNA is thought to activate PKR, a cellular eIF2α kinase whose synthesis is upregulated by interferon α/β (reviewed by Dever *et al.*, 2007). In addition to its kinase domain, PKR contains two canonical double-stranded RNA (dsRNA) binding domains. In the presence of dsRNA, two PKR molecules bind the same dsRNA segment and each molecule phosphorylates the other, generating an activated kinase capable of phosphorylating eIF2α. Finally, the cellular protein PACT, which also binds to dsRNA, can activate PKR in a dsRNA-independent fashion, resulting in apoptosis (Patel and Sen, 1998).

Unchecked, PKR activated by dsRNA in virus-infected cells would effectively deplete active eIF2, inhibit ternary complex formation, and prevent viral and cellular protein synthesis. Should initially-infected cells succeed in inhibiting translation, the viral invader would effectively be stopped in its tracks, denied access to the cellular translational apparatus it needs to complete its life-cycle. This arm of the innate host response then is designed to sacrifice the initially infected cells for the benefit of the larger population. However, numerous viruses, including HSV-1, have captured a variety of functions to counter this cellular response. This struggle for control of the translational machinery is often an integral component of viral pathogenesis (reviewed in Mohr, 2005).

Circumventing disaster: keeping eIF2 functioning in HSV-1 infected cells

To date, two viral gene products, acting at *different stages* of the HSV-1 lifecycle, that facilitate the abundant synthesis of viral proteins by preventing the cellular antiviral response from inactivating the translation initiation factor eIF2 have been described. The first of these was the product of the $\gamma_1$34.5 gene.

After initial reports established the $\gamma_1$34.5 gene was a major HSV-1 neurovirulence determinant and non-essential for growth in cultured monkey kidney cells (Chou *et al.*, 1990; Maclean *et al.*, 1991; Bolovan *et al.*, 1994), further investigation revealed that $\gamma_1$34.5 mutants actually behaved like classical viral host range mutants, exhibiting restricted growth in some lines of cultured cells but not others. Thus, while a standard line of monkey kidney cells were permissive or supported replication of $\gamma_1$34.5 mutant viruses, many human tumour cells were in fact non-permissive, or did not support the growth of $\gamma_1$34.5 mutant derivatives (Chou *et al.*, 1992). Upon infection of non-permissive human tumour cell lines with a $\gamma_1$34.5 mutant strain, all of the events in the viral lifecycle proceeded normally up to and including viral DNA replication and the accumulation of γ_2 late mRNA transcripts. These viral late mRNAs encoding key structural proteins required to complete the viral lifecycle and assemble the next generation of viral progeny, however, were never translated due to a block at the level of protein synthesis, effectively interrupting the viral lifecycle prior to the assembly and release of viral particles (Chou *et al.* 1992). Subsequent biochemical analysis demonstrated that the $\gamma_1$34.5 gene product prevents the accumulation of phosphorylated eIF2α by recruiting a cellular phosphatase, protein phosphatase 1α (PP1α), to remove phosphate from eIF2α (He *et al.*, 1997). Interestingly, the domain of the $\gamma_1$34.5 protein that contains this activity is homologous to a domain in the GADD34 protein, a cellular PP1α-binding protein that promotes eIF2α dephosphorylation in response to other forms of cell stress (Chou and Roizman, 1994). Besides their defect in protein synthesis, $\gamma_1$34.5 mutant derivatives are hypersensitive to interferon α, and therefore more sensitive to this arm of innate host defences that serve to limit viral replication in the host (Cheng *et al.*, 2001; Cerveny *et al.*, 2003). This hypothesis is supported by reports demonstrating that $\gamma_1$34.5 mutant viruses, while neuroattenuated in normal mice and mice with deficiencies in their acquired immune response, exhibit restored neurovirulence in mice with deficiencies in innate immunity (Leib *et al.*, 1999, 2000).

Ensuing genetic studies revealed that HSV-1 actually encodes multiple functions to control eIF2α phosphorylation, as the dsRNA binding protein specified by the true-late Us11 gene prevents PKR activation (Mohr and Gluzman, 1996; Mulvey *et al.*, 1999; Poppers *et al.*, 2000; Khoo *et al.*, 2002). The Us11 protein not only prevents PKR activation in response to dsRNA, but it antagonizes activation mediated by the cellular protein PACT as well (Peters *et al.*, 2002). Analysis of a panel of $\gamma_1$34.5 and Us11 mutants established that both Us11 and $\gamma_1$34.5 gene products act at different times in the productive growth cycle to regulate eIF2α phosphorylation in infected cells (Mulvey *et al.*, 2003; Fig. 7.6). Importantly, Us11 expressed in its natural context as a late γ_2 protein is required to properly regulate PKR activation, eIF2α phosphorylation and viral mRNA translation (Mulvey *et al.*, 2003). Thus, since late proteins are not produced in non-permissive tumour cells infected with a $\gamma_1$34.5 mutant virus, the $\gamma_1$34.5 mutant is actually doubly deficient in that it also fails to translate the γ_2 Us11 mRNA (Chou *et al.*, 1992). Moreover, the interferon sensitivity of $\gamma_1$34.5 mutants, previously attributed solely to the absence of $\gamma_1$34.5 function, likewise results from the absence of the $\gamma_1$34.5 protein and the failure to synthesize the Us11 polypeptide (Mulvey *et al.*, 2004).

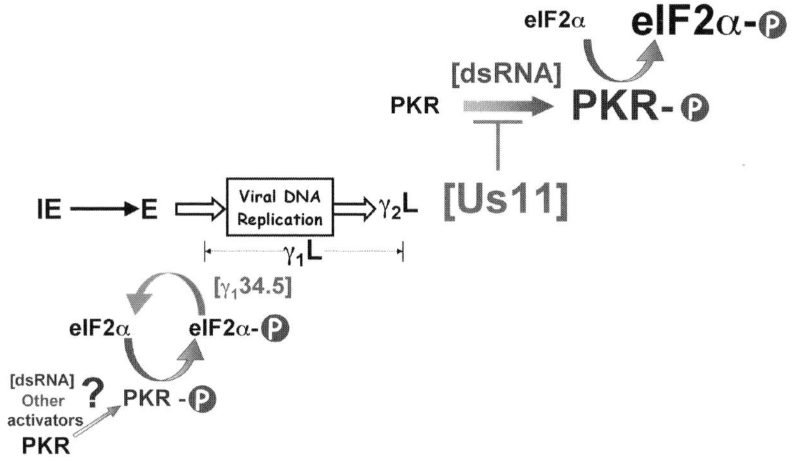

Figure 7.6 Regulation of eIF2α phosphorylation by different functions that act during discrete phases in the HSV-1 lifecycle. Early in the HSV-1 lifecycle, small quantities of dsRNA, or perhaps other effectors that remain to be identified activate the normally dormant cellular PKR kinase. After assembling a dimer of PKR on dsRNA, each subunit of the multimer phosphorylates the other (PKR–P). Since PKR is activated in cells infected with a $\gamma_1$34.5 mutant virus, preventing the translation of γ_2 mRNAs, we propose that the $\gamma_1$34.5 gene product, through its interaction with PP1α, is able to adequately dephosphorylate the quantities of phosphorylated eIF2α (eIF2α–P) produced from the period preceding the initiation of viral DNA synthesis and extending into the initial segment of the late phase (designated as γ_1L). However, late in the viral lifecycle, synthesis of mRNA from γ_2 genes (γ_2 L), many of which are transcribed from ORFs located on opposing DNA strands, results in the production of large quantities of viral dsRNA. In the absence of Us11, the increase in dsRNA concentration generates more activated PKR, which in turn phosphorylates eIF2α. The concentration of phosphorylated eIF2α quickly rises beyond the capacity of the $\gamma_1$34.5–PP1α complex to effectively reverse the reaction, accounting for the observed reduction in viral translation rates in cells infected with a Us11 mutant virus. We suggest that while the $\gamma_1$34.5 protein acts downstream of phosphorylated eIF2α and therefore has the potential to counter a variety of eIF2α kinases, Us11 acts late in infection to specifically antagonize PKR activation in response to the copious levels of dsRNA produced in virus infected cells. In this drawing, relative concentrations of PKR–P, dsRNA, and eIF2α–P at earlier, compared to later times in the viral lifecycle, are represented by character size. (reprinted from Mulvey *et al.*, 2003 with permission.)

As discussed in the preceding section, HSV-1 employs multiple tactics to prevent the inactivation of eIF2 by PKR, an important component of innate host defences; however, at least three other kinases, distinct from PKR, with the capability to phosphorylate eIF2*α* have been identified in mammalian cells (reviewed in Dever, 2002). This family of kinases, by virtue of their intrinsic ability to monitor intracellular conditions, directly transduces stress signals to the translational machinery and thereby globally regulates rates of protein synthesis (Fig. 7.7). By virtue of the fact that the $\gamma_1$34.5 gene product acts downstream of eIF2*α* phosphorylation, it can reverse the consequences of an activated eIF2*α* kinase and should therefore be able to counter the effects of a broad range of eIF2*α* kinases. However, it is important to keep in mind that despite producing wild-type levels of the $\gamma_1$34.5 protein, Us11 mutant viruses do not synthesize proteins at wild-type rates and are sensitive to interferon *α* (Mulvey *et al.* 2003, 2004). Thus, even though they express the $\gamma_1$34.5 phosphatase component, an inhibitor of an upstream kinase is still required for normal rates of translation in HSV-1 infected cells, implying that recruiting PP1*α* may not be adequate to prevent the accumulation of phosphorylated eIF2*α* in HSV-1 infected cells. Just as the HSV-1 Us11 protein specifically antagonizes the dsRNA activated eIF2*α* kinase PKR and

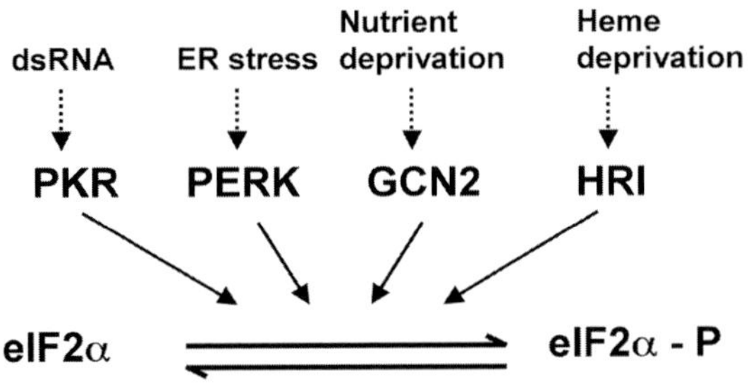

Figure 7.7 Activation of multiple eIF2α kinases by discrete stimuli. The four known eIF2α kinases share a common organization in that each has a kinase domain fused to a ligand binding domain and remains catalytically silent until encountering its respective activating ligand. While the accumulation of dsRNA activates PKR, reduced levels of heme induce HRI activity, nutrient deprivation or ultraviolet light exposure triggers GCN2, and the accrual of unfolded proteins in the ER lumen switches on PERK.

works in conjunction with the $\gamma_1$34.5 protein to preserve supplies of active eIF2, other HSV-1 gene products were hypothesized to target eIF2α kinases other than PKR to specifically prevent the accumulation of phosphorylated eIF2 in response to different stresses, such as the accumulation of misfolded proteins within the ER lumen.

Given the magnitude of the burden that replicating viruses impose on the ER, their potential to produce intracellular stress by challenging ER folding capacity and activating the unfolded protein response (UPR) is not surprising. Once activated, one arm of the UPR represented by IRE1 and ATF6 activates a gene expression program that extends the folding capability of the ER, whereas the other branch, corresponding to the eIF2α kinase PERK, causes a transient translational arrest (Fig. 7.8). Blocking continued production of potential protein clientele alleviates the load on the ER and contributes to maintaining ER homeostasis (reviewed in Ron and Walter, 2007). Remarkably, HSV-1 protein synthesis was maintained in infected cells exposed to acute ER stress, and this did not require either the $\gamma_1$34.5 or Us11 gene products, the two viral functions known to regulate eIF2α phosphorylation (Mulvey *et al.*, 2006). PERK, but not IRE1, resists activation by acute ER stress in HSV-1-infected cells. This required the ER luminal domain of PERK, which associates with the viral glycoprotein gB. Strikingly, gB regulates viral protein accumulation in a PERK-dependent manner illustrating a new strategy whereby a virus-encoded PERK-specific effector contributes to maintaining ER homeostasis (Mulvey *et al.*, 2007; Fig. 7.9).

Double duty for Us11: preventing RNase L activation by inhibiting OAS

To a large extent, the failure of interferon-treated cells to support viral replication reflects activation of latent, interferon-stimulated gene products upon viral infection that subsequently directly curtail viral replication. PKR and oligoadenylate synthetase (OAS) are two key dsRNA-responsive effector proteins that figure prominently in the ability of the interferon-induced proteome to antagonize viral replication. While activated PKR phosphorylates eIF2α to inhibit protein synthesis, oligoadenylate chains with unique 2′ 5′ linkage produced by OAS bind to and activate RNAseL, a latent cellular ribonuclease that accelerates the decay of both mRNA and rRNA. To prevent RNAse L activation in infected cells, the HSV-1 Us11 protein inhibits OAS and thereby prevents the synthesis of the RNase L activator. Inhibition of OAS specifically requires the Us11 dsRNA-binding domain, suggesting a mechanism that, in part, relies on sequestering available dsRNA produced during infection. Thus, in addition to PKR and its protein activator, PACT, the HSV-1 Us11 gene product is able to counteract the activity of OAS, a third cellular protein critical for host defence (Sanchez and Mohr, 2007).

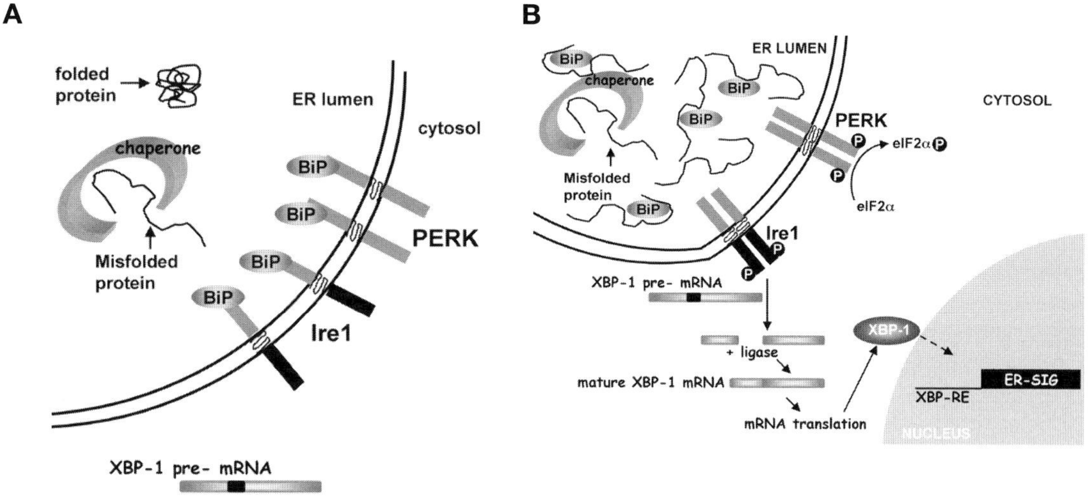

Figure 7.8 Maintenance of ER homeostasis by the cellular unfolded protein response. (A) A non-stressed ER under conditions of chaperone sufficiency is shown. The stress sensors IRE1 and PERK are both inactive, bound to the chaperone BiP in the ER lumen. The XBP-1 precursor mRNA (pre-mRNA) remains unprocessed, unable to direct XBP-1 protein synthesis. (B) ER stress, as measured by chaperone insufficiency, results in the release of BiP from the luminal domains of both IRE1 and PERK, allowing them to form homodimers within the plane of the ER membrane. Once each subunit of the dimer phosphorylates the other, activated PERK phosphorylates eIF2α, whereas the activated endoribonuclease activity of IRE1 cleaves the XBP-1 pre-mRNA to initiate proper processing. Once processing is completed by an RNA ligase, the mature XBP-1 mRNA is translated and the protein product translocates to the nucleus where it activates expression of ER stress-induced genes (SIG). (reprinted from Mulvey *et al.*, 2007 with permission.)

Concluding thoughts

Clearly, HSV-1 goes to great lengths to commandeer the translational machinery within its host cell and preserve its activity during its productive replication cycle. Anything short of absolute success in this endeavour will undoubtedly compromise the production of viral proteins and in turn, the subsequent assembly of viral progeny. To recruit 40S ribosomes to the 5′ end of its capped mRNAs, HSV-1 must successfully engage the translation initiation factor eIF4F, while simultaneously preserving the activity of eIF2, a target of host innate defences and stress sensors. The sheer number of different viral functions required to properly dominate cellular translational control pathways illustrates its fundamental importance to productive viral replication. Out of the approximately 84 HSV-1-encoded ORFs expressed during the productive replication cycle, at least four act to regulate eIF4F-core and associated factors (promote 4E-BP1 inactivation, ICP6, UL47, vhs) while three promote viral protein accumulation by targeting eIF2α kinases ($\gamma_1$34.5, Us11, gB; see Fig. 7.9). Given the complexity of the eukaryotic translation machinery, it would not be the least surprising if further study identified additional viral gene products that target other critical steps in translational control. Indeed, the HSV-1 VP22 RNA binding protein was shown to be required for normal levels of viral protein synthesis on a global scale late in infection, although the underlying mechanism remains unknown (Duffy *et al.*, 2009). In addition, while stimulating host translational control pathways and maintaining their activity is critical for productive viral replication, virus-encoded microRNAs accumulate in latently-infected neurons and repress viral gene expression by inhibiting translation of specific viral mRNA targets (see Chapter 22).

Since translation of viral mRNAs is critical for viral replication and pathogenesis, understanding how cellular signalling pathways that regulate protein synthesis are manipulated by virus-encoded functions in infected cells may expose a new

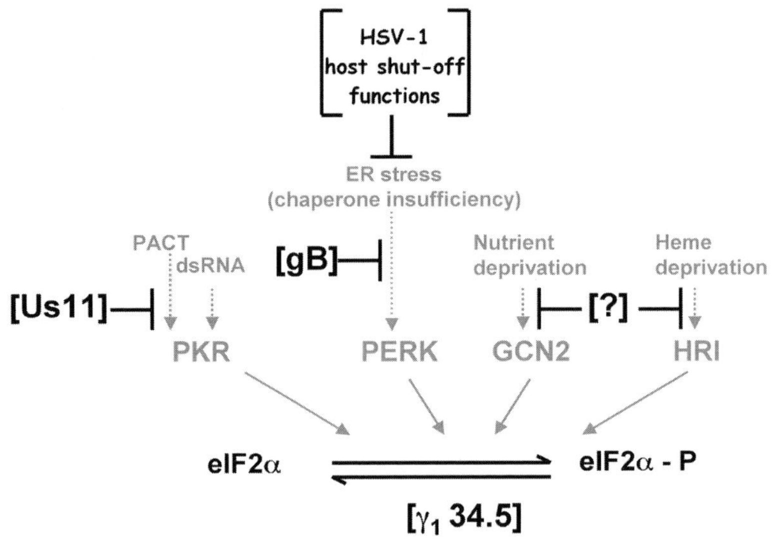

Figure 7.9 Multiple HSV-1-specified functions interact with host eIF2α kinases to regulate viral protein accumulation. The four known mammalian kinases, PKR, PERK, GCN2, and HRI, capable of phosphorylating eIF2α (eIF2α-P) are shown along with their respective activating stimuli. The HSV-1 $\gamma_1$34.5 protein is a GADD34-related polypeptide that binds to PP1 to maintain adequate supplies of phosphorylated, active eIF2. By virtue of acting downstream of eIF2α phosphorylation, it has the potential to counteract many eIF2α kinases. Us11 is a PKR-specific antagonist that physically associates with PKR to prevent its activation by dsRNA and PACT. It also physically associates with dsRNA and PACT. The powerful impairment of host gene expression in HSV-1-infected cells results from the combined action of virion components together with immediate-early genes (collectively termed HSV-1 host shutoff functions). This eliminates the majority of cellular ER clients, and combined with the limited load of viral ER clients early in the productive life cycle, ensures chaperone sufficiency and thereby does not produce ER stress. As most viral glycoprotein production ramps up later in the life cycle, glycoprotein B physically associates with PERK to regulate viral protein accumulation in a PERK-dependent manner, thereby maintaining ER homeostasis. Hypothetical functions targeting the remaining eIF2α-kinases GCN2 and HRI are indicated by a question mark (?). (reprinted from Mulvey *et al.*, 2007 with permission.)

Achilles' heel in herpesvirus biology. Along these lines, virus–host interactions that control translation may prove to be effective targets for the future development of novel antiviral therapies capable of limiting vegetative growth and/or suppressing reactivation. Finally, continued exploration of how protein synthesis is regulated in HSV-1 infected cells will provide further insight into translational control strategies, a basic biological regulatory mechanism. While the tactics discussed herein may appear on the surface to be confined to cells infected with HSV-1, they may ultimately prove to be pivotal in our efforts to understand cellular translational control mechanisms in health and disease, and exactly how these mechanisms are manipulated so effectively by viral pathogens.

Acknowledgements

I would like to thank present (Uyanga Chuluunbaatar, Mariko Kobayashi, Caleb McKinney) and past (Derek Walsh, Matthew Mulvey, Cesar Perez, Carolina Arias) members of my laboratory for many stimulating discussions, for critically reading the manuscript, and the NIH for their financial support of our research. Finally, I apologize to those investigators whose work could not be cited in this review due to space limitations.

References

Algire M.A., Maag, D., and Lorsch, J.R. (2005). Pi release from eIF2, not GTP hydrolysis, is the step controlled by start-site selection during eukaryotic translation initiation. Mol. Cell *20*, 251–262.

Amrani, N., Ghosh, S., Mangus, D.A., and Jacobson, A. (2008).Translation factors promote the formation of two states of the closed-loop mRNP. *Nature 453*, 1276–1280.

Arias, C., Walsh, D., Harbell, J., Wilson, A.C., and Mohr, I. (2009). Activation of a host translational control pathway by a viral developmental switch. PLoS Pathog. 5, e1000334.

Bartkowski, M., and Roizman, B. (1976). RNA synthesis in cells infected with herpes simplex virus XIII. Differences in the methylation patterns of viral RNA during the reproductive cycle. J. Virol. *20*, 583–588.

Bolovan, C.A., Sawtell, N.M., and Thompson, R.L. (1994). ICP34.5 mutants of herpes simplex virus type 1 strain 17syn+ are attenuated for neurovirulence in mice and for replication in confluent primary mouse embryo cell cultures. J. Virol., 68, 48–55.

Buxade, M., Parra-Palau, J.L., and Proud, C., G. (2008). The Mnks: MAP kinase-interacting kinases (MAP kinase signal-integrating kinases). Front. Biosci. *13*, 5359–5373.

Byrd, M.P., Zamora, M., and Lloyd, R.E. (2002). Generation of multiple isoforms of eukaryotic translation initiation factor 4GI by use of alternate translation initiation codons. Mol. Cell *Biol. 22*, 4499–4511.

Caron, S., Charon, M., Cramer, E., Sonenberg, N., and Dusanter-Fourt, I. (2004), Selective modification of eukaryotic initiation factor 4F (eIF4F) at the onset of cell differentiation: recruitment of eIF4GII and long lasting phosphorylation of eIF4E. Mol. Cell Biol. *24*, 4920–4928.

Cerveny, M., Hessefort, S., Yang, K., Cheng, G., Gross, M., and He, B. (2003). Amino acid substitutions in the effector domain of the g134.5 protein of herpes simplex virus 1 have differential effects on viral response to interferon-α. Virology *307*, 290–300.

Cheng, G., Brett, M.E., and He, B. (2001). Val193 and Phe195 of the g134.5 protein of herpes simplex virus 1 are required for viral resistance to interferon a/b. Virology *290*, 115–120.

Chou, J., Kern, E.R., Whitley, R.J., and Roizman, B. (1990). Mapping of herpes simplex virus-1 neurovirulence to gamma (1) 34.5, a gene nonessential for growth in culture. Science *250*, 1262–1266.

Chou, J., and Roizman, B. (1992). The _34.5 gene of Herpes Simplex Virus 1 precludes neuroblastoma cells from triggering total shutoff of protein synthesis characteristic of programmed. cell death in neuronal cells. Proc. Natl. Acad. Sci. U.S.A. *89*, 3266–3270.

Chou, J., and Roizman, B. (1994). Herpes simplex virus 1 gamma(1)34.5 gene function, which blocks the host response to infection, maps in the homologous domain of the genes expressed during growth arrest and DNA damage. Proc. Natl. Acad. Sci. U.S.A. *91*, 5247–55251.

Cohen, J.I., Straus, S.E., and Arvin, A.M. (2007). Varicella-Zoster virus. In Fields' Virology, 5th ed., Knipe, D.M. and Howley, P.M., eds. (Philadelphia, Pa; Lippincott, Williams & Wilkins), pp. 2773–2818.

Connor, J.H., and Lyles, D.S. (2002). Vesicular Stomatitis Virus infection alters the eIF4F translation initiation complex and causes dephosphorylation of the eIF4E binding protein 4E-BP1. J. Virol. *76*, 10177–19187.

Craig, A.W., Haghighat, A., Yu, A.T., and Sonenberg, N. (1998). Interaction of polyadenylate binding protein with the eIF4G homologue PAIP enhances translation. Nature *392*, 520–523.

Cuesta, R., Xi, Q., and Schneider, R.J. (2000). Adenovirus-specific translation by displacement of kinase Mnk1 from cap-initiation complex eIF4F. EMBO J. *19*, 3465–3474.

DeLuca N., McCarthy, A.M., and Schaffer, P. (1985). Isolation and characterrization of deletion mutants of herpes simplex virus type 1 in the gene encoding immediate-early regulatory protein ICP4. J. Virol. *56*, 558–570.

Dever, T.E., Dar, A.C., and Sicheri, F. (2007). The eIF2*α* kinases. In Translational control in biology and medicine, MB Mathews, N Sonenberg, & JWB Hershey, eds., (Cold Spring Harbor Laboratory Press, Cold Spring Harbor, NY), pp. 319–344.

Dobrikova, E., Shveygert, M., Walters, R., and Gromeier, M. (2010). Herpes simplex virus proteins ICP27 and UL47 associate with polyadenylate-binding protein and control its sub-cellular distribution. J. Virol. *84*, 270–279.

Doepker, R.C., Hsu, W.-L., Saffran, H.A., and Smiley, J.R. (2004). Herpes simplex virus virion host shutoff protein is stimulated by translation initiation factors eIF4B and eIF4H. J. Virol. 78, 4684–4699.

Duerst, R.J., and Morrison, L.A. (2004). Herpes simplex virus 2 virion host shutoff protein interferes. with type I interferon production and responsiveness. Virology *322*, 158–167.

Duffy, C., Mbong, E.F., and Baines, J.D. (2009). VP22 of herpes simplex virus 1 promotes protein synthesis at late times in infection and accumulation of a subset of viral mRNAs at early times in infection. J. Virol. *83*, 1009–1017.

Elgadi, M.M., Hayes, C.E., and Smiley, J.R. (1999). The herpes simplex virus vhs protein induces endoribonucleolytic cleavage of target RNAs in cell extracts. J. Virol. 73, 7153–7164.

Ellison, K.S., Maranchuk, R.A., Mottet, K.L., and Smiley, J.R. (2005). Control of VP16 translation by the herpes simplex virus type 1 immediate-early protein ICP27. J. Virol. *79*, 4120–4131.

Esclatine, A., Taddeo, B., Evans, L., and Roizman, B. (2004). The herpes simplex virus 1 UL41 gene-dependent destabilization of cellular RNAs is selective and may be sequence specific. Proc. Natl. Acad. Sci. U.S.A. *101*, 3603–3608.

Everly, Jr, D.N. ., Feng, P., Mian, I.S., and Read, G.S. (2002). mRNA degradation by the virion host shutoff (vhs) protein of herpes simplex virus: genetic and biochemical evidence that Vhs is a nuclease. J. Virol. *76*, 8560–8571.

Feigenblum, D., and Schneider, R.J. (1993). Modification of eukaryotic initiation factor 4F during infection by influenza virus. J. Virol. *67*, 3027–3035.

Feng, P., Everly Jr, D.N., and Read, G.S. (2001). mRNA decay during herpesvirus infections: interaction between a putative viral nuclease and a cellular translation factor. J. Virol. 75, 10272–10280.

Feng, P., Everly Jr, D.N., and Read, G.S. (2005). mRNA decay during herpes simplex virus (HSV) infections: protein–protein interactions involving the HSV virion host shutoff protein and translation factors eIF4H and eIF4A. J. Virol. *79*, 9651–9664.

Fontaine–Rodriguez, E.C., and Knipe, D.M. (2008). Herpes simplex virus ICP27 increases translation of a subset of viral late mRNAs. J. Virol. *82*, 3538–3545.

Fontaine–Rodriguez, E.C., Taylor, T.J., Olesky, M., and Knipe, D.M. (2004). Proteomics of herpes simplex virus infected cell protein 27: association with translation initiation factors. Virology *330*, 487–492.

Fukunaga, R., and Hunter, T. (1997). MNK1, a new MAP kinase-activated protein kinase, isolated by a novel expression screening method for identifying protein kinase substrates. EMBO J. *16*, 1921–1933.

Ganem, D. (2007). Kaposi's sarcoma associated herpesvirus. In Fields' Virology, 5th ed., Knipe, D.M. and Howley, P.M., eds. (Philadelphia, Pa; Lippincott, Williams & Wilkins), pp. 2847–2888.

Goldstein D.J., and Weller, S.K. (1988). Herpes simplex virus type-1 ribonucleotide reductase activity is dispensable for virus growth and DNA synthesis: Isolation and characterization of an ICP6 lacZ insertion mutant. J. Virol. *62*, 196–205.

Gross, J.D., Moerke, N.J., von der Haar, T., Lugovskoy, A.A., Sachs, A.B., McCarthy, J.E., and Wagner, G. (2003). Ribosome loading onto the mRNA cap is driven by conformational coupling between eIF4G and eIF4E. Cell. *115*, 739–750.

Haghighat A., Mader, S., Pause, A., and Sonenberg, N. (1995). Repression of cap-dependent translation by 4E-binding protein 1: competition with p220 for binding to eukaryotic initiation factor-4E. EMBO J. *14*, 5701–5709.

Hardy, W.R., and Sandri-Goldin, R.M. (1994). Herpes simplex virus inhibits host cell splicing and regulatory protein ICP27 is required for this effect. J. Virol. *68*, 7790–7799.

He, B., Gross, M., and Roizman, B. (1997). The gamma(1)34.5 protein of herpes simplex virus 1 complexes with protein phosphatase 1 alpha to dephosphorylate the alpha subunit of eukaryotic initiation factor 2 and preclude the shutoff of protein synthesis by double-stranded RNA-activated protein kinase. Proc. Natl. Acad. Sci. U.S.A., 94, 843–848.

Imataja, H., Gradi, A., and Sonenberg, N. (1998). A newly identified N-terminal amino acid sequence of human eIF4G binds poly(A)-binding protein and functions in poly(A)-dependent translation. *EMBO J. 17*, 7480–7489.

Jacquemont, B., and Roizman, B. (1975). RNA synthesis in cells infected with herpes simplex virus. X. Properties of viral symmetric transcripts and of double-stranded RNA prepared from them. J. Virol. *15*, 707–713.

Kahvejian, A., Roy, G., and Sonenberg, N. (2001). The mRNA closed-loop model: the function of PABP and PABP-interacting proteins in mRNA translation. Cold Spring Harb. Symp. Quant. Biol. *66*, 293–300.

Karim, M.M., Svitkin, Y.V., Kahvejian, A., De Crescenzo, G., Costa-Mattioli, M., and Sonenberg, N. (2006). A mechanism of translational repression by competition of Paip2 with eIF4G for poly(A) binding protein (PABP) binding. Proc. Natl. Acad. Sci. U.S.A. *103*, 9494–9499.

Karr, B.M., and Read, G.S. (1999). The virion host shutoff function of herpes simplex virus degrades the 5′ end of a target mRNA before the 3′ end. Virology *264*, 195–204.

Khaleghpour, K., Svitkin, Y.V., Craig, A.W., DeMaria, C.T., Deo, R.C., Burley, S.K., and Sonenberg, N. (2001). Translational repression by a novel partner of human poly(A) binding protein. Mol. Cell *7*, 205–216.

Khoo, D., Perez, C., and Mohr, I. (2002). Characterization of RNA determinants recognized by the arginine- and proline-rich region of Us11, a herpes simplex virus type 1-encoded double-stranded RNA binding protein that prevents PKR activation. J. Virol., 76, 11971 –11981.

Kieff, E., and Rickinson, A.B. (2007). Epstein–Barr virus and its replication. In Fields virology, 5th ed., Knipe, D.M. and Howley, P.M., eds. (Philadelphia, Pa; Lippincott, Williams & Wilkins), pp. 2603–2654.

Korom, M., Wylie, K.M., and Morrison, L.A. (2008). Selective ablation of virion host shutoff protein RNase activity attenuates herpes simplex virus 2 in mice. J. Virol. *82*, 3642–3653.

Kudchodkar, S.B., Yu, Y., Maguire, T.G., and Alwine, J.C. (2004).Human cytomegalovirus infection induces rapamycin-insensitive phosphorylation of downstream effectors of mTOR kinase. J. Virol. *78*, 11030–11039.

Kwong, A.D., and Frenkel, N. (1987). Herpes simplex virus-infected cells contain a function(s) that destabilizes both host and viral mRNAs. Proc. Natl. Acad. Sci. U.S.A. 84, 1926 –1930.

Larralde, O., Smith, R.W., Wilkie, G.S., Malik, P., Gray, N.K., and Clements, J.B. (2006).Direct stimulation of translation by the multifunctional herpesvirus ICP27 protein. J. Virol. *80*, 1588–1591.

Laurent, A.M., Madjar, J.J., and Greco, A. (1998). Translational control of viral and host protein synthesis during the course of herpes simplex virus type 1 infection: evidence that initiation of translation is the limiting step. J. Gen. Virol. 79, 2765–2775.

Leib, D.A., Harrison, T.E., Laslo, K.M., Machalek, M.A., Moorman, N.J., and Virgin, H.W. (1999). Interferons regulate the phenotype of wild-type and mutant herpes simplex viruses in vivo. J. Exp. Med., 189, 663–672.

Leib, D.A., Machalek, M.A., Williams, B.R., Silverman, R.H., and Virgin, H.W. (2000). Specific phenotypic restoration of an attenuated virus by knockout of a host resistance gene. Proc. Natl. Acad. Sci. U.S.A. *97*, 6097–6101.

Li, Q., Imataka, H., Morino, S., Rogers, Jr, G.W. ., Richter-Cook, N.J., Merrick, W.C., and Sonenberg, N. (1999). Eukaryotic translation initiation factor 4AIII (eIF4AIII) is functionally distinct from eIF4AI and eIF4AII. Mol. Cell Biol. *19*, 7336–7346.

Lloyd, R.E. (2006). Translational control by viral proteinases. Virus Res. *119*, 76–88.

Ma, X.M., and Blenis, J. (2009). Molecular mechanisms of mTOR-mediated translational control. *Nat. Rev.* Mol. Cell Biol. *10*, 307–318.

Maclean, A.R., Ul-Fareed, M., Robertson, L., Harland, J., and Brown, S.M. (1991). Herpes simplex virus type 1 deletion variants 1714 and 1716 pinpoint neurovirulence-related sequences in Glasgow strain 17+ between immediate early gene 1 and the 'a' sequence. J. Gen. Virol. *72*, 631–639.

Mader, S., Lee, H., Pause, A., and Sonenberg, N. (1995). The translation initiation factor eIF-4E binds to a common motif shared by the translation factor eIF-4G and the translational repressors 4E-binding proteins. Mol. Cell Biol. *15*, 4990–4997.

Marcotrigiano, J., Gingras, A.C., Sonenberg, N., and Burley, S.K. (1999). Cap-dependent translation initiation in eukaryotes is regulated by a molecular mimic of eIF4G. Mol. Cell *3*, 707–716.

Mocarski, Jr, E.S. ., Shenk, T., and Pass, R.F. (2007). Cytomegaloviruses, In Fields Virology, 5th ed., Knipe, D.M. and Howley, P.M., eds. (Philadelphia, Pa; Lippincott, Williams & Wilkins), pp. 2701–2772.

Mohr, I. (2005). To replicate or not to replicate: achieving selective oncolytic virus replication in cancer cells through translational control. Oncogene *24*, 7697–7709.

Mohr, I., and Gluzman, Y. (1996). A herpesvirus genetic element which affects translation in the absence of the viral GADD34 function. EMBO J. *15*, 4759–4766.

Mohr, I.J., Pe'ery, T., and Mathews, M.B. (2007). Protein synthesis and translational control during viral infection. In Translational control in biology and medicine, MB Mathews, N Sonenberg, & JWB Hershey, eds., (Cold Spring Harbor Laboratory Press, Cold Spring Harbor, NY), pp. 545–595.

Moorman, N.J., Cristea, I.M., Terhune, S.S., Rout, M.P., Chait, B.T., and Shenk, T. (2008). Human cytomegalovirus protein UL38 inhibits host cell stress responses by antagonizing the tuberous sclerosis protein complex. Cell Host Microbe *3*, 253–262.

Mulvey, M., Arias, C., and Mohr, I. (2006). Resistance of mRNA translation to acute ER stress-inducing agents in HSV-1-infected cells requires. multiple virus-encoded functions. J. Virol. *80*, 7354–7363.

Mulvey, M., Arias, C., and Mohr, I. (2007). Maintenance of ER homeostasis in HSV-1 infected cells through the association of a viral glycoprotein with PERK, a cellular ER stress sensor. J. Virol. *81*, 3377–3390.

Mulvey, M., Camarena, V., and Mohr, I. (2004). Full resistance of HSV-1 infected primary human cells to interferon a require both the Us11 and g134.5 gene products. J. Virol. *78*, 10193–10196.

Mulvey, M., Poppers, J., Ladd, A., and Mohr, I. (1999). A herpesvirus ribosome associated, RNA-binding protein confers a growth advantage upon mutants deficient in a GADD34-related function. J. Virol., 73, 3375–3385.

Mulvey, M., Poppers, J., Sternberg, D., and Mohr, I. (2003). Regulation of eIF2 alpha phosphorylation by different functions that act during discrete phases in the herpes simplex virus type 1 life cycle. J. Virol. 77, 10917–10928.

Murphy, J.A., Duerst, R.J., Smith, T.J., and Morrison, L.A. (2003). Herpes simplex virus type 2 virion host shutoff protein regulates alpha/beta interferon but not adaptive immune responses during primary infection *in vivo*. J. Virol. *77*, 9337–9345.

Oroskar, A.A., and Read, G.S. (1987). A mutant of herpes simplex virus type 1 exhibits increased stability of immediate-early (alpha) mRNAs. J. Virol. *61*, 604–606.

Pasieka, T.J., Lu, B., Crosby, S.D., Wylie, K.M., Morrison, L.A., Alexander, D.E., Menachery, V.D., and Leib, D.A. (2008). Herpes simplex virus virion host shutoff attenuates establishment of the antiviral state. J. Virol. *82*, 5527–5533.

Pasieka, T.J., Lu, B., and Leib, D.A. (2008). Enhanced pathogenesis of an attenuated herpes simplex virus for mice lacking Stat1. J. Virol. *82*, 6052–6055.

Patel, R.C., and Sen, G.C. (1998). PACT, a protein activator of the interferon-induced protein kinase, PKR. EMBO J. *17*, 4379–4390.

Perez-Parada, J., Saffran, H.A., and Smiley, J.R. (2004). RNA degradation induced by the herpes simplex virus vhs protein proceeds 5′ to 3′ *in vitro*. J. Virol. *78*, 13391–13394.

Peters G.A., Khoo, D., Mohr, I., and Sen, G.C. (2002). Inhibition of PACT-mediated activation of PKR by the herpes simplex virus type 1 Us11 protein. J. Virol. *76*, 11054–11064.

Poppers, J., Mulvey, M., Khoo, D., and Mohr, I. (2000). Inhibition of PKR activation by the proline rich RNA binding domain of the HSV-1 Us11 protein. J. Virol. *74*, 112315–11221.

Preston, C.M., and Newton, A.A. (1976). The effects of herpes simplex virus type 1 on cellular DNA-dependent RNA polymerase activities. J. Gen. Virol. *33*, 471–482.

Pyronnet, S., Imataka, H., Gingras, A.-C., Fukunaga, R., Hunter, T., and Sonenberg, N. (1999). Human eukaryotic translation initiation factor 4G (eIF4G) recruits Mnk1 to phosphorylate eIF4E. EMBO J. *18*, 270–279.

Raught, B., Gingras, A.C., Gygy, S.P., Imataka, H., Morino, S., Gradi, A., Aebersold, R., and Sonenberg, N. (2000). Serum-stimulated, rapamycin-sensitive phosphorylation sites in the eukaryotic translation initiation factor 4GI. EMBO J. *19*, 434–444.

Read, G.S., and Frenkel, N. (1983). Herpes simplex virus mutants defective in the virion-associated shutoff of host polypeptide synthesis and exhibiting abnormal synthesis of α (immediate early) viral polypeptides. J. Virol. *46*, 498–512.

Rogers, Jr, G.W., Komar, A.A., and Merrick, W.C. (2002). eIF4A: the godfather. of the DEAD box helicases. Prog. Nucleic Acid Res. Mol. Biol. 72, 307–331.

Roizman, B., Knipe, D.M., and Whitley, R. (2007). Herpes simplex viruses. In Fields' Virology, 5th ed., Knipe, D.M. and Howley, P.M., eds. (Philadelphia, Pa; Lippincott, Williams & Wilkins), pp. 2501–2602.

Ron, D., and Walter, P. (2007). Signal integration in the endoplasmic reticulum unfolded protein response. Nat. Rev. Microbiol. 8, 519–529.

Sandri-Goldin, R.M. (2008). The many roles of the regulatory protein ICP27 during herpes simplex virus infection. Front. Biosci. *13*, 5241–5256.

Sanchez, R., and Mohr, I. (2007). Inhibition of the cellular 2′–5′ oligoadenylate synthetase by the herpes simplex virus Us11 protein. J. Virol. *81*, 3455–3464.

Sarma, N., Agarwal, D., Shiflett, L.A., and Read, G.S. (2008). Small interfering RNAs that deplete the cellular translation factor eIF4H impede mRNA degradation by the virion host shutoff protein of herpes simplex virus. J. Virol. *82*, 6600–6609.

Scheper, G.C., Morrice, N.A., Kleijn, M., and Proud, C.G. (2001). The mitogen-activated protein kinase signal-integrating kinase Mnk2 is a eukaryotic initiation factor 4E kinase with high levels of basal activity in mammalian cells. Mol. Cell Biol. *21*, 743–754.

Scheper, G.C., and Proud, C.G. (2002). Does phosphorylation of the cap-binding protein eIF4E play a role in translation initiation? Eur. J. Biochem. *269*, 5350–5359.

Schneider, R.J., and Mohr, I. (2003). Translation initiation and viral tricks. Trends Biochem. Sci. *28*, 130–136.

Sciabica, K.S., Dai, Q.J., and Sandri-Goldin, R.M. (2003). ICP27 interacts with SRPK1 to mediate HSV splicing inhibition by altering SR protein phosphorylation. *EMBO J. 22*, 1608–1619.

Shibuya, T., Tange, T.O., Sonenberg, N., and Moore, M.J. (2004). eIF4AIII binds spliced mRNA in the exon junction complex and is essential for nonsense-mediated decay. Nat. Struct. Mol. Biol. *11*, 346–351.

Smith, T.J., Morrison, L.A., and Leib, D.A. (2002). Pathogenesis of herpes simplex virus type 2 virion host shutoff (vhs) mutants. J. Virol. *76*, 2054–2061.

Soliman, T., Sandri-Goldin, R.M., and Silverstein, S.J. (1997). Shuttling of the herpes simplex virus type 1 regulatory protein ICP27 between the nucleus and cytoplasm mediates the expression of late proteins. J. Virol. *71*, 9188–9197.

Sonenberg, N., and Hinnebusch, A.G. (2009). Regulation of translation initiation in eukaryotes: mechanisms and biological targets. *Cell 136*, 731–745.

Spencer, C.A., Dahmus, M.E., and Rice, S.A. (1997). Repression of host RNA polymerase II transcription by herpes simplex type 1. J. Virol. *71*, 2031–2040.

Stanners, C.P., and Becker, H. (1971). Control of macromolecular synthesis in proliferating and resting Syrian hamster cells in monolayer culture. I. Ribosome function. J. Cell. Physiol. 77, 31–42.

Strelow, L.I., and Leib, D.A. (1995). Role of the virion host shutoff (vhs) of herpes simplex virus type 1 in latency and pathogenesis. J. Virol. *69*, 6779–6786.

Taddeo, B., Zhang, W., and Roizman, B. (2009). The virion-packaged endoribonuclease of herpes simplex virus 1 cleaves mRNA in polyribosomes. Proc. Natl. Acad. Sci. U.S.A. *106*, 12139–12144.

Thomas, G., and Gordon, J. (1979). Regulation of protein synthesis during the shift of quiescent animal cells into the proliferative state. Cell Biol. Int. Rep. *3*, 307–320.

Thomas, G., and Thomas, G. (1986). Translational control of mRNA expression during the early mitogenic response in Swiss mouse 3T3 cells: identification of specific proteins. J. Cell. Biol. *103*, 2137–2144.

Thomas, G., Thomas, G., and Luther, H. (1981). Transcriptional and translational control of cytoplasmic proteins after serum stimulation of quiescent Swiss 3T3 cells. Proc. Natl. Acad. Sci. U.S.A. *78*, 5712–5716.

Walsh, D., and Mohr, I. (2004). Phosphorylation of eIF4E by Mnk1 enhances HSV-1 translation and replication in quiescent cells. Genes Dev. *18*, 660–672.

Walsh, D., and Mohr, I. (2006). Assembly of an active translation initiation factor complex by a viral protein. Genes Dev. *20*, 461–472.

Walsh, D., Perez, C., Notary, J., and Mohr, I. (2005). Regulation of the translation initiation factor eIF4F by multiple mechanisms in human cytomegalovirus infected cells. J. Virol. *79*, 8057–8064.

Waskiewicz, A.J., Flynn, A., Proud, C.G.. and Cooper, J.A. (1997). Mitogen-activated protein kinases activate the serine/threonine kinases Mnk1 and Mnk2. EMBO J. *16*, 1909–1920.

Waskiewicz, A.J., Johnson, J.C., Penn, B., Mahalingam, M., Kimball, S.R., and Cooper, J.A. (1999). Phosphorylation of the cap-binding protein eukaryotic translation factor 4E by protein kinase Mnk1 *in vivo*. Mol. Cell Biol. *19*, 1871–1880.

Wells, S.E., Hillner, P.E., Vale, R.D., and Sachs, A.B. (1998). Circularization of mRNA by eukaryotic translation initiation factors. Mol. Cell 2, 135–140.

Xi, Q., Cuesta, R., and Schneider, R.J. (2004). Tethering of eIF4G to adenoviral mRNAs by viral 100k protein drives ribosome shunting. Genes Dev. *18*, 1997–2009.

Yamanishi, K., Mori, Y., and Pellet, P.E. (2007). Human herpesviruses 6 and 7. In Fields' Virology, 5th ed., Knipe, D.M. and Howley, P.M. eds. (Philadelphia, Pa; Lippincott, Williams & Wilkins), pp. 2819–2846.

Yoder-Hill, J., Pause, A., Sonenberg, N., and Merrick, W.C. (1993). The p46 subunit of eukaryotic initiation factor (eIF)–4F exchanges with eIF-4A. J. Biol. Chem. *268*, 5566–5573.

Zhou, C., and Knipe, D.M. (2002). Association of herpes simplex virus type 1 ICP8 and ICP27 proteins with cellular RNA polymerase II holoenzyme. J. Virol. *76*, 5893–5904.

Recent Progress in Understanding Herpes Simplex Virus Entry: Relationship of Structure to Function

8

Roselyn J. Eisenberg, Ekaterina E. Heldwein, Gary H. Cohen and Claude Krummenacher

Abstract

Membrane fusion allows exchange of materials between cellular compartments enclosed by lipid membranes. Similarly, entry of enveloped viruses into cells allows the viral contents to be delivered by fusion of the envelope with a target cell membrane. Fusion requires disruption of both layers of the two membranes. For most enveloped viruses, a single surface glycoprotein undergoes conformational changes that bring the bilayer of the virus in proximity with that of the host cell and fusion ensues. In contrast, herpesvirus entry requires three virion glycoproteins, gB and a gH/gL heterodimer, that function as the core fusion machinery. Some herpesviruses require additional proteins, e.g. alphaherpesviruses (with a few exceptions) initiate fusion by binding of glycoprotein gD to a cell receptor. A conformational change then exposes the normally hidden receptor binding residues of gD. This change and/or the exposed residues trigger gB and gH/gL to effect virus-cell and cell–cell fusion. Because of the multiplicity of proteins involved in HSV entry as opposed to entry of enveloped RNA viruses, it has been difficult to unravel the mechanism of how the four entry glycoproteins function. Some favour formation of a multiprotein fusion complex while others suggest it may be more of a stepwise process. Solution of the structures of all four entry proteins, coupled with existing and new information has solved much of this mystery. We now have a much better idea of the outline of the process, but the challenge for the future will be to fill in important details. It is clear that entry of HSV occurs in an exquisitely regulated stepwise process that begins with binding of gD to a receptor, activation of the regulatory protein gH/gL that in turn up-regulates the fusogenic activity of gB. Thus, in some ways, HSV entry is remarkably similar overall to entry by simpler RNA viruses, such as influenza. A single fusion protein gB carries out fusion. What distinguishes HSV entry is the double regulation of this process. This chapter will focus primarily on what we know about the structures of gD, gH/gL and gB along with supporting data for how those proteins function.

Introduction

Herpes simplex virus (HSV) is the prototype member of the diverse herpesvirus family. The two serotypes of the virus (HSV1 and HSV2) are significant human pathogens in spite of effective anti-viral therapy. HSV has a lipid envelope, bearing eleven surface proteins and glycoproteins. To deliver the capsid containing the double-stranded DNA genome into the host cell, HSV must fuse its envelope with a cellular membrane, either at the cell surface or within an endocytic vesicle, depending on the cell type (Delboy *et al.*, 2006; Milne *et al.*, 2005; Nicola *et al.*, 2003; Nicola and Straus, 2000; Fig. 8.1).

Among the 12 envelope glycoproteins, only five, gC, gB, gD, gH, and gL participate in viral cell binding and entry; the last four are required for virus-cell and cell–cell fusion (reviewed in Heldwein and Krummenacher, 2008; Krummenacher *et al.*, 2007). The most important advance in putting all of the previous abundant literature in a new perspective on their roles in entry has been the solution of the structures

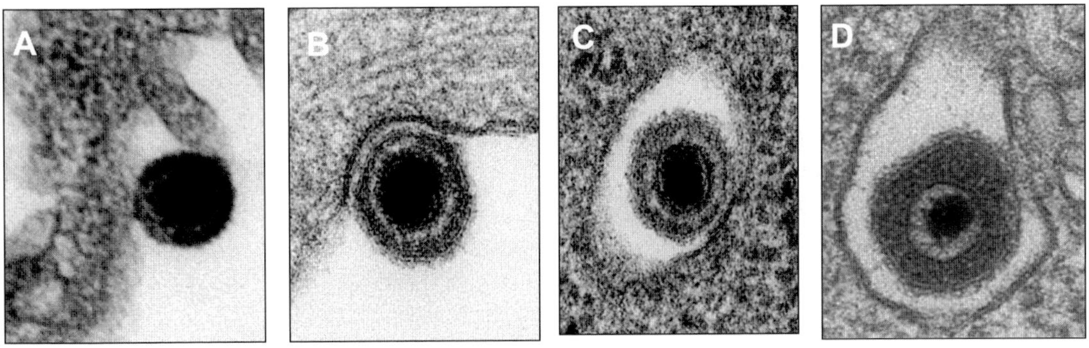

Figure 8.1 HSV entry. (A) Initial attachment (gB, gC to HSPG). (B) Receptor binding by gD. (C,D) Fusion: gB, gH, gL.

of gD, both unbound and bound to one of its receptors, HVEM (herpesvirus entry mediator) (Carfi *et al.*, 2001; Krummenacher *et al.*, 2005) and of gB (Heldwein *et al.*, 2006) and of gH/gL (Chowdary *et al.*, 2010). Hence this chapter will centre around how structural biology has put a new perspective on our understanding of HSV entry. Other important advances have arisen from biochemical and cell-based assays. A number of excellent reviews on this subject have described the state of knowledge prior to solution of the structures of gB and gH/gL (Campadelli-Fiume *et al.*, 2000; Spear, 1993; Spear, 2004; Spear *et al.*, 2000a; Spear and Longnecker, 2003), so we will not attempt to cover much of this older ground in detail except to put some of those remarkable findings in a structural context as appropriate.

All herpesviruses encode genes for gB, gH, and gL, and these three proteins constitute what has been termed their core fusion machinery (Krummenacher *et al.*, 2007; Spear *et al.*, 2000a). HSV attaches to a cell by binding to heparan sulfate proteoglycan (HSPG) via gC, and by binding of gD to one of three cellular receptors: nectin-1, herpesvirus entry mediator (HVEM), or a modified heparan sulfate (Krummenacher *et al.*, 2007; Spear, 2004) (Fig. 8.1). Structural and biochemical studies have shown that gD undergoes conformational changes to permit receptor binding and to potentiate fusion (Krummenacher *et al.*, 2007; Krummenacher *et al.*, 2005). Structural and biochemical studies of gB (Hannah *et al.*, 2009; Hannah *et al.*, 2007; Heldwein *et al.*, 2006) show that it has the features of a trimeric fusion protein but it is important to note that gB cannot function as a fusion protein in the absence of gH/gL. Several studies indicate that gB also binds a non-HSPG cell receptor (Bender *et al.*, 2005; Satoh and Arase, 2008; Satoh *et al.*, 2008), but how that interaction fits into an overall scheme for entry is not clear. Now that the structure of gH/gL has been solved, (Chowdary *et al.*, 2010), it is clear that gH/gL does not resemble any known viral fusion protein, even though motifs present in fusion proteins have been proposed from sequence analysis (Galdiero *et al.*, 2006, 2008, 2007, 2010a,b). One publication suggested that gH/gL might to be capable of carrying out hemifusion when it is expressed in the presence of gD and receptor but in the absence of gB (Subramanian and Geraghty, 2007), and this observation will eventually be resolved by additional studies of gH/gL in light of its structural properties. Indeed, two recent studies have shown that binding of gD to a cell-bound receptor leads to an interaction between gB and gH/gL (Atanasiu *et al.*, 2007; Avitabile *et al.*, 2007) that occurs concurrently with fusion. That this interaction occurs prior to fusion and is a necessary event on the pathway to fusion is also supported by recent studies (Atanasiu *et al.*, 2010b).

In this chapter, we will describe the current art knowledge about HSV entry, focusing primarily on the structures of the HSV glycoproteins and their cell receptors as well as structures of gB and gH/gL. We will also review data regarding key protein–protein interactions that appear to occur during entry by several routes. We will point out those aspects that are well supported by experimental data and we indicate those aspects that still remain mysterious. Finally, we will attempt to place what we know and don't know about the mechanism of HSV entry into the larger context of herpesvirus entry in general. Comparisons with entry by other enveloped viruses will also be made to note similarities and differences.

The structure of gD, its receptors and how gD functions in HSV entry

gD is the receptor binding protein of HSV

HSV-1 gD is a type-I membrane glycoprotein of 369 amino acids. It is an essential glycoprotein entry of HSV and for entry of most alphaherpesviruses (Ligas and Johnson, 1988). Early studies suggested that gD might be the receptor-binding protein of HSV. For instance, when cells were engineered to express gD, they became resistant to infection due to interference with a cellular receptor (previously reviewed in Campadelli-Fiume *et al.*, 2000; Spear, 2004; Spear *et al.*, 2000b; Spear and Longnecker, 2003). This was later confirmed by using soluble forms of gD which could compete with virion-bound gD for receptor binding, thereby preventing virion gD from interacting with the receptor and blocking HSV entry (Johnson *et al.*, 1990; Nicola *et al.*, 1996a,b). Isolation of the first bona fide HSV entry receptor, HVEM (Montgomery *et al.*, 1996), together with solution of the structure of gD bound to HVEM (Carfi *et al.*, 2001), provided firm evidence that this was the case. Additional structures of gD together with biochemical and functional data provided a model for receptor-activated triggering of the fusion machinery by gD (Carfi *et al.*, 2001; Connolly *et al.*, 2002, 2003; Krummenacher *et al.*, 2005; Lazear *et al.*, 2008).

To date, three unrelated molecules have been shown to bind gD and mediate HSV entry: HVEM, nectin-1 and a modified form of heparon sulfate (reviewed in Campadelli-Fiume *et al.*, 2000; Heldwein and Krummenacher, 2008; Spear *et al.*, 2000b). The main receptor for entry into epithelial cells, keratinocytes and neuron is nectin-1 (HveC, PRR1, CD111) (Galen *et al.*, 2006; Huber *et al.*, 2001; Linehan *et al.*, 2004; Simpson *et al.*, 2005; Tiwari *et al.*, 2008). Nectin-1 is a cell adhesion molecule that participates in the organization of adherens junctions and tight junctions of epithelial cells and synapses between neurons (De Regge *et al.*, 2006; Mizoguchi *et al.*, 2002; Takahashi *et al.*, 1999; for review see Takai *et al.*, 2008). Nectin-1 is also a receptor for PRV and BHV-1 (Geraghty *et al.*, 1998; Milne *et al.*, 2001). Other nectins can also be used as entry receptors by other alphaherpesviruses (Geraghty *et al.*, 1998; Warner *et al.*, 1998). Furthermore, nectin-2 may be used by HSV-2 and mutant forms of HSV-1 (Lopez *et al.*, 2000; Warner *et al.*, 1998). Thus, it may turn out that alphaherpesviruses in general use members of the nectin family for entry.

Unlike nectin-1 which fits well with the biology of the virus in its tissue and cellular distribution, HVEM, as a member of the tumour necrosis factor receptor family, is found mostly on lymphocytes, although it is also present at low levels on fibroblasts and other cell types (Krummenacher *et al.*, 2004; Montgomery *et al.*, 1996). HVEM is a key regulator of T-cell responses and has four natural ligands that either up- or down-regulate T-cell proliferation (Kopp *et al.*, 2009; Krummenacher *et al.*, 2004). Are both receptors used for entry *in vivo*, and if so, in what settings? HSV-1 and HSV-2 clinical isolates from various types of lesions all have the ability to use nectin-1 and HVEM, suggesting that both receptors play a role *in vivo* (Krummenacher *et al.*, 2004).

Are both receptors used for entry *in vivo* and if so, in what settings? HSV-1 and HSV-2 clinical isolates all have the ability to use both receptors regardless of what type of lesion they came from (Krummenacher *et al.*, 2004) suggesting that both receptors may be important *in vivo*. Perhaps the best information we have at present comes from studies of receptor knockout mice, particularly those lacking nectin-1 (Murphy *et al.*, 2006; Taylor *et al.*, 2007). When these mice are infected intra-vaginally with HSV, the virus spreads to a limited extent to sensory ganglia but when mice are infected by the intracranial route, the virus does not establish neuronal infection (Kopp *et al.*, 2009; Taylor *et al.*, 2007). These results suggest that nectin-1 is the major entry receptor for HSV *in vivo*. Perhaps, the interaction between virion gD with HVEM enables the virus to maintain itself in a latent state or to limit the extent of reactivation.

The interaction between gD and HVEM is revealed by the structure of the complex

HSV-1 gD is the smallest of the membrane-anchored entry glycoproteins. Despite being dwarfed by gB and gH/gL, gD provides the critical binding of the virus to a cell surface receptor. Crystal structures of the gD ectodomain (either alone or bound to HVEM) show that it is a compact protein organized around a V-like immunoglobulin (IgV) core (Carfi *et al.*, 2001; Krummenacher *et al.*, 2005). The V-domain, possibly acquired from a cellular gene, mostly provides a scaffold for the large N- and C-terminal extensions that wrap around this backbone. The HVEM-binding site on gD is limited to N-terminal residues 1–32 that form a hairpin when gD is bound to HVEM (Fig. 8.2).

This hairpin is supported by a long α-helix α3, which, in turn, is supported by the IgV core (Carfi *et al.*, 2001). The gD/HVEM interface is characterized by three sets of interactions that are critical for function (Carfi *et al.*, 2001; Connolly *et al.*, 2003). At the centre of the interface, the first set involves HVEM Tyr23 that inserts into a pocket encircled by the two arms of the N-terminal hairpin of gD. The second consists of an intermolecular anti-parallel β-sheet, with a short strand on gD supplementing a two-stranded β-sheet of HVEM. The final set comprises hydrogen bonds between gD N15 and HVEM T76 and S74 in the second cysteine-rich domain. Collectively, these interactions ensure stability of the complex. Several laboratory strains of HSV-1 that contain mutations at residues 27 and/or 25 cannot use HVEM as a receptor because the mutated gD cannot bind to it, although these residues do not directly contact HVEM (Campadelli-Fiume *et al.*, 1990; Connolly *et al.*, 2003; Dean *et al.*, 1994; Lopez *et al.*, 2000; Montgomery *et al.*, 1996; Warner *et al.*, 1998). The N-terminal hairpin that composes the HVEM-binding site is flexible in conformation. In unliganded gD, the N-terminus is extended such that residues 1–21 stretch out from the protein core rather than forming a loop; moreover, residues 1–22 are disordered indicating a high degree of flexibility (Fig. 8.2) (Carfi *et al.*, 2001). Thus, HVEM contributes to the formation or stabilization of the N-terminus of gD. The binding site for gD on HVEM is located mostly

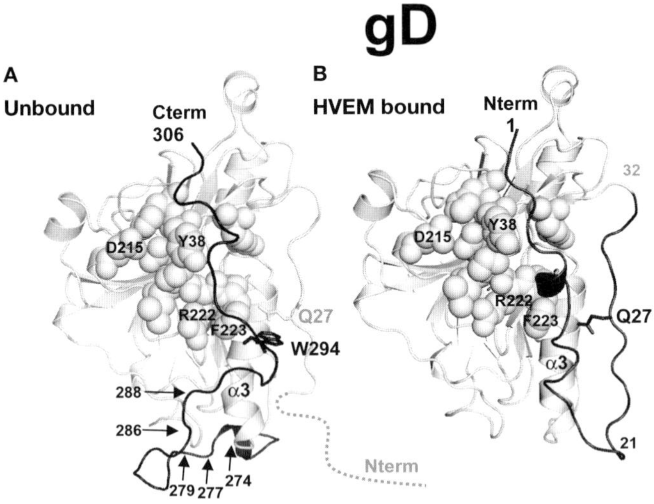

Figure 8.2 Structure of the gD ectodomain either unbound (A) or bound to its receptor HVEM (B). (A) The core of gD is light grey and the C-terminus region (aa 268–306) is black. Residues involved nectin-1 binding are spheres. The key anchoring residue W294 is highlighted. Residues 1–22 of the N-terminus are not resolved in the native form (dotted line). (B) In the HVEM-bound conformation, the C-terminus of gD is not visible and the N-terminus, shown in black is bent at residue 21 to form a hairpin.

in the first and most distal cysteine-rich domain (CRD1) with a critical, albeit limited, involvement of CRD2 (Carfi *et al.*, 2001; Connolly *et al.*, 2002). Interestingly, this binding site overlaps that of a natural cellular ligand for HVEM, BTLA (Compaan *et al.*, 2005). Although BTLA and gD differ in primary sequence and structure, they bind many of the same residues of HVEM including Tyr23 and in both cases, binding involves formation of an intermolecular β-sheet (Compaan *et al.*, 2005).

What do we know about the binding site for nectin-1 on gD?

The information we have about this site comes from studies with mutant forms of both proteins and from studies with monoclonal antibodies. Deletion of the N-terminal hairpin of gD does not significantly affect nectin-1 binding (Spear *et al.*, 2006), indicating that nectin-1 binds gD in a region that is distinct from the HVEM-binding site. Mutagenesis studies showed that several residues from the α3 helix and the N- and C-terminal extensions cooperate in binding to nectin-1 (Fig. 8.2). Three critical binding residues D215, R222, and F223 form a contiguous area on the surface of gD; their simultaneous mutations to alanine prevents nectin-1 binding and usage without affecting HVEM binding (Manoj *et al.*, 2004; Muggeridge *et al.*, 1990; Uchida *et al.*, 2009; Whitbeck *et al.*, 1999). In addition, gD Tyr38 is also critical for nectin-1 binding (Connolly *et al.*, 2005). Part of this putative binding site is masked by the N-terminal hairpin formed when gD is bound to HVEM (Fig. 8.2). When the N-terminal hairpin was locked in place by an intramolecular disulfide bond, as in the gD mutant Y38C-A3C, nectin-1 was no longer able to bind gD, but HVEM usage was unaffected (Connolly *et al.*, 2005). Thus, for gD to bind nectin-1, the gD N-terminus must be extended and flexible as it is when HVEM is not present (Fig. 8.2).

We know that the ectodomain of nectin-1 consists of three Ig domains, a V-fold followed by 2 C-folds (Lopez *et al.*, 1995). There is good evidence that the binding site for gD is located entirely in the V-domain of nectin-1 (Cocchi *et al.*, 1998; Krummenacher *et al.*, 2000, 2007). This domain, when expressed alone, binds to gD with the same affinity as the full nectin-1 ectodomain (Krummenacher *et al.*, 1999). Epitopes of neutralizing antibodies and mutagenesis affecting gD binding map to strands CC′C′ and their connecting loops on the V-domain (Cocchi *et al.*, 2001; Krummenacher *et al.*, 2000; Martinez and Spear, 2002). Functionally, soluble gD prevents nectin-1 mediated cell aggregation and disrupts cell aggregates suggesting that the glycoprotein competes with natural ligands of nectin-1 (Krummenacher *et al.*, 2002; Sakisaka *et al.*, 2001). Therefore, for gD to bind nectin-1 engaged at intercellular junctions, natural nectin/nectin contacts must be disrupted. Indeed, HSV entry is favoured when such junctions are disrupted (Schelhaas *et al.*, 2003; Yoon and Spear, 2002).

The interaction of gD with its third receptor, 3-O-sulfated heparan sulfate (3-O-S HS)

Less information is available about this interaction (Shukla *et al.*, 1999), but it likely involves specific sulfate groups on HS and a well-defined 3-O-S HS-binding site on gD. Mutations and N-terminal deletions in gD that prevent usage of HVEM also impair HSV-1 entry via 3-O-S HS, suggesting that the two binding sites overlap (Yoon and Spear, 2004). Two positively charged pockets on gD could potentially accommodate sulfate groups of 3-O-S-HS (Carfi *et al.*, 2001). One such pocket lies close to the N-terminal hairpin and thus the HVEM-binding site. Structural studies are needed to resolve the interaction with this receptor and to understand how it will activate gD during entry.

Activation of gD

It is remarkable that gD binding to three structurally unrelated receptors using different binding sites activates gD to trigger membrane fusion (Spear *et al.*, 2006). Prior to solving the structure of gD alone or bound to HVEM, a number of mutagenesis studies identified functional regions on the glycoproteins (Chiang *et al.*, 1994; Spear *et al.*, 2006). One of these regions located at the

C-terminus of the ectodomain, residues 261–316, was particularly interesting since it was required for gD function but was not necessary for binding to HVEM or nectin-1 (Chiang *et al.*, 1994; Nicola *et al.*, 1996a; Rux *et al.*, 1998; Willis *et al.*, 1998). Interestingly, deletion of the C-terminus past residue 250 or mutations between 290–299 resulted in a 50–100-fold increase in affinity of gD for HVEM and nectin-1 (Krummenacher *et al.*, 1998; Rux *et al.*, 1998). This increase is caused by a faster rate of complex formation rather than by the stabilization of the gD-receptor complex (Willis *et al.*, 1998). This suggested that the C-terminus was somewhat hindering the initial interaction with HVEM or nectin-1.

Determining the structure of gD in its native conformation, that is, prior to receptor binding, was essential for understanding how this trigger may act at the atomic level (Krummenacher *et al.*, 2005). The absence of the gD C-terminus increased affinity to both receptors, therefore gD truncated at residue 285 was first used to determine the structures of gD alone and bound to HVEM (Carfi *et al.*, 2001). Comparison of these structures identified the conformational change at the N-terminus associated with HVEM binding. Unfortunately, residues past amino acid 259 are disordered in both structures and the flexible C-terminus could not be solved. The C-terminus of the gD ectodomain could be stabilized using a construct in which a cysteine residue was added at position 307 (Handler *et al.*, 1996; Krummenacher *et al.*, 2005). The resulting protein gD306t$_{307Cys}$ formed a disulfide-linked dimer that likely reflects the gD dimer that can be crosslinked on the viral envelope (Handler *et al.*, 1996; Krummenacher *et al.*, 2005). The structure of the gD306t$_{307Cys}$ protomer revealed the location of the C-terminus in the ectodomain of unbound gD and indicated that the N-terminus was extended and flexible (Fig. 8.2a). After wrapping around the IgV core, this segment makes contacts with the α3 helix and the N-terminus (Fig. 8.2a). Consequently, the C-terminus directly precludes access to both nectin-1 and HVEM binding sites, even though each receptor interacts with a different site on gD. Specifically, the gD C-terminus covers key residues involved in nectin-1 binding such as Y38, R222, and F223 (Fig. 8.2a). Thus it is expected that the C-terminus is moved out of the way during nectin-1 binding. In addition, the C-terminus prevents the formation of the N-terminal hairpin required for HVEM binding. Residues 289–306 in the structure of gD306t$_{307Cys}$ occupy the same location as residues 1–16 of the N-terminal hairpin in the structure of gD285 bound to HVEM. The two conformations of gD are mutually exclusive, therefore the formation of the HVEM-binding site requires displacement of the C-terminus.

Assuming that the 3-O-S-HS and HVEM binding sites are closely related (Yoon and Spear, 2004), it is reasonable to assume that movement of the native C-terminus would be required for binding to 3-O-S-HS as well. Thus, binding of any of the three gD receptors would require that the C-terminus be pushed aside (Fusco *et al.*, 2005; Krummenacher *et al.*, 2005). Consistent with this hypothesis, locking the C-terminus in its native conformation in a mutant form of gD with a disulfide bond gD316(A37C-V302C) prevents binding of HVEM and nectin-1 (Krummenacher *et al.*, 2005). To determine the extent of the C-terminal conformational change that is required for gD function, various artificial disulfide bonds were engineered to lock various portions of the C-terminus on the gD core (Lazear *et al.*, 2008). Locking the C terminus at gD residue 274, 277, or 279 did not interfere with receptor binding but impaired the ability of the mutants to promote fusion. Receptor binding and fusion activity were both lost when the C terminus was locked at gD residue 286, 288, or 302. This shows that two portions of the C terminus must move to allow gD function: residues 286 to 306 for binding and residues at or before residue 279 to trigger fusion. The upstream segment between residues 259 and 268 is highly flexible and has not been resolved in any crystal structure to date. This proline-rich region is important for function (Cocchi *et al.*, 2004; Fusco *et al.*, 2005) and could potentially serve as a hinge that dissociates the whole C-terminus from the gD core in response to receptor binding (Zago *et al.*, 2004).

Although the activation of fusion is induced by the displacement of the C-terminus from the gD core, excessive flexibility of the C-terminus maybe deleterious for gD function in entry. One prominent structural feature of the C-terminus is the native configuration of residues W294 and P291 that inserts into the pocket created by the N-terminus and the α3 helix (Krummenacher *et al.*, 2005). Mutation W294A results in destabilization of the C-terminus native conformation as reflected by a faster rate of binding to HVEM and nectin-1, similar to that of gD lacking residues past amino acid 285. Despite this increased affinity full-length gD(W294A) cannot functionally complement a gD-null virus. This suggests that displacement of the gD C-terminus has to occur at the time of receptor binding to properly activate the fusion machinery. It is not yet known whether destabilization of the C-terminus may lead to a premature irreversible activation of membrane fusion or if it may not be properly presented in the absence of receptor. Interestingly, a membrane proximal PxxW motif is conserved in gD from a number of animal alphaherpesviruses (Krummenacher *et al.*, 2007), including PRV and BHV-1, which also use nectins as receptors. Thus, the alphaherpesviruses appear to have developed a sophisticated receptor-induced triggering of gD to activate fusion. This mechanism can nevertheless be subverted by replacing the HVEM-binding site at the N-terminus of gD by an unrelated receptor-binding site (Zhou *et al.*, 2002). One can envisage that gD can be activated by binding to a new cognate receptor mimicking HVEM. It is surprising that artificial constructs lacking most of the core of gD but keeping the C-terminus can allow entry of HSV-based vectors using new cognate receptors (Menotti *et al.*, 2008; Zhou and Roizman, 2007). No structural data are available to determine if the natural activation mechanism is used to allow entry of these vectors. In any case, it is clear that HSV evolved a tightly controlled mechanism to ensure that activation occurs in a timely manner when the virus is apposed to a cell expressing the adequate receptor on its surface.

Entry pathways: endocytosis and fusion at the plasma membrane

The route of entry of HSV and several herpesviruses from all three families, varies according to the type of the cell they infect (reviewed in Heldwein and Krummenacher, 2008). What dictates the choice between endocytosis and fusion at the plasma membrane is not fully understood. Fusion at the plasma membrane was observed directly in primary sensory neurons by electron microscopy (Lycke *et al.*, 1988) and further documented in Vero and Hep2 cells (Nicola *et al.*, 2003; Wittels and Spear, 1991). Thus, the pH-independent fusion at the plasma membrane was long considered the sole route of HSV infection. However, recent studies showed that HSV is endocytosed in many other cell types (Milne *et al.*, 2005; Nicola *et al.*, 2003, 2005). As is the case for many other viruses, HSV endocytosis is thought to be induced by signals generated by one of the multiple interactions between virions and the cells surface (Cheshenko *et al.*, 2007; MacLeod and Minson, 2010). Several pieces of data suggest that gD receptors may play a role in inducing internalization of the virions. The ability of gD to induce internalization of nectin-1 has been characterized in various ways and showed that direct interaction is required (Stiles and Krummenacher, 2010; Stiles *et al.*, 2008). However, the cytoplasmic tail of nectin-1 is not required for gD-induced endocytosis. This suggests that a yet undetermined molecule that associates with nectin-1 may be involved in mediating endocytosis.

Nectin-1 is actively targeted to an endosomal-lysosomal pathway leading to low-pH dependent degradation. The endocytosis of two isoforms of nectin-1 with alternative cytoplasmic tails occurs rapidly, with kinetics similar to that of virus entry. Studies with virions confirmed the importance of the gD–nectin-1 interaction for virion endocytosis (Milne *et al.*, 2005; Stiles and Krummenacher, 2010; Stiles *et al.*, 2008). For instance, in receptor-deficient mouse melanoma B78H1 cells, virions attach to the cell surface. In these cells, expression of HVEM or nectin-1 leads to virus internalization and successful entry (Milne *et al.*, 2005). Furthermore, entering virions and nectin-1 could be localized in the same vesicle 15 minutes post infection of B78H1 cells expressing nectin-1. In contrast, a gD null virus is not endocytosed. Importantly, blocking endocytosis in B78H1-nectin-1 prevents infection. These data show that interaction of gD with a receptor is essential for endocytosis of virions, which in turn is required for entry into these cells. In another model based on CHO cells, the mutant HSV ANG enters CHO cells via endocytosis when nectin-1 is used, but fuses at the plasma membrane when nectin-2 is used (Delboy *et al.*, 2006). These data also point to a critical role for gD and its receptors in the selection of the entry route, but also suggest that other cellular molecules may act in combination with gD and its receptors to direct the virus to an endocytic pathway. The exact pathway used to internalize HSV is not fully determined but most data point to a macropinocytic route (Clement *et al.*, 2006; Nicola and Straus, 2004).

Thus gD binding to its receptor has a dual role during entry. First, gD activates the endocytosis of its receptor to direct the virus to the endocytic pathway; and second the receptor activates gD to trigger the fusion machinery. Both effects are concomitant; however, during endocytic entry, fusion may occur sometime after gD bound to its receptor. The reason for this delay is unclear but can be explained by the need for other cues to initiate fusion. For instance, endocytic entry of HSV is low pH-dependent in some cells and pH-independent in others (Milne *et al.*, 2005; Nicola and Straus, 2004). Endosomal acidification is required for fusion after endocytosis of HSV in HeLa cells, epithelial cells and some keratinocytes (Nicola *et al.*, 2005). Low pH is not required for entry into B78H1-derived cells or CHO-derived cells that express HVEM or nectin-1 (Milne *et al.*, 2005; Nicola *et al.*, 2003). Why low pH is required in some cells and not others remains a mystery, especially given that the same glycoproteins (gB, gD, gH/gL) are required for both entry pathways (Nicola and Straus, 2004). A gD receptor is always needed, and low pH cannot act as a substitute to trigger the fusion machinery (Nicola and Straus, 2004). *In vitro*, acidic pH favours association of virions with liposomes in the presence of a soluble form of the receptor HVEM (Whitbeck *et al.*, 2006). Thus, in some cells, either gD receptor and low pH may be needed to trigger conformational changes in gB and/or gH/gL.

The structure of gB and its role in HSV entry

gB is the most highly conserved entry glycoprotein amongst all herpesviruses. Although necessary, gB alone is insufficient for either viral entry or for virus-free cell–cell fusion. Most herpesviruses require at least gH/gL in addition to gB for both processes. The only exception to this rule was reported for a mutant form of EBV gB. There, a gB mutant with a truncated cytoplasmic tail was able to mediate fusion of epithelial cells in the absence of gH/gL (McShane and Longnecker, 2004). Additionally, it has been observed that gB alone can mediate membrane fusion during nuclear egress of HSV-1, although egress is far less efficient than when gH/gL is also present (Farnsworth *et al.*, 2007).

HSV-1 gB is a 904-residue protein containing a large ectodomain, a single transmembrane anchor, and a cytoplasmic or intraviral domain (Bzik *et al.*, 1986). Solution of the crystal structure of nearly the entire ectodomain of HSV-1 gB (Heldwein *et al.*, 2006) has been a significant breakthrough in understanding the role of gB in entry (Fig. 8.3). Although early observations of gB mutants with hyperfusogenic phenotypes suggested the involvement of gB in fusion (Bzik *et al.*, 1984), the crystal structure provides the most convincing evidence that gB functions as a viral fusion protein. Specifically, the structure of gB is remarkably similar to that of G, the fusion glycoprotein of vesicular stomatitis virus (VSV) (Roche *et al.*, 2006) and the fusion protein gp64 of baculovirus (Kadlec *et al.*, 2008). These proteins show no sequence similarities, and although the VSV G ectodomain is smaller and more compact than gB, the two proteins share a similar overall topology and the conserved folds of the individual domains. This similarity was unexpected given that these proteins belong to unrelated viruses. The structural homology between HSV gB and VSV G strongly suggests that gB is a viral fusion protein. Additional evidenced that this is the case is provided by mutagenesis studies of putative gB fusion loops (Hannah *et al.*, 2007, 2009).

In the structure, the gB ectodomain is a trimer in which three protomers wrap around each other to form a spike-like molecule with the approximate dimensions of 85Åx80Åx160Å. Similarly shaped spikes of gB have been observed on HSV-1 virions using electron microscopy (Stannard *et al.*, 1987). Each protomer consists of five distinct domains: I, the fusion domain; II, the pleckstrin homology (PH) domain; III, the long

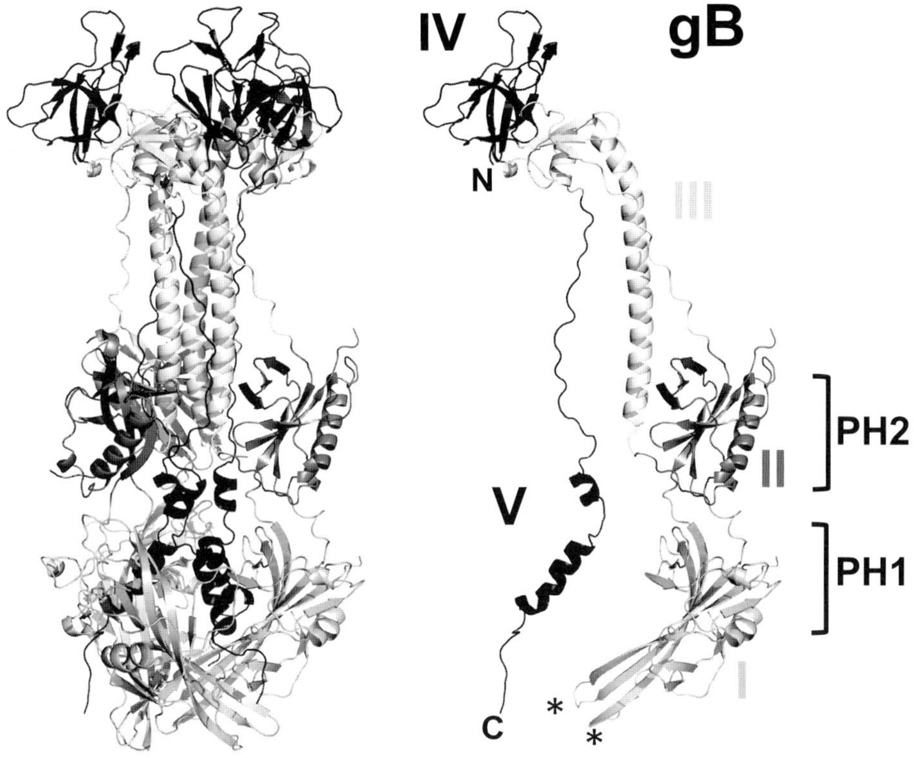

Figure 8.3 Ribbon diagram of the crystal structure of residues 111–725 of the ectodomain of HSV-1 gB. The gB trimer (left) and a single protomer (right) are shown. Each of the 5 structural domains is coloured in a different shade of grey. The two pleckstrin-homology domains are labelled as PH1 and PH2. Asterisks denote the two fusion loops on each protomer.

helical domain; IV, a β-strand domain, and V, a long, linear and partly helical extension. Domain I is located at the base of the gB spike and is proximal to the membrane. Domain I consists of a PH domain and an extended β-sheet containing 2 fusion loops – hydrophobic stretches that can insert into membranes – at the tips of two adjacent finger-like β-hairpins. Domain II is a β-sandwich resembling a PH domain. In cytoplasmic proteins, PH domains bind phospholipids or peptides and function to target these proteins to various cellular compartments (Blomberg *et al.*, 1999; Lemmon and Ferguson, 2000). PH domains in gB have been proposed to serve as protein interaction sites. Domain III consists of a long helix with a globular knob on one end. Three helices within the gB trimer form a triple coiled coil in the protein core. Several β-sheets form a compactly folded domain IV located on the opposite end of the gB spike relative to the putative fusion loops. Three domains IV form a 'crown' on top of the gB molecule. Finally, domain V is an extended ~60-residue-long polypeptide that runs nearly the entire length of the gB molecule and packs into the grooves along the outer side of the central coiled coil. The C-terminus of domain V normally leads into the hydrophobic membrane-proximal region followed by a single transmembrane helix and the cytoplasmic domain. The structure of the gB ectodomain from EBV, another herpesvirus, was subsequently determined (Backovic *et al.*, 2009). Despite the relatively low sequence identity between HSV and EBV gB proteins, the structures look nearly identical, except for relative orientations of domain IV. Thus, the gB fold is likely highly conserved among different herpesviruses and the structures of HSV and EBV gB can be taken to represent gB proteins from other herpesviruses.

HSV gB and VSV G form a new class of viral fusion proteins

Viral fusion proteins enable membrane fusion by undergoing conformational changes that bring the viral envelope and the cell membrane into proximity. Despite differences in architecture, all form trimeric rod-like structures in their postfusion form. Remarkably, all those for which we know both the pre and postfusion form use the same 'fold-back' principle: they bring regions embedded in two prefusion membranes towards one end of a rod-shaped molecule in the postfusion state. This conformational change is thought to drive the fusion of the viral and cell membranes (Harrison, 2008).

Class I fusion proteins are trimers of hairpins that contain a central α-helical coiled coil and N-terminally located fusion peptides (Bullough *et al.*, 1994; Lee *et al.*, 2008; Yin *et al.*, 2005, 2006). Class II fusion proteins are composed of β-structures with internally located fusion loops (Gibbons *et al.*, 2004; Modis *et al.*, 2004). Both HSV gB and VSV G have central α-helical cores like the postfusion trimers of class I fusion proteins and the internal fusion loops of class II fusion proteins. Therefore, together, gB and G define a class III viral fusion proteins (Backovic *et al.*, 2009; Steven and Spear, 2006), which now has another member, gp64 from baculovirus (Kadlec *et al.*, 2008).

Unlike gB, there are now structural data for both prefusion and postfusion forms of VSV G (Roche *et al.*, 2006, 2007). The structure of the gB ectodomain has obvious similarities with the extended postfusion but not the compact prefusion structure of VSV G. It is therefore likely that the available structure of gB represents the postfusion conformation. The prefusion form of gB may be similar to that of VSV G, but so far there hasn't been any confirmation of this model. Importantly, despite obvious similarities in the overall architecture and the folds of the specific domains, the two proteins have significant structural differences that complicate direct modelling of a prefusion conformation of gB, leaving open the possibility that the prefusion form of gB is different from that of VSV G. The two proteins also differ functionally. Unlike VSV G, which binds receptor and accomplishes fusion, gB alone is not sufficient for viral entry; all herpesviruses also require the gH/gL complex and some viruses require additional proteins that bind gH/gL for entry into different cell types (Huber and Compton, 1999; Ryckman *et al.*, 2008). Moreover, the trigger for HSV fusion is, at least in part, the rearrangement of gD induced by receptor binding. In VSV, fusion is triggered by low pH, and the conformational change between the prefusion and the postfusion states is pH-dependent and reversible (Roche *et al.*, 2008). In contrast, it is currently unclear to what extent, the conformation and the function of gB are dependent on pH. For example, the gB ectodomain retains the same overall structure in the crystals obtained under low and high pH (Heldwein *et al.*, 2006). It was recently reported that mildly acidic pH can trigger conformational changes in gB as detected by monoclonal antibodies; an apparent change in the oligomeric state of the protein was also noted (Dollery *et al.*, 2010). One group has proposed a model for a prefusion form of gB, based on that of VSV G (Backovic *et al.*, 2009). However, direct structural information on the prefusion state of gB is needed to understand the fusion-promoting conformational rearrangements in this protein. Understanding the structural basis for the transition of gB from pre- to post-fusion and how it is triggered remains one of the challenges for the future.

Mapping of functional regions on gB with mutants and monoclonal antibodies

gB is the target of many neutralizing antibodies and many investigators have used monoclonal antibodies to map important regions of gB (Atanasiu *et al.*, 2010b; Hannah *et al.*, 2007, 2009; Highlander *et al.*, 1988; Kousoulas *et al.*, 1988; Li *et al.*, 2006; Lin and Spear, 2007; Pereira *et al.*, 1989). Since the structure of gB was solved, epitope mapping efforts were directed at defining functional regions on the structure itself (Bender *et al.*, 2007). In that study, a new expanded panel of monoclonal antibodies to gB was generated and where possible, antibodies from previous studies were incorporated for comparison. Based on this mapping, 4 functional regions (FRs) have been proposed. FR1 encompasses domains I and V and contains the putative fusion loops. FR2 is located on the outward-facing surface of domain II, the PH domain. FR3 maps to the outer ridge of gB 'crown' and contains portions of domains III and IV. Finally, FR4 is composed of the N-terminal unstructured portion of gB, which contains the heparan sulfate binding region. The knowledge of FR locations has simplified the identification of gB regions that are likely involved in interactions with gH/gL as described in more detail further on.

Studies of gB mutants have been a frustrating exercise, which is unsurprising given the complex structure of the trimer (Cai *et al.*, 1987; Kousoulas *et al.*, 1984; Laquerre *et al.*, 1998; Li *et al.*, 2006; Lin and Spear, 2007; Muggeridge, 2000). Once the structure was solved, mutagenesis studies could be carried out in a more rational and targeted fashion. A case in point was construction and analysis of a series of mutants in the putative fusion loops of gB (Hannah *et al.*, 2009; Hannah *et al.*, 2007). In a ribbon rendering of gB (Fig. 8.3a, left), each protomer contains two fusion loops, consisting of three ridges of hydrophobic or non-polar residues (one per protomer) surrounded on each side by charged residues. Each residue in the two loops was changed either to an alanine or to another amino acid residue, e.g. W17R or H263A. Most of the mutants were expressed at wild-type levels on the cell surface. However, in fusion and null-virus complementation assays, substitutions of most polar or charged residues with alanine were functional, while substitutions of non-polar or hydrophobic residues with charged amino acids or alanine were non-functional. The most severely affected non-polar and hydrophobic amino acids were adjacent to each other on the ridge. Their functional importance is clear, but how they accomplish their role in fusion remains to be determined.

It was noted that the ectodomain of wild-type gB but not gD or gH/gL can co-float with rafts on cells (Bender *et al.*, 2003) and with model membranes, e.g. liposomes (Hannah *et al.*, 2009). One presumption was that this interaction might be mediated by the fusion loops of gB as it does with other fusion proteins (Ahn *et al.*, 2002; Hernandez *et al.*, 1997). Three of the four fusion loop mutant proteins failed to do this and the fourth showed an impaired ability to co-float with liposomes (Hannah *et al.*, 2009). The one that showed some activity in the liposome flotation assay also showed some ability to function in cell–cell fusion, though it was null for virus entry (Hannah *et al.*, 2009). Interestingly, the same mutants were recombined back into virus for studies of virion egress from the nucleus. Those mutants that were unable to function in any of the described assays were markedly impaired in egress, at the step of fusion of primary virions with the outer leaflet of the nuclear membrane. They were also severely impaired for cell–cell spread (Wright *et al.*, 2009). Thus, steps of virus entry and egress that involve gB display remarkable similarities. The overall conclusion is that the gB plays a key role in virus–cell and cell–cell fusion events.

gH/gL structure and its role in viral entry

The heterodimeric complex gH/gL is, after gB, the other conserved component of the herpesvirus fusion machinery. gH/gL is also a major target of virus-neutralizing antibodies (Peng *et al.*, 1998), which underscores its importance for virus infection. Until recently, little was known about the role of gH/gL in fusion except that it may have some inherent fusogenic properties. For example, KSHV gH/gL, when expressed in transiently transfected cells, can mediate formation of syncytia in the absence of any other viral proteins, albeit at reduced levels (Pertel, 2002). Similar observations have been made with gH/gL from HCMV and VZV (Cole and Grose, 2003; Kinzler and Compton, 2005). But not all gH/gL proteins have such properties. Also, HSV-1 gH/gL can cause some hemifusion in the absence of gB although the presence of gD and gD receptor is still required for this and gB is needed for full fusion (Subramanian and Geraghty, 2007).

The structure of the ectodomain of HSV-2 gH/gL

HSV-1 gH is an 838 amino-acid glycoprotein with a large ectodomain and a single C-terminal transmembrane anchor; gL is a 224 amino-acid glycoprotein lacking a transmembrane region. Both proteins have signal peptides directing them into the secretory pathway. In HSV-infected cells and on mature virions, gH and gL are always found together, in a stable 1:1 complex (Peng *et al.*, 1998). gL is required for correct folding and trafficking of gH. In HSV, when gH is expressed in the absence of gL, it is retained inside the cell, probably, in the endoplasmic reticulum, lacks important epitopes (Hutchinson *et al.*, 1992; Cairns, 2007) and is not incorporated into the viral envelope (Hutchinson *et al.*, 1992; Roop *et al.*, 1993). Likewise, when gL is expressed in the absence of gH, it has been reported to be secreted from mammalian cells (Dubin and Jiang, 1995; Peng *et al.*, 1998) or retained inside the cells in its immature form (Hutchinson *et al.*, 1992). Recently, the crystal structure of nearly the entire gH ectodomain bound to full-length gL from HSV-2 was determined, revealing a boot-shaped molecule approximately 80 Å × 70 Å (Fig. 8.4 and Chowdary *et al.*, 2010). The C terminus of the gH ectodomain, which would normally lead into its transmembrane region, is located near the 'sole' side of the 'toe'. The unusual shape of the gH/gL structure is reminiscent of the 15–20 nm glycoprotein spikes, observed by cryo electron tomography, which emerge from the HSV-1 envelope at an angle and often appear curved (Grunewald *et al.*, 2003). So, on the virion surface each gH/gL molecule probably appears standing on its 'toe' and leaning to the side.

gH has three domains – the N-terminal domain that binds gL (domain H1), the central helical domain (domain H2), and the C-terminal β-sandwich domain (domain H3) (Fig 8.4). Domain H1 is located in the upper part of the gH/gL 'boot' and contains two subdomains, connected by a linker, missing from the structure. With only ~30% of regular secondary structure, this domain lacks a stable folded core. The central domain H2 is globular and mostly helical. The C-terminal domain H3 is located at the 'toe' end of the 'boot' (Fig 8.4). It is a 10-stranded β-sandwich that looks elongated due to many extended loops. One side of domain H3 has an extension that leads to the transmembrane region. Although gH proteins are conserved among different herpesviruses, the sequence conservation is uneven across the protein (Chowdary *et al.*, 2010). Domain H1 is the most divergent, which is unsurprising given its paucity of regular secondary structure. Domain H2 and especially domain H3 are more conserved and probably have the same fold in different gH proteins. Domain H3 is the most highly conserved and is likely to be functionally important; indeed, several non-functional mutations map to this domain (Galdiero *et al.*, 1997).

The structure reveals extensive interactions between gL and gH. In fact, N-terminal gL interacts almost exclusively with the H1 domain of gH, consistent with previous data (Cairns *et al.*, 2003; Peng *et al.*, 1998). Subdomains H1A and H1B clamp gL like tongs, forming highly complementary interacting surfaces. Like domain H1 of gH, gL lacks a stable structural core, and the two proteins are clearly required for each other's folding, which explains why gH and gL are always found in a complex.

Although gL has been referred to as a chaperone, it is not a chaperone in the true sense of the word. Chaperones assist the non-covalent folding/unfolding and the assembly/disassembly of other macromolecules but do not remain bound to them when the latter perform their normal biological functions (Frydman, 2001). By contrast, the gH/gL complex is very stable, so gL is more of a scaffolding protein for gH, perhaps, with some additional function. Residues Leu166 to Asn224 of gL are missing from the structure (Fig 8.4), but they do not appear critical for fusion in either HSV-1 (Klyachkin *et al.*, 2006) or HSV-2 gL (Chowdary *et al.*, 2010) and are absent from gL proteins in closely related alphaherpesviruses PRV, bovine herpesvirus (BHV), and VZV. Thus, the missing C-terminal region of gL appears to have limited functional importance.

Domain H1 is poorly conserved, and the sequence of gL itself varies substantially among herpesviruses (see Chowdary *et al.*, 2010, for comparisons). It appears that the gH/gL pairs in each herpesvirus have co-evolved to form tight complexes, while presenting quite variable surfaces, perhaps, for binding other viral or cellular proteins. By contrast, domains H2 and H3 of gH are independently folded entities. Being more highly conserved, these domains likely have

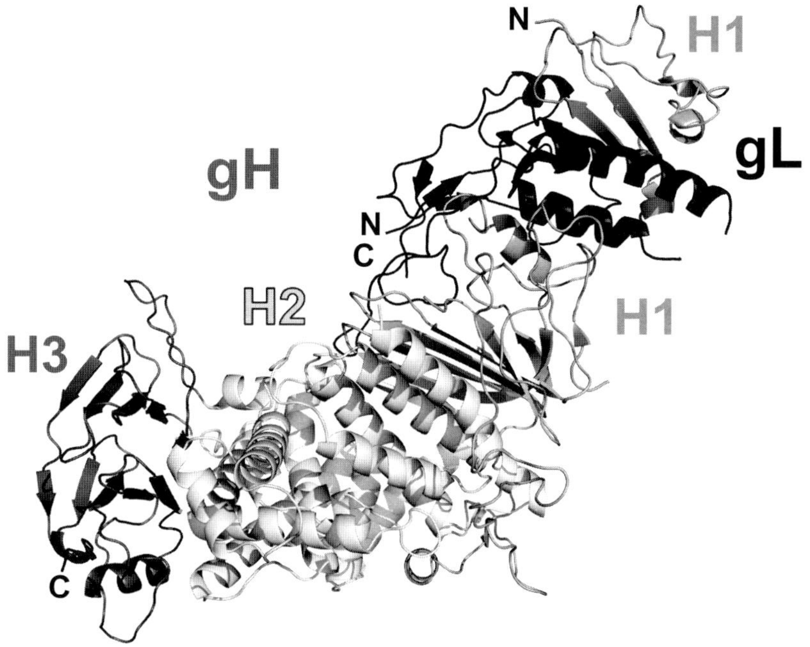

Figure 8.4 Ribbon diagram of the crystal structure of the gH/gL complex from HSV-2. The structure contains nearly the entire ectodomain of gH, residues 48–797, bound to full-length gL. The 3 domains of gH, H1 (light grey), H2 (white) and H3 (dark grey) are labelled. Domain H1 'clamps' gL (black). Chowdary *et al.*, 2010.

similar folds in gH/gL from different herpesviruses and have conserved functional roles.

Several previous reports suggested that gH of a variety of herpesviruses may function as a viral fusion protein (Cole and Grose, 2003; Kinzler and Compton, 2005; Pertel, 2002; Subramanian and Geraghty, 2007). Unexpectedly, the complex revealed by the crystal structure does not resemble any known viral fusion protein. gH has been likened to class I viral fusion proteins based on studies carried out with peptides corresponding to predicted heptad repeats and fusion peptides (Dalla Bona *et al.*, 2006; Galdiero *et al.*, 2005, 2007, 2008b; Gianni *et al.*, 2005, 2006; Lopper and Compton, 2004). Fusion peptides are hydrophobic stretches within viral fusion proteins that are capable of binding to membranes. Several fusion peptides have been proposed in gH, some of which have been shown to bind liposomes as synthetic peptides, with modest affinity (Galdiero *et al.*, 2005, 2007; Gianni *et al.*, 2005). But in the gH/gL structure, these putative fusion peptides are buried helices or β hairpins that participate in multi-stranded sheets and appear unlikely to come out of the hydrophobic core to promote fusion in the context of the entire protein. Moreover, no fusion peptides similar to known fusion peptides of other viral fusion proteins have been identified within the gH/gL structure. As for predicted heptad repeats, these form helices that may stabilize the core of domain H2. But heptad repeats are commonly found in helices that form coiled coils and helical bundles. Although many helices within domain H2 of gH form helical bundles, the trimeric hairpin bundle that is characteristic of the postfusion form of class I and class III fusion proteins (Harrison, 2008) is absent from the gH/gL structure.

An interaction between gB and gH/gL is essential for fusion and entry

Since gB and gH/gL are both needed for fusion, a major puzzle in the field has been to decipher whether they work in tandem as two fusion proteins or as a complex, possibly including gD and receptor, or in some other fashion. Many years ago, it was suggested that entry occurs in a stepwise fashion (Fuller *et al.*, 1989). Later a number of investigators found complexes of gB-gH/gL in lysates of infected and transfected cells (Gianni *et al.*, 2009; Lee and Fuller, 1993), but they may or may not represent the functional unit for fusion.

Recently, a relatively new technique, called bimolecular fluorescence complementation (BiMC) was employed to detect protein interactions in intact transfected cells (Hu and Kerppola, 2003; Kerppola, 2008). In the first studies with HSV glycoproteins it was found that gB and gH/gL can interact with each other, but only when both gD and a gD receptor are also present (Avitabile *et al.*, 2007; Atanasiu *et al.*, 2007). The question then became, is this interaction a by-product of fusion or essential for it to occur? Further investigations led to disparate results and conclusions (Avitabile *et al.*, 2009; Atanasiu *et al.*, 2010b; Chowdary *et al.*, 2010). In one case, the investigators used a mutated form of gB that trafficked in an abnormal manner and showed that this protein interacted with gH/gL in the absence of fusion or gD (Avitabile *et al.*, 2009). Moreover, the cells used for this experiment (293T) were chosen because they over-express proteins after transfection. However, as pointed out by the developers of the assay (Hu and Kerppola, 2003; Kerppola, 2006; Kerppola, 2008), it is essential that the proteins not be over-expressed and that they be expressed in the proper cell compartment to avoid artefacts.

In the second approach, monoclonal antibodies were used to block the interaction between gB and gH/gL that was shown to occur when fusion was triggered by addition of soluble gD to the transfected receptor positive cells (Atanasiu *et al.*, 2010b). The rationale was that if the interaction is essential for fusion/entry, it should be blocked only by antibodies that neutralize HSV infectivity or ones that block fusion. Moreover, if the epitope for that blocking antibody is known, then the interacting site could be mapped. Moreover, some neutralizing antibodies to gB or gH/gL could block fusion at different steps. Using this approach, it was discovered that most of the neutralizing Monoclonal antibodies to gB were able to block fusion. However, only a subset of antibodies also blocked the interaction with gH/gL. These antibodies were previously mapped to FR1 and FR2 of gB which correspond to portions of structural domains II and V (Fig. 8.3 and Bender *et al.*, 2007; Heldwein *et al.*, 2006). Importantly, FR1 contains the fusion loops of gB, while FR2 overlaps structural domain II, one of the two pleckstrin homology domains. Antibodies to FR3 blocked fusion but not the interaction with gH/gL, suggesting that this region, corresponding to structural domain IV (Fig. 8.3), has another function. That FR1 contributes to the interaction is supported by the fact that forms of gB, while with mutations in the fusion loops (Hannah *et al.*, 2007) failed to interact with gH/gL (Atanasiu *et al.*, 2010b). This suggests that insertion of the fusion loops may precede the interaction between gB and gH/gL. On the gH/gL side, fewer reagents were available, particularly ones with virus neutralizing activity. Nevertheless, the potent neutralizing monoclonal antibody LP11 (Gompels and Minson, 1986; Gompels *et al.*, 1991) was able to block both fusion and the interaction with gB (Chowdary *et al.*, 2010). The positions of several monoclonal antibody resistant mutants for LP11 suggest a possible place on gH that interacts with gB, but clearly more data are needed to map this side of the site. Of course, the best way to detail the interaction will be to co-crystallize gB with gH/gL, a challenge for the future. Importantly, in both studies (Atanasiu *et al.*, 2010b; Chowdary *et al.*, 2010), non-neutralizing antibodies failed to block the interaction; only neutralizing antibodies were able to accomplish this. Thus, it is likely that fusion requires an interaction of gB with gH/gL that is dependent on and activated by binding of gD to its receptor. The overall conclusion is that a complex does form, but there is probably a stepwise progression of events from binding of gD to receptor to the end result of fusion.

What is the function of gH/gL?

The overall architecture of the gH/gL complex revealed by the crystal structure does not resemble any known viral fusion proteins. Thus, it has been proposed that instead of having a direct role in fusion, gH/gL plays a role in the transition of gB into its fusion-active state (Chowdary *et al.*, 2010). gH/gL is more likely a positive regulator because gB and gH/gL interact only when gD and a gD receptor are also present (Atanasiu *et al.*, 2007; Avitabile *et al.*, 2007). This implies that gB and gH/gL are not normally associated, which conflicts with the 'negative regulator' model in which gH/gL would function to keep gB inactive. Receptor-bound gD likely acts as a co-activator; it would be interesting to figure out why gD is required in most alphaherpesviruses but not in beta and gammaherpesviruses. Based on its structure, it was hypothesized that gH/gL regulates the transition of gB into its fusion-active state (Chowdary *et al.*, 2010). In a recent study (Atanasiu *et al.*, 2010a) soluble forms of gH/gL plus gD triggered fusion of receptor-bearing cells expressing gB. This is evidence that gB is likely the sole fusogen of HSV and that gH/gL along with gD regulates its activity. Although the mechanism of this activation is yet unknown, it's tempting to speculate that gH/gL could, perhaps, fulfil a role that functionally substitutes for proton binding. Both VSV G and baculovirus gp64 use proton binding to trigger conformational changes involved in fusion (Kadlec *et al.*, 2008; Roche *et al.*, 2006, 2007). If indeed the function of gH/gL is to enable the transition of gB into its proper fusogenic state, it will establish a new paradigm for how viral fusion with cell membranes is accomplished.

Some final thoughts

It is now clear from many studies including structural data about gD, gB and gH/gL (Fig. 8.5) that HSV entry occurs in a stepwise manner that involves a series of highly regulated steps involving protein–protein interactions between the essential glycoproteins gD, gH/gL and gB (Fig. 8.5). So elements of both prevailing models will likely turn out to be correct.

We suggest that first, gD must bind a specific receptor such as nectin-1 or HVEM and undergo a critical set of conformational changes. The changes themselves or residues exposed by those changes then trigger a specific downstream event. Since gH/gL is likely not a fusion protein, we speculate that the activated form of gD binds to a yet undefined region on gH/gL, thereby

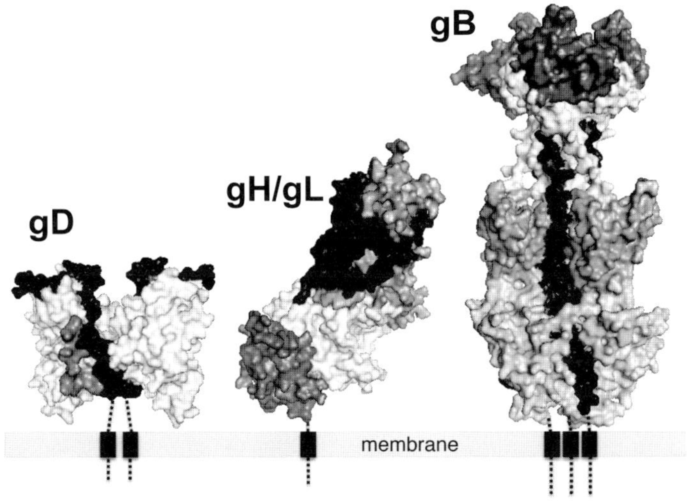

Figure 8.5 Surface representations of the crystal structures of the ectodomains of gD, gH/gL and gB. gD is smaller than gH/gL and gH/gL is more compact than the known structure of gB (likely a post-fusion form). Proteins are coloured by domain using different shades of grey. Each is shown as it may be anchored in the plasma membrane of an infected cell or in the virion envelope. Transmembrane regions and cytoplasmic domains are shown schematically and not to scale with the ectodomains. Relative sizes are consistent with actual size differences, as observed in the crystal structures. Relative glycoprotein orientations may not reflect those on the viral envelope or cell surface.

converting it to a form that binds to gB and activates it into a fusogenic state. This predicts that fusion is carried out exclusively by gB but is doubly regulated. There is good evidence that activation of gB involves an interaction with gH/gL. Is this complex stable in infected cells? Does it include gD? Does it form in an endosome or prior to endocytosis? The interface of the gB–gH/gL complex is a promising target for future efforts to design antiviral compounds. The same is true for the activated form of gD and its viral glycoprotein target. What is the structure of pre-fusion gB and how does it carry out its task of fusion? We suspect that once the structures of pre-fusion gB and of the gB-gH/gL complex are solved, much of the mystery and controversy surrounding the field of HSV entry will be settled. Of course, new questions will undoubtedly emerge from such structural information and from a variety of new studies of entry. However, we posit that the structures that have been solved have already helped to elucidate many aspects of this complex process.

Acknowledgements

The authors thank all colleagues whose names appear on cited articles from previous years. We thank all current members of the laboratories of Drs Heldwein, Krummenacher, Cohen and Eisenberg who have contributed to this chapter in terms of ideas and experiments. Research work from the laboratories of the co-authors of this chapter was supported as follows: U.S. Public Health Service Grants AI-056045 and AI-076231 to RJE; U.S. Public Health Service Grant AI-18289 to GHC; U.S. Public Health Service Grant AI-073384 and The Joseph and Josephine Rabinowitz Award for Research Excellence at the UPENN SDM to CK; and by the NIH grant 1DP20D001996 and Pew Scholar Program in Biomedical Sciences to EEH.

References

Ahn, A., Gibbons, D.L., and Kielian, M. (2002). The fusion peptide of Semliki Forest virus associates with sterol-rich membrane domains. J. Virol. *76*, 3267–3275.

Atanasiu, D., Saw, W.T., Cohen, G.H., and Eisenberg, R.J. (2010a). The cascade of events governing cell–cell fusion induced by HSV glycoproteins gD, gH/gL and gB. J. Virol. 22 September 2010. [Epub ahead of print].

Atanasiu, D., Whitbeck, J., Leon, M.P. d., Lou, H., Hannah, B., Cohen, G., and Eisenberg, R. (2010b). Bimolecular complementation defines functional regions of HSV gB that are involved with gH/gL as necessary steps leading to cell fusion. J. Virol. *84*, 3825–3834.

Atanasiu, D., Whitbeck, J.C., Cairns, T.M., Reilly, B., Cohen, G.H., and Eisenberg, R.J. (2007). Bimolecular complementation reveals that glycoproteins gB and gH/gL of herpes simplex virus interact with each other. during cell fusion. Proc. Natl. Acad. Sci. U.S.A. *104*, 18718–18723.

Avitabile, E., Forghieri, C., and Campadelli-Fiume, G. (2007). Complexes between herpes simplex virus glycoproteins gD, gB, and gH detected in cells by complementation of split enhanced green fluorescent protein. J. Virol. *81*, 11532–11537.

Avitabile, E., Forghieri, C., and Campadelli-Fiume, G. (2009). Cross talking among the glycoproteins involved in herpes simplex virus entry and fusion: the interaction between gB and gH/gL does not necessarily require gD. J. Virol. 83, 10752–10760.

Backovic, M., Longnecker, R., and Jardetzky, T.S. (2009). Structure of a trimeric variant of the Epstein–Barr virus glycoprotein B. Proc. Natl. Acad. Sci. U.S.A. *106*, 2880–2885.

Bender, F.C., Samanta, M., Heldwein, E.E., de Leon, M.P., Bilman, E., Lou, H., Whitbeck, J.C., Eisenberg, R.J., and Cohen, G.H. (2007). Antigenic and mutational analyses of herpes simplex virus glycoprotein B reveal four functional regions. J. Virol. *81*, 3827–3841.

Bender, F.C., Whitbeck, J.C., Leon, M.P. d., Lou, H., Eisenberg, R.J., and Cohen, G.H. (2003). Specific associations of glycoprotein B with lipid rafts during herpes simplex virus entry. J. Virol. *77*, 9542–9552.

Bender, F.C., Whitbeck, J.C., Lou, H., Cohen, G.H., and Eisenberg, R.J. (2005). Herpes simplex virus glycoprotein B binds to cell surfaces independently of heparan sulfate and blocks virus entry. J. Virol. *79*, 11588–11597.

Blomberg, N., Baraldi, E., Nilges, M., and Saraste, M. (1999). The PH superfold: a structural scaffold for multiple functions. Trends Biochem. Sci. *24*, 441–445.

Bullough, P.A., Hughson, F.M., Skehel, J.J., and Wiley, D.C. (1994). Structure of influenza haemagglutinin at the pH of membrane fusion. Nature *371*, 37–43.

Bzik, D.J., DebRoy, C., Fox, B.A., Pederson, N.E., and Person, S. (1986). The nucleotide sequence of the gB glycoprotein gene of HSV-2 and comparison with the corresponding gene of HSV-1. Virology *155*, 322–333.

Bzik, D.J., Fox, B.A., DeLuca, N.A., and Person, S. (1984). Nucleotide sequence of a region of the herpes simplex virus type 1 gB glycoprotein gene: mutations affecting rate of virus entry and cell fusion. Virology *137*, 185–190.

Cai, W., Person, S., Warner, S.C., Zhou, J., and DeLuca, N.A. (1987). Linker-insertion nonsense and restriction-site deletion mutations of the gB glycoprotein gene of herpes simplex virus type 1. J. Virol. *61*, 714–721.

Cairns, T., Milne, R.S.B., Leon, M.P. d., Tobin, D.K., Cohen, G.H., and Eisenberg, R.J. (2003). Structure-function analysis of herpes simplex virus (HSV-1) gD and gH/gL: clues from gD/gH chimeras. J. Virol. *77*, 6731–6742.

Campadelli-Fiume, G., Cocchi, F., Menotti, L., and Lopez, M. (2000). The novel receptors that mediate the entry of herpes simplex viruses and animal alphaherpesviruses into cells. Rev. Med. Virology *10*, 305–319.

Campadelli-Fiume, G., Qi, S., Avitabile, E., Foa-Tomasi, L., Brandimarti, R., and Roizman, B. (1990). Glycoprotein D of herpes simplex virus encodes a domain which precludes penetration of cells expressing the glycoprotein by superinfecting herpes simplex virus. J. Virol. *64*, 6070–6079.

Carfi, A., Willis, S.H., Whitbeck, J.C., Krummenacher, C., Cohen, G.H., Eisenberg, R.J., and Wiley, D.C. (2001). Herpes simplex virus glycoprotein D bound to the human receptor HveA. Molec Cell *8*, 169–179.

Cheshenko, N., Liu, W., Satlin, L.M., and Herold, B.C. (2007). Multiple receptor interactions trigger release of membrane and intracellular calcium stores. critical for herpes simplex virus entry. Mol. Biol. Cell *18*, 3119–3130.

Chiang, H.-Y., Cohen, G.H., and Eisenberg, R.J. (1994). Identification of functional regions of herpes simplex virus glycoprotein gD by using linker-insertion mutagenesis. J. Virol. *68*, 2529–2543.

Chowdary, T., Cairns, T., Atanasiu, D., Cohen, G., Eisenberg, R., and Heldwein, E. (2010). Crystal structure of the gH/gL complex from herpes simplex virus type 2. Submitted for publication. Nat. Struct. Mol. Biol. *17*, 882–888.

Clement, C., Tiwari, V., Scanlan, P.M., Valyi-Nagy, T., Yue, B.Y., and Shukla, D. (2006). A novel role for phagocytosis-like uptake in herpes simplex virus entry. J. Cell Biol. *174*, 1009–1021.

Cocchi, F., Fusco, D., Menotti, L., Gianni, T., Eisenberg, R.J., Cohen, G.H., and Campadelli-Fiume, G. (2004). The soluble ectodomain of herpes simplex virus gD contains a membrane-proximal pro-fusion domain and suffices to mediate virus entry. Proc. Natl. Acad. Sci. U.S.A. *101*, 7445–7450.

Cocchi, F., Lopez, M., Dubreuil, P., Campadelli Fiume, G., and Menotti, L. (2001). Chimeric nectin1-poliovirus receptor molecules identify a nectin1 region functional in herpes simplex virus entry. J. Virol. *75*, 7987–7994.

Cocchi, F., Lopez, M., Menotti, L., Aoubala, M., Dubreuil, P., and Campadelli-Fiume, G. (1998). The V domain of herpesvirus Ig-like receptor (HIgR) contains a major functional region in herpes simplex virus-1 entry into cells and interacts physically with the viral glycoprotein D. Proc. Natl. Acad. Sci. U.S.A. *95*, 15700–15705.

Cole, N.L., and Grose, C. (2003). Membrane fusion mediated by herpesvirus glycoproteins: the paradigm of varicella-zoster virus. Rev. Med. Virology *13*, 207–222.

Compaan, D.M., Gonzalez, L.C., Tom, I., Loyet, K.M., Eaton, D., and Hymowitz, S.G. (2005). Attenuating lymphocyte activity: the crystal structure of the BTLA–HVEM complex. J. Biol. Chem. *280*, 39553–39561.

Connolly, S.A., Landsburg, D.J., Carfi, A., Whitbeck, J.C., Zuo, Y., Wiley, D.C., Cohen, G.H., and Eisenberg, R.J. (2005). Potential nectin-1 binding site on herpes simplex virus glycoprotein d. J. Virol. *79*, 1282–1295.

Connolly, S.A., Landsburg, D.J., Carfi, A., Wiley, D.C., Cohen, G.H., and Eisenberg, R.J. (2003). Structure-based mutagenesis of herpes simplex virus glycoprotein D defines three critical regions at the gD-HveA/HVEM binding interface. J. Virol. *77*, 8127–8140.

Connolly, S.A., Landsburg, D.J., Carfi, A., Wiley, D.C., Eisenberg, R.J., and Cohen, G.H. (2002). Structure-based analysis of the herpes simplex virus glycoprotein D binding site present on herpesvirus entry mediator HveA (HVEM). J. Virol. *76*, 10894–10904.

Dalla Bona, A., Formaggio, F., Peggion, C., Kaptein, B., Broxterman, Q.B., Galdiero, S., Galdiero, M., Vitiello, M., Benedetti, E., and Toniolo, C. (2006). Synthesis, conformation, and bioactivity of novel analogues of the antiviral lipopeptide halovir A. J. Pept Sci. *12*, 748–757.

De Regge, N., Nauwynck, H.J., Geenen, K., Krummenacher, C., Cohen, G.H., Eisenberg, R.J., Mettenleiter, T.C., and Favoreel, H.W. (2006). Alphaherpesvirus gD interaction with sensory neurons triggers formation of varicosities that serve as virus exit sites. J. Cell Biol. *174*, 267–275.

Dean, H.J., Terhune, S.S., Shieh, M.T., Susmarski, N., and Spear, P.G. (1994). Single amino acid substitutions in gD of herpes simplex virus 1 confer resistance to gD-mediated interference and cause cell-type-dependent alterations in infectivity. Virology *199*, 67–80.

Delboy, M.G., Patterson, J.L., Hollander, A.M., and Nicola, A.V. (2006). Nectin-2-mediated entry of a syncytial strain of herpes simplex virus via pH-independent fusion with the plasma membrane of Chinese hamster ovary cells. Virology J. *3*, 105.

Dollery, S.J., Lane, K.D., Delboy, M.G., Roller, D.G., and Nicola, A.V. (2010). Role of the UL45 protein in herpes simplex virus entry via low pH-dependent endocytosis and its relationship to the conformation and function of glycoprotein B. Virus Res. *149*, 115–118.

Dubin, G., and Jiang, H. (1995). Expression of herpes simplex virus type 1 glycoprotein L (gL) in transfected mammalian cells: evidence that gL is not independently anchored to cell membranes. J. Virol. *69*, 4564–4568.

Farnsworth, A., Wisner, T.W., Webb, M., Roller, R., Cohen, G., Eisenberg, R., and Johnson, D.C. (2007). Herpes simplex virus glycoproteins gB and gH function in fusion between the virion envelope and the outer nuclear membrane. Proc. Natl. Acad. Sci. U.S.A. *104*, 10187–10192.

Frydman, J. (2001). Folding of newly translated proteins *in vivo*: the role of molecular chaperones. Annu. Rev. Biochem. *70*, 603–647.

Fuller, A.O., Santos, R.E., and Spear, P.G. (1989). Neutralizing antibodies specific for glycoprotein H of herpes simplex virus permit viral attachment to cells but prevent penetration. J. Virol. *63*, 3435–3443.

Fusco, D., Forghieri, C., and Campadelli-Fiume, G. (2005). The pro-fusion domain of herpes simplex virus glycoprotein D (gD) interacts with the gD N terminus and is displaced by soluble forms of viral receptors. Proc. Natl. Acad. Sci. U.S.A. *102*, 9323–9328.

Galdiero, M., Whiteley, A., Bruun, B., Bell, S., Minson, T., and Browne, H. (1997). Site-directed and linker insertion mutagenesis of herpes simplex virus type 1 glycoprotein H. J. Virol. *71*, 2163–2170.

Galdiero, S., Falanga, A., Vitiello, G., Vitiello, M., Pedone, C., D'Errico, G., and Galdiero, M. (2010a). Role of membranotropic sequences from herpes simplex virus type I glycoproteins B and H in the fusion process. Biochim. Biophys. Acta *1798*, 579–591.

Galdiero, S., Falanga, A., Vitiello, M., Browne, H., Pedone, C., and Galdiero, M. (2005). Fusogenic domains in herpes simplex virus type 1 glycoprotein H. J. Biol. Chem. *280*, 28632–28643.

Galdiero, S., Falanga, A., Vitiello, M., D'Isanto, M., Cantisani, M., Kampanaraki, A., Benedetti, E., Browne, H., and Galdiero, M. (2008a). Peptides containing membrane-interacting motifs inhibit herpes simplex virus type 1 infectivity. Peptides *29*, 1461–1471.

Galdiero, S., Falanga, A., Vitiello, M., D'Isanto, M., Collins, C., Orrei, V., Browne, H., Pedone, C., and Galdiero, M. (2007). Evidence for a role of the membrane-proximal region of herpes simplex virus type 1 glycoprotein h in membrane fusion and virus inhibition. Chem. Biochem. *8*, 885–895.

Galdiero, S., Falanga, A., Vitiello, M., Raiola, L., Fattorusso, R., Browne, H., Pedone, C., Isernia, C., and Galdiero, M. (2008b). Analysis of a membrane interacting region of herpes simplex virus type 1 glycoprotein H. J. Biol. Chem. *283*, 29993–30009.

Galdiero, S., Falanga, A., Vitiello, M., Raiola, L., Russo, L., Pedone, C., Isernia, C., and Galdiero, M. (2010b). The presence of a single N-terminal histidine residue enhances the fusogenic properties of a membranotropic peptide derived from HSV-1 glycoprotein H. J. Biol. Chem. *285*, 17123–17136.

Galdiero, S., Vitiello, M., D'Isanto, M., Falanga, A., Collins, C., Raieta, K., Pedone, C., Browne, H., and Galdiero, M. (2006). Analysis of synthetic peptides from heptad-repeat domains of herpes simplex virus type 1 glycoproteins H and B. J. Gen. Virol. *87*, 1085–1097.

Galen, B., Cheshenko, N., Tuyama, A., Ramratnam, B., and Herold, B.C. (2006). Access to nectin favors herpes simplex virus infection at the apical surface of polarized human epithelial cells. J. Virol. *80*, 12209–12218.

Geraghty, R.J., Krummenacher, C., Cohen, G.H., Eisenberg, R.J., and Spear, P.G. (1998). Entry of alphaherpesviruses mediated by poliovirus receptor-related protein 1 and poliovirus receptor. Science *280*, 1618–1620.

Gianni, T., Amasio, M., and Campadelli-Fiume, G. (2009). Herpes simplex virus gD forms distinct complexes with fusion executors gB and gH/gL in part through the C-terminal profusion domain. J. Biol. Chem. *284*, 17370–17382.

Gianni, T., Menotti, L., and Campadelli-Fiume, G. (2005). A heptad repeat in herpes simplex virus 1 gH, located downstream of the alpha-helix with attributes of a fusion peptide, is critical for virus entry and fusion. J. Virol. *79*, 7042–7049.

Gianni, T., Piccoli, A., Bertucci, C., and Campadelli-Fiume, G. (2006). Heptad repeat 2 in herpes simplex virus 1 gH interacts with heptad repeat 1 and is critical for virus entry and fusion. J. Virol. *80*, 2216–2224.

Gibbons, D.L., Vaney, M.C., Roussel, A., Vigouroux, A., Reilly, B., Lepault, J., Kielian, M., and Rey, F.A. (2004). Conformational change and protein–protein interactions of the fusion protein of Semliki Forest virus. Nature *427*, 320–325.

Gompels, U., and Minson, A. (1986). The properties and sequence of glycoprotein H of herpes simplex virus type 1. Virology *153*, 230–247.

Gompels, U.A., Carss, A.L., Saxby, C., Hancock, D.C., Forrester, A., and Minson, A.C. (1991). Characterization and sequence analyses of antibody-selected antigenic variants of herpes simplex virus show a conformationally complex epitope on glycoprotein H. J. Virol. *65*, 2393–2401.

Grunewald, K., Desai, P., Winkler, D.C., Heymann, J.B., Belnap, D.M., Baumeister, W., and Steven, A.C. (2003). Three-dimensional structure of herpes simplex virus from cryo-electron tomography. Science *302*, 1396–1398.

Handler, C.G., Cohen, G.H., and Eisenberg, R.J. (1996). Cross-linking of glycoprotein oligomers during herpes simplex virus type 1 entry. J. Virol. *70*, 6076–6082.

Hannah, B.P., Cairns, T.M., Bender, F.C., Whitbeck, J.C., Lou, H., Eisenberg, R.J., and Cohen, G.H. (2009). Herpes simplex virus glycoprotein B associates with target membranes via its fusion loops. J. Virol. *83*, 6825–6836.

Hannah, B.P., Heldwein, E.E., Bender, F.C., Cohen, G.H., and Eisenberg, R.J. (2007). Mutational evidence of internal fusion loops in herpes simplex virus glycoprotein B. J. Virol. *81*, 4858–4865.

Harrison, S.C. (2008). Viral membrane fusion. Nat. Struct. Mol. Biol. *15*, 690–698.

Heldwein, E.E., and Krummenacher, C. (2008). Entry of herpesviruses into mammalian cells. Cell. Mol. Life Sci. *65*, 1653–1668.

Heldwein, E.E., Lou, H., Bender, F.C., Cohen, G.H., Eisenberg, R.J., and Harrison, S.C. (2006). Crystal structure of glycoprotein B from herpes simplex virus 1. Science *313*, 217–220.

Hernandez, L.D., Peters, R.J., Delos, S.E., Young, J.A., Agard, D.A., and White, J.M. (1997). Activation of a retroviral membrane fusion protein: soluble receptor-induced liposome binding of the ALSV envelope glycoprotein. J. Cell Biol. *139*, 1455–1464.

Highlander, S.L., Cai, W.H., Person, S., Levine, M., and Glorioso, J.C. (1988). Monoclonal antibodies define a domain on herpes simplex virus glycoprotein B involved in virus penetration. J. Virol. *62*, 1881–1888.

Hu, C.D., and Kerppola, T.K. (2003). Simultaneous visualization of multiple protein interactions in living cells using multicolor fluorescence complementation analysis. Nat. Biotechnol. *21*, 539–545.

Huber, M.T., and Compton, T. (1999). Intracellular formation and processing of the heterotrimeric gH-gL-gO (gCIII) glycoprotein envelope complex of human cytomegalovirus. J. Virol. *73*, 3886–3892.

Huber, M.T., Wisner, T.W., Hegde, N.R., Goldsmith, K.A., Rauch, D.A., Roller, R.J., Krummenacher, C., Eisenberg, R.J., Cohen, G.H., and Johnson, D.C. (2001). Herpes simplex virus with highly reduced gD levels can efficiently enter and spread between human keratinocytes. J. Virol. *75*, 10309–10318.

Hutchinson, L., Browne, H., Wargent, V., Davis-Poynter, N., Primorac, S., Goldsmith, K., Minson, A.C., and Johnson, D.C. (1992). A novel herpes simplex virus glycoprotein, gL, forms a complex with glycoprotein H (gH) and affects normal folding and surface expression of gH. J. Virol. *66*, 2240–2250.

Johnson, D.C., Burke, R.L., and Gregory, T. (1990). Soluble forms of herpes simplex virus glycoprotein D bind to a limited number of cell surface receptors and inhibit virus entry into cells. J. Virol. *64*, 2569–2576.

Kadlec, J., Loureiro, S., Abrescia, N.G., Stuart, D.I., and Jones, I.M. (2008). The postfusion structure of baculovirus gp64 supports a unified view of viral fusion machines. Nat. Struct. Mol. Biol. *15*, 1024–1030.

Kerppola, T.K. (2006). Design and implementation of bimolecular fluorescence complementation (BiFC) assays for the visualization of protein interactions in living cells. Nat. Protoc. *1*, 1278–1286.

Kerppola, T.K. (2008). Bimolecular fluorescence complementation: visualization of molecular interactions in living cells. Methods Cell Biol. *85*, 431–470.

Kinzler, E.R., and Compton, T. (2005). Characterization of human cytomegalovirus glycoprotein-induced cell–cell fusion. J. Virol. *79*, 7827–7837.

Klyachkin, Y.M., Stoops, K.D., and Geraghty, R.J. (2006). Herpes simplex virus type 1 glycoprotein L mutants that fail to promote trafficking of glycoprotein H and fail to function in fusion can induce binding of glycoprotein L-dependent anti-glycoprotein H antibodies. J. Gen. Virol. *87*, 759–767.

Kopp, S.J., Banisadr, G., Glajch, K., Maurer, U.E., Grunewald, K., Miller, R.J., Osten, P., and Spear, P.G. (2009). Infection of neurons and encephalitis after intracranial inoculation of herpes simplex virus requires. the entry receptor nectin-1. Proc. Natl. Acad. Sci. U.S.A. *106*, 17916–17920.

Kousoulas, K.G., Huo, B., and Pereira, L. (1988). Antibody-resistant mutations in cross-reactive and type-specific epitopes of herpes simplex virus 1 glycoprotein B map in separate domains. Virology *166*, 423–431.

Kousoulas, K.G., Pellett, P.E., Pereira, L., and Roizman, B. (1984). Mutations affecting conformation or sequence of neutralizing epitopes identified by reactivity of viable plaques segregate from syn and ts domains of HSV-1(F) gB gene. Virology *135*, 379–394.

Krummenacher, C., Baribaud, F., Ponce De Leon, M., Baribaud, I., Whitbeck, J.C., Xu, R., Cohen, G.H., and Eisenberg, R.J. (2004). Comparative usage of herpesvirus entry mediator A and nectin-1 by laboratory strains and clinical isolates of herpes simplex virus. Virology *322*, 286–299.

Krummenacher, C., Baribaud, I., Ponce de Leon, M., Whitbeck, J.C., Lou, H., Cohen, G.H., and Eisenberg, R.J. (2000). Localization of a binding site for herpes simplex virus glycoprotein D on herpesvirus entry mediator C by using antireceptor monoclonal antibodies. J. Virol. *74*, 10863–10872.

Krummenacher, C., Baribaud, I., Sanzo, J.F., Cohen, G.H., and Eisenberg, R.J. (2002). Effects of herpes simplex virus on structure and function of nectin-1/HveC. J. Virol. *76*, 2424–2433.

Krummenacher, C., Carfi, A., Eisenberg, R.J., and Cohen, G.H. (2007). Entry of herpesviruses into cells: the enigma variations., In Viral Entry into Host Cells, Pohlmann, S., and Simmons, G., eds. (Landes Bioscience).

Krummenacher, C., Nicola, A.V., Whitbeck, J.C., Lou, H., Hou, W., Lambris, J.D., Geraghty, R.J., Spear, P.G., Cohen, G.H., and Eisenberg, R.J. (1998). Herpes simplex virus glycoprotein D can bind to poliovirus receptor-related protein 1 or herpesvirus entry mediator, two structurally unrelated mediators of virus entry. J. Virol. *72*, 7064–7074.

Krummenacher, C., Rux, A.H., Whitbeck, J.C., Ponce-de-Leon, M., Lou, H., Baribaud, I., Hou, W., Zou, C., Geraghty, R.J., Spear, P.G., *et al.* (1999). The first immunoglobulin-like domain of HveC is sufficient to bind herpes simplex virus gD with full affinity, while the third domain is involved in oligomerization of HveC. J. Virol. *73*, 8127–8137.

Krummenacher, C., Supekar, V.M., Whitbeck, J.C., Lazear, E., Connolly, S.A., Eisenberg, R.J., Cohen, G.H., Wiley, D.C., and Carfi, A. (2005). Structure of unliganded HSV gD reveals a mechanism for receptor-mediated activation of virus entry. EMBO J. *24*, 4144–4153.

Laquerre, S., Anderson, D.B., Argnani, R., and Glorioso, J.C. (1998). Herpes simplex virus type 1 glycoprotein B requires. a cysteine residue at position 633 for folding, processing, and incorporation into mature infectious virus particles. J. Virol. *72*, 4940–4949.

Lazear, E., Carfi, A., Whitbeck, J.C., Cairns, T.M., Krummenacher, C., Cohen, G.H., and Eisenberg, R.J. (2008). Engineered disulfide bonds in herpes simplex virus type 1 gD separate receptor binding from fusion initiation and viral entry. J. Virol. *82*, 700–709.

Lee, J.E., Fusco, M.L., Hessell, A.J., Oswald, W.B., Burton, D.R., and Saphire, E.O. (2008). Structure of the Ebola virus glycoprotein bound to an antibody from a human survivor. Nature *454*, 177–182.

Lee, W.C., and Fuller, A.O. (1993). Herpes simplex virus type 1 and pseudorabies virus bind to a common saturable receptor on Vero cells that is not heparan sulfate. J. Virol. *67*, 5088–5097.

Lemmon, M.A., and Ferguson, K.M. (2000). Signal-dependent membrane targeting by pleckstrin homology (PH) domains. Biochem. J. *350 Pt 1*, 1–18.

Li, W., Minova-Foster, T.J., Norton, D.D., and Muggeridge, M.I. (2006). Identification of functional domains in herpes simplex virus 2 glycoprotein B. J. Virol. *80*, 3792–3800.

Ligas, M.W., and Johnson, D.C. (1988). A herpes simplex virus mutant in which glycoprotein D sequences are replaced by β-galactosidase sequences binds to but is unable to penetrate into cells. J. Virol. *62*, 1486–1494.

Lin, E., and Spear, P.G. (2007). Random linker-insertion mutagenesis to identify functional domains of herpes simplex virus type 1 glycoprotein B. Proc. Natl. Acad. Sci. U.S.A. *104*, 13140–13145.

Linehan, M.M., Richman, S., Krummenacher, C., Eisenberg, R.J., Cohen, G.H., and Iwasaki, A. (2004). *In vivo* role of nectin-1 in entry of herpes simplex virus type 1 (HSV-1) and HSV-2 through the vaginal mucosa. J. Virol. *78*, 2530–2536.

Lopez, M., Cocchi, F., Menotti, L., Avitabile, E., Dubreuil, P., and Campadelli-Fiume, G. (2000). Nectin2alpha (PRR2alpha or HveB) and nectin2delta are low-efficiency mediators for entry of herpes simplex virus mutants carrying the Leu25Pro substitution in glycoprotein D. J. Virol. *74*, 1267–1274.

Lopez, M., Eberlé, F., Mattei, M.-G., Gabert, J., Birg, F., Bardin, F., Maroc, C., and Dubreuil, P. (1995). Complementary DNA characterization and chromosomal localization of a human gene related to the poliovirus receptor-encoding gene. Gene *155*, 261–265.

Lopper, M., and Compton, T. (2004). Coiled-coil domains in glycoproteins B and H are involved in human cytomegalovirus membrane fusion. J. Virol. *78*, 8333–8341.

Lycke, E., Hamark, B., Johansson, M., Krotochwil, A., Lycke, J., and Svennerholm, B. (1988). Herpes simplex virus infection of the human sensory neuron. An electron microscopy study. Arch. Virol. *101*, 87–104.

MacLeod, I.J., and Minson, T. (2010). Binding of herpes simplex virus type-1 virions leads to the induction of intracellular signalling in the absence of virus entry. PLoS One 5, e9560.

Manoj, S., Jogger, C.R., Myscofski, D., Yoon, M., and Spear, P.G. (2004). Mutations in herpes simplex virus glycoprotein D that prevent cell entry via nectins and alter cell tropism. Proc. Natl. Acad. Sci. U.S.A. *101*, 12414–12421.

Martinez, W.M., and Spear, P.G. (2002). Amino acid substitutions in the V domain of nectin-1 (HveC) that impair entry activity for herpes simplex virus types 1 and 2 but not for Pseudorabies virus or bovine herpesvirus 1. J. Virol. *76*, 7255–7262.

McShane, M.P., and Longnecker, R. (2004). Cell-surface expression of a mutated Epstein–Barr virus glycoprotein B allows fusion independent of other. viral proteins. Proc. Natl. Acad. Sci. U.S.A. *101*, 17474–17479.

Menotti, L., Cerretani, A., Hengel, H., and Campadelli-Fiume, G. (2008). Construction of a fully retargeted herpes simplex virus 1 recombinant capable of entering cells solely via human epidermal growth factor receptor 2. J. Virol. *82*, 10153–10161.

Milne, R.S., Connolly, S.A., Krummenacher, C., Eisenberg, R.J., and Cohen, G.H. (2001). Porcine HveC, a member of the highly conserved HveC/nectin 1 family, is a functional alphaherpesvirus receptor. Virology *281*, 315–328.

Milne, R.S., Nicola, A.V., Whitbeck, J.C., Eisenberg, R.J., and Cohen, G.H. (2005). Glycoprotein D receptor-dependent, low-pH-independent endocytic entry of herpes simplex virus type 1. J. Virol. *79*, 6655–6663.

Mizoguchi, A., Nakanishi, H., Kimura, K., Matsubara, K., Ozaki-Kuroda, K., Katata, T., Honda, T., Kiyohara, Y., Heo, K., Higashi, M., *et al.* (2002). Nectin: an adhesion molecule involved in formation of synapses. J. Cell Biol. *156*, 555–565.

Modis, Y., Ogata, S., Clements, D., and Harrison, S.C. (2004). Structure of the dengue virus envelope protein after membrane fusion. Nature *427*, 313–319.

Montgomery, R.I., Warner, M.S., Lum, B.J., and Spear, P.G. (1996). Herpes simplex virus-1 entry into cells mediated by a novel member of the TNF/NGF receptor family. Cell *87*, 427–436.

Muggeridge, M.I. (2000). Characterization of cell–cell fusion mediated by herpes simplex virus 2 glycoproteins gB, gD, gH and gL in transfected cells. J. Gen. Virol. *81*, 2017–2027.

Muggeridge, M.I., Wu, T.-T., Johnson, D.C., Glorioso, J.C., Eisenberg, R.J., and Cohen, G.H. (1990). Antigenic and functional analysis of a neutralization site of HSV-1 glycoprotein D. Virology *174*, 375–387.

Murphy, K.M., Nelson, C.A., and Sedy, J.R. (2006). Balancing co-stimulation and inhibition with BTLA and HVEM. Nat. Rev. Immunol. *6*, 671–681.

Nicola, A.V., Cohen, G.H., and Eisenberg, R.J. (1996a). Inhibition of virus entry by soluble HSV gD may be analogous to superinfection restriction by gD-expressing cells., Paper presented at: 15th Annual Meeting, American Society for Virology (London, Ontario, Canada).

Nicola, A.V., Hou, J., Major, E.O., and Straus, S.E. (2005). Herpes simplex virus type 1 enters human epidermal keratinocytes, but not neurons, via a pH-dependent endocytic pathway. J. Virol. *79*, 7609–7616.

Nicola, A.V., McEvoy, A.M., and Straus, S.E. (2003). Roles for endocytosis and low pH in herpes simplex virus entry into HeLa and Chinese hamster ovary cells. J. Virol. *77*, 5324–5332.

Nicola, A.V., and Straus, S.E. (2000). Stable binding of HSV-1 to cells mediated by HveA or HveC., Paper presented at: 25th International Herpesvirus Workshop (Portland, OR, Abst. 2.40).

Nicola, A.V., and Straus, S.E. (2004). Cellular and viral requirements for rapid endocytic entry of herpes simplex virus. J. Virol. *78*, 7508–7517.

Nicola, A.V., Willis, S.H., Naidoo, N.N., Eisenberg, R.J., and Cohen, G.H. (1996b). Structure-function analysis of soluble forms of herpes simplex virus glycoprotein D. J. Virol. *70*, 3815–3822.

Peng, T., Ponce de Leon, M., Jiang, H., Dubin, G., Lubinski, J., Eisenberg, R.J., and Cohen, G.H. (1998). The gH–gL complex of herpes simplex virus (HSV) stimulates neutralizing antibody and protects mice against HSV type 1 challenge. J. Virol. *72*, 65–72.

Pereira, L., Ali, M., Kousoulas, K., Huo, B., and Banks, T. (1989). Domain structure of herpes simplex virus 1 glycoprotein B: neutralizing epitopes map in regions of continuous and discontinuous residues. Virology *172*, 11–24.

Pertel, P.E. (2002). Human herpesvirus 8 glycoprotein B (gB), gH, and gL can mediate cell fusion. J. Virol. *76*, 4390–4400.
Roche, S., Albertini, A.A., Lepault, J., Bressanelli, S., and Gaudin, Y. (2008). Structures. of vesicular stomatitis virus glycoprotein: membrane fusion revisited. Cell. Mol. Life Sci. *65*, 1716–1728.
Roche, S., Bressanelli, S., Rey, F.A., and Gaudin, Y. (2006). Crystal structure of the low-pH form of the vesicular stomatitis virus glycoprotein G. Science *313*, 187–191.
Roche, S., Rey, F.A., Gaudin, Y., and Bressanelli, S. (2007). Structure of the prefusion form of the vesicular stomatitis virus glycoprotein G. Science *315*, 843–848.
Roop, C., Hutchinson, L., and Johnson, D.C. (1993). A mutant herpes simplex virus type 1 unable to express glycoprotein L cannot enter cells, and its particles lack glycoprotein H. J. Virol. *67*, 2285–2297.
Rux, A.H., Willis, S.H., Nicola, A.V., Hou, W., Peng, C., Lou, H., Cohen, G.H., and Eisenberg, R.J. (1998). Functional region IV of glycoprotein D from herpes simplex virus modulates glycoprotein binding to the herpesvirus entry mediator. J. Virol. *72*, 7091–7098.
Ryckman, B.J., Rainish, B.L., Chase, M.C., Borton, J.A., Nelson, J.A., Jarvis, M.A., and Johnson, D.C. (2008). Characterization of the human cytomegalovirus gH/gL/UL128–131 complex that mediates entry into epithelial and endothelial cells. J. Virol. *82*, 60–70.
Sakisaka, T., Taniguchi, T., Nakanishi, H., Takahashi, K., Miyahara, M., Ikeda, W., Yokoyama, S., Peng, Y.F., Yamanishi, K., and Takai, Y. (2001). Requirement of interaction of nectin-1alpha/HveC with afadin for efficient cell–cell spread of herpes simplex virus type 1. J. Virol. *75*, 4734–4743.
Satoh, T., and Arase, H. (2008). HSV-1 infection through inhibitory receptor, PILRalpha. Uirusu *58*, 27–36.
Satoh, T., Arii, J., Suenaga, T., Wang, J., Kogure, A., Uehori, J., Arase, N., Shiratori, I., Tanaka, S., Kawaguchi, Y., *et al.* (2008). PILRalpha is a herpes simplex virus-1 entry coreceptor that associates with glycoprotein B. Cell *132*, 935–944.
Schelhaas, M., Jansen, M., Haase, I., and Knebel-Morsdorf, D. (2003). Herpes simplex virus type 1 exhibits a tropism for basal entry in polarized epithelial cells. J. Gen. Virol. *84*, 2473–2484.
Shukla, D., Liu, J., Blaiklock, P., Shworak, N.W., Bai, X., Esko, J.D., Cohen, G.H., Eisenberg, R.J., Rosenberg, R.D., and Spear, P.G. (1999). A novel role for 3-O-sulfated heparan sulfate in herpes simplex virus 1 entry. Cell *99*, 13–22.
Simpson, S.A., Manchak, M.D., Hager, E.J., Krummenacher, C., Whitbeck, J.C., Levin, M.J., Freed, C.R., Wilcox, C.L., Cohen, G.H., Eisenberg, R.J., and Pizer, L.I. (2005). Nectin-1/HveC Mediates herpes simplex virus type 1 entry into primary human sensory neurons and fibroblasts. J. NeuroVirol. *11*, 208–218.
Spear, P.G. (1993). Membrane fusion induced by herpes simplex virus., In Viral fusion mechanisms, Bentz, J., ed. (Boca Raton, Fla.: CRC Press, Inc.), pp. 201–232.
Spear, P.G. (2004). Herpes simplex virus: receptors and ligands for cell entry. Cell. Microbiol. *6*, 401–410.
Spear, P.G., Eisenberg, R.J., and Cohen, G.H. (2000a). Three classes of cell surface receptors for alphaherpesvirus entry. Virology *275*, 1–8.
Spear, P.G., Eisenberg, R.J., and Cohen, G.H. (2000b). Three classes of cell surface receptors for alphaherpesvirus entry. Virology *275*, 1–8.
Spear, P.G., and Longnecker, R. (2003). Herpesvirus entry: an update. J. Virol. Methods *77*, 10179–10185.
Spear, P.G., Manoj, S., Yoon, M., Jogger, C.R., Zago, A., and Myscofski, D. (2006). Different receptors binding to distinct interfaces on herpes simplex virus gD can trigger events leading to cell fusion and viral entry. Virology *344*, 17–24.
Stannard, L.M., Fuller, A.O., and Spear, P.G. (1987). Herpes simplex virus glycoproteins associated with different morphological entities projecting from the virion envelope. J. Gen. Virol. *68*, 715–725.
Steven, A.C., and Spear, P.G. (2006). Viral glycoproteins and an evolutionary conundrum. Science *313*, 177–178.
Stiles, K.M., and Krummenacher, C. (2010). Glycoprotein D actively induces rapid internalization of two nectin-1 isoforms during herpes simplex virus entry. Virology *399*, 109–119.
Stiles, K.M., Milne, R.S., Cohen, G.H., Eisenberg, R.J., and Krummenacher, C. (2008). The herpes simplex virus receptor nectin-1 is down-regulated after trans-interaction with glycoprotein D. Virology *373*, 98–111.
Subramanian, R.P., and Geraghty, R.J. (2007). Herpes simplex virus type 1 mediates fusion through a hemifusion intermediate by sequential activity of glycoproteins D, H, L, and B. Proc. Natl. Acad. Sci. U.S.A. *104*, 2903–2908.
Takahashi, K., Nakanishi, H., Miyahara, M., Mandai, K., Satoh, K., Satoh, A., Nishioka, H., Aoki, J., Nomoto, A., Mizoguchi, A., and Takai, Y. (1999). Nectin/PRR: an immunoglobulin-like cell adhesion molecule recruited to cadherin-based adherens junctions through interaction with afadin, a PDZ domain-containing protein. J. Cell Biol. *145*, 539–549.
Takai, Y., Ikeda, W., Ogita, H., and Rikitake, Y. (2008). The immunoglobulin-like cell adhesion molecule nectin and its associated protein afadin. Annu Rev. Cell Dev. Biol. *24*, 309–342.
Taylor, J.M., Lin, E., Susmarski, N., Yoon, M., Zago, A., Ware, C.F., Pfeffer, K., Miyoshi, J., Takai, Y., and Spear, P.G. (2007). Alternative entry receptors for herpes simplex virus and their roles in disease. Cell Host Microbe *2*, 19–28.
Tiwari, V., Oh, M.J., Kovacs, M., Shukla, S.Y., Valyi-Nagy, T., and Shukla, D. (2008). Role for nectin-1 in herpes simplex virus 1 entry and spread in human retinal pigment epithelial cells. FEBS J.. 275, 5272–5285.
Uchida, H., Shah, W.A., Ozuer, A., Frampton, A.R., Jr., Goins, W.F., Grandi, P., Cohen, J.B., and Glorioso, J.C. (2009). Generation of HVEM-restricted HSV-1 mutant viruses: Resistance of HVEM-expressing cells and identification of mutations that rescue nectin-1 recognition. J. Virol.

Warner, M.S., Geraghty, R.J., Martinez, W.M., Montgomery, R.I., Whitbeck, J.C., Xu, R., Eisenberg, R.J., Cohen, G.H., and Spear, P.G. (1998). A cell surface protein with herpesvirus entry activity (HveB) confers susceptibility to infection by mutants of herpes simplex virus type 1, herpes simplex virus type 2, and pseudorabies virus. Virology *246*, 179–189.

Whitbeck, J.C., Muggeridge, M.I., Rux, A.H., Hou, W., Krummenacher, C., Lou, H., van Geelen, A., Eisenberg, R.J., and Cohen, G.H. (1999). The major neutralizing antigenic site on herpes simplex virus glycoprotein D overlaps a receptor-binding domain. J. Virol. *73*, 9879–9890.

Whitbeck, J.C., Zuo, Y., Milne, R.S.B., Cohen, G.H., and Eisenberg, R.J. (2006). Stable association of herpes simplex virus with target membrane is triggered by low pH in the presence of the gD receptor HVEM. J. Virol. *80*, 3773–3780.

Willis, S.H., Rux, A.H., Peng, C., Whitbeck, J.C., Nicola, A.V., Lou, H., Hou, W., Salvador, L., Cohen, G.H., and Eisenberg, R.J. (1998). Examination of the kinetics of herpes simplex virus glycoprotein D binding to the herpesvirus entry mediator, using surface plasmon resonance. J. Virol. *72*, 5937–5947.

Wittels, M., and Spear, P.G. (1991). Penetration of cells by herpes simplex virus does not require a low pH-dependent endocytic pathway. Virus Res. *18*, 271–290.

Wright, C.C., Wisner, T.W., Hannah, B.P., Eisenberg, R.J., Cohen, G.H., and Johnson, D.C. (2009). Fusion between perinuclear virions and the outer nuclear membrane requires. the fusogenic activity of herpes simplex virus gB. J. Virol. *83*, 11847–11856.

Yin, H.S., Paterson, R.G., Wen, X., Lamb, R.A., and Jardetzky, T.S. (2005). Structure of the uncleaved ectodomain of the paramyxovirus (hPIV3) fusion protein. Proc. Natl. Acad. Sci. U.S.A. *102*, 9288–9293.

Yin, H.S., Wen, X., Paterson, R.G., Lamb, R.A., and Jardetzky, T.S. (2006). Structure of the parainfluenza virus 5 F protein in its metastable, prefusion conformation. Nature *439*, 38–44.

Yoon, M., and Spear, P.G. (2002). Disruption of adherens junctions liberates nectin-1 to serve as receptor for herpes simplex virus and pseudorabies virus entry. J. Virol. *76*, 7203–7208.

Yoon, M., and Spear, P.G. (2004). Random mutagenesis of the gene encoding a viral ligand for multiple cell entry receptors to obtain viral mutants altered for receptor usage. Proc. Natl. Acad. Sci. U.S.A. *101*, 17252–17257.

Zago, A., Jogger, C.R., and Spear, P.G. (2004). Use of herpes simplex virus and pseudorabies virus chimeric glycoprotein D molecules to identify regions critical for membrane fusion. Proc. Natl Acad. Sci. U.S.A 101, 17498–17503.

Zhou, G., and Roizman, B. (2007). Separation of receptor-binding and profusogenic domains of glycoprotein D of herpes simplex virus 1 into distinct interacting proteins. Proc. Natl. Acad. Sci. U.S.A. *104*, 4142–4146.

Zhou, G., Ye, G.J., Debinski, W., and Roizman, B. (2002). Engineered herpes simplex virus 1 is dependent on IL13Ralpha 2 receptor for cell entry and independent of glycoprotein D receptor interaction. Proc. Natl. Acad. Sci. U.S.A. *99*, 15124–15129.

Structure–Function Profiles of Nine Varicella-zoster Virus Glycoproteins: Endocytosis, Entry and Egress

9

Charles Grose, Susan Vleck, Odd Andre Karlsen and Eduardo A. Montalvo

Abstract

Varicella zoster virus has a smaller genome than herpes simplex virus and therefore encodes fewer glycoproteins. In this chapter, nine VZV glycoproteins are profiled, including gE, gI, gC, gH, gL, gB, gK, gM, and gN. Although all VZV glycoproteins have HSV homologues, functions occasionally have greatly shifted. For example, VZV gE is the predominant VZV glycoprotein and exists as a monomer, dimer and trimer, as well as a gE/gI complex. VZV gE is essential, unlike HSV gE. Even though essential, mutations in gE had been detected in wild type VZV strains that exhibit an accelerated cell-spread phenotype. The VZV gC glycoprotein differs from HSV gC in that both transcription and translation of VZV gC are remarkably delayed in cultured cells; often VZV gC protein is difficult to detect altogether. The VZV gH/gL complex resembles the HSV gH/gL complex is that both are critical for virus induced fusion. Fusion is a prominent feature of VZV infected cells. Neutralization antibody to VZV gH dramatically reduces the spread of virus and limits pathogenesis in the skin. The VZV gB glycoprotein is also involved in virus-induced fusion. Of interest, four VZV glycoproteins (gE, gI, gH and gB) have functional endocytosis motifs in their cytoplasmic tail. Thus, all four are internalized from the cell surface in clathrin coated vesicles. This pathway appears critical for the process of virion envelopment in the assembly compartment. Even though abundant amounts of most glycoproteins are produced in cell culture, assembly of fully enveloped and infectious VZV particles rarely occurs. The particle–plaque forming unit ratio remains an extremely high 40,000:1. Likewise, the aberrant assembly process severely limits any assessment of egress mechanisms.

Introduction

The VZV genome is the smallest among the human herpesviruses. In one sense, therefore, VZV is an evolutionarily minimalist herpesvirus. The genome consists of a linear double stranded DNA molecule 125 kbp in length, which was sequenced in its entirety in 1986 (Davison and Scott, 1986). The genome includes a unique long region flanked by both internal and terminal repeated regions, and a unique short region flanked by repeated regions. Two isomeric forms of predominate by virtue of inversion of the short unique DNA segment and its flanking repeated elements. The viral genome encodes at least 70 open reading frames (ORFs), which are enumerated numerically from 1–70.

The VZV genome specifies several glycoproteins, which are now designated according to the name given to the HSV homologous protein (Grose, 1990a). Since VZV has a smaller genome than other herpesviruses, it has fewer glycoproteins but does code for at least nine glycoproteins (Storlie *et al.*, 2008b). Among these ORFs, only gI and gE are found in the unique short segment, where they are designated as genes 67 and 68, respectively. The genes in the short segment are of particular interest, because this segment is considered the most recent addition in evolutionary time and therefore may represent that portion of the genome which determines the particular niche of a herpesvirus, i.e. the gene products in the unique short segment provide attributes which give VZV its distinguishing properties. Alternatively, in the case of VZV, the niche may be established by the lack of a glycoprotein found in other herpesviruses. Furthermore, we have also uncovered remarkable differences in the expression of some VZV glycoproteins in the long segment, when compared with those of herpes simplex virus (HSV) type 1. This chapter provides an update on the functions of nine VZV glycoproteins.

VZV glycoprotein E (ORF 68)

VZV gE structure

Based on the sequence and structure of gE, it is designated a typical type 1 transmembrane glycoprotein which is 623 amino acids in length with a 24 amino acid cleavable signal sequence (Fig. 9.1). The glycoprotein consists of three major regions: a 544 amino acid hydrophilic extracellular region, a 17 amino acid hydrophobic transmembrane region and a 62 amino acid charged cytoplasmic tail. Examination of the predicted amino acid sequence of gE ectodomain reveals two cysteine-rich regions, three N-linked glycosylation signals and a juxtamembrane domain favouring O-linked glycans (Montalvo *et al.*, 1985). VZV gE is the predominant glycoprotein in an infected cell culture (Weigle and Grose, 1983). As a comparison, VZV gE shows 47% amino acid similarity and 27% amino acid identity with its alpha herpes viral homologue HSV-1 gE. The gE protein has been subjected to

VZV ORF 68 gE

```
  1 MGTVNKPVVG VLMGFGIITG TLRITNPVRA SVLRYDDFHT DEDKLDTNSV
 51 YEPYYHSDHA ESSWVNRGES SRKAYDHNSP YIWPRNDYDG FLENAHEHHG
101 VYNQGRGIDS GERLMQPTQM SAQEDLGDDT GIHVIPTLNG DDRHKIVNVD
151 QRQYGDVFKG DLNPKPQGQR LIEVSVEENH PFTLRAPIQR IYGVRYTETW
201 SFLPSLTCTG DAAPAIQHIC LKHTTCFQDV VVDVDCAENT KEDQLAEISY
251 RFQGKKEADQ PWIVVNTSTL FDELELDPPE IEPGVLKVLR TEKQYLGVYI
301 WNMRGSDGTS TYATFLVTWK GDEKTRNPTP AVTPQPRGAE FHMWNYHSHV
351 FSVGDTFSLA MHLQYKIHEA PFDLLLEWLY VPIDPTCQPM RLYSTCLYHP
401 NAPQCLSHMN SGCTFTSPHL AQRVASTVYQ NCEHADNYTA YCLGISHMEP
451 SFGLILHDGG TTLKFVDTPE SLSGLYVFVV YFNGHVEAVA YTVVSTVDHF
501 VNAIEERGFP PTAGQPPATT KPKEITPVNP GTSPLLRYAA WTGGLAAVVL
551 LCLVIFLICT AKRMRVKAYR VDKSPYNQSM YYAGLPVDDF EDSESTDTEE
601 EFGNAIGGSH GGSSYTVYID KTR
```

VZV ORF 67 gI

```
  1 MFLIHCLISA VIFYIQVTNA LIFKGDHVSL QVNSSLTSIL IPMQNDNYTE
 51 IKGQLVFIGE QLPTGTNYSG TLELLYADTV AFCFRSVQVI RYDGCPRIRT
101 SAFISCRYKH SWHYGNSTDR ISTEPDAGVM LKITKPGIND AGVYVLLVRL
151 DHSRSTDGFI LGVNVYTAGS HHNIHGVIYT SPSLQNGYST RALFQQARLC
201 DLPATPKGSG TSLFQHMLDL RAGKSLEDNP WLHEDVVTTE TKSVVKEGIE
251 NHVYPTDMST LPEKSLNDPP ENLLIIIPIV ASVMILTAMV IVIVISVKRR
301 RIKKHPIYRP NTKTRRGIQN ATPESDVMLE AAIAQLATIR EESPPHSVVN
351 PFVK
```

Figure 9.1 Amino acid sequences of VZV gE and gI. VZV gE is the predominant VZV glycoprotein, which frequently forms a complex with gI. The nucleotide and deduced amino acids sequences were published by Peters *et al.* (2006).

2-dimensional peptide analysis (Fig. 9.2). These gels confirm a prior observation that gE likely exists as monomer, dimer and trimer within infected cells (Olson *et al.*, 1997).

VZV gE glycosylation, palmitoylation and myristoylation

Like other type 1 glycoproteins, the gE molecule is formed through a series of intermediary products as the protein traffics from the endoplasmic reticulum through the Golgi apparatus en route to the plasma membrane. The M_r of the polypeptide backbone has been estimated to be 73 kDa. Formation of mature gE results in a protein with a molecular mass of 98kDa, hence the original designation of gE as gp98. The presence of both high-mannose and complex type glycans on gE, as well as O-linked glycans, has been well documented; thus, the potential N-linked Ser-Xxx-Asp sites and the potential O-linked Ser-Thr-Pro sites on the ectodomain are utilized (Grose, 1990a; Montalvo *et al.*, 1985). Further gE is heavily sialylated and sulfated. Other gE modifications include the attachment of palmitic acid to cysteine residues and myristic acid to an N-terminal glycine residue (Harper and Kangro, 1990).

VZV gE serine/threonine phosphorylation

The gE glycoprotein is heavily phosphorylated (Montalvo and Grose, 1986b). The gE endodomain contains a phosphorylation consensus sequence for casein kinase II, which is critical for the important trafficking event of endocytosis (Grose *et al.*, 1989). Casein kinase II is a serine/threonine protein kinase with properties shared by few other kinases (Tuazon and Traugh, 1991). First, casein kinase II is capable of utilizing GTP almost as well as ATP as a phosphate; secondly, casein kinase II activity is potently inhibited by heparin. The kinase assay was initiated by the addition of either ^{32}P-ATP and ^{32}P-GTP. These results showed that the kinase coprecipitated with its substrate, namely gE; in turn, the same kinase phosphorylated gE in an *in vitro* protein kinase assay (Olson *et al.*, 1997). When heparin was added to the protein kinase assay, phosphorylation of gE was greatly diminished. These results indicate that casein kinase II present in human cells can phosphorylate VZV gE.

The consensus motif for casein kinase II phosphorylation from residues 590–602 contains the following amino acids:

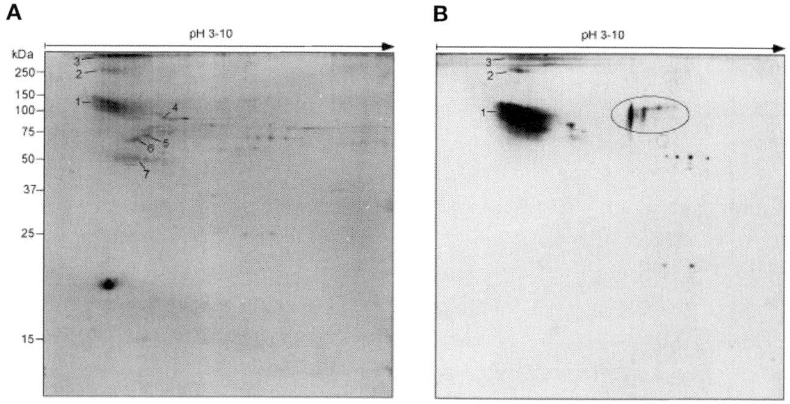

Figure 9.2 Two-dimensional gel electrophoresis of VZV gE purified by immunoaffinity chromatography. (A) Proteins were resolved with isoelectric focusing in a pH 3–10 IPG-strip prior to the second dimension electrophoresis in a 12.5% polyacrylamide gel. Proteins were then visualized with silver staining. (B) Immunoblot of a 2D-gel prepared as in (A), using monoclonal antibody 3B3 directed against gE. The labelled spots in gel A were excised from a parallel SYPRO-Ruby stained 2D-gel and subjected to MALDI-TOF mass spectrometry analysis. They were identified as follows: 1=gE; 2=gE; 3=gE; 4= heat shock protein 70 (Hsp 70); 5=Alpha tubulin (vimentin); 6=Beta tubulin; 7=Actin. The gE protein was detected as a monomer (1), a dimer (2) and possibly also a trimer (3). The encircled spots on the protein immunoblot (B) are most likely precursor forms of gE.

Phe-Glu-Asp-Ser-glu-Ser-thr-Asp-Thr-Glu-glu-Glu-Phe (Fig. 9.3). The notable phosphorylatable serine and threonine residues are surrounded by acidic residues. To determine whether this sequence was the phosphorylation site for casein kinase II, the gE sequence was mutated and the phosphorylation of the gE mutant protein was analysed (Olson *et al.*, 1997; Yao *et al.*, 1992). The two serine residues at 593 and 595 were changed to alanine and phenylalanine respectively and the two threonine residues at 595 and 598 were both changed to alanines. The gE construct with 4 altered amino acids was designated gE-AFAA. When analysed by laser scanning confocal microscopy, the gE-AFAA protein expressed by a recombinant baculovirus was detected both in the cytoplasm and on the plasma membrane; thus, the mutant gE protein was distributed in a similar pattern as wild type gE. Secondly, the biosynthesis of the mutant construct was measured by immunoprecipitation and found to be similar to wild type gE. Thirdly, the phosphorylation of the mutant protein was compared with that of the wild type gE. InstantImager analysis of the two phosphorylated proteins showed a 70% reduction on the mutant gE as compared with wild type gE. Based on these dramatic results, the assumption has been made that gE is phosphorylated primarily by casein kinase II at the canonical phosphorylation consensus sequence in its cytoplasmic tail.

VZV gE tyrosine phosphorylation

In the studies described above, the phosphorylated gene product represents the gE monomer. In further studies with gE, a phosphorylated dimeric

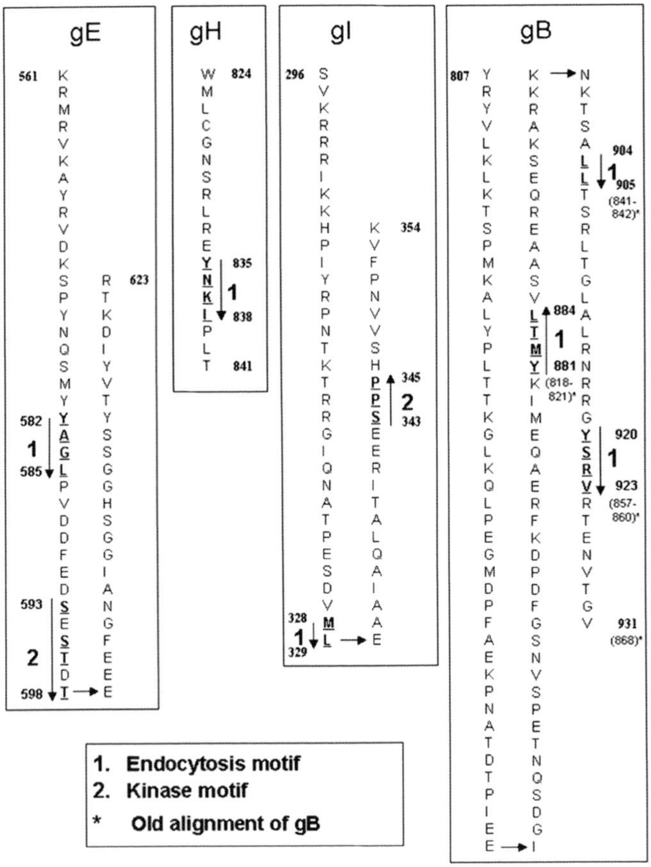

Figure 9.3 Phosphorylation and endocytosis motifs in VZV glycoproteins. The motifs are found in the cytoplasmic tails of the four type 1 viral glycoproteins.

form was also observed. Of interest the phosphorylation of the gE dimer was abolished when GTP was the radiolabel, a result which indicated that the dimer was not primarily phosphorylated by casein kinase II which can utilize GTP nearly as well as ATP as the phosphate source. Again, examination of the gE sequence provided clues as to the nature of the second phosphate acceptor, namely, six tyrosine residues were present in the cytoplasmic tail. To investigate whether tyrosine phosphorylation was involved, insect cells were infected with recombinant baculovirus containing the wild type gE insert (Olson *et al.*, 1997). The precipitated gE protein was transferred to a nitrocellulose membrane and probed with a biotinylated P-tyr-1 antibody, which specifically recognizes phosphotyrosine linkages but not phosphoserine sites. After visualization by chemiluminescence, the anti-phosphotyrosine antibody detected a single protein band at M_r 130 kDa, which was previously identified as the dimeric form of the gE in insect cells.

Because of the importance of the above observation, the experiment was repeated in mammalian cells. Since the transfected mammalian cells did not produce as much gE protein as the baculovirus infected insect cells, the gE transfected cell lysates were concentrated several fold to obtain sufficient gE protein for immunoprecipitation and immunoblot analysis. When the nitrocellulose membrane was probed with the same P-tyr-1 antibody, the phosphotyrosine antibody again detected a protein at Mr 130, which corresponded to the gE protein recognized in the insect cells. Thus, this result documented the presence of phosphotyrosine linkages in gE dimers present in mammalian cells. These dimeric gE forms are present in limited quantities and are not easily detectable by standard electrophoresis in the absence of a highly radioactive ^{32}P-label; however, the gE dimer was easily seen by 2-D gel electrophoresis (Fig. 9.2).

VZV gE endocytosis motif

Endocytosis is the process by which proteins are continually internalized within the cell through specific interactions with clathrin at the plasma membrane (Trowbridge, 1991). Since VZV gE contains potential internalization motifs in its cytoplasmic tail, gE was analysed to determine if it was endocytosed (Olson and Grose, 1997). For the investigation of gE internalization, a confocal microscopy assay was developed (Olson *et al.*, 1998). Namely, Hela cells were transfected with an expression plasmid containing the gE insert. After 16 h, the cells were incubated with primary murine antibody to gE at 4° for 30 min. Then individual monolayers were returned to 37°C for increasing intervals of 0, 10, 15, 30, 30, 45, and 60 min to allow internalization of gE. Thereafter, secondary anti-mouse antibody conjugated to a dye was incubated with the transfected cells to identify the localization of gE. Subsequently the cells were observed by confocal microscopy with optical tomography in micron increments from the surface of the cell to its attachment plane on the plastic surface; the zeta series of gE confocal images were stored in a computer. When the images were subsequently analysed, the following scenario was evident. At the initial time-point at 4°C, antibody bound to gE which remained on the cell surface. At the 5 min time-point, gE remained localized mainly in a clumping pattern on the membrane although some gE was observed within the cell. By 10–15 min at 37°C, however, more gE was detected in the cytoplasm, often within small vesicles. After 30–60 min at 37°C, large amounts of gE were detected in the cytoplasm in a typical pattern consisting of clusters of small vesicles.

To verify that gE internalization was clathrin mediated, a second set of experiments was performed in which gE transfected cells were double stained with an IgG antibody to gE and an IgM antibody to clathrin. When the images of gE and clathrin were merged and observed, they documented that gE was clustering in clathrin coated pits forming the clathrin coated vesicles at the cell membrane. These confocal microscopic analyses demonstrated conclusively that VZV gE is internalized into the cytoplasm from the cell membrane in a pattern consistent with endocytosis of other cell surface receptors, such as the transferrin receptor.

The VZV gE C-tail contains six tyrosine residues but one tyrosine in particular is part of a defined internalization motif, in particular the YAGL sequence (Fig. 9.3). As part of this endocytosis analysis, this tyrosine residue was placed by a glycine residue. As in precious experiments, Hela cells were transfected with the gene encoding the gE-Y582G product and the monolayer was analysed by confocal microscopy. When the gE-Y582G product was first observed at time-point 0, the protein was located at the cell surface. When other cultures were observed at the 15 and 30 min time-points, the mutant protein remained on the cell surface and was not internalized. Therefore, the concluding result of these experiments was that the tyrosine in the gE endodomain YAGL motif was required for efficient internalization of the gE molecule.

Cell spread phenotype of first VZV gE escape mutant virus

Even thought the VZV genome is considered to be very stable, mutations do occur in community outbreaks (Wagenaar *et al.*, 2003). The first mutated VZV isolate was discovered because it was poorly reactive with anti-gE antibodies in a commercial VZV diagnostic kit (Santos *et al.*, 1998). Because the isolate could not be identified, the virus isolate was sent to our laboratory for further analysis. When the isolate (VZV passage 1) was received, the infected cell monolayer was trypsin-dispersed and inoculated onto human melanoma cells for two passages. When CPE covered ~70% of each monolayer, the infected monolayers were probed with mAb 3B3 against gE and with mAb 6B5 against gI and examines by confocal microscopy. In prior studies, we showed that these two mAbs do not cross-react with other viral or cellular proteins (Grose, 1990). As expected from numerous published experiments, mAb 3B3 and mAb 6B5 reacted with the control laboratory strain VZV-32. In marked contrast, the anti-gE mAb 3B3 did not attach to cells infected with the VZV-MSP strain, even though mAb 6B5 did bind the infected cells strongly. As an additional control, the anti-gH mAb 206 was added to cultures individually infected with both VZV strains; all VZV-infected cultures were positive in this assay.

We next immediately addressed the question of whether the VZV-MSP strain fails to express the entire glycoprotein or has lost an epitope on the glycoprotein. In this quest, we were fortunate in that we had begun epitope mapping of gE (Santos *et al.*, 2000). Through this methodology, we established that the MAb 3B3 epitope consisted of at least 11 amino acids (150–161) of gE (Fig. 9.1). The gE B-cell epitope usually includes the following sequence: 150-DQRQYGDVFKGD-161. In the mutated VZV-MSP strain, amino acid 150-D is substituted by 150-N. Thus, the mutated strain has lost the epitope recognized by MAb 3B3. The remainder of the gE sequence is unchanged. Of importance, the entire VZV-MSP genome has now been sequenced (Grose *et al.*, 2004; Peters *et al.*, 2006). The VZV-MSP strain also exhibits a new phenotype, in that the virus spreads faster in cultured cells than other strains and also the viral particles emerge uniformly over the cell surface rather than along viral highways as is typical for all other wild type strains. Since we isolated the first gE mutant strain, similar gE mutant strains have been isolated in Canada, Sweden and Italy (Tipples *et al.*, 2002).

VZV glycoprotein I (ORF 67)

VZV gI structure

The gI protein is typically found in a gE/gI complex (Yao *et al.*, 1993). Based on the sequence and structure of gI, it is designated a typical type 1 transmembrane glycoprotein comprised of a total of 354 amino acids (Fig. 9.1). The ectodomain includes 279 amino acids, the transmembrane domain 17 amino acids and the ectodomain 58 amino acids. The current terminology gI for ORF 67 is based on its homology to the HSV-1 gI glycoprotein. Like the other VZV glycoproteins, VZV gI had historical designations such as gp 64 and gpIV. Unlike the situation in HSV-1 infected cells, VZV gE and gI are not produced in equivalent amounts; rather, VZV gE is more abundant than gI.

VZV gI glycosylation

The ectodomain contains four potential N-linked glycosylation sites as well as a juxtamembrane region compatible with O-linkages (Yao *et al.*, 1993). Furthermore, studies with radiolabelled glycans and inhibitors of glycosylation have confirmed the presence of both complex type

N-linked glycans and O-linked glycans on VZV gI.

VZV gI serine phosphorylation

Like VZV gE, gI is phosphorylated during infection in cell culture. Initial studies in a transient transfection system documented that gI expressed in the absence of any other VZV protein was phosphorylated on its cytoplasmic tail by a serine protein kinase present in cultured cells (Yao and Grose, 1994). Yet, examination of the gI sequence quickly established that no casein kinase II phosphorylation consensus motif was present in the endodomain. Therefore, others serine residues were the phosphoacceptors. One sequence on the gI endodomain Ser-Pro–Pro was of special interest, because of its semblance to the phosphorylation consensus motif for cyclin dependent kinases (Fig. 9.3). To investigate this possibility, the three residues in the gI endodomain were individually mutated. Three expression plasmids containing the following mutations were studied in detail: gI-S343A, gI-P344A and gI-P345A. After 32P-radiolabeling, phosphorylation of each gI mutant protein was compared with wild type protein by phosphorimaging analysis. The phosphorylation signals of gI-S343A and gI-P344A were about 10% of that seen with wild type gI while that of gI-P345A was about 30% of with type gI. These results indicated that both serine residue 343 and proline residue 344 were essential for gI phosphorylation. In contrast, proline 345 modulated but did not abrogate the phosphorylation of gI. In other words, the principal gI phosphorylation site is a serine-proline sequence in the cytoplasmic tail.

Since the above sequence is usually associated with phosphorylation by a cyclin dependent kinase (CDK), an *in vitro* kinase assay was performed with gI as the substrate and CDK1 and CDK2 as the protein kinases (Ye *et al.*, 1999). Of note, the gI product was phosphorylated by both kinases. When the gI phosphorylation signals of CDK1 and CDK2 were compared, a 4-fold higher phosphorylation signal accompanied CDK1 in the protein kinase assay. As a specificity control, CDK6 was included in the kinase assay but no phosphorylation of gI was observed.

VZV gI endocytosis motif

VZV gI expressed in a transient transfection system was investigated in

a similar manner as previously described for gE. To this end, gI transfected cells were incubated with antibody to gI and analysed for endocytosis by laser scanning confocal microscopy and optical tomography at increasing time intervals (Olson and Grose, 1998). As seen with gE, gI was detectable on the surface at time-point 0, while some gI was found within the cytoplasm within 15 min. By 30–60 min, large amounts of gI were observed within the cytoplasm in small vesicle-filled clusters. These results clearly indicated that VZV gI underwent endocytosis from the cell membrane in a similar pattern to gE under the same transfection conditions.

Endocytosis of cell surface receptors is known to be dependent on specific amino acid sequence within the cytoplasmic tail. As described above, the gE endodomain contains a YAGL internalization motif. Upon examination of the gI cytoplasmic tail, no tyrosine containing internalization motif was identified. However, the gI endodomain did contain another potential endocytosis motif, namely, a variation on the dileucine motif. Several receptors, such as the invariant chain associated with the major histocompatibility complex class II molecule as well as the CD4 molecule, contain a dileucine motif which effects efficient internalization of the receptor from the plasma membrane. The cytoplasmic tail of gI contains a Met-Leu sequence at residues 328–329, which corresponds to a dileucine motif (Fig. 9.3). In order to determine if the Met-Leu sequence was responsible for gI endocytosis, the sequence was changed to Ala–Ala by site directed mutagenesis. Subsequently, Hela cells were transfected with the wild type gI gene or the mutant gI gene designated gI-AA. When gI-AA was expressed in cells and analysed by confocal microscopy, the mutant protein was present on the cell surface at time-point 0. After 30 min, gI-AA was still localized mainly to the cell membrane. In contrast, wild type gI was internalized within the cytoplasm after 30 min. These results showed that the dileucine motif in the gI endodomain was functioning as a signal for endocytosis.

VZV glycoprotein gC (ORF 14)

VZV gC structure

The VZV gC protein exhibits remarkable differences form his HSV-1 counterpart (Stingley *et al.*, 2000). Shortly after the publication of the sequence of the complete VZV genome, investigations established the basic characteristics of VZV gC or gpV as it was first known (Kinchington *et al.*, 1990a,b). The gC protein is encoded by ORF14 and is a type 1 glycosylated polypeptide around 105 kDa in mass that traffics to the plasma membrane. ORF14 contains the variable repeat region R2 that differs in length in strains of VZV (Tyler *et al.*, 2007). Examination of its amino sequence shows 560 amino acids, including the repeated region from amino acids 33 to 140 that is expressed as approximately 8 repeated sequences of TSAATRKPDPAVAP residues (Fig. 9.4).

The unglycosylated backbone has a mass of 61 kDa while the observed mass is approximately 105 kDa, a difference that indicates over 40 kDa of post-translational additions to the protein. Using prediction algorithms, four N-linked glycosylation sites and 25 O-linked mucin type sites exclusively on the repeated pattern were identified (Fig. 9.4). The observed molecular weight and glycosylation prediction indicates that more that one third of VZV gC molecular weight is due is oligosaccharide chains. The membrane spanning helical domain is predicated to be near the C-terminus at amino acids 531 to 554, leaving only six amino acids extending into the cytosol. Genomic analysis showed that VZV gC is approximately 30% similar to HSV gC.

Differences between HSV gC and VZV gC expression

HSV-1 gC is a typical true late (gamma-2) gene product, which is expressed towards the end of the replication cycle in cultured cells (Levine *et al.*, 1990). As an alphaherpesvirus, VZV is also considered to have a 12–18 hr replication cycle. In most VZV experiments, the inoculum consists of trypsin-dispersed infected cells from a monolayer with advanced CPE (Grose and

VZV ORF 14 gC

```
  1 MKRIQINLIL TIACIQLSTE SQPTPVSITE LYTSAATRKP DPAVAPTSAA
 51 SRKPDPAVAP TSAASRKPDP AVAPTSAASR KPDPAVAPTS AATRKPDPAV
101 APTSAASRKP DPAVAPTSAA TRKPDPAVAP TSAASRKPDP AANTQHSQPP
151 FLYENIQCVH GGIQSIPYFH TFIMPCYMRL TTGQQAAFKQ QQKTYEQYSL
201 DPEGSNITRW KSLIRPDLHI EVWFTRHLID PHRQLGNALI RMPDLPVMLY
251 SNSADLNLIN NPEIFTHAKE NYVIPDVKTT SDFSVTILSM DATTEGTYIW
301 RVVNTKTKNV ISEHSITVTT YYRPNITVVG DPVLTGQTYA AYCNVSKYYP
351 PHSVRVRWTS RFGNIGKNFI TDAIQEYANG LFSYVSAVRI PQQKQMDYPP
401 PAIQCNVLWI RDGVSNMKYS AVVTPDVYPF PNVSIGIIDG HIVCTAKCVP
451 RGVVHFVWWV NDSPINHENS EITGVCDQNK RFVNMQSSCP TSELDGPITY
501 SCHLDGYPKK FPPFSAVYTY DASTYATTFS VVAVIIGVIS ILGTLGLIAV
551 IATLCIRCCS
```

Figure 9.4 Amino acid sequence of VZV gC. This glycoprotein contains a repeated sequence (R2) within its ectodomain. The nucleotide and deduced amino acids sequences were published by Peters *et al.* (2006).

Brunel, 1978). Because the titre of input virus is so low, however, several replication cycles are required before cytopathology is visible in the newly infected monolayers. While examining glycoprotein expression at increasing intervals after infection, we observed that all syncytia at early time points (24 and 48 hpi) were clearly labelled with a monoclonal antibody against gE, but only an occasional VZV induced syncytium was labelled with anti-gC antibody (Storlie *et al.*, 2006). Even though conditions of cell associated VZV infection are asynchronous, by 48 hpi all input viral populations would have undergone 3–4 replication cycles. VZV gE and gH were easily detectable; yet, VZV gC was invariably difficult to detect by confocal microscopy at 48 hpi.

To further investigate the expression of gC as compared with gE by another method, we infected replicate monolayers and harvested them over a 96 hr time period. When examined by immunoblotting, this experiment confirmed that gC protein expression was markedly delayed when compared with gE, namely, abundant amounts of gC protein were not observed until 72 hpi. Between 72 and 96 hpi, increasing amounts of gC expression were finally detected in the infected monolayers. In contrast, western blotting showed that gE was present in abundance at 24 hpi.

Furthermore, two other explanations for the above gC effect were excluded. First, we investigated whether the effect was specific to one VZV strain. Our initial experiments were performed with the VZV-32 strain used in this laboratory for many years. This strain has now been completely sequenced and contains an intact genome with no major genetic changes when compared with VZV Dumas (Peters *et al.*, 2006; Tyler *et al.*, 2007). We also examined gC expression after infection of cells with 2 additional strains, including VZV-MSP and VZV-Ellen. VZV-MSP has also been completely sequenced and found to have an intact gC gene. The results with the latter two strains demonstrated a delayed gC biosynthesis similar to VZV-32 strain. Secondly, we investigated whether the effect was specific to one cell line: Our initial experiments were performed in human melanoma cells, a substrate highly susceptible to VZV infection. To this end, we infected simian derived Vero cells and found that gC production was similarly delayed. In short, the marked delay in gC biosynthesis in VZV infected cells was not restricted to one VZV strain or one cell line (Storlie *et al.*, 2006).

Delayed VZV gC transcription in cultured cells

Since our protein studies suggested an effect on gC transcription, we tested to see if there was delay in the gC transcription that could account for the delay in protein expression. To further investigate, we used real-time Real Time RT-PCR to determine if this delay in VZV gC expression resulted from a parallel delay in gC transcription (Storlie *et al.*, 2008a). In order to reduce the effects of heterogeneity on infections resulting from the presence of inoculum cells, we selected cell-free inoculum to infect the monolayers that we analysed (Fig. 9.5). Real Time RT-PCR revealed a delay in the appearance of gC transcripts that mirrored a similar delay in the appearance of gC protein. Glycoprotein C transcript levels rose in untreated infected cells by 72 hpi, and reached a peak at 96 hpi. In contrast to the gC mRNA results, gE transcript levels were easily detectable in the first 24 hpi, when gE protein was also detectable.

Thus, there are marked differences between the kinetics of VZV and HSV gC transcription (Ling *et al.*, 1991). Easily detectable HSV gC transcription certainly occurs within the first replication cycle in cultured cells, with the subsequent production of HSV gC protein (Stingley *et al.*, 2000). That process is markedly delayed in VZV infected cells. Early investigations found that the nucleotide binding domains in the 5′ and 3′ UTR regions of VZV ORF14 were different from those in the nucleotide sequence of HSV gC, suggesting a fundamental difference in transcriptional control between the two alphaherpesviruses (Ruyechan *et al.*, 2003).

Absence of gC in aberrant viral particles

The role of VZV gC is confounded by the fact that earlier studies were performed before we knew that very little gC protein was produced in infected cells. By comparison, HSV gC appears

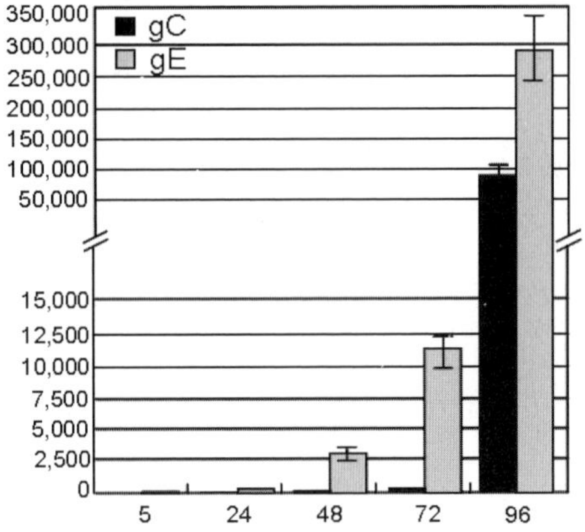

Figure 9.5 Delayed transcription of VZV gC as compared to gE in cultured cells. After inoculation of monolayers, Real Time RT-PCR was performed in quadruplicate to detect mRNA in the monolayers at increasing intervals after infection. Data were analysed with Applied Biosystems RQ Manager software. Error bars indicate 95% confidence intervals. Note minimal gC transcription until 72 hpi. X axis = hpi; Y axis = glycoprotein-specific mRNA levels.

to have at least two functions as a member of the viral envelope: one, attaching to cells via heparan sulfate and two, binding to the complement component C3b and preventing the generation of C5 convertase and thus stopping the complement cascade from destroying the infected cell (Spear and Longnecker, 2003; Spear *et al.*, 1992). Spear has suggested that the latter function is the more important function of HSV gC. In fact, researchers have identified a shortened version of HSV gC that is secreted from the infected cell, apparently to engage the complement response extracellularly.

Recently, we have discovered that VZV grown in cultured cells has an extremely high particle:PFU ratio of 40, 000:1 (Carpenter *et al.*, 2009). In other words, many viral particles are assembled in cell culture, but very few are infectious. In fact, most viral particles resemble light (L) particles, in that they have an outer envelope but lack an internal capsid (Carpenter *et al.*, 2008). Since very little gC is produced during the first 48–72 hpi, we suggest that many envelopes (although assembled) also lack gC. Further studies are required to assess whether a lack of gC during VZV assembly is a contributing factor towards the invariably low titre and high number of aberrant particles.

VZV gC protein in human zoster vesicles

The skin vesicle is the final site of virion assembly during chickenpox and zoster. Since complete viral particles are easily detectable in the human vesicle (Fig. 9.6), we investigated whether gC was expressed in these vesicles. We immunolabelled a biopsy specimen removed from a human zoster vesicle with antibodies to various viral proteins; then we attached fluorescently labelled secondary antibodies to those primary antibodies in order to identify location and approximate concentration of glycoproteins. We had previously documented that gE and gH were present in abundance in human skin vesicles (Weigle and Grose, 1983). In a recent VZV gC study (Storlie *et al.*, 2008b), we also observed a similar abundance of intense spherical gC immunofluorescent labelling in zoster vesicles (Fig. 9.6). This vesicle study also implies that the delay in gC expression seen in cultured cells is not present in the infected human.

VZV glycoproteins gH (ORF37) and gL (ORF60)

VZV gH and gL structure

VZV glycoproteins gH (118 kDa) and gL (18 kDa) are encoded by open reading frames 37

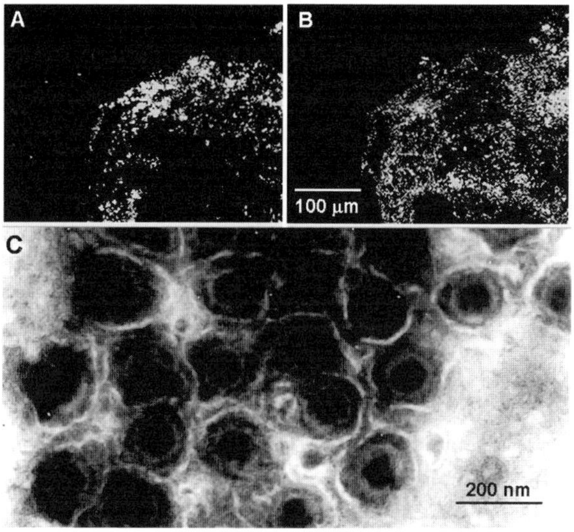

Figure 9.6 Human zoster skin vesicle immunostained with anti-gC antibody. Two sections of a vesicle biopsy (A and B) were immunolabelled with mouse anti-gC monoclonal antibody and linked with the Alexa 488 fluorescent antibody. They were viewed by confocal microscopy. Another section of a zoster vesicle was examined by transmission electron microscopy (C), in order to illustrate the presence of numerous enveloped VZV particles.

and 60, respectively (Forghani *et al.*, 1994; Grose, 1990b). The gH sequence is highly conserved among herpesviruses, whereas function but not sequence is conserved among the herpesvirus gL proteins, many of which were initially identified by genome position rather than sequence homology (Kaye *et al.*, 1992; Spaete *et al.*, 1993). Similar to gE and in contrast to gC, abundant amounts of the type 1 VZV gH are produced in infected cells. However, early attempts to study herpes viral gH function independent of infection in cell culture were largely unsuccessful until it was discovered that gH associated with a second glycoprotein called gL (Maresova *et al.*, 2000). Subsequent coexpression, mutation, and trafficking studies revealed that VZV gH also requires a chaperone activity provided by the gL glycoprotein for efficient cell membrane expression of fusion-competent gH molecules (Cole and Grose, 2003). More recent studies demonstrated that the fusogenic activity of VZV gH and ultimately the incorporation of this important glycoprotein into the virion envelope is further regulated by endocytosis of the mature protein once it has reached the cell membrane (Maresova *et al.*, 2005). The sequences of gH and gL are included in Fig. 9.7.

Interaction and trafficking of the VZV gH/gL complex

Interactions between gH and gL have been extensively investigated (Duus and Grose, 1996; Duus *et al.*, 1995). Immunoprecipitation of the gL protein from cells coexpressing gH with polyclonal anti-gL serum also pulled down pre-gH molecules, indicating that a pre-gH:gL complex was formed. Only pre-gH:mature gL complexes were immunoprecipitated from cell lysates expressing both proteins; no complexes with mature gH molecules or immature gL (pre-gL) molecules were detected. Disrupting the intramolecular cysteine disulfide bonds of the gL molecule by mutating different cysteine residues to alanines resulted in loss of the pre-gH:gL complex formation, and loss of mature, functional gH molecule expression in immunoprecipitation, immunofluorescence and fusion assays, providing an additional line of evidence for a gL chaperone function. Monensin treatment of gL-expressing cells did not alter the glycosylation of gL, indicating that the processing of gL is completed in the *cis*/medial Golgi, and does not require trafficking to the *trans* Golgi network (TGN).

VZV ORF 37 gH

```
  1 MFALVLAVVI LPLWTTANKS YVTPTPATRS IGHMSALLRE YSDRNMSLKL
 51 EAFYPTGFDE ELIKSLHWGN DRKHVFLVIV KVNPTTHEGD VGLVIFPKYL
101 LSPYHFKAEH RAPFPAGRFG FLSHPVTPDV SFFDSSFAPY LTTQHLVAFT
151 TFPPNPLVWH LERAETAATA ERPFGVSLLP ARPTVPKNTI LEHKAHFATW
201 DALARHTFFS AEAIITNSTL RIHVPLFGSV WPIRYWATGS VLLTSDSGRV
251 EVNIGVGFMS SLISLSSGLP IELIVVPHTV KLNAVTSDTT WFQLNPPGPD
301 PGPSYRVYLL GRGLDMNFSK HATVDICAYP EESLDYRYHL SMAHTEALRM
351 TTKADQHDIN EESYYHIAAR IATSIFALSE MGRTTEYFLL DEIVDVQYQL
401 KFLNYILMRI GAGAHPNTIS GTSDLIFADP SQLHDELSLL FGQVKPANVD
451 YFISYDEARD QLKTAYALSR GQDHVNALSL ARRVIMSIYK GLLVKQNLNA
501 TERQALFFAS MILLNFREGL ENSSRVLDGR TTLLLMTSMC TAAHATQAAL
551 NIQEGLAYLN PSKHMFTIPN VYSPCMGSLR TDLTEEIHVM NLLSAIPTRP
601 GLNEVLHTQL DESEIFDAAF KTMMIFTTWT AKDLHILHTH VPEVFTCQDA
651 AARNGEYVLI LPAVQGHSYV ITRNKPQRGL VYSLADVDVY NPISVVYLSR
701 DTCVSEHGVI ETVALPHPDN LKECLYCGSV FLRYLTTGAI MDIIIIDSKD
751 TERQLAAMGN STIPPFNPDM HGDDSKAVLL FPNGTVVTLL GFERRQAIRM
801 SGQYLGASLG GAFLAVVGFG IIGWMLCGNS RLREYNKIPL T
```

VZV ORF 60 gL

```
  1 MASHKWLLQI VFLKTITIAY CLHLQDDTPL FFGAKPLSDV SLIITEPCVS
 51 SVYEAWDYAA PPVSNLSEAL SGIVVKTKCP VPEVILWFKD KQMAYWTNPY
101 VTLKGLAQSV GEEHKSGDIR DALLDALSGV WVDSTPSSTN IPENGCVWGA
151 DRLFQRVCQ
```

Figure 9.7 Amino acid sequences of VZV gH and gL. VZV gH and gL are usually found as a complex. The nucleotide and deduced amino acids sequences were published by Peters *et al.* (2006).

A mechanism for the retrograde trafficking of gL from the Golgi apparatus back to the ER has been proposed, whereby the hydrophobic ER-targeting motif in gL is masked by association with gH and unmasked when gH is mature and the gH:gL complex dissociates in the Golgi. Collectively, these studies lead to a model of gH:gL processing and trafficking: the immature gL traffics as far as the *cis*/medial Golgi, where it matures and then returns to the ER, associates with pre-gH, and escorts it to the *cis*/medial and *trans* Golgi. Somewhere between the *cis*/medial and *trans* Golgi, the mature gH and gL dissociate and gL either returns to the ER or is targeted to a lysosome, where it is degraded. Meanwhile, mature gH moves on to the plasma membrane.

Fusion mediated by VZV gH and gL coexpression

VZV gH transfected alone in cell lines does not result in detectable levels of the protein on outer cell membranes, and only a 97 kDa immature, endoH-sensitive precursor gH (pre-gH) is immunoprecipitated with monoclonal antibodies against gH. However, in the presence of coexpressed gL, high levels of mature, endo-H resistant 118 kDa gH molecules are detected at the; Pasieka *et al.*, 2003b, 2004). Furthermore, the cells in the gH/gL expressing monolayer form polykaryons containing 6–24 nuclei in confocal microscopy fusion assays. This gH-mediated fusion was inhibited by the same monoclonal antibody shown to inhibit syncytia formation in

VZV-infected melanoma monolayers, providing the first direct evidence that the mechanism of cell fusion in VZV infection was mediated by coexpression of the gH and gL glycoproteins (Rodriguez *et al.*, 1993).

VZV gL is also glycosylated, presumably on its one predicted N-linked glycosylation site. Although processing of gH was shown to be dependent upon gL expression, gL processing does not appear to require coexpression of gH, as gL expressed alone was shown to be endo-H resistant and endo-F sensitive. The endo-H sensitive pre-gL is 14 kDa, while the endo-F sensitive mature gL is 18 kDa. VZV gL is unique among the VZV glycoproteins in its ability to facilitate gH maturation and cell membrane expression. Coexpression of either of two other VZV glycoproteins, gE or gI, with gH did not support the complete maturation of gH, although the presence of either of these glycoproteins did allow aggregates of pre-gH molecules to reach cell membrane, where they were detected by immunofluorescence as a cap-like structure. Collectively this set of studies suggested that gL provides a chaperone function to facilitate gH-mediated fusion (Duus and Grose, 1996; Duus *et al.*, 1995). Consistent with this idea, the gL protein is known to contain a peptide sequence closely matching a 16-residue hydrophobic ER-targeting motif encoded by many cellular ER resident proteins known to be involved in glycoprotein processing.

VZV gH endocytosis motif

Endocytosis motifs are located in the cytoplasmic tails of VZV gE and gI (Fig. 9.3). Of note, the short gC cytoplasmic tail also contains a functional YNKI motif, which mediates clathrin-dependent and antibody-independent gH endocytosis in the presence of gL (Fig. 9.3). This clathrin-dependent endocytosis event occurs in both gH/gL-transfected and VZV-infected cells, where the majority of the endocytosis occurred within 24 hours post-infection (Pasieka *et al.*, 2003a). A subsequent analysis of gH endocytosis on virus-mediated cell–cell fusion revealed that regulation of fusion occurs by this mechanism. In addition, coexpression of gE further modulates gH-mediated fusion activity by coendocytosis of a gH:gE complex from the cell membrane (Pasieka *et al.*, 2004). Thus, endocytosis is certainly an important component of gH trafficking within an infected cell.

Role of VZV gH in virus entry and egress

All monoclonal antibodies produced in this laboratory have been tested in a complement independent neutralization assay (Grose *et al.*, 1983). The titre in a complement independent neutralization assay correlates with a titre is a fluorescent antibody to membrane antigen (FAMA) assay, which in turn correlates with a protective antibody response (Grose *et al.*, 1979). The VZV monoclonal antibody that has the highest neutralization titre is that against gH (Montalvo and Grose, 1986a). In a subsequent experiment, the effect of anti-gH antibody on the spread of infection was evaluated by high resolution scanning electron microscopy (Rodriguez *et al.*, 1993). Two antibodies were initially selected, the first being an anti-gH antibody clone MAb 206 with a 1:40,000 titre of complement-independent neutralization activity in a plaque reduction assay. The second control MAb called 5C6 precipitates the VZV ORF 62 regulatory protein, the homologue of the HSV ICP4 protein, and has no neutralization activity. No cytopathology developed in the cultures treated with anti-gH antibody, while those incubated with MAb 5C6 developed syncytia with viral egress as seen in untreated infected monolayers. When viewed by scanning electron microscopy, cultures treated with anti-gH antibody lacked visual evidence of syncytia and had almost no viral particles on the cell surface (Fig. 9.8). When the anti-gH antibody was removed from the culture medium, viral egress was easily seen in another 48 h_r by scanning electron microscopy (Fig. 9.8).

The severe combined immunodeficient mouse (SCID) has become the best animal model for studying VZV infection (Moffat *et al.*, 1998). In order to determine the contribution of gH to and the effect of anti-gH antibody on VZV pathogenesis in skin in this model, MAb 206 was administered to SCID mice with pOka-inoculated skin xenografts starting either 6 hours post infect (hpi) or 4 days post infection (dpi) and continued every 4 days through 12 dpi (Vleck

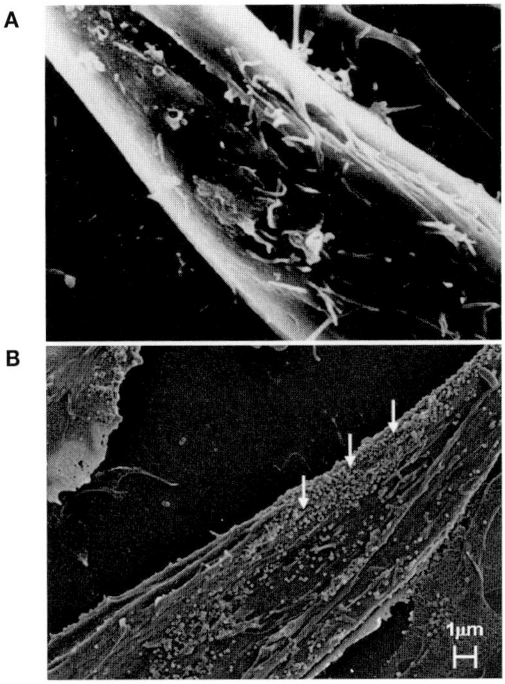

Figure 9.8 Inhibition of viral egress by neutralization antibody to VZV gH. Monolayers were inoculated with VZV cell free virus and then incubated in the presence of anti-gH monoclonal antibody 206 in the culture medium. After 48 hours, no viral egress was evident (A). However, when the antibody was removed from the culture medium, viral egress (arrows) was evident after another 48 hr of incubation (B). The cells in monolayer in panel B also show signs of extensive virus-induced fusion.

et al., 2009). Virus growth was analysed every 7 days until 42dpi. Among the 70% mock-treated xenografts that were infected, VZV titre and genome copy number and lesion size increased over time. Anti-gH MAb 206 administered at 6 hpi significantly reduced the frequency of skin xenograft infection by 42%. Virus titres, genome copies and lesion size were decreased during the period in which antibody was administered in xenografts that became infected, and normal growth kinetics lagged behind the mock-treated xenografts by 14–21 days. In contrast, administering anti-gH antibody four days post infection suppressed VZV replication during the period of administration but did not reduce the frequency of infection, and normal growth kinetics lagged by 7–14 days. The neutralizing anti-gH MAb 206 blocked virus entry, cell fusion or both in human skin in the SCID model.

In cells, MAb 206 bound to plasma membranes and to surface virus particles, blocking spread of the virus. Antibody-gH and antibody–virion complexes were internalized into endosomes within infected cells. VZV gH endocytosis is antibody-independent but clathrin-dependent (Pasieka *et al.*, 2003b), and this report demonstrated that MAb 206 binds and is internalized with gH. MAb 206 and gH colocalized with EEA1, a marker for early endosomes, and Vps4, a marker for multivesicular bodies (MVB), indicating that MAb 206 and gH underwent endocytosis and sorting via the MVB pathway (Babst *et al.*, 1997; Rusten *et al.*, 2007). Vps4 is required for transport of endocytosed proteins between pre-vacuolar endosomes and vacuoles, and for autophagy, likely for autophagosome–endolysosome fusion. The gH-Vps4 colocalization could occur during sorting steps as gH is targeted to the TGN. Alternatively, this association might indicate that VZV uses the MVB pathway for assembly and egress of virus particles, as has been suggested for HSV (Calistri *et al.*, 2007). The MAb 206–gH complexes travelled through endosomes and the MVB pathway but were not targeted to the TGN.

The trafficking of these complexes was disrupted, and they could be sorted through late endosomes to lysosomes or autophagosomes, which are induced during VZV infection (Takahashi *et al.*, 2009). This could occur if the antibody disrupted gH interaction with virus or cellular proteins, or through antibody interaction with Fc receptors on the surface of infected fibroblasts. The MAb 206-labelled virions were identified within cells, suggesting that labelled particles on the surface of infected cells were also internalized and possibly targeted for degradation.

This report demonstrated for the first time that gH contributes to VZV pathogenicity in skin within the SCIDhu model and that the administration of anti-gH antibody was effective at preventing or reducing VZV pathogenicity in human skin (Vleck *et al.*, 2009). The prevention of infection might be attributed to a block on virus attachment or entry into cells. The reduced pathogenicity in xenografts that became infected could be attributed to the block in spread of virus from cell-to-cell, inhibition of canonical gH trafficking or potential virus particle degradation.

VZV glycoprotein gB (ORF31)

VZV gB structure

VZV gB is a type 1 membrane protein that is proteolytically cleaved into disulfide-linked subunits (Montalvo and Grose, 1987). An endoproteolytic cleavage event analogous to that

VZV ORF 31 gB

```
  1 MSPCGYYSKW RNRDRPEYRR NLRFRRFFSS IHPNAAAGSG FNGPGVFITS
 51 VTGVWLCFLC IFSMFVTAVV SVSPSSFYES LQVEPTQSED ITRSAHLGDG
101 DEIREAIHKS QDAETKPTFY VCPPPTGSTI VRLEPTRTCP DYHLGKNFTE
151 GIAVVYKENI AAYKFKATVY YKDVIVSTAW AGSSYTQITN RYADRVPIPV
201 SEITDTIDKF GKCSSKATYV RNNHKVEAFN EDKNPQDMPL IASKYNSVGS
251 KAWHTTNDTY MVAGTPGTYR TGTSVNCIIE EVEARSIFPY DSFGLSTGDI
301 IYMSPFFGLR DGAYREHSNY AMDRFHQFEG YRQRDLDTRA LLEPAARNFL
351 VTPHLTVGWN WKPKRTEVCS LVKWREVEDV VRDEYAHNFR FTMKTLSTTF
401 ISETNEFNLN QIHLSQCVKE EARAIINRIY TTRYNSSHVR TGDIQTYLAR
451 GGFVVVFQPL LSNSLARLYL QELVRENTNH SPQKHPTRNT RSRRSVPVEL
501 RANRTITTTS SVEFAMLQFT YDHIQEHVNE MLARISSSWC QLQNRERALW
551 SGLFPINPSA LASTILDQRV KARILGDVIS VSNCPELGSD TRIILQNSMR
601 VSGSTTRCYS RPLISIVSLN GSGTVEGQLG TDNELIMSRD LLEPCVANHK
651 RYFLFGHHYV YYEDYRYVRE IAVHDVGMIS TYVDLNLTLL KDREFMPLQV
701 YTRDELRDTG LLDYSEIQRR NQMHSLRFYD IDKVVQYDSG TAIMQGMAQF
751 FQGLGTAGQA VGHVVLGATG ALLSTVHGFT TFLSNPFGAL AVGLLVLAGL
801 VAAFFAYRYV LKLKTSPMKA LYPLTTKGLK QLPEGMDPFA EKPNATDTPI
851 EEIGDSQNTE PSVNSGFDPD KFREAQEMIK YMTLVSAAER QESKARKKNK
901 TSALLTSRLT GLALRNRRGY SRVRTENVTG V
```

Figure 9.9 Amino acid sequences of VZV gB. The nucleotide and deduced amino acids sequences were published by Peters *et al.* (2006).

in VZV gB does not occur in the processing of HSV-1 gB. VZV gB isolated from virus consists mainly of a 120– 140-kDa full-length product. The reduced products include two polypeptides of similar Mr (approximately 60 kDa). According to Davison and Scott (1986), the corresponding DNA sequence specifies a polypeptide of 868 amino acid residues, comprised of an unusually short eight amino acid signal peptide, a large ectodomain, a hydrophobic transmembrane region (aa 699–743) followed by a positively charged C-terminal domain containing 125 residues (Fig. 9.9). As outlined below, we now know that the gB ORF is larger than originally predicted.

Reassessment of the original VZV gB signal sequence

Typical signal sequences of a class 1 glycoprotein are between 16 and 30 residues in length. Because of previous inefficient attempts to detect gB expression starting from nucleotide 190–192 ATG codon, we analysed this issue in more detail (Maresova *et al.*, 2003). In the original publication of the entire VZV sequence, an unusually short signal sequence was projected for the gB gene 31. Subsequently, we analysed the signal sequences of HSV-1 gB, EHV-1 gB, PRV gB. After these investigations, we postulated that the authentic initiating methionine codon was located 189 bp upstream of the original projection. This realignment indicates that the signal sequence cleavage site is present between residues Ser and Val (numbers 71 and 72). In the case of VZV gB, therefore, the designation of the initiating methionine 63-aa upstream of the original methionine produces an unusually long signal sequence of 71 residues.

In short, the VZV gB sequence was re-examined in light of recent knowledge about unusually long gB signal peptides in other herpes viral gB homologs. Through mutational analysis, the discovery was made that the authentic initiating methionine for VZV gB is a codon beginning at genome nucleotide 56,819. The total length for the VZV gB primary translation product was 931 amino acids with a 71-aa signal sequence (Fig. 9.8). Considering the likely signal sequence cleavage site to be located between Ser 71 and Val 72, the length of the mature VZV gB polypeptide would then be 860 amino acids prior to further internal cleavage between amino acids Arg 494 and Ser 495.

VZV gB endocytosis motif

Endocytosis is a common trafficking feature of VZV glycoproteins. The VZV gB glycoprotein contains three potential endocytosis motifs in its long cytoplasmic tail (Heineman and Hall, 2001). The three motifs include the following: YMTL (aa 818–821), YSRV (aa 857–868) and LL (aa 841–842) (Fig. 9.3). Mutational analyses demonstrated that the two tyrosine-based motifs influenced VZV gB endocytosis from the plasma membrane and its subsequent trafficking within the cytoplasm.

VZV glycoprotein gK (ORF 5)

VZV gK structure

VZV gK is a protein consisting of 340 amino acids (Fig. 9.9). The sequence is predicted to contain four membrane-spanning domains and two N-linked glycosylation sites (Mo *et al.*, 1999). The mature glycoprotein migrates at M_r 40 kDa. The VZV gK protein is 28% identical to HSV-1 gK.

VZV gK endocytosis motif

HSV-1 gK is internalized from the plasma membrane by clathrin-mediated endocytosis. An initial characterization of VZV gK has been published. VZV gK contains a consensus tyrosine-based endocytosis motif and likely is internalized in a similar pathway (Govero *et al.*, 2007; Hall *et al.*, 2007). Thus, VZV gK joins four other VZV glycoproteins (gE, gI, gB and gH) with endocytosis motifs (Fig. 9.3).

VZV glycoproteins gM (ORF 50) and gN (ORF 9A)

VZV gM structure

VZV gM is a protein consisting of 435 amino acids (Fig. 9.10). VZV gM is predicted to have eight membrane-spanning domains, as well as one N-linked glycosylation site (Yamagishi *et al.*, 2008). VZV gM also has a cysteine residue in the ectodomain, which forms a disulfide bond with its partner, gN. The glycoprotein complex functions

VZV ORF 5 gK

1 MQALGIKTEH FIIMCLLSGH AVFTLWYTAR VKFEHECVYA TTVINGGPVV

51 WGSYNNSLIY VTFVNHSTFL DGLSGYDYSC RENLLSGDTM VKTAISTPLH

101 DKIRIVLGTR NCHAYFWCVQ LKMIFFAWFV YGMYLQFRRI RRMFGPFRSS

151 CELISPTSYS LNYVTRVISN ILLGYPYTKL ARLLCDVSMR RDGMSKVFNA

201 DPISFLYMHK GVTLLMLLEV IAHISSGCIV LLTLGVAYTP CALLYPTYIR

251 ILAWVVVCTL AIVELISYVR PKPTKDNHLN HINTGGIRGI CTTCCATVMS

301 GLAIKCFYIV IFAIAVVIFM HYEQRVQVSL FGESENSQKH

VZV ORF 50 gM

1 MGTQKKGPRS EKVSPYDTTT PEVEALDHQM DTLNWRIWII QVMMFTLGAV

51 MLLATLIAAS SEYTGIPCFY AAVVDYELFN ATLDGGVWSG NRGGYSAPVL

101 FLEPHSVVAF TYYTALTAMA MAVYTLITAA IIHRETKNQR VRQSSGVAWL

151 VVDPTTLFWG LLSLWLLNAV VLLLAYKQIG VAATLYLGHF ATSVIFTTYF

201 CGRGKLDETN IKAVANLRQQ SVFLYRLAGP TRAVFVNLMA ALMAICILFV

251 SLMLELVVAN HLHTGLWSSV SVAMSTFSTL SVVYLIVSEL ILAHYIHVLI

301 GPSLGTLVAC ATLGTAAHSY MDRLYDPISV QSPRLIPTTR GTLACLAVFS

351 VVMLLLRLMR AYVYHRQKRS RFYGAVRRVP ERVRGYIRKV KPAHRNSRRT

401 NYPSQGYGYV YENDSTYETD REDELLYERS NSGWE

VZV ORF 9A gN

1 MGSITASFIL ITMQILFFCE DSSGEPNFAE RNFWHASCSA RGVYIDGSMI

51 TTLFFYASLL GVCVALISLA YHACFRLFTR SVLRSTW

Figure 9.10 Amino acid sequences of VZV gK, gM and gN. These latter two glycoproteins are found as a complex in infected cells. The nucleotide and deduced amino acids sequences were published by Peters *et al.* (2006).

during cell-to-cell spread, but is not essential for virus growth.

VZV gN structure

VZV gN is a protein consisting of 87 amino acids (Fig. 9.10). In the closely related alphaherpesvirus pseudorabies virus, gN is typically found as a disulfide-linked complex together with gM on the virion (Jons *et al.*, 1998). In mutated pseudorabies virus, gN is absent from gM-negative virions whereas gM is detectable in gN-negative virions. Thus, gM appears to be required for proper trafficking of gN during virion assembly.

VZV receptor and VZV gE

Insulin degradation enzyme

VZV does not attach to any of the known receptors for HSV (Spear, 2004). In 2006, insulin degradation enzyme (IDE) was reported to be a cellular receptor for VZV; IDE mediated infection and cell-to-cell spread of the virus via its proposed ligand gE (Li *et al.*, 2006). IDE is a zinc metalloprotease that is known to degrade a number of small proteins, including insulin and beta-amyloid (Duckworth *et al.*, 1998). Data in the VZV receptor report included the following observations. Initially, a truncated VZV gE protein was immobilized onto protein A-Sepharose beads with anti-gE monoclonal antibody. When the beads were incubated with cell lysates, a protein identified as IDE was attached to the complex. The authors also performed co-immunoprecipitation assays. When gE was precipitated from VZV-infected melanoma cells, they observed that IDE was also detectable in the precipitate, and vice-versa. When a rabbit polyclonal anti-IDE Ab was added to a monolayer prior to VZV infection, VZV spread was subsequently inhibited by 30–45%. Knockdown of IDE by siRNA inhibited VZV infection and cell-to-cell spread. Finally, these deficiencies were corrected by expression of exogenous IDE. In a subsequent publication, the IDE binding domain was localized to the gE N-terminal amino acids 32 to 71 (Ali *et al.*, 2009).

In a subsequent contradictory report, the authors re-assessed the data about the VZV gE/IDE interaction (Carpenter, 2010). They first confirmed the interaction. But they found that the interaction occurred predominantly in the cytosol with a precursor non-glycosylated gE form (73 kDa) rather than at the cell surface with the mature gE glycoprotein (98 kDa). In fact, the gE/IDE interaction occurred when VZV-infected cells were treated with tunicamycin, a chemical which prevents the biosynthesis of mature glycosylated gE in the ER/Golgi and thus prevents transport of gE to the cell surface (Montalvo *et al.*, 1985; Tkacz and Lampen, 1975). The above results suggested that the VZV gE/IDE interaction may occur either in the ER or adjacent to the ER, a site where IDE is known to interact with other proteins during ER stress and retrotranslocation. In the same report, the authors demonstrated that IDE antibody in culture medium inhibited the growth of both infected and well as uninfected cultured cells. Thus, IDE is unlikely to be the cell surface receptor for VZV.

Cellular binding partners of gE

In order to define the cellular binding partners of gE, the viral glycoprotein was purified by affinity chromatography and then subjected to Orbitrap mass spectrometry (Carpenter, 2010). Components of the cytoskeleton such as vimentin, actin and tubulin were detected in the purified gE fraction. In addition, four members of the unfolded protein response were detected: heat shock proteins HSPA5 (BiP), HSPA8, and HSPD1, as well as PPIA (peptidyl-propyl *cis–trans* isomer catalyst). The discovery of the latter proteins suggests that abundant viral glycoprotein production leads to ER stress during the virus growth cycle in cultured cells.

Conclusion

VZV has fewer glycoproteins that HSV. The functions of the homologous glycoproteins have exhibited both similarities and differences, but some differences are striking. Notable among the differences is the essential and predominant role of VZV gE. VZV gE has sometimes been called the VZV navigator protein because of its important role in cell-to-cell spread. Even a single missense mutation in gE can lead to an accelerated cell spread phenotype. VZV gE also is heavily phosphorylated by both cellular and viral kinases, and these phosphorylation events facilitate its endocytosis from the infected cell surface. VZV gE is often found as a complex partner with gI, another phosphorylated glycoprotein with an endocytosis motif in its endodomain. Although not essential like gE, absence of gI profoundly inhibits cell spread.

A second glycoprotein with striking differences from HSV is VZV gC, a gene found in the true late kinetic phase of replication. In contrast to HSV gC, both VZV gC transcription and protein production are markedly delayed in cultured cells where transcripts are first detected around 72 h_r post infection. Since infected cultures are often harvested between 72 to 96 h_r post infection, many infected monolayers in the past may have been harvested when there was little or no gC production. This limited gC production was documented in several cell lines infected with different VZV laboratory strains.

VZV gH and gL form a complex in a similar manner to HSV gH and gL. VZV gH is an important fusogen, and this property is influenced by the endocytosis of the gH from the cell surface. VZV gH also harbours a complement independent neutralization epitope. Monoclonal antibodies directed against this epitope have a dramatic effect on progression of infection in cultured cells, where egress of particles is blocked. Similarly, in the SCID-hu model of VZV infection, anti-gH monoclonal antibody inhibits progression of infection and thereby limits pathogenesis. The ability of a single VZV anti-gH antibody to halt progression of infection and egress of viral particles is another striking difference from HSV.

VZV gB has similarities with HSV gB, in that both participate in virus-induced cell fusion. As with gE gI and gH, gB also has an endocytosis motif in its endodomain. The other three VZV glycoproteins gK, gM and gN have not yet been extensively characterized but studies to date have not uncovered remarkable differences from their HSV counterparts.

Of note, the VZV genome lacks the HSV gD glycoprotein, a prominent component of the HSV fusion quartet of glycoproteins gH, gL, gB and gD. In contrast also to HSV, VZV does not attach to the same receptors. Although IDE has been proposed as a VZV receptor, recent data suggest that gE interacts with precursor gE in the cytosol and not as a cell surface receptor.

Finally, glycoproteins are obviously the essential components of the virion envelope. Yet, high resolution imaging studies of VZV grown in cell culture have documented that the vast majority of progeny are aberrant particles containing an envelope but lacking an internal capsid. These particles have been termed light or L particles. These particles are assembled before VZV gC is produced and therefore likely contain an envelope lacking one of the major structural glycoproteins. Furthermore, enumeration of the numbers of L particles and complete particles by high resolution imaging has documented an extremely high particle: PFU ratio of 40,000:1.Thus, studies of VZV entry and egress are confounded by the large number of aberrant particles in cultured cells.

Acknowledgements

The research described in this chapter was supported by grants from the National Institutes of Health, Bethesda, MD.

References

Ali, M.A., Li, Q., Fischer, E.R., and Cohen, J.I. (2009). The insulin degrading enzyme binding domain of varicella-zoster virus (VZV) glycoprotein E is important for cell-to-cell spread and VZV infectivity, while a glycoprotein I binding domain is essential for infection. Virology *386*, 270–279.

Babst, M., Sato, T.K., Banta, L.M., and Emr, S.D. (1997). Endosomal transport function in yeast requires. a novel AAA-type ATPase, Vps4p. EMBO J. *16*, 1820–1831.

Calistri, A., Sette, P., Salata, C., Cancellotti, E., Forghieri, C., Comin, A., Gottlinger, H., Campadelli-Fiume, G., Palu, G., and Parolin, C. (2007). Intracellular trafficking and maturation of herpes simplex virus type 1 gB and virus egress require functional biogenesis of multivesicular bodies. J. Virol. *81*, 11468–11478.

Carpenter, J.E., Henderson, E.P., and Grose, C. (2009). Enumeration of an extremely high particle to pfu ratio for varicella zoster virus. J. Virol. *83*, 6917–6921.

Carpenter, J.E., Hutchinson, J.A., Jackson, W., and Grose, C. (2008). Egress of light particles among filopodia on the surface of Varicella-Zoster virus-infected cells. J. Virol. *82*, 2821–2835.

Carpenter, J.E., Jackson, W., deSouza, G.A., Haarr, L., and Grose, C. (2010). Insulin degrading enzyme binds to the nonglycosylated precursor of varicella zoster virus gE protein found in the endoplasmic reticulum. J. Virol. *84*, 847–855.

Cole, N.L., and Grose, C. (2003). Membrane fusion mediated by herpesvirus glycoproteins: the paradigm of varicella-zoster virus. Rev. Med. Virology *13*, 207–222.

Davison, A.J., and Scott, J.E. (1986). The complete DNA sequence of varicella-zoster virus. J. Gen. Virol. *67*, 1759–1816.

Duckworth, W.C., Bennett, R.G., and Hamel, F.G. (1998). Insulin degradation: progress and potential. Endocr. Rev. *19*, 608–624.

Duus, K.M., and Grose, C. (1996). Multiple regulatory effects of varicella-zoster virus (VZV) gL on trafficking patterns and fusogenic properties of VZV gH. J. Virol. *70*, 8961–8971.

Duus, K.M., Hatfield, C., and Grose, C. (1995). Cell surface expression and fusion by the varicella-zoster virus gH:gL glycoprotein complex: analysis by laser scanning confocal microscopy. Virology *210*, 429–440.

Forghani, B., Ni, L., and Grose, C. (1994). Neutralization epitope of the varicella-zoster virus gH:gL glycoprotein complex. Virology *199*, 458–462.

Govero, J., Hall, S., and Heineman, T.C. (2007). Intracellular localization of varicella-zoster virus ORF39 protein and its functional relationship to glycoprotein K. Virology *358*, 291–302.

Grose, C. (1990a). Glycoproteins encoded by varicella-zoster virus: biosynthesis, phosphorylation, and intracellular trafficking. Annu. Rev. Microbiol. *44*, 59–80.

Grose, C. (1990b). Glycoproteins of varicella-zoster virus and their herpes simplex homologs. Rev. Infect. Dis. *13*, S960–963.

Grose, C., and Brunel, P.A. (1978). Varicella-zoster virus: isolation and propagation in human melanoma cells at 36 and 32 degrees C. Infect. Immun. *19*, 199–203.

Grose, C., Edmond, B.J., and Brunell, P.A. (1979). Complement-enhanced neutralizing antibody response to varicella-zoster virus. J. Infect. Dis. *139*, 432–437.

Grose, C., Edwards, D.P., Friedrichs, W.E., Weigle, K.A., and McGuire, W.L. (1983). Monoclonal antibodies against three major glycoproteins of varicella-zoster virus. Infect. Immun. *40*, 381–388.

Grose, C., Jackson, W., and Traugh, J.A. (1989). Phosphorylation of varicella-zoster virus glycoprotein gpI by mammalian casein kinase II and casein kinase I. J. Virol. *63*, 3912–3918.

Grose, C., Tyler, S., Peters, G., Hiebert, J., Stephens, G.M., Ruyechan, W.T., Jackson, W., Storlie, J., and Tipples, G.A. (2004). Complete DNA sequence analyses of the first two varicella-zoster virus glycoprotein E (D150N) mutant viruses found in North America: evolution of genotypes with an accelerated cell spread phenotype. J. Virol. *78*, 6799–6807.

Hall, S.L., Govero, J.L., and Heineman, T.C. (2007). Intracellular transport and stability of varicella-zoster virus glycoprotein K. Virology *358*, 283–290.

Harper, D.R., and Kangro, H.O. (1990). Lipoproteins of varicella-zoster virus. J. Gen. Virol. *71*, 459–463.

Heineman, T.C., and Hall, S.L. (2001). VZV gB endocytosis and Golgi localization are mediated by YXXphi motifs in its cytoplasmic domain. Virology *285*, 42–49.

Jons, A., Dijkstra, J.M., and Mettenleiter, T.C. (1998). Glycoproteins M and N of pseudorabies virus form a disulfide-linked complex. J. Virol. *72*, 550–557.

Kaye, J.F., Gompels, U.A., and Minson, A.C. (1992). Glycoprotein H of human cytomegalovirus (HCMV) forms a stable complex with the HCMV UL115 gene product. J. Gen. Virol. *73*, 2693–2698.

Kinchington, P.R., Ling, P., Pensiero, M., Gershon, A., Hay, J., and Ruyechan, W.T. (1990a). A possible role for glycoprotein gpV in the pathogenesis of varicella-zoster virus. Adv. Exp. Med. Biol. *278*, 83–91.

Kinchington, P.R., Ling, P., Pensiero, M., Moss, B., Ruyechan, W.T., and Hay, J. (1990b). The glycoprotein products of varicella-zoster virus gene 14 and their defective accumulation in a vaccine strain (Oka). J. Virol. *64*, 4540–4548.

Levine, M., Krikos, A., Glorioso, J.C., and Homa, F.L. (1990). Regulation of expression of the glycoprotein genes of herpes simplex virus type 1 (HSV-1). Adv. Exp. Med. Biol. *278*, 151–164.

Li, Q., Ali, M.A., and Cohen, J.I. (2006). Insulin degrading enzyme is a cellular receptor mediating varicella-zoster virus infection and cell-to-cell spread. Cell *127*, 305–316.

Ling, P., Kinchington, P.R., Ruyechan, W.T., and Hay, J. (1991). A detailed analysis of transcripts mapping to varicella zoster virus gene 14 (glycoprotein V). Virology *184*, 625–635.

Maresova, L., Kutinova, L., Ludvikova, V., Zak, R., Mares, M., and Nemeckova, S. (2000). Characterization of interaction of gH and gL glycoproteins of varicella-zoster virus: their processing and trafficking. J. Gen. Virol. *81*, 1545–1552.

Maresova, L., Pasieka, T., Wagenaar, T., Jackson, W., and Grose, C. (2003). Identification of the authentic varicella-zoster virus gB (gene 31) initiating methionine overlapping the 3′ end of gene 30. J. Med. Virol. *70*, S64–70.

Maresova, L., Pasieka, T.J., Homan, E., Gerday, E., and Grose, C. (2005). Incorporation of three endocytosed varicella-zoster virus glycoproteins, gE, gH, and gB, into the virion envelope. J. Virol. *79*, 997–1007.

Mo, C., Suen, J., Sommer, M., and Arvin, A. (1999). Characterization of Varicella-Zoster virus glycoprotein K (open reading frame 5) and its role in virus growth. J. Virol. *73*, 4197–4207.

Moffat, J.F., Zerboni, L., Kinchington, P.R., Grose, C., Kaneshima, H., and Arvin, A.M. (1998). Attenuation of the vaccine Oka strain of varicella-zoster virus and role of glycoprotein C in alphaherpesvirus virulence demonstrated in the SCID-hu mouse. J. Virol. *72*, 965–974.

Montalvo, E.A., and Grose, C. (1986a). Neutralization epitope of varicella zoster virus on native viral glycoprotein gp118 (VZV glycoprotein gpIII). Virology *149*, 230–241.

Montalvo, E.A., and Grose, C. (1986b). Varicella zoster virus glycoprotein gpI is selectively phosphorylated by a virus-induced protein kinase. Proc. Natl. Acad. Sci. U.S.A. *83*, 8967–8971.

Montalvo, E.A., and Grose, C. (1987). Assembly and processing of the disulfide-linked varicella-zoster virus glycoprotein gpII(140). J. Virol. *61*, 2877–2884.

Montalvo, E.A., Parmley, R.T., and Grose, C. (1985). Structural analysis of the varicella-zoster virus gp98–gp62 complex: posttranslational addition of N-linked and O-linked oligosaccharide moieties. J. Virol. *53*, 761–770.

Olson, J.K., Bishop, G.A., and Grose, C. (1997). Varicella-zoster virus Fc receptor gE glycoprotein: serine/threonine and tyrosine phosphorylation of monomeric and dimeric forms. J. Virol. *71*, 110–119.

Olson, J.K., and Grose, C. (1997). Endocytosis and recycling of varicella-zoster virus Fc receptor glycoprotein gE: internalization mediated by a YXXL motif in the cytoplasmic tail. J. Virol. *71*, 4042–4054.

Olson, J.K., and Grose, C. (1998). Complex formation facilitates endocytosis of the varicella-zoster virus gE:gI Fc receptor. J. Virol. *72*, 1542–1551.

Olson, J.K., Santos, R.A., and Grose, C. (1998). Varicella-zoster virus glycoprotein gE: endocytosis and trafficking of the Fc receptor. J. Infect. Dis. *178 Suppl 1*, S2–6.

Pasieka, T.J., Maresova, L., and Grose, C. (2003a). A functional YNKI motif in the short cytoplasmic tail of varicella-zoster virus glycoprotein gH mediates

clathrin-dependent and antibody-independent endocytosis. J. Virol. *77*, 4191–4204.
Pasieka, T.J., Maresova, L., Shiraki, K., and Grose, C. (2004). Regulation of varicella-zoster virus-induced cell-to-cell fusion by the endocytosis-competent glycoproteins gH and gE. J. Virol. *78*, 2884–2896.
Pasieka, T.J., Woolson, R.F., and Grose, C. (2003b). Viral induced fusion and syncytium formation: measurement by the Kolmogorov-Smirnov statistical test. J. Virol. Methods *111*, 157–161.
Peters, G.A., Tyler, S.D., Grose, C., Severini, A., Gray, M.J., Upton, C., and Tipples, G.A. (2006). A full-genome phylogenetic analysis of varicella-zoster virus reveals a novel origin of replication-based genotyping scheme and evidence of recombination between major circulating clades. J. Virol. *80*, 9850–9860.
Rodriguez, J.E., Moninger, T., and Grose, C. (1993). Entry and egress of varicella virus blocked by same anti-gH monoclonal antibody. Virology *196*, 840–844.
Rusten, T.E., Vaccari, T., Lindmo, K., Rodahl, L.M., Nezis, I.P., Sem-Jacobsen, C., Wendler, F., Vincent, J.P., Brech, A., Bilder, D., and Stenmark, H. (2007). ESCRTs and Fab1 regulate distinct steps of autophagy. Curr. Biol. *17*, 1817–1825.
Ruyechan, W.T., Peng, H., Yang, M., and Hay, J. (2003). Cellular factors and IE62 activation of VZV promoters. J. Med. Virol. *70* Suppl *1*, S90–94.
Santos, R.A., Hatfield, C.C., Cole, N.L., Padilla, J.A., Moffat, J.F., Arvin, A.M., Ruyechan, W.T., Hay, J., and Grose, C. (2000). Varicella-zoster virus gE escape mutant VZV-MSP exhibits an accelerated cell-to-cell spread phenotype in both infected cell cultures. and SCID-hu mice. Virology *275*, 306–317.
Santos, R.A., Padilla, J.A., Hatfield, C., and Grose, C. (1998). Antigenic variation of varicella zoster virus Fc receptor gE: loss of a major B cell epitope in the ectodomain. Virology *249*, 21–31.
Spaete, R.R., Perot, K., Scott, P.I., Nelson, J.A., Stinski, M.F., and Pachl, C. (1993). Coexpression of truncated human cytomegalovirus gH with the UL115 gene product or the truncated human fibroblast growth factor receptor results in transport of gH to the cell surface. Virology *193*, 853–861.
Spear, P.G. (2004). Herpes simplex virus: receptors and ligands for cell entry. Cell. Microbiol. *6*, 401–410.
Spear, P.G., and Longnecker, R. (2003). Herpesvirus Entry: an Update. J. Virol. *77*, 10179–10185.
Spear, P.G., Shieh, M.T., Herold, B.C., WuDunn, D., and Koshy, T.I. (1992). Heparan sulfate glycosaminoglycans as primary cell surface receptors for herpes simplex virus. Adv. Exp. Med. Biol. *313*, 341–353.
Stingley, S.W., Ramirez, J.J.G., Aguilar, S.A., Simmen, K., Sandri-Goldin, R.M., Ghazal, P., and Wagner, E.K. (2000). Global Analysis of Herpes Simplex Virus Type 1 Transcription Using an Oligonucleotide-Based DNA Microarray. J. Virol. *74*, 9916–9927.
Storlie, J., Carpenter, J.E., Jackson, W., and Grose, C. (2008a). Discordant varicella-zoster virus glycoprotein C expression and localization between cultured cells and human skin vesicles. Virology *382*, 171–181.
Storlie, J., Jackson, W., Hutchinson, J., and Grose, C. (2006). Delayed biosynthesis of varicella-zoster virus glycoprotein C: upregulation by hexamethylene bisacetamide and retinoic acid treatment of infected cells. J. Virol. *80*, 9544–9556.
Storlie, J., Maresova, L., Jackson, W., and Grose, C. (2008b). Comparative analyses of the 9 glycoprotein genes found in wild-type and vaccine strains of varicella-zoster virus. J. Infect. Dis. *197* Suppl 2, S49–53.
Takahashi, M.N., Jackson, W., Laird, D.T., Culp, T.D., Grose, C., Haynes, J.I., 2nd, and Benetti, L. (2009). Varicella-Zoster Virus Infection Induces Autophagy in Both Cultured Cells and Human Skin Vesicles. J. Virol. *83*, 5466–5476.
Tipples, G.A., Stephens, G.M., Sherlock, C., Bowler, M., Hoy, B., Cook, D., and Grose, C. (2002). New variant of varicella-zoster virus. Emerg. Infect. Dis. *8*, 1504–1505.
Tkacz, J.S., and Lampen, O. (1975). Tunicamycin inhibition of polyisoprenyl N-acetylglucosaminyl pyrophosphate formation in calf-liver microsomes. Biochim. Biophys. Res. Commun. *65*, 248–257.
Trowbridge, I.S. (1991). Endocytosis and signals for internalization. Curr. Opin. Cell Biol. *3*, 634–641.
Tuazon, P.T., and Traugh, J.A. (1991). Casein kinase I and II–multipotential serine protein kinases: structure, function, and regulation. Adv Second Messenger Phosphoprotein Res. *23*, 123–164.
Tyler, S.D., Peters, G.A., Grose, C., Severini, A., Gray, M.J., Upton, C., and Tipples, G.A. (2007). Genomic cartography of varicella-zoster virus: a complete genome-based analysis of strain variability with implications for attenuation and phenotypic differences. Virology *359*, 447–458.
Vleck, S.E., Oliver, S.L., Reichelt, M., Rajamani, J., Zerboni, L., Jones, C., Zehnder, J., Grose, C., and Arvin, A.M. (2009). Anti-Glycoprotein H Antibody Impairs the Pathogenicity of Varicella-Zoster Virus in Skin Xenografts in the SCID Mouse Model. J. Virol. *84*, 141–152.
Wagenaar, T.R., Chow, V.T., Buranathai, C., Thawatsupha, P., and Grose, C. (2003). The out of Africa model of varicella-zoster virus evolution: single nucleotide polymorphisms and private alleles distinguish Asian clades from European/North American clades. Vaccine *21*, 1072–1081.
Weigle, K.A., and Grose, C. (1983). Common expression of varicella-zoster viral glycoprotein antigens *in vitro* and in chickenpox and zoster vesicles. J. Infect. Dis. *148*, 630–638.
Yamagishi, Y., Sadaoka, T., Yoshii, H., Somboonthum, P., Imazawa, T., Nagaike, K., Ozono, K., Yamanishi, K., and Mori, Y. (2008). Varicella-zoster virus glycoprotein M homolog is glycosylated, is expressed on the viral envelope, and functions in virus cell-to-cell spread. J. Virol. *82*, 795–804.
Yao, Z., and Grose, C. (1994). Unusual phosphorylation sequence in the gpIV (gI) component of the varicella-zoster virus gpI-gpIV glycoprotein complex (VZV gE–gI complex). J. Virol. *68*, 4204–4211.

Yao, Z., Jackson, W., Forghani, B., and Grose, C. (1993). Varicella-zoster virus glycoprotein gpI/gpIV receptor: expression, complex formation, and antigenicity within the vaccinia virus-T7 RNA polymerase transfection system. J. Virol. *67*, 305–314.

Yao, Z., Jones, D.H., and Grose, C. (1992). Site-directed mutagenesis of herpesvirus glycoprotein phosphorylation sites by recombination polymerase chain reaction. PCR Methods Appl *1*, 205–207.

Ye, M., Duus, K.M., Peng, J., Price, D.H., and Grose, C. (1999). Varicella-zoster virus Fc receptor component gI is phosphorylated on its endodomain by a cyclin-dependent kinase. J. Virol. *73*, 1320–1330.

Nucleocapsid Structure, Assembly and DNA Packaging of Herpes Simplex Virus

10

James F. Conway and Fred L. Homa

Abstract

The herpes simplex virion consists of an external membrane envelope, a proteinaceous layer called the tegument, and an icosahedral capsid containing the double-stranded linear DNA genome. The capsid shell is 125 nm in diameter and consists of 162 capsomers (150 hexons, 11 pentons and a portal) which lie on a T=16 icosahedral lattice. The capsid shell consists of four major structural proteins VP5, VP19C, VP23 and VP26 which are the products of the HSV UL19, UL38, UL18 and UL35 genes. In addition to the four major structural proteins the HSV-1 capsid contains a number of minor capsid proteins. These include the UL6, UL15, UL17, UL25, UL28 and UL33 proteins, all of which (along with the HSV-1 UL32 protein) are required for the processing and packaging of replicated viral DNA into preformed capsid shells. The UL6, UL17, UL25 and UL33 proteins remain associated with DNA containing capsids while UL15 and UL28 do not. This review is a summary of our present knowledge with respect to how the capsid is assembled, how DNA is packaged and what is known about the role of the seven packaging proteins in this process. In addition, recent advances in our understanding the structure of the four distinct types of capsids that are present in HSV infected cells as determined by three-dimensional image reconstructions from cryo-electron microscopy (cryoEM) will be presented and discussed.

Introduction

The herpesviruses comprise a large family of double stranded DNA viruses. Eight of the herpesviruses infect humans: herpes simplex virus types 1 and 2 (HSV-1 and HSV-2), varicella zoster virus (VZV), human cytomegalovirus (HCMV), Epstein–Barr virus (EBV), and human herpesviruses 6, 7, and 8 (HHV-6, HHV-7, and HHV-8). Several of these viruses are important human pathogens and all remain with the host for life. Herpesviruses are capable of residing in a latent or quiescent state within the host where they avoid immune clearance. Primary or recurrent infections can be life threatening in immunocompromised patients such as AIDS or transplant patients where infection with HCMV can result in retinitis, pneumonia, and gastrointestinal disease (Dykewicz, 2001; Field, 1999; Griffiths, 1996; Smith *et al.*, 2001). Current treatments against HSV-1 and HSV-2 and the other human herpesviruses rely primarily on blocking viral DNA replication. With the exception of acyclovir (and its prodrug valacyclovir) nearly all drugs that block viral DNA replication, including ganciclovir and cidofovir, have significant toxic side effects (mainly because these drugs inhibit host polymerases) that include neutropenia, CNS symptoms and liver toxicity (Crumpacker, 1996; Emery and Hassan-Walker, 2002; McGavin and Goa, 2001; Villarreal, 2003; Wagstaff, 2001). Experience with anti-bacterials has taught us that it is rarely satisfactory to have a single class of drug to treat infectious agents that can readily generate resistant mutants. The success of mechanism-based antiviral research that led to the discovery of novel HIV antivirals, such as non-nucleoside reverse transcriptase inhibitors and HIV protease inhibitors, has generated interest in the discovery of new herpesvirus antivirals that exploit novel processes such as virus entry into host cells and virus assembly. Capsid formation and DNA packaging represent attractive targets for new anti-herpesvirus therapeutics. Capsid assembly involves highly specific interactions among at least five different proteins and seven additional proteins are involved in preparation of DNA for encapsidation and packaging. Most of the proteins involved in capsid assembly and DNA packaging are conserved in other herpesviruses suggesting that these mechanisms will also be similar between them. The discovery of small organic molecules that inhibit HSV replication by blocking processing and encapsidation demonstrates the validity of targeting the cleavage/packaging complex for therapeutic intervention. These efforts have lead to the discovery of three types of compounds, benzimidazole ribosides, substituted thioureas and BAY 38–4766, that act by inhibiting steps in DNA processing and encapsidation (Buerger *et al.*, 2001; Krosky *et al.*, 1998; Newcomb and Brown, 2002; van Zeijl *et al.*, 2000; Zacny *et al.*, 1999). Isolation of drug resistant mutants demonstrated that two of these drugs target the terminase subunits of HCMV (Krosky *et al.*, 1998; Underwood *et al.*, 1998; van Zeijl *et al.*, 2000; Zacny *et al.*, 1999) and the third compound acts by blocking the incorporation of the portal protein into HSV-1 capsids (Newcomb and Brown, 2002).

Herpesvirus capsid structure

Both the structure and the assembly pathway of herpesvirus capsids resemble those of dsDNA bacteriophages such as T4, T7, P22, HK97 and λ (Black, 1994; Casjens, 1988; Prevelige and King, 1993; Steven *et al.*, 2005). For example, in both HSV-1 and dsDNA phages, a closed DNA-free procapsid is first formed and later filled with DNA; capsid assembly involves interaction of the major capsid protein with a scaffolding protein or domain not found in the mature virion; DNA is packaged by an ATP-driven pump located at a specialized vertex of the capsid containing a dodecameric portal structure; and packaging triggers conformational changes in the capsid termed expansion. Recent evidence suggests that the herpesvirus major capsid protein fold is similar to that of bacteriophage HK97, and probably most or all other dsDNA-tailed phages, (Baker *et al.*, 2005a; Steven *et al.*, 2005) and we have visualized several dsDNA phages with the distinctive T=16 capsid morphology that was previously considered diagnostic of herpesvirus (Duda *et al.*, 2006; Weigele *et al.*, 2007). Other animal viruses do not resemble dsDNA phage as closely as the herpesviruses in capsid structure or in the process of capsid formation and maturation.

Electron microscopy and biochemical studies provide insight into the structure of the HSV-1 capsid and how it is assembled in the infected cell nucleus (Brown, 2002; Homa and Brown, 1997; Rixon, 1993; Steven, 1996). Infected cells contain four types of capsids. The procapsid is the first closed capsid structure formed in the nucleus and is difficult to isolate due to its fragile nature. The procapsid is the precursor for the three stable capsid types (A-, B-, and C-capsids) that accumulate in the nucleus (Fig. 10.1). The stable capsid types can be distinguished morphologically in electron micrographs and can be separated from each on sucrose density gradients (Gibson and Roizman, 1972). The shell structure is essentially the same for all four capsids but they differ with respect to the minor capsid proteins that are found on the capsid exterior and in the material present inside the capsid cavity. A- and B-capsids lack DNA. The cavity of the procapsid and the B-capsid contain the cleaved form (VP22a) of the scaffolding protein, while in A-capsids the cavity lacks significant amounts of either DNA or protein (Newcomb and Brown, 1991). C-capsids contain the viral DNA and a heterodimer of viral proteins, UL25 and UL17, which form what is referred to as the C-capsid specific component (CCSC) that binds to five sites around each of the 12 capsid vertexes to stabilize the C-capsid after DNA is packaged. The C-capsid is similar, if not identical, in structure and composition to the capsids present in infectious virions (Booy *et al.*, 1991). Studies with HSV mutants demonstrate that procapsids are able to package DNA and mature into infectious virions (Booy *et al.*, 1991; Brown, 2002; Gibson and Roizman, 1972; Homa and Brown, 1997; Sheaffer *et al.*, 2001; Trus *et al.*, 1996). However, A-capsids are considered to be abortive forms that result from failed attempts to

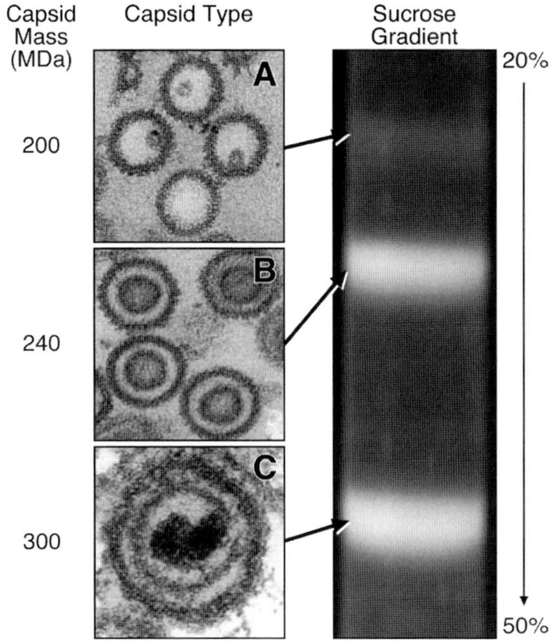

Figure 10.1 Three types of HSV capsids (A, B and C) that are found in HSV-infected cells. A-capsids are empty while B-capsids contain a remnant of the scaffold core and C-capsids enclose one copy of dsDNA genome of the virus. The left panels are thin-section electron micrographs and the right panels show capsid bands from cells infected with HSV that were separated on 20 to 50% sucrose gradients.

package DNA and B-capsids are angularized capsids that never enter the packaging pathway. The mass of each capsid type has been determined by dark-field scanning transmission electron microscopy indicating that B-capsids (M_r ~245×10^6) can be regarded as A-capsids (M_r ~203×10^6) that contain about Mr 40×10^6 of VP22a while C-capsids (M_r~303×10^6) can be considered as A-capsids that contain the Mr 101×10^6 viral genome (Newcomb *et al.*, 1993).

The structure of the HSV-1 capsid has been determined by three-dimensional image reconstructions calculated to resolutions of 3–0.84 nm from cryo-electron microscopy (cryoEM) of all three capsid types (Baker *et al.*, 2003; Booy *et al.*, 1994; Cardone *et al.*, 2007; Cheng *et al.*, 2002; Conway *et al.*, 1993, 1996; Grunewald *et al.*, 2003; Heymann *et al.*, 2003; Newcomb *et al.*, 1993; Schrag *et al.*, 1989; Spencer *et al.*, 1997; Trus *et al.*, 1995, 1999, 2001, 2004; Zhou *et al.*, 1994, 1995, 2000). Icosahedral reconstructions show that the capsid is a protein shell approximately 15 nm thick and 125 nm in diameter (Fig. 10.2). Its major structural features are 162 capsomers (150 hexons, 11 pentons and a portal) that lie on a T=16 icosahedral lattice. The pentons each have five-fold rotational symmetry and are located at all but one capsid vertex, which is occupied by the dodecameric portal complex. The hexons have six-fold symmetry and occupy capsid edges and faces. All capsomers have an axial channel that extends through the shell into the capsid interior. The capsomers are connected by asymmetric structures called triplexes (320 in all) that lie on the capsid floor at the base of the capsomer protrusions. One triplex is found at each local 3-fold position where it makes contacts with all three adjacent capsomers.

The capsid shell is composed predominantly of four proteins, a major capsid protein, VP5, and three less abundant proteins, VP19C, VP23 and VP26 (Fig. 10.2). The protruding capsomers are all made of VP5, hexons each contain six copies of VP5 while pentons have five. The capsid, therefore, contains a total of 955 copies of VP5 (one vertex being occupied by the portal structure). Formation of the capsid from hexamers and pentamers of VP5 conforms to the quasi-equivalence

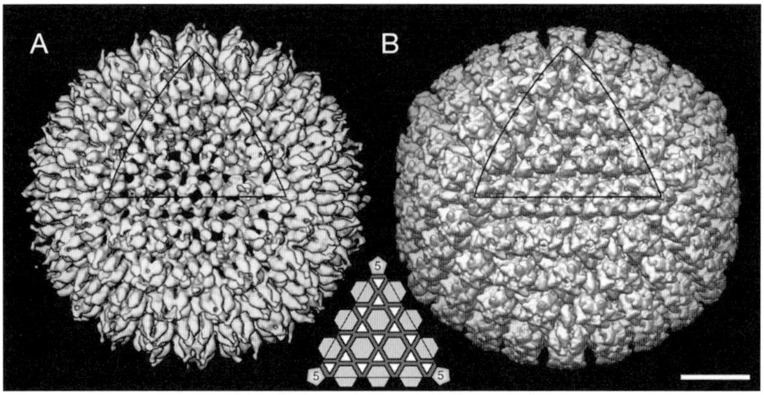

Figure 10.2 Surface views of HSV procapsid (A) (Cheng *et al.*, 2002) and capsid (B) structures calculated from cryoEM data. A triangular face has been outlined on each capsid and capsomer positions are indicated in the inset where '5' corresponds to vertices occupied by pentons, hexons represented as hexagons, and triplex molecules by triangles. Bar = 25nm. The procapsid image was used with permission.

theory for the architecture of icosahedral virus capsids (Caspar and Klug, 1962). VP19C and VP23 together make up the triplexes (Newcomb *et al.*, 1993; Spencer *et al.*, 1998). On average, triplexes contain one copy of VP19C and two of VP23, although triplex composition may prove to vary with position in the capsid shell. VP26 is located at the distal tips of the hexons, but not pentons, with one VP26 molecule bound to each hexon-associated VP5 (Trus *et al.*, 1995; Wingfield *et al.*, 1997; Zhou *et al.*, 1995). In comparison to the alpha- and betaherpesviruses, the gammaherpesviruses bind the VP26 analogue (ORF65 in KSHV) to pentons as well as hexons (Trus *et al.*, 2001).

In addition to the shell proteins, HSV-1 B-capsids contain a large amount of the scaffolding protein (the product of the UL26.5 gene), and smaller amounts of the two products, VP24 and VP21, encoded by the UL26 gene (Liu and Roizman, 1991, 1993). During capsid assembly the scaffolding protein forms a core (the scaffold) internal to the capsid shell and interacts directly with VP5 in the shell (Desai and Person, 1999; Huang *et al.*, 2007; Person and Desai, 1998; Walters *et al.*, 2003; Warner *et al.*, 2001). The site recognized by VP5 has been localized to a discrete region within the C-terminal 14 amino acids of the scaffold that forms a hydrophobic amphipathic helix (Hong *et al.*, 1996; Oien *et al.*, 1997; Thomsen *et al.*, 1995). Genetic analysis of second-site revertant mutants demonstrated that the N-terminus of VP5 is required for the interaction with the scaffold protein (Walters *et al.*, 2003). Results from alanine-scanning mutagenesis of hydrophobic residues in the N-terminus of VP5 revealed seven hydrophobic residues (I27, L35, F39, L58, L65, L67, and L71) that reside in two predicted alpha helices are important for this bimolecular interaction (Huang *et al.*, 2007). The scaffolding protein is lost after capsid assembly, but before or during the time DNA enters, and is not present in the mature HSV-1 virion. Intermediates in the capsid assembly pathway have been identified using a cell-free assembly system (Newcomb *et al.*, 1994, 1996, 1999). Using a panel of recombinant baculoviruses (rBV) encoding capsid proteins, the system is constituted by mixing either extracts of rBV-infected insect cells containing HSV-1 proteins or by mixing the purified capsid proteins together. The *in vitro* assembly system has demonstrated that mature capsids are formed by way of a partial procapsid which grows by addition of capsid and scaffold subunits to form the spherical procapsid (Trus *et al.*, 1996) (Fig. 10.2). The procapsid then angularizes to form the mature, icosahedral capsid (Heymann *et al.*, 2003). This capsid assembly and maturation process is similar to the morphogenesis of dsDNA bacteriophages such as P22 and λ (Bjornsti *et al.*, 1983; Black, 1994; Casjens, 1988; Earnshaw *et al.*, 1978; Hendrix, 1985; Hohn,

1983; Prevelige and King, 1993) and particularly well characterized for phage HK97 (Conway *et al.*, 1995, 2001; Lata *et al.*, 2000; Steven *et al.*, 2005).

C-capsids and capsids of intact virions contain a single copy of the herpesvirus genome. The packaged DNA is free of other proteins, and the DNA is very tightly condensed inside the capsid. Electron micrographs of HSV-1 C-capsids, for example, show an interduplex spacing of 2.6 nm, a value comparable to that found in dsDNA bacteriophages and near the theoretical limit for close packing of DNA duplexes (Booy *et al.*, 1991). Although electron micrographs reveal local areas where packaged HSV-1 DNA strands are aligned, there does not appear to be any pattern of longer range organization. Packaged HSV-1 DNA is considered to be partly or mostly disordered similar to the liquid crystalline state described for packaged DNA in dsDNA bacteriophages such as T4, λ, and T7 (Cerritelli *et al.*, 1997; Earnshaw and Harrison, 1977; Earnshaw *et al.*, 1978).

Capsid assembly and DNA packaging pathway

During the herpesvirus cleavage/packaging reaction (Fig. 10.3) the procapsid assembles from several structural proteins including a scaffold that is subsequently digested by the viral

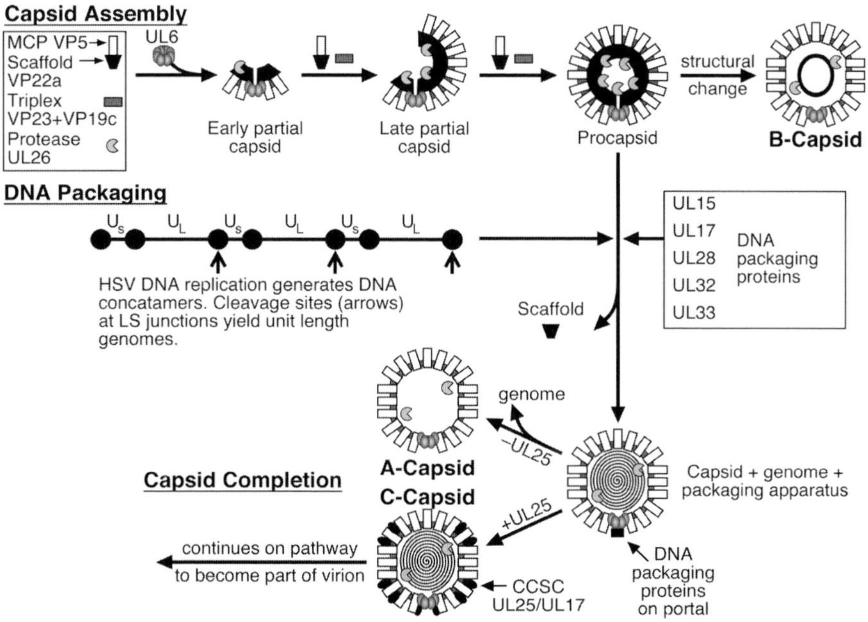

Figure 10.3 Capsid assembly and DNA-packaging pathway for HSV-1. Early events in assembly appear to be controlled when capsid proteins are transported to the nucleus. The scaffold protein (preVP22a) chaperons the major capsid protein (VP5) with the small capsid protein (VP26) bound to VP5 into the nucleus. The triplex protein VP19C, chaperons the second triplex protein VP23 into the nucleus. The scaffold/VP5 complex comes together by virtue of the scaffold protein's ability to self-interact. The triplex complex (VP23/VP19C) is added to the capsid during the formation of the hexons and pentons. The growing capsid shell assembles into an unstable procapsid that undergoes a structural transformation into the stable polyhedral B-capsid, which is a dead-end product. During DNA packaging the portal (UL6) of the procapsid interacts with the terminase composed of UL15, UL28, and UL33 which is an ATP dependent packaging motor that also cleaves the packaged unit length DNA. The scaffold is removed from the capsid during the packaging reaction through the action of the UL26 encoded protease (which is found in all capsid forms and in virions). The DNA containing capsid is stabilized by addition of the CCSC to the capsid vertices and the resulting C-capsid exits the nucleus. If the CCSC is not added the DNA is not retained in the capsid and an empty A-capsid is generated that is retained in the nucleus. The cleavage/packaging and capsid completion reactions can be viewed as separate steps in the overall process of generating a stable DNA-containing capsid.

protease encoded by the UL26 gene and expelled. The DNA chromosome is incorporated into the procapsid through the ring-shaped portal present at a unique vertex and requiring the action of six cleavage/packaging proteins that interact with the procapsid either during capsid assembly or during DNA packaging. Each capsid contains a single portal, located at a unique capsid vertex, that is composed of 12 UL6 protein molecules. The position of the portal requires that capsid formation takes place in such a way that a portal is incorporated into one of the 12 capsid vertices. The location of the portal at a unique vertex in the capsid suggests that there is a mechanism to control portal incorporation (Cardone *et al.*, 2007; Chang *et al.*, 2007). Direct interaction of the scaffold with the portal has been demonstrated by *in vitro* experiments with purified proteins (Newcomb *et al.*, 2003). In studies done using an *in vitro* capsid assembly assay, the interaction of the scaffold protein and portal proteins was found to be essential for incorporation of the portal into the capsid (Newcomb *et al.*, 2005). More importantly, these studies found that in order for the portal to be incorporated into the capsid, it needs to be present when the assembly reaction is started. Delaying addition of the portal to the assembly reaction resulted in assembly of capsids that lacked a portal complex. Studies with deletion mutants demonstrate that amino acids 143 to 151 of the scaffold protein are critical for the formation of portal-containing capsids (Huffman *et al.*, 2008; Singer *et al.*, 2005; Yang and Baines, 2008).

DNA packaging and capsid stabilization

The HSV-1 genome has two segments of unique sequence, designated as long (U_L) and short (U_S), that are flanked by inverted repeats, denoted *ba* and *ca*, respectively (Fig. 10.4). Generation of linear unit-length genomes from concatemeric viral DNA occurs through specific DNA cleavage within the terminally repeated *a* sequences, which contain all of the *cis*-acting signals required for genome maturation (Deiss *et al.*, 1986; Deiss and Frenkel, 1986; Mocarski and Roizman, 1982; Spaete and Frenkel, 1982; Spaete and Frenkel, 1985; Stow *et al.*, 1983; Varmuza and Smiley, 1985). The *a* sequence is composed of a copy of a 20-bp direct repeat (DR1) at each end and two segments of unique sequence U_c and U_b that are separated by a series of directly repeated elements (the 12 bp DR2 sequences – see Fig. 10.4). The specific signals for DNA cleavage and packaging, termed *pac1* and *pac2*, are located within the Ub and Uc regions, respectively. *pac1* is characterized primarily by two stretches of five to eight G nucleotides separated by a 3 to 7-nt T-rich region, whereas *pac2* contains a conserved CGCCGCG motif located near a run of 5 to 10 T's (Brown, 2002; Deiss and Frenkel, 1986; McVoy *et al.*,

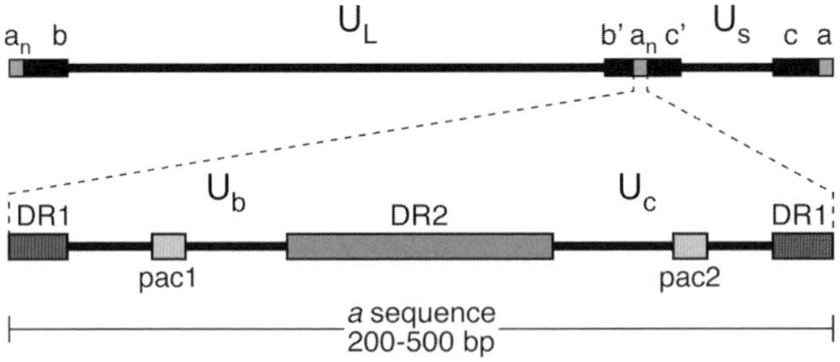

Figure 10.4 The cis–acting cleavage and packaging signals are located within the terminal *a* sequences of the HSV genome. The unique L (long) and S (short) components of the HSV genome are flanked by inverted repeats (b and c) with multiple copies of the *a* sequence found at the L termini and at the L/S junction while only a single a sequence is found at the S termini. The *a* sequence is bordered by two copies of a 20 bp direct repeat called the DR1. The interior of the *a* sequence is composed of a 12 bp repeat sequence, DR2. The DR1 and DR2 sequence are flanked by the terminal U_b and U_c sequences that include the two packaging sequences pac1 and pac2 that are recognized by the HSV cleavage/packaging proteins. In the linear HSV genome the *pac*1 site lies 30 bp from the S terminus and the *pac*2 site lies 32 bp from the L terminus.

1998; Nasseri and Mocarski, 1988; Smiley *et al.*, 1990; Smiley *et al.*, 1992). The DR1 element contains the site of DNA cleavage, although this sequence itself is not required for double-stranded break formation (Varmuza and Smiley, 1985). The DR2 repeats are also dispensable for cleavage and packaging (Smiley *et al.*, 1990).

HSV DNA replication generates a DNA concatemer where only the L-end of the genome is exposed. The S-end of the genomes are found in junctions with the L-end (Martinez *et al.*, 1996; Zhang *et al.*, 1994). The exposure of one genome end in the concatemer is observed in other herpesviruses including human cytomegalovirus, equine herpesvirus 1, murine cytomegalovirus, guinea pig cytomegalovirus, and bovine herpesvirus 1 (Brown, 2002). The exposure of L ends in HSV has led to the assumption that DNA encapsidation begins at the L end and proceeds towards S. Since the S-end is the last end to be packaged, uncoating must begin at the S-end. Using an *in vitro* uncoating assay it has recently been demonstrated that the S segment end does indeed exit the capsid first (Newcomb *et al.*, 2009).

Based on studies with HSV-1 mutants, at least seven viral genes have been shown essential for DNA packaging (Abbotts *et al.*, 2000; al-Kobaisi *et al.*, 1991; Baines *et al.*, 1994; Baines and Roizman, 1992; Cavalcoli *et al.*, 1993; Chang *et al.*, 1996; Lamberti and Weller, 1996, 1998; McNab *et al.*, 1998; Patel and MacLean, 1995; Patel *et al.*, 1996; Poon and Roizman, 1993; Preston *et al.*, 1983; Reynolds *et al.*, 2000; Salmon and Baines, 1998; Salmon *et al.*, 1998; Sheaffer *et al.*, 2001; Taus and Baines, 1998; Taus *et al.*, 1998; Tengelsen *et al.*, 1993; Yu *et al.*, 1997; Yu and Weller, 1998ab). DNA and unfilled capsids accumulate, but no packaging takes place in cells infected with these mutants, which map to the UL6, UL15, UL17, UL25, UL28, UL32, and UL33 genes (Fig. 10.5). DNA packaging takes place in the infected cell nucleus, but active DNA replication is not required at the time encapsidation takes place (Church *et al.*, 1998). There does appear to be an energy requiring step in this process since depletion of ATP blocks DNA packaging (Dasgupta and Wilson, 1999). The proteins encoded by the UL6, UL15, UL17, UL25, UL28 and UL33 genes have been shown to reside as minor components of the capsid (Beard and Baines, 2004; Cockrell *et al.*, 2009; Homa and Brown, 1997; McNab *et al.*, 1998; Mettenleiter, 2004, 2006;

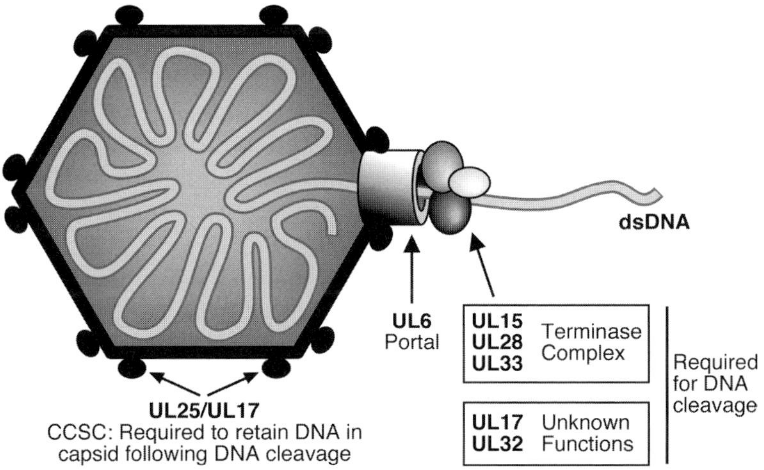

Figure 10.5 Model of the HSV capsid/DNA packaging complex and the proposed function of the cleavage/packaging proteins. HSV terminase complex (UL15, UL28, UL33) binds the U_L-terminus of replicated DNA, cleaves at the *pac*2 site and docks on the portal-vertex (UL6) of the capsid. The U_L-end of the genome is packaged followed by the U_S-end via the terminase/portal complex. The second cleavage at the *pac*1 site generates a unit length HSV genome that forms a stable DNA containing C-capsid following the addition of 5 CCSCs to each of the portal vertices.

Ogasawara *et al.*, 2001; Salmon and Baines, 1998; Sheaffer *et al.*, 2001; Thurlow *et al.*, 2005) and are found in procapsids, polyhedral capsids (mature B-capsids) and in C-capsids although the level of each protein differs with the different capsid types (Sheaffer *et al.*, 2001). Suggested functions for each of the six proteins are based on similarity to DNA packaging in dsDNA bacteriophages. In phage T4 the large subunit of the T4 terminase has amino acid sequence homology to the UL15 protein suggesting that it may function in DNA cleavage or translocation into the capsid (Davison, 1992). The HSV-1 UL28 protein and its HCMV homologue (UL56) both bind specifically to DNA packaging sequences within the viral genome, suggesting they also may be part of the terminase function (Adelman *et al.*, 2001; Bogner, 2002). The dodecameric portal, composed of the UL6 protein, is located at a single capsid vertex (Cardone *et al.*, 2007; Chang *et al.*, 2007; Newcomb *et al.*, 2001b; Newcomb *et al.*, 2003). Electron microscopy and image reconstruction of purified UL6 demonstrated that the complex consists of a dodecameric ring that has an axial channel and peripheral flanges, and resembles known bacteriophage portals in shape (Cardone *et al.*, 2007; Chang *et al.*, 2007; Trus *et al.*, 2004). Docking experiments with a capsid model from which one penton had been computationally excised showed that the portal complex fits snugly into the vacant vertex (Trus *et al.*, 2004) although the precise orientation is disputed (Cardone *et al.*, 2007; Chang *et al.*, 2007). Tailed bacteriophage assembly proceeds through the formation of spherical precursor capsids (procapsids) with a unique portal vertex occupied by a dodecameric ring through which the DNA enters and exits the capsid (Bazinet and King, 1985; Valpuesta and Carrascosa, 1994). The portal vertex is the docking site for the packaging proteins such as terminase, the protein responsible for translocation of DNA into the capsid in an ATP-dependent fashion and cleavage of monomeric units from concatemeric DNA. Portal proteins are found as integral capsid components in both procapsids and mature capsids while terminase proteins are only associated with capsids that are in the process of packaging DNA and are not present in mature capsids or virions. In this respect, UL6 is identical to the phage portal proteins since the level of UL6 protein associated with procapsids, B-capsids and C-capsids is invariant (Sheaffer *et al.*, 2001). In contrast, the amount of the two putative HSV-1 terminase subunits (UL15 and UL28) are reduced in C-capsids relative to the amount found in the HSV-1 procapsids, again fitting with what is found with phage terminase proteins (Sheaffer *et al.*, 2001).

The HSV-1 UL28 gene encodes an 85-kDa protein that has been expressed in bacteria (Adelman *et al.*, 2001) and shown to recognize conserved HSV-1 DNA packaging sequences. Specifically, it was found to bind to the *pac*1 site but only when the DNA was heat denatured. The *pac*1 site appears to adopt a novel DNA conformation when denatured and UL28 binds to only one strand of the *pac*1 motif. The novel DNA binding activity of the UL28 protein and its interaction with the putative nuclease, UL15 protein, suggest a complex process for recognizing and cleaving viral DNA at the proper cleavage site (Jacobson *et al.*, 2006; Yang and Baines, 2006; Yang *et al.*, 2007).

The HSV-1 UL25 gene encodes a 62-kDa protein that acts to maintain stable incorporation of DNA in capsids. In cells infected with a UL25 null virus, DNA appears to enter the capsid and is cleaved by the cleavage/packaging machinery but the DNA is subsequently lost resulting in the accumulation of A-capsids that lack both DNA and scaffold proteins, suggesting that the function of the UL25 protein is to retain DNA in the capsid following the cleavage event (Cockrell *et al.*, 2009; McNab *et al.*, 1998; Ogasawara *et al.*, 2001; Stow, 2001). A similar phenomenon has been reported in phage P22 mutants defective in the gene products gp4, gp10, and gp26 (Black, 1988; Casjens, 1988). In cells infected with these mutants, the filled capsids are unstable and lose mature DNA resulting in empty capsids. The role of these proteins is thought to be 'head completion' the closing of the channel at the unique vertex where DNA enters along with stabilizing the DNA inside the capsid.

The crystal structure of the UL25 gene product showed that it contains a flexible N-terminal tail and a more rigid core of packed α-helices (Bowman *et al.*, 2006). UL25 is a capsid-associated protein present on all capsid forms in increasing amounts from procapsid to B-, A-, C-capsids, and virions (Newcomb *et al.*, 2006; Sheaffer *et al.*, 2001). UL25 attaches to the exterior capsid vertices in a proposed heterodimer with UL17 (Cockrell *et al.*, 2009; McNab *et al.*, 1998; Newcomb *et al.*, 2006; Trus *et al.*, 2007) and has been shown to interact with the triplex protein, VP19c, and the major capsid protein, VP5. These observations have been corroborated by cryoEM–based capsid reconstructions predicting that UL25 contacts the triplexes and hexons surrounding the penton (Conway *et al.*, 2010; Trus *et al.*, 2007). Conformational changes of the capsid proteins induced by DNA packaging are proposed to expose UL25 binding sites at the capsid surface. Much like the accessory proteins of the dsDNA phages, accumulation of UL25 protein may reinforce the capsid to prevent loss of DNA (Conway *et al.*, 2010; Trus *et al.*, 2007). The conformation of UL25 around the portal vertex has not yet been visualized, however, and it may have a more active role in retaining packaged DNA. Finally, two recent studies have shown that UL25 is required for attachment of the tegument to the capsid through its interaction with the large tegument protein UL36 (Coller *et al.*, 2007) and that UL25 is required at a very early stage in infection when the viral genome uncoats (Preston *et al.*, 2008). The functions of the UL17 and UL32 proteins in the cleavage/packaging reaction are unclear, although they appear to serve more of an accessory function in the DNA cleavage/packaging reaction (Chang *et al.*, 1996; Lamberti and Weller, 1998; Salmon *et al.*, 1998; Taus *et al.*, 1998; Thurlow *et al.*, 2005, 2006).

Capsid structure

The herpesvirus capsid, comprising a number of different protein species in hundreds of copies, is relatively complex compared to many other virus capsids. Consequently, crystallization of capsids and capsid subunits has proven difficult and the atomic resolution information available on structural proteins is limited to parts of just two proteins from HSV-1: the 604-amino acid upper domain of the major capsid protein VP5 (PDB ID: 1NO7), and all but the 133 N-terminal residues of the minor capsid protein UL25 (PDB ID: 2F5U). Nonetheless, much is known about herpesvirus capsid structure and assembly, primarily from electron microscopy. Classical methods, such as negative staining and metal shadowing, have provided important morphological data on virions and purified nucleocapsids. Micrographs reveal the external shape of the capsid, aided by its large size and extensive capsomer 'turrets', including the icosahedral nature of the capsid geometry and the arrangement of capsomers (Johnson and Speir, 1997) according to triangulation number T=16 (Fig. 10.6).

Quantitative descriptions of herpesvirus capsid structure have come from cryo-electron microscopy (cryoEM), where biological samples are protected from dehydration by being embedded in a thin layer of vitrified buffer, and images are treated as mass-projections of the native or near-native structures. Three-dimensional image reconstruction techniques can then reconstitute a model of the original structure from these cryoEM micrographs (Crowther, 1971). Herpesvirus capsids were amongst the early samples studied by these methods, including equine herpesvirus 1 (Baker *et al.*, 1990) and human HSV-1 (Booy *et al.*, 1991; Schrag *et al.*, 1989). The structures calculated were the first to reveal the shapes of penton and hexon capsomers, corresponding to pentamers and hexamers of the major capsid protein (mcp), respectively, and the 'triplex' molecules interspersed between capsomers on the exterior surface. Internal features such as channels along the central axis of capsomers were also described, but the packaged dsDNA was not well resolved due in part to lack of long-range order as well as the limited resolutions of 3.5–4.0 nm achieved for those early structures. Subsequent structural and biochemical experiments in the following 2–3 years identified locations of HSV-1 major capsid proteins VP5 (Trus *et al.*, 1992) and VP26 (Booy *et al.*, 1994; Trus *et al.*, 1995; Zhou *et al.*, 1995), and determined that both VP19C and VP23 constitute the triplex molecules (Newcomb *et al.*, 1993; Zhou *et al.*, 1994). More recently, subnanometer resolution has been achieved for the HSV-1 B-capsid, (Zhou *et al.*, 2000) sufficient for tentative assignment of helices to tubular regions of the density map and description of the spatial arrangement of subunits in the triplex. In

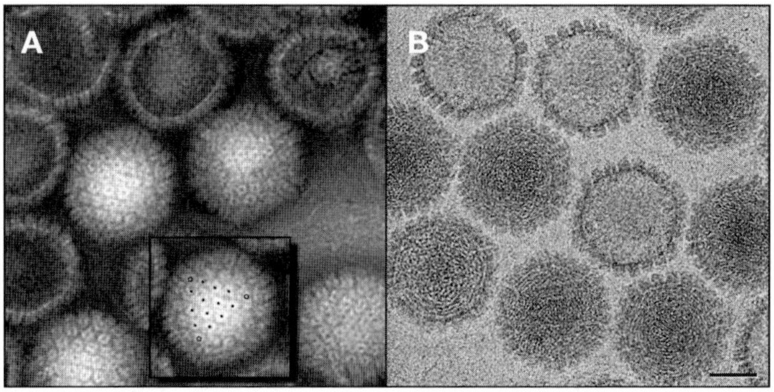

Figure 10.6 Herpesvirus nucleocapsids visualized by electron microscopy. (A) Capsid morphology is readily visible by negative stain EM (the contrast agent used here is 1% phosphotungstic acid) although many particles have collapsed (darker interiors). Capsomers evident on the top surface of the intact whiter capsids can be assigned (e.g. inset) to hexavalent and pentavalent positions (dots and open pentagons, respectively) from which the triangulation number T=16 can be determined. (B) By cryoEM, capsids appear dark on a lighter background and most particles appear intact with strong capsomer extensions visible at the peripheries. Differences in interior density readily distinguish empty A-capsids from DNA-filled C-capsids, for which the dsDNA inter-duplex spacing of ~2.5 nm is evident as swirls and spots. Images such as these are used to reconstruct three-dimensional models of the capsid at resolutions beyond 1 nm. Bar = 50 nm.

addition, the correct handedness of the capsid has been determined – a critical feature to resolve in the absence of absolute chirality in the EM density (Cheng *et al.*, 2002). The locations of the major structural proteins of the herpesvirus capsid, as demonstrated by EM, are shown in Fig. 10.7.

This description of the herpesvirus capsid architecture bears striking resemblances to those of dsDNA tailed phages such as T4 and λ. For example, pentameric and hexameric capsomers are arranged according to quasi-equivalent icosahedral geometry (Caspar and Klug, 1962), and the capsids are decorated by smaller accessory molecules at positions of local 3-fold symmetry (T4 Soc, λ gpD, and HSV triplex). Several other important structural similarities have surfaced in recent years. For one, the herpesvirus capsid subunits assemble around a scaffolding core into a precursor form (Trus *et al.*, 1996), analogous to a phage procapsid or prohead, and the scaffold structure is subsequently degraded by a viral protease. The procapsid includes the main structural proteins – the major capsid protein, VP5; the triplex proteins VP19C and VP23; the hexon tip-binding VP26 – as well as another strikingly phage-like structure, the recently discovered UL6 portal (Newcomb *et al.*, 2001a; Trus *et al.*, 2004), a dodecamer that replaces a penton at one vertex. Portal–scaffold complexes appear likely to nucleate capsid assembly, and the portal is where the terminase–dsDNA complex binds to initiate packaging of the genome (White *et al.*, 2003), most probably as an ATP-driven pump pushing the dsDNA through the central channel of the portal. Yet another common feature is the capsid 'expansion' transformation concurrent with packaging that is observed in λ-like phages (Conway *et al.*, 1995, 2001; Dokland and Murialdo, 1993). Although the herpesvirus capsid does not undergo the large-scale remodelling of phages, the capsid surface is nonetheless reshaped by forming more extensive contacts between capsomers that close the many holes penetrating the surface and presumably confer additional stability to withstand the internal pressure of the densely packed genome (Heymann *et al.*, 2003). Unlike the lambdoid phages, herpesvirus capsids do not bind a connector-tail structure, and instead are coated with a deep layer of partially ordered 'tegument' proteins that is further enveloped with a membrane. However, the earlier steps of capsid assembly, expansion and DNA packaging are strongly reminiscent of the well-known pathway of dsDNA tailed phages.

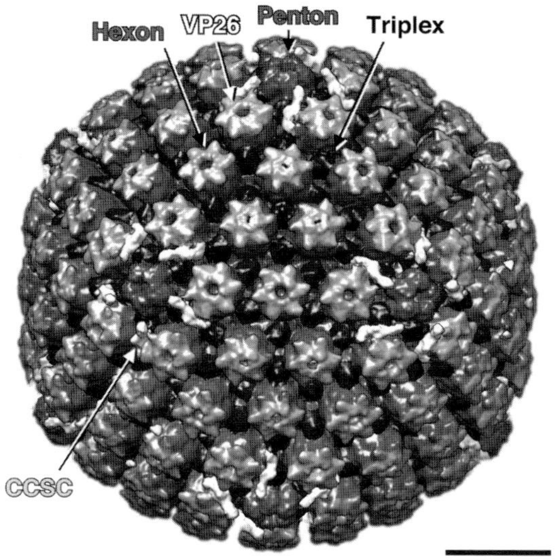

Figure 10.7 Surface rendition of the Herpesvirus C-capsid including the essential minor molecule, CCSC. Pentons comprise 5 copies of the VP5 major capsid protein and are located at the capsid vertices. Hexons include 6 copies of the VP5 major capsid protein as well as six copies of the minor capsid protein, VP26, bound to the tips (but not to pentons). Triplex molecules bind between the capsomer turrets, and support the CCSC density that binds around pentamers. CCSC is composed of UL25 in the region distal to the adjacent penton, and putatively UL17 in the proximal part. Bar = 25 nm. A colour version of this figure can be found at: http://www.pitt.edu~jxc100/2010-WellerBook-Ch10-Fig7/Fig7_WebVersion.htm

Similarities between the herpesvirus and phage capsids appear to go even deeper. The mature capsid of the lambdoid phage, HK97, has been solved crystallographically (Helgstrand *et al.*, 2003) revealing a capsid protein fold that has since been seen in other related phage capsids, including T4 (Fokine *et al.*, 2005), and by high resolution cryoEM in ε15 (Jiang *et al.*, 2008), P22 (Jiang *et al.*, 2003), phi29 (Morais *et al.*, 2005), T7 (Agirrezabala *et al.*, 2007), T5 (Effantin *et al.*, 2006), as well as λ (Lander *et al.*, 2008), despite little to no detectable sequence similarities. Indeed, this fold looks likely to be common to all tailed phages, but remarkably it is also consistent with the lower domain of the VP5 mcp subunit in the 0.85 nm cryoEM structure of HSV-1 B-capsid (Baker *et al.*, 2005b). The herpesvirus VP5 is considerably larger at 150 kDa than the mature HK97 gp5 protein of 31 kDa, but nonetheless the region proposed to share a common fold provides much of the connected capsid density with the remainder of VP5 arranged as extensive loops piled on top. Even the distinctive T=16 capsid geometry of herpesvirus has recently been seen in two phages – SPO1 (Duda *et al.*, 2006) and Syn9 (Weigele *et al.*, 2007) – further extending the commonalities between herpesvirus capsid structure and assembly with bacteriophages.

These similarities are significant in suggesting not only a common lineage between herpesviruses and dsDNA tailed phages of at least the capsid proteins (Bamford *et al.*, 2005; Hendrix, 2002; Steven *et al.*, 2005) but also indicate how aspects of the herpesvirus capsid assembly pathway may work based on the simpler and more tractable phage systems. For example, the proposed existence of a herpesvirus portal that functions in a manner analogous to phage portals was tested by *in situ* immunogold labelling of a candidate subunit, UL6, and visualization of over-expressed rings (Newcomb *et al.*, 2001a), and the portal assembly was later characterized structurally by cryoEM of isolated particles (Trus *et al.*, 2004) and by tomographic reconstruction of capsids (Cardone *et al.*, 2007; Chang *et al.*, 2007). The overall structure of the herpesvirus portal at 1.6 nm resolution is seen to resemble phage portals (Trus *et al.*, 2004). In hindsight, the portals were not previously detected on herpesvirus capsids since the dodecameric portal mass

(12 × 74 kDa) is only 20% higher than that of a regular penton (5 × 150 kDa) and the dimensions and location are also similar. This is in contrast to bacteriophages which have similar sized portal subunits but much smaller mcps, generally of 30–50 kDa.

The herpesvirus capsid forms, A, B and C, are termed from the distinct bands of purified nuclear extract that form on a sucrose gradient. However, at current resolutions the capsid structures are largely indistinguishable, and run separately on the gradient due to the different contents of the interior volume – tightly-packaged dsDNA for C-capsids; scaffold remnants for B-capsids; and empty (i.e. buffer) for A-capsids. Their positions on the capsid assembly pathway as currently understood are shown in Fig. 10.3. In this scheme, A and B capsids are off-pathway dead-ends – B-capsids retain much of the proteolysed scaffold but failed to package DNA, whereas A capsids likely did embark on the packaging reaction, expelling remnants of the scaffold from the capsid interior, but did not retain the genome.

The roles of minor capsid proteins are important in the production of C-capsids and one molecule, the HSV-1 C-capsid-specific component (CCSC) comprising the UL25 protein and possibly also UL17, has recently been studied by cryoEM (Conway *et al.*, 2010; Trus *et al.*, 2007). CCSC was observed to be absent on A- and B-capsids, but present in C-capsids at ~50% occupancy. At some stage during the packaging process, CCSC molecules bind around the vertices, apparently functioning to maintain the DNA within the capsid. Although a number of alternative quasi-equivalent binding sites are available, the CCSC molecule is found in discernible quantities adjacent to pentons, bridging the peri-pentonal triplex molecule and the next proximal-most triplex. Modelling with the UL25 crystal structure and mutational studies both suggest that UL25 is located within the CCSC at the end distal to the penton vertex. Several lines of evidence suggest that UL17 and UL25 dimerize, hence the remaining CCSC density is proposed to be UL17. The UL17 portion of the CCSC molecule appears to approach pentons in the cryoEM reconstructions, although it apparently does not make contact. One of several roles attributed to UL25 is to aid in retaining the internal dsDNA after packaging, since UL25-null virus produces abundant A-capsids and DNA processed to genome-length units, presumably as a consequence of packaging and 'leakage' (Cockrell *et al.*, 2009; McNab *et al.*, 1998). This suggests that the CCSC may have a specific conformation around the portal vertex different from the icosahedrally averaged density visualized so far that is representative only of penton vertices. There is a clear need for lower-symmetry capsid structures that reveal the portal vertex separately from the other 11 vertices so that more details of the assembly pathway may be studied.

Several studies using cryo-electron tomography have visualized non-icosahedral aspects of herpesvirus structure, although at considerably lower resolutions than achieved with 'single particle' cryoEM. The capsid, for example, was found to be located off-centre within the virion with one vertex closely approaching the viral envelope, correlating with a region of low glycoprotein density, while the opposite vertex was separated from the envelope by a ~35 nm deep layer of tegument protein (Grunewald *et al.*, 2003), raising questions on both the final stages of assembling capsids into virions, and the process of cell entry. The portal was not observed in this study, but more recent tomographic work has targeted it (Cardone *et al.*, 2007; Chang *et al.*, 2007; Deng *et al.*, 2007). Portals are not convincingly detectable in tomograms and different approaches were used to locate portal vertices to allow tomograms to be combined so that resolution would improve. However, two of these studies placed the mass of the portal towards the interior surface of the capsid, although one used 6 M urea to strip pentons from the capsid to enhance detection of the portal vertex. The third put the portal more towards the exterior of the capsid, much like a capsomer, and added support for this localization by visualizing capsids with freeze-fracture electron microscopy without finding any vertices lacking external protrusions. Given the limited resolution of cryo-electron tomography, a generally accepted verdict on the portal may need to await methodological advances.

Much of the structural work described above was done with HSV-1 particles, but other herpesvirus capsids have also been examined by cryoEM, including simian cytomegalovirus (Trus *et al.*, 1999), and the gammaherpesvirus Kaposi's sarcoma-associated virus (HHV8) (Deng *et al.*, 2008; Trus *et al.*, 2001). The cytomegalovirus study, together with another on HSV-1 (Chen *et al.*, 2001), visualized partially tegumented capsids, showing the innermost-layers interacting with specific capsid subunits including the triplex and penton. The cytomegalovirus B-capsid was not observed to be different from that of HSV-1 at the resolution achieved. Kaposi's sarcoma-associated virus capsids also differed little from HSV-1 except that in one study the analogue of the small decoration protein, VP26, that binds HSV-1 hexons, was found to bind both hexons and pentons in HHV8 (Trus *et al.*, 2001) and is in fact essential for capsid assembly, unlike in the alphaherpesviruses (Perkins *et al.*, 2008).

Capsid structures have also been solved for distant cousins of the animal herpesviruses, including those of a fish (Booy *et al.*, 1996) and an oyster (Davison *et al.*, 2005). The capsid of the channel catfish virus is remarkably similar to those of human herpesviruses, including icosahedral geometry with T=16 built from hexameric and pentameric capsomers interspaced with triplex-like density on the exterior surface. The capsid diameter is smaller owing to the shorter extents of capsomer protrusions, but also from lack of a VP26-analogue, which while dispensable for capsid formation in HSV-1 is apparently completely dispensed with in channel catfish virus. Likewise, the only invertebrate herpesvirus known to date, OsHV (of oyster), has similar size and geometry to HSV-1 but shorter capsomers (Davison *et al.*, 2005).

Acknowledgements

The research in the laboratories of F.L.H. and J.F.C. was supported by National Institutes of Health awards (AI060836) and by the Commonwealth of Pennsylvania (SAP 4100031302).

References

Abbotts, A.P., Preston, V.G., Hughes, M., Patel, A.H., and Stow, N.D. (2000). Interaction of the herpes simplex virus type 1 packaging protein UL15 with full-length and deleted forms of the UL28 protein. J. Gen. Virol. *81*, 2999–3009.

Adelman, K., Salmon, B., and Baines, J.D. (2001). Herpes simplex virus DNA packaging sequences adopt novel structures. that are specifically recognized by a component of the cleavage and packaging machinery. Proc. Natl. Acad. Sci. U.S.A. *98*, 3086–3091.

Agirrezabala, X., Velazquez-Muriel, J.A., Gomez-Puertas, P., Scheres, S.H., Carazo, J.M., and Carrascosa, J.L. (2007). Quasi-atomic model of bacteriophage t7 procapsid shell: insights into the structure and evolution of a basic fold. Structure *15*, 461–472.

al-Kobaisi, M.F., Rixon, F.J., McDougall, I., and Preston, V.G. (1991). The herpes simplex virus UL33 gene product is required for the assembly of full capsids. Virology *180*, 380–388.

Baines, J.D., Poon, A.P., Rovnak, J., and Roizman, B. (1994). The herpes simplex virus 1 UL15 gene encodes two proteins and is required for cleavage of genomic viral DNA. J. Virol. *68*, 8118–8124.

Baines, J.D., and Roizman, B. (1992). The cDNA of UL15, a highly conserved herpes simplex virus 1 gene, effectively replaces the two exons of the wild-type virus. J. Virol. *66*, 5621–5626.

Baker, M.L., Jiang, W., Bowman, B.R., Zhou, Z.H., Quiocho, F.A., Rixon, F.J., and Chiu, W. (2003). Architecture of the herpes simplex virus major capsid protein derived from structural bioinformatics. J. Mol. Biol. *331*, 447–456.

Baker, M.L., Jiang, W., Rixon, F.J., and Chiu, W. (2005a). Common ancestry of herpesviruses and tailed DNA bacteriophages. J. Virol. *79*, 14967–14970.

Baker, M.L., Jiang, W., Rixon, F.J., and Chiu, W. (2005b). Common ancestry of herpesviruses and tailed DNA bacteriophages. J. Virol. *79*, 14967–14970.

Baker, T.S., Newcomb, W.W., Booy, F.P., Brown, J.C., and Steven, A.C. (1990). Three-dimensional structures. of maturable and abortive capsids of equine herpesvirus 1 from cryoelectron microscopy. J. Virol. *64*, 563–573.

Bamford, D.H., Grimes, J.M., and Stuart, D.I. (2005). What does structure tell us about virus evolution? Curr. Opin. Struct. Biol.

Bazinet, C., and King, J. (1985). The DNA translocating vertex of dsDNA bacteriophage. Annu. Rev. Microbiol. *39*, 109–129.

Beard, P.M., and Baines, J.D. (2004). The DNA cleavage and packaging protein encoded by the U(L)33 gene of herpes simplex virus 1 associates with capsids. Virology *324*, 475–482.

Bjornsti, M.A., Reilly, B.E., and Anderson, D.L. (1983). Morphogenesis of bacteriophage phi 29 of Bacillus subtilis: oriented and quantized *in vitro* packaging of DNA protein gp3. J. Virol. *45*, 383–396.

Black, L. (1988). DNA packaging in dsDNA bacteriophages. In The Bacteriophages, Calendar, R., ed. (New York, Plenum Press), pp. 321–373.

Black, L., M.K Showe, and A.C Steven (1994). Morphogenesis of the T4 head. In 'Bacteriophage T4 (Washington, D.C., ASM Press).

Bogner, E. (2002). Human cytomegalovirus terminase as a target for antiviral chemotherapy. Rev. Med. Virology *12*, 115–127.

Booy, F.P., Newcomb, W.W., Trus, B.L., Brown, J.C., Baker, T.S., and Steven, A.C. (1991). Liquid-crystalline, phage-like packing of encapsidated DNA in herpes simplex virus. Cell *64*, 1007–1015.

Booy, F.P., Trus, B.L., Davison, A.J., and Steven, A.C. (1996). The capsid architecture of channel catfish virus, an evolutionarily distant herpesvirus, is largely conserved in the absence of discernible sequence homology with herpes simplex virus. Virology *215*, 134–141.

Booy, F.P., Trus, B.L., Newcomb, W.W., Brown, J.C., Conway, J.F., and Steven, A.C. (1994). Finding a needle in a haystack: detection of a small protein (the 12-kDa VP26) in a large complex (the 200-MDa capsid of herpes simplex virus). Proc. Natl. Acad. Sci. U.S.A. *91*, 5652–5656.

Bowman, B.R., Welschhans, R.L., Jayaram, H., Stow, N.D., Preston, V.G., and Quiocho, F.A. (2006). Structural characterization of the UL25 DNA-packaging protein from herpes simplex virus type 1. J. Virol. *80*, 2309–2317.

Brown, J., M.A McVoy, and F.L Homa. (2002). Packaging DNA into herpesvirus capsids. In Structure–function relationships of human pathogenic viruses, Bogner, A.H.E., ed. (New York, Kluwer Acacdemic/Plenum Publishers), pp. 111–155.

Buerger, I., Reefschlaeger, J., Bender, W., Eckenberg, P., Popp, A., Weber, O., Graeper, S., Klenk, H.D., Ruebsamen-Waigmann, H., and Hallenberger, S. (2001). A novel nonnucleoside inhibitor specifically targets cytomegalovirus DNA maturation via the UL89 and UL56 gene products. J. Virol. *75*, 9077–9086.

Cardone, G., Winkler, D.C., Trus, B.L., Cheng, N., Heuser, J.E., Newcomb, W.W., Brown, J.C., and Steven, A.C. (2007). Visualization of the herpes simplex virus portal *in situ* by cryo-electron tomography. Virology *361*, 426–434.

Casjens, S., and R Hendrix (1988). Control mechanisms in dsDNA bacteriophage assembly. In 'The Bacteriophages', Vol 1 (New York, Plenum Press).

Caspar, D.L., and Klug, A. (1962). Physical principles in the construction of regular viruses. Cold Spring Harb. Symp. Quant. Biol. *27*, 1–24.

Cavalcoli, J.D., Baghian, A., Homa, F.L., and Kousoulas, K.G. (1993). Resolution of genotypic and phenotypic properties of herpes simplex virus type 1 temperature-sensitive mutant (KOS) tsZ47: evidence for allelic complementation in the UL28 gene. Virology *197*, 23–34.

Cerritelli, M.E., Cheng, N., Rosenberg, A.H., McPherson, C.E., Booy, F.P., and Steven, A.C. (1997). Encapsidated conformation of bacteriophage T7 DNA. Cell *91*, 271–280.

Chang, J.T., Schmid, M.F., Rixon, F.J., and Chiu, W. (2007). Electron cryotomography reveals the portal in the herpesvirus capsid. J. Virol. *81*, 2065–2068.

Chang, Y.E., Poon, A.P., and Roizman, B. (1996). Properties of the protein encoded by the UL32 open reading frame of herpes simplex virus 1. J. Virol. *70*, 3938–3946.

Chen, D.H., Jakana, J., McNab, D., Mitchell, J., Zhou, Z.H., Dougherty, M., Chiu, W., and Rixon, F.J. (2001). The pattern of tegument–capsid interaction in the herpes simplex virus type 1 virion is not influenced by the small hexon-associated protein VP26. J. Virol. *75*, 11863–11867.

Cheng, N., Trus, B.L., Belnap, D.M., Newcomb, W.W., Brown, J.C., and Steven, A.C. (2002). Handedness of the herpes simplex virus capsid and procapsid. J. Virol. *76*, 7855–7859.

Church, G.A., Dasgupta, A., and Wilson, D.W. (1998). Herpes simplex virus DNA packaging without measurable DNA synthesis. J. Virol. *72*, 2745–2751.

Cockrell, S.K., Sanchez, M.E., Erazo, A., and Homa, F.L. (2009). Role of the UL25 protein in herpes simplex virus DNA encapsidation. J. Virol. *83*, 47–57.

Coller, K.E., Lee, J.I., Ueda, A., and Smith, G.A. (2007). The capsid and tegument of the alphaherpesviruses are linked by an interaction between the UL25 and VP1/2 proteins. J. Virol. *81*, 11790–11797.

Conway, J.F., Cockrell, S.K., Copeland, A.M., Newcomb, W.W., Brown, J.C., and Homa, F.L. (2010). Labeling and localization of the herpes simplex virus capsid protein UL25 and its interaction with the two triplexes closest to the penton. J. Mol. Biol,. 397, 575–586.

Conway, J.F., Duda, R.L., Cheng, N., Hendrix, R.W., and Steven, A.C. (1995). Proteolytic and conformational control of virus capsid maturation: the bacteriophage HK97 system. J. Mol. Biol. *253*, 86–99.

Conway, J.F., Trus, B.L., Booy, F.P., Newcomb, W.W., Brown, J.C., and Steven, A.C. (1993). The effects of radiation damage on the structure of frozen hydrated HSV-1 capsids. J. Struct. Biol. *111*, 222–233.

Conway, J.F., Trus, B.L., Booy, F.P., Newcomb, W.W., Brown, J.C., and Steven, A.C. (1996). Visualization of three-dimensional density maps reconstructed from cryoelectron micrographs of viral capsids. J. Struct. Biol. *116*, 200–208.

Conway, J.F., Wikoff, W.R., Cheng, N., Duda, R.L., Hendrix, R.W., Johnson, J.E., and Steven, A.C. (2001). Virus maturation involving large subunit rotations and local refolding. Science *292*, 744–748.

Crowther, R.A. (1971). Procedures for three-dimensional reconstruction of spherical viruses by Fourier synthesis from electron micrographs. Philos. Trans. R. Soc. Lond. B Biol. Sci. *261*, 221–230.

Crumpacker, C.S. (1996). Ganciclovir. N. Engl. J. Med. *335*, 721–729.

Dasgupta, A., and Wilson, D.W. (1999). ATP depletion blocks herpes simplex virus DNA packaging and capsid maturation. J. Virol. *73*, 2006–2015.

Davison, A.J. (1992). Channel catfish virus: a new type of herpesvirus. Virology *186*, 9–14.

Davison, A.J., Trus, B.L., Cheng, N., Steven, A.C., Watson, M.S., Cunningham, C., Le Deuff, R.M., and Renault, T. (2005). A novel class of herpesvirus with bivalve hosts. J. Gen. Virol. *86*, 41–53.

Deiss, L.P., Chou, J., and Frenkel, N. (1986). Functional domains within the a sequence involved in the cleavage-packaging of herpes simplex virus DNA. J. Virol. *59*, 605–618.

Deiss, L.P., and Frenkel, N. (1986). Herpes simplex virus amplicon: cleavage of concatemeric DNA is linked to packaging and involves amplification of the terminally reiterated a sequence. J. Virol. *57*, 933–941.

Deng, B., O'Connor, C.M., Kedes, D.H., and Zhou, Z.H. (2007). Direct visualization of the putative portal in the Kaposi's sarcoma-associated herpesvirus capsid by cryoelectron tomography. J. Virol. *81*, 3640–3644.

Deng, B., O'Connor, C.M., Kedes, D.H., and Zhou, Z.H. (2008). Cryo-electron tomography of Kaposi's sarcoma-associated herpesvirus capsids reveals dynamic scaffolding structures. essential to capsid assembly and maturation. J. Struct. Biol. *161*, 419–427.

Desai, P., and Person, S. (1999). Second site mutations in the N-terminus of the major capsid protein (VP5) overcome a block at the maturation cleavage site of the capsid scaffold proteins of herpes simplex virus type 1. Virology *261*, 357–366.

Dokland, T., and Murialdo, H. (1993). Structural transitions during maturation of bacteriophage lambda capsids. J. Mol. Biol. *233*, 682–694.

Duda, R.L., Hendrix, R.W., Huang, W.M., and Conway, J.F. (2006). Shared architecture of bacteriophage SPO1 and herpesvirus capsids. Curr. Biol. *16*, R11–13.

Dykewicz, C.A. (2001). Summary of the Guidelines for Preventing Opportunistic Infections among Hematopoietic Stem Cell Transplant Recipients. Clin. Infect. Dis. *33*, 139–144.

Earnshaw, W.C., and Harrison, S.C. (1977). DNA arrangement in isometric phage heads. Nature *268*, 598–602.

Earnshaw, W.C., King, J., Harrison, S.C., and Eiserling, F.A. (1978). The structural organization of DNA packaged within the heads of T4 wild-type, isometric and giant bacteriophages. Cell *14*, 559–568.

Effantin, G., Boulanger, P., Neumann, E., Letellier, L., and Conway, J.F. (2006). Bacteriophage T5 Structure Reveals Similarities with HK97 and T4 Suggesting Evolutionary Relationships. J. Mol. Biol. *361*, 993–1002.

Emery, V.C., and Hassan-Walker, A.F. (2002). Focus on new drugs in development against human cytomegalovirus. Drugs *62*, 1853–1858.

Field, A.K. (1999). Human cytomegalovirus: challenges, opportunities and new drug development. Antivir. Chem. Chemother. *10*, 219–232.

Fokine, A., Leiman, P.G., Shneider, M.M., Ahvazi, B., Boeshans, K.M., Steven, A.C., Black, L.W., Mesyanzhinov, V.V., and Rossmann, M.G. (2005). Structural and functional similarities between the capsid proteins of bacteriophages T4 and HK97 point to a common ancestry. Proc. Natl. Acad. Sci. U.S.A. *102*, 7163–7168.

Gibson, W., and Roizman, B. (1972). Proteins specified by herpes simplex virus. 8. Characterization and composition of multiple capsid forms of subtypes 1 and 2. J. Virol. *10*, 1044–1052.

Griffiths, P.D. (1996). Herpesviruses and AIDS. J. Antimicrob. Chemother. *37 Suppl B*, 87–95.

Grunewald, K., Desai, P., Winkler, D.C., Heymann, J.B., Belnap, D.M., Baumeister, W., and Steven, A.C. (2003). Three-dimensional structure of herpes simplex virus from cryo-electron tomography. Science *302*, 1396–1398.

Helgstrand, C., Wikoff, W.R., Duda, R.L., Hendrix, R.W., Johnson, J.E., and Liljas, L. (2003). The refined structure of a protein catenane: the HK97 bacteriophage capsid at 3.44 A resolution. J. Mol. Biol. *334*, 885–899.

Hendrix, R. (1985). Shape determination in virus assembly: the bacteriophage example. In Virus Structure and Assembly., Casjens, S., ed. (Boston, Jones and Bartlett Publishers), pp. 169–203.

Hendrix, R.W. (2002). Bacteriophages: evolution of the majority. Theor Popul Biol. *61*, 471–480.

Heymann, J.B., Cheng, N., Newcomb, W.W., Trus, B.L., Brown, J.C., and Steven, A.C. (2003). Dynamics of herpes simplex virus capsid maturation visualized by time-lapse cryo-electron microscopy. Nat. Struct. Biol. *10*, 334–341.

Hohn, B. (1983). DNA sequences necessary for packaging of bacteriophage lambda DNA. Proc. Natl. Acad. Sci. U.S.A. *80*, 7456–7460.

Homa, F.L., and Brown, J.C. (1997). Capsid assembly and DNA packaging in herpes simplex virus. Rev. Med. Virology *7*, 107–122.

Hong, Z., Beaudet-Miller, M., Durkin, J., Zhang, R., and Kwong, A.D. (1996). Identification of a minimal hydrophobic domain in the herpes simplex virus type 1 scaffolding protein which is required for interaction with the major capsid protein. J. Virol. *70*, 533–540.

Huang, E., Perkins, E.M., and Desai, P. (2007). Structural features. of the scaffold interaction domain at the N terminus of the major capsid protein (VP5) of herpes simplex virus type 1. J. Virol. *81*, 9396–9407.

Huffman, J.B., Newcomb, W.W., Brown, J.C., and Homa, F.L. (2008). Amino acids 143 to 150 of the herpes simplex virus type 1 scaffold protein are required for the formation of portal-containing capsids. J. Virol. *82*, 6778–6781.

Jacobson, J.G., Yang, K., Baines, J.D., and Homa, F.L. (2006). Linker insertion mutations in the herpes simplex virus type 1 UL28 gene: effects on UL28 interaction with UL15 and UL33 and identification of a second-site mutation in the UL15 gene that suppresses a lethal UL28 mutation. J. Virol. *80*, 12312–12323.

Jiang, W., Baker, M.L., Jakana, J., Weigele, P.R., King, J., and Chiu, W. (2008). Backbone structure of the infectious epsilon15 virus capsid revealed by electron cryomicroscopy. Nature *451*, 1130–1134.

Jiang, W., Li, Z., Zhang, Z., Baker, M.L., Prevelige, P.E., Jr., and Chiu, W. (2003). Coat protein fold and maturation transition of bacteriophage P22 seen at subnanometer resolutions. Nat. Struct. Biol. *10*, 131–135.

Johnson, J.E., and Speir, J.A. (1997). Quasi-equivalent viruses: a paradigm for protein assemblies. J. Mol. Biol. *269*, 665–675.

Krosky, P.M., Underwood, M.R., Turk, S.R., Feng, K.W., Jain, R.K., Ptak, R.G., Westerman, A.C., Biron, K.K., Townsend, L.B., and Drach, J.C. (1998). Resistance of human cytomegalovirus to benzimidazole ribonucleosides maps to two open reading frames: UL89 and UL56. J. Virol. *72*, 4721–4728.

Lamberti, C., and Weller, S.K. (1996). The herpes simplex virus type 1 UL6 protein is essential for cleavage and packaging but not for genomic inversion. Virology *226*, 403–407.

Lamberti, C., and Weller, S.K. (1998). The herpes simplex virus type 1 cleavage/packaging protein, UL32, is involved in efficient localization of capsids to replication compartments. J. Virol. *72*, 2463–2473.

Lander, G.C., Evilevitch, A., Jeembaeva, M., Potter, C.S., Carragher, B., and Johnson, J.E. (2008). Bacteriophage lambda stabilization by auxiliary protein gpD: timing, location, and mechanism of attachment determined by cryo-EM. Structure *16*, 1399–1406.

Lata, R., Conway, J.F., Cheng, N., Duda, R.L., Hendrix, R.W., Wikoff, W.R., Johnson, J.E., Tsuruta, H., and Steven, A.C. (2000). Maturation dynamics of a viral capsid: visualization of transitional intermediate states. Cell *100*, 253–263.

Liu, F., and Roizman, B. (1993). Characterization of the protease and other. products of amino-terminus-proximal cleavage of the herpes simplex virus 1 UL26 protein. J. Virol. *67*, 1300–1309.

Liu, F.Y., and Roizman, B. (1991). The herpes simplex virus 1 gene encoding a protease also contains within its coding domain the gene encoding the more abundant substrate. J. Virol. *65*, 5149–5156.

Martinez, R., Sarisky, R.T., Weber, P.C., and Weller, S.K. (1996). Herpes simplex virus type 1 alkaline nuclease is required for efficient processing of viral DNA replication intermediates. J. Virol. *70*, 2075–2085.

McGavin, J.K., and Goa, K.L. (2001). Ganciclovir: an update of its use in the prevention of cytomegalovirus infection and disease in transplant recipients. Drugs *61*, 1153–1183.

McNab, A.R., Desai, P., Person, S., Roof, L.L., Thomsen, D.R., Newcomb, W.W., Brown, J.C., and Homa, F.L. (1998). The product of the herpes simplex virus type 1 UL25 gene is required for encapsidation but not for cleavage of replicated viral DNA. J. Virol. *72*, 1060–1070.

McVoy, M.A., Nixon, D.E., Adler, S.P., and Mocarski, E.S. (1998). Sequences within the herpesvirus-conserved pac1 and pac2 motifs are required for cleavage and packaging of the murine cytomegalovirus genome. J. Virol. 72, 48–56.

Mettenleiter, T.C. (2004). Budding events in herpesvirus morphogenesis. Virus Res. *106*, 167–180.

Mettenleiter, T.C. (2006). Intriguing interplay between viral proteins during herpesvirus assembly or: the herpesvirus assembly puzzle. Vet. Microbiol. *113*, 163–169.

Mocarski, E.S., and Roizman, B. (1982). Structure and role of the herpes simplex virus DNA termini in inversion, circularization and generation of virion DNA. Cell *31*, 89–97.

Morais, M.C., Choi, K.H., Koti, J.S., Chipman, P.R., Anderson, D.L., and Rossmann, M.G. (2005). Conservation of the capsid structure in tailed dsDNA bacteriophages: the pseudoatomic structure of phi29. Mol. Cell *18*, 149–159.

Nasseri, M., and Mocarski, E.S. (1988). The cleavage recognition signal is contained within sequences surrounding an a–a junction in herpes simplex virus DNA. Virology *167*, 25–30.

Newcomb, W.W., and Brown, J.C. (1991). Structure of the herpes simplex virus capsid: effects of extraction with guanidine hydrochloride and partial reconstitution of extracted capsids. J. Virol. *65*, 613–620.

Newcomb, W.W., and Brown, J.C. (2002). Inhibition of herpes simplex virus replication by WAY-150138: assembly of capsids depleted of the portal and terminase proteins involved in DNA encapsidation. J. Virol. *76*, 10084–10088.

Newcomb, W.W., Cockrell, S.K., Homa, F.L., and Brown, J.C. (2009). Polarized DNA ejection from the herpesvirus capsid. J. Mol. Biol. *392*, 885–894.

Newcomb, W.W., Homa, F.L., and Brown, J.C. (2005). Involvement of the portal at an early step in herpes simplex virus capsid assembly. J. Virol. *79*, 10540–10546.

Newcomb, W.W., Homa, F.L., and Brown, J.C. (2006). Herpes simplex virus capsid structure: DNA packaging protein UL25 is located on the external surface of the capsid near the vertices. J. Virol. *80*, 6286–6294.

Newcomb, W.W., Homa, F.L., Thomsen, D.R., Booy, F.P., Trus, B.L., Steven, A.C., Spencer, J.V., and Brown, J.C. (1996). Assembly of the herpes simplex virus capsid: characterization of intermediates observed during cell-free capsid formation. J. Mol. Biol. *263*, 432–446.

Newcomb, W.W., Homa, F.L., Thomsen, D.R., and Brown, J.C. (2001a). *In vitro* assembly of the herpes simplex virus procapsid: formation of small procapsids at reduced scaffolding protein concentration. J. Struct. Biol. *133*, 23–31.

Newcomb, W.W., Homa, F.L., Thomsen, D.R., Trus, B.L., Cheng, N., Steven, A., Booy, F., and Brown, J.C. (1999). Assembly of the herpes simplex virus procapsid from purified components and identification of small complexes containing the major capsid and scaffolding proteins. J. Virol. 73, 4239–4250.

Newcomb, W.W., Homa, F.L., Thomsen, D.R., Ye, Z., and Brown, J.C. (1994). Cell-free assembly of the herpes simplex virus capsid. J. Virol. *68*, 6059–6063.

Newcomb, W.W., Juhas, R.M., Thomsen, D.R., Homa, F.L., Burch, A.D., Weller, S.K., and Brown, J.C. (2001b). The UL6 gene product forms the portal for entry of DNA into the herpes simplex virus capsid. J. Virol. 75, 10923–10932.

Newcomb, W.W., Thomsen, D.R., Homa, F.L., and Brown, J.C. (2003). Assembly of the herpes simplex virus capsid: identification of soluble scaffold–portal complexes and their role in formation of portal-containing capsids. J. Virol. 77, 9862–9871.

Newcomb, W.W., Trus, B.L., Booy, F.P., Steven, A.C., Wall, J.S., and Brown, J.C. (1993). Structure of the herpes simplex virus capsid. Molecular composition of the pentons and the triplexes. J. Mol. Biol. *232*, 499–511.

Ogasawara, M., Suzutani, T., Yoshida, I., and Azuma, M. (2001). Role of the UL25 gene product in packaging DNA into the herpes simplex virus capsid: location of

UL25 product in the capsid and demonstration that it binds DNA. J. Virol. *75*, 1427–1436.

Oien, N.L., Thomsen, D.R., Wathen, M.W., Newcomb, W.W., Brown, J.C., and Homa, F.L. (1997). Assembly of herpes simplex virus capsids using the human cytomegalovirus scaffold protein: critical role of the C terminus. J. Virol. *71*, 1281–1291.

Patel, A.H., and MacLean, J.B. (1995). The product of the UL6 gene of herpes simplex virus type 1 is associated with virus capsids. Virology *206*, 465–478.

Patel, A.H., Rixon, F.J., Cunningham, C., and Davison, A.J. (1996). Isolation and characterization of herpes simplex virus type 1 mutants defective in the UL6 gene. Virology *217*, 111–123.

Perkins, E.M., Anacker, D., Davis, A., Sankar, V., Ambinder, R.F., and Desai, P. (2008). Small capsid protein pORF65 is essential for assembly of Kaposi's sarcoma-associated herpesvirus capsids. J. Virol. *82*, 7201–7211.

Person, S., and Desai, P. (1998). Capsids are formed. in a mutant virus blocked at the maturation site of the UL26 and UL26.5 open reading frames of herpes simplex virus type 1 but are not formed. in a null mutant of UL38 (VP19C). Virology *242*, 193–203.

Poon, A.P., and Roizman, B. (1993). Characterization of a temperature-sensitive mutant of the UL15 open reading frame of herpes simplex virus 1. J. Virol. *67*, 4497–4503.

Preston, V.G., Coates, J.A., and Rixon, F.J. (1983). Identification and characterization of a herpes simplex virus gene product required for encapsidation of virus DNA. J. Virol. *45*, 1056–1064.

Preston, V.G., Murray, J., Preston, C.M., McDougall, I.M., and Stow, N.D. (2008). The UL25 gene product of herpes simplex virus type 1 is involved in uncoating of the viral genome. J. Virol. *82*, 6654–6666.

Prevelige, P.E., Jr., and King, J. (1993). Assembly of bacteriophage P22: a model for ds-DNA virus assembly. Prog. Med. Virology *40*, 206–221.

Reynolds, A.E., Fan, Y., and Baines, J.D. (2000). Characterization of the U(L)33 gene product of herpes simplex virus 1. Virology *266*, 310–318.

Rixon, F. (1993). Structure and assembly of herpesviruses. Seminars in Virology *4*, 135–144.

Salmon, B., and Baines, J.D. (1998). Herpes simplex virus DNA cleavage and packaging: association of multiple forms of U(L)15-encoded proteins with B capsids requires. at least the U(L)6, U(L)17, and U(L)28 genes. J. Virol. *72*, 3045–3050.

Salmon, B., Cunningham, C., Davison, A.J., Harris, W.J., and Baines, J.D. (1998). The herpes simplex virus type 1 U(L)17 gene encodes virion tegument proteins that are required for cleavage and packaging of viral DNA. J. Virol. *72*, 3779–3788.

Schrag, J.D., Prasad, B.V., Rixon, F.J., and Chiu, W. (1989). Three-dimensional structure of the HSV1 nucleocapsid. Cell *56*, 651–660.

Sheaffer, A.K., Newcomb, W.W., Gao, M., Yu, D., Weller, S.K., Brown, J.C., and Tenney, D.J. (2001). Herpes simplex virus DNA cleavage and packaging proteins associate with the procapsid prior to its maturation. J. Virol. *75*, 687–698.

Singer, G.P., Newcomb, W.W., Thomsen, D.R., Homa, F.L., and Brown, J.C. (2005). Identification of a region in the herpes simplex virus scaffolding protein required for interaction with the portal. J. Virol. *79*, 132–139.

Smiley, J.R., Duncan, J., and Howes, M. (1990). Sequence requirements for DNA rearrangements induced by the terminal repeat of herpes simplex virus type 1 KOS DNA. J. Virol. *64*, 5036–5050.

Smiley, J.R., Lavery, C., and Howes, M. (1992). The herpes simplex virus type 1 (HSV-1) a sequence serves as a cleavage/packaging signal but does not drive recombinational genome isomerization when it is inserted into the HSV-2 genome. J. Virol. *66*, 7505–7510.

Smith, S.R., Butterly, D.W., Alexander, B.D., and Greenberg, A. (2001). Viral infections after renal transplantation. Am J. Kidney Dis. *37*, 659–676.

Spaete, R.R., and Frenkel, N. (1982). The herpes simplex virus amplicon: a new eucaryotic defective-virus cloning-amplifying vector. Cell *30*, 295–304.

Spaete, R.R., and Frenkel, N. (1985). The herpes simplex virus amplicon: analyses of cis-acting replication functions. Proc. Natl. Acad. Sci. U.S.A. *82*, 694–698.

Spencer, J.V., Newcomb, W.W., Thomsen, D.R., Homa, F.L., and Brown, J.C. (1998). Assembly of the herpes simplex virus capsid: preformed. triplexes bind to the nascent capsid. J. Virol. *72*, 3944–3951.

Spencer, J.V., Trus, B.L., Booy, F.P., Steven, A.C., Newcomb, W.W., and Brown, J.C. (1997). Structure of the herpes simplex virus capsid: peptide A862-H880 of the major capsid protein is displayed on the rim of the capsomer protrusions. Virology *228*, 229–235.

Steven, A., and P.G Spear (1996). Herpesvirus capsid assembly and envelopment. In Structural Biology of Viruses, Burnett, W.C.R., and Garcea, R., ed. (New York, Oxford University Press), pp. Eds.), pp. 312–351.

Steven, A.C., Heymann, J.B., Cheng, N., Trus, B.L., and Conway, J.F. (2005). Virus maturation: dynamics and mechanism of a stabilizing structural transition that leads to infectivity. Curr. . Opin. Struct. Biol. *15*, 227–236.

Stow, N.D. (2001). Packaging of genomic and amplicon DNA by the herpes simplex virus type 1 UL25-null mutant KUL25NS. J. Virol. *75*, 10755–10765.

Stow, N.D., McMonagle, E.C., and Davison, A.J. (1983). Fragments from both termini of the herpes simplex virus type 1 genome contain signals required for the encapsidation of viral DNA. Nucleic Acids Res. *11*, 8205–8220.

Taus, N.S., and Baines, J.D. (1998). Herpes simplex virus 1 DNA cleavage/packaging: the UL28 gene encodes a minor component of B capsids. Virology *252*, 443–449.

Taus, N.S., Salmon, B., and Baines, J.D. (1998). The herpes simplex virus 1 UL 17 gene is required for localization of capsids and major and minor capsid proteins to intranuclear sites where viral DNA is cleaved and packaged. Virology *252*, 115–125.

Tengelsen, L.A., Pederson, N.E., Shaver, P.R., Wathen, M.W., and Homa, F.L. (1993). Herpes simplex virus type 1 DNA cleavage and encapsidation require the product of the UL28 gene: isolation and characterization of two UL28 deletion mutants. J. Virol. *67*, 3470–3480.

Thomsen, D.R., Newcomb, W.W., Brown, J.C., and Homa, F.L. (1995). Assembly of the herpes simplex virus capsid: requirement for the carboxyl-terminal twenty-five amino acids of the proteins encoded by the UL26 and UL26.5 genes. J. Virol. *69*, 3690–3703.

Thurlow, J.K., Murphy, M., Stow, N.D., and Preston, V.G. (2006). Herpes simplex virus type 1 DNA-packaging protein UL17 is required for efficient binding of UL25 to capsids. J. Virol. *80*, 2118–2126.

Thurlow, J.K., Rixon, F.J., Murphy, M., Targett-Adams, P., Hughes, M., and Preston, V.G. (2005). The herpes simplex virus type 1 DNA packaging protein UL17 is a virion protein that is present in both the capsid and the tegument compartments. J. Virol. *79*, 150–158.

Trus, B.L., Booy, F.P., Newcomb, W.W., Brown, J.C., Homa, F.L., Thomsen, D.R., and Steven, A.C. (1996). The herpes simplex virus procapsid: structure, conformational changes upon maturation, and roles of the triplex proteins VP19c and VP23 in assembly. J. Mol. Biol. *263*, 447–462.

Trus, B.L., Cheng, N., Newcomb, W.W., Homa, F.L., Brown, J.C., and Steven, A.C. (2004). Structure and polymorphism of the UL6 portal protein of herpes simplex virus type 1. J. Virol. *78*, 12668–12671.

Trus, B.L., Gibson, W., Cheng, N., and Steven, A.C. (1999). Capsid structure of simian cytomegalovirus from cryoelectron microscopy: evidence for tegument attachment sites. J. Virol. *73*, 2181–2192.

Trus, B.L., Heymann, J.B., Nealon, K., Cheng, N., Newcomb, W.W., Brown, J.C., Kedes, D.H., and Steven, A.C. (2001). Capsid structure of Kaposi's sarcoma-associated herpesvirus, a gammaherpesvirus, compared to those of an alphaherpesvirus, herpes simplex virus type 1, and a betaherpesvirus, cytomegalovirus. J. Virol. *75*, 2879–2890.

Trus, B.L., Homa, F.L., Booy, F.P., Newcomb, W.W., Thomsen, D.R., Cheng, N., Brown, J.C., and Steven, A.C. (1995). Herpes simplex virus capsids assembled in insect cells infected with recombinant baculoviruses: structural authenticity and localization of VP26. J. Virol. *69*, 7362–7366.

Trus, B.L., Newcomb, W.W., Booy, F.P., Brown, J.C., and Steven, A.C. (1992). Distinct monoclonal antibodies separately label the hexons or the pentons of herpes simplex virus capsid. Proc. Natl. Acad. Sci. U.S.A. *89*, 11508–11512.

Trus, B.L., Newcomb, W.W., Cheng, N., Cardone, G., Marekov, L., Homa, F.L., Brown, J.C., and Steven, A.C. (2007). Allosteric signaling and a nuclear exit strategy: binding of UL25/UL17 heterodimers to DNA-Filled HSV-1 capsids. Mol. Cell *26*, 479–489.

Underwood, M.R., Harvey, R.J., Stanat, S.C., Hemphill, M.L., Miller, T., Drach, J.C., Townsend, L.B., and Biron, K.K. (1998). Inhibition of human cytomegalovirus DNA maturation by a benzimidazole ribonucleoside is mediated through the UL89 gene product. J. Virol. *72*, 717–725.

Valpuesta, J.M., and Carrascosa, J.L. (1994). Structure of viral connectors and their function in bacteriophage assembly and DNA packaging. Q. Rev. Biophy. *27*, 107–155.

van Zeijl, M., Fairhurst, J., Jones, T.R., Vernon, S.K., Morin, J., LaRocque, J., Feld, B., O'Hara, B., Bloom, J.D., and Johann, S.V. (2000). Novel class of thiourea compounds that inhibit herpes simplex virus type 1 DNA cleavage and encapsidation: resistance maps to the UL6 gene. J. Virol. *74*, 9054–9061.

Varmuza, S.L., and Smiley, J.R. (1985). Signals for site–specific cleavage of HSV DNA: maturation involves two separate cleavage events at sites distal to the recognition sequences. Cell *41*, 793–802.

Villarreal, E.C. (2003). Curr. ent and potential therapies for the treatment of herpes-virus infections. Prog. Drug Res. *60*, 263–307.

Wagstaff, A., and H.M Bryson (2001). Foscarnet: a reappraisal of its antiviral activity, pharmacokinetic properties and therapeutic use in immunocompromised patients with viral infections. Drugs *48*, 199–226.

Walters, J.N., Sexton, G.L., McCaffery, J.M., and Desai, P. (2003). Mutation of single hydrophobic residue I27, L35, F39, L58, L65, L67, or L71 in the N terminus of VP5 abolishes interaction with the scaffold protein and prevents closure of herpes simplex virus type 1 capsid shells. J. Virol. *77*, 4043–4059.

Warner, S.C., Chytrova, G., Desai, P., and Person, S. (2001). Mutations in the N-terminus of VP5 alter its interaction with the scaffold proteins of herpes simplex virus type 1. Virology *284*, 308–316.

Weigele, P.R., Pope, W.H., Pedulla, M.L., Houtz, J.M., Smith, A.L., Conway, J.F., King, J., Hatfull, G.F., Lawrence, J.G., and Hendrix, R.W. (2007). Genomic and structural analysis of Syn9, a cyanophage infecting marine *Prochlorococcus* and *Synechococcus*. Environ Microbiol. *9*, 1675–1695.

White, C.A., Stow, N.D., Patel, A.H., Hughes, M., and Preston, V.G. (2003). Herpes simplex virus type 1 portal protein UL6 interacts with the putative terminase subunits UL15 and UL28. J. Virol. *77*, 6351–6358.

Wingfield, P.T., Stahl, S.J., Thomsen, D.R., Homa, F.L., Booy, F.P., Trus, B.L., and Steven, A.C. (1997). Hexon-only binding of VP26 reflects differences between the hexon and penton conformations of VP5, the major capsid protein of herpes simplex virus. J. Virol. *71*, 8955–8961.

Yang, K., and Baines, J.D. (2006). The Putative Terminase Subunit of Herpes Simplex Virus 1 Encoded by UL28 Is Necessary and Sufficient To Mediate Interaction between pUL15 and pUL33. J. Virol. *80*, 5733–5739.

Yang, K., and Baines, J.D. (2008). Domain within herpes simplex virus 1 scaffold proteins required for interaction with portal protein in infected cells and incorporation of the portal vertex into capsids. J. Virol. *82*, 5021–5030.

Yang, K., Homa, F., and Baines, J.D. (2007). Putative terminase subunits of herpes simplex virus 1 form a complex in the cytoplasm and interact with portal protein in the nucleus. J. Virol. *81*, 6419–6433.

Yu, D., Sheaffer, A.K., Tenney, D.J., and Weller, S.K. (1997). Characterization of ICP6::lacZ insertion mutants of the UL15 gene of herpes simplex virus type 1 reveals the translation of two proteins. J. Virol. *71*, 2656–2665.

Yu, D., and Weller, S.K. (1998a). Genetic analysis of the UL 15 gene locus for the putative terminase of herpes simplex virus type 1. Virology *243*, 32–44.

Yu, D., and Weller, S.K. (1998b). Herpes simplex virus type 1 cleavage and packaging proteins UL15 and UL28 are associated with B but not C capsids during packaging. J. Virol. *72*, 7428–7439.

Zacny, V.L., Gershburg, E., Davis, M.G., Biron, K.K., and Pagano, J.S. (1999). Inhibition of Epstein–Barr virus replication by a benzimidazole L-riboside: novel antiviral mechanism of 5, 6-dichloro-2-(isopropylamino)–1-beta-L-ribofuranosyl-1H-benzimidazole. J. Virol. *73*, 7271–7277.

Zhang, X., Efstathiou, S., and Simmons, A. (1994). Identification of novel herpes simplex virus replicative intermediates by field inversion gel electrophoresis: implications for viral DNA amplification strategies. Virology *202*, 530–539.

Zhou, Z.H., Dougherty, M., Jakana, J., He, J., Rixon, F.J., and Chiu, W. (2000). Seeing the herpesvirus capsid at 8.5 A. Science *288*, 877–880.

Zhou, Z.H., He, J., Jakana, J., Tatman, J.D., Rixon, F.J., and Chiu, W. (1995). Assembly of VP26 in herpes simplex virus-1 inferred from structures. of wild-type and recombinant capsids. Nat. Struct. Biol. *2*, 1026–1030.

Zhou, Z.H., Prasad, B.V., Jakana, J., Rixon, F.J., and Chiu, W. (1994). Protein subunit structures. in the herpes simplex virus A-capsid determined from 400 kV spot-scan electron cryomicroscopy. J. Mol. Biol. *242*, 456–469.

Nuclear Egress and Envelopment of HSV

11

Joel D. Baines and Kari L. Roberts

Abstract

In a process unique in biology, all herpesviruses obtain their initial virion envelope by budding through the inner nuclear membrane. In the most prominent model of virion egress, the envelope of the perinuclear virion then fuses with the luminal surface of the outer nuclear membrane, releasing the de-enveloped capsid into the cytosol for subsequent budding events. The pUL31/pUL34 protein complex is a major player in the initial budding event, and mediates several relevant functions including disruption of the nuclear lamina, recruitment of other viral proteins to the inner nuclear membrane and perinuclear virion, and budding of the nucleocapsid through the inner nuclear membrane. This review focuses on pUL31, pUL34, and other proteins known or believed to be involved in the initial budding event and egress from the perinuclear space.

Introduction

Herpesvirus nucleocapsids, like those of all herpesviruses, are assembled in the nucleus and undergo an initial or primary envelopment step at the inner nuclear membrane (INM). Subsequent to this step, the envelope of perinuclear virions is believed to fuse with the inner face of the outer nuclear membrane (ONM), releasing the de-enveloped nucleocapsid into the cytosol which subsequently buds through Golgi or *trans*-Golgi membranes in a final envelopment step. Through the process of exocytosis, vesicles bearing these virions make their way to the plasma membrane where they fuse, releasing the virion into the extracellular space. This review will explore the process by which nucleocapsids exit the nucleus (Fig. 11.1).

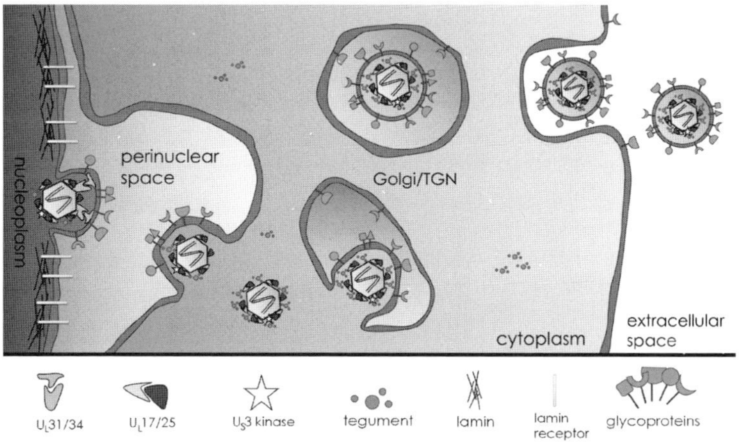

Figure 11.1 Nuclear and cytoplasmic egress of HSV-1.

Anatomy of the nuclear membrane

In a process unique to all herpesviruses, nucleocapsids obtain their initial virion envelope by budding through the inner nuclear membrane (INM) into the perinuclear space. The anatomy of the nuclear membrane (NM) poses a number of molecular problems that must be overcome for successful virion budding and egress.

The NM completely encloses the nucleus and consists of two lipid bilayers: the INM and the ONM. These are separated by a lumen that is continuous with the lumen of the endoplasmic reticulum (ER), while the ONM is analogous to the ER membrane. The nuclear pore membrane abuts nuclear pore complexes and is continuous with both the INM and ONM (Fig. 11.2).

The nuclear lamina lines the entire nucleoplasmic surface of the INM, and is responsible for maintaining the shape and structural integrity of the nucleus (Belmont *et al.*, 1993; Gruenbaum *et al.*, 2000). The nuclear lamina comprises a number of diverse proteins that are interlinked through a complex series of interactions. A major component of the nuclear lamina in mammalian cells consists of a meshwork of type V microfilaments composed of lamin types A, B and C (Fisher *et al.*, 1986; Hoger *et al.*, 1988; Hoger *et al.*, 1990). Type A and C lamins are splice variants from a single gene, *LMNA*, causing lamin C to have a unique C-terminus lacking the final 80 amino acids of mature lamin A, and containing 6 unique amino acids. Lamins B1 and B2 are derived from two different different genes, *LMNB1* and *LMNB2*, respectively. Like all intermediate filament subunits, lamins form homodimers that interact through their globular head and tail domains to form long filaments (Fisher *et al.*, 1986). Although linked, the Lamin B and Lamin A/C networks can be disassembled separately from one another (Izumi *et al.*, 2000).

Although some lamin is detectable in the nucleoplasm, most lamin molecules are retained at the nuclear periphery by interaction with lamin receptors that are integral membrane proteins targeted to and retained at the INM. There are many different types of lamin receptors including those that interact with B type lamins (lamin B receptor, lamin associated polypeptide (LAP)2beta), A type lamins (LEM2), or both (emerin, MAN1) (reviewed in Wagner and Krohne, 2007). Moreover, many of these lamin receptors interact with one another. Both lamins and their receptors have a number of functions, including mechanosensing activities, transcriptional regulation, chromatin organization, chromosome localization, DNA replication, and DNA repair (reviewed in Dechat *et al.*, 2008).

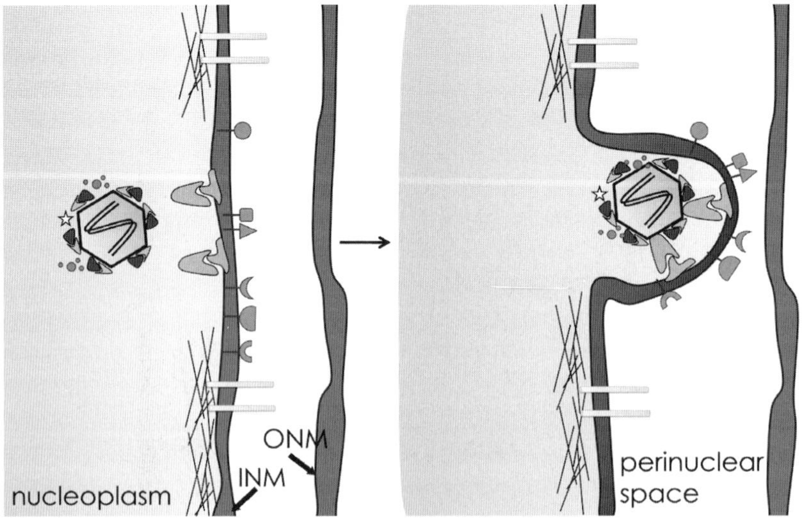

Figure 11.2 HSV-1 budding into the inner nuclear membrane.

The nuclear lamina is disrupted early in mitosis through phosphorylation of A and B type lamins through the activities of p34cdc2/cyclin B kinase and protein kinase C (Peter *et al.*, 1990; Collas *et al.*, 1997). Dephosphorylation correlates with reassembly of the nuclear membrane as both lamin types become incorporated into the nuclear lamina of daughter nuclei. The lamina is also disrupted in the late stages of apoptosis when it is correlated with phosphorylation of lamin B1 by protein kinase C (PKC) beta II, and PKC delta, and cleavage by caspases 3 and 6 (Cross *et al.*, 2000; Chiarini *et al.*, 2002). Nuclear membrane herniations and/or fragmentation result from the resulting disruptions in the lamina.

Viral proteins targeted to the INM of HSV infected cells

Targeting of viral glycoproteins to the INM is believed to involve exploitation of cellular targeting pathways. The INM targeting mechanism of integral membrane proteins likely involves translation of a given protein in association with the ER membrane, followed by retrograde trafficking past the contiguous nuclear pore membrane, culminating in delivery to the INM (Fig. 11.3). Cytosolic/nucleosolic domains greater than 70 kDa severely hamper delivery to the INM, presumably because they impede movement of this bulky domain through the nuclear pore and past its associated membrane (Soullam and Worman, 1995). It is of interest that all herpes simplex virus glycoproteins contain predicted cytosolic domains smaller than this upper limit suggesting that, at least theoretically, all could localize in the INM (McGeoch *et al.*, 1988; Dolan *et al.*, 1998).

It has recently become apparent that nuclear importins help mediate INM targeting of integral membrane proteins, just as they are involved in nuclear importation of cytosolic proteins (Lusk *et al.*, 2007). The targeting is also believed to be mechanistically similar, except for the obvious difference that integral membrane proteins remain tethered to the membrane by their hydrophobic transmembrane domains during transit to the INM. In baculoviruses, a largely hydrophobic domain of 33 amino acids has been shown to contain sufficient information to target chimeric proteins to the INM (Braunagel *et al.*, 2004). This domain interacts with importin alpha 16 to help mediate INM targeting (Saksena *et al.*, 2006). In yeast, sequences very similar to canonical nuclear localization signals recognized by importin beta 1 have been shown to be required for INM targeting (King *et al.*, 2006).

Herpes simplex virus proteins that are targeted to the INM include the myristoylated protein encoded by UL11, the major glycoproteins gB, gD, gM, and gH, the integral membrane phosphoproteins encoded by the UL34 gene (pUL34),

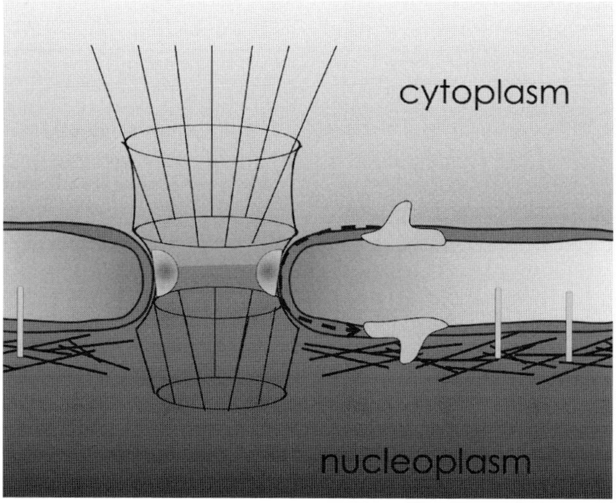

Figure 11.3 Nuclear pore complex and retrograde trafficking of pUL34 from the outer nuclear membrane to the inner nuclear membrane.

and its interaction partner pUL31 (Baines *et al.*, 1995, 2007; Gilbert *et al.*, 1994; Jensen and Norrild, 1998a; Torrisi *et al.*, 1992; Reynolds *et al.*, 2002; Farnsworth *et al.*, 2007). Glycoprotein C is also detectable in the perinuclear virion, but not the INM, presumably because the virion associated gC is at a higher concentration or exhibits a conformation more amenable to antibody binding (Gilbert *et al.*, 1994; Jensen and Norrild, 1998a,b). Although little is known concerning the mechanism by which these proteins are targeted to the INM, it has recently become clear that gD, gM and gH localization at the INM is enhanced significantly by pUL34 (Wills *et al.*, 2009; Wills and Baines, unpublished data). Moreover, pUL34 interacts with the cytosolic domain of gD in the absence of other proteins and coimmunoprecipitates with immature gD from infected cell lysates, suggesting that the gD/pUL34 interaction is involved in targeting ER-resident gD to the INM (Wills *et al.*, 2009). While pUL34 may be a key component involved in targeting gD to the INM by a direct interaction, other proteins could interact with either gD or pUL34 as they transit retrograde to the INM, and thereby become targeted to the INM indirectly. Of interest in this regard are the observations that gD has been shown to interact with gB and gH/gL, at least transiently (Avitabile *et al.*, 2007; Atanasiu *et al.*, 2007). Other signals for INM targeting of herpesvirus glycoproteins have been identified including the transmembrane domain of HSV-1 gB, and a basic peptide in human cytomegalovirus gB which is distinct from canonical nuclear localization signals (Gilbert *et al.*, 1994; Meyer and Radsak, 2000). These data suggest that there may be more than one pathway to target integral membrane proteins to the INM of HSV infected cells.

Some HSV proteins destined to reside in the nascent perinuclear virion are cytosolic proteins targeted to the INM by diffusion or active transport through the nuclear pore, followed by association with ligands anchored to the INM. This is the most likely pathway by which pUL31 associates with the INM, because it contains an essential nuclear localization signal and most pUL31 in the infected cell associates with pUL34 or lamins (Lotzerich *et al.*, 2006b; McGeoch *et al.*, 1988). In the absence of pUL31 the ratios of gD and gM at the INM versus that at the ONM decrease, suggesting that pUL31 plays a role in anchoring viral membrane proteins to the INM.

Still other viral proteins become incorporated into the primary virion by association with the budding nucleocapsid. Caveats for assigning members to this group includes the possibilities that (i) proteins less tightly associated with the capsid may or may not be seen in purified capsid preparations when different purification protocols and detection methods are used and (ii) immunogold labelling, while definitive when significantly above background, is also notoriously insensitive and wholly dependent on the activity of the primary antibody used in the experiment. Proteins likely to become incorporated into the perinuclear virion by association with the nucleocapsid include the major capsid proteins, the portal vertex encoded by UL6, and the pUL17/pUL25 complex which decorates the capsid surface (Newcomb *et al.*, 2001; Thurlow *et al.*, 2005; Trus *et al.*, 2007; Wills *et al.*, 2006). A small number of copies of virion host shutoff (VHS) also associate with the capsid and likely become incorporated into the nascent virion during INM budding (Read and Patterson, 2007). The large tegument protein VP1/2 and its interaction partner encoded by UL37 have also been noted in purified capsid preparations, suggesting that they might eventually become incorporated into the nascent virion by association with the nucleocapsid (Bucks *et al.*, 2007). Along these lines, pUL37 may facilitate nuclear egress (Desai *et al.*, 2001) (see below).

Still other proteins are targeted to the perinuclear virion in unknown ways. For example, the abundant tegument proteins VP16 and VP22 have been identified in perinuclear virions (Naldinho-Souto *et al.*, 2006; Padula *et al.*, 2009), but which mechanism accounts for their incorporation is unclear at this time.

Selection of DNA-containing (C) capsids for envelopment

That DNA containing (type C) capsids are preferentially enveloped at the INM is a long standing observation (Roizman and Furlong, 1974). The lead hypothesis to explain this phenomenon is based on the observation that pUL25 is enriched in C capsids and interacts with pUL17 to form a C capsid-specific complex (CCSC) (Trus *et al.*, 2007; Sheaffer *et al.*, 2001). The CCSC is located at the capsid surface atop triplexes linking the 20 pentons within the capsid to neighbouring hexons. Because it is specifically located on the surface of C capsids, or is at least more abundant in C capsids, the CCSC has been proposed to directly or indirectly engage the envelopment machinery at the INM, thus accounting for the selection of C capsids during envelopment at the INM (Trus *et al.*, 2007). Although this model is consistent with the absence of envelopment of pUL17 and pUL25 null capsids (Salmon *et al.*, 1998; McNab *et al.*, 1998), it does not fully account for the observation that B capsids (that lack DNA) are readily enveloped in cells infected with viruses lacking terminase components pUL15 and pUL28 (Tengelsen *et al.*, 1993; Baines *et al.*, 1994). Because most terminase molecules are lost from C capsids (Yu and Weller, 1998) it is logical to propose that the association of terminase with B capsids may act to inhibit envelopment of capsids lacking DNA. Thus both positive selection of C capsids by the CCSC, and negative selection against B capsids by the associated terminase may play roles in selecting which capsids are to become enveloped. Direct evidence of whether either of these possibilities is involved awaits identification of envelopment proteins that interact with the CCSC and whether the CCSC associates with capsids lacking terminase components.

The pUL31/pUL34 nuclear envelopment complex

The proteins most required for nuclear egress are encoded by UL31 and UL34. pUL34 orients within the INM as a type 2 integral membrane protein, with the bulk of the protein located in the nucleoplasm and only the terminal 5 amino acids located in the perinuclear space (Shiba *et al.*, 2000; McGeoch *et al.*, 1988; Fig. 11.3). In contrast, the primary sequence of pUL31 predicts a largely hydrophobic protein lacking a transmembrane domain (McGeoch *et al.*, 1988). The UL31 and UL34 proteins interact and this interaction is critical to their proper localization in the INM (Reynolds *et al.*, 2001; Reynolds *et al.*, 2002). Amino acids 137–181 of pUL34 within an emerin chimera are sufficient to target pUL31 to sites containing the chimera (Liang and Baines, 2005). Using a TEM-1 beta-lactamase-based two hybrid technique, amino acids 59–127 of the mouse cytomegalovirus orthologue of UL31 were sufficient to complement a pUL34 chimera (Schnee *et al.*, 2006). This region represents the first of four conserved within UL31, and the corresponding region of HSV-1 pUL31 was the only conserved domain to mediate association with the HSV-1 and pseudorabies virus UL34 homologues (Schnee *et al.*, 2006).

Deletion of UL34 precludes primary envelopment of nucleocapsids in most cell lines (Roller *et al.*, 2000). Whereas deletion of UL31 inhibits envelopment in most cells, primary envelopment is less affected in rabbit skin cells, suggesting these cells partially complement UL31 (Liang *et al.*, 2004). Defects in primary envelopment have been noted for the UL31 and UL34 orthologues in other herpesviruses (Farina *et al.*, 2005; Granato *et al.*, 2008; Fuchs *et al.*, 2002; Bubeck *et al.*, 2004; Lotzerich *et al.*, 2006a; Wagenaar *et al.*, 1995). Deletion of UL31 causes UL34 to localize less tightly to the nuclear rim, probably because it becomes redistributed throughout the ER (Reynolds *et al.*, 2001). In turn, deletion of pUL34 causes pUL31 to localize in the nucleoplasm, rather than in association with the nuclear rim.

Both pUL31 and pUL34 have been shown to interact with lamin A *in vitro* in the absence of other cellular or viral proteins, and pUL34 was shown to pull down lamin A and lamin B1 from cellular lysates (Reynolds *et al.*, 2004; Mou *et al.*, 2008). The normal smooth distribution of UL34 throughout the nuclear rim is partly dependent on several functions including:

1. pUL31 as mentioned above.
2. The product of US3 (pUS3), a kinase that phosphorylates both pUL31 and pUL34 (Poon and Roizman, 2005; Purves *et al.*, 1991; Kato *et al.*, 2005). In the absence of US3 kinase activity, pUL31 and pUL34 cluster aberrantly within the NM, whereas these proteins are normally evenly distributed throughout the NM (Klupp *et al.*, 2001;

Reynolds *et al.*, 2002; Ryckman and Roller, 2004), and

3 Lamin A/C inasmuch as pUL34 clusters aberrantly in the NM of LMNA$^{-/-}$ mouse embryonic fibroblasts (Mou *et al.*, 2008). It is unclear whether the dependence of pUL34 localization on lamin A is a direct or indirect effect inasmuch as removal of lamin A/C might have several deleterious effects on the lamina and nuclear rim.

Negotiating the nuclear lamina

During infection, herpes simplex virus induces conformational changes in lamin A, mislocalization of lamins A and B1, and small or large perforations in the nuclear lamina depending on the cell line examined (Scott and O'Hare, 2001; Reynolds *et al.*, 2004; Simpson-Holley *et al.*, 2005; Bjerke and Roller, 2006). All of the HSV-induced changes in the nuclear lamina are dependent on the products of UL31 and UL34. Perhaps secondary to HSV-mediated lamina disruption during HSV infection, the size of the nucleus increases, and this increase is also dependent on pUL31 and pUL34 (Monier *et al.*, 2000; Simpson-Holley *et al.*, 2005). The nuclear expansion is dependent on an intact actin cytoskeleton but is not essential for viral replication (Simpson-Holley *et al.*, 2005).

Several mechanisms are likely responsible for lamina disruption. In the first of these, protein kinase C (PKC) is activated during HSV infection, and PKC delta and alpha are recruited to the nuclear membrane in a ul31 and UL34 dependent manner (Park and Baines, 2006). The recruitment of PKC to the nuclear rim correlates with an HSV-induced increase in phosphorylation of lamin B (Park and Baines, 2006). The observation that PKC delta is partly responsible for phosphorylation of lamins during apoptosis leading to their eventual degradation invites the hypothesis that PKC phosphorylation may have something to do with the perforation of the nuclear lamina in HSV-infected cells. In the second potential mechanism, the US3 encoded kinase has been shown to directly phosphorylate lamin A at multiple sites and to be required for the high level of lamin A phosphorylation noted in HSV infected cells (Mou *et al.*, 2007). Moreover, US3 kinase was sufficient to induce some solubilization of lamin A from permeabilized nuclei suggesting that it is sufficient to depolymerize lamin A (Mou *et al.*, 2007). Finally, and despite the previously mentioned observation, the lamina remains perforated as revealed by indirect immunofluorescence, even in the absence of US3, albeit in a different distribution pattern (Bjerke and Roller, 2006; Mou *et al.*, 2007). This observation, together with observations that overexpression of pUL34 and especially pUL31, can disrupt the nuclear lamina dramatically in the absence of infection (Reynolds *et al.*, 2004; Bjerke and Roller, 2006), suggest that the nuclear envelopment complex has lamin disruption capabilities of its own.

Disruption of the lamina also corresponds to displacement of chromatin from certain regions of the nuclear rim (Simpson-Holley *et al.*, 2004), and displacement of lamin receptors like emerin from the nuclear rim (Leach *et al.*, 2007). It has been proposed that the displacement of chromatin, as well as disruption of the nuclear lamina, is important for nucleocapsids to gain access to the INM (Simpson-Holley *et al.*, 2004; Reynolds *et al.*, 2004; Scott and O'Hare, 2001). On the other hand, there is only limited evidence that the nuclear lamina imposes a barrier to infection, because it is difficult to parse the effects of pUL31 and pUL34 on envelopment from their effects on lamina disruption. In one study, replication of a US3 kinase dead mutant at high multiplicities of infection was enhanced slightly in *LmnA*$^{-/-}$ cells as opposed to *LmnA*$^{+/+}$ cells (Mou *et al.*, 2008). This result supports the proposition that one role of US3 kinase activity is to overcome a lamin A–dependent barrier.

Given its many effects on cell biology, it is logical that, while HSV must perforate the lamina, it must also limit its destruction to avoid toxic effects that might impede virus replication. This may be one reason why the lamins are disrupted in specific regions rather than globally, and may explain the only limited solubilization of lamins (i.e. dissociation from the INM) noted upon HSV infection (Simpson-Holley *et al.*, 2004; Reynolds *et al.*, 2004).

US3, a kinase that regulates nuclear envelopment and egress from the nucleus

In the absence of US3 kinase activity, enveloped nucleocapsids accumulate in the perinuclear space of alphaherpesvirus infected cells (Wagenaar *et al.*, 1995; Klupp *et al.*, 2001; Reynolds *et al.*, 2002; Ryckman and Roller, 2004; Schumacher *et al.*, 2005). Perhaps consequential to the accumulation of virions in the perinuclear space, the intranuclear space expands, causing the nuclear membrane to herniate inwardly towards the nucleoplasm. Moreover, the distribution of the nuclear envelopment complex is dramatically altered in the absence of US3 kinase activity, from smooth to very punctate (Bjerke and Roller, 2006; Mou *et al.*, 2007; Ryckman and Roller, 2004). At least some of the sites of pUL31/pUL34 accumulation seen in cells infected with US3 kinase dead mutants coincide with the NM herniations containing perinuclear virions (Reynolds *et al.*, 2002). The various manifestations of the absence of US3 kinase activity on the morphology of the nuclear membrane can be reproduced by ablating US3 phosphorylation sites located at the N-terminus of pUL31 (Mou *et al.*, 2009). Moreover, the US3 kinase dead phenotypes at the NM are largely precluded in cells infected with a double mutant lacking US3 kinase activity and containing an acidic pUL31 N-terminus (Mou *et al.*, 2009). In contrast, alteration of the known UL34 phosphorylation site does not alter the morphology of the nuclear membrane or the localization of the nuclear envelopment complex substantially (Ryckman and Roller, 2004). Thus, phosphorylation of pUL31 by US3 kinase is critical for the proper localization of the nuclear envelopment complex, and efficient egress of virions from the nucleus.

As in cells infected with HSV lacking US3 kinase activity, perinuclear virions accumulate aberrantly in cells infected with viruses simultaneously lacking gH and gB (Farnsworth *et al.*, 2007). Because gB and gH are essential components of the HSV membrane fusion machinery during entry (Manservigi and Spear, 1977; Sarmiento *et al.*, 1979; Cai *et al.*, 1988; Davis-Poynter *et al.*, 1994; Forrester *et al.*, 1992), it has been proposed that these proteins are also involved in fusion of the perinuclear virion envelope with the ONM during the process of de-envelopment (Farnsworth *et al.*, 2007). Along these lines, it is of interest that US3 has been shown to phosphorylate the cytosolic/nucleosolic tail of gB (Wisner *et al.*, 2009; Imai *et al.*, 2010), perhaps representing a US3-mediated mechanism to alter the conformation of gB and thereby initiate fusion between the perinuclear virion envelope and ONM. Phosphorylation of gB may also be involved indirectly in the fusion reaction because most phosphorylated gB is localized at the NM, as opposed to the substantial amounts of unphosphorylated gB located at the plasma membrane and cytoplasm (Imai *et al.*, 2010). Thus, US3 phosphorylation of gB may also be involved in the targeting of the potential fusion apparatus to the NM and perinuclear virion.

Other HSV proteins implicated in primary envelopment and nuclear egress of nucleocapsids

Analysis of viral null mutants have identified several gene products that may be involved in primary envelopment or nuclear egress, but none of these genes promote exit from the nucleus to the extent of UL31 and UL34. The growing list is as follows.

- **UL37:** More nucleocapsids than normal accumulate in the nuclei of cells infected with null mutants of UL37, indicating that this protein facilitates, but is ultimately dispensable for primary envelopment or egress from the nucleus (Desai *et al.*, 2001).
- **UL51:** Analysis of a UL51 null mutant revealed abundant nucleocapsids in the perinuclear space but relatively few enveloped particles in the cytoplasm, suggesting a role for UL51 in exiting the perinuclear space (Nozawa *et al.*, 2005).
- **gK:** A virus lacking glycoprotein K displays a predominance of nucleocapsids in the nucleus in tightly packed infected cells, suggesting that at least under some conditions gK plays a role in envelopment at the INM (Jayachandra *et al.*, 1997).

- **VP16 and vhs:** Cells infected with a virus lacking UL48 encoding VP16, and to a greater extent, a double mutant lacking VP16 and UL41 encoding vhs contained large numbers of enveloped particles in evaginations of the perinuclear space, and virtually no enveloped particles in the cytoplasm (Mossman *et al.*, 2000). The VP16 null phenotype was partially relieved by secondary mutations in UL53, encoding gK, and a deletion of UL41. These data further implicate VP16, vhs, and gK in a step involved after primary envelopment, but how they might facilitate the process is unknown. It is intriguing, however, that VP16 and vhs have been shown to interact (Smibert *et al.*, 1994; Lam *et al.*, 1996), have been identified in perinuclear virions (Read and Patterson, 2007; Naldinho-Souto *et al.*, 2006), and at least VP16 has been shown to interact with viral glycoprotein H (Gross *et al.*, 2003), an important component of the HSV membrane fusion machine. These observations suggest the possibility that the VP16 and vhs tegument proteins are somehow involved in triggering glycoprotein-mediated fusion between the perinuclear virion envelope and the ONM.

Conclusions

The extensive amount of experimental data has contributed to the following model of primary envelopment and nuclear egress. pUL31 enters the nuclear pore through its nuclear localization signal and enters the nucleus. pUL34 is translated in association with the ER membrane and through a nuclear localization signal at its C terminus engages nuclear import machinery in the cytosol. The nuclear transport machinery then acts to move pUL34, which is embedded in the ONM past the nuclear pore membrane, retrograde to the INM where it engages pUL31. The two proteins also interact with lamin A and directly or indirectly with lamin B1, essentially anchoring them at the INM. Us3 also associates with the pUL31/pUL34 complex. Immature gD interacts with pUL34 in the ER, and follows this protein on its path to the INM. Other interactions with gD may cause immature gH, and gM in the ER to be recruited to the INM through retrograde trafficking, whereas gB trafficking to the NM is facilitated through its phosphorylation by pUS3. The nuclear lamina is disrupted by a combination of localized US3 kinase activity to phosphorylate lamin A, phosphorylation of lamin B by PKC delta and alpha recruited to the INM, and competition for lamins and lamin receptors by the pUL31/pUL34 complex. The nuclear envelopment complex and associated glycoproteins and kinase are now in positions to engage the nucleocapsid.

Nucleocapsids are assembled in the nucleoplasm. In a final assembly step, the terminase disengages from the nucleocapsid as it completes DNA packaging and the pUL17/pUL25 complex adopts a conformation uniquely localized on the nucleocapsid surface. This pUL17/pUL25 complex then interacts directly or indirectly with the pUL31/pUL34 complex and primary envelopment ensues. The resulting perinuclear virion contains at least pUS3, VP22, VP16, vhs, gB, gC, gD, gH, gM, and possibly gK. US3-mediated phosphorylation of pUL31 and gB triggers gB/gH mediated fusion of the perinuclear virion envelope with the luminal side of the ONM. The phosphorylation of pUL31 and possibly pUL34 may also help disengage the nucleocapsid from the pUL31/pUL34 complex. Thus, the nucleocapsid with limited tegument on its surface is released into the cytosol and the pUL31/pUL34 complex and viral glycoproteins remain associated with the ONM.

This model poses both well supported and highly speculative steps. Its purpose is to invite research that will promote both molecular understanding of the egress pathway and additional molecular detail.

References

Atanasiu, D., Whitbeck, J.C., Cairns, T.M., Reilly, B., Cohen, G.H., and Eisenberg, R.J. (2007). Bimolecular complementation reveals that glycoproteins gB and gH/gL of herpes simplex virus interact with each other. during cell fusion. Proc. Natl. Acad. Sci. U.S.A. *104*, 18718–18723.

Avitabile, E., Forghieri, C., and Campadelli-Fiume, G. (2007). Complexes between herpes simplex Virus Glycoproteins gD, gB, and gH Detected in Cells by Complementation of Split Enhanced Green Fluorescent Protein. J. Virol. *81*, 11532–11537.

Baines, J.D., Jacob, R.J., Simmerman, L., and Roizman, B. (1995). The U_L11 gene products of herpes simplex virus 1 are present in the nuclear and cytoplasmic membranes, and intranuclear dense bodies of infected cells. J. Virol. *69*, 825–833.

Baines, J.D., Poon, A.P.W., Rovnak, J., and Roizman, B. (1994). The U_L15 gene of herpes simplex virus encodes two proteins and is required for cleavage of viral DNA. J. Virol. *68*, 8118–8124.

Baines, J.D., Wills, E.G., Pennington, J., Jacob, R.J., and Roizman, B. (2007). Glycoprotein M of herpes simplex virus type 1 is incorporated into virions during budding at the inner nuclear membrane. J. Virol. *81*, 800–812.

Belmont, A.S., Zhai, Y., and Thilenius, A. (1993). Lamin B distribution and association with peripheral chromatin revealed by optical sectioning and electron microscopy tomography. J. Cell Biol. *123*, 1671–1685.

Bjerke, S.L., and Roller, R.J. (2006). Roles for herpes simplex virus type 1 U(L)34 and U(S)3 proteins in disrupting the nuclear lamina during herpes simplex virus type 1 egress. Virology. *347*, 261–276.

Braunagel, S.C., Williamson, S.T., Saksena, S., Zhong, Z., Russell, W.K., Russell, D.H., and Summers, M.D. (2004). Trafficking of ODV-E66 is mediated via a sorting motif and other. viral proteins: Facilitated trafficking to the inner nuclear membrane. Proc. Natl. Acad. Sci. U.S.A. *101*, 8372–8377.

Bubeck, A., Wagner, M., Ruzsics, Z., Lotzerich, M., Iglesias, M., Singh, I.R., and Koszinowski, U.H. (2004). Comprehensive mutational analysis of a herpesvirus gene in the viral genome context reveals a region essential for virus replication. J. Virol. *78*, 8026–8035.

Bucks, M.A., O'Regan, K.J., Murphy, M.A., Wills, J.W., and Courtney, R.J. (2007). Herpes simplex virus type 1 tegument proteins VP1/2 and UL37 are associated with intranuclear capsids. Virology *361*, 316–324.

Cai, W., Gu, B., and Person, S. (1988). Role of glycoprotein B of herpes simplex virus type 1 in viral entry and cell fusion. J. Virol. *62*, 2596–2604.

Chiarini, A., Whitfield, J.F., Armato, U., and Dal Pra, I. (2002). Protein kinase C-beta II is an apoptotic lamin kinase in polyomavirus-transformed, etoposide-treated pyF111 rat fibroblasts. J. Biol. Chem. *277*, 18827–18839.

Collas, P., Thompson, L., Fields, A.P., Poccia, D.L., and Courvalin, J.C. (1997). Protein kinase C-mediated interphase lamin B phosphorylation and solubilization. J. Biol. Chem. *272*, 21274–21280.

Cross, T., Griffiths, G., Deacon, E., Sallis, R., Gough, M., Watters, D., and Lord, J.M. (2000). PKC-delta is an apoptotic lamin kinase. Oncogene *19*, 2331–2337.

Davis-Poynter, N., B.E.L.L., S., Minson, T., and Browne, H. (1994). Analysis of the contributions of herpes simplex virus type 1 membrane proteins to the induction of cell–cell fusion. J. Virol. *68*, 7586–7590.

Dechat, T., Pfleghaar, K., Sengupta, K., Shimi, T., Shumaker, D.K., Solimando, L., and Goldman, R.D. (2008). Nuclear lamins: major factors in the structural organization and function of the nucleus and chromatin. Genes Dev. 22, 832–853.

Desai, P., Sexton, G.L., McCaffery, J.M., and Person, S. (2001). A null mutation in the gene encoding the herpes simplex virus type 1 UL37 polypeptide abrogates virus maturation. J. Virol. *75*, 10259–10271.

Dolan, A., Jamieson, F.E., Cunningham, C., Barnett, B.C., and McGeoch, D.J. (1998). The genome of herpes simplex virus type 2. J. Virol. *72*, 2010–2021.

Farina, A., Feederle, R., Raffa, S., Gonnella, R., Santarelli, R., Frati, L., Angeloni, A., Torrisi, M.R., Faggioni, A., and Delecluse, H.J. (2005). BFRF1 of epstein–barr virus is essential for efficient primary viral envelopment and egress. J. Virol. *79*, 3703–3712.

Farnsworth, A., Wisner, T.W., Webb, M., Roller, R., Cohen, G., Eisenberg, R., and Johnson, D.C. (2007). Herpes simplex virus glycoproteins gB and gH function in fusion between the virion envelope and the outer nuclear membrane. Proc. Natl. Acad. Sci. U.S.A. *104*, 10187–10192.

Fisher, D.Z., Chaudhary, N., and Blobel, G. (1986). cDNA sequencing of nuclear lamins A and C reveals primary and secondary structural homology to intermediate filament proteins. Proc. Natl. Acad. Sci. U.S.A. *83*, 6450–6454.

Forrester, A., Farrell, H., Wilkinson, G., Kaye, J., Davis-Poynter, N., and Minson, T. (1992). Construction and properties of a mutant of herpes simplex virus type 1 with glycoprotein H coding sequences deleted. J. Virol. *66*, 341–348.

Fuchs, W., Klupp, B.G., Granzow, H., Osterrieder, N., and Mettenleiter, T.C. (2002). The interacting UL31 and UL34 gene products of pseudorabies virus are involved in egress from the host-cell nucleus and represent components of primary enveloped but not mature virions. J. Virol. *76*, 364–378.

Gilbert, R., Ghosh, K., Rasile, L., and Ghosh, H.P. (1994). Membrane anchoring domain of herpes simplex virus glycoprotein gB is sufficient for nuclear envelope localization. J. Virol. *68*, 2272–2285.

Granato, M., Feederle, R., Farina, A., Gonnella, R., Santarelli, R., Hub, B., Faggioni, A., and Delecluse, H.J. (2008). Deletion of Epstein–Barr virus BFLF2 leads to impaired viral DNA packaging and primary egress as well as to the production of defective viral particles. J. Virol. *82*, 4042–4051.

Gross, S.T., Harley, C.A., and Wilson, D.W. (2003). The cytoplasmic tail of herpes simplex virus glycoprotein H binds to the tegument protein VP16 *in vitro* and *in vivo*. Virology *317*, 1–12.

Gruenbaum, Y., Wilson, K.L., Harel, A., Goldberg, M., and Cohen, M. (2000). Review: Nuclear lamins-structural proteins with fundamental functions. Journal of Structural Biology *129*, 313–323.

Hoger, T.H., Krohne, G., and Franke, W.W. (1988). Amino acid sequence and molecular characterization of murine lamin B as deduced from cDNA clones. Eur. J. Cell Biol. *47*, 283–290.

Hoger, T.H., Zatloukal, K., Waizenegger, I., and Krohne, G. (1990). Characterization of a second highly conserved B-type lamin present in cells previously thought to contain only a single B-type lamin. Chromosoma *100*, 67–69.

Imai, T., Sagou, K., Arii, J., and Kawaguchi, Y. (2010). Effects of phosphorylation of herpes simplex virus 1 envelope glycoprotein B by Us3 kinase *in vivo* and *in vitro*. J. Virol. *84*, 153–162.

Izumi, M., Vaughan, O.A., Hutchison, C.J., and Gilbert, D.M. (2000). Head and/or CaaX domain deletions of lamin proteins disrupt preformed. lamin A and C but not lamin B structure in mammalian cells. Mol. Biol. Cell *11*, 4323–4337.

Jayachandra, S., Baghian, A., and Kousoulas, K.G. (1997). Herpes simplex virus type 1 glycoprotein K is not essential for infectious virus production in actively replicating cells but is required for efficient envelopment and translocation of infectious virions from the cytoplasm to the extracellular space. J. Virol. *71*, 5012–5024.

Jensen, H.L., and Norrild, B. (1998a). Herpes simplex virus type 1-infected human embryonic lung cells studied by optimized immunogold cryosection electron microscopy. J. Histochem. Cytochem. *46*, 487–496.

Jensen, H.L., and Norrild, B. (1998b). Herpes simplex Virus Type 1-infected Human Embryonic Lung Cells Studied by Optimized Immunogold Cryosection Electron Microscopy. J. Histochem. Cytochem. 46, 487–496.

Kato, A., Yamamoto, M., Ohno, T., Kodaira, H., Nishiyama, Y., and Kawaguchi, Y. (2005). Identification of proteins phosphorylated directly by the Us3 protein kinase encoded by herpes simplex virus 1. J. Virol. *79*, 9325–9331.

King, M.C., Lusk, C., and Blobel, G. (2006). Karyopherin-mediated import of integral inner nuclear membrane proteins. Nature *442*, 1003–1007.

Klupp, B.G., Granzow, H., and Mettenleiter, T.C. (2001). Effect of the pseudorabies virus US3 protein on nuclear membrane localization of the UL34 protein and virus egress from the nucleus. J. Gen. Virol. *82*, 2363–2371.

Lam, Q., Smibert, C.A., Koop, K.E., Lavery, C., Capone, J.P., Weinheimer, S.P., and Smiley, J.R. (1996). Herpes simplex virus VP16 rescues viral mRNA from destruction by the virion host shutoff function. EMBO J. *15*, 2575–2581.

Leach, N., Bjerke, S.L., Christensen, D.K., Bouchard, J.M., Mou, F., Park, R., Baines, J., Haraguchi, T., and Roller, R.J. (2007). Emerin Is hyperphosphorylated and redistributed in herpes simplex virus type 1-infected cells in a manner dependent on both UL34 and US3. J. Virol. *81*, 10792–10803.

Liang, L., and Baines, J.D. (2005). Identification of an essential domain in the Herpes Simplex Virus 1 U(L)34 protein that is necessary and sufficient to interact with U(L)31 protein. J. Virol. *79*, 3797–3806.

Liang, L., Tanaka, M., Kawaguchi, Y., and Baines, J.D. (2004). Cell lines that support replication of a novel herpes simplex 1 UL31 deletion mutant can properly target UL34 protein to the nuclear rim in the absence of UL31. Virology *329*, 68–76.

Lotzerich, M., Ruzsics, Z., and Koszinowski, U.H. (2006a). Functional domains of murine cytomegalovirus nuclear egress protein M53/p38. J. Virol. *80*, 73–84.

Lotzerich, M., Ruzsics, Z., and Koszinowski, U.H. (2006b). Functional Domains of Murine Cytomegalovirus Nuclear Egress Protein M53/p38. J. Virol. *80*, 73–84.

Lusk, C.P., Blobel, G., and King, M.C. (2007). Highway to the inner nuclear membrane: rules for the road. Nat. Rev. Microbiol. 8, 414–420.

Manservigi, R., and Spear, P.G. (1977). Cell fusion induced by herpes simplex virus is promoted and suppressed by different viral glycoproteins. Proc. Natl. Acad. Sci. U.S.A. *74*, 3913–3917.

McGeoch, D.J., Dalrymple, M.A., Davison, A.J., Dolan, A., Frame, M.C., McNab, D., Perry, L.J., Scott, J.E., and Taylor, P. (1988). The complete DNA sequence of the long unique region in the genome of herpes simplex virus type 1. J. Gen. Virol. *69*, 1531–1574.

McNab, A.R., Desai, P., Person, S., Roof, L.L., Thomsen, D.R., Newcomb, W.W., Brown, J.C., and Homa, F.L. (1998). The product of the herpes simplex virus type 1 UL25 gene is required for encapsidation but not for cleavage of replicated DNA. J. Virol. *72*, 1060–1070.

Meyer, G.A., and Radsak, K.D. (2000). Identification of a novel signal sequence that targets transmembrane proteins to the nuclear envelope inner membrane. J. Biol. Chem. *275*, 3857–3866.

Monier, K., Armas, J.C., Etteldorf, S., Ghazal, P., and Sullivan, K.F. (2000). Annexation of the interchromosomal space during viral infection. Nat. Cell Biol. *2*, 661–665.

Mossman, K.L., Sherburne, R., Lavery, C., Duncan, J., and Smiley, J.R. (2000). Evidence that herpes simplex virus VP16 is required for viral egress downstream of the initial envelopment event. J. Virol. *74*, 6287–6299.

Mou, F., Forest, T., and Baines, J.D. (2007). US3 of herpes simplex virus type 1 encodes a promiscuous protein kinase that phosphorylates and alters localization of lamin A/C in infected cells. J. Virol. *81*, 6459–6470.

Mou, F., Wills, E., and Baines, J.D. (2009). Phosphorylation of the Herpes simplex virus 1 UL31 protein by the US3 encoded kinase regulates localization of the nuclear envelopment complex and egress of nucleocapsids. J. Virol. *83*, 5181–5191.

Mou, F., Wills, E.G., Park, R., and Baines, J.D. (2008). Effects of Lamin A/C, Lamin B1, and viral US3 kinase activity on viral infectivity, virion egress, and the targeting of herpes simplex virus UL34-encoded protein to the inner nuclear membrane. J. Virol. *82*, 8094–8104.

Naldinho-Souto, R., Browne, H., and Minson, T. (2006). Herpes simplex virus tegument protein VP16 is a component of primary enveloped virions. J. Virol. *80*, 2582–2584.

Newcomb, W.W., Juhas, R.M., Thomsen, D.R., Homa, F.L., Burch, A.D., Weller, S.K., and Brown, J.C. (2001). The UL6 gene product forms the portal for entry of DNA into the herpes simplex virus capsid. J. Virol. *75*, 10923–10932.

Nozawa, N., Kawaguchi, Y., Tanaka, M., Kato, A., Kato, A., Kimura, H., and Nishiyama, Y. (2005). Herpes simplex virus type 1 UL51 protein is involved in maturation and egress of virus particles. J. Virol. *79*, 6947–6956.

Padula, M.E., Sydnor, M.L., and Wilson, D.W. (2009). Isolation and preliminary characterization of herpes simplex virus 1 primary enveloped virions from the perinuclear space. J. Virol. *83*, 4757–4765.

Park, R., and Baines, J.D. (2006). Herpes simplex virus type 1 infection induces activation and recruitment of protein kinase C to the nuclear membrane and increased phosphorylation of lamin B. J. Virol. *80*, 494–504.

Peter, M., Nakagawa, J., Doree, M., Labbe, J.C., and Nigg, E.A. (1990). *In vitro* disassembly of the nuclear lamina and M phase-specific phosphorylation of lamins by cdc2 kinase. Cell *61*, 591–602.

Poon, A.P., and Roizman, B. (2005). Herpes simplex virus 1 ICP22 regulates the accumulation of a shorter mRNA and of a truncated US3 protein kinase that exhibits altered functions. J. Virol. *79*, 8470–8479.

Purves, F.C., Spector, D., and Roizman, B. (1991). The herpes simplex virus protein kinase encoded by the U_S3 gene mediates posttranslational modification of the phosphoprotein encoded by the U_L34 gene. J. Virol. *65*, 5757–5764.

Read, G.S., and Patterson, M. (2007). Packaging of the virion host shutoff (Vhs) protein of herpes simplex virus: two forms of the Vhs polypeptide are associated with intranuclear B and C capsids, but only one is associated with enveloped virions. J. Virol. *81*, 1148–1161.

Reynolds, A.E., Liang, L., and Baines, J.D. (2004). Conformational changes in the nuclear lamina of cells infected with herpes simplex virus 1 requires. genes UL31 and UL34. J. Virol. *78*, 5564–5575.

Reynolds, A.E., Ryckman, B., Baines, J.D., Zhou, Y., Liang, L., and Roller, R.J. (2001). U_L31 and U_L34 proteins of herpes simplex virus type 1 form a complex that accumulates at the nuclear rim and is required for envelopment of nucleocapsids. J. Virol. *75*, 8803–8817.

Reynolds, A.E., Wills, E.G., Roller, R.J., Ryckman, B.J., and Baines, J.D. (2002). Ultrastructural localization of the HSV-1 U_L31, U_L34, and U_S3 proteins suggests specific roles in primary envelopment and egress of nucleocapsids. J. Virol. *76*, 8939–8952.

Roizman, B., and Furlong, D. (1974). The replication of herpesviruses. In Comprehensive Virology, Fraenkel-Conrat, H. and Wagner, R.R., eds. (New York: Plenum Press), pp. 229–403.

Roller, R., Zhou, Y., Schnetzer, R., Ferguson, J., and Desalvo, D. (2000). Herpes simplex virus type 1 U_L34 gene product is required for viral envelopment. J. Virol. *74*, 117–129.

Ryckman, B.J., and Roller, R.J. (2004). Herpes simplex virus type 1 primary envelopment: UL34 protein modification and the US3-UL34 catalytic relationship. J. Virol. *78*, 399–412.

Saksena, S., Summers, M.D., Burks, J.K., Johnson, A.E., and Braunagel, S.C. (2006). Importin-[alpha]-16 is a translocon-associated protein involved in sorting membrane proteins to the nuclear envelope. Nat. Struct. Mol. Biol. *13*, 500–508.

Salmon, B., Cunningham, C., Davison, A.J., Harris, W.J., and Baines, J.D. (1998). The herpes simplex virus 1 U_L17 gene encodes virion tegument proteins that are required for cleavage and packaging of viral DNA. J. Virol. *72*, 3779–3788.

Sarmiento, M., Haffey, M., and Spear, P.G. (1979). Membrane proteins specified by herpes simplex virus type 1. III. Role of glycoprotein VP7(B2) in virion infectivity. J. Virol. *29*, 1149–1158.

Schnee, M., Ruzsics, Z., Bubeck, A., and Koszinowski, U.H. (2006). Common and specific properties of herpesvirus UL34/UL31 protein family members revealed by protein complementation assay. J. Virol. *80*, 11658–11666.

Schumacher, D., Tischer, B.K., Trapp, S., and Osterrieder, N. (2005). The protein encoded by the US3 orthologue of Marek's disease virus is required for efficient de-envelopment of perinuclear virions and involved in actin stress fiber breakdown. J. Virol. *79*, 3987–3997.

Scott, E.S., and O'Hare, P. (2001). Fate of the inner nuclear membrane protein lamin b receptor and nuclear lamins in herpes simplex virus type 1 infection. J. Virol. *75*, 8818–8830.

Sheaffer, A.K., Newcomb, W.W., Gao, M., Yu, D., Weller, S.K., Brown, J.C., and Tenney, D.J. (2001). Herpes simplex virus DNA cleavage and packaging proteins associate with the procapsid prior to its maturation. J. Virol. *75*, 687–698.

Shiba, C., Daikoku, T., Goshima, F., Takakuwa, H., Yamauchi, Y., Koiwai, O., and Nishiyama, Y. (2000). The UL34 gene product of herpes simplex virus type 2 is a tail-anchored type II membrane protein that is significant for virus envelopment. J. Gen. Virol. *81*, 2397–2405.

Simpson-Holley, M., Baines, J., Roller, R., and Knipe, D.M. (2004). Herpes simplex virus 1 U(L)31 and U(L)34 gene products promote the late maturation of viral replication compartments to the nuclear periphery. J. Virol. *78*, 5591–5600.

Simpson-Holley, M., Colgrove, R.C., Nalepa, G., Harper, J.W., and Knipe, D.M. (2005). Identification and functional evaluation of cellular and viral factors involved in the alteration of nuclear architecture during herpes simplex virus 1 infection. J. Virol. *79*, 12840–12851.

Smibert, C.A., Popova, B., Xiao, P., Capone, J.P., and Smiley, J.R. (1994). Herpes simplex virus VP16 forms a complex with the virion host shutoff protein vhs. J. Virol. *68*, 2339–2346.

Soullam, B., and Worman, H.J. (1995). Signals and structural features. involved in integral membrane protein targeting to the inner nuclear membrane. J. Cell Biol. *130*, 15–27.

Tengelsen, L.A., Pedersen, N.E., Shaver, P.R., Wathen, M.W., and Homa, F.L. (1993). Herpes simplex virus type 1 DNA cleavage and capsidation require the product of the UL28 gene: isolation and characterization of two UL28 deletion mutants. J. Virol. *67*, 3470–3480.

Thurlow, J.K., Rixon, F.J., Murphy, M., Targett-Adams, P., Hughes, M., and Preston, V.G. (2005). The herpes simplex virus type 1 DNA packaging protein UL17 is a virion protein that is present in both the capsid and the tegument compartments. J. Virol. *79*, 150–158.

Torrisi, M.R., Di Lazzaro, C., Pavan, A., Pereira, L., and Campadelli-Fiume, G. (1992). Herpes simplex virus envelopment and maturation studies by fracture label. J. Virol. *66*, 554–561.

Trus, B.L., Newcomb, W.W., Cheng, N., Cardone, G., Marekov, L., Homa, F.L., Brown, J.C., and Steven, A.C. (2007). Allosteric Signaling and a Nuclear Exit Strategy: Binding of UL25/UL17 Heterodimers to DNA-Filled HSV-1 Capsids. Mol. Cell *26*, 479–489.

Wagenaar, F., Pol, J.M., Peeters, B., Gielkens, A.L., de Wind, N., and Kimman, T.G. (1995). The US3-encoded protein kinase from pseudorabies virus affects egress of virions from the nucleus. J. Gen. Virol. *76*, 1851–1859.

Wagner, N., and Krohne, G. (2007). LEM-Domain Proteins: New Insights into Lamin-Interacting Proteins. In Int. Rev. Cytol.A Survey of Cell Biology, Kwang, W. J., ed. Academic Press), pp. 1–46.

Wills, E., Scholtes, L., and Baines, J.D. (2006). Herpes Simplex Virus 1 DNA Packaging Proteins encoded by UL6, UL15, UL17, UL28, and UL33 are located on the external surface of the viral capsid. J. Virol. *80*, 10894–10899.

Wills, E., Mou, F., and Baines, J.D. (2009). The UL31 and UL34 Gene Products of Herpes Simplex Virus 1 are required for Optimal Localization of Viral Glycoproteins D and M to the inner nuclear membrane of infected cells. J. Virol. *83*, 4800–4809.

Wisner, T.W., Wright, C.C., Kato, A., Kawaguchi, Y., Mou, F., Baines, J.D., Roller, R.J., and Johnson, D.C. (2009). Herpesvirus gB-induced fusion between the virion envelope and outer nuclear membrane during virus egress is regulated by the viral US3 kinase. J. Virol. *83*, 3115–3126.

Yu, D., and Weller, S.K. (1998). Herpes simplex virus type 1 cleavage and packaging proteins UL15 and UL28 are associated with B but not C capsids during packaging. J. Virol. *72*, 7428–7439.

Apoptosis Modulation During Herpes Simplex Virus Replication

12

Christopher R. Cotter and John A. Blaho

Abstract

Consequences of human herpes simplex virus (HSV) infection include the induction of apoptosis and the concomitant synthesis of proteins which act to prevent this process from killing the infected cell. Recent data have clarified our current understanding of the mechanisms of induction and prevention of apoptosis by HSV; which ultimately establishes a delicate balance of pro- and antiapoptotic modulating factors in infected cells. These findings emphasize the fact that modulation of apoptosis by HSV during infection is a multicomponent phenomenon involving a combination of viral and cellular factors. We review recent evidence showing how this important human pathogen modulates the fundamental cell death process.

Introduction

Apoptosis is a highly regulated form of cell suicide which is critical to ensure homeostatic balance in multicellular organisms. This phenomenon is an energy-dependent process that differs from other forms of cell death based on a number of distinguishing biochemical and morphological changes occurring in the apoptotic cell. Understanding what these changes are and how they affect the cell affords researchers important tools to investigate apoptosis in a variety of model systems.

Hallmarks of apoptosis

In the nucleus of a cell undergoing apoptosis, genomic DNA is cleaved into multiple nucleosomal fragments that can be detected as either a ladder or a smear using gel electrophoresis (Wyllie *et al.*, 1980). The fragmented DNA can also be end-labelled using the TUNEL assay (terminal deoxynucleotide transferase-mediated incorporation of labelled nucleotides at DNA breakage sites) (Gavrieli *et al.*, 1992). Chromatin undergoes condensation, which may be stained, visualized, and subject to quantification by phase-contrast microscopy (Kerr *et al.*, 1972; Wyllie *et al.*, 1980). Cytosolic proteins that mediate apoptosis also undergo modifications, which may be detected via immunoblotting. These include cleavage of pro-caspases into caspases, as well as degradation of specific caspase substrates, such as the DNA repair enzyme PARP (poly(ADP-ribose) polymerase) (Alnemri *et al.*, 1996; Kaufmann *et al.*, 1993; Lazebnik *et al.*, 1994). Both the plasma and mitochondrial membranes are altered to reflect the apoptotic state of the cell. Phosphatidylserine (PS) moieties become exposed on the extracellular surface of the cell and can be detected by Annexin V staining and subsequent flow cytometric analyses (Fadok *et al.*, 1998; Koopman *et al.*, 1994; Vanags *et al.*, 1996). The membrane permeability of the mitochondria increases supporting an influx of hydrogen ions and a disruption of the potential between the inner and outer mitochondrial membranes. Finally, the appearance of small blebs on the plasma membrane is indicative of an apoptotic phenotype and can be visualized by microscopy.

Apoptotic pathways

Multiple 'death' signals drive cells to undergo apoptosis. Depending on the nature of the signal, one of two molecular pathways is activated

(Green, 1998; Nagata, 1997; Sun *et al.*, 1999). The extrinsic pathway, also known as the death receptor pathway, is typically triggered by binding of certain ligands (Fas ligand or TNF-alpha) to their cognate cell surface death receptors (Fas/CD95 or TNF-R1) (Ashkenazi and Dixit, 1998). Ligand–receptor binding complexes serve as docking stations for adaptor proteins to bind and through a series of protein–protein interactions ultimately result in the activation of effector caspases which engender the apoptotic phenotype. The upstream caspase of the extrinsic pathway is caspase-8, present in the steady-state in the form of an inactive pro-caspase.

The second pathway involves the transmission of the apoptotic signal through the mitochondria and, as such, is often referred to as the mitochondrial or intrinsic pathway (Green and Evan, 2002). Inducers of the intrinsic pathway include exposure to Ultraviolet (UV) radiation, growth factor withdrawal, DNA damage, oxidative stress, and viral infection. The distribution and abundance of mitochondrial proteins of the Bcl-2 family are altered following induction of the intrinsic pathway resulting in disruption of the mitochondrial membrane potential and subsequent release of cytochrome c into the cytoplasm (Green, 1998; Petit *et al.*, 1996). Cytoplasmic cytochrome c associates with Apaf-1 (Green, 1998; Sun *et al.*, 1999). The complex between these two proteins recruits and activates the upstream caspase-9 (Sun *et al.*, 1999).

Both caspase-8 and caspase-9 serve to activate caspase-3. As such, caspase-3 activation represents a convergence point for both the extrinsic and intrinsic pathways. Caspase-3 is an effector caspase that cleaves (and renders non-functional) multiple cellular proteins important for maintaining the integrity of the cell. Examples of caspase-3 targets are the DNAse inhibitor of DNA fragmentation factor 45 (DFF-45) and the DNA repair enzyme, PARP (Kaufmann *et al.*, 1993; Lazebnik *et al.*, 1994; Liu *et al.*, 1997). Structural elements, such as lamin B and actin, are also disrupted following caspase-3 activation (Lazebnik *et al.*, 1995; Mashima *et al.*, 1995). These changes result in the aforementioned biochemical and morphological changes characteristic of an apoptotic cell.

Viral infection and apoptosis

Viral infection, by definition, is a calculated hijacking of the host's cellular machinery by the virus for the sole purpose of replicating its own genome. Apoptosis induction is one escape mechanism by which the infected cell attempts to interfere with this hijacking (Koyama *et al.*, 2000). In doing so, the virus infected cell attempts to limit virus replication, prevent progeny virus production, and ultimately protect other cells from being infected. Not surprisingly, many viruses have evolved mechanisms to actively block this premature death of infected cells. Members of the Herpesviridae family have been well characterized to interfere with apoptosis modulation. This family of viruses contains three subfamilies (α, β, γ) broadly based on the duration of their replication cycle and host range. Kaposi's sarcoma associated virus, a gammaherpesvirus, synthesizes anti-apoptotic proteins that interfere with the enzymatic activities of caspases (Belanger *et al.*, 2001; Djerbi *et al.*, 1999; Thome *et al.*, 1997). The betaherpesvirus, human cytomegalovirus, also has anti-apoptotic genes that interfere with caspase function, in addition to viral proteins that inhibit cellular pro-apoptotic proteins (Arnoult *et al.*, 2004; Poncet *et al.*, 2004; Skaletskaya *et al.*, 2001). The remainder of this review will focus on the modulation of apoptosis, from induction to inhibition by the prototypical alphaherpesvirus, Herpes simplex virus type 1 (HSV-1).

Herpes simplex virus life cycle

Herpes simplex virus type 1 (HSV-1) is a neurotropic human virus that primarily infects epithelial cells during its lytic replication. The typical presentation of HSV-1 infection is in the form of orofacial lesions, usually cold sores, though cases of genital HSV-1 infections in young women appear to be on the rise (Pena *et al.*, 2009). In healthy individuals, the host immune system is quite efficient in clearing this initial infection. However, the virus gains access to the neuronal ganglia that innervate the epithelium and establishes a latent infection there with very limited gene expression. Recurrent infections of the epithelium at or near the initial site of infection can occur at any time as a result of the virus reactivating from its latent stronghold in ganglia. By adulthood, it is estimated that close to 80% of the population has encountered the virus at some point in their lives and are seropositive for HSV (Schillinger *et al.*, 2004).

Lytic HSV-1 replication begins by virus binding to the surface of the host cell through an interaction between a subset of viral glycoproteins (gB and gC) and cellular heparin sulfate moieties (Shieh *et al.*, 1992; WuDunn and Spear, 1989). A second interaction between another viral glycoprotein, gD, and the cell surface receptor HVEM occurs followed by fusion of the viral envelope with the host cell plasma membrane (Geraghty *et al.*, 1998; Montgomery *et al.*, 1996; Warner *et al.*, 1998; Whitbeck *et al.*, 1999). Immediately thereafter, the viral nucleocapsid and tegument proteins are deposited into the cytoplasm (Morgan *et al.*, 1968). Once the viral DNA enters the nucleus a temporal, cascade of viral gene expression follows consisting of three discrete phases – immediate early, early, and late.

The first class of viral genes to be transcribed is the immediate early (IE), or alpha (α) genes. An incoming tegument protein, VP16, complexes with a series of host cellular proteins to drive expression of this class in an RNA polymerase II-dependent fashion (Campbell *et al.*, 1984; Gerster and Roeder, 1988; Kristie and Sharp, 1993). A subset of IE genes including ICP (infected cell protein) 4 and 27 are critical for regulating the later phases of the infection cycle (DeLuca *et al.*, 1985; DeLuca and Schaffer, 1985; Dixon and Schaffer, 1980; McCarthy *et al.*, 1989; Preston, 1979; Sacks *et al.*, 1985; Watson and Clements, 1980). As such, deletion viruses for these genes are impaired in early and late gene expression, in addition to being replication defective. The next class to be transcribed is the early (E) or beta (β) genes, which encode viral proteins involved in viral DNA synthesis (Boehmer and Lehman, 1997). Lastly, the late (L) or gamma (γ) genes are expressed. Members of this class include tegument components, such as VP16 and virion host shut-off (vhs), as well as viral surface glycoproteins. A complete viral replication cycle typically requires between 12–18 hours of infection.

HSV-1 infected cells undergo multiple alterations at both the structural and biochemical level. Collectively, these changes are referred to as cytopathic effect, or CPE (Roizman, 1974). CPE as a result of HSV-1 infection includes the following: (1) enlargement and disaggregation of the nucleolus; (2) chromatin damage; (3) destabilization of the cytoskeleton resulting in the rounding up of cells; (4) folding and insertion of viral glycoproteins into organelle membranes; (5) cell fusion (polykaryocytosis) and (6) a global decrease in macromolecular synthesis (Avitabile *et al.*, 1995; Hampar and Ellison, 1961; Heeg *et al.*, 1986; Roizman, 1962; Roizman and Roane, 1964; Schwartz and Roizman, 1969; Shipkey *et al.*, 1967). Initially, it was believed that HSV-1 induced CPE was the result of necrosis (and subsequent cell lysis). However extensive work over the last decade has revealed that modulation of the apoptotic pathway (from induction to prevention) is a defining feature HSV-1 infection.

Apoptotic modulation by herpes simplex virus

In 1997, a report by Koyama and Adachi showed that HSV-1 infected cells display classic apoptotic features (membrane blebbing, DNA fragmentation, and chromatin condensation) when *de novo* protein synthesis is blocked using the translational inhibitor cycloheximide (CHX) (Koyama and Adachi, 1997). A similar phenomenon was later observed with HSV-2 (Pena *et al.*, 2009; Yedowitz and Blaho, 2005). Additional studies extended this finding to show that other features of apoptosis were detectable in CHX-treated HSV infected cells such as caspase cleavage, mitochondrial membrane potential changes, and PS flipping (Aubert and Blaho, 1999; Gautier *et al.*, 2003; Jerome *et al.*, 2001; Kraft *et al.*, 2006; Nguyen *et al.*, 2005). An important implication from these studies is that apoptosis modulation during infection occurs in two phases. The first phase (induction) occurs in the absence of *de novo* synthesized viral proteins. The second phase (prevention) absolutely requires the gene products of viral replication. Attempts to understand the mechanism and viral trigger of HSV-1 induced apoptosis, as well as which viral proteins block apoptosis has been an active area of research for many years.

Apoptosis induction

As previous reports had demonstrated that the triggering event occurred without viral protein synthesis, a reductionist approach was taken to focus on key early events surrounding the initial phases of virus infection (Aubert *et al.*, 1999; Koyama and Adachi, 1997). Possible trigger(s) included the following: virus binding to the host cell; viral envelope/host membrane fusion; translocation of the capsid to the nucleus; or immediate early (IE)/α viral gene expression.

Studies utilizing UV-irradiated viruses were important in that they revealed that the trigger was due, not to virus binding or entry, but rather to expression of IE genes (Aubert *et al.*, 1999; Sanfilippo *et al.*, 2004). The initial series of experiments that pointed to IE gene expression as the apoptotic triggering event involved a dose–response, time course of UV irradiation treatment to determine an amount of UV light sufficient to abrogate viral gene transcription, yet leave in tact (and thus functional) the virion associated protein, VP16 (Sanfilippo *et al.*, 2004). Exposure of virus stocks to UV light for a minimum of 4 minutes was shown to be enough time for virions to become replication defective (no viral gene expression detected and less than 10 PFU/ml progeny virus produced). Importantly, at this 4-minute treatment UV light, incoming VP16 remained functional in its capacity to act as a transcriptional activator. HSV-1 viral stocks UV-inactivated in this manner failed to induce apoptosis showing that IE gene expression was a requirement, yet binding and entry were not (Aubert *et al.*, 1999; Sanfilippo *et al.*, 2004).

To directly assess the contribution of IE gene expression to HSV-1 induced apoptosis, the use of a series of recombinant viruses deficient in IE gene expression was employed (Sanfilippo and Blaho, 2006). Consistent with work using UV inactivated viruses and temperature sensitive HSV-1 mutants, infection of HEp-2 cells (a human epithelial cell line) with HSV-1 (*d109*), a virus lacking all five immediate early genes, failed to trigger apoptosis (Samaniego *et al.*, 1998). Using recombinant viruses deficient in single IE genes (either ICP4, ICP22, ICP27, or ICP0), it was found that ICP0 expression was critical for apoptosis induction, whereas ICPs 4, 22, and 27 were dispensable (Cai and Schaffer, 1992; Johnson *et al.*, 1992; Ogle and Roizman, 1999; Soliman *et al.*, 1997). Further, infection with the recombinant virus, *d106*, which only expresses ICP0 also induced apoptosis similar to wild-type HSV-1 (Samaniego *et al.*, 1998). A plasmid encoding ICP0 was also shown to trigger apoptosis extending these findings using a virus-free system (Sanfilippo and Blaho, 2006).

The use of one last recombinant virus was used to confirm that ICP0 mRNA was the apoptotic trigger. HSV-1 (*n212*) contains a non-sense mutation in the ICP0 orf producing a complete ICP0 transcript, yet no protein (Cai and Schaffer, 1992). Infection with *n212* resulted in classical apoptotic morphologies of infected cells showing that the ICP0 transcript, but not the protein was the apoptotic trigger during HSV-1 infection. To date, this remains the sole example of viral induced apoptosis being triggered during infection as the result of the expression of a single viral gene. The mechanism of ICP0 mRNA induction of apoptosis remains elusive. Some interesting possibilities include ICP0 RNA processing via the cellular microRNA machinery or interaction with some pro-apoptotic protein or transcription factor serving to drive apoptosis.

Apoptosis prevention

The use of metabolic inhibitors that interfere with normal cell function and a knowledge of the well-defined temporal nature of viral gene expression during HSV-1 infection has led to insight into the virus's modulation of apoptosis. Treatment of HEp-2 cells with the translational inhibitor cycloheximide (CHX) up to 3 hours post infection (hpi) with HSV-1 resulted in apoptosis (Aubert *et al.*, 1999). However, CHX treatment 6 hpi or later resulted in no apoptosis. These CHX treatment experiments highlighted key features of HSV-1 infection, apoptosis induction, and prevention. First, the trigger of apoptosis during HSV-1 infection is produced within the first 3 hours following infection. This is consistent with all data described earlier describing the requirement for IE gene ICP0 to induce cell death during

HSV-1 infection (Sanfilippo and Blaho, 2006). A second important implication is that infected cell anti-apoptotic proteins are produced within a specific time frame between 3 and 6 hpi. This time frame of anti-apoptotic protein production was given the name the 'prevention window' to emphasize that factors produced during this 'window' of time are capable of blocking both extrinsic and intrinsic apoptotic stimuli (Aubert *et al.*, 1999; Goodkin *et al.*, 2003). Again, a similar apoptotic prevention window was also observed with HSV-2 (Yedowitz and Blaho, 2005). In addition, the anti-apoptotic state generated in infected cells by wild type HSV-1 is quite ubiquitous in its capacity to block apoptosis driven by multiple stimuli including: staurosporine (protein kinase inhibitor), ceramide, etoposide, granzyme B, UV irradiation, hyperthermia, heat, as well as treatment with either TNF alpha or Fas antibody (Aubert *et al.*, 1999; Galvan and Roizman, 1998; Goodkin *et al.*, 2003; Javouhey *et al.*, 2008; Jerome *et al.*, 2001; Leopardi and Roizman, 1996; Perng *et al.*, 2000).

Extensive research to uncover which viral proteins have anti-apoptotic functions has revealed ten candidate genes. Classification of viral anti-apoptotic proteins can be divided into two classes – essential viral genes (required for productive viral replication) and non-essential viral genes. The IE gene products, ICP4 and ICP27 are essential viral genes produced in the early stages following lytic HSV-1 infection. Infection of either monkey kidney Vero cells or human epithelial HEp-2 cells with recombinant viruses deficient in these genes display hallmark characteristics of apoptosis (Aubert and Blaho, 1999; Leopardi and Roizman, 1996). The exogenous expression of ICP4 during infection with HSV-1 (*d*120), the ICP4 deletion mutant, restored the observed suppression of apoptosis observed during wild-type infection supporting the data that this viral protein is a critical anti-apoptotic protein (Leopardi and Roizman, 1996). An extensive study using multiple ICP27 domain mutants revealed that the C-terminus region is critical for apoptosis inhibition (Aubert *et al.*, 2001). Interestingly, a terminally differentiated cell line of neuronal origin appears to be resistant to apoptosis induced by *d*120 infection (Aiamkitsumrit *et al.*, 2007).

The remaining anti-apoptotic viral proteins are non-essential viral genes whose capacity to inhibit apoptosis is necessary but oftentimes not sufficient to do so. ICP22 is a non-essential immediate early protein that becomes degraded during apoptosis at late times post-infection (Aubert and Blaho, 1999). Furthermore, a deletion virus for this gene shows elevated apoptosis when compared to wild-type infection suggesting that one role during infection is the inhibition of virus-induced cell death (Aubert *et al.*, 1999). The ICP10 protein of HSV-2 has also been implicated in apoptosis prevention (Han *et al.*, 2009; Langelier *et al.*, 2002; Perkins *et al.*, 2003). The latency-associated transcript of HSV-1 (LAT) is the sole viral gene expressed during latent HSV-1 infection (Deatly *et al.*, 1988; Stevens *et al.*, 1987). The trigeminal ganglia of both mice and rabbits show increased apoptosis, as measured by TUNEL staining and PARP cleavage, for LAT-null infections compared to wild type (Branco and Fraser, 2005; Perng *et al.*, 2000). Interestingly, this anti-apoptotic capacity appears to show cell type specificity, as $CD8^+$ T-cells are unaffected by the LAT defect (Branco and Fraser, 2005). Multiple *in vitro* studies have strengthened the anti-apoptotic role of LAT. One study suggests that a potential mechanism of LAT-induced inhibition of apoptosis is through impaired caspase-3 activation (Carpenter *et al.*, 2007). Of the anti-apoptotic HSV-1 genes, the role of LAT in blocking viral-induced apoptosis is the most controversial. One report suggests that the function of LAT as a pro-survival viral factor may be due to blockage of other cellular death pathways, not apoptosis (Thompson and Sawtell, 2001). In addition, the metabolic processing of the LAT gene via the cellular microRNA machinery (Umbach *et al.*, 2008; Umbach *et al.*, 2009) to elicit it's anti-apoptotic function has been proposed.

Two viral glycoproteins have also been described to block apoptosis during infection (Jerome *et al.*, 2001; Jerome *et al.*, 1999; Zhou *et al.*, 2000; Zhou and Roizman, 2001). The U_S6 gene product, Glycoprotein D (gD), binding to the HVEM receptor is a critical mediator of host cell entry (Geraghty *et al.*, 1998; Montgomery *et al.*, 1996; Warner *et al.*, 1998; Whitbeck *et al.*, 1999). As such, studying the role of gD during infection with a null virus requires high MOIs which results in viral intake by the endocytic pathway (Zhou *et al.*, 2000). Under these experimental conditions, infection of cells with a gD-null virus revealed an apoptotic phenotype, which could be reversed by exogenous expression of the protein in a baculoviral system (Zhou *et al.*, 2000). The domain responsible for apoptosis inhibition is distinct from those essential for receptor binding (Zhou *et al.*, 2003). Glycoprotein J (the U_S5 gene product) has also been shown to be capable of blocking apoptosis during infection with a null virus, as well as in virus-free systems when supplied *in trans* following Fas and Granzyme B treatment (Jerome *et al.*, 2001; Jerome *et al.*, 1999). Another protein component of mature virion particles, UL14, has also been suggested to possess anti-apoptotic activity (Yamauchi *et al.*, 2003). In this report, cell lines stably expressing this minor tegument protein were more resistant to environmental apoptosis triggers. Another viral protein shown to impair partial resistance to cells following environmental stress (heat-induced apoptosis) is U_S11 (Javouhey *et al.*, 2008).

The viral protein kinase U_S3 has been extensively studied with respect to apoptosis modulation. U_S3-deficient viruses show elevated apoptosis upon Vero cell infection (Leopardi *et al.*, 1997). Interestingly, baculoviral expression of U_S3 can block cytochrome c release from the mitochondria and pro–caspase-3 cleavage following infection with the ICP4-null virus *d*120. This capacity to block *d*120-induced apoptosis, in addition to virus-free over-expression experiments revealed insight into the mechanism of U_S3-mediated inhibition of apoptosis. When U_S3 and the pro-apoptotic cellular protein Bad are co-expressed, there is less apoptosis and less Bad cleavage (Benetti *et al.*, 2003; Munger and Roizman, 2001). As such, a possible mechanism of U_S3-mediated inhibition of apoptosis may be the result of a block to Bad-cleavage. Importantly, the anti-apoptotic function of U_S3 appears to be a function of its kinase activity (Benetti and Roizman, 2007).

Cellular factors

Understanding the functional interplay between viral proteins and host proteins involved in apoptotic modulation has revealed important insight into mechanisms of HSV-1's pro- and anti-apoptotic activities. One such example is the previously described association of the viral kinase U_S3 with pro-apoptotic Bcl-2 family members (Benetti *et al.*, 2003; Munger and Roizman, 2001). Further, HSV-1 has a well-demonstrated impact on caspase activities within an infected cell (Benetti and Roizman, 2007; Munger *et al.*, 2003).

NF-κB

In addition to apoptosis, the host response to virus infection includes the production of pro-inflammatory cytokines (TNF-α, IL-6, and IL-12) and the induction of antiviral genes through the Interferon pathway (Liou, 2002). Binding of the transcription factor, NF-κB to target sites of the host genome is a critical first step for the induction of both of these events. Nuclear translocation of NF-κB has been observed during HSV-1 infection, and the appearance of NF-κB in the nucleus coincides with the viral block to apoptosis (Goodkin *et al.*, 2003; Gregory *et al.*, 2004; Patel *et al.*, 1998). Studies performed in cell lines engineered to express a dominant negative form of IκBα (the protein bound to NF-κB and rendering it inactive in the steady-state) underwent apoptosis when infected with wild type HSV-1 (Goodkin *et al.*, 2003). An implication from this work, as well as another report that showed the failure of a pro-apoptotic HSV-1 recombinant virus to induce NF-κB translocation to the nucleus support a model of apoptosis prevention requiring NF-κB (Goodkin *et al.*, 2003). Others, however, have reported that the role of NF-κB during HSV-1 infection may be associated primarily with immune regulation (PKR-dependent induction), as opposed to apoptosis prevention (Taddeo *et al.*, 2003; Taddeo *et al.*, 2004).

p53

The tumour-suppressor gene, p53, functions within cells to appropriately respond to multiple types of environmental insults, such as DNA damage, which compromise the integrity of the cell (Levine, 1997; Prives and Hall, 1999). Apoptosis, as a result of p53 activation, is critical for homeostatic maintenance in the face of such stresses. It is, therefore, not surprising that mutations in this gene are present in over 50% of all human tumours (Hainaut *et al.*, 1998; Hainaut and Hollstein, 2000). Unfortunately, the same mutations that inactivate the function of p53 also render those same cells resistant to anticancer therapeutics. Novel approaches to deal with aberrant p53 signalling are crucial.

Recent reports showed that transformed cells can be driven to apoptose following infection with recombinant HSV-1 viruses (Aubert and Blaho, 1999, 2003; Nguyen *et al.*, 2005). Such recombinant viruses are generally both pro-apoptotic and replication-defective. The status of p53 (modification and distribution within the cell) was a determinant of whether or not these viruses were capable of killing the cells (Aubert and Blaho, 2003). Tumour cells that contained under-modified p53 were highly susceptible to HSV-1 induced apoptosis, a phenomenon referred to as oncoapoptosis. In contrast, primary fibroblast cells of both human and murine origin that contain healthy levels of the p53 protein were resistant to apoptosis following the same infections (Aubert and Blaho, 2003). A clear distinction in the reaction of different cell types to HSV-1 induced apoptosis was apparent from these studies. A follow-up to this study surveyed the susceptibility of multiple tumour cell lines to undergo oncoapoptosis (Nguyen *et al.*, 2007). Interestingly, certain tumour cells were resistant to HSV-1-induced apoptosis. The conclusion from this work was that in order for a tumour cell to be susceptible to oncoapoptosis it must possess intact, functional intrinsic apoptotic machinery. Determining which cellular factors result in resistance remains an important area of this research.

Perspective

An intricate balance exists between pro- and anti-apoptotic factors in a cell during the steady-state. This homeostatic balance is no doubt disrupted following infection with virus. Apoptosis that results due to HSV-1 infection is a complex, tightly regulated process. Unravelling the mechanisms that underlie HSV-1's modulation of apoptosis remains an active area of research. By-products of the viral replication cycle both induce and block apoptosis during infection. Further, this induction and prevention is the result of multiple interactions between viral RNA/proteins with host RNA/proteins. The recent findings that certain cell types show a greater disruption of this apoptotic balance following HSV-1 infection highlights the importance of continued research in this field with the discovery of potential novel therapeutics as a desired endpoint.

Acknowledgements

We wish to thank all of the individuals in our laboratory whose hard work has set the basis for developing this interesting research project. Individuals who played significant roles in generating the data that served as the basis of this review include Martine Aubert, Jennifer O'Toole, Renee Baranin, Louis Serico, Christine Sanfilippo, Renzo Lambardozzi, Natalie Chirimuuta, Margot Goodkin, Elise Morton, Jamie Yedowitz, Marie Nguyen, Rachel Kraft, Kristen Peña, Leah Kang, and Elisabeth Gennis. These studies were supported in part by grants from the United States Public Health Service (AI38873 and AI48582 to J.A.B.) and the American Cancer Society (JFRA 634 to J.A.B.). J.A.B. thanks the Lucille P. Markey Charitable Trust and the National Foundation for Infectious Diseases for their support. C.R.C. is a predoctoral trainee and was supported in part by a United States Public Health Service Institutional Research Training Award (AI07647).

References

Aiamkitsumrit, B., Zhang, X., Block, T.M., Norton, P., Fraser, N.W., and Su, Y.H. (2007). Herpes simplex virus type 1 ICP4 deletion mutant virus d120 infection failed to induce apoptosis in nerve growth factor-differentiated PC12 cells. J. NeuroVirol. *13*, 305–314.

Alnemri, E.S., Livingston, D.J., Nicholson, D.W., Salvesen, G., Thornberry, N.A., Wong, W.W., and Yuan, J. (1996). Human ICE/CED-3 protease nomenclature. Cell *87*, 171.

Arnoult, D., Bartle, L.M., Skaletskaya, A., Poncet, D., Zamzami, N., Park, P.U., Sharpe, J., Youle, R.J., and Goldmacher, V.S. (2004). Cytomegalovirus cell death suppressor vMIA blocks Bax – but not Bak-mediated apoptosis by binding and sequestering Bax at mitochondria. Proc. Natl. Acad. Sci. U.S.A. *101*, 7988–7993.

Ashkenazi, A., and Dixit, V.M. (1998). Death receptors: signaling and modulation. Science *281*, 1305–1308.

Aubert, M., and Blaho, J.A. (1999). The herpes simplex virus type 1 regulatory protein ICP27 is required for the prevention of apoptosis in infected human cells. J. Virol. 73, 2803–2813.

Aubert, M., and Blaho, J.A. (2003). Viral oncoapoptosis of human tumor cells. Gene Ther. *10*, 1437–1445.

Aubert, M., O'Toole, J., and Blaho, J.A. (1999). Induction and prevention of apoptosis in human HEp-2 cells by herpes simplex virus type 1. J. Virol. 73, 10359–10370.

Aubert, M., Rice, S.A., and Blaho, J.A. (2001). Accumulation of herpes simplex virus type 1 early and leaky-late proteins correlates with apoptosis prevention in infected human HEp-2 cells. J. Virol. 75, 1013–1030.

Avitabile, E., Di Gaeta, S., Torrisi, M.R., Ward, P.L., Roizman, B., and Campadelli-Fiume, G. (1995). Redistribution of microtubules and Golgi apparatus in herpes simplex virus-infected cells and their role in viral exocytosis. J. Virol. *69*, 7472–7482.

Belanger, C., Gravel, A., Tomoiu, A., Janelle, M.E., Gosselin, J., Tremblay, M.J., and Flamand, L. (2001). Human herpesvirus 8 viral FLICE-inhibitory protein inhibits Fas-mediated apoptosis through binding and prevention of procaspase-8 maturation. J. Hum Virology *4*, 62–73.

Benetti, L., Munger, J., and Roizman, B. (2003). The herpes simplex virus 1 US3 protein kinase blocks caspase-dependent double cleavage and activation of the proapoptotic protein BAD. J. Virol. *77*, 6567–6573.

Benetti, L., and Roizman, B. (2007). In transduced cells, the US3 protein kinase of herpes simplex virus 1 precludes activation and induction of apoptosis by transfected procaspase 3. J. Virol. *81*, 10242–10248.

Boehmer, P.E., and Lehman, I.R. (1997). Herpes simplex virus DNA replication. Annu. Rev. Biochem. *66*, 347–384.

Branco, F.J., and Fraser, N.W. (2005). Herpes simplex virus type 1 latency-associated transcript expression protects trigeminal ganglion neurons from apoptosis. J. Virol. *79*, 9019–9025.

Cai, W., and Schaffer, P.A. (1992). Herpes simplex virus type 1 ICP0 regulates expression of immediate-early, early, and late genes in productively infected cells. J. Virol. *66*, 2904–2915.

Campbell, M.E., Palfreyman, J.W., and Preston, C.M. (1984). Identification of herpes simplex virus DNA sequences which encode a trans-acting polypeptide responsible for stimulation of immediate early transcription. J. Mol. Biol. *180*, 1–19.

Carpenter, D., Hsiang, C., Brown, D.J., Jin, L., Osorio, N., BenMohamed, L., Jones, C., and Wechsler, S.L. (2007). Stable cell lines expressing high levels of the herpes simplex virus type 1 LAT are refractory to caspase 3 activation and DNA laddering following cold shock induced apoptosis. Virology *369*, 12–18.

Deatly, A.M., Spivack, J.G., Lavi, E., O'Boyle, D.R., 2nd, and Fraser, N.W. (1988). Latent herpes simplex virus type 1 transcripts in peripheral and central nervous system tissues of mice map to similar regions of the viral genome. J. Virol. *62*, 749–756.

DeLuca, N.A., McCarthy, A.M., and Schaffer, P.A. (1985). Isolation and characterization of deletion mutants of herpes simplex virus type 1 in the gene encoding immediate-early regulatory protein ICP4. J. Virol. *56*, 558–570.

DeLuca, N.A., and Schaffer, P.A. (1985). Activation of immediate-early, early, and late promoters by temperature-sensitive and wild-type forms of herpes simplex virus type 1 protein ICP4. Mol. Cell Biol. *5*, 1997–2008.

Dixon, R.A., and Schaffer, P.A. (1980). Fine-structure mapping and functional analysis of temperature-sensitive mutants in the gene encoding the herpes simplex virus type 1 immediate early protein VP175. J. Virol. *36*, 189–203.

Djerbi, M., Screpanti, V., Catrina, A.I., Bogen, B., Biberfeld, P., and Grandien, A. (1999). The inhibitor of death receptor signaling, FLICE-inhibitory protein defines a new class of tumor progression factors. J. Exp. Med. *190*, 1025–1032.

Fadok, V.A., Bratton, D.L., Frasch, S.C., Warner, M.L., and Henson, P.M. (1998). The role of phosphatidylserine in recognition of apoptotic cells by phagocytes. Cell Death Differ. *5*, 551–562.

Galvan, V., and Roizman, B. (1998). Herpes simplex virus 1 induces and blocks apoptosis at multiple steps during infection and protects cells from exogenous inducers in a cell-type-dependent manner. Proc. Natl. Acad. Sci. U.S.A. *95*, 3931–3936.

Gautier, I., Coppey, J., and Durieux, C. (2003). Early apoptosis-related changes triggered by HSV-1 in individual neuronlike cells. Exp. Cell. Res. *289*, 174–183.

Gavrieli, Y., Sherman, Y., and Ben-Sasson, S.A. (1992). Identification of programmed. cell death *in situ* via specific labeling of nuclear DNA fragmentation. J. Cell Biol. *119*, 493–501.

Geraghty, R.J., Krummenacher, C., Cohen, G.H., Eisenberg, R.J., and Spear, P.G. (1998). Entry of alphaherpesviruses mediated by poliovirus receptor-related protein 1 and poliovirus receptor. Science *280*, 1618–1620.

Gerster, T., and Roeder, R.G. (1988). A herpesvirus trans-activating protein interacts with transcription factor OTF-1 and other. cellular proteins. Proc. Natl. Acad. Sci. U.S.A. *85*, 6347–6351.

Goodkin, M.L., Ting, A.T., and Blaho, J.A. (2003). NF-kappaB is required for apoptosis prevention during herpes simplex virus type 1 infection. J. Virol. *77*, 7261–7280.

Green, D.R. (1998). Apoptotic pathways: the roads to ruin. Cell *94*, 695–698.

Green, D.R., and Evan, G.I. (2002). A matter of life and death. Cancer Cell *1*, 19–30.

Gregory, D., Hargett, D., Holmes, D., Money, E., and Bachenheimer, S.L. (2004). Efficient replication by herpes simplex virus type 1 involves activation of the IkappaB kinase-IkappaB-p65 pathway. J. Virol. *78*, 13582–13590.

Hainaut, P., Hernandez, T., Robinson, A., Rodriguez-Tome, P., Flores, T., Hollstein, M., Harris, C.C., and Montesano, R. (1998). IARC Database of p53 gene mutations in human tumors and cell lines: updated compilation, revised formats and new visualisation tools. Nucleic Acids Res. *26*, 205–213.

Hainaut, P., and Hollstein, M. (2000). p53 and human cancer: the first ten thousand mutations. Adv. Cancer Res. 77, 81–137.

Hampar, B., and Ellison, S.A. (1961). Chromosomal aberrations induced by an animal virus. Nature *192*, 145–147.

Han, J.Y., Miller, S.A., Wolfe, T.M., Pourhassan, H., and Jerome, K.R. (2009). Cell type-specific induction and inhibition of apoptosis by Herpes Simplex virus type 2 ICP10. J. Virol. *83*, 2765–2769.

Heeg, U., Dienes, H.P., Muller, S., and Falke, D. (1986). Involvement of actin-containing microfilaments in HSV-induced cytopathology and the influence of inhibitors of glycosylation. Arch. Virol. *91*, 257–270.

Javouhey, E., Gibert, B., Arrigo, A.P., Diaz, J.J., and Diaz-Latoud, C. (2008). Protection against heat and staurosporine mediated apoptosis by the HSV-1 US11 protein. Virology *376*, 31–41.

Jerome, K.R., Chen, Z., Lang, R., Torres, M.R., Hofmeister, J., Smith, S., Fox, R., Froelich, C.J., and Corey, L. (2001). HSV and glycoprotein J. inhibit caspase activation and apoptosis induced by granzyme B or Fas. J. Immunol. *167*, 3928–3935.

Jerome, K.R., Fox, R., Chen, Z., Sears, A.E., Lee, H., and Corey, L. (1999). Herpes simplex virus inhibits apoptosis through the action of two genes, Us5 and Us3. J. Virol. *73*, 8950–8957.

Johnson, P.A., Miyanohara, A., Levine, F., Cahill, T., and Friedmann, T. (1992). Cytotoxicity of a replication-defective mutant of herpes simplex virus type 1. J. Virol. *66*, 2952–2965.

Kaufmann, S.H., Desnoyers, S., Ottaviano, Y., Davidson, N.E., and Poirier, G.G. (1993). Specific proteolytic cleavage of poly(ADP-ribose) polymerase: an early marker of chemotherapy-induced apoptosis. Cancer Res. *53*, 3976–3985.

Kerr, J.F., Wyllie, A.H., and Curr. ie, A.R. (1972). Apoptosis: a basic biological phenomenon with wide-ranging implications in tissue kinetics. Br J. Cancer *26*, 239–257.

Koopman, G., Reutelingsperger, C.P., Kuijten, G.A., Keehnen, R.M., Pals, S.T., and van Oers, M.H. (1994). Annexin V for flow cytometric detection of phosphatidylserine expression on B cells undergoing apoptosis. Blood *84*, 1415–1420.

Koyama, A.H., and Adachi, A. (1997). Induction of apoptosis by herpes simplex virus type 1. J. Gen. Virol. *78*, 2909–2912.

Koyama, A.H., Fukumori, T., Fujita, M., Irie, H., and Adachi, A. (2000). Physiological significance of apoptosis in animal virus infection. Microbes Infect. *2*, 1111–1117.

Kraft, R.M., Nguyen, M.L., Yang, X.H., Thor, A.D., and Blaho, J.A. (2006). Caspase 3 activation during herpes simplex virus 1 infection. Virus Res. *120*, 163–175.

Kristie, T.M., and Sharp, P.A. (1993). Purification of the cellular C1 factor required for the stable recognition of the Oct-1 homeodomain by the herpes simplex virus alpha-trans-induction factor (VP16). J. Biol. Chem. *268*, 6525–6534.

Langelier, Y., Bergeron, S., Chabaud, S., Lippens, J., Guilbault, C., Sasseville, A.M., Denis, S., Mosser, D.D., and Massie, B. (2002). The R1 subunit of herpes simplex virus ribonucleotide reductase protects cells against apoptosis at, or upstream of, caspase-8 activation. J. Gen. Virol. *83*, 2779–2789.

Lazebnik, Y.A., Kaufmann, S.H., Desnoyers, S., Poirier, G.G., and Earnshaw, W.C. (1994). Cleavage of poly(ADP-ribose) polymerase by a proteinase with properties like ICE. Nature *371*, 346–347.

Lazebnik, Y.A., Takahashi, A., Moir, R.D., Goldman, R.D., Poirier, G.G., Kaufmann, S.H., and Earnshaw, W.C. (1995). Studies of the lamin proteinase reveal multiple parallel biochemical pathways during apoptotic execution. Proc. Natl. Acad. Sci. U.S.A. *92*, 9042–9046.

Leopardi, R., and Roizman, B. (1996). The herpes simplex virus major regulatory protein ICP4 blocks apoptosis induced by the virus or by hyperthermia. Proc. Natl. Acad. Sci. U.S.A. *93*, 9583–9587.

Leopardi, R., Van Sant, C., and Roizman, B. (1997). The herpes simplex virus 1 protein kinase US3 is required for protection from apoptosis induced by the virus. Proc. Natl. Acad. Sci. U.S.A. *94*, 7891–7896.

Levine, A.J. (1997). p53, the cellular gatekeeper for growth and division. Cell *88*, 323–331.

Liou, H.C. (2002). Regulation of the immune system by NF-kappaB and IkappaB. J. Biochem. Mol. Biol. *35*, 537–546.

Liu, X., Zou, H., Slaughter, C., and Wang, X. (1997). DFF, a heterodimeric protein that functions downstream of caspase-3 to trigger DNA fragmentation during apoptosis. Cell *89*, 175–184.

Mashima, T., Naito, M., Fujita, N., Noguchi, K., and Tsuruo, T. (1995). Identification of actin as a substrate of ICE and an ICE-like protease and involvement of an ICE-like protease but not ICE in VP-16-induced U937 apoptosis. Biochim. Biophys. Res. Commun. *217*, 1185–1192.

McCarthy, A.M., McMahan, L., and Schaffer, P.A. (1989). Herpes simplex virus type 1 ICP27 deletion mutants exhibit altered patterns of transcription and are DNA deficient. J. Virol. *63*, 18–27.

Montgomery, R.I., Warner, M.S., Lum, B.J., and Spear, P.G. (1996). Herpes simplex virus-1 entry into cells mediated by a novel member of the TNF/NGF receptor family. Cell *87*, 427–436.

Morgan, C., Rose, H.M., and Mednis, B. (1968). Electron microscopy of herpes simplex virus. I. Entry. J. Virol. *2*, 507–516.

Munger, J., Hagglund, R., and Roizman, B. (2003). Infected cell protein No. 22 is subject to proteolytic cleavage by caspases activated by a mutant that induces apoptosis. Virology *305*, 364–370.

Munger, J., and Roizman, B. (2001). The US3 protein kinase of herpes simplex virus 1 mediates the post-translational modification of BAD and prevents

BAD-induced programmed. cell death in the absence of other. viral proteins. Proc. Natl. Acad. Sci. U.S.A. *98*, 10410–10415.

Nagata, S. (1997). Apoptosis by death factor. Cell *88*, 355–365.

Nguyen, M.L., Kraft, R.M., and Blaho, J.A. (2005). African green monkey kidney Vero cells require de novo protein synthesis for efficient herpes simplex virus 1-dependent apoptosis. Virology *336*, 274–290.

Nguyen, M.L., Kraft, R.M., and Blaho, J.A. (2007). Susceptibility of cancer cells to herpes simplex virus-dependent apoptosis. J. Gen. Virol. *88*, 1866–1875.

Ogle, W.O., and Roizman, B. (1999). Functional anatomy of herpes simplex virus 1 overlapping genes encoding infected-cell protein 22 and US1.5 protein. J. Virol. *73*, 4305–4315.

Patel, A., Hanson, J., McLean, T.I., Olgiate, J., Hilton, M., Miller, W.E., and Bachenheimer, S.L. (1998). Herpes simplex type 1 induction of persistent NF-kappa B nuclear translocation increases the efficiency of virus replication. Virology *247*, 212–222.

Peña, K.C., Adelson, M.E., Mordechai, E., and Blaho, J.A. (2009). Genital herpes simplex virus type 1 in women: Detection in cervicovaginal specimens from gynecological practices in the United States. J. Clin. Microbiol. *48*, 150–153.

Perkins, D., Pereira, E.F., and Aurelian, L. (2003). The herpes simplex virus type 2 R1 protein kinase (ICP10 PK) functions as a dominant regulator of apoptosis in hippocampal neurons involving activation of the ERK survival pathway and upregulation of the antiapoptotic protein Bag-1. J. Virol. *77*, 1292–1305.

Perng, G.C., Jones, C., Ciacci-Zanella, J., Stone, M., Henderson, G., Yukht, A., Slanina, S.M., Hofman, F.M., Ghiasi, H., Nesburn, A.B., *et al.* (2000). Virus-induced neuronal apoptosis blocked by the herpes simplex virus latency-associated transcript. Science *287*, 1500–1503.

Petit, P.X., Susin, S.A., Zamzami, N., Mignotte, B., and Kroemer, G. (1996). Mitochondria and programmed. cell death: back to the future. FEBS Lett. *396*, 7–13.

Poncet, D., Larochette, N., Pauleau, A.L., Boya, P., Jalil, A.A., Cartron, P.F., Vallette, F., Schnebelen, C., Bartle, L.M., Skaletskaya, A., *et al.* (2004). An anti-apoptotic viral protein that recruits Bax to mitochondria. J. Biol. Chem. *279*, 22605–22614.

Preston, C.M. (1979). Control of herpes simplex virus type 1 mRNA synthesis in cells infected with wild-type virus or the temperature-sensitive mutant tsK. J. Virol. *29*, 275–284.

Prives, C., and Hall, P.A. (1999). The p53 pathway. J. Pathol. *187*, 112–126.

Roizman, B. (1962). Polykaryocytosis induced by viruses. Proc. Natl. Acad. Sci. U.S.A. *48*, 228–234.

Roizman, B. (1974). Herpesviruses, latency and cancer: a biochemical approach. J. Reticuloendothel. Soc. *15*, 312–321.

Roizman, B., and Roane, P.R., Jr. (1964). The multiplication of herpes simplex virus. II. The relation between protein synthesis and the duplication of viral DNA in infected Hep-2 cells. Virology *22*, 262–269.

Sacks, W.R., Greene, C.C., Aschman, D.P., and Schaffer, P.A. (1985). Herpes simplex virus type 1 ICP27 is an essential regulatory protein. J. Virol. *55*, 796–805.

Samaniego, L.A., Neiderhiser, L., and DeLuca, N.A. (1998). Persistence and expression of the herpes simplex virus genome in the absence of immediate-early proteins. J. Virol. *72*, 3307–3320.

Sanfilippo, C.M., and Blaho, J.A. (2006). ICP0 gene expression is a herpes simplex virus type 1 apoptotic trigger. J. Virol. *80*, 6810–6821.

Sanfilippo, C.M., Chirimuuta, F.N., and Blaho, J.A. (2004). Herpes simplex virus type 1 immediate-early gene expression is required for the induction of apoptosis in human epithelial HEp-2 cells. J. Virol. *78*, 224–239.

Schillinger, J.A., Xu, F., Sternberg, M.R., Armstrong, G.L., Lee, F.K., Nahmias, A.J., McQuillan, G.M., Louis, M.E., and Markowitz, L.E. (2004). National seroprevalence and trends in herpes simplex virus type 1 in the United States, 1976–1994. Sex. Transm. Dis. *31*, 753–760.

Schwartz, J., and Roizman, B. (1969). Concerning the egress of herpes simplex virus from infected cells: electron and light microscope observations. Virology *38*, 42–49.

Shieh, M.T., WuDunn, D., Montgomery, R.I., Esko, J.D., and Spear, P.G. (1992). Cell surface receptors for herpes simplex virus are heparan sulfate proteoglycans. J. Cell Biol. *116*, 1273–1281.

Shipkey, F.H., Erlandson, R.A., Bailey, R.B., Babcock, V.I., and Southam, C.M. (1967). Virus biographies. II. Growth of herpes simplex virus in tissue culture. Exp. Mol. Pathol *6*, 39–67.

Skaletskaya, A., Bartle, L.M., Chittenden, T., McCormick, A.L., Mocarski, E.S., and Goldmacher, V.S. (2001). A cytomegalovirus-encoded inhibitor of apoptosis that suppresses caspase-8 activation. Proc. Natl. Acad. Sci. U.S.A. *98*, 7829–7834.

Soliman, T.M., Sandri-Goldin, R.M., and Silverstein, S.J. (1997). Shuttling of the herpes simplex virus type 1 regulatory protein ICP27 between the nucleus and cytoplasm mediates the expression of late proteins. J. Virol. *71*, 9188–9197.

Stevens, J.G., Wagner, E.K., Devi-Rao, G.B., Cook, M.L., and Feldman, L.T. (1987). RNA complementary to a herpesvirus alpha gene mRNA is prominent in latently infected neurons. Science *235*, 1056–1059.

Sun, X.M., MacFarlane, M., Zhuang, J., Wolf, B.B., Green, D.R., and Cohen, G.M. (1999). Distinct caspase cascades are initiated in receptor-mediated and chemical-induced apoptosis. J. Biol. Chem. *274*, 5053–5060.

Taddeo, B., Luo, T.R., Zhang, W., and Roizman, B. (2003). Activation of NF-kappaB in cells productively infected with HSV-1 depends on activated protein kinase R and plays no apparent role in blocking apoptosis. Proc. Natl. Acad. Sci. U.S.A. *100*, 12408–12413.

Taddeo, B., Zhang, W., Lakeman, F., and Roizman, B. (2004). Cells lacking NF-kappaB or in which NF-kappaB is not activated vary with respect to ability to sustain herpes simplex virus 1 replication and are not susceptible to apoptosis induced by a replication-incompetent mutant virus. J. Virol. *78*, 11615–11621.

Thome, M., Schneider, P., Hofmann, K., Fickenscher, H., Meinl, E., Neipel, F., Mattmann, C., Burns, K., Bodmer, J.L., Schroter, M., *et al.* (1997). Viral FLICE-inhibitory proteins (FLIPs) prevent apoptosis induced by death receptors. Nature *386*, 517–521.

Thompson, R.L., and Sawtell, N.M. (2001). Herpes simplex virus type 1 latency-associated transcript gene promotes neuronal survival. J. Virol. *75*, 6660–6675.

Umbach, J.L., Kramer, M.F., Jurak, I., Karnowski, H.W., Coen, D.M., and Cullen, B.R. (2008). MicroRNAs expressed by herpes simplex virus 1 during latent infection regulate viral mRNAs. Nature *454*, 780–783.

Umbach, J.L., Nagel, M.A., Cohrs, R.J., Gilden, D.H., and Cullen, B.R. (2009). Analysis of human alphaherpesvirus microRNA expression in latently infected human trigeminal ganglia. J. Virol. *83*, 10677–10683.

Vanags, D.M., Porn-Ares, M.I., Coppola, S., Burgess, D.H., and Orrenius, S. (1996). Protease involvement in fodrin cleavage and phosphatidylserine exposure in apoptosis. J. Biol. Chem. *271*, 31075–31085.

Warner, M.S., Geraghty, R.J., Martinez, W.M., Montgomery, R.I., Whitbeck, J.C., Xu, R., Eisenberg, R.J., Cohen, G.H., and Spear, P.G. (1998). A cell surface protein with herpesvirus entry activity (HveB) confers susceptibility to infection by mutants of herpes simplex virus type 1, herpes simplex virus type 2, and pseudorabies virus. Virology *246*, 179–189.

Watson, R.J., and Clements, J.B. (1980). A herpes simplex virus type 1 function continuously required for early and late virus RNA synthesis. Nature *285*, 329–330.

Whitbeck, J.C., Muggeridge, M.I., Rux, A.H., Hou, W., Krummenacher, C., Lou, H., van Geelen, A., Eisenberg, R.J., and Cohen, G.H. (1999). The major neutralizing antigenic site on herpes simplex virus glycoprotein D overlaps a receptor-binding domain. J. Virol. *73*, 9879–9890.

WuDunn, D., and Spear, P.G. (1989). Initial interaction of herpes simplex virus with cells is binding to heparan sulfate. J. Virol. *63*, 52–58.

Wyllie, A.H., Kerr, J.F., and Curr. ie, A.R. (1980). Cell death: the significance of apoptosis. Int. Rev. Cytol *68*, 251–306.

Yamauchi, Y., Daikoku, T., Goshima, F., and Nishiyama, Y. (2003). Herpes simplex virus UL14 protein blocks apoptosis. Microbiol. Immunol. *47*, 685–689.

Yedowitz, J.C., and Blaho, J.A. (2005). Herpes simplex virus 2 modulates apoptosis and stimulates NF-kappaB nuclear translocation during infection in human epithelial HEp-2 cells. Virology *342*, 297–310.

Zhou, G., Avitabile, E., Campadelli-Fiume, G., and Roizman, B. (2003). The domains of glycoprotein D required to block apoptosis induced by herpes simplex virus 1 are largely distinct from those involved in cell–cell fusion and binding to nectin1. J. Virol. *77*, 3759–3767.

Zhou, G., Galvan, V., Campadelli-Fiume, G., and Roizman, B. (2000). Glycoprotein D or J. delivered in trans blocks apoptosis in SK-N-SH cells induced by a herpes simplex virus 1 mutant lacking intact genes expressing both glycoproteins. J. Virol. *74*, 11782–11791.

Zhou, G., and Roizman, B. (2001). The domains of glycoprotein D required to block apoptosis depend on whether. glycoprotein D is present in the virions carrying herpes simplex virus 1 genome lacking the gene encoding the glycoprotein. J. Virol. *75*, 6166–6172.

Mechanisms of Subversion of Type I Interferon Responses by Alphaherpesviruses

13

Paul T. Sobol and Karen L. Mossman

Abstract

Key to the innate immune response to alphaherpesvirus infection is the expression and secretion of type I interferons (IFNs). This family of cytokines bolsters a host offensive to invading pathogens by inducing IFN stimulated genes (ISGs). Not surprisingly, the evolutionary pressure faced by alphaherpesviruses to adapt to the type I IFN response has shaped alphaherpesvirus evolution at the very interface of the virus–host interaction. The cumulative effects of type I IFN expression on alphaherpesvirus replication *in vitro* and dissemination *in vivo* are discussed in this chapter, along with mechanisms employed by these viruses to subvert the type I IFN response. Specifically, this chapter will summarize how alphaherpesviruses block type I IFN production, inhibit the effects of type I IFN signal transduction and suppress downstream IFN-dependent effector pathways with the aims of augmenting viral replication and dissemination.

Introduction

Central to the innate immune response to viral infection is the expression of IFNs, a family of pleiotropic and immunomodulatory cytokines possessing antiviral and antiproliferative functions (Stark *et al.*, 1998). The IFN family of cytokines is divided into three groups depending on which receptor complex is utilized: type I IFNs bind IFN-α receptor 1 and 2 (IFNAR1/2), type II IFN binds the IFN-γ receptor (IFNGR) and type III IFNs bind the interleukin (IL)–28 receptor α (IL-28Rα) and IL-10 receptor β (IL-10Rβ). There are five type I IFNs (IFN-α, -β, -ε, -κ, and -ω), one type II IFN (IFN-γ) and three type III IFNs (IFN-λ1–3 or IL-28A/B and IL-29) (Kotenko *et al.*, 2003; Schroder *et al.*, 2004; Platanias, 2005). Although no structural homology exists across the IFN subtypes, there are 13 homologous subspecies of IFN-α (IFN-α1–13). IFN types I and III are expressed by most nucleated cells, primarily in response to the presence of virally-derived nucleic acids. Epithelial cells and fibroblasts have been identified as the main cell types responsible for the production of IFN-β, while macrophages, leucocytes and dendritic cells (DCs) primarily produce IFN-α (Fitzgerald-Bocarsly, 1993; Bjorck, 2001; Vollstedt *et al.*, 2003). Among the most important immune cell types involved in controlling virus infection are plasmacytoid DCs (pDCs), a specialized subset of the DC compartment, which produce the majority of type I IFN *in vivo* (Hochrein *et al.*, 2004; Rasmussen *et al.*, 2007). In contrast, type II IFN is predominantly expressed by NK cells and activated $CD8^+$ T-lymphocytes (Biron and Brossay, 2001; Platanias, 2005).

While effective immunity against alphaherpesviruses relies on both innate and adaptive mechanisms, the innate immune response has been shown to be of paramount importance in both murine models of HSV infection and in human studies of clinically manifested complications of HSV-1/2 lytic reactivation (Gresser *et al.*, 1976; Zawatzky *et al.*, 1982a–c Pedersen *et al.*, 1983; Ellermann-Eriksen *et al.*, 1986; Shupack *et al.*, 1992; Vollstedt *et al.*, 2004). Initial studies of mouse strains that were susceptible (BALB/c) or resistant (C57Bl/6) to HSV-1 infection illustrated that rapid induction of type I IFNs positively correlated with a resistant phenotype (Mossman, 2002; Halford *et al.*, 2004). Subsequent studies showed that depletion of IFN-α/β abrogated HSV-1 resistance (Lopez, 1985; Su *et al.*, 1990; Kunder *et al.*, 1993; Halford *et al.*, 1997) and that type I IFN induction inhibited HSV-1 spread from the inoculation site to the peripheral nervous system (Halford *et al.*, 2004). Topical administration of IFN-α was also shown to greatly ameliorate HSV-2 shedding and subsequent lytic reactivation in humans (Shupack *et al.*, 1992). More recent studies have shown that exogenous expression of type I IFNs via adenoviral and plasmid DNA constructs could suppress viral replication and increase the survival of mice infected intraocularly with HSV-1 (Cui and Carr, 2000; Al-Khatib *et al.*, 2004, 2005; Gill *et al.*, 2006). Several groups have also shown increased HSV-2 dissemination and pathogenesis in $IFNAR^{-/-}$ mice after intravaginal HSV-2 administration (Murphy *et al.*, 2003; Gill *et al.*, 2006; Svensson *et al.*, 2007). In addition, depletion of accessory proteins involved in the type I IFN response substantially abrogated control of HSV-1 replication and pathogenesis in acute and corneal challenge models of infection (Vollstedt *et al.*, 2004; Pasieka *et al.*, 2009).

As with HSV-1 and -2, type I IFNs attenuate Varicella-zoster virus (VZV) replication *in vitro* (Ito *et al.*, 1992; Balachandra *et al.*, 1994; Tovey and Maury, 1999; Desloges *et al.*, 2005a). The importance of type I IFNs in controlling VZV pathogenesis is also illustrated by recent *in vivo* studies of VZV infection of human skin xenografts, where IFN-α was downregulated in VZV-infected cells but was potently induced in surrounding epidermal cells (Ku *et al.*, 2004; Ku *et al.*, 2005). Furthermore, a clinical study showed that therapeutic IFN-α administration to immunocompromised children with VZV ameliorated the extent of VZV pathology (Arvin, 1996). Taken together, these studies underscore the manifold contributions of type I IFNs to establishing immunity against alphaherpesviruses. Although the recently discovered type III IFNs are emerging as important regulators of innate immune responses at epithelial surfaces, where many alphaherpesviruses replicate, few studies have yet addressed their role in controlling alphaherpesvirus infection. This chapter will thus focus on the induction of type I IFN in response to alphaherpesvirus infection, along with strategies used by these viruses to counter type I IFN mediated responses. Given the similarities between type I and III IFNs, it is likely that many of the issues discussed in this chapter will be relevant to these new IFN family members as well.

Cellular recognition of alphaherpesviruses and the type I IFN response

Induction of type I IFN in response to alphaherpesvirus infection

The type I IFN antiviral response occurs in a biphasic manner, consisting of IFN production (Fig. 13.1) and subsequent signal transduction (Fig. 13.2) (Juang *et al.*, 1998; Yoneyama *et al.*, 1998; Sato *et al.*, 2000; Mossman and Ashkar, 2005; Honda *et al.*, 2006; Honda and Taniguchi, 2006). In the first phase, germline-encoded pattern recognition receptors (PRRs) sense and bind to evolutionarily conserved viral motifs inherent within viral particles (e.g. viral glycoproteins) or produced during viral replication (e.g. genomic intermediates). The main inducing factor in generating type I IFN expression is dsRNA, which is produced during the life cycles of not only RNA viruses, but also of DNA viruses through symmetrical transcription (Jacquemont and Roizman, 1975). Engagement of PRRs with their viral ligands leads to the recruitment of various adapter proteins and kinases, culminating in multiple IFN regulatory factor (IRF)3 and/or IRF7 phosphorylation events in the C-terminus by either TANK-binding kinase 1 (TBK-1) in fibroblasts or by the inhibitor of NF-κB kinase epsilon (IKK-ε) in immune cells such as pDCs (Fitzgerald *et al.*, 2003; Lees-Miller and Meek, 2003; Sharma *et al.*, 2003; McWhirter *et al.*, 2004; Meek *et al.*, 2004). Once phosphorylated, IRF3 and IRF7 homo- or hetero-dimerize and translocate to the nucleus where they bind to response elements in the promoter region of IFN genes, often in association with additional transcription factors, such as NF-κB and AP-1, and activators, such as cAMP response element-binding (creb) binding

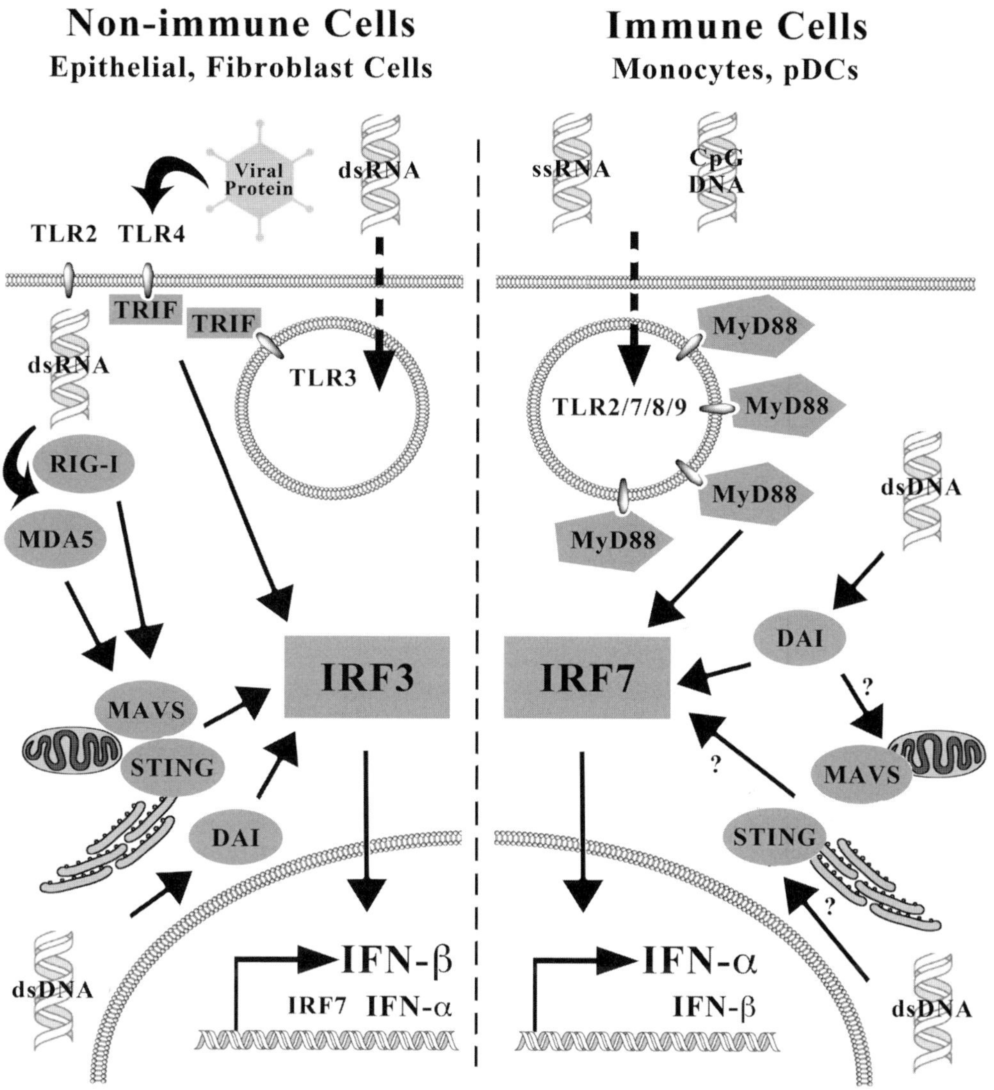

Figure 13.1 The model of alphaherpesvirus-mediated type I interferon production. Virus infection activates latently expressed transcription factors following recognition of conserved viral motifs, causing induction of IFN-α and IFN–β in a cell-type-dependent context. Non-immune cells such as epithelial and fibroblast cells predominantly utilize RLRs and TLR3 to recognize dsRNA, and require IRF3 for the induction of IFN-β. Immune cells such as inflammatory monocytes and pDCs utilize TLR2/7/8/9 to recognize ssRNA and CpG DNA species, and require IRF7 for the induction of IFN-α.

protein (CBP) and p300. In fibroblasts and epithelial cells, IRF3 is constitutively expressed and is a key mediator of IFN-β production while in immune cells, including pDCs, IRF7 is constitutively expressed, and is critical for production of IFN-α (Fig. 13.1) (Honda *et al.*, 2006; Honda and Taniguchi, 2006).

Perhaps the best characterized PRRs are Toll-like receptors (TLRs), which recognize a diversity of bacterially and virally derived lipids, polysaccharides, lipopeptides and nucleic acids, and localize to the plasma membrane or within endosomes (Takeuchi and Akira, 2009). TLR2 and TLR4 bind viral glycoproteins and are localized to the cell membrane while TLR3, TLR7 and

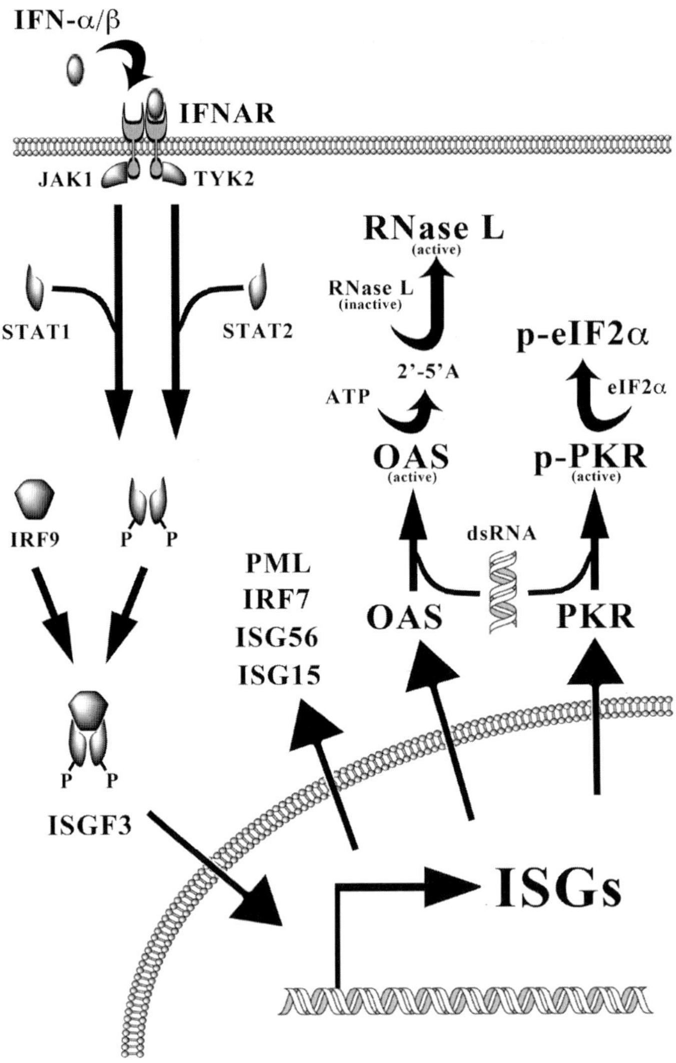

Figure 13.2 Induction of interferon-stimulated genes following type I interferon production. Binding of type I IFNs to their cognate receptor leads to the expression of ISGs through activation of the JAK-STAT pathway. Although latently expressed, OAS and PKR levels are increased significantly by IFN signalling. Both enzymes become activated after binding to dsRNA, a by-product of virus infection. Activation of the OAS-RNase L pathway leads to the degradation of RNA species, while activation of PKR leads to phosphorylation of eIF2α and translational repression. The cumulative effects of ISG induction leads to a classic antiviral state, whereby subsequent viral replication is halted and cells are primed for apoptosis.

TLR9 bind viral dsRNA, ssRNA and CpG-rich DNA, respectively, and are localized within the endosome (Hemmi *et al.*, 2000; Alexopoulou *et al.*, 2001; Lund *et al.*, 2003). Although several alphaherpesvirus components have been shown to stimulate type I IFN expression through TLRs, very few studies have identified essential roles for individual receptors. For example, TLR2 and TLR9 have been shown to recognize both an unidentified virion glycoprotein and HSV-2 genomic DNA, respectively (Lund *et al.*, 2003; Krug *et al.*, 2004; Kurt-Jones *et al.*, 2004; Aravalli *et al.*, 2005; Cheng *et al.*, 2007). Murine models of HSV infection have illustrated that while pDCs mediate early recognition of HSV-1 in a TLR9-dependent manner, and that cells lacking TLR9 fail to produce type I IFN early after HSV-2 infection (Lund *et al.*, 2003; Rasmussen *et al.*, 2007; Sorensen *et al.*, 2008), TLR9-independent IFN production by other cell types occurs (Hochrein *et al.*, 2004; Rasmussen *et al.*, 2007). Other studies have shown that mice deficient for TLR2 are resistant to lethal HSV-1 encephalitis and express less pro-inflammatory cytokines following HSV-1 infection (Kurt-Jones *et al.*, 2004; Mansur *et al.*, 2005), suggesting that TLR2 mainly functions to elicit pro-inflammatory cytokine production. Recently, TLR2 was shown to be important in producing type I IFN in bone marrow-derived monocytes in response to vaccinia virus, another dsDNA virus (Barbalat *et al.*, 2009), suggesting that TLR2 may also mediate IFN induction in response to alphaherpesviruses. Of interest, a recent study reported that humans possessing a dominant-negative TLR3 allele were prone to HSV-1 encephalitis, presupposing the importance of TLR3 in establishing innate immune responses against HSV-1 in the central nervous system (Zhang *et al.*, 2007).

While TLRs recognize viral motifs at the plasma membrane or within endosomes, additional PRRs recognize viral motifs within the cytoplasm and endoplasmic reticulum, particularly viral genomic material (Fig. 13.1). The cytoplasmic RNA helicases retinoic acid inducible gene I (RIG-I) and melanoma differentiation-associated protein 5 (MDA5) bind dsRNA and, through the adaptor protein MAVS (mitochondrial antiviral signalling), promote type I IFN expression (Kawai *et al.*, 2005; Li *et al.*, 2009). Murine fibroblasts deficient for RIG-I fail to mount a type I IFN response at early times after HSV-2 infection, and recent studies have shown that TLR9 and RIG-I cooperate to induce the type I IFN response against HSV-2 (Rasmussen *et al.*, 2007; Rasmussen *et al.*, 2009). In support of this observation, IFN-β expression by infected fibroblasts has been shown to be dependent on MAVS, further implicating the RNA helicases in the type I IFN response to alphaherpesviruses (Kawai *et al.*, 2005; Meylan *et al.*, 2005; Seth *et al.*, 2005; Xu *et al.*, 2005; Ishii *et al.*, 2006; Cheng *et al.*, 2007).

Recently, PRRs that recognize cytoplasmic DNA have been shown to be crucial in inducing type I IFN production and pro-inflammatory cytokines. These sensors include the DNA dependent activator of IRFs (DAI), NOD-like receptor (NLR)/inflammasome activator Nalp3, absent in melanoma 2 (AIM2) and other, currently unknown intracellular DNA sensors (Muruve *et al.*, 2008; Fernandes-Alnemri *et al.*, 2009; Hornung *et al.*, 2009). Additionally, stimulator of interferon genes (STING) functions as a DNA sensor within the endoplasmic reticulum. DAI was recently shown to stimulate type I IFN responses following HSV-1 infection, since the DAI inhibitor adenosine deaminase acting on RNA 1 (ADAR1) inhibits HSV-1-induced type I IFN expression (Wang *et al.*, 2008). However, in knockout fibroblasts and mice, DAI was shown to be redundant for type I IFN responses following HSV-1 infection (Takaoka *et al.*, 2007; Ishii *et al.*, 2008). In contrast, STING has been shown to be essential for type I IFN responses following HSV-1 infection of fibroblasts and mice (Ishikawa and Barber, 2008; Ishikawa *et al.*, 2009) as STING knockout mice are highly susceptible to lethal HSV-1 challenge. Of interest, STING was also shown to be essential for plasmid DNA vaccine- and vaccinia viral vector-mediated antibody production and $CD8^+$ T-cell responses (Ishikawa *et al.*, 2009). Studies of AIM2 and Nalp3 have shown that while these intracellular DNA sensors are required for inflammasome-mediated IL-1β expression and caspase 1-dependent apoptosis, they are not required for the type I IFN response to foreign DNA (Muruve *et al.*, 2008; Fernandes-Alnemri *et al.*, 2009; Hornung *et al.*, 2009).

In addition to an IFN-dependent antiviral response to alphaherpesvirus replication, recent research has underscored the importance of an upstream, IFN-independent antiviral response to incoming viral particles which induces a particular subset of ISGs capable of eliciting a full antiviral state (Nicholl *et al.*, 2000; Mossman *et al.*, 2001; Preston *et al.*, 2001; Collins *et al.*, 2004). While IRF3 is necessary for this response, the mechanism involved in sensing virus particles has yet to be determined, but is independent of RNA helicases and TLRs (Paladino *et al.*, 2006). This response to herpesvirus particle entry is not only MOI-dependent, but is also reliant on a currently unidentified member of the phosphoinositide

3-kinase family of proteins (Noyce *et al.*, 2006). Taken together, it appears that several features of alphaherpesviruses, either inherent to the physical viral particle or produced during its replication cycle, are recognized by multiple PRRs which then induce IFN-dependent and independent responses in an effort to quell viral dissemination through the induction of an antiviral state.

Inhibition of alphaherpesvirus replication by type I IFN

In a generalized context, induction of the type I IFN response by alphaherpesvirus-infected cell types promotes the activation of immune cells, such as NK cells, macrophages, DCs, and B and T lymphocytes to halt viral dissemination, generate durable adaptive immunity towards alphaherpesvirus immunological determinants and reduce viral load (Borysiewicz and Sissons, 1994; Tanaka *et al.*, 1998; Biron *et al.*, 1999; Deshpande *et al.*, 2000; Goodbourne *et al.*, 2000). While inhibition of the immediate early and early classes of viral gene products by IFN have been described (Mittnacht *et al.*, 1988; Oberman and Panet, 1988; Chatterjee and Burns, 1990; DeStasio and Taylor, 1990; Maheshwari *et al.*, 1994; Holberg-Petersen *et al.*, 1996; Wintergerst *et al.*, 1996), it has become apparent that the specific means by which IFN mediates anti-alphaherpesvirus innate immunity is poorly characterized. This lack of information is due in part to the complexity of alphaherpesviruses, the cell-type-specific effects of IFN expression and responsiveness, the multifactorial nature of what constitutes an effective antiviral state and, perhaps most importantly, the *in vivo* context of how the type I IFN response interplays with humoral and cellular arms of the innate and adaptive immune systems.

In the second phase of the type I IFN innate immune response, secreted IFNs bind to their cognate receptor in an autocrine or paracrine fashion, inducing the phosphorylation of signal transducer and activator of transcription-1 and -2 (STAT-1/2) in a Janus-activated kinase 1- (JAK1) and tyrosine kinase 2- (TYK2)-dependent manner, respectively (Platanias, 2005) (Fig. 13.2). Phosphorylated STAT–1/2 heterodimers associate with constitutively expressed IRF9 to form the ISGF3 complex, which then translocates to the nucleus and stimulates the production of ISGs. While as many as 300 ISGs have been identified (de Veer *et al.*, 2001), the three best-characterized ISGs are the dsRNA-dependent protein kinase (PKR), oligoadenylate synthetase (OAS) and the Mx family of dynamin-like GTPases. The N-terminus of PKR binds dsRNA generated by viral replication, resulting in C-terminal autophosphorylation, homodimerization and activation (Sadler and Williams, 2008). PKR then phosphorylates the eukaryotic initiation factor 2 (eIF2) subunit α, which is the regulatory subunit of the eIF2 trimeric complex required for protein synthesis initiation (Garcia *et al.*, 2007; Sadler and Williams, 2008). Like PKR, OAS is upregulated by IFN and binds dsRNA, which results in the synthesis of 2′–5′-linked oligoadenylates (2′–5′ A) from ATP (Hovanessian *et al.*, 1977). 2′–5′ A in turn bind to and activate ribonuclease L (RNase L), an endoribonuclease which degrades viral and host RNA, including ribosomal RNA (rRNA), resulting in protein synthesis arrest and the induction of caspase 9-dependent apoptosis via the pro-apoptotic c-Jun N-terminal kinase pathway (Sadler and Williams, 2008). Induction of apoptosis has been shown to be an antiviral defence mechanism utilized to abrogate viral spread (Barber, 2001).

Despite the contributions of PKR, OAS-RNase L and Mx1 towards inhibition of RNA virus replication, only PKR has been definitively shown to affect alphaherpesvirus replication. Studies investigating the effects of the OAS-RNase L pathway on HSV-1 replication remain divided (Fujihara *et al.*, 1989; Taylor *et al.*, 1995; Khabar *et al.*, 2000; Zheng *et al.*, 2001; Harle *et al.*, 2002; Al-khatib *et al.*, 2003, 2004, 2005) and Mx1 expression has been shown to have no effect on HSV-1 replication (Pavlovic *et al.*, 1990). However, it has been shown that mice deficient for PKR have a pronounced susceptibility to HSV-1 and HSV-2 (Leib *et al.*, 2000; Al-Khatib *et al.*, 2004; Carr *et al.*, 2005) and, in the context of an HSV-2 infection, that the PKR pathway is the dominant means by which type I IFN exerts its antiviral activities (Carr *et al.*, 2005). Similarly, studies of VZV have illustrated that PKR, but not the OAS-RNase L pathway, mediates IFN-β-dependent inhibition of replication (Desloges *et al.*, 2004, 2005a).

Underscoring the importance of multiple ISGs in mounting a complete antiviral response, mice triply deficient for PKR, RNase L and Mx1 elicit IFN-induced antiviral effects (Zhou *et al.*, 1999). Recent studies have shown that ISG15, a ubiquitin-like molecule, is important in mediating the antiviral effects of IFN *in vivo* (Lenschow *et al.*, 2007). The survival of mice deficient for ISG15 was found to be significantly decreased relative to wild-type mice after intracranial or corneal administration of HSV-1. In addition to ISG15, other proteins involved in the type I IFN response to alphaherpesvirus infection include ISG56, a protein that binds to eIF3 and inhibits protein synthesis (Guo *et al.*, 2000), and IP-10, a C-X-C chemokine responsible for monocyte, NK, DC and T-cell migration to sites of viral infection (Neville *et al.*, 1997). These ISGs are rapidly induced in response to HSV-1 entry and replication (Nicholl *et al.*, 2000; Mossman *et al.*, 2001). In addition, a considerable amount of research has ascertained the anti-herpesvirus properties of the promyelocytic leukaemia protein (PML), a multifunctional ISG and tumour suppressor that regulates p53-dependent apoptosis (de Stanchina *et al.*, 2004), MHC-I expression (Zheng *et al.*, 1998; Zhang *et al.*, 2003), transcriptional regulation, and antiviral immunity (Dellaire and Bazett-Jones, 2004; Everett and Chelbi-Alix, 2007).

Alphaherpesvirus-mediated subversion of the type I interferon response

Given the importance of the type I IFN system in the innate immune response, the redundancy of PRRs in monitoring virus entry and replication and the numerous downstream effector pathways of the IFN response, it is not surprising that alphaherpesviruses have developed means to counteract the type I IFN response over the course of co-evolution with their hosts. As DNA viruses with large coding capacities, alphaherpesviruses employ multiple means to inhibit various aspects of the type I IFN response simultaneously, from dampening IFN production to abrogating IFN signalling to inhibiting effector ISG function (Table 13.1).

Inhibition of interferon production

Regardless of the mode of recognition, the induction of type I IFN and/or ISGs in response to alphaherpesvirus infection requires IRF3 and/or IRF7 (Fig. 13.1). As with other DNA and RNA viruses, alphaherpesviruses such as HSV-1 target IRF3 and IRF7 activation. In the absence of replication, HSV-1 particles induce robust IRF3 activation and ISG production; however, induction of the type I IFN response is abolished following HSV-1 gene expression (Nicholl *et al.*, 2000; Mossman *et al.*, 2001). The HSV-1 immediate early protein ICP0 was shown to be responsible for blocking ISG induction, via inhibition of IRF3 and IRF7 (Eidson *et al.*, 2002; Lin *et al.*, 2004). ICP0 is a promiscuous viral and cellular transactivator that modulates cell cycle progression, promotes the proteasome-dependent degradation of several cellular proteins and is involved in modulating HSV latency (Everett, 2000; Hagglund and Roizman, 2004). The majority of the functions ascribed to ICP0 are dependent on its RING finger domain, which possesses E3 ubiquitin ligase activity (Everett, 2000; Boutell *et al.*, 2002). While various studies have shown that inhibition of IRF3-mediated IFN expression is dependent on the RING finger domain of ICP0, neither IRF3 nor upstream activators of IRF3 are degraded following infection with HSV-1 (Eidson *et al.*, 2002; Lin *et al.*, 2004). Despite several studies, the mechanism of ICP0-mediated repression of IRF3 activation is currently unknown. One study has shown that following Sendai virus (SeV)-induced IRF3 activation, ICP0 sequesters IRF3 in the cytoplasm (Melroe *et al.*, 2004). Another study suggests that ICP0 inhibits IFN-β expression and ISG induction in the nucleus by preventing the association of IRF3 and the co-activator CBP/p300 (Melroe *et al.*, 2007). A recent study, however, found that overexpression of ICP0, which induces its nuclear localization, fails to block dsRNA-mediated ISG induction (Everett and Orr, 2009). Thus further work is required to elucidate the mechanism of IRF inhibition by ICP0.

Table 13.1 Alphaherpesvirus proteins involved in subverting the type I IFN response

Inhibition of	Mechanism of evasion	Virus	Viral protein(s)
IFN production	Inhibition of IRF activation	HSV-1, HSV-2	ICP0, ICP27
	Inhibition of TLR3 expression	HSV-1	Us3
	Disruption of TBK-1 activity	HSV-1	ICP34.5
	Unknown	HSV-1	UL13
		VZV	IE63
IFN signalling	Inhibition of STAT-1 phosphorylation	HSV-1, HSV-2, VZV	vhs, UL13, ICP0, ICP27
	Disruption of ISGF3 complex formation	HSV-1	vhs
	Increased expression of SOCS-3	HSV-1	vhs, UL13
ISG production and function	Degradation of ISG mRNA	HSV-1	vhs
	Degradation of ISGs (e.g. PML, Sp100)	HSV-1, HSV-2	ICP0
	Inhibition of PKR activity	HSV-1	ICP34.5, Us11
		VZV	IE63
	Inhibition of 2′–5′ A production	HSV-1	Us11

In addition to ICP0, several other HSV proteins have been shown to inhibit the activation of IRF3, including ICP27, an immediate early gene product involved in viral mRNA nuclear export, host transcriptional repression, mRNA stability and mRNA splicing (Smith *et al.*, 2005; Melchjorsen *et al.*, 2006; Sandri-Goldin, 2008). HSV-1 recombinants lacking ICP27 were shown to substantially increase IRF3 activation and induce significantly greater levels of type I and III IFNs compared with wild-type HSV-1 in macrophages and DCs (Melchjorsen *et al.*, 2006), indicating ICP27 plays an early role in inhibiting either IRF3 activation or an upstream activator/kinase of IRF3. In contrast, deletion of ICP27 did not affect ISG levels in primary human fibroblasts (Lin *et al.*, 2004) or IRF3 activation in response to SeV infection (Melroe *et al.*, 2004), indicating the ability of ICP27 to inhibit IRF3 activation and ISG induction may be cell type- and context-dependent.

At least three late gene products contribute to the dampening of type I IFN production in response to alphaherpesvirus infection. Us3 is a protein kinase that is conserved in all alphaherpesviruses (McGeoch and Davison, 1986) and has been shown to have anti-apoptotic activity (Nishiyama, 2004). In human histiocytic lymphoma cells, infection with HSV-1 deficient for Us3 resulted in increased IRF3 activation and accumulation of IFN-β and TLR3 transcripts compared to infection with wild-type HSV-1 (Peri *et al.*, 2008). How Us3 functions in this capacity is currently unknown. ICP34.5, a major neurovirulence factor and regulatory subunit of protein phosphatase 1 (PP1), interacts with and prevents TBK-1 from phosphorylating IRF3, preventing downstream induction of ISGs and IFN expression (Chou and Roizman, 1992; Chou *et al.*, 1995; Verpooten *et al.*, 2009). Finally, the virion host shutoff (vhs) protein, a tegument-derived gene product and endoribonuclease with broad specificities for viral and host mRNA species, has been implicated in blocking type I IFN production (Kwong and Frenkel, 1987; Smiley, 2004). Infection of murine cells with vhs-deficient HSV-1 and HSV-2 has been shown to produce significantly more type I IFNs relative to wild-type HSV (Suzutani *et al.*, 2000; Duerst and Morrison, 2004; Pasieka *et al.*, 2008).

Inhibition of interferon signalling

In addition to dampening production of type I IFNs, alphaherpesviruses have evolved strategies to abrogate IFN signalling. In the initial stages of HSV-1 infection, various viral proteins act to inhibit IFN-induced JAK1, TYK2 and STAT1/2 phosphorylation, while at later times of infection, a substantial decrease in JAK1 and STAT2 protein levels ensues (Yokota *et al.*, 2001). Correlative with these observations was the induction of suppressor of cytokines 3 (SOCS-3), a negative regulator of the type I IFN response, at 2 hours after HSV-1 infection (Yokota *et al.*, 2004). Subsequent experimentation confirmed these observations and infection of a variety of cell types with gene deletion mutants of HSV-1 suggested that multiple viral protein(s) were involved.

Vhs is one of the viral proteins involved in counteracting various components of the type I IFN signalling cascade. Vhs-null HSV-1 is exquisitely sensitive to the effects of type I IFN (Pasieka *et al.*, 2008) and infection with vhs-null HSV-1 results in higher levels of STAT-1 phosphorylation levels and a sustained induction of ISG transcripts compared to infection with WT HSV-1 (Overton *et al.*, 1994; Smiley *et al.*, 2001; Lin *et al.*, 2004; Eisemann *et al.*, 2007; Loret *et al.*, 2008). Additionally, vhs was shown to disrupt the formation of the ISGF3 complex (Chee and Roizman, 2004). The importance of vhs in blocking IFN-mediated innate responses was recapitulated *in vivo*, where vhs-null HSV-2 replicated to wild-type HSV-2 levels and produced substantial genital disease and hind limb paralysis in IFNAR knockout mice, but not in wild-type mice (Murphy *et al.*, 2003).

The UL13 protein has also been shown to be involved in suppressing the JAK-STAT pathway during the second phase of the type I IFN response. UL13 is a tegument-derived viral serine/threonine protein kinase involved in temporally regulating HSV genome replication and the subcellular localization of a variety of viral proteins, including ICP0 (Smith and Smith, 1989; Overton *et al.*, 1994; Ogle *et al.*, 1997). Infection with UL13-deficient HSV-1 leads to an accumulation of phosphorylated STAT1, and UL13 modulates the function of vhs (Overton *et al.*, 1994). Taken together, it appears that tegument-derived vhs and UL13 play a role in inhibiting STAT1 phosphorylation at early times after HSV infection (Overton *et al.*, 1994; Yokota *et al.*, 2001; Yokota *et al.*, 2004). In addition, a recent study by Johnson and colleagues illustrated that infection with ICP27-deficient HSV-1, but not with wild-type HSV-1, resulted in STAT1 phosphorylation and nuclear localization (Johnson *et al.*, 2008).

Inhibition of interferon stimulated gene function

To ensure maximum suppression of the type I IFN response, alphaherpesviruses also express an array of proteins involved in directly negating the effects of ISGs designed to inhibit viral transcriptional, translational and replication programmes.

While ICP0 has been characterized in the context of inhibiting IRF activation and IFN production, it was originally shown to counteract an IFN-mediated block to virus transcription (Mossman *et al.*, 2000; Mossman and Smiley, 2002). While neither the mechanism of IFN mediated inhibition of transcription nor the counter-response by ICP0 have been fully elucidated, it is likely that the cellular protein PML is involved. Once expressed, ICP0 translocates to the nucleus where it promotes the proteasome-dependent degradation of SUMO-1 modified forms of PML (Maul *et al.*, 1993; Everett and Maul, 1994). Although various studies have clearly delineated a role for PML in inhibiting HSVs (Everett *et al.*, 2008a; Sobol *et al.*, 2009), whether PML-mediated antiviral responses are an integral component of the type I IFN response remains controversial (Chee *et al.*, 2003; Everett *et al.*, 2008b; Sobol *et al.*, 2009). Specifically, exogenous IFN treatment was shown to have a significantly lower antiviral effect against ICP0-deficient HSV-1 in PML null murine fibroblasts than in wild-type fibroblasts (Chee *et al.*, 2003). However, deregulation of PML in human prostate adenocarcinoma cells failed to alter type I IFN production or responsiveness relative to cells with intact PML signalling (Sobol *et al.*, 2009). Of interest, while VZV encodes a homologue of ICP0, ORF61p, a recent study found that ORF61p fails to induce PML degradation. Instead, components of PML nuclear bodies were found to accelerate VZV replication (Kyratsous and Silverstein, 2009), indicating that HSV and VZV have evolved different strategies to manipulate their host environment to ensure successful replication and spread.

The specific role the OAS-RNase L pathway plays in abrogating alphaherpesvirus replication is unclear, as neither HSV nor VZV significantly activate this pathway (Smith *et al.*, 2003; Desloges *et al.*, 2005b). However, pathogenesis mediated by HSV-2 is reduced in mice lacking RNase L (Duerst and Morrison, 2007) and the OAS-RNase L pathway has been shown to be required for IFN-mediated resistance to HSV-1 *in vitro* and *in vivo* (Carr *et al.*, 2003; Austin *et al.*, 2005; Rasmussen *et al.*, 2009). Furthermore, studies have shown that HSV-1 infection reduces the cellular levels of 2′–5′ A and RNase L-dependent rRNA degradation following type I IFN treatment (Cayley *et al.*, 1984). Consistent with a role of this cellular pathway in dampening alphaherpesvirus replication, Us11, a tegument-derived late gene product of HSV-1, was found to be both necessary and sufficient to inhibit OAS activity (Sanchez and Mohr, 2007). Since the dsRNA-binding domain within Us11 was required for its activity and the association of Us11 with OAS was disrupted by nuclease treatment, it was hypothesized that Us11 functions, in part, by sequestering dsRNA from OAS.

The ability of Us11 to bind and sequester dsRNA also dampens the activation of the dsRNA dependent kinase PKR. Indeed, expression of Us11 at early times post-infection was shown to drastically inhibit PKR autophosphorylation and kinase activity; a 68 amino acid N-terminal fragment containing the dsRNA binding domain was sufficient for this inhibitory activity (Poppers *et al.*, 2000; Cassady and Gross, 2002). Us11 was also shown to inhibit PKR activation by PACT, a protein activator of PKR, through interaction with its C-terminal domain with the N-terminal domain of PKR (Peters *et al.*, 2002). Furthermore, expression of Us11 strongly increased the replication of VZV, confirming that the PKR pathway has evolved to modulate the replication of multiple alphaherpesviruses (Desloges *et al.*, 2005a). Since repression of PKR function required Us11 expression prior to PKR activation in several studies (Cassady *et al.*, 1998a,b; Cassady and Gross, 2002), the role of naturally expressed Us11 was unclear. However, following infection with Us11 deletion mutants of HSV-1, enhanced PKR activation and eIF2α phosphorylation were observed (Mulvey *et al.*, 2003). In addition to Us11, the HSV-1 ICP34.5 protein is required to completely inhibit eIF2α phosphorylation and ensure the maintenance of cap-dependent translation in HSV-1 infected cells. ICP34.5 is involved in recruiting PP1α to phosphorylated eIF2α, resulting in its dephosphorylation and the release of translational repression on HSV-1 cap-dependent translation (He *et al.*, 1997). Importantly, deletion of ICP34.5 in the HSV-1 genome was fully complemented in PKR-deficient mice (Leib *et al.*, 2000).

While much is less is known about the mechanisms VZV utilizes to counteract the type I IFN response, a recent study by Ambagala *et al.* (2007) illustrated that deletion of the immediate early gene product expressed through ORF63 (IE63) rendered VZV hypersensitive to the effects of IFN-α; IE63 gene deletions were also complemented by growth on U2OS cells, which also complement the growth of ICP0-null HSV-1 (Ambagala and Cohen, 2007). Subsequent experimentation illustrated that IE63 was involved in counteracting eIF2α phosphorylation, although the specific mechanism involved remains unknown. Taken together, it appears that alphaherpesviruses have evolved multiple, redundant means of counteracting all phases of the type I IFN response in an effort to affect unimpeded replication and spread.

Conclusions

The type I IFN response plays a paramount role in the earliest phases of the immune response to inhibit alphaherpesvirus replication and dissemination. Given the inherent complexity of alphaherpesviruses, as well as the delicate co-evolution created as a result of the symbiotic relationship alphaherpesviruses have with their hosts, viral countermeasures to the host type I IFN response are many, varied and redundant. As more research in the innate immunology and virology communities continues, so too will our understanding of the evolutionary forces at work that balance virus and host.

Acknowledgements

The Canadian Institutes of Health Research and the Canadian Breast Cancer Foundation funded work from the Mossman laboratory.

References

Al-khatib, K., Williams, B.R., Silverman, R.H., Halford, W., and Carr, D.J. (2003). The murine double-stranded RNA-dependent protein kinase PKR and the murine 2′, 5′-oligoadenylate synthetase-dependent RNase L are required for IFN-beta-mediated resistance against herpes simplex virus type 1 in primary trigeminal ganglion culture. Virology *313*, 126–135.

Al-Khatib, K., Williams, B.R., Silverman, R.H., Halford, W., and Carr, D.J. (2004). Distinctive roles for 2′, 5′-oligoadenylate synthetases and double-stranded RNA-dependent protein kinase R in the *in vivo* antiviral effect of an adenoviral vector expressing murine IFN-beta. J. Immunol. *172*, 5638–5647.

Al-Khatib, K., Williams, B.R., Silverman, R.H., Halford, W., and Carr, D.J. (2005). Dichotomy between survival and lytic gene expression in RNase L- and PKR-deficient mice transduced with an adenoviral vector expressing murine IFN-beta following ocular HSV-1 infection. Exp. Eye Res. *80*, 167–173.

Alexopoulou, L., Holt, A.C., Medzhitov, R., and Flavell, R.A. (2001). Recognition of double-stranded RNA and activation of NF-[kappa]B by Toll-like receptor 3. Nature *413*, 732–738.

Ambagala, A.P., and Cohen, J.I. (2007). Varicella-Zoster virus IE63, a major viral latency protein, is required to inhibit the alpha interferon-induced antiviral response. J. Virol. *81*, 7844–7851.

Aravalli, R.N., Hu, S., Rowen, T.N., Palmquist, J.M., and Lokensgard, J.R. (2005). Cutting edge: TLR2-mediated proinflammatory cytokine and chemokine production by microglial cells in response to herpes simplex virus. J. Immunol. *175*, 4189–4193.

Arvin, A.M. (1996). Immune responses to varicella-zoster virus. Infect. Dis. Clin. North Am. *10*, 529–570.

Austin, B.A., James, C., Silverman, R.H., and Carr, D.J. (2005). Critical role for the oligoadenylate synthetase/RNase L pathway in response to IFN-beta during acute ocular herpes simplex virus type 1 infection. J. Immunol. *175*, 1100–1106.

Balachandra, K., Thawaranantha, D., Ayuthaya, P.I., Bhumisawasdi, J., Shiraki, K., and Yamanishi, K. (1994). Effects of human alpha, beta and gamma interferons on varicella zoster virus *in vitro*. Southeast Asian J. Trop. Med. Public Health *25*, 252–257.

Barbalat, R., Lau, L., Locksley, R.M., and Barton, G.M. (2009). Toll-like receptor 2 on inflammatory monocytes induces type I interferon in response to viral but not bacterial ligands. Nat. Immunol. *10*, 1200–1207.

Barber, G.N. (2001). Host defense, viruses and apoptosis. Cell Death Differ. *8*, 113–126.

Biron, C.A., and Brossay, L. (2001). NK cells and NKT cells in innate defense against viral infections. Curr. Opin. Immunol. *13*, 458–464.

Biron, C.A., Nguyen, K.B., Pien, G.C., Cousens, L.P., and Salazar-Mather, T.P. (1999). Natural killer cells in antiviral defense: function and regulation by innate cytokines. Annu. Rev. Immunol. *17*, 189–220.

Bjorck, P. (2001). Isolation and characterization of plasmacytoid dendritic cells from Flt3 ligand and granulocyte-macrophage colony-stimulating factor-treated mice. Blood *98*, 3520–3526.

Borysiewicz, L.K., and Sissons, J.G. (1994). Cytotoxic T cells and human herpes virus infections. Curr. . Top. Microbiol. Immunol. *189*, 123–150.

Boutell, C., Sadis, S., and Everett, R.D. (2002). Herpes simplex virus type 1 immediate-early protein ICP0 and its isolated ring finger domain act as ubiquitin E3 ligases *in vitro*. J. Virol. *76*, 841–850.

Carr, D.J., Al-khatib, K., James, C.M., and Silverman, R. (2003). Interferon-beta suppresses herpes simplex virus type 1 replication in trigeminal ganglion cells through an RNase L-dependent pathway. J. Neuroimmunol. *141*, 40–46.

Carr, D.J., Tomanek, L., Silverman, R.H., Campbell, I.L., and Williams, B.R. (2005). RNA-dependent protein kinase is required for alpha-1 interferon transgene-induced resistance to genital herpes simplex virus type 2. J. Virol. *79*, 9341–9345.

Cassady, K.A., and Gross, M. (2002). The herpes simplex virus type 1 U(S)11 protein interacts with protein kinase R in infected cells and requires. a 30-amino-acid sequence adjacent to a kinase substrate domain. J. Virol. 76, 2029–2035.

Cassady, K.A., Gross, M., and Roizman, B. (1998a). The herpes simplex virus US11 protein effectively compensates for the gamma1(34.5) gene if present before activation of protein kinase R by precluding its phosphorylation and that of the alpha subunit of eukaryotic translation initiation factor 2. J. Virol. *72*, 8620–8626.

Cassady, K.A., Gross, M., and Roizman, B. (1998b). The second-site mutation in the herpes simplex virus recombinants lacking the $g_1$34.5 genes precludes shutoff of protein synthesis by blocking the phosphorylation of eIF-2a. J. Virol. *72*, 7005–7011.

Cayley, P.J., Davies, J.A., McCullagh, K.G., and Kerr, I.M. (1984). Activation of the ppp(A2′p)nA system in interferon-treated, herpes simplex virus-infected cells and evidence for novel inhibitors of the ppp(A2′p) nA-dependent RNase. Eur. J. Biochem. *143*, 165–174.

Chatterjee, S., and Burns, P. (1990). Expression of herpes simplex virus type 1 glycoproteins in interferon-treated human neuroblastoma cells. J. Virol. *64*, 5209–5213.

Chee, A.V., Lopez, P., Pandolfi, P.P., and Roizman, B. (2003). Promyelocytic leukemia protein mediates interferon-based anti-herpes simplex virus 1 effects. J. Virol. 77, 7101–7105.

Chee, A.V., and Roizman, B. (2004). Herpes simplex virus 1 gene products occlude the interferon signaling pathway at multiple sites. J. Virol. *78*, 4185–4196.

Cheng, G., Zhong, J., Chung, J., and Chisari, F.V. (2007). Double-stranded DNA and double-stranded RNA induce a common antiviral signaling pathway in human cells. Proc. Natl. Acad. Sci. U.S.A. *104*, 9035–9040.

Chou, J., Chen, J.J., Gross, M., and Roizman, B. (1995). Association of a M(r) 90, 000 phosphoprotein with protein kinase PKR in cells exhibiting enhanced phosphorylation of translation initiation factor eIF-2 alpha and premature shutoff of protein synthesis after infection with gamma 134.5- mutants of herpes simplex virus 1. Proc. Natl. Acad. Sci. U.S.A. *92*, 10516–10520.

Chou, J., and Roizman, B. (1992). The gamma 1(34.5) gene of herpes simplex virus 1 precludes neuroblastoma cells from triggering total shutoff of protein synthesis characteristic of programed. cell death in neuronal cells. Proc. Natl. Acad. Sci. U.S.A. *89*, 3266–3270.

Collins, S.E., Noyce, R.S., and Mossman, K.L. (2004). Innate cellular response to virus particle entry requires. IRF3 but not virus replication. J. Virol. *78*, 1706–1717.

Cui, B., and Carr, D.J. (2000). A plasmid construct encoding murine interferon beta antagonizes the replication of herpes simplex virus type I *in vitro* and *in vivo*. J. Neuroimmunol. *108*, 92–102.

de Stanchina, E., Querido, E., Narita, M., Davuluri, R.V., Pandolfi, P.P., Ferbeyre, G., and Lowe, S.W. (2004). PML is a direct p53 target that modulates p53 effector functions. Mol. Cell *13*, 523–535.

de Veer, M.J., Holko, M., Frevel, M., Walker, E., Der, S., Paranjape, J.M., Silverman, R.H., and Williams, B.R. (2001). Functional classification of interferon-stimulated genes identified using microarrays. J. Leukoc. Biol. *69*, 912–920.

Dellaire, G., and Bazett-Jones, D.P. (2004). PML nuclear bodies: dynamic sensors of DNA damage and cellular stress. Bioessays *26*, 963–977.

Deshpande, S.P., Kumaraguru, U., and Rouse, B.T. (2000). Dual role of B cells in mediating innate and acquired immunity to herpes virus infections. Cell. Immunol. *202*, 79–87.

Desloges, N., Rahaus, M., and Wolff, M.H. (2005a). Role of the protein kinase PKR in the inhibition of varicella-zoster virus replication by beta interferon and gamma interferon. J. Gen. Virol. *86*, 1–6.

Desloges, N., Rahaus, M., and Wolff, M.H. (2005b). Varicella-zoster virus does not significantly induce cell defence mechanism mediated by the 2–5A/RNase L pathway during its replication cycle. Med. Microbiol. Immunol. *194*, 25–31.

DeStasio, R.P., and Taylor, M.W. (1990). Specific effect of interferon on the herpes simplex virus type 1 transactivation event. J. Virol. *64*, 2588–2593.

Duerst, R.J., and Morrison, L.A. (2004). Herpes simplex virus 2 virion host shutoff protein interferes. with type I interferon production and responsiveness. Virology *322*, 158–167.

Duerst, R.J., and Morrison, L.A. (2007). Herpes simplex virus type 2-mediated disease is reduced in mice lacking RNase L. Virology *360*, 322–328.

Eidson, K.M., Hobbs, W.E., Manning, B.J., Carlson, P., and DeLuca, N.A. (2002). Expression of herpes simplex virus ICP0 inhibits the induction of interferon-stimulated genes by viral infection. J. Virol. *76*, 2180–2191.

Eisemann, J., Muhl-Zurbes, P., Steinkasserer, A., and Kummer, M. (2007). Infection of mature dendritic cells with herpes simplex virus type 1 interferes. with the interferon signaling pathway. Immunobiology *212*, 877–886.

Ellermann-Eriksen, S., Liberto, M.C., Iannello, D., and Mogensen, S.C. (1986). X-linkage of the early *in vitro* alpha/beta interferon response of mouse peritoneal macrophages to herpes simplex virus type 2. J. Gen. Virol. *67*, 1025–1033.

Everett, R.D. (2000). ICP0, a regulator of herpes simplex virus during lytic and latent infection. Bioessays *22*, 761–770.

Everett, R.D., and Chelbi-Alix, M.K. (2007). PML and PML nuclear bodies: implications in antiviral defence. Biochimie *89*, 819–830.

Everett, R.D., and Maul, G.G. (1994). HSV-1 IE protein Vmw110 causes redistribution of PML. EMBO J. *13*, 5062–5069.

Everett, R.D., and Orr, A. (2009). Herpes simplex virus type 1 regulatory protein ICP0 aids infection in cells with a preinduced interferon response but does not impede interferon-induced gene induction. J. Virol. *83*, 4978–4983.

Everett, R.D., Parada, C., Gripon, P., Sirma, H., and Orr, A. (2008a). Replication of ICP0-null mutant herpes simplex virus type 1 is restricted by both PML and Sp100. J. Virol. *82*, 2661–2672.

Everett, R.D., Young, D.F., Randall, R.E., and Orr, A. (2008b). STAT-1- and IRF-3-dependent pathways are not essential for repression of ICP0-null mutant herpes simplex virus type 1 in human fibroblasts. J. Virol. *82*, 8871–8881.

Fernandes-Alnemri, T., Yu, J.W., Datta, P., Wu, J., and Alnemri, E.S. (2009). AIM2 activates the inflammasome and cell death in response to cytoplasmic DNA. Nature *458*, 509–513.

Fitzgerald, K.A., McWhirter, S.M., Faia, K.L., Rowe, D.C., Latz, E., Golenbock, D.T., Coyle, A.J., Liao, S.M., and Maniatis, T. (2003). IKKepsilon and TBK1 are essential components of the IRF3 signaling pathway. Nat. Immunol. *4*, 491–496.

Fitzgerald-Bocarsly, P. (1993). Human natural interferon-a producing cells. Pharmacol. Ther. *60*, 39–62.

Fujihara, M., Milligan, J.R., and Kaji, A. (1989). Effect of 2′, 5′-oligoadenylate on herpes simplex virus-infected cells and preventive action of 2′, 5′-oligoadenylate on the lethal effect of HSV-2. J. Interferon Res. *9*, 691–707.

Garcia, M.A., Meurs, E.F., and Esteban, M. (2007). The dsRNA protein kinase PKR: virus and cell control. Biochimie *89*, 799–811.

Gill, N., Deacon, P.M., Lichty, B., Mossman, K.L., and Ashkar, A.A. (2006). Induction of innate immunity against herpes simplex virus type 2 infection via local delivery of Toll-like receptor ligands correlates with beta interferon production. J. Virol. *80*, 9943–9950.

Goodbourne, S., Didcock, L., and Randall, R.E. (2000). Interferons: cell signalling, immune modulation, antiviral responses and virus countermeasures. J. Gen. Virol. *81*, 2341–2364.

Gresser, I., Tovey, M.G., Maury, C., and Bandu, M.T. (1976). Role of interferon in the pathogenesis of virus diseases in mice as demonstrated by the use of anti-interferon serum. II. Studies with herpes simplex, Moloney sarcoma, vesicular stomatitis, Newcastle disease, and influenza viruses. J. Exp. Med. *144*, 1316–1323.

Guo, J., Hui, D.J., Merrick, W.C., and Sen, G.C. (2000). A new pathway of translational regulation mediated by eukaryotic initiation factor 3. EMBO J. *19*, 6891–6899.

Hagglund, R., and Roizman, B. (2004). Role of ICP0 in the strategy of conquest of the host cell by herpes simplex virus 1. J. Virol. *78*, 2169–2178.

Halford, W.P., Balliet, J.W., and Gebhardt, B.M. (2004). Re-evaluating natural resistance to herpes simplex virus type 1. J. Virol. *78*, 10086–10095.

Halford, W.P., veress, L.A., Gebhardt, B.M., and Carr, D.J.J. (1997). Innate and acquired immunity to herpes simplex virus type 1. Virology *236*, 328–337.

Harle, P., Cull, V., Agbaga, M.P., Silverman, R., Williams, B.R., James, C., and Carr, D.J. (2002). Differential effect of murine alpha/beta interferon transgenes on antagonization of herpes simplex virus type 1 replication. J. Virol. *76*, 6558–6567.

He, B., Gross, M., and Roizman, B. (1997). The $g_1$34.5 protein of herpes simplex virus 1 complexes with protein phosphatase 1 a to dephosphorylate the a subunit of the eukaryotic translation initiation factor 2 and preclude the shutoff of protein synthesis by double-stranded RNA-activated protein kinase. Proc. Natl. Acad. Sci. U.S.A. *94*, 843–848.

Hemmi, H., Takeuchi, O., Kawai, T., Kaisho, T., Sato, S., Sanjo, H., Matsumoto, M., Hoshino, K., Wagner, H., Takeda, K., and Akira, S. (2000). A Toll-like receptor recognizes bacterial DNA. Nature *408*, 740–745.

Hochrein, H., Schlatter, B., O'Keeffe, M., Wagner, C., Schmitz, F., Schiemann, M., Bauer, S., Suter, M., and Wagner, H. (2004). Herpes simplex virus type-1 induces IFN-alpha production via Toll-like receptor 9-dependent and -independent pathways. Proc. Natl. Acad. Sci. U.S.A. *101*, 11416–11421.

Holberg-Petersen, M., Bukholm, G., Haarr, L., Langeland, N., and Degre, M. (1996). Human interferon reduces surface expression but not total production of herpes simplex virus type 1 glycoproteins gC and gE in heterologous hamster cells. J. Interferon Cytokine Res. *16*, 717–724.

Honda, K., Takaoka, A., and Taniguchi, T. (2006). Type I interferon gene induction by the interferon regulatory factor family of transcription factors. Immunity *25*, 349–360.

Honda, K., and Taniguchi, T. (2006). IRFs: master regulators of signalling by Toll-like receptors and cytosolic pattern-recognition receptors. Nat. Rev. Immunol. *6*, 644–658.

Hornung, V., Ablasser, A., Charrel-Dennis, M., Bauernfeind, F., Horvath, G., Caffrey, D.R., Latz, E., and Fitzgerald, K.A. (2009). AIM2 recognizes cytosolic dsDNA and forms a caspase-1-activating inflammasome with ASC. Nature *458*, 514–518.

Hovanessian, A.G., Brown, R.E., and Kerr, I.M. (1977). Synthesis of low molecular weight inhibitor of protein synthesis with enzyme from interferon-treated cells. Nature *268*, 537–540.

Ishii, K.J., Coban, C., Kato, H., Takahashi, K., Torii, Y., Takeshita, F., Ludwig, H., Sutter, G., Suzuki, K., Hemmi, H., Sato, S., Yamamoto, M., Uematsu, S., Kawai, T., Takeuchi, O., and Akira, S. (2006). A Toll-like receptor-independent antiviral response induced by double-stranded B-form DNA. Nat. Immunol. *7*, 40–48.

Ishii, K.J., Kawagoe, T., Koyama, S., Matsui, K., Kumar, H., Kawai, T., Uematsu, S., Takeuchi, O., Takeshita, F., Coban, C., and Akira, S. (2008). TANK-binding kinase-1 delineates innate and adaptive immune responses to DNA vaccines. Nature *451*, 725–729.

Ishikawa, H., and Barber, G.N. (2008). STING is an endoplasmic reticulum adaptor that facilitates innate immune signalling. Nature *455*, 674–678.

Ishikawa, H., Ma, Z., and Barber, G.N. (2009). STING regulates intracellular DNA-mediated, type I interferon-dependent innate immunity. Nature *461*, 788–792.

Ito, M., Mizutani, K., Nakano, T., Yasuda, N., Kamiya, T., Kitamura, K., Torigoe, S., Ihara, T., Kamiya, H., and Sakurai, M. (1992). Flow cytometric analysis of effects of cytokines on the expression of varicella-zoster virus glycoproteins. Antiviral Res. *18*, 317–326.

Jacquemont, B., and Roizman, B. (1975). RNA synthesis in cells infected with herpes simplex virus. X. Properties of viral symmetric transcripts and of double-stranded RNA prepared from them. J. Virol. *15*, 707–713.

Johnson, K.E., Song, B., and Knipe, D.M. (2008). Role for herpes simplex virus 1 ICP27 in the inhibition of type I interferon signaling. Virology *374*, 487–494.

Juang, Y., Lowther, W., Kellum, M., Au, W.C., Lin, R., Hiscott, J., and Pitha, P.M. (1998). Primary activation of interferon A and interferon B gene transcription by interferon regulatory factor 3. Proc. Natl. Acad. Sci. U.S.A. *95*, 9837–9842.

Kawai, T., Takahashi, K., Sato, S., Coban, C., Kumar, H., Kato, H., Ishii, K.J., Takeuchi, O., and Akira, S. (2005). IPS-1, an adaptor triggering RIG-I- and Mda5-mediated type I interferon induction. Nat. Immunol. *6*, 981–988.

Khabar, K.S.A., Dhalla, M., Siddiqui, Y., Zhou, A., Al-Ahdal, M.N., Der, S.D., Silverman, R.H., and Williams, B.R.G. (2000). Effect of deficiency of the double-stranded RNA dependent protein kinase, PKR, on antiviral resistance in the presence or absence of ribonuclease L: HSV-1 replication is particularly sensitive to deficiency of the major IFN-mediated enzymes. J. Interferon Cytokine Res. *20*, 653–659.

Kotenko, S.V., Gallagher, G., Baurin, V.V., Lewis-Antes, A., Shen, M., Shah, N.K., Langer, J.A., Sheikh, F., Dickensheets, H., and Donnelly, R.P. (2003). IFN-lambdas mediate antiviral protection through a distinct class II cytokine receptor complex. Nature Immunol. *4*, 69–77.

Krug, A., Luker, G.D., Barchet, W., Leib, D.A., Akira, S., and Colonna, M. (2004). Herpes simplex virus type 1 activates murine natural interferon-producing cells through toll-like receptor 9. Blood *103*, 1433–1437.

Ku, C.C., Besser, J., Abendroth, A., Grose, C., and Arvin, A.M. (2005). Varicella-Zoster virus pathogenesis and immunobiology: new concepts emerging from investigations with the SCIDhu mouse model. J. Virol. *79*, 2651–2658.

Ku, C.C., Zerboni, L., Ito, H., Graham, B.S., Wallace, M., and Arvin, A.M. (2004). Varicella-zoster virus transfer to skin by T Cells and modulation of viral replication by epidermal cell interferon-alpha. J. Exp. Med. *200*, 917–925.

Kunder, S.C., Kelly, K.M., and Morahan, P.S. (1993). Biological response modifier-mediated resistance to herpesvirus infections requires. induction of alpha/beta interferon. Antiviral Res. *21*, 129–139.

Kurt-Jones, E.A., Chan, M., Zhou, S., Wang, J., Reed, G., Bronson, R., Arnold, M.M., Knipe, D.M., and Finberg, R.W. (2004). Herpes simplex virus 1 interaction with Toll-like receptor 2 contributes to lethal encephalitis. Proc. Natl. Acad. Sci. U.S.A. *101*, 1315–1320.

Kwong, A.D., and Frenkel, N. (1987). Herpes simplex virus-infected cells contain a function(s) that destabilizes both host and viral mRNAs. Proc. Natl. Acad. Sci. U.S.A. *84*, 1926–1930.

Kyratsous, C.A., and Silverstein, S.J. (2009). Components of nuclear domain 10 bodies regulate varicella-zoster virus replication. J. Virol. *83*, 4262–4274.

Lees-Miller, S.P., and Meek, K. (2003). Repair of DNA double strand breaks by non-homologous end joining. Biochimie *85*, 1161–1173.

Leib, D.A., Machalek, M.A., Williams, B.R., Silverman, R.H., and Virgin, H.W. (2000). Specific phenotypic restoration of an attenuated virus by knockout of a host resistance gene. Proc. Natl. Acad. Sci. U.S.A. *97*, 6097–6101.

Lenschow, D.J., Lai, C., Frias-Staheli, N., Giannakopoulos, N.V., Lutz, A., Wolff, T., Osiak, A., Levine, B., Schmidt, R.E., Garcia-Sastre, A., Leib, D.A., Pekosz, A., Knobeloch, K.P., Horak, I., and Virgin, H.W.t. (2007). IFN-stimulated gene 15 functions as a critical antiviral molecule against influenza, herpes, and Sindbis viruses. Proc. Natl. Acad. Sci. U.S.A. *104*, 1371–1376.

Li, X., Ranjith-Kumar, C.T., Brooks, M.T., Dharmaiah, S., Herr, A.B., Kao, C., and Li, P. (2009). The RIG-I-like receptor LGP2 recognizes the termini of double-stranded RNA. J. Biol. Chem. *284*, 13881–13891.

Lin, R., Noyce, R.S., Collins, S.E., Everett, R.D., and Mossman, K.L. (2004). The herpes simplex virus ICP0 RING finger domain inhibits IRF3- and IRF7-mediated activation of interferon-stimulated genes. J. Virol. *78*, 1675–1684.

Lopez, C. (1985). Natural resistance mechanisms in herpes simplex virus infections. In The Herpesviruses. Roizman, B. and Lopez, C., eds. (New York, USA: Plenum Press), pp. 37–68.

Loret, S., Guay, G., and Lippe, R. (2008). Comprehensive characterization of extracellular herpes simplex virus type 1 virions. J. Virol. *82*, 8605–8618.

Lund, J., Sato, A., Akira, S., Medzhitov, R., and Iwasaki, A. (2003). Toll-like receptor 9-mediated recognition of Herpes simplex virus-2 by plasmacytoid dendritic cells. J. Exp. Med. *198*, 513–520.

Maheshwari, R.K., Sidhu, G., Singh, A.K., Sivaram, S.S., Kinchington, P.R., Hay, J., and Friedman, R.M. (1994). Defective transport of herpes simplex virus glycoprotein in interferon-treated cells: role of intracellular pH. J. Interferon Res. *14*, 319–324.

Mansur, D.S., Kroon, E.G., Nogueira, M.L., Arantes, R.M., Rodrigues, S.C., Akira, S., Gazzinelli, R.T., and Campos, M.A. (2005). Lethal encephalitis in myeloid differentiation factor 88-deficient mice infected with herpes simplex virus 1. Am. J. Pathol. *166*, 1419–1426.

Maul, G.G., Guldner, H.H., and Spivack, J.G. (1993). Modification of discrete nuclear domains induced by herpes simplex virus type 1 immediate early gene 1 product (ICP0). J. Gen. Virol. *74*, 2679–2690.

McGeoch, D.J., and Davison, A.J. (1986). Alphaherpesviruses possess a gene homologous to the protein kinase gene family of eukaryotes and retroviruses. Nucleic Acids Res. *14*, 1765–1777.

McWhirter, S.M., Fitzgerald, K.A., Rosains, J., Rowe, D.C., Golenbock, D.T., and Maniatis, T. (2004). IFN-regulatory factor 3-dependent gene expression is defective in Tbk1-deficient mouse embryonic fibroblasts. Proc. Natl. Acad. Sci. U.S.A. *101*, 233–238.

Meek, K., Gupta, S., Ramsden, D.A., and Lees-Miller, S.P. (2004). The DNA-dependent protein kinase: the director at the end. Immunol. Rev. *200*, 132–141.

Melchjorsen, J., Siren, J., Julkunen, I., Paludan, S.R., and Matikainen, S. (2006). Induction of cytokine expression by herpes simplex virus in human monocyte-derived macrophages and dendritic cells is dependent on virus replication and is counteracted by ICP27 targeting NF-kappaB and IRF-3. J. Gen. Virol. *87*, 1099–1108.

Melroe, G.T., DeLuca, N.A., and Knipe, D.M. (2004). Herpes simplex virus 1 has multiple mechanisms for blocking virus-induced interferon production. J. Virol. *78*, 8411–8420.

Melroe, G.T., Silva, L., Schaffer, P.A., and Knipe, D.M. (2007). Recruitment of activated IRF-3 and CBP/p300 to herpes simplex virus ICP0 nuclear foci: Potential role in blocking IFN-beta induction. Virology *360*, 305–321.

Meylan, E., Curr. an, J., Hofmann, K., Moradpour, D., Binder, M., Bartenschlager, R., and Tschopp, J. (2005). Cardif is an adaptor protein in the RIG-I antiviral pathway and is targeted by hepatitis C virus. Nature *437*, 1167–1172.

Mittnacht, S., Straub, P., Kirchner, H., and Jacobsen, H. (1988). Interferon treatment inhibits onset of herpes simplex virus immediate-early transcription. Virology *164*, 201–210.

Mossman, K.L. (2002). Activation and inhibition of virus and interferon: the herpesvirus story. Viral Immunol. *15*, 3–15.

Mossman, K.L., and Ashkar, A.A. (2005). Herpesviruses and the innate immune response. Viral Immunol. *18*, 267–281.

Mossman, K.L., Macgregor, P.F., Rozmus, J.J., Goryachev, A.B., Edwards, A.M., and Smiley, J.R. (2001). Herpes simplex virus triggers and then disarms a host antiviral response. J. Virol. *75*, 750–758.

Mossman, K.L., Saffran, H.A., and Smiley, J.R. (2000). Herpes simplex virus ICP0 mutants are hypersensitive to interferon. J. Virol. *74*, 2052–2056.

Mossman, K.L., and Smiley, J.R. (2002). Herpes simplex virus ICP0 and ICP34.5 counteract distinct interferon–induced barriers to virus replication. J. Virol. *76*, 1995–1998.

Mulvey, M., Poppers, J., Sternberg, D., and Mohr, I. (2003). Regulation of eIF2alpha phosphorylation by different functions that act during discrete phases in the herpes simplex virus type 1 life cycle. J. Virol. *77*, 10917–10928.

Murphy, J.A., Duerst, R.J., Smith, T.J., and Morrison, L.A. (2003). Herpes simplex virus type 2 virion host shutoff protein regulates alpha/beta interferon but not adaptive immune responses during primary infection *in vivo*. J. Virol. *77*, 9337–9345.

Muruve, D.A., Pétrilli, V., Zaiss, A.K., White, L.R., Clark, S.A., Ross, P.J., Parks, R.J., and Tschopp, J. (2008). The inflammasome recognizes cytosolic microbial and host DNA and triggers an innate immune response. Nature *452*, 103–107.

Neville, L.F., Mathiak, G., and Bagasra, O. (1997). The Immunobiology of Interferon-gamma Inducible Protein 10 kD (IP-10): a novel, pleiotropic member of the C-X-C chemokine superfamily. Cytokine Growth Factor Rev. *8*, 207–219.

Nicholl, M.J., Robinson, L.H., and Preston, C.M. (2000). Activation of cellular interferon-responsive genes after infection of human cells with herpes simplex virus type 1. J. Gen. Virol. *81*, 2215–2218.

Nishiyama, Y. (2004). Herpes simplex virus gene products: the accessories reflect her lifestyle well. Rev. Med. Virol. *14*, 33–46.

Noyce, R.S., Collins, S.E., and Mossman, K.L. (2006). Identification of a novel pathway essential for the immediate-early, interferon-independent antiviral response to enveloped virions. J. Virol. *80*, 226–235.

Oberman, F., and Panet, A. (1988). Inhibition of transcription of herpes simplex virus immediate early genes in interferon-treated human cells. J. Gen. Virol. *69*, 1167–1177.

Ogle, W.O., Ng, T.I., Carter, K.L., and Roizman, B. (1997). The UL13 protein kinase and the infected cell type are determinants of posttranslational modification of ICP0. Virology *235*, 406–413.

Overton, H., McMillan, D., Hope, L., and Wong-Kai-In, P. (1994). Production of host shutoff-defective mutants of herpes simplex virus type 1 by inactivation of the UL13 gene. Virology *202*, 97–106.

Paladino, P., Cummings, D.T., Noyce, R.S., and Mossman, K.L. (2006). The IFN-independent response to virus particle entry provides a first line of antiviral defense that is independent of TLRs and retinoic acid-inducible gene I. J. Immunol. *177*, 8008–8016.

Pasieka, T.J., Cilloniz, C., Lu, B., Teal, T.H., Proll, S.C., Katze, M.G., and Leib, D.A. (2009). Host responses to wild-type and attenuated herpes simplex virus infection in the absence of Stat1. J. Virol. *83*, 2075–2087.

Pasieka, T.J., Lu, B., Crosby, S.D., Wylie, K.M., Morrison, L.A., Alexander, D.E., Menachery, V.D., and Leib, D.A. (2008). Herpes simplex virus virion host shutoff attenuates establishment of the antiviral state. J. Virol. *82*, 5527–5535.

Pavlovic, J., Zurcher, T., Haller, O., and Staeheli, P. (1990). Resistance to influenza virus and vesicular stomatitis virus conferred by expression of human MxA protein. J. Virol. *64*, 3370–3375.

Pedersen, E.B., Haahr, S., and Mogensen, S.C. (1983). X-linked resistance of mice to high doses of herpes simplex virus type 2 correlates with early interferon production. Infect Immun. *42*, 740–746.

Peri, P., Mattila, R.K., Kantola, H., Broberg, E., Karttunen, H.S., Waris, M., Vuorinen, T., and Hukkanen, V. (2008). Herpes simplex virus type 1 Us3 gene deletion influences toll-like receptor responses in cultured monocytic cells. Virol. J. *5*, 140.

Peters, G.A., Khoo, D., Mohr, I., and Sen, G.C. (2002). Inhibition of PACT-mediated activation of PKR by the herpes simplex virus type 1 Us11 protein. J. Virol. *76*, 11054–11064.

Platanias, L.C. (2005). Mechanisms of type-I- and type-II-interferon-mediated signalling. Nat. Rev. Immunol. *5*, 375–386.

Poppers, J., Mulvey, M., Khoo, D., and Mohr, I. (2000). Inhibition of PKR activation by the proline-rich RNA binding domain of the herpes simplex virus type 1 Us11 protein. J. Virol. *74*, 11215–11221.

Preston, C.M., Harman, A.N., and Nicholl, M.J. (2001). Activation of interferon response factor-3 in human cells infected with herpes simplex virus type 1 or human cytomegalovirus. J. Virol. *75*, 8909–8916.

Rasmussen, S.B., Jensen, S.B., Nielsen, C., Quartin, E., Kato, H., Chen, Z.J., Silverman, R.H., Akira, S., and Paludan, S.R. (2009). Herpes simplex virus infection is sensed by both Toll-like receptors and retinoic acid-inducible gene- like receptors, which synergize to induce type I interferon production. J. Gen. Virol. *90*, 74–78.

Rasmussen, S.B., Sorensen, L.N., Malmgaard, L., Ank, N., Baines, J.D., Chen, Z.J., and Paludan, S.R. (2007). Type I interferon production during herpes simplex virus infection is controlled by cell-type-specific viral recognition through Toll-like receptor 9, the mitochondrial antiviral signaling protein pathway, and novel recognition systems. J. Virol. *81*, 13315–13324.

Sadler, A.J., and Williams, B.R. (2008). Interferon-inducible antiviral effectors. Nat. Rev. Immunol. *8*, 559–568.

Sanchez, R., and Mohr, I. (2007). Inhibition of cellular 2–5′ oligoadenylate synthetase by the herpes simplex virus type 1 Us11 protein. J. Virol. *81*, 3455–3464.

Sandri-Goldin, R.M. (2008). The many roles of the regulatory protein ICP27 during herpes simplex virus infection. Front Biosci. 13, 5241–5256.

Sato, M., Suemori, H., Hata, N., Asagiri, M., Ogasawara, K., Nakao, K., Nakay, T., Katsuki, M., Noguchi, S., Tanaka, N., and Taniguchi, T. (2000). Distinct and essential roles of transcription factors IRF-3 and IRF-7 in response to viruses for IFN-a/b gene induction. Immunity *13*, 539–548.

Schroder, K., Hertzog, P.J., Ravasi, T., and Hume, D.A. (2004). Interferon-gamma: an overview of signals, mechanisms and functions. J. Leuko. Biol. *75*, 163–189.

Seth, R.B., Sun, L., Ea, C.K., and Chen, Z.J. (2005). Identification and characterization of MAVS, a mitochondrial antiviral signaling protein that activates NF-kappaB and IRF 3. Cell *122*, 669–682.

Sharma, S., tenOever, B.R., Grandvaux, N., Zhou, G.P., Lin, R., and Hiscott, J. (2003). Triggering the interferon antiviral response through an IKK-related pathway. Science *300*, 1148–1151.

Shupack, J., Stiller, M., Davis, I., Kenny, C., and Jondreau, L. (1992). Topical alpha-interferon ointment with dimethyl sulfoxide in the treatment of recurrent genital herpes simplex. Dermatology *184*, 40–44.

Smiley, J.R. (2004). Herpes simplex virus virion host shutoff protein: immune evasion mediated by a viral RNase? J. Virol. *78*, 1063–1068.

Smiley, J.R., Elgadi, M.M., and Saffran, H.A. (2001). Herpes simplex virus vhs protein. Methods Enzymol. *342*, 440–451.

Smith, R.F., and Smith, T.F. (1989). Identification of new protein kinase-related genes in three herpesviruses, herpes simplex virus, varicella-zoster virus, and Epstein–Barr virus. J. Virol. *63*, 450–455.

Smith, R.W., Malik, P., and Clements, J.B. (2005). The herpes simplex virus ICP27 protein: a multifunctional post-transcriptional regulator of gene expression. Biochem. Soc. Trans. *33*, 499–501.

Smith, T.J., Silverman, R.H., and Leib, D.A. (2003). RNase L activity does not contribute to host RNA degradation induced by herpes simplex virus infection. J. Gen. Virol. *84*, 925–928.

Sobol, P.T., Hummel, J.L., Rodrigues, R.M., and Mossman, K.L. (2009). PML has a predictive role in tumor cell permissiveness to interferon-sensitive oncolytic viruses. Gene Ther. *16*, 1077–1087.

Sorensen, L.N., Reinert, L.S., Malmgaard, L., Bartholdy, C., Thomsen, A.R., and Paludan, S.R. (2008). TLR2 and TLR9 synergistically control herpes simplex virus infection in the brain. J. Immunol. *181*, 8604–8612.

Stark, G.R., Kerr, I.M., Williams, B.R., Silverman, R.H., and Schreiber, R.D. (1998). How cells respond to interferons. Annu. Rev. Biochem. *67*, 227–264.

Su, Y.H., Oakes, J.E., and Lausch, R.N. (1990). Ocular avirulence of a herpes simplex virus type 1 strain is associated with heightened sensitivity to alpha/beta interferon. J. Virol. *64*, 2187–2192.

Suzutani, T., Nagamine, M., Shibaki, T., Ogasawara, M., Yoshida, I., Daikoku, T., Nishiyama, Y., and Azuma, M. (2000). The role of the UL41 gene of herpes simplex virus type 1 in evasion of non-specific host defence mechanisms during primary infection. J. Gen. Virol. *81*, 1763–1771.

Svensson, A., Bellner, L., Magnusson, M., and Eriksson, K. (2007). Role of IFN-alpha/beta signaling in the prevention of genital herpes virus type 2 infection. J. Reprod. Immunol. *74*, 114–123.

Takaoka, A., Wang, Z., Choi, M.K., Yanai, H., Negishi, H., Ban, T., Lu, Y., Miyagishi, M., Kodama, T., Honda, K., Ohba, Y., and Taniguchi, T. (2007). DAI (DLM-1/ZBP1) is a cytosolic DNA sensor and an activator of innate immune response. Nature *448*, 501–505.

Takeuchi, O., and Akira, S. (2009). Innate immunity to virus infection. Immunol. Rev. *227*, 75–86.

Tanaka, N., Sato, M., Lamphier, M.S., Nozawa, H., Oda, E., Noguchi, S., Schreiber, R.D., Tsujimoto, Y., and Taniguchi, T. (1998). Type I interferons are essential mediators of apoptotic death in virally infected cells. Genes Cells *3*, 29–37.

Taylor, J.L., Witt, P.L., Irizarry, A., Tom, P., and O'Brien, W.J. (1995). 2′, 5′-oligoadenylate synthetase in interferon-alpha- and acyclovir-treated herpes simplex virus-infected cells. J. Interferon Cytokine Res. *15*, 27–30.

Tovey, M.G., and Maury, C. (1999). Oromucosal interferon therapy: marked antiviral and antitumor activity. J. Interferon Cytokine Res. *19*, 145–155.

Verpooten, D., Ma, Y., Hou, S., Yan, Z., and He, B. (2009). Control of TANK-binding kinase 1-mediated signaling by the gamma(1)34.5 protein of herpes simplex virus 1. J. Biol. Chem. *284*, 1097–1105.

Vollstedt, S., Arnold, S., Schwerdel, C., Franchini, M., Alber, G., Di Santo, J.P., Ackermann, M., and Suter, M. (2004). Interplay between alpha/beta and gamma interferons with B, T, and natural killer cells in the defense against herpes simplex virus type 1. J. Virol. *78*, 3846–3850.

Vollstedt, S., Franchini, M., Hefti, H.P., Odermatt, B., O'Keeffe, M., Alber, G., Glanzmann, B., Riesen, M., Ackermann, M., and Suter, M. (2003). Flt3 ligand-treated neonatal mice have increased innate immunity against intracellular pathogens and efficiently control virus infections. J. Exp. Med. *197*, 575–584.

Wang, Z., Choi, M.K., Ban, T., Yanai, H., Negishi, H., Lu, Y., Tamura, T., Takaoka, A., Nishikura, K., and Taniguchi, T. (2008). Regulation of innate immune responses by DAI (DLM-1/ZBP1) and other. DNA-sensing molecules. Proc. Natl. Acad. Sci. U.S.A. *105*, 5477–5482.

Wintergerst, U., Gangemi, J.D., Kern, E.R., Whitley, R.J., and Chatterjee, S. (1996). Differential effect of recombinant human and mouse interferons on replication of herpes simplex virus type 1 in mouse cells. Antiviral Res. *32*, 27–34.

Xu, L.G., Wang, Y.Y., Han, K.J., Li, L.Y., Zhai, Z., and Shu, H.B. (2005). VISA is an adapter protein required for virus-triggered IFN-beta signaling. Mol. Cell *19*, 727–740.

Yokota, S., Yokosawa, N., Kubota, T., Suzutani, T., Yoshida, I., Miura, S., Jimbow, K., and Fujii, N. (2001). Herpes simplex virus type 1 suppresses the interferon signaling pathway by inhibiting phosphorylation of STATs and janus kinases during an early infection stage. Virology *286*, 119–124.

Yokota, S.-i., Yokosawa, N., Okabayashi, T., Suzutani, T., Miura, S., Jimbow, K., and Fujii, N. (2004). Induction of Suppressor of Cytokine Signaling-3 by Herpes Simplex Virus Type 1 Contributes to Inhibition of the Interferon Signaling Pathway. J. Virol. *78*, 6282–6286.

Yoneyama, M., Suhara, W., Fukuhara, Y., Fukuda, M., Nishida, E., and Fujita, T. (1998). Direct triggering of the type I interferon system by virus infection: activation of a transcription factor complex containing IRF-3 and CBP/p300. EMBO J. *17*, 1087–1095.

Zawatzky, R., Engler, H., and Kirchner, H. (1982a). Experimental infection of inbred mice with herpes simplex virus. III. Comparison between newborn and adult C57BL/6 mice. J. Gen. Virol. *60*, 25–29.

Zawatzky, R., Gresser, I., DeMaeyer, E., and Kirchner, H. (1982b). The role of interferon in the resistance of C57BL/6 mice to various doses of herpes simplex virus type 1. J. Infect. Dis. *146*, 405–410.

Zawatzky, R., Kirchner, H., DeMaeyer-Guignard, J., and DeMaeyer, E. (1982c). An X-linked locus influences the amount of circulating interferon induced in the mouse by herpes simplex virus type 1. J. Gen. Virol. *63*, 325–332.

Zhang, H., Melamed, J., Wei, P., Cox, K., Frankel, W., Bahnson, R.R., Robinson, N., Pyka, R., Liu, Y., and Zheng, P. (2003). Concordant down-regulation of proto-oncogene PML and major histocompatibility antigen HLA class I expression in high-grade prostate cancer. Cancer Immunity 3, 2.

Zhang, S.Y., Jouanguy, E., Ugolini, S., Smahi, A., Elain, G., Romero, P., Segal, D., Sancho-Shimizu, V., Lorenzo, L., Puel, A., *et al.* (2007). TLR3 deficiency in patients with herpes simplex encephalitis. Science *317*, 1522–1527.

Zheng, P., Guo, Y., Niu, Q., Levy, D.E., Dyck, J.A., Lu, S., Sheiman, L.A., and Liu, Y. (1998). Proto-oncogene PML controls genes devoted to MHC class I antigen presentation. Nature *396*, 373–376.

Zheng, X., Silverman, R.H., Zhou, A., Goto, T., Kwon, B.S., Kaufman, H.E., and Hill, J.M. (2001). Increased severity of HSV-1 keratitis and mortality in mice lacking the 2–5A-dependent RNase L gene. Invest. Ophthalmol. Vis. Sci. *42*, 120–126.

Zhou, A., Paranjape, J.M., Der, S.D., Williams, B.R.G., and Silverman, R.H. (1999). Interferon action in triply deficient mice reveals the existence of alternative antiviral pathways. Virology *258*, 435–440.

Molecular Chaperones and Alphaherpesvirus Infection

14

Christine M. Livingston, Christos Kyratsous, Saul Silverstein and Sandra K. Weller

Abstract

Molecular chaperone proteins have long been recognized to play diverse and important roles in the life cycles of viruses from bacteriophage to SV40 to herpesviruses. The alphaherpesviruses HSV-1 and VZV not only interact with and reorganize cellular chaperones and co-chaperones but alphaherpesviruses also encode their own molecular chaperones. Cellular chaperones such as Hsp70, Hsc70 and Hsp90 are required for efficient production of infectious virus in that they play essential roles in nuclear transport of viral proteins, protein quality control and maintenance of cellular homeostasis during viral infection. These findings raise the possibility that molecular chaperones could be utilized as effective targets for antiviral therapy. Here we review the evidence that replication of the human alphaherpesviruses herpes simplex virus type 1 and 2 (HSV-1 and 2) and varicella zoster virus (VZV) requires the activities of cellular and viral molecular chaperones.

Introduction

The cellular machinery responsible for protein homeostasis and integrity of the cellular proteome is referred to as protein quality control (PQC). Cellular PQC machinery is responsible for maintaining a functional cell proteome by mediating protein folding and refolding, proteolytic degradation of terminally misfolded or damaged proteins and protection from toxic effects of aggregation. PQC is maintained by protein refolding machinery (molecular chaperone proteins) and protein degradation machinery (26S proteasome, ubiquitin, ubiquitin ligases and deubiquitinating enzymes) (reviewed in McClellan *et al.*, 2005). Misfolded, damaged or abnormal proteins can arise in the cell following exposure to environmental cell stressors such as heat shock, oxidative stress and proteotoxic chemical treatment. Mutations in cellular genes and infection by viruses with robust gene expression patterns can also result in challenges to PQC. An imbalance of protein quality control due to an inability to efficiently detect, degrade or refold abnormal proteins is the basis for many protein misfolding disorders in which loss-of-function or gain-of-function from misfolding of specific proteins cause cellular toxicity. Cell survival relies on a functional PQC system, and it is now clear that accumulation of misfolded or abnormal proteins is a characteristic of many human diseases in which PQC mechanisms are defective or overwhelmed.

Viral infection can lead to proteotoxic stress in cells because of massive overproduction of viral proteins in a short period of time, and viruses have evolved to make use of cellular chaperones during infection. Hsp70 family members have long been known to play essential roles in the lifecycles of many viruses. For example, dnaK was discovered in *E. coli* when it was learned that dnaK mutants could not support bacteriophage λ DNA replication (Georgopoulos, 1977). During λ phage replication, the viral origin binding protein O recognizes and binds the viral DNA origin. The λ P viral protein interacts with O and the host helicase dnaB at the origin which results in recruitment of dnaK. The interaction between dnaK and λP liberates dnaB from the complex to unwind

DNA and commence replication (Liberek *et al.*, 1988). In addition, purified dnaK was shown to complement *in vitro* replication of single-stranded M13 bacteriophage DNA in the presence of λ phage O and P replication proteins (Zylicz and Georgopoulos, 1984). Hsc70 is required for initiation of polyomavirus SV40 replication via its interaction with SV40 large T antigen (for review see (Sullivan and Pipas, 2002). Hsp70 and Hsp40 are also required for initiation of papillomavirus replication *in vitro*. Purified Hsp70 and Hsp40 were reported to bind papillomaviral helicase protein E1 and facilitate its binding to the viral origin resulting in stimulation of cell-free DNA replication (Liu *et al.*, 1998). Furthermore, Hsc70 has been implicated in entry of rotavirus into host cells (Guerrero *et al.*, 2002; Perez-Vargas *et al.*, 2006; Zarate *et al.*, 2003), the capsid disassembly reaction of reovirus (Ivanovic *et al.*, 2007) and replication of influenza virus (Melville *et al.*, 1999; Watanabe *et al.*, 2006). A recent report suggests that picornavirus replication is also dependent on the Hsp90 chaperone machine; maturation of the large capsid protein P1 can be inhibited by inhibition of Hsp90 (Geller *et al.*, 2007). Thus viruses that infect species from *E. coli* to human share a requirement for the highly conserved molecular chaperone systems.

Cellular molecular chaperones

Molecular chaperones can mediate either refolding or promote degradation of abnormal proteins through their association with specific co-factors. Molecular chaperones were first referred to as heat shock proteins because of their high levels of expression during thermal stress. Subsequently, it was shown that molecular chaperones are expressed at high levels in the presence of numerous stressors that induce proteotoxic stress (reviewed in Georgopoulos and Welch, 1993). Some molecular chaperones guide protein folding by preventing and disrupting improper protein interactions in an energy-dependent manner, and a defining characteristic of chaperones is that they are not associated with the final product (Ellis, 1987; Frydman, 2001; Hartl, 1996; Tavaria *et al.*, 1996). Protein folding is a challenge in the crowded environment of cellular subcompartments. Chaperones are necessary to prevent improper interactions of exposed hydrophobic protein surfaces that will lead to protein aggregation, and to mediate disaggregation and refolding of abnormal proteins. Rather than directly mediating refolding of a polypeptide, some chaperone proteins such as Hsp90 have been referred to as 'holdases' because they hold a polypeptide so that it can fold into its native conformation (Buchner, 1999).

Molecular chaperone families

Molecular chaperones are grouped into superfamilies of functionally similar proteins. The Hsp70 chaperone family (DnaK in *E. coli* and Ssa1–4 in *Saccharomyces cerevisiae*) consists of the highly inducible Hsp70 protein and its constitutively expressed counterpart Hsc70. Hsp70 chaperones generally have a molecular weight of about 70kDa and consist of an N-terminal ATPase domain and a C-terminal substrate-binding domain (reviewed in Hartl, 1996). Hydrolysis of ATP fuels substrate binding and release cycles in which Hsp70 has low affinity for a substrate polypeptide in the ATP bound state but high affinity for substrate in the ADP bound state. Hsp70 chaperones guide proper folding of *de novo* polypeptides by binding to hydrophobic stretches of amino acids in the N terminus of extended polypeptide chains as they emerge from the ribosome. The Hsp70 chaperone system works in cooperation with other chaperone systems such as the large cytosolic chaperonin TriC and the dnaJ protein Hsp40 (Frydman *et al.*, 1994) as well as cochaperones to mediate folding of de novo polypeptides. Cofactors are important for coupling Hsc70 and Hsp70 to the proteasome system, for example, CHIP (C terminus of Hsc70-interacting protein) an E3 ubiquitin ligase and Bag1 are cofactors of Hsc70 and Hsp70 that couple the chaperone system to the proteasome system.

Hsp70 is highly inducible during cell stress while Hsc70 is constitutively and abundantly expressed but only somewhat inducible (Hightower and White, 1981; Ingolia and Craig, 1982; Lowe and Moran, 1984; Palter *et al.*, 1986). Hsp70 and Hsc70 chaperones share considerable (81%) amino acid homology (Dworniczak and Mirault, 1987) and play somewhat redundant functions in the cell. Both operate on *de novo* polypeptides, and yeast strains carrying mutations in the Hsc70 gene were able to grow at elevated temperatures due to increased levels of Hsp70 (Craig and Jacobsen, 1985). During heat shock at 42°C, both Hsc70 and Hsp70 relocalize to nucleoli (Ellis *et al.*, 2000; Hattori *et al.*, 1993; Pelham, 1984). These two proteins, however, do play distinct roles in cells. Hsc70 regulates formation and dissociation of protein complexes such as clathrin-coated endocytic pits (Chappell *et al.*, 1986; Rothman and Schmid, 1986; Ungewickell, 1985), reovirus outer capsids during viral entry into host cells (Ivanovic *et al.*, 2007) and aggregates of misfolded protein (Wacker *et al.*, 2004).

Hsp90, an abundant and highly conserved member of the Hsp90 chaperone family, is a specialized chaperone that is essential for maintaining the activities of signal transduction proteins such as steroid hormone receptors and protein kinases. It's role is to mediate both activation and proteasomal degradation of these targets (reviewed in Mayer *et al.*, 2009). Hsp90 generally does not mediate folding of nascent polypeptides, rather, Hsp90 interacts with proteins that are in near-native state conformations. In the cytosol, Hsp90 often works cooperatively with other chaperone systems such as Hsp70 and cochaperones such as Hop. Hop binds both Hsc70 and Hsp90 simultaneously via two distinct tricopeptide repeat (TPR) motifs to mediate transfer of substrates between the two chaperone systems. The activity of Hsp90 is inhibited by ansamycin antibiotics such as geldanamycin that bind to the N-terminal ATPase domain of Hsp90 with nanomolar affinity (Stebbins *et al.*, 1997; Whitesell *et al.*, 1994). Hsp90 is a central regulator of many signal transduction pathways and drugs, such as derivatives of geldanamycin that target the activity of Hsp90, are currently in clinical trials for cancer treatment. Geldanamycin is also a potent inhibitor of herpes viral infection in part because Hsp90 is required for nuclear transport of essential viral proteins (Burch and Weller, 2005; Szczepaniak, Nellissery and Weller, unpublished). The requirements of Hsp90 for efficient herpesviral infection will be discussed later in this review.

Roles of host chaperones in herpesvirus infection

Herpesviruses are large dsDNA viruses that replicate in the nucleus of mammalian cells. The human alphaherpesviruses, herpes simplex virus type 1 (HSV-1), herpes simplex virus type 2 (HSV-2) and varicella zoster virus (VZV), share the ability to establish latent infections in sensory ganglia. The life cycle of HSV is relatively short in that progeny virus is produced within 12 hours of infection and gene expression is robust. In contrast, the lifecycle of VZV is much longer and viral spreading is inefficient (Weller, 1953). In fact, it is very difficult to obtain high titre stocks of VZV. Both HSV and VZV appear to recruit and require the activities of cellular chaperones. In addition both HSV and VZV also appear to encode putative viral chaperones for intracellular transport, efficient protein translation and perhaps even capsid assembly.

Regulation of heat shock protein expression during HSV-1 infection

As described above, cells have evolved homeostatic mechanisms to respond to various stressors including heat, oxidative, chemical, mutational stress and viral infection. The interaction between HSV-1/HSV-2 infection and stress response pathways is likely to be quite complex. Some stress responses may be detrimental to virus infection while other components of the stress response pathways may be beneficial. HSV has been reported to induce aspects of the cellular heat shock response including increasing mRNA levels but not necessarily protein synthesis of Hsp70 family members. The Preston lab reported that expression of three cellular stress proteins (25, 70, 90 kDa) was induced in chick embryo fibroblasts infected at non-permissive temperature with a ts ICP4 mutant (tsK), but not wt HSV-1 (Notarianni and Preston, 1982). These results imply that HSV-1 infection induces production of heat shock proteins but that ICP4 can prevent this induction. Two other studies have suggested that Hsp70 mRNA levels increase following HSV-1 infection (Kobayashi *et al.*, 1994; Phillips *et al.*, 1991). On the other hand, total protein levels of Hsp70, Hsc70 and Hsp40 do not appear to change significantly during infection (Burch and Weller, 2004). In a recent proteomic study that measured global changes in host cell gene expression during HSV-1 infection, Hsp70 and Hsc70 were reportedly downregulated (Antrobus *et al.*, 2009). These experiments suggest that wt HSV infection can induce some elements of the cell stress response such as upregulation of mRNA levels of heat shock proteins but does not increase protein levels. It will be of interest to determine how and why HSV only induces only a partial stress response.

Spatial reorganization of cellular chaperone proteins in HSV-1-infected cells

Although heat shock protein levels do not appear to be upregulated significantly during HSV-1 infection, their subcellular localization is altered dramatically. During HSV-1 infection, the cellular chaperone proteins Hsp70, Hsc70, Hsp40 and Hsp90 are spatially redistributed to nuclear foci called virus-induced chaperone-enriched (VICE) domains (Fig. 14.1) (Burch and Weller, 2004, 2005). VICE domains form adjacent to viral replication compartments, sites of genome replication and encapsidation. In addition to chaperone proteins, VICE domains contain several components of the protein remodelling machinery such as the 20S proteasome, polyubiquitinated proteins and putative misfolded proteins (Burch and Weller, 2004; Livingston *et al.*, 2009). Hsc70 is specifically reorganized to VICE domains whereas other chaperones such as Hsp90 are also abundant in replication compartments.

VICE domains form during the earliest stages of infection adjacent to prereplicative viral foci (Livingston *et al.*, 2008). Replication compartment formation and virus production are impaired in cells overexpressing dominant-negative Hsc70 suggesting that Hsc70's ATPase activity is required for productive infection (Li *et al.*, 2008). Relocalization of Hsc70 to VICE domains may facilitate formation of replication compartments by concentrating cellular chaperones at a location adjacent to developing replication compartments. Several non-mutually exclusive possibilities have been proposed for how VICE domains function to promote viral infection: 1. VICE domains function as nuclear protein quality control centres in which abnormal proteins are refolded, sequestered or degraded by cellular chaperone and proteasomal machinery. 2. VICE domains sequester 'dangerous signals' such as misfolded protein and or other cellular proteins that might otherwise trigger additional cell stress or apoptosis. 3. The proximity of VICE domains to replication compartments provides access to cellular chaperones needed to respond to events such as damage to proteins during stress (Livingston *et al.*, 2009) or the presence of stalled transcription complexes (Li *et al.*, 2008).

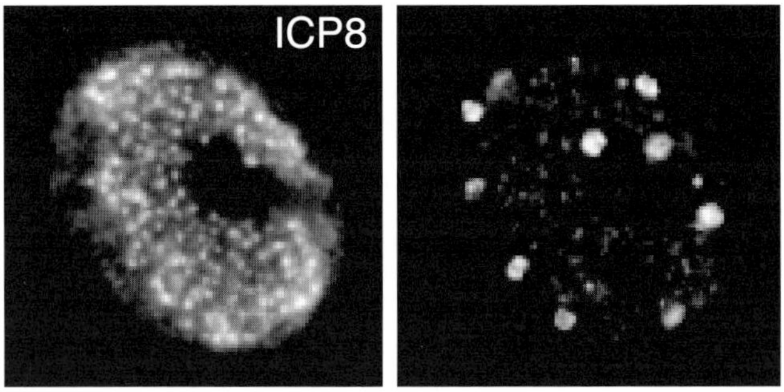

Figure 14.1 Hsc70 is detected in virus-induced chaperone-enriched (VICE) domains during HSV-1 infection. Vero cells (African Green Monkey kidney fibroblasts) were infected with HSV-1 strain KOS at an MOI of 10 for 6 hours. Infected cells were prepared for immunofluorescence imaging of ICP8 (left) and Hsc70 (right) as described in (Livingston *et al.*, 2009). The panels shown represent the same cell. The image was captured using a Zeiss LSM 510 Meta confocal microscope and a Zeiss plan-apochromat 63x1.4 objective lens.

VICE domains as centres of nuclear protein quality control

VICE domains appear to share several features with nuclear bodies, known as intranuclear inclusion bodies, which form in protein misfolding disease states. Intranuclear inclusion bodies form in the presence of excess misfolded protein such as mutant huntingtin and ataxin-1 protein in cells from patients afflicted with Huntington's disease or spinal cerebellar ataxia. In these neurodegenerative disease states, the protein processing capabilities of the proteasome and refolding pathways are exceeded or impaired. Intranuclear inclusion bodies containing mutant huntingtin or ataxin-1 protein resemble VICE domains in that both types of nuclear bodies contain molecular chaperones, the proteasome, ubiquitin and sometimes PML (Chai *et al.*, 1999; Davies *et al.*, 1997; Fu *et al.*, 2005b; Matsumoto *et al.*, 2006; Paulson *et al.*, 1997; Ross, 1997; Skinner *et al.*, 1997; Stenoien *et al.*, 2002; Woulfe, 2007).

Consistent with the notion that VICE domains act as centres for protein quality control, their formation appears to correlate with the amount of infectious virus. They form in 100% of cells infected at a relatively high MOI of 10 PFU/cell at 6 hours post infection; whereas, only 20% of cells infected at an MOI of 0.1 PFU/cell at that time point contain VICE domains (Livingston *et al.*, 2009). HSV-1 infection at high MOI would be expected to produce large amounts of viral proteins that may overwhelm the protein processing capabilities of the cell. We have also shown that putative misfolded proteins are detected in VICE domains. For example, the HSV-1 capsid portal protein UL6 normally forms a 12-mer homooligomeric ring that sits at the base of the viral capsid (Newcomb *et al.*, 2001). UL6 is generally detected in replication compartments (Taus *et al.*, 1998) (Lamberti, Lukonis and Weller, unpublished observations) in which genomic cleavage and packaging are thought to take place. Interestingly, a subpopulation of UL6 is detected by a unique monoclonal antibody (4G9) in VICE domains (Burch and Weller, 2004). Immunoprecipitation of UL6 with the 4G9 antibody revealed that this subpopulation is also ubiquitinated. It is possible that this subpopulation of UL6 represents misfolded protein or improper folding intermediates of the oligomeric ring. A model misfolded protein, GFP170* (Fu *et al.*, 2005a; Garcia-Mata *et al.*, 1999), also localized to VICE domains during infection, providing additional evidence that misfolded proteins are sequestered in VICE domains (Livingston *et al.*, 2009).

Sequestration of misfolded proteins and other signals that might otherwise induce apoptosis

We have proposed that VICE domains play a cytoprotective role during HSV-1 infection by sequestering misfolded protein (Livingston *et al.*, 2009). It is possible that formation of VICE domains represents an attempt by the virus to delay or prevent apoptosis (Burch and Weller, 2004). Intranuclear inclusion bodies have also been shown to be cytoprotective (Tanaka *et al.*, 2004; Taylor *et al.*, 2003). Misfolded proteins can be highly toxic to the cell if they are not sequestered, refolded or degraded. Thus it may be beneficial for the virus to sequester proteins that may otherwise trigger premature apoptosis during infection. Interestingly, we have also shown that other proteins appear to be sequestered in VICE domains. Wilkinson and Weller reported in 2006 that DNA repair proteins ATRIP (obligate interaction partner of ATR) and an endogenous species of phosphorylated RPA are sequestered in VICE domains during infection (Wilkinson and Weller, 2006). Because phosphorylated RPA is associated with the DNA damage response, it is possible that sequestration of this form of RPA in VICE domains may prevent or delay induction of premature apoptosis in infected cells.

Cellular chaperones are relocalized to VICE domains so that they are accessible when needed for replication/transcription or repair of viral proteins in replication compartments

Hsc70 is rapidly exchanged between VICE domains and the surrounding nucleoplasm (Livingston *et al.*, 2009) allowing it to sample other nuclear environments such as replication compartments. The proximity of VICE domains to replication compartments may provide access to chaperone proteins that might be needed during stress. For instance, we have shown that Hsc70 is recruited to replication compartments in infected cells that have been heat shocked by incubation at 42°C for 1 hour (Livingston *et al.*, 2009). Heat shock causes viral proteins to become insoluble indicating they are damaged or misfolded. The availability of Hsc70 in VICE domains may facilitate recruitment of Hsc70 into replication compartments.

VICE domains and Hsc70 have also been proposed to play a role in resolution of stalled transcription complexes that are expected to occur in replication compartments. The ATPase activity of Hsc70 is required for VICE domain formation, replication compartment formation and efficient virus production (Li *et al.*, 2008) (Livingston and Weller, unpublished results). The Sandri-Goldin laboratory has reported direct protein–protein interactions between ICP27 and Hsc70 and between ICP27 and RNA Pol II (Li *et al.*, 2008; Rojas *et al.*, 2009). It is possible that the proximity of VICE domains to replication compartments may facilitate recruitment of Hsc70 to the transcriptional apparatus allowing proteasomal degradation of stalled RNAP II complexes.

The proposed roles of VICE domains appear, at first glance diverse, however, each of the proposed functions involves disaggregation of proteins or prevention of improper protein–protein interactions consistent with roles in protein quality control.

Requirements for formation of VICE domains

Formation of VICE domains requires active protein processing machinery and infection with live HSV as it does not occur when the proteasome is inhibited by MG132, when the protein translational machinery is inhibited by cycloheximide (Burch and Weller, 2004) or during infection with UV-inactivated HSV-1 (Livingston and Weller, unpublished). VICE domains form, albeit at lower frequency, in the presence of drugs that inhibit viral DNA replication (Livingston *et al.*, 2008). Three HSV-1 immediate-early proteins have been implicated in formation of VICE domains: ICP0, ICP27 and ICP22. ICP0 is a RING finger E3 ubiquitin ligase that mediates proteasomal degradation of numerous cellular innate immune/anti-viral proteins (see Chapters 4 and 13). Growth of an ICP0 null virus is severely inhibited in partially restrictive cell lines such as Vero cells (African Green Monkey kidney fibroblast) at low MOIs; however, at higher MOIs, ICP0 null viruses replicate to almost wild-type levels (Cai and Schaffer, 1992). VICE domain formation was reported to be severely compromised in cells infected in the absence of ICP0 at low MOI (Burch and Weller, 2004); however, Bastian *et al.* recently reported that even under these conditions, VICE domains could be detected at late times post infection (Bastian *et al.*, 2009). The defect in VICE domain formation at early times can be explained by the observations that ICP0 facilitates efficient viral gene expression especially at low MOI and that VICE domain formation correlates with robust gene expression. The fact that VICE domains do eventually form in cells infected with ICP0 null mutants indicates that ICP0 is not obligatorily required for VICE domain formation.

Similarly, Li *et al.* reported that ICP27 is required for efficient VICE domain formation within the first 6 hours after infection (Li *et al.*, 2008); however, it has subsequently been shown that VICE domains eventually form in the absence of ICP27 at later times post infection (24 hours) (Bastian *et al.*, 2009). The only viral protein, thus far, that appears to be essential for reorganization of Hsc70 into VICE domains in Vero cells is ICP22. ICP22 is a phosphoprotein that is required for productive infection in some cell types (HEL fibroblasts) but dispensable in others (Vero) (Poffenberger *et al.*, 1993; Sears *et al.*, 1985). ICP22 affects the phosphorylation state of RNA pol II (Fraser and Rice, 2007; Rice *et al.*, 1995). In the absence of ICP22, Hsc70 was not recruited to VICE domains even at high MOIs at late times after infection. ICP22 also plays a direct role in reorganization of Hsc70 in transiently transfected cells in which Hsc70 colocalized with ICP22 nuclear inclusions. The observation that ICP22 is essential for Hsc70 recruitment to VICE domains and that ICP22 is not required for viral growth in some cell lines suggests that localization of Hsc70 to VICE domains may be essential for viral growth in some but not all cell types.

Small heat shock proteins and HSV-1 replication

In addition to the members of the Hsp90 and Hsp70 family members, human cells encode at least 10 small heat shock proteins that are characterized by a low molecular weight of 12–40 kDa and the presence of a C terminal α crystallin domain. The α crystallin domain is important for oligomerization and cytoprotective activities. Small heat shock proteins recognize and bind destabilized proteins independently of ATP (reviewed in McHaourab *et al.*, 2009). The small heat shock protein Hsp27 (HspB1 in human) is a constitutively-expressed and heat-inducible multifunctional chaperone that prevents protein aggregation and protects cells against apoptosis during heat and oxidative stress (reviewed in Arrigo, 2007). Hsp27 can form a variety of higher order oligomeric complexes up to 1000 kDa in weight; and its activities appear to depend on its oligomerization status. Heat stress results in induction of Hsp27 expression, localization of Hsp27 to nuclear foci called stress granules and phosphorylation of Hsp27 on three serine residues (ser15, 78 and 82) by MAPKAP kinases 2 and 3 (mitogen-activated protein kinase-activated protein). Phosphorylation of Hsp27 results in dissociation of large Hsp27 oligomers into smaller oligomeric complexes and correlates with loss of chaperone activity (reviewed in Kostenko and Moens, 2009). Small oligomeric complexes of phosphorylated Hsp27 were not able to mediate refolding or prevent thermal aggregation of citrate synthase *in vitro* (Rogalla *et al.*, 1999). MAPKAP kinase 2/3 is activated during heat stress by the p38 kinase (Rouse *et al.*, 1994). In other words, p38 activation during stress results in phosphorylation of Hsp27 via MAPKAP kinase 2/3.

The p38 kinase is activated by phosphorylation in HSV-1 infected cells in a manner that is dependent on the presence of the multifunctional immediate-early protein ICP27 (Gillis *et al.*, 2009; Hargett *et al.*, 2005; McLean and Bachenheimer, 1999). Treatment of infected cells with MAPK/p38 kinase inhibitors had no effect on HSV-1 transcription but reduced production of infectious virus by as much as 85–90% (almost 1 log) in Vero cells (Karaca *et al.*, 2004) suggesting that the activities of MAPK/p38 contribute to efficient viral production. Recently it was shown that Hsp27 is reorganized from a diffuse cytoplasmic staining pattern in uninfected cells to nuclear foci in HSV-1-infected cells (Mathew *et al.*, 2009). Whether or not these nuclear foci are related to nuclear stress granules or VICE domains has not been directly tested. Depletion of Hsp27 resulted in a decrease in virus yield suggesting that Hsp27 plays a beneficial role in HSV-1 infection (Mathew *et al.*, 2009) however specific roles of Hsp27 during HSV-1 infection have not been identified. A role for Hsp27 has been described for other viral lifecycles such as adenoviral infection during which phosphorylated Hsp27 is required for efficient transport of incoming adenoviral capsids to the nucleus (Suomalainen *et al.*, 2001). It will be interesting to determine if Hsp27 functions similarly during HSV-1 infection. Considering the cytoprotective properties of Hsp27, it is also possible that Hsp27 plays a role in inhibition of apoptosis triggered by viral infection.

Specific roles of cellular chaperone proteins in the HSV-1 life cycle

By analogy with other viruses such as phage lambda and SV40 we suggest that HSV may have evolved to utilize cellular chaperones in specific viral processes including DNA replication, nuclear transport, and capsid assembly. In this section we will review the literature supporting roles for the cellular PQC machinery in these three essential viral processes.

Cellular heat shock proteins have been reported to enhance the binding activity of the homodimeric HSV-1 origin binding protein, UL9, to the viral origin (oriS). An interaction between UL9 and the human homologue of bacterial dnaJ chaperone protein, hTid1 was reported to enhance binding of UL9 to oriS *in vitro* (Eom and Lehman, 2002). Another study showed that an interaction between UL9 and Hsp70 and Hsp40 enhanced the oriS-binding and DNA-dependent ATPase activities of UL9 (Tanguy Le Gac and Boehmer, 2002). These studies raise the possibility that cellular chaperones play a role in initiation of HSV-1 DNA synthesis in infected cells; however, additional experiments are needed to confirm this hypothesis.

The observation that viral yields are diminished in cells in the presence of geldanamycin (GA) indicates that Hsp90 is important for HSV-1 infection (Burch and Weller, 2005; Li *et al.*, 2004). GA is a benzoquinone ansamycin antibiotic that inhibits Hsp90 function by binding the ADP/ATP binding pocket of Hsp90 (Stebbins *et al.*, 1997; Whitesell *et al.*, 1994). In GA-treated cells, viral DNA synthesis is impaired, and nuclear transport of the viral polymerase, UL30, is abrogated (Burch and Weller, 2005; Li *et al.*, 2004). In addition, Hsp90 appears to be required for nuclear import of the major capsid protein VP5 (Szczepaniak, Nellissery and Weller, unpublished observation). It has been reported that client proteins of Hsp90 are destabilized in the presence of Hsp90 inhibitors. GA-bound Hsp90 recruits the E3 ubiquitin ligase CHIP which then mediates proteasomal degradation of Hsp90 client proteins (Xu *et al.*, 2002). UL30 was destabilized by GA in a proteasome-dependent manner (Burch and Weller, 2005), suggesting that UL30 may be a direct client of Hsp90; however, VP5 was not degraded under these conditions. Hsp90-facilitated nuclear import of VP5 may require another viral protein that acts as an adaptor protein (Szczepaniak, Nellissery and Weller, unpublished observation).

Hsp90 has also been reported to facilitate the nuclear transport or function of polymerases from two other mammalian viruses. Hsp90 activity is required for nuclear import of Influenza viral RNA polymerase subunits PB1 and PB2 (Naito *et al.*, 2007). Hsp90 has also been shown to bind to the reverse transcriptase of Hepatitis C and facilitate formation of a ribonucleoprotein complex consisting of the polymerase and viral RNA (Hu and Seeger, 1996). It will be of interest to determine whether Hsp90 also plays a direct role in HSV polymerase function. Indeed, Hsp90 colocalizes with UL30 in replication compartments (Burch and Weller, 2005), and it is possible that Hsp90 plays a functional role at the replication fork in addition to its role in nuclear transport. Hsc70 also facilitates nuclear import of some HSV proteins. For example, the ATPase activity of Hsc70 is required for efficient nuclear import of the HSV-1 capsid portal protein UL6 (Livingston, Nellissery, Szczepaniak and Weller, unpublished results).

Cellular chaperones may play a role during HSV capsid assembly. The minor capsid protein UL6 forms a portal ring structure composed of 12 subunits through which DNA enters the capsid. Although UL6 can be detected in replication compartments (Patel *et al.*, 1996; Taus *et al.*, 1998), a subpopulation of ubiquitinylated UL6 can be detected in VICE domains (as described above). Cell chaperones such as Hsc70 may be needed to maintain UL6 in a properly folded state for oligomerization. Formation of the UL6 12-mer portal ring itself or the process by which portal rings are incorporated into procapsids may also require assistance by the PQC machinery. The tegument protein VP22 has also been reported to localize in structures that resemble VICE domains (Hutchinson *et al.*, 2002). Other capsid proteins such as VP5, VP19c, VP23, UL26 are not found in VICE domains and instead are localized to replication compartments where capsid assembly and encapsidation are thought to take place (Burch and Weller, 2004). One intriguing possibility is that VICE domains represent specialized centres for specific aspects of capsid assembly or quality control of capsid proteins that require chaperone assistance such as UL6. Additional experiments will be needed to ascertain whether host chaperones play a direct role in portal ring assembly or incorporation of portal rings into procapsids.

In summary, it is likely that we have only scratched the surface of potential interactions between cellular chaperone systems and viral proteins and that other examples in which cellular chaperones play direct and indirect roles in intracellular transport, gene expression, DNA replication, capsid assembly and genome encapsidation will be revealed in the future.

Putative viral chaperones

In addition to utilizing cellular chaperones, many viruses encode proteins that play roles in protein folding, protein trafficking and assembly of multiprotein complexes. HSV-1 and HSV-2 encode a ribonucleotide reductase (RR) consisting of subunits $R1_2R2_2$ whose primary function has been thought to convert ribonucleotides to deoxyribonucleotides that are needed during viral DNA synthesis. Several lines of evidence suggest that the R1 subunit may also exhibit chaperone activities. The N-terminal 400 aa of R1 from HSV-2 is dispensable for RR activity and encodes an α crystallin domain similar to the small heat shock protein Hsp27 (Chabaud *et al.*, 2003). Interestingly, the HSV-1 RR subunit (ICP6) has also been reported to exhibit chaperone activity in that it associates with eIF4G to promote formation of eIF4F complex assembly in order to control protein translation during infection (Walsh and Mohr, 2006).

The N-terminus of the HSV-1 and HSV-2 tegument protein UL14 and the substrate-binding domain of Hsp70 share significant homology (Yamauchi *et al.*, 2002) suggesting that UL14 may also function as a viral chaperone. Overexpression of UL14 in Hep-2 cells was reported to correlate with increased folding and activity of newly-synthesized luciferase compared with overexpression of mutant UL14 (Yamauchi *et al.*, 2002). UL14 also blocks the apoptotic response in the presence of drugs which generally induce apoptosis (Yamauchi *et al.*, 2003). Although UL14 normally resides in the cytoplasm of transfected cells, UL14 can translocate to the nucleus/nucleolus and to nuclear stress granules following heat stress (Yamauchi *et al.*, 2002) similar to how Hsp70 and Hsc70 behave in heat-stressed cells (Ellis *et al.*, 2000; Hattori *et al.*, 1993; Pelham, 1984). Furthermore, UL14 plays a role in nuclear transport of at least three viral proteins: minor capsid protein VP26, the cleavage and packaging protein UL33 and the alpha transactivating factor VP16 in transformed cells (Yamauchi *et al.*, 2001). The ability of UL14 to mediate nuclear translocation of viral proteins, to refold firefly luciferase, and prevent apoptosis combined with its homology to the substrate-binding domain of Hsp70 suggests that UL14 is an HSV-1/HSV-2-encoded chaperone.

Another putative HSV-1 chaperone, UL32, is essential for encapsidation of viral genomes despite not being a constituent of either procapsids or mature virions (Lamberti and Weller, 1998) (Kosinski and Weller, unpublished results). The amino acid sequence of UL32 reveals the presence of several CXXC motifs that resemble those found in the bacterial chaperone Hsp33 and other redox sensitive chaperone proteins such as thioredoxin (Kosinski and Weller, unpublished). In cells infected with a UL32 null mutant virus, it appears that one or more capsid proteins are misfolded (Kosinski, Szczepaniak and Weller, unpublished). These observations suggest that UL32 may be a chaperone needed for assembly of an encapsidation-competent procapsid. Although the precise function of HSV-1 UL32 is unknown, it is absolutely essential for cleavage and packaging of viral genomes but does not appear to be a component of either procapsids or mature virions. The amino acid sequence of UL32 resembles that of Hsp33, a bacterial chaperone that acts as a redox sensitive chaperone (Kosinski and Weller, unpublished). Like the cellular redox protein thioredoxin, UL32 contains several CXXC motifs raising the possibility that UL32 plays a role in disfulfide bond formation (Kosinski and Weller, unpublished). It is tempting to speculate that UL32 may mediate formation of disulfide bonds during assembly of an encapsidation-competent procapsid.

Effects of thermal stress on HSV-1/HSV-2 infection

Proteotoxic stress of uninfected cells results in activation of heat shock factor-1 (HSF-1) that, in turn, mediates induction of the cellular heat shock response by binding DNA sequences in the promoter regions of heat shock-responsive genes causing their upregulation (reviewed in Whitesell and Lindquist, 2009). A prototypical consequence of HSF-1 induction is an increase in Hsp70 mRNA and protein levels. An early study from the Morimoto laboratory demonstrated that HSV-1 infection causes a transient (3–7 hpi) increase in hsp70 mRNA levels followed by downregulation of transcription from the hsp70 gene later (10 hpi) during infection. Interestingly, HSF-1 did not bind heat shock elements in HSV-1-infected HeLa cell lysates suggesting that

HSF-1 was not active during infection (Phillips *et al.*, 1991). It was proposed that the hsp70 gene promoter could be selectively activated in an HSF-1-independent fashion by viral immediate-early transactivating proteins such as ICP4.

The ability of HSV-1 and HSV-2 to withstand temperature stress appears to be dependent on cell type, suggesting that temperature sensitive cellular factors or pathways may be required for viral infection. HSV-1 can replicated in mammalian cells at 39°C while growth of HSV-2 was inhibited at this temperature (Crouch and Rapp, 1972b) suggesting that there may be differences between HSV-1 and HSV-2 in terms of reliance on temperature sensitive cellular factors. The apparent temperature sensitivity of wild type HSV-2 appears to be dependent on cell type. HSV-1 replicated similarly in rabbit and hamster cells at 33°C; however, virus growth at 39°C was inhibited in hamster but not rabbit cells (Crouch and Rapp, 1972a). In addition, it appears that prior heat treatment could reduce the ability of the HSV-1 strain HFEM to form plaques in infected cell monolayers (Gharpure, 1965). Taken together, these results suggest that in certain cell types a temperature-sensitive function required for efficient HSV infection may be destroyed during heat treatment. Interestingly, our laboratory has shown that insertion and deletion mutants in the ICP6 (RR1) gene appear to be temperature-sensitive (Goldstein and Weller, 1988). Although ICP6 is not required for growth in dividing cells, growth of a null mutant virus (ΔICP6) was compromised in growth-arrested (serum starved) cells or in cells cultured at a non-permissive temperature of 39.5°C. We hypothesized that an alternative pathway(s) for obtaining deoxyribonucleotides such as the cellular ribonucleotide reductase can compensate for defects in viral RR but that this alternative source is not available in cells that are non-dividing or grown at 39.5°C.

Counteracting negative aspects of the heat shock response

Apoptosis may be induced if protein quality control machinery is overwhelmed during cell stress. Since induction of apoptosis might be expected to be detrimental to lytic infection, HSV-1 has apparently evolved to encode proteins that can prevent or delay the apoptotic response (see Chapter 12). One example is Us11 which is an RNA binding protein that interacts with cellular protein kinases, and it has been shown to inhibit the activities of antiviral proteins such as RNA-dependent protein kinase (PKR) and 2′–5′ oligoadenylate synthetase (OAS). Us11 was reported to accelerate recovery of protein synthesis and inhibit apoptosis in 43°C heat shocked HeLa cells (Diaz-Latoud *et al.*, 1997; Javouhey *et al.*, 2008). Thus, Us11 appears to exhibit cytoprotective activities in stressed cells and may be a part of the mechanism by which apoptosis is inhibited during HSV infection.

Positive aspects of the heat shock response for HSV-1 infection

Although some aspects of heat stress can be deleterious to viral productivity due to inactivation of essential cellular or viral factors, it appears that heat stress activates other pathways that are beneficial for viral infection. The alphaherpesviruses establish lifelong latent infections in the nuclei of trigeminal ganglia during which latency-associated RNA transcripts (LATs) are abundantly expressed from a circularized, non-integrated viral genome while little to no transcription of lytic genes takes place (reviewed in Bloom *et al.*). Interestingly, it was recently shown that LATs may play a cytoprotective role during exposure of latently infected cells to cold shock by upregulating heat shock protein expression (Atanasiu *et al.*, 2006; Moriya *et al.*, 1994).

Reactivation from latency has been reported to require the immediate-early protein ICP0 in a RING finger-dependent manner (Halford and Schaffer, 2001; Harris *et al.*, 1989) and involves complex epigenetic regulation of the genome (Bloom *et al.*). Temperature and chemical inducers of the heat shock response such as sodium benzoate and heat shock at 42–44°C have been reported to induce reactivation of latent/quiescent HSV-1 infection in animal models and cultured cells (Halford *et al.*, 1996; Moriya *et al.*, 1994; Sawtell and Thompson, 1992). The requirement for ICP0 for reactivation from latency could be overcome by heat stressing quiescently infected tissue culture cells and latently infected mice (Preston and Nicholl, 2008; Thompson and Sawtell, 2006). Heat stress can also be beneficial to lytic infection in the absence of ICP0. HSV-1 lytic infection in the absence of ICP0 is severely compromised at low MOI. However, it was found that heat shock of cells prior to infection with an ICP0 null virus resulted in wild type levels of virus production (Bringhurst and Schaffer, 2006). Thus, cellular factors capable of promoting reactivation from latency or promoting growth of an ICP0 deficient mutant may be induced by heat shock. It will be interesting to determine which aspects of the heat shock response are required.

In a recent study, a panel of HSV-1 promoter–luciferase constructs representing each kinetic class (immediate-early, early and late) was assayed for transcriptional upregulation in DNA transformed and infected cells during heat stress (Kushnir *et al.*). Luciferase assay and microarray analysis showed that the ICP0 and ICP22 promoters were most strongly upregulated following heat shock at 43°C. Further analysis of the ICP0 promoter revealed binding sites for two nuclear factor Y (NF-Y) sites, CCAAT, required for heat shock-induced transcriptional upregulation of ICP0. Interestingly, NF-Y is a heterotrimeric transcription factor that is necessary for both basal and induced activities of multiple stress responsive genes including Hsp70 (reviewed in Mantovani, 1999). Kushnir *et al.* concluded that ICP0 is upregulated in response to stress and that binding of NF-Y protein to NF-Y sites in the ICP0 promoter may play a role in reactivation from latency. Given the role of NF-Y in inducing expression of cellular heat stress responsive genes, it is intriguing to speculate that the ICP0 gene is stress-responsive. The specific roles of ICP0 in heat stressed cells have yet to be elucidated. In any case, induction of cellular heat stress in these systems was necessary to overcome the requirement of ICP0 for reactivation from latency.

Reorganization of chaperone proteins during VZV infection

Chaperone proteins, including Hsp70, Hsc70 and Hsp90, and their co-chaperone partners, like BAG3, are reorganized during lytic infection with varicella zoster virus (VZV) (Kyratsous and

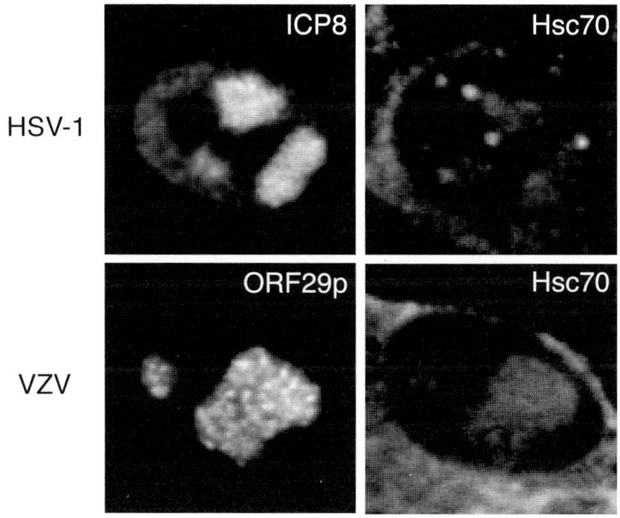

Figure 14.2 Hsc70 is detected in replication compartments during VZV infection. MeWo (human melanoma) cells were infected with HSV-1 Strain 17 for 6 hours (top panels) or VZV at an MOI of 0.05 for 24 hours (bottom panels) to account for differences in productivity between the two viruses. Infected cells were fixed and prepared for immunofluorescence analysis of the HSV-1 single stranded DNA binding protein ICP8 or the VZV single stranded DNA binding protein ORF29p and Hsc70. Images were captured on a Zeiss LSM 410 confocal microscope using a plan-apochromat 63x1.4 objective lens. Each set of images in the top and bottom panels represent the same cell.

Silverstein, 2007). These proteins are predominantly cytoplasmic in uninfected cells. However, in infected cells a fraction of these proteins concentrates inside the nucleus and co-localizes with virus proteins such as ORF29p and ORF62, that function in virus DNA replication and transcription respectively.

Although chaperones are reorganized during infection with both HSV-1 and VZV, the precise localization of these host proteins differs significantly. During HSV-1 infection VICE domains decorate the periphery of replication compartments (Burch and Weller, 2004) (Figs. 14.1 and 14.2 top panels). In contrast, in VZV infected cells, these same proteins are found to co-localize with viral replication/transcription machinery (Kyratsous and Silverstein, 2007) (Fig. 14.2 bottom panels). There are several non-mutually exclusive ways to explain the lack of VICE formation during VZV infection:

1. Expression of HSV-1 ICP0 is important for efficient formation of VICE domains (Burch and Weller, 2004). The VZV orthologue, ORF61p, lacks specific activities of ICP0; in fact VZV itself can be considered a phenocopy of an ICP0 deficient HSV-1 virus (Kyratsous and Silverstein, 2008; Kyratsous *et al.*, 2009b). It is thus possible that expression of ORF61p during VZV infection fails to provide activities of ICP0 that are required for proper localization of chaperone proteins to the periphery of replication compartments.
2. As mentioned above, efficient VICE domain formation during HSV-1 infection correlates with multiplicity of infection and is linked to host 'stress' that is induced during robust gene expression at very early stages of virus replication. VICE domains appear to develop as part of the host protein quality control mechanisms (Livingston *et al.*, 2009). At low multiplicities, VICE domains are only detected in a small percentage of infected cells. In contrast, at high multiplicities, when production of large amounts of virus gene products constitutes a burden on host cell metabolic machinery, chaperone proteins redistribute in VICE in all cells. Therefore, it is possible that HSV-1 has evolved to form VICE domains to facilitate its replication by regulating host response to increased production of virus proteins. Unlike HSV-1, VZV grown in tissue culture does not reach high titres and its replication is less robust than that of HSV-1. As a result, at least under experimental conditions, VZV replication might be less deleterious to host metabolic machinery and the virus has evolved unique mechanisms to utilize chaperone proteins. It should be noted that VICE domains do not form in VZV infected cells even when infection is allowed to progress for several rounds and spreads to neighbouring cells.

A role for the co-chaperone BAG3 in VZV and HSV replication

Members of the BAG family of proteins are unique regulators of Hsp/Hsc70 activities. The <u>b</u>cl-2-<u>a</u>ssociate athanogene proteins were originally identified by their association with Bcl-2 an anti-apoptotic protein (Lee *et al.*, 1999; Takayama *et al.*, 1999). Subsequently these proteins were also shown to interact with Hsp/Hsc70 and regulate their function (Takayama *et al.*, 1999). They were thus characterized as co-chaperones. To date, six human BAG proteins have been identified. Their unique BAG domain is located at their C-terminus and is necessary and sufficient for interaction with their chaperone partners (Doong *et al.*, 2003; Takayama *et al.*, 1997; Takayama and Reed, 2001; Takayama *et al.*, 1999). This domain is required for BAG3 mediated regulation of chaperone activity. At least for BAG-1 it was shown that binding induces conformational changes to the ATPase domain of Hsp70 resulting in a reduction of nucleotide affinity, in a manner antagonistic to Hip (Hohfeld, 1998; Hohfeld and Jentsch, 1997; Sondermann *et al.*, 2001). While eukaryotic cytosol lacks a sequence homologue of bacterial GrpE, the structural and biological similarities between BAG domains' interaction with Hsp70s have raised an interesting possibility: that BAG family members are the eukaryotic equivalents of bacterial nucleotide exchange factors (Briknarova *et al.*, 2001; Hohfeld, 1998;

Hohfeld and Jentsch, 1997; Sondermann *et al.*, 2001).

BAG3 is a protein partner of VZV ORF29p

The relationship between alphaherpesviruses and BAG3 was first described when it was discovered in a protein–protein interaction screen for partners of VZV ORF29p (Kyratsous and Silverstein, 2007). Association of BAG3 with ORF29p, a protein required for virus replication, and the involvement of BAG3-regulated pathways in stabilization and proper nuclear localization of the virus protein raised the possibility that co-chaperone proteins might be required for completion of the virus replication cycle. Consistent with this hypothesis, siRNA mediated depletion of BAG3 reduced VZV plaquing efficiency and greatly diminished its ability to replicate and spread to neighbouring cells.

After infection and spread of VZV in cells with wild-type endogenous levels of BAG3, nuclei of neighbouring infected cells form ring-shaped structures within the polykaryocytes that form, and virus proteins stain as a continuous layer, which is indicative of VZV-directed syncytia formation (Tzanck, 1947). Strikingly, when levels of BAG3 are decreased these ring-shaped structures are never seen, even when infection is allowed to progress for several days. This demonstrates that virus spread is inefficient and defective.

Nevertheless, in BAG3 depleted cells chaperones, such as Hsp70/Hsc70 and Hsp90, and virus proteins still reorganize and co-localize in replication/transcription foci in a manner similar to what is described above. Thus, nuclear localization of these chaperone proteins is not sufficient for successful replication of the viral genome. Regulation of heat shock protein activity by BAG3 and/or other co-chaperone proteins is also required.

Importantly, it should be noted that cells stably silenced for BAG3 do not exhibit defects in their morphology or growth rate when compared to their parental cells. Therefore, any slow down in virus proliferation results from loss of the co-chaperone protein and is not a result of gross defects in host metabolic machinery.

During these studies it was reported that unlike VZV, wild-type HSV plaquing efficiency and spread are not affected when levels of BAG3 are decreased (Kyratsous and Silverstein, 2007). It was posited that ICP0 compensated for lack of BAG3. A follow-up study showed that expression of ICP0, an immediate early protein encoded by HSV, is required for efficient virus replication when BAG3 levels are decreased (Kyratsous and Silverstein, 2008). Therefore, a novel function for ICP0, compensation for loss of co-chaperone activity, was first described in these studies.

Curiously, VZV encodes an orthologue of ICP0, called ORF61p. The two proteins were considered to have similar functions, as they both activate virus promoters and enhance infectivity of virus DNA (Moriuchi *et al.*, 1992; Moriuchi *et al.*, 1993). However, unlike ICP0, expression of ORF61p is not sufficient to compensate and bypass loss of BAG3 and thus does not mimic ICP0's function in this setting (Kyratsous and Silverstein, 2008). This was the first indication that the two virus proteins possess distinct activities, an observation that was later verified by directly comparing their functions (Kyratsous and Silverstein, 2009; Kyratsous *et al.*, 2009a).

Analysis of gene expression in the presence and absence of BAG3 revealed that it is required to boost immediate early gene expression. However, transcription fusions of promoters from affected genes did not reveal a direct requirement for BAG3 for increased reporter activity. It is thus possible, that BAG3 enhances accumulation of immediate early gene products via a post-transcriptional mechanism. The precise molecular details of BAG3's contribution to accumulation of immediate early proteins remain unknown. However, it is clear that BAG3 is a positive regulator of these processes. Two different scenarios are consistent with current studies. Perhaps, similar to other host proteins such as HCF and Oct1, or virus proteins such as VP16 and ICP0, BAG3 can activate gene expression. This might only be important in the context of virus replication and is therefore possibly not observed in reporter assays. Alternatively, BAG3 might be required for release of a block that leads to gene silencing. The latter hypothesis suggests that BAG3 is an anti-repressor and thus an upstream positive regulator of virus replication. In the presence of ICP0, requirements for host proteins are different and BAG3 levels do not affect virus yield although they alter viral immediate early protein abundance. Promoter fusion assays suggest that BAG3 is not involved in transcription of virus genes. Analysis of BAG3 localization during HSV infection revealed that it accumulated in the cytoplasm at the same time as ICP0. Therefore, it is likely that there is a link between the co-chaperone and the proposed cytoplasmic regulatory functions of ICP0 that interfere with host post-transcriptional regulation to increase virus protein accumulation (Kawaguchi *et al.*, 1997; Lopez *et al.*, 2001).

The role of Hsp90 during VZV infection

Heat shock proteins can regulate the balance between protein folding and protein turnover (McClellan and Frydman, 2001; McClellan *et al.*, 2005). Hsp90 is thought to be a key regulator of this process, depending on its association with other chaperone and co-chaperone proteins (Neckers *et al.*, 1999; Prodromou *et al.*, 1997). Because BAG3 abrogates protein degradation mediated by heat shock proteins (Doong *et al.*, 2003), it was hypothesized that its interaction partner ORF29p (see above) is a client for Hsp90 directed proteasomal degradation.

Geldanamycin and other ansamycin antibiotics interact with and inhibit the ATPase domain of Hsp90, shifting the balance of Hsp90 activity from protein folding to degradation of its clients (Schneider *et al.*, 1996; Whitesell *et al.*, 1994). Therefore, it is expected that when cells are treated with ansamycin antibiotics, Hsp90 is locked in its ADP bound conformation (Grenert *et al.*, 1997; Sullivan *et al.*, 1997) and its clients are rapidly degraded by the proteasome. Such treatment of cells expressing ORF29p resulted in a decrease in protein half-life; ORF29p's t1/2 is more than 6h in mock treated cells, but it is degraded in less than 1h in ansamycin treated cells (Kyratsous and Silverstein, 2007).

Treatment of ORF29p expressing cells with ansamycin not only destabilizes the virus protein, but it also prevents proper nuclear localization (Kyratsous and Silverstein, 2007). This was consistent with other studies showing that when ORF29p is restricted to the cytoplasm it is significantly less stable compared to when it is nuclear (Stallings *et al.*, 2006).

ORF29p is the VZV ssDNA binding protein and it is assumed to function during replication of the virus genome (Kinchington *et al.*, 1988). As such, it is indispensable for viral growth (Cohen *et al.*, 2007). It was hypothesized that destabilization of the protein and restriction to the cytoplasm will prohibit virus replication. In support of this prediction it was demonstrated that drug attenuation of Hsp90 ATPase activity with ansamycin antibiotics efficiently inhibited VZV plaque formation and spread in tissue culture (Kyratsous and Silverstein, 2007). It is likely that inhibition of virus replication is a result of interfering with both cell and virus targets, as ansamycin antibiotics are already known to affect a plethora of host cell proteins (Neckers and Neckers, 2005). As far as VZV targets are concerned, the drug affects at least ORF29p localization and stability. Therefore, ansamycin antibiotics can be used as potent inhibitors of VZV replication.

Summary

In this chapter we have reviewed the evidence that alphaherpesviruses HSV-1 and VZV have evolved to commandeer the host protein quality control mechanisms. Cellular chaperones such as Hsp70, Hsc70 and Hsp90 are required for efficient production of infectious virus playing essential roles in nuclear transport of viral proteins, protein quality control and maintenance of cellular homeostasis during viral infection. These viruses also encode proteins that play important roles in inhibition of apoptosis, transport of viral proteins to the nucleus, translation, and virus assembly. Knowledge of how viruses utilize viral and host proteins to ensure that the cellular environment is conducive to viral infection is important for a more complete understanding of these pathogenic viruses. Moreover, these findings raise the possibility that molecular chaperones could be utilized as effective targets for antiviral therapy.

References

Antrobus, R., Grant, K., Gangadharan, B., Chittenden, D., Everett, R.D., Zitzmann, N., and Boutell, C. (2009).

Proteomic analysis of cells in the early stages of herpes simplex virus type-1 infection reveals widespread changes in the host cell proteome. Proteomics *9*, 3913–3927.
Arrigo, A.P. (2007). The cellular 'networking' of mammalian Hsp27 and its functions in the control of protein folding, redox state and apoptosis. Adv. Exp. Med. Biol. *594*, 14–26.
Atanasiu, D., Kent, J.R., Gartner, J.J., and Fraser, N.W. (2006). The stable 2-kb LAT intron of herpes simplex stimulates the expression of heat shock proteins and protects cells from stress. Virology *350*, 26–33.
Bastian, T.W., Livingston, C.M., Weller, S.K., and Rice, S.A. (2010). Herpes simplex virus type 1 immediate-early protein ICP22 is required for VICE domain formation during productive viral infection. J. Virol. *84*, 2384–2394.
Bloom, D.C., Giordani, N.V., and Kwiatkowski, D.L. (2010). Epigenetic regulation of latent HSV-1 gene expression. Biochim Biophys Acta *1799*, 246–256.
Briknarova, K., Takayama, S., Brive, L., Havert, M.L., Knee, D.A., Velasco, J., Homma, S., Cabezas, E., Stuart, J., Hoyt, D.W., *et al.* (2001). Structural analysis of BAG1 cochaperone and its interactions with Hsc70 heat shock protein. Nat. Struct. Biol. *8*, 349–352.
Bringhurst, R.M., and Schaffer, P.A. (2006). Cellular stress rather. than stage of the cell cycle enhances the replication and plating efficiencies of herpes simplex virus type 1 ICP0- viruses. J. Virol. *80*, 4528–4537.
Buchner, J. (1999). Hsp90 & Co. – a holding for folding. Trends Biochem. Sci. *24*, 136–141.
Burch, A.D., and Weller, S.K. (2004). Nuclear sequestration of cellular chaperone and proteasomal machinery during herpes simplex virus type 1 infection. J. Virol. *78*, 7175–7185.
Burch, A.D., and Weller, S.K. (2005). Herpes simplex virus type 1 DNA polymerase requires. the mammalian chaperone hsp90 for proper localization to the nucleus. J. Virol. *79*, 10740–10749.
Cai, W., and Schaffer, P.A. (1992). Herpes simplex virus type 1 ICP0 regulates expression of immediate-early, early, and late genes in productively infected cells. J. Virol. *66*, 2904–2915.
Chabaud, S., Lambert, H., Sasseville, A.M., Lavoie, H., Guilbault, C., Massie, B., Landry, J., and Langelier, Y. (2003). The R1 subunit of herpes simplex virus ribonucleotide reductase has chaperone-like activity similar to Hsp27. FEBS Lett. *545*, 213–218.
Chai, Y., Koppenhafer, S.L., Shoesmith, S.J., Perez, M.K., and Paulson, H.L. (1999). Evidence for proteasome involvement in polyglutamine disease: localization to nuclear inclusions in SCA3/MJD and suppression of polyglutamine aggregation *in vitro*. Hum. Mol. Genet. *8*, 673–682.
Chappell, T.G., Welch, W.J., Schlossman, D.M., Palter, K.B., Schlesinger, M.J., and Rothman, J.E. (1986). Uncoating ATPase is a member of the 70 kilodalton family of stress proteins. Cell *45*, 3–13.
Cohen, J.I., Krogmann, T., Pesnicak, L., and Ali, M.A. (2007). Absence or overexpression of the Varicella-Zoster Virus (VZV) ORF29 latency-associated protein impairs late gene expression and reduces VZV latency in a rodent model. J. Virol. *81*, 1586–1591.
Craig, E.A., and Jacobsen, K. (1985). Mutations in cognate genes of Saccharomyces cerevisiae hsp70 result in reduced growth rates at low temperatures. Mol. Cell Biol. *5*, 3517–3524.
Crouch, N.A., and Rapp, F. (1972a). Cell-dependent differences in the production of infectious herpes simplex virus at a supraoptimal temperature. J. Virol. *9*, 223–230.
Crouch, N.A., and Rapp, F. (1972b). Differential effect of temperature on the replication of herpes simplex virus type 1 and type 2. Virology *50*, 939–941.
Davies, S.W., Turmaine, M., Cozens, B.A., DiFiglia, M., Sharp, A.H., Ross, C.A., Scherzinger, E., Wanker, E.E., Mangiarini, L., and Bates, G.P. (1997). Formation of neuronal intranuclear inclusions underlies the neurol. ogical dysfunction in mice transgenic for the HD mutation. Cell *90*, 537–548.
Diaz-Latoud, C., Diaz, J.J., Fabre-Jonca, N., Kindbeiter, K., Madjar, J.J., and Arrigo, A.P. (1997). Herpes simplex virus Us11 protein enhances recovery of protein synthesis and survival in heat shock treated HeLa cells. Cell Stress Chaperones *2*, 119–131.
Doong, H., Rizzo, K., Fang, S., Kulpa, V., Weissman, A.M., and Kohn, E.C. (2003). CAIR-1/BAG-3 abrogates heat shock protein-70 chaperone complex-mediated protein degradation: accumulation of poly-ubiquitinated Hsp90 client proteins. J. Biol. Chem. *278*, 28490–28500.
Dworniczak, B., and Mirault, M.E. (1987). Structure and expression of a human gene coding for a 71 kd heat shock 'cognate' protein. Nucleic Acids Res. *15*, 5181–5197.
Ellis, J. (1987). Proteins as molecular chaperones. Nature *328*, 378–379.
Ellis, S., Killender, M., and Anderson, R.L. (2000). Heat-induced alterations in the localization of HSP72 and HSP73 as measured by indirect immunohistochemistry and immunogold electron microscopy. J. Histochem. Cytochem. *48*, 321–332.
Eom, C.Y., and Lehman, I.R. (2002). The human DnaJ. protein, hTid-1, enhances binding of a multimer of the herpes simplex virus type 1 UL9 protein to oris, an origin of viral DNA replication. Proc. Natl. Acad. Sci. U.S.A. *99*, 1894–1898.
Fraser, K.A., and Rice, S.A. (2007). Herpes simplex virus immediate-early protein ICP22 triggers loss of serine 2-phosphorylated RNA polymerase II. J. Virol. *81*, 5091–5101.
Frydman, J. (2001). Folding of newly translated proteins *in vivo*: the role of molecular chaperones. Annu. Rev. Biochem. *70*, 603–647.
Frydman, J., Nimmesgern, E., Ohtsuka, K., and Hartl, F.U. (1994). Folding of nascent polypeptide chains in a high molecular mass assembly with molecular chaperones. Nature *370*, 111–117.
Fu, L., Gao, Y.S., and Sztul, E. (2005a). Transcriptional repression and cell death induced by nuclear aggregates of non-polyglutamine protein. Neurobiol. Dis. *20*, 656–665.

Fu, L., Gao, Y.S., Tousson, A., Shah, A., Chen, T.L., Vertel, B.M., and Sztul, E. (2005b). Nuclear aggresomes form by fusion of PML-associated aggregates. Mol. Biol. Cell *16*, 4905–4917.

Garcia-Mata, R., Bebok, Z., Sorscher, E.J., and Sztul, E.S. (1999). Characterization and dynamics of aggresome formation by a cytosolic GFP-chimera. J. Cell Biol. *146*, 1239–1254.

Geller, R., Vignuzzi, M., Andino, R., and Frydman, J. (2010). Evolutionary constraints on chaperone-mediated folding provide an antiviral approach refractory to development of drug resistance. Genes Dev. *21*, 195–205.

Georgopoulos, C., and Welch, W.J. (1993). Role of the major heat shock proteins as molecular chaperones. Annu Rev. Cell Biol. *9*, 601–634.

Georgopoulos, C.P. (1977). A new bacterial gene (groPC) which affects lambda DNA replication. Mol. Gen. Genet. *151*, 35–39.

Gharpure, M.A. (1965). A heat-sensitive cellular function required for the replication of DNA viruses but not RNA viruses. Virology *27*, 308–319.

Gillis, P.A., Okagaki, L.H., and Rice, S.A. (2009). Herpes simplex virus type 1 ICP27 induces p38 mitogen-activated protein kinase signaling and apoptosis in HeLa cells. J. Virol. *83*, 1767–1777.

Goldstein, D.J., and Weller, S.K. (1988). Factor(s) present in herpes simplex virus type 1-infected cells can compensate for the loss of the large subunit of the viral ribonucleotide reductase: characterization of an ICP6 deletion mutant. Virology *166*, 41–51.

Grenert, J.P., Sullivan, W.P., Fadden, P., Haystead, T.A., Clark, J., Mimnaugh, E., Krutzsch, H., Ochel, H.J., Schulte, T.W., Sausville, E., *et al.* (1997). The amino-terminal domain of heat shock protein 90 (hsp90) that binds geldanamycin is an ATP/ADP switch domain that regulates hsp90 conformation. J. Biol. Chem. *272*, 23843–23850.

Guerrero, C.A., Bouyssounade, D., Zarate, S., Isa, P., Lopez, T., Espinosa, R., Romero, P., Mendez, E., Lopez, S., and Arias, C.F. (2002). Heat shock cognate protein 70 is involved in rotavirus cell entry. J. Virol. *76*, 4096–4102.

Halford, W.P., Gebhardt, B.M., and Carr, D.J. (1996). Mechanisms of herpes simplex virus type 1 reactivation. J. Virol. *70*, 5051–5060.

Halford, W.P., and Schaffer, P.A. (2001). ICP0 is required for efficient reactivation of herpes simplex virus type 1 from neuronal latency. J. Virol. *75*, 3240–3249.

Hargett, D., McLean, T., and Bachenheimer, S.L. (2005). Herpes simplex virus ICP27 activation of stress kinases JNK and p38. J. Virol. *79*, 8348–8360.

Harris, R.A., Everett, R.D., Zhu, X.X., Silverstein, S., and Preston, C.M. (1989). Herpes simplex virus type 1 immediate-early protein Vmw110 reactivates latent herpes simplex virus type 2 in an *in vitro* latency system. J. Virol. *63*, 3513–3515.

Hartl, F.U. (1996). Molecular chaperones in cellular protein folding. Nature *381*, 571–579.

Hattori, H., Kaneda, T., Lokeshwar, B., Laszlo, A., and Ohtsuka, K. (1993). A stress-inducible 40kDa protein (hsp40): purification by modified two-dimensional gel electrophoresis and co-localization with hsc70(p73) in heat-shocked HeLa cells. J. Cell Sci. *104*, 629–638.

Hightower, L.E., and White, F.P. (1981). Cellular responses to stress: comparison of a family of 71-–73-kilodalton proteins rapidly synthesized in rat tissue slices and canavanine-treated cells in culture. J. Cell Physiol *108*, 261–275.

Hohfeld, J. (1998). Regulation of the heat shock conjugate Hsc70 in the mammalian cell: the characterization of the anti-apoptotic protein BAG-1 provides novel insights. Biol. Chem. *379*, 269–274.

Hohfeld, J., and Jentsch, S. (1997). GrpE-like regulation of the hsc70 chaperone by the anti-apoptotic protein BAG-1. EMBO J. *16*, 6209–6216.

Hu, J., and Seeger, C. (1996). Hsp90 is required for the activity of a hepatitis B virus reverse transcriptase. Proc. Natl. Acad. Sci. U.S.A. *93*, 1060–1064.

Hutchinson, I., Whiteley, A., Browne, H., and Elliott, G. (2002). Sequential localization of two herpes simplex virus tegument proteins to punctate nuclear dots adjacent to ICP0 domains. J. Virol. *76*, 10365–10373.

Ingolia, T.D., and Craig, E.A. (1982). Drosophila gene related to the major heat shock-induced gene is transcribed at normal temperatures. and not induced by heat shock. Proc. Natl. Acad. Sci. U.S.A. *79*, 525–529.

Ivanovic, T., Agosto, M.A., Chandran, K., and Nibert, M.L. (2007). A role for molecular chaperone Hsc70 in reovirus outer capsid disassembly. J. Biol. Chem. *282*, 12210–12219.

Javouhey, E., Gibert, B., Arrigo, A.P., Diaz, J.J., and Diaz-Latoud, C. (2008). Protection against heat and staurosporine mediated apoptosis by the HSV-1 US11 protein. Virology *376*, 31–41.

Karaca, G., Hargett, D., McLean, T.I., Aguilar, J.S., Ghazal, P., Wagner, E.K., and Bachenheimer, S.L. (2004). Inhibition of the stress-activated kinase, p38, does not affect the virus transcriptional program of herpes simplex virus type 1. Virology *329*, 142–156.

Kawaguchi, Y., Bruni, R., and Roizman, B. (1997). Interaction of herpes simplex virus 1 alpha regulatory protein ICP0 with elongation factor 1delta: ICP0 affects translational machinery. J. Virol. *71*, 1019–1024.

Kinchington, P.R., Inchauspe, G., Subak-Sharpe, J.H., Robey, F., Hay, J., and Ruyechan, W.T. (1988). Identification and characterization of a varicella-zoster virus DNA-binding protein by using antisera directed against a predicted synthetic oligopeptide. J. Virol. *62*, 802–809.

Kobayashi, K., Ohgitani, E., Tanaka, Y., Kita, M., and Imanishi, J. (1994). Herpes simplex virus-induced expression of 70kDa heat shock protein (HSP70) requires. early protein synthesis but not viral DNA replication. Microbiol. Immunol. *38*, 321–325.

Kostenko, S., and Moens, U. (2009). Heat shock protein 27 phosphorylation: kinases, phosphatases, functions and pathology. Cell. Mol. Life Sci. *66*, 3289–3307.

Kushnir, A.S., Davido, D.J., and Schaffer, P.A. Role of Nuclear Factor Y in Stress-Induced Activation of the

Herpes Simplex Virus Type 1 ICP0 Promoter. J. Virol. *84*, 188–200.
Kyratsous, C.A., and Silverstein, S.J. (2007). BAG3, a host cochaperone, facilitates varicella-zoster virus replication. J. Virol. *81*, 7491–7503.
Kyratsous, C.A., and Silverstein, S.J. (2008). The co-chaperone BAG3 regulates Herpes Simplex Virus replication. Proc. Natl. Acad. Sci. U.S.A. *105*, 20912–20917.
Kyratsous, C.A., and Silverstein, S.J. (2009). Components of nuclear domain 10 bodies regulate varicella-zoster virus replication. J. Virol. *83*, 4262–4274.
Kyratsous, C.A., Silverstein, S.J., DeLong, C.R., and Panagiotidis, C.A. (2009a). Chaperone-fusion expression plasmid vectors for improved solubility of recombinant proteins in *Escherichia coli*. Gene *440*, 9–15.
Kyratsous, C.A., Walters, M.S., Panagiotidis, C.A., and Silverstein, S.J. (2009b). Complementation of a herpes simplex virus ICP0 null mutant by varicella-zoster virus ORF61p. J. Virol. *83*, 10637–10643.
Lamberti, C., and Weller, S.K. (1998). The herpes simplex virus type 1 cleavage/packaging protein, UL32, is involved in efficient localization of capsids to replication compartments. J. Virol. *72*, 2463–2473.
Lee, J.H., Takahashi, T., Yasuhara, N., Inazawa, J., Kamada, S., and Tsujimoto, Y. (1999). Bis, a Bcl-2-binding protein that synergizes with Bcl-2 in preventing cell death. Oncogene *18*, 6183–6190.
Li, L., Johnson, L.A., Dai-Ju, J.Q., and Sandri-Goldin, R.M. (2008). Hsc70 focus formation at the periphery of HSV-1 transcription sites requires. ICP27. PLoS ONE 3, e1491.
Li, Y.H., Tao, P.Z., Liu, Y.Z., and Jiang, J.D. (2004). Geldanamycin, a ligand of heat shock protein 90, inhibits the replication of herpes simplex virus type 1 *in vitro*. Antimicrob. Agents Chemother. *48*, 867–872.
Liberek, K., Georgopoulos, C., and Zylicz, M. (1988). Role of the Escherichia coli DnaK and DnaJ. heat shock proteins in the initiation of bacteriophage lambda DNA replication. Proc. Natl. Acad. Sci. U.S.A. *85*, 6632–6636.
Liu, J.S., Kuo, S.R., Makhov, A.M., Cyr, D.M., Griffith, J.D., Broker, T.R., and Chow, L.T. (1998). Human Hsp70 and Hsp40 chaperone proteins facilitate human papillomavirus-11 E1 protein binding to the origin and stimulate cell-free DNA replication. J. Biol. Chem. *273*, 30704–30712.
Livingston, C.M., DeLuca, N.A., Wilkinson, D.E., and Weller, S.K. (2008). Oligomerization of ICP4 and rearrangement of heat shock proteins may be important for herpes simplex virus type 1 prereplicative site formation. J. Virol. *82*, 6324–6336.
Livingston, C.M., Ifrim, M.F., Cowan, A.E., and Weller, S.K. (2009). Virus-Induced Chaperone-Enriched (VICE) domains function as nuclear protein quality control centers during HSV-1 infection. PLoS Pathog. *5*, e1000619.
Lopez, P., Van Sant, C., and Roizman, B. (2001). Requirements for the nuclear-cytoplasmic translocation of infected-cell protein 0 of herpes simplex virus 1. J. Virol. *75*, 3832–3840.
Lowe, D.G., and Moran, L.A. (1984). Proteins related to the mouse L-cell major heat shock protein are synthesized in the absence of heat shock gene expression. Proc. Natl. Acad. Sci. U.S.A. *81*, 2317–2321.
Mantovani, R. (1999). The molecular biology of the CCAAT-binding factor NF-Y. Gene *239*, 15–27.
Mathew, S.S., Della Selva, M.P., and Burch, A.D. (2009). Modification and reorganization of the cytoprotective cellular chaperone Hsp27 during herpes simplex virus type 1 infection. J. Virol. *83*, 9304–9312.
Matsumoto, G., Kim, S., and Morimoto, R.I. (2006). Huntingtin and mutant SOD1 form aggregate structures. with distinct molecular properties in human cells. J. Biol. Chem. *281*, 4477–4485.
Mayer, M.P., Prodromou, C., and Frydman, J. (2009). The Hsp90 mosaic: a picture emerges. Nat. Struct. Mol. Biol. *16*, 2–6.
McClellan, A.J., and Frydman, J. (2001). Molecular chaperones and the art of recognizing a lost cause. Nat. Cell Biol. 3, E51–53.
McClellan, A.J., Tam, S., Kaganovich, D., and Frydman, J. (2005). Protein quality control: chaperones culling corrupt conformations. Nat. Cell Biol. *7*, 736–741.
McHaourab, H.S., Godar, J.A., and Stewart, P.L. (2009). Structure and mechanism of protein stability sensors: chaperone activity of small heat shock proteins. Biochemistry *48*, 3828–3837.
McLean, T.I., and Bachenheimer, S.L. (1999). Activation of cJUN N-terminal kinase by herpes simplex virus type 1 enhances viral replication. J. Virol. *73*, 8415–8426.
Melville, M.W., Tan, S.L., Wambach, M., Song, J., Morimoto, R.I., and Katze, M.G. (1999). The cellular inhibitor of the PKR protein kinase, P58(IPK), is an influenza virus-activated co-chaperone that modulates heat shock protein 70 activity. J. Biol. Chem. *274*, 3797–3803.
Moriuchi, H., Moriuchi, M., Smith, H.A., Straus, S.E., and Cohen, J.I. (1992). Varicella-zoster virus open reading frame 61 protein is functionally homologous to herpes simplex virus type 1 ICP0. J. Virol. *66*, 7303–7308.
Moriuchi, H., Moriuchi, M., Straus, S.E., and Cohen, J.I. (1993). Varicella-zoster virus (VZV) open reading frame 61 protein transactivates VZV gene promoters and enhances the infectivity of VZV DNA. J. Virol. *67*, 4290–4295.
Moriya, A., Yoshiki, A., Kita, M., Fushiki, S., and Imanishi, J. (1994). Heat shock-induced reactivation of herpes simplex virus type 1 in latently infected mouse trigeminal ganglion cells in dissociated culture. Arch. Virol. *135*, 419–425.
Naito, T., Momose, F., Kawaguchi, A., and Nagata, K. (2007). Involvement of Hsp90 in assembly and nuclear import of influenza virus RNA polymerase subunits. J. Virol. *81*, 1339–1349.
Neckers, L., and Neckers, K. (2005). Heat-shock protein 90 inhibitors as novel cancer chemotherapeutics – an update. Expert Opin. Emerg Drugs *10*, 137–149.
Neckers, L., Schulte, T.W., and Mimnaugh, E. (1999). Geldanamycin as a potential anti-cancer agent: its

molecular target and biochemical activity. Invest New Drugs *17*, 361–373.

Newcomb, W.W., Juhas, R.M., Thomsen, D.R., Homa, F.L., Burch, A.D., Weller, S.K., and Brown, J.C. (2001). The UL6 gene product forms the portal for entry of DNA into the herpes simplex virus capsid. J. Virol. *75*, 10923–10932.

Notarianni, E.L., and Preston, C.M. (1982). Activation of cellular stress protein genes by herpes simplex virus temperature-sensitive mutants which overproduce immediate early polypeptides. Virology *123*, 113–122.

Palter, K.B., Watanabe, M., Stinson, L., Mahowald, A.P., and Craig, E.A. (1986). Expression and localization of Drosophila melanogaster hsp70 cognate proteins. Mol. Cell Biol. *6*, 1187–1203.

Patel, A.H., Rixon, F.J., Cunningham, C., and Davison, A.J. (1996). Isolation and characterization of herpes simplex virus type 1 mutants defective in the UL6 gene. Virology *217*, 111–123.

Paulson, H.L., Perez, M.K., Trottier, Y., Trojanowski, J.Q., Subramony, S.H., Das, S.S., Vig, P., Mandel, J.L., Fischbeck, K.H., and Pittman, R.N. (1997). Intranuclear inclusions of expanded polyglutamine protein in spinocerebellar ataxia type 3. Neuron *19*, 333–344.

Pelham, H.R. (1984). Hsp70 accelerates the recovery of nucleolar morphology after heat shock. EMBO J. *3*, 3095–3100.

Perez-Vargas, J., Romero, P., Lopez, S., and Arias, C.F. (2006). The peptide-binding and ATPase domains of recombinant hsc70 are required to interact with rotavirus and reduce its infectivity. J. Virol. *80*, 3322–3331.

Phillips, B., Abravaya, K., and Morimoto, R.I. (1991). Analysis of the specificity and mechanism of transcriptional activation of the human hsp70 gene during infection by DNA viruses. J. Virol. *65*, 5680–5692.

Poffenberger, K.L., Raichlen, P.E., and Herman, R.C. (1993). *In vitro* characterization of a herpes simplex virus type 1 ICP22 deletion mutant. Virus Genes *7*, 171–186.

Preston, C.M., and Nicholl, M.J. (2008). Induction of cellular stress overcomes the requirement of herpes simplex virus type 1 for immediate-early protein ICP0 and reactivates expression from quiescent viral genomes. J. Virol. *82*, 11775–11783.

Prodromou, C., Roe, S.M., O'Brien, R., Ladbury, J.E., Piper, P.W., and Pearl, L.H. (1997). Identification and structural characterization of the ATP/ADP-binding site in the Hsp90 molecular chaperone. Cell *90*, 65–75.

Rice, S.A., Long, M.C., Lam, V., Schaffer, P.A., and Spencer, C.A. (1995). Herpes simplex virus immediate-early protein ICP22 is required for viral modification of host RNA polymerase II and establishment of the normal viral transcription program. J. Virol. *69*, 5550–5559.

Rojas, S., Corbin-Lickfett, K.A., Escudero-Paunetto, L., and Sandri-Goldin, R.M. (2010). ICP27 phosphorylation site mutants are defective in herpes simplex virus 1 replication and gene expression. J. Virol. *84*, 2200–2211.

Ross, C.A. (1997). Intranuclear neuronal inclusions: a common pathogenic mechanism for glutamine-repeat neurodegenerative diseases? Neuron *19*, 1147–1150.

Rothman, J.E., and Schmid, S.L. (1986). Enzymatic recycling of clathrin from coated vesicles. Cell *46*, 5–9.

Rouse, J., Cohen, P., Trigon, S., Morange, M., Alonso-Llamazares, A., Zamanillo, D., Hunt, T., and Nebreda, A.R. (1994). A novel kinase cascade triggered by stress and heat shock that stimulates MAPKAP kinase-2 and phosphorylation of the small heat shock proteins. Cell *78*, 1027–1037.

Sawtell, N.M., and Thompson, R.L. (1992). Rapid *in vivo* reactivation of herpes simplex virus in latently infected murine ganglionic neurons after transient hyperthermia. J. Virol. *66*, 2150–2156.

Schneider, C., Sepp-Lorenzino, L., Nimmesgern, E., Ouerfelli, O., Danishefsky, S., Rosen, N., and Hartl, F.U. (1996). Pharmacologic shifting of a balance between protein refolding and degradation mediated by Hsp90. Proc. Natl. Acad. Sci. U.S.A. *93*, 14536–14541.

Sears, A.E., Halliburton, I.W., Meignier, B., Silver, S., and Roizman, B. (1985). Herpes simplex virus 1 mutant deleted in the alpha 22 gene: growth and gene expression in permissive and restrictive cells and establishment of latency in mice. J. Virol. *55*, 338–346.

Skinner, P.J., Koshy, B.T., Cummings, C.J., Klement, I.A., Helin, K., Servadio, A., Zoghbi, H.Y., and Orr, H.T. (1997). Ataxin-1 with an expanded glutamine tract alters nuclear matrix-associated structures. Nature *389*, 971–974.

Sondermann, H., Scheufler, C., Schneider, C., Hohfeld, J., Hartl, F.U., and Moarefi, I. (2001). Structure of a Bag/Hsc70 complex: convergent functional evolution of Hsp70 nucleotide exchange factors. Science *291*, 1553–1557.

Stallings, C.L., Duigou, G.J., Gershon, A.A., Gershon, M.D., and Silverstein, S.J. (2006). The cellular localization pattern of Varicella-Zoster virus ORF29p is influenced by proteasome-mediated degradation. J. Virol. *80*, 1497–1512.

Stebbins, C.E., Russo, A.A., Schneider, C., Rosen, N., Hartl, F.U., and Pavletich, N.P. (1997). Crystal structure of an Hsp90–geldanamycin complex: targeting of a protein chaperone by an antitumor agent. Cell *89*, 239–250.

Stenoien, D.L., Mielke, M., and Mancini, M.A. (2002). Intranuclear ataxin1 inclusions contain both fast- and slow-exchanging components. Nat. Cell Biol. *4*, 806–810.

Sullivan, C.S., and Pipas, J.M. (2002). T antigens of simian virus 40: molecular chaperones for viral replication and tumorigenesis. Microbiol. Mol. Biol. Rev. *66*, 179–202.

Sullivan, W., Stensgard, B., Caucutt, G., Bartha, B., McMahon, N., Alnemri, E.S., Litwack, G., and Toft, D. (1997). Nucleotides and two functional states of hsp90. J. Biol. Chem. *272*, 8007–8012.

Suomalainen, M., Nakano, M.Y., Boucke, K., Keller, S., and Greber, U.F. (2001). Adenovirus-activated PKA and p38/MAPK pathways boost microtubule-mediated nuclear targeting of virus. EMBO J. *20*, 1310–1319.

Takayama, S., Bimston, D.N., Matsuzawa, S., Freeman, B.C., Aime-Sempe, C., Xie, Z., Morimoto, R.I., and Reed, J.C. (1997). BAG-1 modulates the chaperone activity of Hsp70/Hsc70. EMBO J. *16*, 4887–4896.

Takayama, S., and Reed, J.C. (2001). Molecular chaperone targeting and regulation by BAG family proteins. Nat. Cell Biol. *3*, E237–241.

Takayama, S., Xie, Z., and Reed, J.C. (1999). An evolutionarily conserved family of Hsp70/Hsc70 molecular chaperone regulators. J. Biol. Chem. *274*, 781–786.

Tanaka, M., Kim, Y.M., Lee, G., Junn, E., Iwatsubo, T., and Mouradian, M.M. (2004). Aggresomes formed. by alpha-synuclein and synphilin-1 are cytoprotective. J. Biol. Chem. *279*, 4625–4631.

Tanguy Le Gac, N., and Boehmer, P.E. (2002). Activation of the herpes simplex virus type-1 origin-binding protein (UL9) by heat shock proteins. J. Biol. Chem. *277*, 5660–5666.

Taus, N.S., Salmon, B., and Baines, J.D. (1998). The herpes simplex virus 1 UL 17 gene is required for localization of capsids and major and minor capsid proteins to intranuclear sites where viral DNA is cleaved and packaged. Virology *252*, 115–125.

Tavaria, M., Gabriele, T., Kola, I., and Anderson, R.L. (1996). A hitchhiker's guide to the human Hsp70 family. Cell Stress Chaperones *1*, 23–28.

Taylor, J.P., Tanaka, F., Robitschek, J., Sandoval, C.M., Taye, A., Markovic-Plese, S., and Fischbeck, K.H. (2003). Aggresomes protect cells by enhancing the degradation of toxic polyglutamine-containing protein. Hum. Mol. Genet. *12*, 749–757.

Thompson, R.L., and Sawtell, N.M. (2006). Evidence that the herpes simplex virus type 1 ICP0 protein does not initiate reactivation from latency *in vivo*. J. Virol. *80*, 10919–10930.

Tzanck, A. (1947). Le cytodiagnostic immediat en dermatologie. Bull Soc. Fr Dermatol Syph *7*, 68.

Ungewickell, E. (1985). The 70-kd mammalian heat shock proteins are structurally and functionally related to the uncoating protein that releases clathrin triskelia from coated vesicles. EMBO J. *4*, 3385–3391.

Wacker, J.L., Zareie, M.H., Fong, H., Sarikaya, M., and Muchowski, P.J. (2004). Hsp70 and Hsp40 attenuate formation of spherical and annular polyglutamine oligomers by partitioning monomer. Nat. Struct. Mol. Biol. *11*, 1215–1222.

Walsh, D., and Mohr, I. (2006). Assembly of an active translation initiation factor complex by a viral protein. Genes Dev. *20*, 461–472.

Watanabe, K., Fuse, T., Asano, I., Tsukahara, F., Maru, Y., Nagata, K., Kitazato, K., and Kobayashi, N. (2006). Identification of Hsc70 as an influenza virus matrix protein (M1) binding factor involved in the virus life cycle. FEBS Lett. *580*, 5785–5790.

Weller, T.H. (1953). Serial propagation *in vitro* of agents producing inclusion bodies derived from varicella and herpes zoster. Proc. Soc. Exp. Biol. Med. *83*, 340–346.

Whitesell, L., and Lindquist, S. (2009). Inhibiting the transcription factor HSF1 as an anticancer strategy. Expert Opin. Ther. Targets *13*, 469–478.

Whitesell, L., Mimnaugh, E.G., De Costa, B., Myers, C.E., and Neckers, L.M. (1994). Inhibition of heat shock protein HSP90-pp60v-src heteroprotein complex formation by benzoquinone ansamycins: essential role for stress proteins in oncogenic transformation. Proc. Natl. Acad. Sci. U.S.A. *91*, 8324–8328.

Wilkinson, D.E., and Weller, S.K. (2006). Herpes simplex virus type I disrupts the ATR-dependent DNA-damage response during lytic infection. J. Cell Sci. *119*, 2695–2703.

Woulfe, J.M. (2007). Abnormalities of the nucleus and nuclear inclusions in neurodegenerative disease: a work in progress. Neuropathol. Appl. Neurobiol. *33*, 2–42.

Xu, W., Marcu, M., Yuan, X., Mimnaugh, E., Patterson, C., and Neckers, L. (2002). Chaperone-dependent E3 ubiquitin ligase CHIP mediates a degradative pathway for c-ErbB2/Neu. Proc. Natl. Acad. Sci. U.S.A. *99*, 12847–12852.

Yamauchi, Y., Daikoku, T., Goshima, F., and Nishiyama, Y. (2003). Herpes simplex virus UL14 protein blocks apoptosis. Microbiol. Immunol. *47*, 685–689.

Yamauchi, Y., Wada, K., Goshima, F., Daikoku, T., Ohtsuka, K., and Nishiyama, Y. (2002). Herpes simplex virus type 2 UL14 gene product has heat shock protein (HSP)-like functions. J. Cell Sci. *115*, 2517–2527.

Yamauchi, Y., Wada, K., Goshima, F., Takakuwa, H., Daikoku, T., Yamada, M., and Nishiyama, Y. (2001). The UL14 protein of herpes simplex virus type 2 translocates the minor capsid protein VP26 and the DNA cleavage and packaging UL33 protein into the nucleus of coexpressing cells. J. Gen. Virol. *82*, 321–330.

Zarate, S., Cuadras, M.A., Espinosa, R., Romero, P., Juarez, K.O., Camacho-Nuez, M., Arias, C.F., and Lopez, S. (2003). Interaction of rotaviruses with Hsc70 during cell entry is mediated by VP5. J. Virol. *77*, 7254–7260.

Zylicz, M., and Georgopoulos, C. (1984). Purification and properties of the *Escherichia coli* dnaK replication protein. J. Biol. Chem. *259*, 8820–8825.

Interactions Between HSV-1 and the DNA Damage Response

15

Matthew D. Weitzman and Sandra K. Weller

Abstract

The cellular DNA damage machinery responds to virus infection and the foreign genomes that accumulate in the nuclei of infected cells. Many DNA viruses have been shown to manipulate the cellular DNA damage response pathways in order to create environments conducive to their own replication. Some cellular factors are activated during infection while others are inactivated. In this chapter we will focus on the complex cellular responses triggered by infection with HSV-1 and the consequences of its manipulation by viral factors.

Introduction

Cells possess a complex network of machinery to sense and repair different forms of DNA damage. Viral infection presents cells with foreign DNA, providing model systems to study recognition and processing of damaged DNA (Lilley *et al.*, 2007). Many viruses are known to interact with components of the cellular DNA damage response network and they can help to pinpoint the cellular proteins that regulate specific processes. HSV-1 infection activates and exploits aspects of the cellular DNA damage signalling cascade while inhibiting others (Lilley *et al.*, 2005, 2010; Parkinson *et al.*, 1999; Shirata *et al.*, 2005; Wilkinson and Weller, 2004, 2005; Wilkinson and Weller, 2006). During lytic infection, replication compartments are formed within the infected cell nucleus, and cellular DNA repair proteins are recruited into these structures (Lilley *et al.*, 2005; Shirata *et al.*, 2005; Taylor and Knipe, 2004; Wilkinson and Weller, 2004). Some components of the repair machinery are inactivated or degraded during HSV-1 infection. It is possible that inactivation of some DNA repair pathways may free components of these pathways for utilization by HSV-1. Alternatively, some parts of the cellular DNA damage pathways may be intrinsically antiviral and therefore inactivation is necessary for efficient viral replication. In this chapter we review the known interactions between HSV and the cellular DNA damage response, and discuss their impact on different stages of the virus lifecycle, as well as the consequences for the infected cell.

The HSV replication cycle

HSV carries out much of its life cycle in the cell nucleus. After entry at the cell membrane, viral capsids travel through the cytoplasm, dock at the nuclear pore and release viral genomes into the nucleus (Ojala *et al.*, 2000). Although the fate of viral genomes once they enter the nucleus is still not well understood, it appears that some factors such as PML and other proteins are recruited in order to repress viral gene expression (discussed below and in Chapters 4 and 6). The viral immediate early protein ICP0 disrupts repressive structures on the viral genomes paving the way for three waves of gene expression, immediate early, early and late genes and lytic replication.

The appearance of viral genomes in the nucleus might be expected to induce a cellular DNA damage response because the incoming linear genomes possess double strand ends that resemble double strand breaks and also contain nicks and gaps (reviewed in Wilkinson and Weller, 2003). It is possible that cellular proteins are recruited to incoming viral genomes for the purpose of repairing the nicks and gaps. Repair of nicks and gaps followed by end-ligation might be expected to result in the formation of covalently closed circular DNA molecules. This is supported by observations that suggest that the incoming viral genomes circularize (Garber *et al.*, 1993; Poffenberger and Roizman, 1985; Strang and Stow, 2005). This model has, however, recently been called into question after an unsuccessful attempt to directly visualize covalently closed circular viral DNA in cells infected with wild-type HSV-1 using Gardella gels (Jackson and DeLuca, 2003). In contrast, circular forms could be detected in cells infected with an ICP0 mutant virus. Additional experiments will be required to resolve the question of whether incoming viral genomes can circularize and whether circular or linear genomes function as templates for viral DNA replication. It is possible that inter- or intra-molecular recombination events not detected by the Gardella gel protocol generate novel DNA structures used to initiate viral DNA synthesis.

The *cis*- and *trans*- acting elements required for viral DNA replication are discussed in detail in Chapter 6. The products of DNA replication are larger-than-unit length concatemers consisting of tandem repeats of the viral genome (Ben-Porat and Rixon, 1979). Concatemer production is an essential step for the generation of progeny virus, since the packaging machinery must recognize longer-than-unit-length concatemers during the encapsidation step.

The overall mechanism of HSV DNA replication is poorly understood. Although, it was previously thought that linear virion DNA circularized in infected cells and that concatemers were generated by rolling circle replication (reviewed in Boehmer and Lehman, 1997), this model has never been directly confirmed. Simple rolling circle replication does not explain the observation that genomic inversions occur as soon as viral DNA synthesis can be detected (Lamberti and Weller, 1996; Severini *et al.*, 1994; Zhang *et al.*, 1994). We and others have proposed that by analogy with the well-studied DNA bacteriophages, HSV uses a recombination-dependent mechanism to generate head-to-tail concatemers (reviewed in Wilkinson and Weller, 2003 and discussed in Chapter 6).

Cellular DNA repair pathways

Double-strand breaks in DNA (DSBs) are generated in chromosomal DNA by collapse of replication forks, during programmed rearrangements, by physical stress, and by damaging agents such as ionizing radiation (IR). These lesions threaten cell viability and can lead to chromosomal translocations and genomic instability. DSBs can be repaired via non-homologous end-joining (NHEJ) or homologous recombination (HR), depending on availability of repair templates and the stage of the cell cycle (Branzei and Foiani, 2008; Shrivastav *et al.*, 2008). HR requires functions of the Rad51-family of proteins, while NHEJ is initiated by association of Ku70/80 proteins with DNA ends and recruitment of the DNA-dependent protein kinase catalytic subunit (DNA-PKcs) (Sancar *et al.*, 2004). Auto-phosphorylation of DNA-PKcs leads to its release from DNA, allowing processing of the ends for repair via the functions of DNA ligase IV and X-ray repair cross-complementing group (XRCC4).

The cellular DNA damage response

Damage at cellular chromatin is recognized by sensor molecules that initiate a network of signal transduction cascades, collectively referred to as the DNA damage response. The MRN complex, consisting of Mre11, Rad50 and Nbs1, acts as a sensor of DSBs and is recruited to the site of DNA damage (Lavin, 2007; Petrini and Stracker, 2003; Williams *et al.*, 2007). The PI3 kinase-like kinases (PIKKs) called ATM (ataxia-telangiectasia mutated), ATR (ATM and Rad3-related) and DNA-dependent protein kinase (DNA-PK) are the central signal transducers for the DNA damage response (Harper and Elledge, 2007; Lavin and Kozlov, 2007; Shiloh, 2001). ATM and DNA-PK control the responses to DSBs, while ATR responds primarily to single-strand DNA damage and stalled replication forks. The MRN complex recruits and activates the ATM kinase at the damage site, which then undergoes autophosphorylation on residue S1981 (Bakkenist and Kastan, 2003; Falck *et al.*, 2005; Lavin, 2007; Lee and Paull, 2004). Activated ATM phosphorylates a large number of substrates that mediate checkpoints and repair (Matsuoka *et al.*, 2007). Among the proteins phosphorylated in the DNA damage cascade are mediators (53BP1 and Mdc1) and effectors (Chk2) of the checkpoint responses. One of the first proteins to be phosphorylated upon DNA damage is the histone variant H2AX (Rogakou *et al.*, 1998), which spreads over an extensive region of chromatin flanking the break (Rogakou *et al.*, 1999) and acts as a signal for recruitment of DNA damage proteins to DSBs (Huen and Chen, 2008). This produces subnuclear structures termed irradiation-induced foci (IRIF) (Fernandez-Capetillo *et al.*, 2003a). The recruitment of factors to damage sites occurs by an ordered assembly process, and various signals for the recruitment and retention of repair proteins to damaged chromatin have now been elucidated (Huen and Chen, 2008). For example, phosphorylation, methylation, sumoylation and ubiquitination of repair factors now appear to play important roles in the formation of IRIF (Huen and Chen, 2008; Panier and Durocher, 2009). ATR is activated during S phase to regulate origin firing and repair of damaged replication forks (Cimprich and Cortez, 2008). Processing and resection at DSBs generates single-stranded DNA (ssDNA) involved in strand-invasion and homologous recombination (Branzei and Foiani, 2008; Mimitou and Symington, 2009). ATR is recruited to ssDNA–dsDNA junctions through association with its cofactor the ATR-interacting-protein (ATRIP) that recognizes RPA-coated ssDNA (Cimprich and Cortez, 2008). ATR activation leads to phosphorylation of a number of targets including Chk1 for checkpoint activation and RPA32, Smc1, and Rad9 to mediate repair (Cimprich and Cortez, 2008). Many targets of the DNA-PKcs kinase activity have been identified *in vitro*, although biological relevance is unclear in many cases. There is tremendous cross-talk between the PIKKs, and together they form a complex regulatory network that coordinates NHEJ, HR, and the checkpoint pathways (Branzei and Foiani, 2008; Shrivastav *et al.*, 2008).

ND10/PODs and the cellular DNA damage response

The nuclear domains known as ND10, promyelocytic leukaemia nuclear bodies (PML-NB) or PML oncogenic domains (PODs), are dynamic nuclear bodies ranging in size between 0.2 and 1 mm present in most mammalian cells. ND10 have been implicated in various cellular functions including regulation of gene expression, senescence, the interferon response, apoptosis, tumour suppression, alternative telomere lengthening and DNA damage responses (reviewed in Bernardi and Pandolfi, 2007; Borden and Culjkovic, 2009). ND10 have also been implicated in cellular responses to DNA damage and stress (Dellaire and Bazett-Jones, 2004). A number of proteins associated with DNA repair and checkpoint signalling are localized within ND10 even in the absence of DNA damage, such as RPA, ATR, Chk2, MRN and BLM (Barr *et al.*, 2003; Bischof *et al.*, 2001; Mirzoeva and Petrini, 2001). In response to IR, components of the ND10 including PML are recruited to sites of DSBs (Carbone *et al.*, 2002; Dellaire *et al.*, 2006; Xu *et al.*, 2003), and PML has been reported as a target for the checkpoint kinases Chk2 and ATR (Bernardi and Pandolfi, 2003; Yang *et al.*, 2002). These and other studies have suggested that ND10 may play a role in the recognition of DNA damage and the recruitment of cellular repair proteins to sites of damage (Boe *et al.*, 2006; Carbone *et al.*, 2002; Dellaire and Bazett-Jones, 2004, 2007; Dellaire *et al.*, 2006).

ND10/PODs and the response to virus infection

ND10 have also been associated with the genomes of several DNA viruses, and they have been suggested to play an intrinsic antiviral defence role (reviewed in Everett and Chelbi-Alix, 2007; Tavalai and Stamminger, 2008) and discussed in detail in Chapters 4 and 13. The incoming viral genomes and early replication compartments are associated with ND10 for several DNA viruses including HSV-1 (Ishov and Maul, 1996; Maul *et al.*, 1996). Many viruses also encode early proteins that initially co-localize with ND10 and then redistribute components of these structures (reviewed in Everett, 2006; Tavalai and Stamminger, 2008). This may be required to counteract repression of viral gene expression by ND10 proteins. Repressive components of the ND10, such as PML, Sp100 and hDaxx, are recruited to sites adjacent to HSV-1 viral genomes at the early stages of virus infection (Everett and Murray, 2005; Everett *et al.*, 2007; Everett *et al.*, 2004; Everett *et al.*, 2003). Although the molecular nature of the trigger for this recruitment is not clear, it does not require viral protein expression and may therefore reflect the ability of PML and other ND10 proteins to recognize some property of the incoming genome itself. It is possible that cellular factors associated with the DNA damage response are involved in sensing these incoming genomes. In support of this idea, we have observed phosphorylation of H2AX at sites adjacent to incoming HSV-1 viral genomes in the nucleus detected at the edge of an emerging plaque (MDW, unpublished observation). In the absence of ICP0, DNA repair proteins such as 53BP1 are also rapidly relocated in response to virus infection (MDW, unpublished observation). This suggests DNA repair proteins are part of an intrinsic host defence activated at early times of infection.

Emerging observations suggest that the intrinsic antiviral defences lead to silencing of viral genomes, and that both ND10 factors and DNA damage repair pathways are involved. HSV-1 has evolved a mechanism to counteract both of these intrinsic antiviral responses through ICP0-mediated targeting of key cellular regulatory proteins (see Chapter 4). This includes degradation of PML and Sp100 in the case of ND10 disruption (Chelbi-Alix and de The, 1999; Everett *et al.*, 1998), and degradation of RNF8 and RNF168 to block the DNA damage response (Lilley *et al.*, 2010). Interestingly, several reports have provided evidence for a link between transcriptional silencing and the recruitment of DNA damage proteins. For instance, during meiotic sex chromosome inactivation (MSCI), the X and Y chromosomes are transcriptionally silenced by a mechanism that requires the DNA-repair protein ATR (Turner, 2007; Turner *et al.*, 2005). Furthermore, ATR has been shown to colocalize with γH2AX on regions of heterochromatin or 'silenced' chromosomes (Fernandez-Capetillo *et al.*, 2003b; Turner, 2007; Turner *et al.*, 2005). The recent observations of Lilley *et al.* (Lilley *et al.*, 2010) suggest that manipulation of the DNA damage response by ICP0 (as described below) counteracts silencing of viral genomes by DNA repair proteins. This may be important during the early stages of lytic infection and/or for efficient escape from latency.

Viral replication compartments and recruitment of cellular proteins

HSV-1 infection induces the formation of large globular replication compartments (RCs) in the nuclei of infected cells. Viral gene expression, DNA replication, capsid assembly and genome encapsidation are thought to occur within these domains (de Bruyn Kops and Knipe, 1988; Lamberti and Weller, 1996; Quinlan *et al.*, 1984). Replication compartments form via a series of intermediate structures which require the ordered assembly of viral and cellular proteins (Burkham *et al.*, 2001; Carrington-Lawrence and Weller, 2003; Liptak *et al.*, 1996; Livingston *et al.*, 2008; Lukonis and Weller, 1997; Taylor *et al.*, 2003). Replication compartments can be detected with antibodies to viral DNA replication proteins such as ICP8 and the polymerase (Bush *et al.*, 1991; Quinlan *et al.*, 1984), as well as cellular factors such as PML (Burkham *et al.*, 2001).

Several cellular proteins with functions in replication and repair have been reported to associate with HSV replication compartments. These have been observed by co-localization in immunofluorescence experiments (Gregory and Bachenheimer, 2008; Lilley *et al.*, 2005; Taylor and Knipe, 2004; Wilcock and Lane, 1991; Wilkinson and Weller, 2004) and also identified by proteomic analysis of proteins associated with the viral ICP8 protein (Taylor and Knipe, 2004). Proteins shown to co-localize with replication compartments include ATM, Mre11, Rad50, Nbs1, RPA, MSH2, BRCA1, BLM, WRN, DNA ligase I, RAD51 and Ku86 (Gregory and Bachenheimer, 2008; Lilley *et al.*, 2005; Taylor and Knipe, 2004; Wilcock and Lane, 1991; Wilkinson and Weller, 2004). Recent work from our laboratory suggests that ATRIP is also recruited to replication compartments (Mohni and Weller, unpublished results).

Activation of ATM-dependent damage signalling during HSV-1 infection

HSV-1 and HSV-2 infections activate the ATM signalling pathway. This has been demonstrated by immunoblotting with an antibody generated to ATM auto-phosphorylated at residue S1981, a marker of ATM activation (Gregory and Bachenheimer, 2008; Lilley *et al.*, 2005; Shirata *et al.*, 2005). Downstream signalling has also been observed using phosphospecific antibodies or detected as a shift in the migration of substrates including Nbs1, Chk2, 53BP1, and p53 (Gregory and Bachenheimer, 2008; Lilley *et al.*, 2005; Shirata *et al.*, 2005; Takaoka *et al.*, 2003; Wilkinson and Weller, 2004). These phosphorylation events are dependent upon the ATM kinase and a functional MRN complex (Lilley *et al.*, 2005; Shirata *et al.*, 2005). Proteins not known to be involved in DNA repair may also be phosphorylated in an ATM-dependent fashion, such as the transcription factor Sp1 (Iwahori *et al.*, 2007). Many other proteins are substrates for ATM in response to DNA damage (Matsuoka *et al.*, 2007) and it remains to be seen how many of these are also phosphorylated during virus infection. The damage marker γH2AX accumulates in areas surrounding viral replication compartments, which may represent breaks in the cellular chromatin that is peripherally displaced during infection (Wilkinson and Weller, 2006). It has also been suggested that expression of ICP0 alone in the absence of virus is sufficient to activate some DNA damage signalling (Li *et al.*, 2008), although other studies suggest that ICP0 is not sufficient to activate ATM/MRN signalling (Lilley *et al.*, 2005; Lilley *et al.*, 2010; Wilkinson and Weller, 2004; Wilkinson and Weller, 2006).

Manipulating the cellular damage response

Although it is clear that HSV-1 replication activates features of the cellular DNA damage pathway, it is also evident that there is selective inhibition of other facets of the cellular response. For example, ICP0 induces proteasome-mediated degradation of the catalytic subunit of DNA-PK at least in some cell types, and virus replication is more efficient in cells lacking DNA-PKcs, especially at low multiplicities of infection (Lees-Miller *et al.*, 1996; Parkinson *et al.*, 1999). Together, these observations imply that DNA-PK may be an antiviral factor that the virus degrades in order to promote lytic infection. There is also a report that loss of Mre11 occurs at late times of infection in human foreskin keratinocytes and human embryonic fibroblasts (Gregory and Bachenheimer, 2008). It is unlikely that Mre11 is antiviral, however, since it appears to benefit virus growth (Lilley *et al.*, 2005). The ability of the virus to manipulate the levels of DNA-PKcs and Mre11 appear to be somewhat cell type specific. Additional experiments will be required to determine the consequences of these manipulations in protein levels.

The activation of the ATM kinase by HSV infection would normally be expected to activate the ATR kinase; however, two proteins that are normally targeted by ATR, Chk1 and RPA32, are not phosphorylated during wild-type HSV infection (Lilley *et al.*, 2005; Shirata *et al.*, 2005; Wilkinson and Weller, 2004, 2005). Although it was originally suggested that the lack of activation was a result of the spatial uncoupling of ATR from its cofactor ATRIP (Wilkinson and Weller, 2006), it now appears that the interaction between ATR and ATRIP is not disrupted by viral infection and that ATR and ATRIP are recruited to viral replication compartments and are required for efficient HSV replication (Mohni and Weller, unpublished results). It is intriguing that, although ATR and ATRIP may play a beneficial role in infection, HSV has apparently evolved a mechanism that prevents phosphorylation of at least two downstream ATR targets.

In addition to phosphorylation, ubiquitination of H2A and H2AX is emerging as a central coordinator of the DNA damage response (Huen and Chen, 2008; Panier and Durocher, 2009). We discovered that ICP0 induced the proteasome-mediated degradation of two cellular ubiquitin ligases, RNF8 and RNF168 (Lilley *et al.*, 2010). These cellular ligases normally serve to maintain the ubiquitinated mark on H2A and H2AX and thus mediate the accumulation of downstream repair proteins at the sites of double-strand breaks (Doil *et al.*, 2009; Kolas *et al.*, 2007; Mailand *et al.*, 2007; Stewart *et al.*, 2009). ICP0 expression therefore leads to a decrease in ubiquitination of the cellular histones and prevents the retention of downstream DNA repair proteins at sites of cellular damage (Lilley *et al.*, 2010).

Consequences for virus and host cell

Herpesviruses manipulate cellular DNA damage response pathways in order to create an environment conducive to their own replication. This is likely to involve a combination of selective exploitation and inactivation of cellular factors.

Activation of the ATM-dependent damage response may be beneficial to viral growth. Mutant cell lines and drug treatments demonstrate that viral DNA replication is reduced in conditions where there is no ATM signalling (Lilley *et al.*, 2005). HSV infection causes the accumulation of some cellular repair factors to viral compartments (e.g. RPA, WRN, MRN, Rad51, and ATM), and it is possible that these proteins play direct or indirect roles in viral DNA replication. For example, virus growth is reduced in cells lacking functional Mre11 and ATM, and there is an extended eclipse phase in the absence of these cellular proteins (Lilley *et al.*, 2005).

Conversely, the virus appears to inhibit components of the DNA damage response, such as degradation of DNA-PKcs. As discussed above, the role of NHEJ and end-joining in the HSV lifecycle is still unclear. Evidence for and against the circularization of viral genomes during infection has been reported (Garber *et al.*, 1993; Jackson and DeLuca, 2003; Poffenberger and Roizman, 1985). It has also been suggested that the cellular NHEJ pathway inhibits herpesvirus replication, as evidenced by increased yields in cells deficient for Ku70 (Taylor and Knipe, 2004), and DNA-PKcs (Parkinson *et al.*, 1999). In contrast to these observations, RNAi-mediated knockdown of cellular DNA ligases suggested that DNA ligase IV and its co-factor XRCC4 may be required for efficient HSV-1 replication (Muylaert and Elias, 2007). Further work is required to define the requirements for end-joining and its role in replication.

As described in detail in Chapters 4 and 13, ICP0 plays an important function in counteracting intrinsic host defences to virus infection and abrogating silencing. This may be achieved by degradation of cellular factors implicated in ND10 and/or the DNA damage response. Mutant viruses unable to express ICP0 are transcriptionally silenced until reactivation by superinfection with wild-type HSV or by providing ICP0 *in trans* (Hobbs *et al.*, 2001; Preston, 2007; Preston and Nicholl, 1997; Samaniego *et al.*, 1998). The mechanisms underlying repression and ICP0-induced reactivation of quiescent viral genomes are not fully understood (reviewed in Efstathiou and Preston, 2005; Knipe and Cliffe, 2008). We discovered that ICP0 induces degradation of histone ubiquitin ligases RNF8 and RNF168, and that this results in a striking loss of ubiquitinated forms of H2A and H2AX (Lilley *et al.*, 2010). In addition to its role in DNA damage, ubiquitinated H2A is also known to be involved in transcriptional repression (Zhou *et al.*, 2009). It is implicated in silencing the inactive X chromosome (Xi) in mice (Fang *et al.*, 2004) and the promoters of polycomb-target genes such as Hox (Wang *et al.*, 2004). Therefore, targeting of cellular H2A ubiquitin ligases provides a potential functional link between the role of ICP0 in the viral lifecycle and the cellular mechanisms of transcriptional silencing. We have speculated that loss of repressive ubiquitinated H2A marks on viral DNA may play a role in promoting the initial stages of lytic infection and in achieving efficient reactivation from latency (Lilley *et al.*, 2010). This hypothesis would predict that growth and silencing of ICP0 deficient viruses would be altered in cells that lack RNF8 and/or RNF168. This is supported by the observation that the plaque-forming efficiency of an ICP0-null virus was enhanced in RNF8 deficient cells (Lilley *et al.*, 2010), suggesting that degradation of this cellular ligase is beneficial for HSV-1 replication. It will be interesting to examine silencing of genes during quiescent viral infections in cells that lack aspects of the damage signalling pathway.

A predicted cellular consequence of manipulation of the DNA damage response and repair would be increased sensitivity of ICP0-expressing cells to DNA damaging agents. Indeed, enhanced radiosensitivity has been reported for glioblastoma multiforme cell lines expressing ICP0 (Hadjipanayis and DeLuca, 2005). Upon irradiation cells expressing ICP0 showed persistent γH2AX foci, increased apoptosis and decreased survival. It is difficult to say whether this results from degradation of DNA-PKcs, RNF8 and RNF168, or additional unidentified substrates. Inactivation of DNA repair could contribute to the enhanced tumouricidal effect observed when oncolytic HSV is combined with ionizing radiation and other DNA-damaging agents (Jiang *et al.*, 2006).

Interactions of other herpesviruses with the DNA damage response

Human cytomegalovirus (HCMV) is a representative betaherpesvirus that like HSV-1 activates aspects of the cellular DNA damage response. HCMV manipulates the cell cycle to create an environment suitable for viral DNA replication without competition with cell replication (Kalejta and Shenk, 2002). Aspects of the cellular DNA response are induced by HCMV infection and there is evidence of ATM and ATR activation (Gaspar and Shenk, 2006; Luo *et al.*, 2007). Although some of the expected substrates are phosphorylated, there is incomplete activation of the full pathway and there is a block to downstream events (Gaspar and Shenk, 2006; Luo *et al.*, 2007). Expression of IE1–72 is sufficient to induce ATM signalling, possibly through its perturbation of normal cell cycle progression (Castillo *et al.*, 2005). Precisely how HCMV subverts parts of the damage signalling pathways is unclear but it has been proposed that there is selective recruitment of cellular factors at viral replication centres (Luo *et al.*, 2007), while another report suggested mislocalization of cellular checkpoint proteins to the cytoplasm (Gaspar and Shenk, 2006). The impact of DNA damage and repair proteins on the virus has yet to be completely demonstrated, but it may contribute to efficient virus replication (T. Kowalik, personal communication). One possibility is that ATM signalling contributes towards the pathology of HCMV infection, for example in endothelial injury (Shen *et al.*, 2004). In addition to its affects on DSB damage signalling, HCMV infection can also modulate other DNA repair pathways such as base excision repair (Ranneberg-Nilsen *et al.*, 2006), which could promote genetic variability in progeny genomes.

Herpesviruses that lead to tumourigenesis have been linked to DNA damage pathways. Epstein–Barr virus (EBV) undergoes productive replication in B cells and epithelial cells, and is maintained as an extra-chromosomal episome. Lytic infection activates ATM signalling and phosphorylation of downstream substrates such as Chk2, p53 and H2AX (Kudoh *et al.*, 2005). MRN and recombination factors are recruited to EBV compartments and are loaded onto the newly synthesized viral genome (Kudoh *et al.*, 2005; Kudoh *et al.*, 2009). These observations suggest that recombination could aid viral genome DNA synthesis during the lytic phase. EBV has also been implicated in the pathogenesis of human malignancies. Multiple EBV proteins have been shown to induce DNA damage, while at the same time inactivating cell cycle checkpoints (Gruhne *et al.*, 2009b). The replication protein EBNA1 can induce DSBs, damage signalling and genomic instability, possibly via activation of reactive oxygen species (Gruhne *et al.*, 2009a). The Kaposi sarcoma herpesvirus (KSHV) expresses a number of proteins with interactions with DNA damage pathways. The v-cyclin protein induced senescence and a robust damage response, which may contribute to malignant transformation (Koopal *et al.*, 2007). In addition, KSHV encodes a viral interferon regulatory factor (vIRF1) that was shown to interact directly with ATM kinase and inhibit its activity (Shin *et al.*, 2006). The murine gammaherpesvirus 68 (γHV68) encodes a kinase protein orf36 that is sufficient to induce γH2AX phosphorylation (Tarakanova *et al.*, 2007)

Comparisons with other viral systems

The cellular DNA repair machinery is manipulated in a variety of ways by many viruses (Lilley *et al.*, 2007). Different strategies are employed to exploit or inactivate the cellular pathways. For example, in the case of Adenovirus the wild-type virus prevents activation of damage signalling through mislocalization and degradation of the MRN sensor, ensuring that it is kept away from viral replication centres (Stracker *et al.*, 2002). Activation of ATM and ATR is observed only with mutant Adenoviruses that cannot inactivate the MRN complex (Carson *et al.*, 2003). There may be commonalities between viruses in the need to inhibit aspects of the cellular signalling response.

It is interesting to note that in addition to HSV-1, a number of other viruses inactivate cellular factors involve in the NHEJ pathway of repair, presumably to prevent processing and end-joining of viral genomes. For example, Adenoviral proteins inactivate NHEJ through degradation and mislocalization of the MRN complex (Stracker *et al.*, 2002), and binding to DNA-PKcs (Boyer *et al.*, 1999), dissociation of the DNA ligase IV/XRCC4 complex (Jayaram *et al.*, 2008) and degradation of DNA ligase IV (Baker *et al.*, 2007). In contrast, other viruses may utilize the NHEJ machinery. For example, adeno-associated virus (AAV) replication leads to activation of a DNA damage response in which DNA-PKcs is the primary kinase (Schwartz *et al.*, 2009) and utilizes NHEJ for genome integration (Daya *et al.*, 2009).

Summary and future directions

Clearly further investigation is required to unravel the complex relationship between the cellular DNA damage network and HSV-1 recombination and replication in order to understand how signalling and repair pathways contribute towards creating an intracellular environment conducive to viral infection. An ordered series of steps is required to initiate replication and recruit protein complexes to the viral origin. DNA recombination and repair have intrinsic roles in HSV-1 DNA replication, and may be required to generate viral replication intermediates. Repair mechanisms may contribute positively to viral DNA replication by recruiting the cellular proteins involved in homologous recombination. HSV-1 infection, thus, activates and exploits components of the cellular DNA damage signalling cascades. On the other hand, HSV-1 infection appears to inhibit other components of damage signalling pathways. HSV infection preferentially activates the ATM kinase but then inhibits the formation of damage foci on cellular DNA. HSV also inactivates the other damage signalling kinases: ATR through an unknown mechanism and DNA-PKcs through degradation. The DNA damage response may need to be modulated because it contributes to host intrinsic antiviral defences. Damage signalling could lead to silencing of viral genomes or other negative consequences of activation, such as establishment of cell cycle checkpoints or induction of apoptosis. Inhibition of repair pathways would therefore serve to counteract these intrinsically antiviral roles. The ICP0 protein plays an important role in counteracting cellular resistance to virus infection (see Chapters 4 and 13). Identifying the cellular targets of ICP0 will pinpoint key regulatory factors of fundamental cellular processes and will reveal how the virus overcomes intracellular barriers. Some of the most important questions remaining to be deciphered in our understanding of HSV and the damage response are: (i) what is the trigger for the DNA damage response activated during virus infection, (ii) how does recruitment of cellular repair and recombination factors regulate virus replication, (iii) what is the mechanism for recombination and the contribution of viral/cellular proteins, and (iv) how do host repair pathways limit virus infection. The answers to these questions will go a long way towards understanding the mechanism of viral DNA replication, the role of cellular proteins, and the complex interactions that determine the outcome of the virus–host interaction.

Acknowledgements

This article is an overview of HSV-1 and DNA repair interactions, and is not intended to be an exhaustive review of the topic. We apologize for primary work not covered here either from lack of space or inadvertent oversight. We thank members of the Weller and Weitzman labs for helpful discussions and comments on the manuscript. Work on DNA repair in the Weitzman lab has been supported by an NIH grant (CA097093) and a Pioneer Developmental Chair from the Salk Institute. Work on DNA replication and recombination in the Weller lab has been supported by NIH grants (AI21747 and AI069136).

References

Baker, A., Rohleder, K.J., Hanakahi, L.A., and Ketner, G. (2007). Adenovirus E4, 34k and E1b 55k oncoproteins target host DNA ligase IV for proteasomal degradation. J. Virol. *81*, 7034–7040.

Bakkenist, C.J., and Kastan, M.B. (2003). DNA damage activates ATM through intermolecular autophosphorylation and dimer dissociation. Nature *421*, 499–506.

Barr, S.M., Leung, C.G., Chang, E.E., and Cimprich, K.A. (2003). ATR kinase activity regulates the intranuclear translocation of ATR and RPA following ionizing radiation. Curr. Biol. *13*, 1047–1051.

Ben-Porat, T., and Rixon, F.J. (1979). Replication of herpesvirus DNA. IV: analysis of concatemers. Virology *94*, 61–70.

Bernardi, R., and Pandolfi, P.P. (2003). Role of PML and the PML-nuclear body in the control of programmed. cell death. Oncogene *22*, 9048–9057.

Bernardi, R., and Pandolfi, P.P. (2007). Structure, dynamics and functions of promyelocytic leukaemia nuclear bodies. Nat. Rev. Microbiol. 8, 1006–1016.

Bischof, O., Kim, S.H., Irving, J., Beresten, S., Ellis, N.A., and Campisi, J. (2001). Regulation and localization of the Bloom syndrome protein in response to DNA damage. J. Cell Biol. *153*, 367–380.

Boe, S.O., Haave, M., Jul-Larsen, A., Grudic, A., Bjerkvig, R., and Lonning, P.E. (2006). Promyelocytic leukemia nuclear bodies are predetermined processing sites for damaged DNA. J. Cell Sci. *119*, 3284–3295.

Boehmer, P.E., and Lehman, I.R. (1997). Herpes simplex virus DNA replication. Annu. Rev. Biochem. *66*, 347–384.

Borden, K.L., and Culjkovic, B. (2009). Perspectives in PML: a unifying framework for PML function. Front Biosci. 14, 497–509.

Boyer, J., Rohleder, K., and Ketner, G. (1999). Adenovirus E4, 34k and E4, 11k inhibit double strand break repair and are physically associated with the cellular DNA-dependent protein kinase. Virology *263*, 307–312.

Branzei, D., and Foiani, M. (2008). Regulation of DNA repair throughout the cell cycle. Nat. Rev. Microbiol. 9, 297–308.

Burkham, J., Coen, D.M., Hwang, C.B., and Weller, S.K. (2001). Interactions of herpes simplex virus type 1 with ND10 and recruitment of PML to replication compartments. J. Virol. *75*, 2353–2367.

Bush, M., Yager, D.R., Gao, M., Weisshart, K., Marcy, A.I., Coen, D.M., and Knipe, D.M. (1991). Correct intranuclear localization of herpes simplex virus DNA polymerase requires. the viral ICP8 DNA-binding protein. J. Virol. *65*, 1082–1089.

Carbone, R., Pearson, M., Minucci, S., and Pelicci, P.G. (2002). PML NBs associate with the hMre11 complex and p53 at sites of irradiation induced DNA damage. Oncogene *21*, 1633–1640.

Carrington-Lawrence, S.D., and Weller, S.K. (2003). Recruitment of polymerase to herpes simplex virus type 1 replication foci in cells expressing mutant primase (UL52) proteins. J. Virol. *77*, 4237–4247.

Carson, C.T., Schwartz, R.A., Stracker, T.H., Lilley, C.E., Lee, D.V., and Weitzman, M.D. (2003). The Mre11 complex is required for ATM activation and the G2/M checkpoint. EMBO J. *22*, 6610–6620.

Castillo, J.P., Frame, F.M., Rogoff, H.A., Pickering, M.T., Yurochko, A.D., and Kowalik, T.F. (2005). Human cytomegalovirus IE1–72 activates ataxia telangiectasia mutated kinase and a p53/p21-mediated growth arrest response. J. Virol. *79*, 11467–11475.

Chelbi-Alix, M.K., and de The, H. (1999). Herpes virus induced proteasome-dependent degradation of the nuclear bodies-associated PML and Sp100 proteins. Oncogene *18*, 935–941.

Cimprich, K.A., and Cortez, D. (2008). ATR: an essential regulator of genome integrity. Nat. Rev. Microbiol. 9, 616–627.

Daya, S., Cortez, N., and Berns, K.I. (2009). Adeno-associated virus site-specific integration is mediated by proteins of the nonhomologous end-joining pathway. J. Virol. *83*, 11655–11664.

de Bruyn Kops, A., and Knipe, D.M. (1988). Formation of DNA replication structures. in herpes virus-infected cells requires. a viral DNA binding protein. Cell *55*, 857–868.

Dellaire, G., and Bazett-Jones, D.P. (2004). PML nuclear bodies: dynamic sensors of DNA damage and cellular stress. Bioessays *26*, 963–977.

Dellaire, G., and Bazett-Jones, D.P. (2007). Beyond repair foci: subnuclear domains and the cellular response to DNA damage. Cell Cycle *6*, 1864–1872.

Dellaire, G., Ching, R.W., Ahmed, K., Jalali, F., Tse, K.C., Bristow, R.G., and Bazett-Jones, D.P. (2006). Promyelocytic leukemia nuclear bodies behave as DNA damage sensors whose response to DNA double-strand breaks is regulated by NBS1 and the kinases ATM, Chk2, and ATR. J. Cell Biol. *175*, 55–66.

Doil, C., Mailand, N., Bekker-Jensen, S., Menard, P., Larsen, D.H., Pepperkok, R., Ellenberg, J., Panier, S., Durocher, D., Bartek, J., *et al.* (2009). RNF168 binds and amplifies ubiquitin conjugates on damaged chromosomes to allow accumulation of repair proteins. Cell *136*, 435–446.

Efstathiou, S., and Preston, C.M. (2005). Towards an understanding of the molecular basis of herpes simplex virus latency. Virus Res. *111*, 108–119.

Everett, R.D. (2006). Interactions between DNA viruses, ND10 and the DNA damage response. Cell. Microbiol. *8*, 365–374.

Everett, R.D., and Chelbi-Alix, M.K. (2007). PML and PML nuclear bodies: implications in antiviral defence. Biochimie *89*, 819–830.

Everett, R.D., Freemont, P., Saitoh, H., Dasso, M., Orr, A., Kathoria, M., and Parkinson, J. (1998). The disruption of ND10 during herpes simplex virus infection correlates with the Vmw110- and proteasome-dependent loss of several PML isoforms. J. Virol. *72*, 6581–6591.

Everett, R.D., and Murray, J. (2005). ND10 components relocate to sites associated with herpes simplex virus type 1 nucleoprotein complexes during virus infection. J. Virol. *79*, 5078–5089.

Everett, R.D., Murray, J., Orr, A., and Preston, C.M. (2007). Herpes simplex virus type 1 genomes are associated with ND10 nuclear substructures. in quiescently infected human fibroblasts. J. Virol. *81*, 10991–11004.

Everett, R.D., Sourvinos, G., Leiper, C., Clements, J.B., and Orr, A. (2004). Formation of nuclear foci of the herpes simplex virus type 1 regulatory protein ICP4 at early times of infection: localization, dynamics, recruitment of ICP27, and evidence for the de novo induction of ND10-like complexes. J. Virol. *78*, 1903–1917.

Everett, R.D., Sourvinos, G., and Orr, A. (2003). Recruitment of herpes simplex virus type 1 transcriptional regulatory protein ICP4 into foci juxtaposed to ND10 in live, infected cells. J. Virol. *77*, 3680–3689.

Falck, J., Coates, J., and Jackson, S.P. (2005). Conserved modes of recruitment of ATM, ATR and DNA-PKcs to sites of DNA damage. Nature *434*, 605–611.

Fang, J., Chen, T., Chadwick, B., Li, E., and Zhang, Y. (2004). Ring1b-mediated H2A ubiquitination associates with inactive X chromosomes and is involved in initiation of X inactivation. J. Biol. Chem. *279*, 52812–52815.

Fernandez-Capetillo, O., Celeste, A., and Nussenzweig, A. (2003a). Focusing on foci: H2AX and the recruitment of DNA-damage response factors. Cell Cycle *2*, 426–427.

Fernandez-Capetillo, O., Mahadevaiah, S.K., Celeste, A., Romanienko, P.J., Camerini-Otero, R.D., Bonner, W.M., Manova, K., Burgoyne, P., and Nussenzweig, A. (2003b). H2AX is required for chromatin remodeling and inactivation of sex chromosomes in male mouse meiosis. Dev. Cell *4*, 497–508.

Garber, D.A., Beverley, S.M., and Coen, D.M. (1993). Demonstration of circularization of herpes simplex virus DNA following infection using pulsed field gel electrophoresis. Virology *197*, 459–462.

Gaspar, M., and Shenk, T. (2006). Human cytomegalovirus inhibits a DNA damage response by mislocalizing checkpoint proteins. Proc. Natl. Acad. Sci. U.S.A. *103*, 2821–2826.

Gregory, D.A., and Bachenheimer, S.L. (2008). Characterization of mre11 loss following HSV-1 infection. Virology *373*, 124–136.

Gruhne, B., Sompallae, R., Marescotti, D., Kamranvar, S.A., Gastaldello, S., and Masucci, M.G. (2009a). The Epstein–Barr virus nuclear antigen-1 promotes genomic instability via induction of reactive oxygen species. Proc. Natl. Acad. Sci. U.S.A. *106*, 2313–2318.

Gruhne, B., Sompallae, R., and Masucci, M.G. (2009b). Three Epstein–Barr virus latency proteins independently promote genomic instability by inducing DNA damage, inhibiting DNA repair and inactivating cell cycle checkpoints. Oncogene.

Hadjipanayis, C.G., and DeLuca, N.A. (2005). Inhibition of DNA repair by a herpes simplex virus vector enhances the radiosensitivity of human glioblastoma cells. Cancer Res. *65*, 5310–5316.

Harper, J.W., and Elledge, S.J. (2007). The DNA damage response: ten years after. Mol. Cell *28*, 739–745.

Hobbs, W.E., Brough, D.E., Kovesdi, I., and DeLuca, N.A. (2001). Efficient activation of viral genomes by levels of herpes simplex virus ICP0 insufficient to affect cellular gene expression or cell survival. J. Virol. *75*, 3391–3403.

Huen, M.S., and Chen, J. (2008). The DNA damage response pathways: at the crossroad of protein modifications. Cell Res. *18*, 8–16.

Ishov, A.M., and Maul, G.G. (1996). The periphery of nuclear domain 10 (ND10) as site of DNA virus deposition. J. Cell Biol. *134*, 815–826.

Iwahori, S., Shirata, N., Kawaguchi, Y., Weller, S.K., Sato, Y., Kudoh, A., Nakayama, S., Isomura, H., and Tsurumi, T. (2007). Enhanced phosphorylation of transcription factor sp1 in response to herpes simplex virus type 1 infection is dependent on the ataxia telangiectasia-mutated protein. J. Virol. *81*, 9653–9664.

Jackson, S.A., and DeLuca, N.A. (2003). Relationship of herpes simplex virus genome configuration to productive and persistent infections. Proc. Natl. Acad. Sci. U.S.A. *100*, 7871–7876.

Jayaram, S., Gilson, T., Ehrlich, E.S., Yu, X.F., Ketner, G., and Hanakahi, L. (2008). E1B 55k-independent dissociation of the DNA ligase IV/XRCC4 complex by E4, 34k during adenovirus infection. Virology *382*, 163–170.

Jiang, H., Alonso, M.M., Gomez-Manzano, C., Piao, Y., and Fueyo, J. (2006). Oncolytic viruses and DNA-repair machinery: overcoming chemoresistance of gliomas. Expert Rev. Anticancer Ther. *6*, 1585–1592.

Kalejta, R.F., and Shenk, T. (2002). Manipulation of the cell cycle by human cytomegalovirus. Front Biosci. *7*, d295–306.

Knipe, D.M., and Cliffe, A. (2008). Chromatin control of herpes simplex virus lytic and latent infection. Nat. Rev. Microbiol. *6*, 211–221.

Kolas, N.K., Chapman, J.R., Nakada, S., Ylanko, J., Chahwan, R., Sweeney, F.D., Panier, S., Mendez, M., Wildenhain, J., Thomson, T.M., *et al.* (2007). Orchestration of the DNA-damage response by the RNF8 ubiquitin ligase. Science *318*, 1637–1640.

Koopal, S., Furuhjelm, J., Jarviluoma, A., Jaamaa, S., Pyakurel, P., Pussinen, C., Wirzenius, M., Biberfeld, P., Alitalo, K., Laiho, M., *et al.* (2007). Viral Oncogene-Induced DNA Damage Response Is Activate in Kaposi Sarcoma Tumorigenesis. PLoS Pathog. 3, e140.

Kudoh, A., Fujita, M., Zhang, L., Shirata, N., Daikoku, T., Sugaya, Y., Isomura, H., Nishiyama, Y., and Tsurumi, T. (2005). Epstein–Barr virus lytic replication elicits ATM checkpoint signal transduction while providing an S-phase-like cellular environment. J. Biol. Chem. *280*, 8156–8163.

Kudoh, A., Iwahori, S., Sato, Y., Nakayama, S., Isomura, H., Murata, T., and Tsurumi, T. (2009). Homologous recombinational repair factors are recruited and loaded onto the viral DNA genome in Epstein–Barr virus replication compartments. J. Virol. *83*, 6641–6651.

Lamberti, C., and Weller, S.K. (1996). The herpes simplex virus type 1 UL6 protein is essential for cleavage and packaging but not for genomic inversion. Virology *226*, 403–407.

Lavin, M.F. (2007). ATM and the Mre11 complex combine to recognize and signal DNA double-strand breaks. Oncogene *26*, 7749–7758.

Lavin, M.F., and Kozlov, S. (2007). ATM activation and DNA damage response. Cell Cycle *6*, 931–942.

Lee, J.H., and Paull, T.T. (2004). Direct activation of the ATM protein kinase by the Mre11/Rad50/Nbs1 complex. Science *304*, 93–96.

Lees-Miller, S.P., Long, M.C., Kilvert, M.A., Lam, V., Rice, S.A., and Spencer, C.A. (1996). Attenuation of DNA-dependent protein kinase activity and its catalytic subunit by the herpes simplex virus type 1 transactivator ICP0. J. Virol. *70*, 7471–7477.

Li, H., Baskaran, R., Krisky, D.M., Bein, K., Grandi, P., Cohen, J.B., and Glorioso, J.C. (2008). Chk2 is required for HSV-1 ICP0-mediated G2/M arrest and enhancement of virus growth. Virology *375*, 13–23.

Lilley, C.E., Carson, C.T., Muotri, A.R., Gage, F.H., and Weitzman, M.D. (2005). DNA repair proteins affect the lifecycle of herpes simplex virus 1. Proc. Natl. Acad. Sci. U.S.A. *102*, 5844–5849.

Lilley, C.E., Chaurushiya, M.S., Boutell, C., Landry, S., Suh, J., Panier, S., Everett, R.D., Stewart, G.S., Durocher, D., and Weitzman, M.D. (2010). A viral E3 ligase targets RNF8 and RNF168 to control histone ubiquitination and DNA damage responses. EMBO J. *29*, 943–955.

Lilley, C.E., Schwartz, R.A., and Weitzman, M.D. (2007). Using or abusing: viruses and the cellular DNA damage response. Trends Microbiol. *15*, 119–126.

Liptak, L.M., Uprichard, S.L., and Knipe, D.M. (1996). Functional order of assembly of herpes simplex virus DNA replication proteins into prereplicative site structures. J. Virol. *70*, 1759–1767.

Livingston, C.M., DeLuca, N.A., Wilkinson, D.E., and Weller, S.K. (2008). Oligomerization of ICP4 and rearrangement of heat shock proteins may be important for herpes simplex virus type 1 prereplicative site formation. J. Virol. *82*, 6324–6336.

Lukonis, C.J., and Weller, S.K. (1997). Formation of herpes simplex virus type 1 replication compartments by transfection: requirements and localization to nuclear domain 10. J. Virol. *71*, 2390–2399.

Luo, M.H., Rosenke, K., Czornak, K., and Fortunato, E.A. (2007). Human cytomegalovirus disrupts both ataxia telangiectasia mutated protein (ATM)- and ATM-Rad3-related kinase-mediated DNA damage responses during lytic infection. J. Virol. *81*, 1934–1950.

Mailand, N., Bekker-Jensen, S., Faustrup, H., Melander, F., Bartek, J., Lukas, C., and Lukas, J. (2007). RNF8 ubiquitylates histones at DNA double-strand breaks and promotes assembly of repair proteins. Cell *131*, 887–900.

Matsuoka, S., Ballif, B.A., Smogorzewska, A., McDonald, E.R., 3rd, Hurov, K.E., Luo, J., Bakalarski, C.E., Zhao, Z., Solimini, N., Lerenthal, Y., *et al.* (2007). ATM and ATR substrate analysis reveals extensive protein networks responsive to DNA damage. Science *316*, 1160–1166.

Maul, G.G., Ishov, A.M., and Everett, R.D. (1996). Nuclear domain 10 as preexisting potential replication start sites of herpes simplex virus type-1. Virology *217*, 67–75.

Mimitou, E.P., and Symington, L.S. (2009). DNA end resection: many nucleases make light work. DNA Repair (Amst) 8, 983–995.

Mirzoeva, O.K., and Petrini, J.H. (2001). DNA damage-dependent nuclear dynamics of the Mre11 complex. Mol. Cell Biol. *21*, 281–288.

Muylaert, I., and Elias, P. (2007). Knockdown of DNA ligase IV/XRCC4 by RNA interference inhibits herpes simplex virus type I DNA replication. J. Biol. Chem. *282*, 10865–10872.

Ojala, P.M., Sodeik, B., Ebersold, M.W., Kutay, U., and Helenius, A. (2000). Herpes simplex virus type 1 entry into host cells: reconstitution of capsid binding and uncoating at the nuclear pore complex *in vitro*. Mol. Cell Biol. *20*, 4922–4931.

Panier, S., and Durocher, D. (2009). Regulatory ubiquitylation in response to DNA double-strand breaks. DNA Repair (Amst) 8, 436–443.

Parkinson, J., Lees-Miller, S.P., and Everett, R.D. (1999). Herpes simplex virus type 1 immediate-early protein vmw110 induces the proteasome-dependent degradation of the catalytic subunit of DNA-dependent protein kinase. J. Virol. *73*, 650–657.

Petrini, J.H., and Stracker, T.H. (2003). The cellular response to DNA double-strand breaks: defining the sensors and mediators. Trends Cell Biol. *13*, 458–462.

Poffenberger, K.L., and Roizman, B. (1985). A noninverting genome of a viable herpes simplex virus 1: presence of head-to-tail linkages in packaged genomes and requirements for circularization after infection. J. Virol. *53*, 587–595.

Preston, C.M. (2007). Reactivation of expression from quiescent herpes simplex virus type 1 genomes in the absence of immediate-early protein ICP0. J. Virol. *81*, 11781–11789.

Preston, C.M., and Nicholl, M.J. (1997). Repression of gene expression upon infection of cells with herpes simplex virus type 1 mutants impaired for immediate-early protein synthesis. J. Virol. *71*, 7807–7813.

Quinlan, M.P., Chen, L.B., and Knipe, D.M. (1984). The intranuclear location of a herpes simplex virus DNA-binding protein is determined by the status of viral DNA replication. Cell *36*, 857–868.

Ranneberg-Nilsen, T., Bjoras, M., Luna, L., Slettebakk, R., Dale, H.A., Seeberg, E., and Rollag, H. (2006). Human cytomegalovirus infection modulates DNA base excision repair in fibroblast cells. Virology *348*, 389–397.

Rogakou, E.P., Boon, C., Redon, C., and Bonner, W.M. (1999). Megabase chromatin domains involved in DNA double-strand breaks *in vivo*. J. Cell Biol. *146*, 905–916.

Rogakou, E.P., Pilch, D.R., Orr, A.H., Ivanova, V.S., and Bonner, W.M. (1998). DNA double-stranded breaks induce histone H2AX phosphorylation on serine 139. J. Biol. Chem. *273*, 5858–5868.

Samaniego, L.A., Neiderhiser, L., and DeLuca, N.A. (1998). Persistence and expression of the herpes simplex virus genome in the absence of immediate-early proteins. J. Virol. *72*, 3307–3320.

Sancar, A., Lindsey-Boltz, L.A., Unsal-Kacmaz, K., and Linn, S. (2004). Molecular mechanisms of mammalian

DNA repair and the DNA damage checkpoints. Annu. Rev. Biochem. *73*, 39–85.

Schwartz, R.A., Carson, C.T., Schuberth, C., and Weitzman, M.D. (2009). Adeno-associated virus replication induces a DNA damage response coordinated by DNA-dependent protein kinase. J. Virol. *83*, 6269–6278.

Severini, A., Morgan, A.R., Tovell, D.R., and Tyrrell, D.L. (1994). Study of the structure of replicative intermediates of HSV-1 DNA by pulsed-field gel electrophoresis. Virology *200*, 428–435.

Shen, Y.H., Utama, B., Wang, J., Raveendran, M., Senthil, D., Waldman, W.J., Belcher, J.D., Vercellotti, G., Martin, D., Mitchelle, B.M., *et al.* (2004). Human cytomegalovirus causes endothelial injury through the ataxia telangiectasia mutant and p53 DNA damage signaling pathways. Circ Res. *94*, 1310–1317.

Shiloh, Y. (2001). ATM and ATR: networking cellular responses to DNA damage. Curr. Opin. Genet Dev. *11*, 71–77.

Shin, Y.C., Nakamura, H., Liang, X., Feng, P., Chang, H., Kowalik, T.F., and Jung, J.U. (2006). Inhibition of the ATM/p53 signal transduction pathway by Kaposi's sarcoma-associated herpesvirus interferon regulatory factor 1. J. Virol. *80*, 2257–2266.

Shirata, N., Kudoh, A., Daikoku, T., Tatsumi, Y., Fujita, M., Kiyono, T., Sugaya, Y., Isomura, H., Ishizaki, K., and Tsurumi, T. (2005). Activation of ataxia telangiectasia-mutated DNA damage checkpoint signal transduction elicited by herpes simplex virus infection. J. Biol. Chem. *280*, 30336–30341.

Shrivastav, M., De Haro, L.P., and Nickoloff, J.A. (2008). Regulation of DNA double-strand break repair pathway choice. Cell Res. *18*, 134–147.

Stewart, G.S., Panier, S., Townsend, K., Al-Hakim, A.K., Kolas, N.K., Miller, E.S., Nakada, S., Ylanko, J., Olivarius, S., Mendez, M., *et al.* (2009). The RIDDLE syndrome protein mediates a ubiquitin-dependent signaling cascade at sites of DNA damage. Cell *136*, 420–434.

Stracker, T.H., Carson, C.T., and Weitzman, M.D. (2002). Adenovirus oncoproteins inactivate the Mre11-Rad50-NBS1 DNA repair complex. Nature *418*, 348–352.

Strang, B.L., and Stow, N.D. (2005). Circularization of the herpes simplex virus type 1 genome upon lytic infection. J. Virol. *79*, 12487–12494.

Takaoka, A., Hayakawa, S., Yanai, H., Stoiber, D., Negishi, H., Kikuchi, H., Sasaki, S., Imai, K., Shibue, T., Honda, K., *et al.* (2003). Integration of interferon-alpha/beta signalling to p53 responses in tumour suppression and antiviral defence. Nature *424*, 516–523.

Tarakanova, V.L., Leung-Pineda, V., Hwang, S., Yang, C.W., Matatall, K., Basson, M., Sun, R., Piwnica-Worms, H., Sleckman, B.P., and Virgin, H.W.t. (2007). Gamma-herpesvirus kinase actively initiates a DNA damage response by inducing phosphorylation of H2AX to foster viral replication. Cell Host Microbe *1*, 275–286.

Tavalai, N., and Stamminger, T. (2008). New insights into the role of the subnuclear structure ND10 for viral infection. Biochim. Biophys. Acta *1783*, 2207–2221.

Taylor, T.J., and Knipe, D.M. (2004). Proteomics of herpes simplex virus replication compartments: association of cellular DNA replication, repair, recombination, and chromatin remodeling proteins with ICP8. J. Virol. *78*, 5856–5866.

Taylor, T.J., McNamee, E.E., Day, C., and Knipe, D.M. (2003). Herpes simplex virus replication compartments can form by coalescence of smaller compartments. Virology *309*, 232–247.

Turner, J.M. (2007). Meiotic sex chromosome inactivation. Development *134*, 1823–1831.

Turner, J.M., Mahadevaiah, S.K., Fernandez-Capetillo, O., Nussenzweig, A., Xu, X., Deng, C.X., and Burgoyne, P.S. (2005). Silencing of unsynapsed meiotic chromosomes in the mouse. Nat. Genet. *37*, 41–47.

Wang, H., Wang, L., Erdjument-Bromage, H., Vidal, M., Tempst, P., Jones, R.S., and Zhang, Y. (2004). Role of histone H2A ubiquitination in Polycomb silencing. Nature *431*, 873–878.

Wilcock, D., and Lane, D.P. (1991). Localization of p53, retinoblastoma and host replication proteins at sites of viral replication in herpes-infected cells. Nature *349*, 429–431.

Wilkinson, D.E., and Weller, S.K. (2003). The role of DNA recombination in herpes simplex virus DNA replication. IUBMB Life *55*, 451–458.

Wilkinson, D.E., and Weller, S.K. (2004). Recruitment of cellular recombination and repair proteins to sites of herpes simplex virus type 1 DNA replication is dependent on the composition of viral proteins within prereplicative sites and correlates with the induction of the DNA damage response. J. Virol. *78*, 4783–4796.

Wilkinson, D.E., and Weller, S.K. (2005). Inhibition of the herpes simplex virus type 1 DNA polymerase induces hyperphosphorylation of replication protein A and its accumulation at S-phase-specific sites of DNA damage during infection. J. Virol. *79*, 7162–7171.

Wilkinson, D.E., and Weller, S.K. (2006). Herpes simplex virus type I disrupts the ATR-dependent DNA-damage response during lytic infection. J. Cell Sci. *119*, 2695–2703.

Williams, R.S., Williams, J.S., and Tainer, J.A. (2007). Mre11-Rad50-Nbs1 is a keystone complex connecting DNA repair machinery, double-strand break signaling, and the chromatin template. Biochem. Cell Biol. *85*, 509–520.

Xu, Z.X., Timanova-Atanasova, A., Zhao, R.X., and Chang, K.S. (2003). PML colocalizes with and stabilizes the DNA damage response protein TopBP1. Mol. Cell Biol. *23*, 4247–4256.

Yang, S., Kuo, C., Bisi, J.E., and Kim, M.K. (2002). PML-dependent apoptosis after DNA damage is regulated by the checkpoint kinase hCds1/Chk2. Nat. Cell Biol. *4*, 865–870.

Zhang, X., Efstathiou, S., and Simmons, A. (1994). Identification of novel herpes simplex virus replicative intermediates by field inversion gel electrophoresis: implications for viral DNA amplification strategies. Virology *202*, 530–539.

Zhou, W., Wang, X., and Rosenfeld, M.G. (2009). Histone H2A ubiquitination in transcriptional regulation and DNA damage repair. Int. J. Biochem. Cell. Biol. *41*, 12–15.

Varicella-zoster Virus Pathogenesis and Latency

16

Leigh Zerboni and Ann M. Arvin

Abstract

Varicella-zoster virus (VZV) is a human alphaherpesvirus and the aetiological agent of varicella during primary infection and herpes zoster upon reactivation from latency. In this chapter we describe clinical observations and experience with VZV disease in the natural human host. We present a refined paradigm for VZV pathogenesis that emphasizes the importance of VZV lymphotropism. We also present key findings in our investigations of VZV pathogenesis and latency using SCID mouse-human tissue xenografts. Experimental infection with over 100 recombinant VZV viruses containing mutations in open reading frames encoding viral glycoproteins, viral kinases, regulatory proteins and as well as promoter elements has revealed tissue-specific requirements for VZV gene motifs and promoter elements for replication in skin, T-lymphocytes and neurons. We have repeatedly observed that replication phenotypes of VZV mutants in cultured cells are not predictive of behaviour within differentiated intact tissues under physiological (*in vivo*) conditions. These observations underscore the importance of *in vivo* investigations in identifying molecular mechanisms of VZV pathogenesis. These experiments provide insight into VZV interactions with human host cells and have potential relevance in the design of 'second generation' recombinant VZV vaccines.

Introduction

Varicella-zoster virus (VZV) is the causative agent of varicella, during primary infection, and herpes zoster upon reactivation from latency. The 125 kilobase VZV genome contains open reading frames (ORFs) encoding at least 70 genes; although the kinetics of VZV gene transcription has not been reported, these genes have been classified as immediate early (IE), early (E) and late (L) genes based on homologues in herpes simplex virus 1 (HSV-1) (Davison and Scott, 1986; Cohen *et al.*, 2007). We have recently reported the kinetics of VZV protein expression using novel methods to identify newly infected cells in culture (Reichelt *et al.*, 2008). VZV was the first human herpesvirus to be fully sequenced and information on the genetic diversity of VZV isolates continues to emerge; VZV is now thought to represent at least five distinct geographic clades (reviewed in Breuer *et al.*, 2010). Two intriguing growth characteristics distinguish VZV from other human alphaherpesviruses, a marked host cell restriction and cell-association of infectious viral progeny (Cohen *et al.*, 2007). VZV also possesses molecular properties that distinguish the virus from other alphaherpesviruses, including lack of an orthologous gene encoding glycoprotein D (Davison and Scott, 1986), the requirement for glycoprotein E for viral replication in cell culture (Mallory *et al.*, 1997; Mo *et al.*, 2002) and the absence of the latency-associated transcript (LAT) found in HSV and related alphaherpesviruses.

In the first part of this chapter, we describe knowledge of VZV disease in the natural human host gained through clinical observation. In the second part of this chapter, we summarize our investigations of VZV pathogenesis through experimental infection of VZV clinical isolates and recombinant viruses in severe compromised immunodeficient (SCID) mice xenotransplanted with human tissue xenografts. Since the development of a cosmid-based mutagenesis system for construction of recombinant VZV mutants (Cohen and Seidel, 1993; Kemble *et al.*, 2000), we have tested over 100 recombinant VZV viruses containing mutations in open reading frames encoding viral glycoproteins, viral kinases, regulatory proteins and promoter elements. Our experiments have revealed tissue-specific virulence determinants within VZV gene motifs and promoter elements. One recurring theme in these investigations has been that *in vitro* replication phenotypes are not predictive of behaviour within differentiated intact tissues within an *in vivo* microenvironment. This observation draws attention to the importance of *in vivo* infection in identifying tissue-specific molecular requirements for VZV pathogenesis. We hope that these pursuits will have relevance in the design of 'second generation' recombinant VZV vaccines and novel therapeutic approaches to the treatment of VZV-related diseases.

Clinical observation and experience with VZV disease in the human host

VZV is highly infectious, with an 80–90% attack rate following close contact exposure; initial exposure almost always results in a clinically apparent primary infection, which is commonly called 'chickenpox' (Cohen *et al.*, 2007). VZV is the only human herpesvirus for which vaccines are licensed and widely distributed and so the incidence of chickenpox has declined dramatically in nations where the vaccine is given as a routine childhood immunization (Arvin, 2001). However, VZV remains a medically important human alphaherpesvirus in that the live, attenuated vaccine is not appropriate for all patient populations, the vaccine virus has the capacity to establish latency and the long-term consequences of vaccination on the incidence of herpes zoster are undetermined.

Primary varicella disease

Whereas the closely related simplex viruses (HSV-1, -2) spread among humans by direct contact with cutaneous lesions (Roizman *et al.*, 2007), primary VZV infection is hypothesized to be initiated by inoculation of pharyngeal epithelial cells upon exposure to aerosolized infected cell droplets (Grose, 1981; Sawyer *et al.*, 1992). Spread from direct contact with skin lesions reinforces the efficiency of VZV transmission. A 10–21 day incubation period followed by early symptoms such as fever, malaise and headache precede onset of an erythematous maculopapular rash. These eruptions rapidly progress to a vesicular exanthem that is the signature characteristic of 'chickenpox' (Fig. 16.1A). Histopathological examination of skin biopsy specimens demonstrates

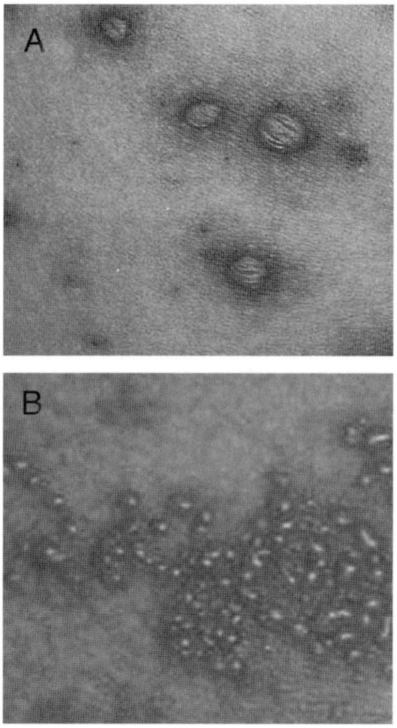

Figure 16.1 Skin lesions associated with varicella and herpes zoster. (A) Photograph of the acute vesicular exanthem that is characteristic of primary varicella infection. Erythematous maculopapular lesions evolve into fluid-filled vesicles containing high titre infectious virus. Lesions will usually crust and heal after 24–48 hours without scarring. (B) Reactivation of latent VZV is associated with an erythematous exanthem corresponding to a cutaneous dermatome, which is innervated by sensory fibres from a single spinal ganglia. Cutaneous lesions form small clusters and may be a source of transmissible virus.

the formation of vesicles and bullae in the superficial epithelium and the presence of cytopathic effects that can extend into the dermis (Miliauskas and Webber, 1984). Epidermal changes include focal necrosis, acantholysis, multinucleation, giant 'ballooning' cells and Cowdry type A intranuclear inclusions (Annunziato *et al.*, 2000; Miliauskas and Webber, 1984).

The initial crop of vesicular skin lesions will crust and heal within 24–48 hours, to be followed by successive waves of lesions over a period of 7–10 days. Ulcerative lesions may also appear on mucous membranes of the upper respiratory tract and genitals. In total, 100–300 lesions typically arise during the course of infection and occur as successive crops; overall lesion number appears to increase with age (Arvin, 2001; Arvin, 2005). These fluid-filled lesions are intensely pruritic and contain high titres of infectious virus, enhancing spread of the contagion to susceptibles with close contact. Differential diagnoses include impetigo, contact dermatitis or other skin infections. Virus isolation by direct culture of vesicular fluid is the definitive method for diagnosing VZV infection, but polymerase chain reaction (PCR) methods are now used more commonly.

In healthy individuals, a robust humoral and cell-mediated immune response will resolve primary infection within 10–14 days without complication (Arvin, 2005). Initial lesions often penetrate more deeply, resulting in characteristic shallow scars, often on the face and along the scalp hairline. Skin previously damaged by acne, impetigo or eczema may also have scarring. Superinfection with group A beta-haemolytic *Streptococcus* or *Staphylococcus aureus* is a rare but serious complication. Older adolescents and adults are more prone to extensive rash, and at higher risk of haemorrhagic or neurological complications, as well as other sequelae such as varicella pneumonia, myocarditis and nephritis (Arvin, 2005). Patients with compromised cellular immunity resulting from T cell immunodeficiency or chemotherapy are at high risk of disseminated varicella, the most severe consequence of which is interstitial pneumonitis resulting in respiratory failure (Feldman *et al.*, 1975).

VZV latency

VZV establishes a lifelong persistence within the cell bodies of sensory nerve ganglia known as latency. VZV is presumed to reach ganglia via retrograde transport of VZ virions along nerve axons enervating skin lesions, although recent studies suggest that circulating VZV-infected lymphocytes may traffic to sensory nerve ganglia during the viremic incubation period providing an alternative route for VZV neurotropism (Asano *et al.*, 1985; Zerboni *et al.*, 2005b; Gershon *et al.*, 2009). It is not known if VZV requires lytic replication in neuronal ganglia to establish latent infection, but infectious VZV has been recovered from ganglia in patients who died of disseminated varicella (Bastian *et al.*, 1974).

Taking HSV as the prototype, alphaherpesviral latency has been described as the persistence of the viral genome in a non-replicating state without productive gene transcription or translation; this quiescent state facilitates lack of recognition by host immune factors. In recent years a more dynamic view of HSV-1 latency has emerged from animal models which indicates that frequent low-level viral gene expression may occur but complete reactivation, i.e. infectious viral progeny transiting axons, is repressed by resident virus-specific T cells through interferon-mediated antiviral responses (reviewed in Khanna *et al.*, 2004). Competent host immune surveillance appears to be critical for control of HSV-1 reactivation (Khanna *et al.*, 2004). Persistent T cell infiltrates have been observed in human trigeminal ganglia co-infected with latent HSV-1 and VZV (Theil *et al.*, 2003).

In the absence of small laboratory animal models, analyses of VZV latency have relied on histological examination of cadaver ganglia obtained at autopsy from patients who died without evidence of clinical VZV reactivation. However, there are significant discrepancies among these reports as described below. This approach is encumbered by the postmortem interval between death and autopsy during which physiological changes occur within cells that may alter the true state of VZV during latency. Variance associated with sampling error is another concern in that autopsies are most routinely performed on patients with a medical cause of death, such as patients who died under intensive care or circumstances in which host T cell immune responses are impaired. Whether as a consequence of these factors, or the use of divergent methods, results using cadaver ganglia have been inconsistent. The detection of incipient or abortive reactivation, as observed in HSV-1 animal models, would explain some of these inconsistencies.

Localization of latent VZV genomes in sensory nerve ganglia

VZV DNA can be readily detected by polymerase chain reaction (PCR) in human sensory ganglia, with the highest prevalence rate in trigeminal ganglia (Gilden *et al.*, 1983, 2001; Inoue *et al.*, 2010; Mahalingam *et al.*, 1992). Recently, Inoue and colleagues examined 414 trigeminal ganglia from 207 cadavers of mixed age and gender (Inoue *et al.*, 2010). VZV DNA was detected in 94.4% of trigeminal ganglia whereas HSV-1 was detected in 60.6%; there were no significant differences in detection of VZV with respect to age or gender (Inoue *et al.*, 2010).

Early attempts to determine the ganglionic cell type harbouring latent VZV DNA employed *in situ* histochemical (ISH) detection methods but results were inconsistent. Whereas one study reported a VZV-specific signal in 3–30% of neurons and 5–30% of satellite cells (Lungu *et al.*, 1995), another study (using PCR amplification prior to ISH detection) detected VZV DNA in 2–5% of neurons and in only a small fraction (<0.1%) of non-neuronal cells (Kennedy *et al.*, 1998). Given that supportive satellite cells completely invest individual ganglionic neurons, reliably discriminating the source of the ISH detection signal is challenging. More recent data employing fractionation methods to separate neuronal and non-neuronal cell populations prior to analysis implicate neurons as the primary cell type in which VZV genomes persist during latency with only rare detection in satellite cells (LaGuardia *et al.*, 1999; Levin *et al.*, 2003). Using fractionation methods, Levin *et al.* (Levin *et al.*, 2003) detected VZV DNA in 1.5% of neurons and none in 20,700 satellite cells.

Prevalence and abundance of latent VZV genomes

Various methods have been employed to quantify the latent viral load (genome copies per ganglion cell) in sensory nerve ganglia. An early study, performed when methods for determining DNA copy number were less precise, reported 6 to 31 VZV genomes per 10^5 unfractionated ganglion cells (Mahalingam *et al.*, 1993). More recent estimates of viral DNA burden using quantitative real time PCR have reported 258 ± 38 VZV genomes per 10^5 unfractionated ganglion cells (Pevenstein *et al.*, 1999) and 577–55,543 VZV genomes per 10^5 unfractionated ganglion cells (Cohrs *et al.*, 2000). Wang *et al.* (2005) employed laser-capture microdissection to separate neuronal and non-neuronal populations followed by quantitative real-time PCR and detected VZV DNA in 1.0–6.9% of neurons per ganglion with ~6.9 genomes per VZV-positive cell, only nine VZV genomes were detected in 14,200 non-neuronal cells (Wang *et al.*, 2005).

VZV gene transcription in latently infected neurons

VZV does not possess an obvious counterpart to the latency-associated transcript (LAT) locus present in many alphaherpesviral genomes (Davison and Scott, 1986). In contrast, transcripts corresponding to a wide array of VZV genes have been reported in cadaver ganglia. Early work using radiolabeled DNA probes detected VZV RNA in a small fraction of neurons in trigeminal ganglia (Hyman *et al.*, 1983; Gilden *et al.*, 1987). A subsequent study using radiolabeled RNA probes and *in situ* autoradiography, detected RNAs corresponding to VZV ORFs 29 and 62, but not 10, 28 or 61; the signal was localized to non-neuronal cells (Meier *et al.*, 1993). Kennedy and colleagues (Kennedy *et al.*, 1999) employed a DIG-labelled probe and detected a VZV-specific RNA signal corresponding to ORF29 and 63 in most specimens, whereas transcripts corresponding to ORFs 4 and 21 were less frequently detected. Cohrs and colleagues screened a cDNA library prepared from cadaver ganglia and detected polyadenylated RNA corresponding to VZV ORFs 21, 29, 62 and 63 but not 4, 10, 51 and 61 (Cohrs *et al.*, 1994, 1996).

More recently, quantitative real-time RT-PCR demonstrated mRNAs for ORFs 63 (19/22 ganglia), 21 (5/22 ganglia) and 29 (3/22 ganglia); ORF63 transcripts were significantly more abundant (>2000-fold) than transcripts corresponding to ORFs 21 and 29 (Cohrs *et al.*, 2000). RNA corresponding to ORF66 has been described in one report (Cohrs *et al.*, 2003). With the possible exception of ORF63, which is the most consistently detected, the collective data indicates that viral transcripts may be expressed in some latently infected ganglia; however, the low abundance and inconsistent detection could be explained by incipient reactivation or transcriptional aberrancies related to postmortem changes, rather than representing a defining characteristic of VZV latency. Further work is needed to clarify the VZV transcriptome in latently infected ganglia.

VZV protein expression in latently infected neurons

Mahalingam and colleagues (Mahalingam *et al.*, 1996) were the first to report detection of VZV protein in cadaver ganglia. IE63 protein was detected in five ganglia from two of nine (22%) seropositive adults using a standard immunohistochemical detection method with a rabbit polyclonal antibody. No IE63 protein was detected in 16 ganglia from seven seropositive adults (Mahalingam *et al.*, 1996). In those ganglia that yielded a positive signal (5 of 21), IE63 protein was observed in 2–4 neurons per section; five sections per ganglia were examined. Kennedy and colleagues (Kennedy *et al.*, 2000), noting the requirement to screen multiple sections from each patient to detect positive staining, reported IE63 protein in trigeminal ganglia from 5 of 20 (25%) patients.

Whereas in the previous reports cited, VZV IE63 protein was observed in a small proportion of the total number of ganglia studied, Lungu and colleagues (Lungu *et al.*, 1998) detected IE63 protein in one ganglia examined (100%) from each of three individuals; the IE63-specific signal was detected in 3–9% of neurons. VZV proteins corresponding to IE62 and ORFs 4, 21 and 29 were also detected in 6–24% of neurons; all VZV proteins were described as having a cytoplasmic localization (Lungu *et al.*, 1998). Subsequently, other groups reported detection of cytoplasmic VZV proteins in cadaver ganglia using immunohistochemical detection methods (Cohrs *et al.*, 2003; Grinfeld and Kennedy, 2004; Hufner *et al.*, 2006; Theil *et al.*, 2003). Grinfeld and Kennedy (Grinfeld and Kennedy, 2004) reported cytoplasmic localization of IE62, IE63, ORF21 and ORF29 proteins in a significant number of neurons (5–10%) in nearly all ganglia examined; IE63 protein was detected in all of 10 ganglia, from 10 individuals (one ganglion examined per individual).

Cytoplasmic restriction of VZV transactivating proteins has been the basis for proposing a mechanism for the maintenance of VZV latency through exclusion of these proteins from the neuronal nucleus. However, a need for further investigation was indicated by discordance in the reported prevalence and abundance of IE63 protein expression, which ranged from a rare event (2–4 neurons per section) in ~25% of ganglia examined (Mahalingam *et al.*, 1996; Kennedy *et al.*, 2000) to abundant expression in 100% of ganglia examined (Lungu *et al.*, 1998; Grinfeld and Kennedy, 2004). Of note, although transcripts corresponding to ORF63 have been detected in cadaver ganglia by different methods, full length ORF63 RNA has not been demonstrated.

We examined IE63 expression in cadaver ganglia by evaluating a large number of sections from 27 ganglia recovered from 18 individuals (Zerboni *et al.*, 2010). Using standard immunohistochemical (IHC) detection methods as previously applied, we determined that reliable discrimination of the chromogen detection signal was confounded by an abundance of cytoplasmic neuronal pigments, in particular neuromelanin (Zerboni *et al.*, 2010). Neuromelanin accumulates in the cytoplasm of catecholaminergic neurons, which comprise a subpopulation of neurons in sensory ganglia (Katz *et al.*, 1983; Price and Mudge, 1983) and has been described in human trigeminal ganglia and dorsal root ganglia (Kishikawa *et al.*, 1988; Sulzer *et al.*, 2008). We demonstrated the presence of neuromelanin using the method of Lillie (Lillie, 1956) to differentiate melanin and lipofuscins. Neuronal pigments pose well-documented technical challenges to IHC detection due to the accumulation of iron within pigment granules that can interfere with IHC oxidation-reduction reactions (Nadji, 1986; Ramos-Vara, 2005). We modified our IHC protocol, adapting a method used in the histopathological examination of melanomas to exclude pigment interference (Kamino and Tam, 1991; Kligora *et al.*, 1999; Ramos-Vara, 2005). We detected IE63 protein in ganglia from only one of 18 patients; expression of glycoprotein E and an immune cell infiltrate was observed in the same ganglion, which suggests that, in this case, IE63 protein expression may indicate abortive or incipient reactivation (Zerboni *et al.*, 2010).

Our observations support the earlier reports by Mahalingam and colleagues (Mahalingam *et al.*, 1996) and Kennedy and colleagues (Kennedy *et al.*, 2000) and suggest that IE63 protein expression, which has been called a latency-associated protein, is not a defining characteristic of latency. The same caveats about neuromelanin interference in IHC detection that we observed in staining for IE63 protein would apply to detection of other VZV proteins. Further investigation to resolve the extent of VZV protein expression during persistence is important for developing a hypothesis about how VZV latency is maintained.

VZV reactivation

Abortive or incomplete reactivation of latent VZV genomes may be limited to an individual neuron and its invested satellite cells, but complete reactivation resulting in newly synthesized infectious particles transiting from the neuronal cell body via anterograde axonal transport is more likely to involve multiple neurons within a single ganglion or a few neighbouring ganglia. Egress of infectious particles from nerve endings to the involved skin tissue results in replication within epidermal cells but the extent of cell–cell spread and rash extension is probably dependent on host cell immune factors (Arvin, 2005; Cohen *et al.*, 2007). The mechanism by which VZV latency is subverted and/or replication of the latent genome is initiated is unknown. VZV cannot be recovered from explanted trigeminal or dorsal root ganglia whereas the same methods are effective at recovery of infectious HSV (Plotkin *et al.*, 1989). This suggests that the 'trigger' required for VZV reactivation in human neurons is different than those requirements of HSV-1 and HSV-2.

Herpes zoster

Herpes zoster, also known as shingles, occurs in a distribution corresponding to the cutaneous dermatome that is innervated by neurons in the ganglion in which VZV is reactivating (Fig. 16.1B). Of seropositive individuals living to 85 years and beyond, 50% will experience a clinical episode of herpes zoster, most commonly along a thoracic dermatome (Cohen *et al.*, 2007). Intense unilateral pain is the typical prodrome of herpes zoster. In the earliest stages, erythematous maculopapular patches appear on the skin surface, which may gradually evolve into vesicular lesions that resemble those of primary varicella infection (Annunziato *et al.*, 2000). As with 'chickenpox', these lesions contain infectious virus and so herpes zoster is source of transmissible virus. Resolution of herpes zoster is associated with an increase of VZV-specific T cell responses (Arvin, 2005). Healthy individuals generally recover fully from herpes zoster although post-herpetic neuralgia can cause significant morbidity; unlike HSV reactivation, clinical recurrence is rare (Arvin, 2005). Motor function can be affected as in cases when herpes zoster involving cranial nerve ganglia cause facial palsies associated with Ramsay Hunt syndrome. Immunocompromised patients are at increased risk for dissemination as well as meningoencephalitis. VZV DNA can be detected in cerebrospinal fluid samples of most herpes zoster patients, as well as increased protein and lymphocytes (Arvin, 2005).

Other VZV-related morbidities

Post-herpetic neuralgia, the most common complication of VZV, is persistence of pain after resolution of the zoster rash. Post-herpetic neuralgia is more common in the elderly and immunocompromised patients. 'Zoster sine herpete' is another condition in which VZV reactivation does not progress to cutaneous lesion formation but is associated with pain and paresis. The aetiology of post-herpetic neuralgia and 'zoster sine herpete' are unknown but may be consequences of ganglionitis or damage to the involved sensory neuron.

Investigation of tissue-specific virulence and VZV pathogenesis using the SCID mouse–human xenograft model

Despite the wealth of clinical observation, critical events in VZV pathogenesis such as the state of the VZV genome during latency and the mechanism by which reactivation is initiated, remain unresolved. Unlike other alphaherpesviruses that are capable of replicating within common laboratory animal species, VZV exhibits a marked host cell restriction such that critical aspects of disease are not recapitulated upon infection of non-human primates (Cohen *et al.*, 2007). The lack of a small permissive animal model and the perplexing highly cell-associated nature of the virus in cell culture have been formidable obstacles to investigation of VZV pathogenesis. These have, in part, been overcome by the capacity to establish xenografts of human tissues relevant to VZV pathogenesis in mice with severe combined immunodeficiency (SCID) (Fig. 16.2) and recombinant DNA techniques which permit construction of stable single-clone VZV recombinant viruses using VZV cosmid-based genomic constructs or bacterial artificial chromosomes (Fig. 16.3).

In this section, we summarize our major findings using experimental infection of SCID mouse-human xenografts with clinical isolates

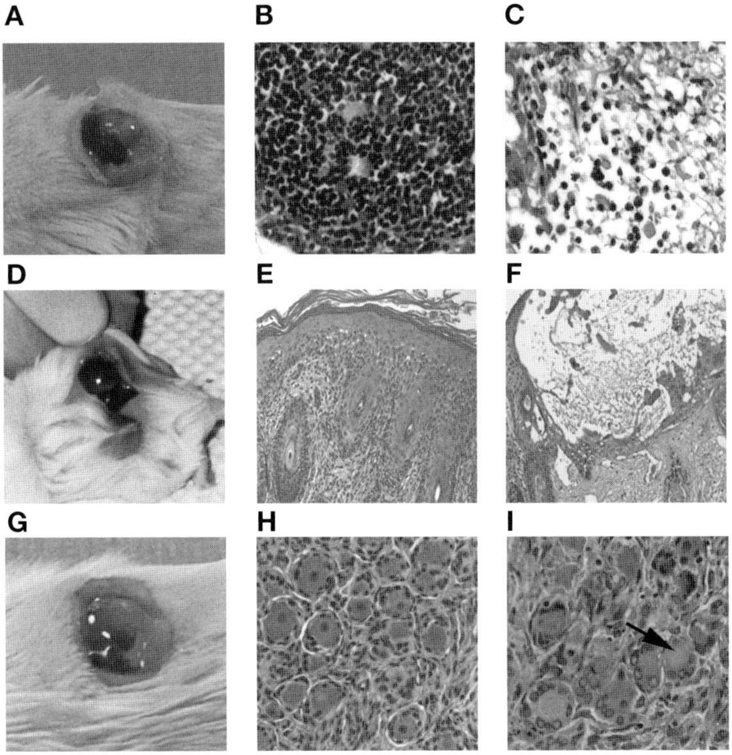

Figure 16.2 The SCID mouse-human model of VZV pathogenesis. Human tissue xenografts have served as a valuable alternative for *in vivo* studies in the absence of opportunities to investigate VZV pathogenesis in a small animal model. (A) Photograph of a T-cell xenograft attached to the murine renal capsule. (B) Histological section of uninfected T-cell xenograft. (C) T cell xenograft 21 days after VZV-infection exhibiting significant lymphocyte depletion. (D) Photograph of bilateral subcutaneous skin xenografts, 6 weeks after transplantation. (E) Uninfected skin xenograft, showing organotypic skin architecture. (F) Skin xenograft 21 days after VZV-infection with a large vesicular lesion and widespread acantholysis. (G) DRG xenograft attached to the murine renal capsule. (H). Uninfected DRG xenograft exhibiting organotypic architecture (I) DRG xenograft 14 days after VZV-infection with significant cytopathic effect in sensory ganglionic cells (black arrow). Histological panels are stained with haematoxylin and eosin. Reprinted from Zerboni and Arvin (2008).

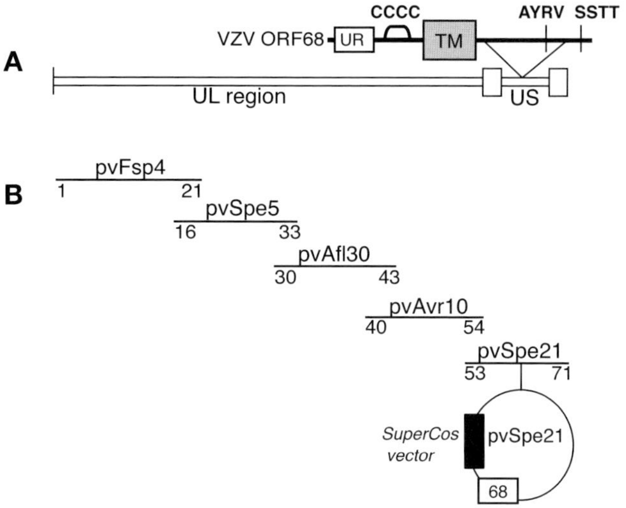

Figure 16.3 Generation of VZV recombinant virus using cosmid-based mutagenesis. The 124 kb VZV genome (A) features a unique long region (UL) and unique short region (US) flanked by repeat elements. VZV genomic cosmids (B) contain overlapping fragments that span the genome and are subcloned into SuperCos vectors. Genetic motifs and promoter elements can be modified by PCR-directed mutagenesis creating a mutant cosmid, which is then cotransfected with the intact cosmids to yield a stable single-clone recombinant genome. Shown is an example of gene motifs in ORF68 encoding gE: UR = unique region, CCCC = cysteine-rich repeat region, TM = transmembrane domain, AYRV = endocytosis motif, SSTT = phosphorylation motif.

as well as recombinant VZV viruses containing mutations in open reading frames encoding viral glycoproteins, viral kinases, regulatory proteins and promoter elements. This approach has facilitated investigation of molecular determinants of virulence in human tissues with relevance to VZV pathogenesis.

Generation of human skin and T-cell xenografts in SCID mice

Human tissue xenografts have served as a valuable alternative for *in vivo* studies in the absence of opportunities to investigate VZV pathogenesis in a small animal model. SCID mice support foreign tissues over a long period of time and are not susceptible to the consequences of VZV infection of the human tissue xenograft because VZV is not infectious for mice, even those with deficiencies of adaptive immunity.

T cell xenografts (also known as thy/liv xenografts) are constructed by xenotransplantation of human fetal thymus and liver tissues (gestational age 14–21 weeks) under the renal capsule in 6–8 week old homozygous *scid/scid* CB.17 mice (Fig. 16.2A–C). Engraftment for more than 6 months will result in a fused organ resembling an age-matched human thymus, with well-demarcated thymic cortical and medullar areas. T cell xenografts are infected by direct inoculation with VZV-infected fibroblasts (~50 µl, 10^4–10^5 PFU/ml) and are typically examined within 7–21 days of infection.

Skin xenografts are constructed by bi-lateral subcutaneous xenotransplantation of full-thickness human fetal skin (gestational age 18–24 weeks) in 6–8 week old homozygous *scid/scid* CB.17 mice (Fig. 16.2D–F). At 5 weeks post xenotransplantation, >100% of human skin xenografts exhibit organotypic features of differentiated human skin. Skin xenografts are infected by direct inoculation with VZV-infected fibroblasts (~50 µl, 10^4–10^5 PFU/ml) and complete vesicular lesions are typically observed within 21 days of infection. Except for the absence of a lymphocytic inflammatory response, inoculation of human skin xenografts with VZV-infected cells results in epidermal changes that are indistinguishable from biopsies of natural varicella lesions.

T cell infectivity and skin virulence are assessed using an array of methods including viral titration of tissue homogenates, histological examination of tissue sections, analyses of virion morphogenesis by transmission electron microscopy (EM), protein expression profiles, quantitative DNA/RNA analysis and flow cytometric analysis (T cells).

VZV T-cell tropism facilitates skin pathogenesis

Unlike other alphaherpesviruses, VZV exhibits a marked T cell tropism (Koropchak *et al.*, 1989; Moffat *et al.*, 1995; Zerboni *et al.*, 2000). The capacity of VZV to replicate in T-lymphocytes provides a mechanism for virus amplification and spread that is unique among the human alphaherpesviruses. Primary VZV infection was originally hypothesized to resemble the Fenner model for infectious ectromelia (mousepox) pathogenesis, in which inoculation via the respiratory route was followed by replication in peripheral blood mononuclear cells within regional lymphoid tissue and subsequent viremic spread to reticuloendothelial organs (Grose, 1981). Viral amplification occurring during a 10–21 day incubation period within reticuloendothelial organs was presumed to result in a secondary viraemia that delivered the virus to cutaneous sites of infection (Grose, 1981).

Through experimental co-culture of tonsil T-lymphocytes with VZV-infected cell monolayers, we observed that VZV preferentially infects activated $CD4^+$ memory T cells and especially those that express skin homing markers such as cutaneous leucocyte antigen (CLA) and chemokine receptor-4 (Ku *et al.*, 2002). VZV-infected tonsil $CD4^+$ memory T cells that are introduced into the circulation of SCID mice with human skin xenografts traffic to human skin and initiate replication within the xenograft (Ku *et al.*, 2004). VZV replication in primary human fibroblast and keratinocyte cultures is rapidly lytic, however VZV replication in differentiated full-thickness skin xenografts progresses slowly. Although VZV-infected tonsil T cells can be isolated from skin xenografts within 24 hours of transfer, spread of VZV takes 10–21 days to form cutaneous lesions extending to the skin surface (Ku *et al.*, 2004). Histopathological observations of infected skin tissue sections after T cell transfer indicates that lesion formation occurs via gradual cell–cell spread and formation of polykaryocytes by fusion of VZV-infected skin cells similar to observations in skin biopsy sections during natural varicella infection (Ku *et al.*, 2004).

The observation that lesion extension to the skin surface takes 10–21 days in skin xenografts, which mimics the 10–21 day incubation period during natural infection of the human host, suggests that intrinsic immune responses might modulate the pathogenic process of vesicle formation. Since SCID mice lack adaptive immune responses, we investigated the role of innate immune responses in skin tissues to modulate cell–cell spread and polykaryon formation. We observed markers of interferon pathway activation including phosphorylation and nuclear translocation of Stat1 protein, activation of the NF-κB pathway and extensive IFN-α expression in uninfected epidermal cells surrounding foci of VZV-infected cells in skin xenografts (Ku *et al.*, 2004). In a separate experiment, skin xenografts in SCID mice were infected with VZV and animals were subsequently administered antibody that blocked the IFN-α/β receptor (Ku *et al.*, 2004). Skin lesions in antibody treated mice were more extensive and virus yields were significantly increased compared with untreated mice (Ku *et al.*, 2004).

Taken together, these data suggest that VZV lymphotropism may have evolved as a mechanism to exploit the process of programmed immune surveillance and to overcome innate immune barriers in skin. We proposed a refinement to the mousepox model wherein inoculation of epithelial cells via the respiratory route facilitates infection of lymphoid tissues of the Waldeyer's tonsillar ring, preferentially infecting $CD4^+$ memory T cells that carry out immune surveillance and express skin homing markers (Fig. 16.4) (Ku *et al.*, 2004, 2005). Migration of VZV-infected skin-homing memory T cells facilitates direct transport of VZV to skin, where innate immune barriers impede progression of cell–cell spread to the skin surface during the 10–21 day incubation period. Viral amplification within skin

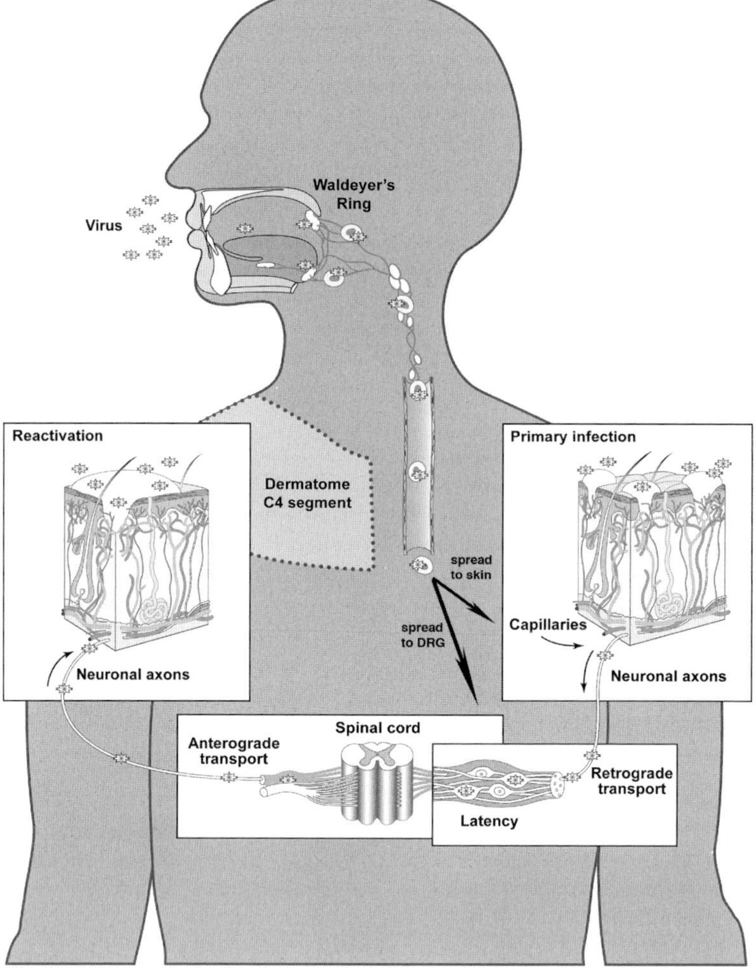

Figure 16.4 Schema for VZV pathogenesis. Varicella infection is acquired by inoculation of pharyngeal epithelial cells via the respiratory route. A 10–21 day incubation period allows for infection of T cells and viral amplification in regional lymphoid tissues (Waldeyer's ring), the transport of virus to skin by infected T cells and gradual progression of lesion formation against the innate antiviral responses of epidermal cells. VZV may reach sensory nerve ganglia by the haematogenous route or by retrograde axonal transport from cutaneous nerve endings to establish a lifelong latent infection. Reactivation of latent VZV may result in anterograde transport of infectious particles to reinfect skin, causing herpes zoster. Reprinted from Zerboni and Arvin (2008).

cells also provides the opportunity for infection of additional skin memory T cells as they migrate through sites where skin lesions are forming. Newly infected T cells could then re-enter the circulation and deliver VZV to distant skin sites, which would account for the viraemia detected at the onset of varicella symptoms and the successive crops of lesions that appear during primary VZV infection (Ku *et al.*, 2004, 2005).

Investigation of vaccine attenuation in skin xenografts

The VZV vaccine (Oka-Merck; VARIVAX) was licensed in the United States in 1995 and is the only vaccine for prevention of a human herpesvirus infection. The live, attenuated vaccine (vaccine Oka) was generated by serial passage of a clinical isolate (parental Oka) in human and guinea pig embryo fibroblasts (Takahashi *et al.*, 1974).

Clinical evidence for vaccine Oka attenuation is that vaccine-associated rashes occur in fewer than 5% of healthy children and adults (Arvin, 2001). However, vaccine Oka retains the capacity to cause varicella in immunocompromised patients and to establish latency and reactivate in high-risk patients. Moffat *et al.* (1998) demonstrated that the vaccine Oka is reduced in the capacity to infect skin xenografts, but not T cells, when compared to parental Oka (Moffat *et al.*, 1998a). Infection of skin xenografts *in vivo* was associated with decreased virus yields and slower development of cutaneous lesions (Moffat *et al.*, 1998a).

We hypothesized that vaccine Oka attenuation results from a reduced capacity to replicate in skin. Altered skin virulence would provide the adaptive immune response a longer interval to develop and to potentially clear infectious virus from the inoculation site before the formation of skin lesions (Arvin, 2001). We investigated the molecular basis for attenuation of vaccine Oka by generation of cosmid-derived recombinant parental Oka (pOka), recombinant vaccine Oka (vOka) and six pOka/vOka chimeric viruses (Niizuma *et al.*, 2003; Zerboni *et al.*, 2005a). All recombinant viruses (pOka, vOka and chimeric viruses) exhibited similar growth phenotypes and plaque morphology *in vitro* (Zerboni *et al.*, 2005a). Recombinant vOka retained altered skin virulence *in vivo*, and evaluation of the chimeric viruses demonstrated that various combinations of VZV genome fragments were associated with an attenuation phenotype (Zerboni *et al.*, 2005a). These experiments indicated that vaccine Oka attenuation is multi-factorial and is not the result of a single mutation but likely reflects mutations in several genes that cumulatively alter skin virulence. Recently, molecular analysis of infectious virus recovered from individuals with vaccine-related skin lesions demonstrated that the Oka vaccine exhibits a marked genetic heterogeneity (Quinlivan *et al.*, 2006). Vaccine preparations contain mixtures of VZV genomes with varying genetic alterations that may have accumulated during serial passage in cultured cells, some of which exhibit skin virulence (Quinlivan *et al.*, 2006) as well as parent genomes that carry no mutations. Some selection for VZV genomes that have more capacity to replicate in skin may contribute to cutaneous lesion formation and transmission of the virus from vaccine recipients which is known to occur (Quinlivan *et al.*, 2007).

Investigation of molecular determinants of virulence and pathogenesis in human skin and T cell xenografts

In this section, we briefly summarize our analysis of recombinant VZV viruses in skin and T cell xenografts (Fig. 16.3). These experiments have demonstrated that VZV skin virulence can be diminished by even single amino acid substitutions, such as those that alter glycoprotein intracellular trafficking or disrupt viral kinase function. The differential requirement for VZV protein functional motifs, dependent upon the target cell type, is a recurring observation in our investigation of VZV pathogenesis *in vivo*. Changes in nucleotide sequences in VZV gene promoters that disrupt sites for host cell transactivators can also be attenuating in the tissue microenvironment. These observations lay the foundation for the development of a 'second generation' varicella vaccine that represents a single VZV genome in which the molecular basis for attenuation is defined.

Viral glycoproteins

The role of VZV glycoproteins in virus entry, endocytosis and egress are covered in Chapter 9. We will focus on the potential role of VZV glycoproteins in tissue-specific pathogenesis. Histopathological observation of infected skin xenografts indicates that lesion formation occurs via cell–cell spread and formation of polykaryons by fusion of VZV-infected epidermal cells; whereas VZV-infected T cells do not undergo fusion. Infection in T cell xenografts depends on the release of VZV particles. Therefore, VZV glycoproteins that are associated with syncytia formation *in vitro* are likely to contribute to the process of cell–cell spread in skin *in vivo* and may be less important for T cell infectivity.

Glycoprotein E VZV gE is a type I membrane protein and the most abundant VZV protein in infected cell cultures. Whereas orthologous alphaherpesviral gE species are dispensable for virus replication, VZV gE is essential (Mallory *et al.*, 1997; Mo *et al.*, 2002). VZV gE forms non-covalent heterodimers with glycoprotein I (Yao *et al.*, 1993); gE, gI and the gE–gI complex play critical roles in cell–cell spread and infection of skin and T-cell xenografts (Berarducci *et al.*, 2006; Berarducci *et al.*, 2009; Berarducci *et al.*, 2010). We examined the requirement for a large non-conserved region of the VZV gE endodomain for skin and T-cell infection (Berarducci 2006, 2009, 2010), including a region which disrupts binding of gE to insulin degrading enzyme (IDE) (Berarducci *et al.*, 2010), which has been proposed to be a VZV receptor (Li *et al.*, 2006; Li *et al.*, 2007).

Deletion of the unique 188 amino acid portion of the gE ectodomain (minus the 27 aa leader sequence) was lethal for replication in cell culture (Berarducci *et al.*, 2006). Partial deletions removing residues Y51-P187 restored replication in culture but the mutant was unable to replicate in skin or T cell xenografts; a ΔP27-Y51 mutant retained the capacity to replicate in skin xenografts but cell–cell spread was significantly impaired (Berarducci *et al.*, 2006) (Table 16.1). A ΔP27-G90 mutant, which was defective in binding to cellular IDE, also exhibited reduced skin virulence; however, gE–IDE interaction is retained in the ΔP27-Y51 mutant (Berarducci *et al.*, 2010). These observations indicate that the region of aa90–187 is critical for both skin and T cell pathogenesis and gE–IDE interaction is not required for viral entry and cell–cell spread in skin, but a contribution to virulence cannot be excluded. In another set of experiments, the serine in position 31 was shown to be important for skin infectivity whereas alanine substitution of the serine in position 49 had no effect (Berarducci *et al.*, 2006).

Disruption of the first cysteine-rich region, which is required for gE–gI complex formation, altered gE expression on plasma membranes and significantly decreased VZV cell–cell spread in culture and skin but did not alter entry into T cells (Berarducci *et al.*, 2009). The differential requirement for a functional gE–gI heterodimer in skin and T cells may reflect the need for the gE–gI complex in cell–cell spread and polykaryon formation, which is a component of skin pathogenesis but does not occur in T cells. Deleting gI has similar negative consequences on skin virulence but also impairs T cell infectivity *in vivo*, which may reflect a more general requirement for gI in pathogenesis *in vivo* (Moffat *et al.*, 2002).

The gE cytosolic domain contains functional motifs which are conserved among alphaherpesvirus gE orthologues and may function in tissue-specific tropism of VZV, these include an AYRV motif (aa568–571) that mediates gE trafficking to the *trans*-Golgi network and a phosphorylated acid cluster domain, SSTT (aa588–601) (Moffat *et al.*, 2004). Alanine-substitution in the AYRV motif and SSTT phosphorylation motif were associated with higher viral titres *in vitro*; gE expression was increased but aberrant in the AYRV mutant (Moffat *et al.*, 2004). The AYRV mutant exhibited impaired skin virulence but altered gE trafficking had a less significant effect on T cell virulence (Moffat *et al.*, 2004). Mutation of the SSTT phosphorylation had no effect on replication in skin and T cell xenografts *in vivo* (Moffat *et al.*, 2004).

Glycoprotein H VZV gH is a highly conserved, essential type 1 membrane protein and one of the core herpesviral fusogens. VZV gH is an important target of humoral immune responses and anti-gH antibodies are capable of complement-independent neutralization of VZV infection *in vitro* (Vleck *et al.*, 2010). We investigated the effects of antibody-mediated interference on cell–cell spread of VZV in skin xenografts by administration of monoclonal antibody to gH (Vleck *et al.*, 2010). Antibody given at 6 hours post infection significantly reduced cell–cell spread and VZV genome copies were reduced when examined by quantitative real-time PCR (Vleck *et al.*, 2010). In that SCID mice lack a polyclonal B-cell response, the inhibitory effect might be attributed to a block on viral entry in that anti-gH antibodies may mask functional domains required to mediate fusion upon viral attachment. The observation that passive antibody administration reduces VZV skin virulence may have clinical relevance in the development of an alternative prophylaxis in patients at risk for severe primary infection.

Table 16.1 Cell culture and *in vivo* replication kinetics of select recombinant VZV mutants that have been examined in the SCID mouse-human model for VZV pathogenesis

		Cell culture *in vitro*		SCIDhu mice *in vivo*	
	Virus	Growth kinetics	Plaque size	Skin xenografts	T cell xenografts
Recombinant Oka	+	+++	**Normal**	++	++
Glycoproteins					
ΔgE	–				
ΔgE P27-P187	–				
ΔgE P27-G90	+	++	Small	+	++
ΔgE Y51-G90	+	+++	Small	++	
ΔgE_Y51-P187	+	+	Small	–	–
ΔgE P27-Y51	+	++	Small	+/–	
ΔgE S49A	+	++	Normal	++	
ΔgE S31A	+	+	Normal	+/–	
ΔgE cysteine208–236	+	++	Small	+/–	
gE YAGL (internalization motif)	–				
gE AYRV (endocytosis motif)	+	++++	Normal	+	+
gE SSTT (phosphorylation motif)	+	++++	Normal	++	++
ΔgI (complete deletion)	+	++	Small	–	–
ΔgB 491-RSSR-494	+	+++	Normal	+	
gB 428-GSGG-431	+	+++	Normal	+	
Viral kinases					
ORF47- ΔC (C-terminal deletion)	+	++	Normal	+	–
ORF47P-S1 (kinase motif mutant)	+	++	Normal	++	
ORF47P-S2 (kinase motif mutant)	+	++	Normal	++	
ORF47D-N (kinase motif mutant)	+	++	Normal	+	–
ORF66S (stop codon mutant)	+	+			+
ORF66 S48A (kinase domain mutant)	+				
ORF66 S331A (kinase domain mutant)	+				
ORF66 G102A (kinase domain mutant)	+	++		+	+/–
ORF66 S250P (kinase domain mutant)	+	++		++	++
Regulatory proteins					
ΔORF62 (complete deletion)	+	+++	Normal	++	
ΔORF71 (complete deletion)	+	+++	Normal	++	
ΔORF62/71 (double deletion)	–				
Δ63/70 (double deletion)	–				
Δ63 (single copy deletion)	+	++	Normal	++	++
ORF63 T171 (phosphorylation mutant)	+	+	Small	+	++
ORF63 S181 (phosphorylation mutant)	+	+	Small	+	++
ORF63 S185 (phosphorylation mutant)	+	+	Small	+	++

Table 16.1 continued

		Cell culture *in vitro*		SCIDhu mice *in vivo*	
	Virus	Growth kinetics	Plaque size	Skin xenografts	T cell xenografts
Tegument proteins					
ΔORF10 (complete deletion)	+	+++	Normal	+	++
ORF10 P28A (acidic domain mutant)	+	+++	Normal	++	
ORF10 P28S (acidic domain mutant)	+	+++	Normal	++	
ΔORF11	+	+++	Normal	+/–	
ΔORF12	+	+++	Normal	++	
ΔORF10/11	+	+++	Normal	+/–	
ΔORF10/11/12	+	+++	Normal	+/–	
ΔORF11/12	+	+++	Normal	+/–	
Promoter mutants					
gI-SP1 (trans factor binding mutant)	+	++	Normal	+/–	+
gI-USF (trans factor binding mutant)	+	++	Normal	++	++
gI-SP1/USF	+	++	Normal	–	+/–
gE-proΔI	+	++	Normal	++	
gE-proΔII	+	++	Normal	++	
ORF10-Δpro	+	+++	Normal	+	
ORF10-proΔ+39/+63	+	+++	Normal	++	
ORF10-proΔUSF	+	+++	Normal	+	
ORF61-proΔSp1a	+	++	Normal	++	
ORF61-proΔSp1ac	+	+	Small	+	

Glycoprotein B VZV gB is a highly conserved membrane protein and one of the core herpesviral fusogens. Our investigation of gB using structure-based mutation demonstrated a requirement for residues in two sites within the putative fusion loop (Oliver *et al.*, 2009). We also investigated the requirement for a putative subtilisin-like proprotein convertase furin recognition motif, in that mature gB is known to be cleaved into two polypeptides of 66 and 68 kDa (Oliver *et al.*, 2009). Deletion of the furin recognition motif eliminated the 66 kDa cleavage product but the mutation had no effect on replication or plaque morphology *in vitro* (Oliver *et al.*, 2009). However, viral titres recovered from infected skin xenografts were significantly reduced with the furin deletion mutant (pOka-gBΔ^{491}RSSR^{494}) compared to the parental virus (pOka) at 10 and 21 days after infection (Oliver *et al.*, 2009). The consequences of disrupting gB furin cleavage on the VZV fusion machinery is difficult to assess in a cell-associated virus, but detailed EM analysis did not reveal any effect on virion morphogenesis. In that HSV gB does not possess a furin cleavage motif, our results may indicate a difference in the requirement or function of VZV gB in skin virulence compared to HSV gB.

Viral kinases

The VZV genome encodes two serine/threonine kinases that are common to all alphaherpesviruses, ORF47 protein (p) and ORF66p. ORF47p and ORF66p are localized to the virion tegument and are dispensable for replication in cell culture (Heineman and Cohen, 1995; Heineman *et al.*, 1996). Targets of ORF47-mediated phosphorylation include IE62 and gE. Cytoplasmic localization of IE62 in infected cell cultures is associated with ORF66-mediated phosphorylation (Kinchington *et al.*, 2000). Our investigations in differentiated skin and T cells have demonstrated tissue-specific requirements for both viral kinases in VZV pathogenesis *in vivo*.

ORF47 protein VZV ORF47 is orthologous to HSV U_L13 and is highly conserved in all herpesviruses. Insertion of a stop codon in ORF47 has no effect on replication in cell culture but completely abrogated infection in skin and T cell xenografts (Moffat *et al.*, 1998b). This suggests that ORF47 functions may be complemented by related cellular kinases, such as CKII family members, or the ORF66 kinase, *in vitro* but not *in vivo*. We investigated the requirement for ORF47 kinase activity *in vivo* by C-terminal truncation and by targeting two putative C-terminal kinase motifs, DYS and PPE (Besser *et al.*, 2003). Disruption of the DYS motif (rOka47D-N mutant) eliminated ORF47 kinase activity but ORF47 homodimer formation and IE62 interaction was preserved (Besser *et al.*, 2003). Replication in skin xenografts was significantly reduced, but eliminating kinase function was not lethal to infection as observed with the stop codon mutant (Besser *et al.*, 2003). Although ORF47 kinase function was not required for replication in skin xenografts, the kinase defective mutant was unable to replicate in T cell xenografts (Besser *et al.*, 2004).

Histological and EM analysis of rOka47D-N-infected tissue sections revealed alterations in virion assembly and egress; however, polykaryon formation in skin was intact (Besser *et al.*, 2004). Expression of VZV gE on the cellular membranes of infected skin cells was limited, which is consistent with the requirement for ORF47-directed phosphorylation of gE in appropriate accumulation in the *trans* Golgi network (TGN) and membrane localization (Besser *et al.*, 2003). In that reduced gE membrane expression did not completely prevent syncytia formation suggests that other VZV proteins may facilitate polykaryon formation in skin when gE intracellular trafficking is disrupted. This observation suggests that preservation of syncytia formation may allow for limited cell–cell spread despite reduced replication and altered virion formation, whereas VZV T cell infection has a stricter requirement for unimpaired virion assembly and egress because infected T cells do not fuse.

ORF66 protein We investigated the requirement for ORF66p kinase activity by stop codon mutation and mutations in putative protein kinase subdomains at G102 and S250 (Schaap *et al.*, 2005; Schaap-Nutt *et al.*, 2006). G102A substitution blocked ORF66 kinase activity and resulted in nuclear retention of IE62 at late times post infection; S250P substitution had no effect on kinase activity (Schaap-Nutt *et al.*, 2006). The ORF66p stop codon mutant and the G102A mutant were only moderately attenuated for replication in skin xenografts but were significantly diminished in T cell xenografts *in vivo* (Schaap *et al.*, 2005; Schaap-Nutt *et al.*, 2006). Virion morphogenesis in T cells was impaired, with few intact virions observed by EM analysis (Schaap *et al.*, 2005; Schaap-Nutt *et al.*, 2006).

Further analysis showed that the ORF66 stop codon mutant and the kinase defective mutant had lost the capacity to block apoptosis, which is typically observed upon VZV-infection of T cells (Schaap *et al.*, 2005). Mechanisms to prolong T cell survival upon infection may be critical for spread of VZV during viraemia. Our observations with the ORF66 stop codon mutant suggest that VZV ORF66p may be required to block T cell apoptosis during virion assembly. In addition to enhancing infection by blocking apoptosis, we have identified immune evasion functions that may further enhance T cell tropism during VZV pathogenesis (Schaap *et al.*, 2005). Blocking ORF66 expression or kinase activity reduced the capacity of VZV to interfere with interferon signalling pathways and MHC class I expression in VZV-infected cells (Schaap *et al.*, 2005).

Regulatory proteins

VZV encodes several regulatory proteins with homologues common to other alphaherpesviruses. Progress towards identification of the specific molecular requirements for regulatory proteins in viral pathogenesis is difficult because introducing major mutations into these ORFs is often incompatible with replication.

IE62 protein IE62 is a large phosphoprotein that is encoded by duplicate genes, ORF62/71. As the major viral transactivator, IE62 functional domains include a nuclear localization signal, an N-terminal activation domain and a DNA-binding domain. Although only one copy of the duplicated gene is required for replication, serial passage in cultured cells and skin xenografts *in vivo* is associated with restoring the deleted copy to yield complete genomes (Sato *et al.*, 2003). Single-copy expression of IE62 from a non-native site reduces plaque size and titre *in vitro* and blocks replication in skin xenografts *in vivo* (Sato *et al.*, 2003).

IE63 protein IE63 is a small regulatory phosphoprotein that is encoded by duplicate genes, ORF63/70. Single copy deletion of either ORF63 or ORF70 at their native sites, or single copy expression at an ectopic site, allows for normal replication in cell culture and skin xenografts *in vivo* (Sommer *et al.*, 2001). We also constructed single-copy IE63 mutants, expressed from an ectopic site, with alanine substitution in T171, S181 and S185 (Baiker *et al.*, 2004a). These putative phosphorylation mutants grew poorly in cultured cells, with reduced viral titres and a small plaque phenotype. Decreased ORF47 and gE synthesis was observed, suggesting that IE63 phosphorylation may be required for IE62-induced expression of these gene products. Of interest, the IE63 serine/threonine mutants were significantly attenuated in skin but not in T cell xenografts (Baiker *et al.*, 2004a). Based on our previous observations, the reduction in ORF47 and gE expression is likely to contribute to reduced virulence of ORF63-deficient mutants in skin xenografts.

Tegument proteins

ORFs 9–12 ORFs 9 through 12 comprise a highly conserved gene cluster that encode for four putative tegument proteins. ORF9 is a member of the herpesvirus UL49 gene family and interacts with IE62 (Cilloniz *et al.*, 2007). ORF10, the HSV-1 VP16 homologue, is known to enhance IE62-mediated transactivation of VZV genes (Moriuchi *et al.*, 1995). We investigated the requirements for ORF10 by constructing a complete deletion of ORF10 and targeted mutations in the acidic activation domain and putative binding site for human cellular factor-1 (HCF-1) (Che *et al.*, 2006). Whereas VP16 is required for HSV-1 replication, ORF10 was dispensable *in vitro* despite significant effects on virion morphogenesis (Che *et al.*, 2006). Absence of ORF10 expression resulted in reduced viral yields and smaller lesions in skin xenografts; replication in T cells was unaffected (Che *et al.*, 2006). Site-directed mutations in the acidic activation domain and HCF-1 binding site had no effect on skin virulence.

Deletion of ORF9 was incompatible with VZV replication *in vitro* whereas ORFs 11 and 12 were dispensable; ORF9 was shown to be a VZV tegument protein (Che *et al.*, 2008). Deletion of ORF12 had no effect on skin virulence but deletion of ORF11 alone, or ORF11 in combination with ORF10 or ORF12, significantly reduced viral titres (Che *et al.*, 2008). Deletion of ORFs 10–12 (pOkaΔ10/11/12) severely restricted skin infectivity (Che *et al.*, 2008). Taken together, these results indicate that individual genes within the ORF10 to ORF12 cluster can be deleted without any deleterious effects on VZV replication *in vitro*; however, the ORF10-ORF12 cluster is required for VZV virulence in skin xenografts *in vivo*. ORF11 appears to be the critical gene product necessary for skin pathogenesis.

VZV promoters

Consensus binding sites for cellular proteins that function in the transactivation of host genes have been mapped to many VZV promoters. Functional analysis of these sequences using reporter-based promoter assays demonstrated that some consensus binding sites enhance viral gene transactivation by IE62 and other VZV regulatory proteins. We have investigated the role of consensus binding sites for host cell transcription factors within the promoters of ORF67 (encoding gI), ORF68 (encoding gE), ORF10 and ORF61. Taken together, these experiments demonstrate that VZV tissue-specific pathogenesis is dependent upon the presence and function of cellular transcription factors that work in concert with viral regulatory proteins.

gI promoter. Consensus binding sites for specificity factor 1 (Sp1) and upstream regulatory factor (USF) are found within the ORF67 promoter region (Ito *et al.*, 2003). Mutation of the Sp1 and USF sites individually did not alter VZV replication *in vitro*, but a dual Sp1/USF mutant exhibited reduced gI expression, smaller plaque size and lower viral titres (Ito *et al.*, 2003). Disrupting the Sp1 site alone significantly reduced virulence in skin and T cells *in vivo* (Ito *et al.*, 2003). The dual Sp1/USF mutant was unable to replicate in skin but retained some infectivity in T cells xenografts (Ito *et al.*, 2003). These experiments reflect a differential requirement for high levels of gI expression in skin and T cell xenografts in that less gI was required for productive infection of T cells than was required for skin infection.

gE promoter The VZV gE promoter contains consensus binding sites for several cellular transcription factors, including two Sp1 sites (Berarducci *et al.*, 2007). We constructed three stepwise deletions of the gE promoter (ΔI, ΔII and ΔIII) as well as point mutations of two Sp1 binding sites located in region III (Berarducci *et al.*, 2007). Deletion or point mutation of a non-canonical Sp1 binding site was incompatible for viral replication; whereas mutation of a canonical Sp1 binding site was compatible for virus recovery but viral titres and plaque size were significantly reduced. Using an *in vitro* reporter assay, we determined that IE62 exerts the most regulatory control on the gE promoter compared with ORF61, ORF4 or IE63; Sp1 binding was critical for IE62-mediated transactivation of the gE promoter *in vitro*. In contrast to the dramatic consequences of altering Sp1 binding, deletion of consensus binding sites for 13 and 6 cellular transcription factors with the ΔI and ΔII mutations, respectively, were compatible with viral replication and did not alter infectivity of cultured cells or in tonsil T cells *in vitro*, or skin xenografts *in vivo*. These results indicate that the major cell protein involved in gE promoter regulation is Sp1. Cellular factors other than Sp1 do not appear to play a significant role in IE62-mediated transactivation of the gE promoter in skin or human T cells.

ORF10 promoter. Consensus binding sites for octamer transcription factor 1 (Oct-1) and USF are found within the ORF10 promoter region (Che *et al.*, 2007). We identified the minimal ORF10 promoter and constructed mutations in the Oct-1 and USF binding sites, as well as putative TATA boxes (Che *et al.*, 2007). We demonstrated a requirement for the USF binding site for IE62-mediated transactivation of the ORF10 promoter; mutation of the site demonstrated its requirement for ORF10 transcription (Che *et al.*, 2007). Consistent with our observations showing a requirement for ORF10 in normal skin virulence, a promoter deletion mutant and a site-specific USF mutant were deficient for skin infectivity (Che *et al.*, 2007).

ORF61 promoter VZV ORF61 encodes a regulatory protein with transactivating capacity that is orthologous to HSV-1 ICP0. We mapped the minimal ORF61 promoter and generated mutants in three putative Sp1 binding sites, designated Sp1a, Sp1b, and Sp1c. Sp1a and Sp1c were shown to function synergistically *in vitro* and the dual mutant (pOka-61proΔSp1ac) exhibited reduced ORF61, ORF66 and gE expression (Wang *et al.*, 2009). The pOka-61proΔSp1ac replicated poorly in skin xenografts, indicating that the Sp1 binding sites were important for ORF61 expression *in vivo* (Wang *et al.*, 2009).

Infection of human dorsal root ganglia (DRG) xenografts in SCID mice

Tissues acquired postmortem from individuals who died during primary varicella infection or reactivation are extremely rare and as a consequence very little is known about acute VZV infection in sensory ganglia and the molecular requirements for VZV neurovirulence. Investigations using cadaver ganglia acquired from patients without acute VZV infection at the time of death, and so presumed to represent latent infection, have not yielded a consistent picture of the state of VZV during latent infection. We developed a SCID mouse-human neural model by xenotransplantation of intact human dorsal root ganglia (Fig. 16.3G–I) (Zerboni *et al.*, 2005b). Individual

DRG (1–2 mm^3) are dissected from human fetal spinal tissues (gestational age 18–24 weeks) and are implanted under the renal capsule in 6–8 week old homozygous *scid/scid* CB.17 mice (Zerboni *et al.*, 2005b). DRG xenografts survive up to 8 months post xenotransplantation and maintain all ganglionic cell populations that are potentially relevant to VZV neuropathobiology within their normal tissue architecture under physiological (*in vivo*) conditions.

At 4 weeks post transplantation, >95% of human DRG xenografts exhibit organotypic features of human adult ganglia including clusters of small and large diameter neuronal cell bodies in association with satellite cells (neuron-satellite cell complexes), within a dense supportive matrix comprised of fibroblasts and other cells types, collagen fibres and extending axons (Fig. 16.3H) (Zerboni *et al.*, 2005b). Human endothelial cells (expressing PECAM-1) line blood vessels supporting the graft and resident neurons express human neuron-specific cell markers such as synaptophysin and neural cell adhesion molecule (NCAM) (Zerboni *et al.*, 2005b). DRG xenografts can be directly injected with VZV-infected fibroblasts (~10 μl, 10^{2}–10^{4} PFU) or passively infected by transfer of VZV-infected T lymphocytes into the tail vein of SCID-DRG mice (Zerboni *et al.*, 2005b). VZV neurovirulence is assessed by histological and TEM analysis, quantitative real-time PCR for VZV nucleic acids, protein expression and infectious release assay. Murine tissues within SCID-DRG mice are not permissive for VZV infection and so this model is suitable for examination of acute VZV replication in human DRG as well as long-term persistence of latent VZV.

In this section, we summarize our investigation of VZV neuropathogenesis using the SCID DRG model. Our results demonstrate that VZV DRG infection is a biphasic process in which an acute replicative phase transitions into long-term viral persistence with low-level differential viral gene transcription in the absence of viral protein expression. Of particular interest, the transition to viral persistence does not require adaptive immune responses, which SCID mice lack.

Acute infection of human sensory ganglia in SCID-DRG mice

Acute infection of human sensory ganglia in SCID-DRG mice is associated with marked cytopathic changes in the tissue within 10–14 days of infection; these changes include the presence of misshapen nuclei or denucleated neuronal cell bodies, cytoplasmic and intranuclear inclusions (Fig. 16.3I) (Zerboni *et al.*, 2005b). Detection of high copies of VZV genomic DNA by quantitative real-time PCR (10^{7}–10^{9}/10^{5} cells) correlates with recovery of infectious virus from mechanically disrupted xenografts (Fig. 16.5a). VZV mRNA and proteins of all kinetic classes are expressed and can be detected by quantitative real-time RT-PCR, Western blot and immunostaining methods (Zerboni *et al.*, 2005b). Neuronal pigments, which accumulate with age in human neurons, are absent.

Production of infectious VZV occurs in neurons and satellite cells Single-cell histological analysis of acute VZV infection by correlative immunofluorescence-electron microscopy (IF-EM) revealed that assembly and packaging of infectious VZV may occur in satellite cells as well as neurons (Reichelt *et al.*, 2008). Intranuclear and cytoplasmic virions were observed within satellite cells as well as neurons; extracellular particles were observed external to the cellular membranes of both cell types (Reichelt *et al.*, 2008). This finding is interesting considering the preponderance of data from investigations using latently infected cadaver ganglia that neurons, and not satellite cells, harbour the latent viral genome. However, it is consistent with the evidence that VZV reactivation is associated with infection of satellite cells as well as neurons in cadaver ganglia from patients with herpes zoster at the time of death (Esiri and Tomlinson, 1972).

Polykaryon formation is a characteristic of VZV infection of DRG Another interesting observation revealed through ultrastructural analysis of acutely-infected human DRG xenografts is the presence of polykaryon formation in neuron-satellite cell complexes, that is, fusion of membranes separating an individual neuronal cell body from

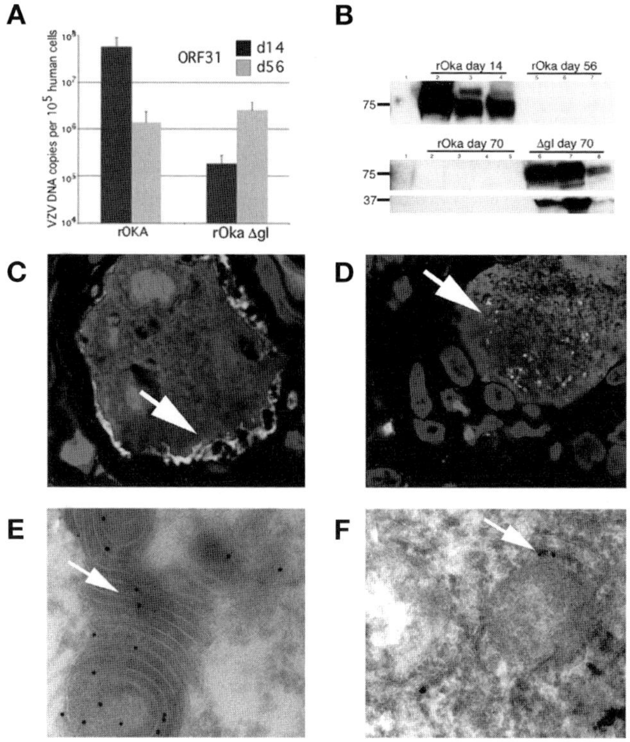

Figure 16.5 Replication of glycoprotein I deletion (ΔgI) in human DRG xenografts. (A) Quantitative real time PCR for VZV genomes using a probe for ORF31 in rOka and ΔgI-infected DRG at 14 and 56 days after infection; unlike rOka-infected DRG, the ΔgI mutant does not exhibit the expected transition to viral persistence. (B) Western blot using an anti-VZV human polyclonal antibody (B1 and $B2_a$) and anti-IE63 ($B2_b$), in DRG infected with rOka at day 14 and day 56 (B1) or day 70 (B2), or ΔgI at day 70 after infection. Viral persistence rOka-infected DRG occurs in the absence of viral gene expression whereas protein is expressed in ΔgI-infected DRG at 70 days after infection. (C and D) rOka-infected DRG at 14 days after infection, bright staining on neuronal membranes using a mouse monoclonal antibody to VZV gE (C, white arrow), which is retained in the cytoplasm in ΔgI-infected neurons (D, white arrow). (E–F) ImmunoEM of ΔgI-infected neurons with gold-labelled gE (black dots), white arrow denotes strange convolutions in Golgi structures (E) and rough ER (F).

its surrounding supportive satellite cells (Reichelt *et al.*, 2008). Polykaryon formation facilitates cell–cell spread and is a hallmark of VZV skin lesions and infection of cultured cells. Polykaryon formation within individual neuron-satellite cell complexes may facilitate recruitment of satellite cells as sites for virion assembly, thus providing a mechanism for VZV amplification during reactivation. Polykaryon formation may also facilitate spread of VZV to adjacent neuron satellite–cell complexes within individual ganglia, which could potentiate the neuropathic consequences of acute VZV infection, which is associated with widespread ganglionitis. Although ultrastructural analysis of such tissues is rare, a similar observation of incomplete boundaries between neurons and satellite cells was reported in a trigeminal ganglion obtained at autopsy from patient who died with recurrent VZV (Esiri and Tomlinson, 1972).

Persistence of VZV in human neuronal ganglia in SCID-DRG mice

Within 8 weeks of acute infection, the marked cytopathic effect that characterizes productive infection of DRG xenografts has resolved (Zerboni *et al.*, 2005b). In some areas, clusters of neurons and satellite cells retain organotypic features of human

DRG whereas in others histopathic indicators of a prior infectious process, such as Nageotte nodules and fibrosis, are observed (Zerboni *et al.*, 2005b). Infectious virus cannot be recovered from disrupted tissues and virus particles are not observed by ultrastructrual TEM analysis. However, VZV genomes can be detected by quantitative real-time PCR. The defining criteria for herpesvirus 'latency' is the persistence of the HSV genome in the absence of infectious virus production, and the capacity to reactivate and cause disease in the periphery (Roizman *et al.*, 2007). Presumably, the latent genomes that persist in DRG neurons after clearance of productive VZV infection retain the capacity to reactivate; however, in the absence of any neural connection to the mouse periphery we refer to this stage as 'persistence' rather than latency.

Latent viral load during VZV persistence in DRG We have characterized the state of VZV during persistence in human DRG xenografts and compared this with current knowledge about the state of VZV during latency in the human host (Zerboni *et al.*, 2005b). We routinely detect 10^5–10^7 VZV genomes per 10^5 ganglion cells at 56 days after infection by quantitative real-time PCR, which represents a 100-fold decrease when compared with acutely infected DRG xenografts (Fig. 16.5A). These values are significantly higher than estimates of the VZV DNA burden in latently infected cadaver ganglia. This is not unexpected in that we evaluate DRG only 8 weeks following acute infection whereas cadaver ganglia are usually acquired decades following primary varicella infection.

Transcriptional regulation during VZV persistence in DRG Several groups have reported detection of VZV RNAs in latently infected cadaver ganglia, in particular transcripts corresponding to ORF63. At 8 weeks post infection, we have observed accumulation of VZV RNAs corresponding to ORFs 62 and 63, but not 31 (encoding gB) by quantitative real-time RT PCR (Zerboni *et al.*, 2005b). A similar block of late gene expression was reported during VZV infection of human neural stem cells xenotransplanted into the brain of non-obese diabetic SCID mice (Baiker *et al.*, 2004b). Further investigation is needed, but these results suggest that differential regulation of VZV gene expression is characteristic of VZV latency.

VZV protein expression during persistence in DRG. Whereas RNAs are likely to deteriorate during the postmortem interval, proteins are relatively stable (Hynd *et al.*, 2003). However, discrepancies among reports of VZV protein abundance in cadaver ganglia are confounding. We have examined numerous tissue sections from persistently infected DRG xenografts and have not detected any VZV proteins by IHC methods or immunoblot using antibodies to IE63, IE62 and viral glycoprotein (Fig. 16.5B). We employed live animal imaging to further assess VZV protein expression during the course of VZV infection in DRG xenografts. DRG were inoculated with pOka-63/70-lucifease, in which ORF63 and ORF70 have a luciferase fusion cassette (Oliver *et al.*, 2008). The progress of VZV infection was evaluated over a 70-day interval, with luminescence values corresponding to IE63 protein expression assessed every 3–7 days. We observed increasing IE63 protein expression from days 3–15, which correlates with the acute phase of VZV infection in DRG xenografts. After day 15, values declined sharply and by day 56 no IE63 protein expression was detected (Oliver *et al.*, 2008). At the conclusion of the experiment, DRG were analysed to confirm the persistence of VZV genomes despite the absence of IE63 protein expression (Oliver *et al.*, 2008).

Demonstration of neurovirulence in VZV Oka vaccine

We investigated the capacity of vaccine Oka to enter and replicate in human neurons in DRG xenografts. DRG xenografts were directly inoculated with vaccine Oka and the clinical (parental) Oka strain of VZV and assessed at 14 and 28 days after infection (Zerboni *et al.*, 2005b). VZV genome copies, recovery of infectious virus and cytopathic effect were comparable at both timepoints (Zerboni *et al.*, 2005b). Like parental Oka, vaccine Oka replicated efficiently in neurons and satellite cells causing significantly cytopathic effect and producing infectious virus. Within 28 days of infection, infectious virus production had ceased and no cytopathic effect was evident. The preservation of neurovirulence suggests that the accumulation of mutations that attenuate vaccine Oka infection in skin do not alter neurotropism or neurovirulence. These observations are significant in that a recent study by Gershon and colleagues (Gershon *et al.*, 2009) documents VZV latency in multiple sensory ganglia recovered from vaccinated children who did not develop a vaccine-related rash. Efforts to eliminate the capacity of vaccine Oka to establish latency and reactivate will likely require development of second generation varicella vaccines with a molecular basis for altered neurovirulence. Experiments to identify specific molecular determinants of neurovirulence are ongoing.

VZV T-cell tropism facilitates neurotropism

The work of Ku and colleagues (Ku *et al.*, 2005) led to new insights about VZV pathogenesis and immunobiology that emphasize the importance of VZV tropism for $CD4^+$ memory T cells in direct haematogenous spread of VZV to skin. Similarly, early spread of VZV to sensory ganglia by circulating VZV-infected T cells is supported by detection of VZV DNA in ganglia recovered from immunocompromised children who died during the incubation period of varicella and before the appearance of the rash. We hypothesized that circulating VZV-infected T cells could directly infect DRG xenografts (Zerboni *et al.*, 2005b). VZV-infected T cells were injected into the tail vein of mice with human DRG xenografts; VZV DNA was detected in ganglion xenografts and infectious virus was recovered from homogenized tissues (Zerboni *et al.*, 2005b). These experiments demonstrate that VZV-infected T cells can traffic to and productively infect human sensory ganglia *in vivo*. These observations support the hypothesis that VZV may spread directly to sensory ganglia through a haematogenous route, providing a second mechanism by which the virus can reach neuronal sites for persistence. The recent study by Gershon and colleagues (Gershon *et al.*, 2009) documenting VZV latency among vaccines in the absence of a vaccine-associated rash is evidence that VZV may reach ganglion neurons in healthy children by viraemia.

Molecular determinants of VZV neuropathogenesis

Experimental infection of human DRG xenografts with recombinant VZV mutants enables analysis of gene products or motifs within genes with the potential to alter neuropathogenesis or neurovirulence. Using this approach, our aim is to define the contributions of VZV genes and genetic elements and identify specific molecular determinants of neurovirulence. These findings may have direct relevance towards the development of neuroattenuated varicella vaccines.

Whereas glycoprotein I was strictly required for replication in skin and T cell xenografts *in vivo* (Moffat *et al.*, 2002), we observed a markedly different phenotype in DRG xenografts (Zerboni *et al.*, 2007). Unlike the expected biphasic cycle observed with recombinant Oka (rOka), infection with rOkaΔgI, which lacks ORF67, significantly delayed onset of productive infection and showed no transition to persistence up to 70 days after infection (Zerboni *et al.*, 2007). VZV genome copies increased significantly from day 14 to 28, and then from day 28 to 56 (Fig. 16.5A). Infectious virus was recovered from only 1/11 DRG xenografts at day 14, but from 2/4 DRG xenografts at day 70 (by which time rOka exhibits persistence with no release of infectious virus). Histopathic effect, even at day 70 when infectious virus was recovered, was greatly reduced with only rare DRG neurons expressing viral proteins (Zerboni *et al.*, 2007). In contrast to the prominent membrane localization of gE which is typically observed in rOka-infected DRG, in rOkaΔgI-infected DRG gE expression was greatly reduced with only marginal punctate foci within the neuronal cytoplasm (Fig. 16.5C and D). Ultrastructural immuno-EM revealed the presence of unusual Golgi stacks in the cytoplasm of those few neurons that contained evidence of virion particle assembly (Fig. 16.5E) and retention of gE in the rough-endoplasmic reticulum (Fig. 16.5F) (Zerboni *et al.*, 2007). These observations suggest that genetic requirements for VZV replication and spread in neurons are different than in skin and T cells. Moreover, these results demonstrate that mutations which attenuate viral replication in cultured cells may have unpredictable consequences in terminally differentiated neuronal cells *in vivo*.

We also examined the contributions of consensus binding sites in the gI promoter for Sp1 and USF cellular transcription factors (Zerboni *et al.*, 2007). Neither the rOka-gI-Sp1 mutant nor the rOka-gI-Sp1/USF mutant were impaired for replication in DRG xenografts compared with rOka (Zerboni *et al.*, 2007). The two promoter mutants replicated equally well and exhibited the expected shift to VZV persistence following acute infection (Zerboni *et al.*, 2007). These results are of interest in that rOka-gI-Sp1 and rOka-gI-Sp1/USF were impaired for VZV replication in cell culture and in T cell xenografts (Ito *et al.*, 2003). The dual Sp1/USF promoter mutant was unable to replicate in skin xenografts (Ito *et al.*, 2003). As we have previously noted, these results demonstrate tissue-specific requirements for host cell transcription factors in facilitating VZV gene expression and suggest that the requirements for intact VZV gene and promoter functions are in fact less stringent in DRG than skin and T cells *in vivo*.

Conclusion

Advances in basic molecular tools for construction of recombinant viruses and the development of novel approaches for *in vivo* investigation have provided the opportunity to define genetic modifications that may improve upon the current varicella vaccine. While the experiments described in this chapter have contributed to our knowledge about VZV pathogenesis and latency, progress is slow and further investigation is needed. Our perspective is that development of a recombinant VZV vaccine that retains the capacity to elicit a robust antiviral immune response through attenuated replication in skin, but is deficient for T cell infection and neurotropism, would significantly advance control of varicella and herpes zoster in the human population.

Acknowledgements

This chapter summarizes the dedication and achievements of the graduate students, postdoctoral fellows and research staff who have worked on VZV molecular virology and pathogenesis over the past ten years, including (alphabetically) Allison Abendroth, Armin Baiker, Barbara Berarducci, Jaya Besser, Xibing Che, Stuart Hinchliffe, Hideki Ito, Jeremy Jones, Chi-Chi Ku, Suzanne Mallory, Chengjun Mo, Jennifer Moffat, Takahiro Niizumi, Stefan Oliver, Jaya Rajamani, Vasavi Ramachandran, Mike Reichelt, Bunji Sato, Anne Schaap-Nutt, Marvin Sommer, Shaye Stamatis, Susan Vleck, and Li Wang. This work was supported by AI053846, AI20459 and CA49605.

References

Annunziato, P.W., Lungu, O., Panagiotidis, C., Zhang, J.H., Silvers, D.N., Gershon, A.A., and Silverstein, S.J. (2000). Varicella-zoster virus proteins in skin lesions: implications for a novel role of ORF29p in chickenpox. J. Virol. *74*, 2005–2010.

Arvin, A.M. (2001). Varicella vaccine: genesis, efficacy, and attenuation. Virology *284*, 153–158.

Arvin, A. (2005). Aging, immunity, and the varicella-zoster virus. N. Engl. J. Med. *352*, 2266–2267.

Asano, Y., Itakura, N., Hiroishi, Y., Hirose, S., Nagai, T., Ozaki, T., Yazaki, T., Yamanishi, K., and Takahashi, M. (1985). Viremia is present in incubation period in nonimmunocompromised children with varicella. J. Pediatr *106*, 69–71.

Baiker, A., Bagowski, C., Ito, H., Sommer, M., Zerboni, L., Fabel, K., Hay, J., Ruyechan, W., and Arvin, A.M. (2004a). The immediate-early 63 protein of Varicella-Zoster virus: analysis of functional domains required for replication *in vitro* and for T-cell and skin tropism in the SCIDhu model *in vivo*. J. Virol. *78*, 1181–1194.

Baiker, A., Fabel, K., Cozzio, A., Zerboni, L., Fabel, K., Sommer, M., Uchida, N., He, D., Weissman, I., and Arvin, A.M. (2004b). Varicella-zoster virus infection of human neural cells *in vivo*. Proc. Natl. Acad. Sci. U.S.A. *101*, 10792–10797.

Bastian, F.O., Rabson, A.S., Yee, C.L., and Tralka, T.S. (1974). Herpesvirus varicellae isolated from human dorsal root ganglia. Arch Pathol 97, 331–333.

Berarducci, B., Ikoma, M., Stamatis, S., Sommer, M., Grose, C., and Arvin, A.M. (2006). Essential functions of the unique N-terminal region of the varicella-zoster virus glycoprotein E ectodomain in viral replication and in the pathogenesis of skin infection. J. Virol. *80*, 9481–9496.

Berarducci, B., Sommer, M., Zerboni, L., Rajamani, J., and Arvin, A.M. (2007). Cellular and viral factors regulate the varicella-zoster virus gE promoter during viral replication. J. Virol. *81*, 10258–10267.

Berarducci, B., Rajamani, J., Reichelt, M., Sommer, M., Zerboni, L., and Arvin, A.M. (2009). Deletion of the first cysteine-rich region of the varicella-zoster virus glycoprotein E ectodomain abolishes the gE and gI interaction and differentially affects cell–cell spread and viral entry. J. Virol. *83*, 228–240.

Berarducci, B., Rajamani, J., Zerboni, L., Che, X., Sommer, M., and Arvin, A.M. (2010). Functions of the unique N-terminal region of glycoprotein E in the pathogenesis of varicella-zoster virus infection. Proc. Natl. Acad. Sci. U.S.A. *107*, 282–287.

Besser, J., Sommer, M.H., Zerboni, L., Bagowski, C.P., Ito, H., Moffat, J., Ku, C.C., and Arvin, A.M. (2003). Differentiation of varicella-zoster virus ORF47 protein kinase and IE62 protein binding domains and their contributions to replication in human skin xenografts in the SCID-hu mouse. J. Virol. *77*, 5964–5974.

Besser, J., Ikoma, M., Fabel, K., Sommer, M.H., Zerboni, L., Grose, C., and Arvin, A.M. (2004). Differential requirement for cell fusion and virion formation in the pathogenesis of varicella-zoster virus infection in skin and T cells. J. Virol. *78*, 13293–13305.

Breuer, J., Grose, C., Norberg, P., Tipples, G., and Schmid, D.S. (2010). A proposal for a common nomenclature for viral clades that form the genus varicella-zoster virus. J. Gen. Virol. *91*, 821–828.

Che, X., Zerboni, L., Sommer, M.H., and Arvin, A.M. (2006). Varicella-zoster virus open reading frame 10 is a virulence determinant in skin cells but not in T cells *in vivo*. J. Virol. *80*, 3238–3248.

Che, X., Berarducci, B., Sommer, M., Ruyechan, W.T., and Arvin, A.M. (2007). The ubiquitous cellular transcriptional factor USF targets the varicella-zoster virus open reading frame 10 promoter and determines virulence in human skin xenografts in SCIDhu mice *in vivo*. J. Virol. *81*, 3229–3239.

Che, X., Reichelt, M., Sommer, M.H., Rajamani, J., Zerboni, L., and Arvin, A.M. (2008). Functions of the ORF9-to-ORF12 gene cluster in varicella-zoster virus replication and in the pathogenesis of skin infection. J. Virol. *82*, 5825–5834.

Cilloniz, C., Jackson, W., Grose, C., Czechowski, D., Hay, J., and Ruyechan, W.T. (2007). The varicella-zoster virus (VZV) ORF9 protein interacts with the IE62 major VZV transactivator. J. Virol. *81*, 761–774.

Cohen, J.I., and Seidel, K.E. (1993). Generation of varicella-zoster virus (VZV) and viral mutants from cosmid DNAs: VZV thymidylate synthetase is not essential for replication *in vitro*. Proc. Natl. Acad. Sci. U.S.A. *90*, 7376–7380.

Cohen, J.I., Straus, S.E., and Arvin, A. (2007). Varicella-zoster virus replication, pathogenesis and management. In Fields' virology, Fields, B.N., Knipe, D.M., and Howley, P.M., eds. (Philadelphia, Wolters Kluwer Health/Lippincott Williams & Wilkins).

Cohrs, R.J., Srock, K., Barbour, M.B., Owens, G., Mahalingam, R., Devlin, M.E., Wellish, M., and Gilden, D.H. (1994). Varicella-zoster virus (VZV) transcription during latency in human ganglia: construction of a cDNA library from latently infected human trigeminal ganglia and detection of a VZV transcript. J. Virol. *68*, 7900–7908.

Cohrs, R.J., Barbour, M., and Gilden, D.H. (1996). Varicella-zoster virus (VZV) transcription during latency in human ganglia: detection of transcripts mapping to genes 21, 29, 62, and 63 in a cDNA library enriched for VZV RNA. J. Virol. *70*, 2789–2796.

Cohrs, R.J., Randall, J., Smith, J., Gilden, D.H., Dabrowski, C., van Der Keyl, H., and Tal-Singer, R. (2000). Analysis of individual human trigeminal ganglia for latent herpes simplex virus type 1 and varicella-zoster virus nucleic acids using real-time PCR. J. Virol. *74*, 11464–11471.

Cohrs, R.J., Gilden, D.H., Kinchington, P.R., Grinfeld, E., and Kennedy, P.G. (2003). Varicella-zoster virus gene 66 transcription and translation in latently infected human Ganglia. J. Virol. *77*, 6660–6665.

Davison, A.J., and Scott, J.E. (1986). The complete DNA sequence of varicella-zoster virus. J. Gen. Virol. *67*, 1759–1816.

Esiri, M.M., and Tomlinson, A.H. (1972). Herpes zoster: demonstration of virus in trigeminal nerve and ganglion by immunofluorescence and electron microscopy. J. Neurol. Sci. *15*, 35–48.

Feldman, S., Hughes, W.T., and Daniel, C.B. (1975). Varicella in children with cancer: Seventy-seven cases. Pediatrics *56*, 388–397.

Gershon, A., Chen, J.J., Davis, L., Krinsky, C., Cowles, R., and Gershon, M.D. (2009). Distribution of latent varicella-zoster virus in sensory ganglia and gut after vaccination and wildtype infection: evidence for viremic spread. 34th International Herpesvirus Workshop Ithaca, New York.

Gilden, D.H., Vafai, A., Shtram, Y., Becker, Y., Devlin, M., and Wellish, M. (1983). Varicella-zoster virus DNA in human sensory ganglia. Nature *306*, 478–480.

Gilden, D.H., Rozenman, Y., Murray, R., Devlin, M., and Vafai, A. (1987). Detection of varicella-zoster virus nucleic acid in neurons of normal human thoracic ganglia. Ann. Neurol. 22, 377–380.

Gilden, D.H., Gesser, R., Smith, J., Wellish, M., Laguardia, J.J., Cohrs, R.J., and Mahalingam, R. (2001). Presence of VZV and HSV-1 DNA in human nodose and celiac ganglia. Virus Genes *23*, 145–147.

Grinfeld, E., and Kennedy, P.G. (2004). Translation of varicella-zoster virus genes during human ganglionic latency. Virus Genes *29*, 317–319.

Grose, C. (1981). Variation on a theme by Fenner: the pathogenesis of chickenpox. Pediatrics *68*, 735–737.

Heineman, T.C., and Cohen, J.I. (1995). The varicella-zoster virus (VZV) open reading frame 47 (ORF47) protein kinase is dispensable for viral replication and is not required for phosphorylation of ORF63 protein, the VZV homolog of herpes simplex virus ICP22. J. Virol. *69*, 7367–7370.

Heineman, T.C., Seidel, K., and Cohen, J.I. (1996). The varicella-zoster virus ORF66 protein induces kinase activity and is dispensable for viral replication. J. Virol. *70*, 7312–7317.

Hufner, K., Derfuss, T., Herberger, S., Sunami, K., Russell, S., Sinicina, I., Arbusow, V., Strupp, M., Brandt, T., and Theil, D. (2006). Latency of alpha-herpes viruses is accompanied by a chronic inflammation in human trigeminal ganglia but not in dorsal root ganglia. J. Neuropathol. Exp. Neurol. 65, 1022–1030.

Hyman, R.W., Ecker, J.R., and Tenser, R.B. (1983). Varicella-zoster virus RNA in human trigeminal ganglia. Lancet 2, 814–816.

Hynd, M.R., Lewohl, J.M., Scott, H.L., and Dodd, P.R. (2003). Biochemical and molecular studies using human autopsy brain tissue. J. Neurochem. *85*, 543–562.

Inoue, H., Motani-Saitoh, H., Sakurada, K., Ikegaya, H., Yajima, D., Hayakawa, M., Sato, Y., Otsuka, K., Kobayashi, K., Nagasawa, S., *et al.* (2010). Detection of varicella-zoster virus DNA in 414 human trigeminal ganglia from cadavers by the polymerase chain reaction: a comparison of the detection rate of varicella-zoster virus and herpes simplex virus type 1. J. Med. Virol. *82*, 345–349.

Ito, H., Sommer, M.H., Zerboni, L., He, H., Boucaud, D., Hay, J., Ruyechan, W., and Arvin, A.M. (2003). Promoter sequences of varicella-zoster virus glycoprotein I targeted by cellular transactivating factors Sp1 and USF determine virulence in skin and T cells in SCIDhu mice *in vivo*. J. Virol. *77*, 489–498.

Kamino, H., and Tam, S.T. (1991). Immunoperoxidase technique modified by counterstain with azure B as a diagnostic aid in evaluating heavily pigmented melanocytic neoplasms. J. Cutan Pathol *18*, 436–439.

Katz, D.M., Markey, K.A., Goldstein, M., and Black, I.B. (1983). Expression of catecholaminergic characteristics by primary sensory neurons in the normal adult rat *in vivo*. Proc. Natl. Acad. Sci. U.S.A. *80*, 3526–3530.

Kemble, G.W., Annunziato, P., Lungu, O., Winter, R.E., Cha, T.A., Silverstein, S.J., and Spaete, R.R. (2000). Open reading frame S/L of varicella-zoster virus encodes a cytoplasmic protein expressed in infected cells. J. Virol. *74*, 11311–11321.

Kennedy, P.G., Grinfeld, E., and Gow, J.W. (1998). Latent varicella-zoster virus is located predominantly in neurons in human trigeminal ganglia. Proc. Natl. Acad. Sci. U.S.A. *95*, 4658–4662.

Kennedy, P.G., Grinfeld, E., and Gow, J.W. (1999). Latent Varicella-zoster virus in human dorsal root ganglia. Virology *258*, 451–454.

Kennedy, P.G., Grinfeld, E., and Bell, J.E. (2000). Varicella-zoster virus gene expression in latently infected and explanted human ganglia. J. Virol. *74*, 11893–11898.

Khanna, K.M., Lepisto, A.J., Decman, V., and Hendricks, R.L. (2004). Immune control of herpes simplex virus during latency. Curr. . Opin. Immunol. *16*, 463–469.

Kinchington, P.R., Fite, K., and Turse, S.E. (2000). Nuclear accumulation of IE62, the varicella-zoster virur (VZV) major transcriptional regulatory protein, is inhibited by phosphorylation mediated by the VZV open reading frame 66 protein kinase. J. Virol. *74*, 2265–2277.

Kishikawa, M., Matsumoto, M., Iseki, M., Kondo, H., Shimada, O., Tsuda, N., and Fujii, H. (1988). Quantitative study of neurons with intracytoplasmic pigments in dorsal root ganglia: atomic bomb and aging. Acta Med. Nagasaki 33, 58–62.

Kligora, C.J., Fair, K.P., Clem, M.S., and Patterson, J.W. (1999). A comparison of melanin bleaching and azure blue counterstaining in the immunohistochemical diagnosis of malignant melanoma. Mod Pathol *12*, 1143–1147.

Koropchak, C.M., Solem, S.M., Diaz, P.S., and Arvin, A.M. (1989). Investigation of varicella-zoster virus infection of lymphocytes by *in situ* hybridization. J. Virol. *63*, 2392–2395.

Ku, C.C., Padilla, J.A., Grose, C., Butcher, E.C., and Arvin, A.M. (2002). Tropism of varicella-zoster virus for human tonsillar CD4(+) T lymphocytes that express activation, memory, and skin homing markers. J. Virol. *76*, 11425–11433.

Ku, C.C., Zerboni, L., Ito, H., Graham, B.S., Wallace, M., and Arvin, A.M. (2004). Varicella-zoster virus transfer to skin by T Cells and modulation of viral replication by epidermal cell interferon-alpha. J. Exp. Med. *200*, 917–925.

Ku, C.C., Besser, J., Abendroth, A., Grose, C., and Arvin, A.M. (2005). Varicella-Zoster virus pathogenesis and immunobiology: new concepts emerging from investigations with the SCIDhu mouse model. J. Virol. *79*, 2651–2658.

LaGuardia, J.J., Cohrs, R.J., and Gilden, D.H. (1999). Prevalence of varicella-zoster virus DNA in dissociated human trigeminal ganglion neurons and nonneuronal cells. J. Virol. *73*, 8571–8577.

Levin, M.J., Cai, G.Y., Manchak, M.D., and Pizer, L.I. (2003). Varicella-zoster virus DNA in cells isolated from human trigeminal ganglia. J. Virol. *77*, 6979–6987.

Li, Q., Ali, M.A., and Cohen, J.I. (2006). Insulin degrading enzyme is a cellular receptor mediating varicella-zoster virus infection and cell-to-cell spread. Cell *127*, 305–316.

Li, Q., Krogmann, T., Ali, M.A., Tang, W.J., and Cohen, J.I. (2007). The amino terminus of varicella-zoster virus (VZV) glycoprotein E is required for binding to insulin-degrading enzyme, a VZV receptor. J. Virol. *81*, 8525–8532.

Lillie, R.D. (1956). A Nile blue staining technic for the differentiation of melanin and lipofuscins. Stain Technol *31*, 151–153.

Lungu, O., Annunziato, P.W., Gershon, A., Staugaitis, S.M., Josefson, D., LaRussa, P., and Silverstein, S.J. (1995). Reactivated and latent varicella-zoster virus in human dorsal root ganglia. Proc. Natl. Acad. Sci. U.S.A. *92*, 10980–10984.

Lungu, O., Panagiotidis, C.A., Annunziato, P.W., Gershon, A.A., and Silverstein, S.J. (1998). Aberrant intracellular localization of Varicella-Zoster virus regulatory proteins during latency. Proc. Natl. Acad. Sci. U.S.A. *95*, 7080–7085.

Mahalingam, R., Wellish, M.C., Dueland, A.N., Cohrs, R.J., and Gilden, D.H. (1992). Localization of herpes simplex virus and varicella zoster virus DNA in human ganglia. Ann. Neurol. 31, 444–448.

Mahalingam, R., Wellish, M., Lederer, D., Forghani, B., Cohrs, R., and Gilden, D. (1993). Quantitation of latent varicella-zoster virus DNA in human trigeminal ganglia by polymerase chain reaction. J. Virol. *67*, 2381–2384.

Mahalingam, R., Wellish, M., Cohrs, R., Debrus, S., Piette, J., Rentier, B., and Gilden, D.H. (1996). Expression of protein encoded by varicella-zoster virus open reading frame 63 in latently infected human ganglionic neurons. Proc. Natl. Acad. Sci. U.S.A. 93, 2122–2124.

Mallory, S., Sommer, M., and Arvin, A.M. (1997). Mutational analysis of the role of glycoprotein I in varicella-zoster virus replication and its effects on glycoprotein E conformation and trafficking. J. Virol. *71*, 8279–8288.

Meier, J.L., Holman, R.P., Croen, K.D., Smialek, J.E., and Straus, S.E. (1993). Varicella-zoster virus transcription in human trigeminal ganglia. Virology *193*, 193–200.

Miliauskas, J.R., and Webber, B.L. (1984). Disseminated varicella at autopsy in children with cancer. Cancer *53*, 1518–1525.

Mo, C., Lee, J., Sommer, M., Grose, C., and Arvin, A.M. (2002). The requirement of varicella zoster virus glycoprotein E (gE) for viral replication and effects of glycoprotein I on gE in melanoma cells. Virology *304*, 176–186.

Moffat, J.F., Stein, M.D., Kaneshima, H., and Arvin, A.M. (1995). Tropism of varicella-zoster virus for human $CD4^+$ and $CD8^+$ T lymphocytes and epidermal cells in SCID-hu mice. J. Virol. *69*, 5236–5242.

Moffat, J.F., Zerboni, L., Kinchington, P.R., Grose, C., Kaneshima, H., and Arvin, A.M. (1998a). Attenuation of the vaccine Oka strain of varicella-zoster virus and role of glycoprotein C in alphaherpesvirus virulence demonstrated in the SCID-hu mouse. J. Virol. *72*, 965–974.

Moffat, J.F., Zerboni, L., Sommer, M.H., Heineman, T.C., Cohen, J.I., Kaneshima, H., and Arvin, A.M. (1998b). The ORF47 and ORF66 putative protein kinases of varicella-zoster virus determine tropism for human T cells and skin in the SCID-hu mouse. Proc. Natl. Acad. Sci. U.S.A. *95*, 11969–11974.

Moffat, J., Ito, H., Sommer, M., Taylor, S., and Arvin, A.M. (2002). Glycoprotein I of varicella-zoster virus is required for viral replication in skin and T cells. J. Virol. *76*, 8468–8471.

Moffat, J., Mo, C., Cheng, J.J., Sommer, M., Zerboni, L., Stamatis, S., and Arvin, A.M. (2004). Functions of the C-terminal domain of varicella-zoster virus glycoprotein E in viral replication *in vitro* and skin and T-cell tropism *in vivo*. J. Virol. *78*, 12406–12415.

Moriuchi, H., Moriuchi, M., and Cohen, J.I. (1995). Proteins and cis-acting elements associated with transactivation of the varicella-zoster virus (VZV) immediate-early gene 62 promoter by VZV open reading frame 10 protein. J. Virol. *69*, 4693–4701.

Nadji, M. (1986). Immunoperoxidase techniques. I. Facts and artifacts. Am J. Dermatopathol *8*, 32–36.

Niizuma, T., Zerboni, L., Sommer, M.H., Ito, H., Hinchliffe, S., and Arvin, A.M. (2003). Construction of varicella-zoster virus recombinants from parent Oka cosmids and demonstration that ORF65 protein is dispensable for infection of human skin and T cells in the SCID-hu mouse model. J. Virol. *77*, 6062–6065.

Oliver, S.L., Zerboni, L., Sommer, M., Rajamani, J., and Arvin, A.M. (2008). Development of recombinant varicella-zoster viruses expressing luciferase fusion proteins for live *in vivo* imaging in human skin and dorsal root ganglia xenografts. J. Virol. Methods *154*, 182–193.

Oliver, S.L., Sommer, M., Zerboni, L., Rajamani, J., Grose, C., and Arvin, A.M. (2009). Mutagenesis of varicella-zoster virus glycoprotein B: putative fusion loop residues are essential for viral replication, and the furin cleavage motif contributes to pathogenesis in skin tissue *in vivo*. J. Virol. *83*, 7495–7506.

Pevenstein, S.R., Williams, R.K., McChesney, D., Mont, E.K., Smialek, J.E., and Straus, S.E. (1999). Quantitation of latent varicella-zoster virus and herpes simplex virus genomes in human trigeminal ganglia. J. Virol. *73*, 10514–10518.

Plotkin, S.A., Starr, S.E., Connor, K., and Morton, D. (1989). Zoster in normal children after varicella vaccine. J. Infect. Dis. *159*, 1000–1001.

Price, J., and Mudge, A.W. (1983). A subpopulation of rat dorsal root ganglion neurones is catecholaminergic. Nature *301*, 241–243.

Quinlivan, M.A., Gershon, A.A., Nichols, R.A., La Russa, P., Steinberg, S.P., and Breuer, J. (2006). Vaccine Oka varicella-zoster virus genotypes are monomorphic in single vesicles and polymorphic in respiratory tract secretions. J. Infect. Dis. *193*, 927–930.

Quinlivan, M.L., Gershon, A.A., Al Bassam, M.M., Steinberg, S.P., LaRussa, P., Nichols, R.A., and Breuer, J. (2007). Natural selection for rash-forming genotypes of the varicella-zoster vaccine virus detected within immunized human hosts. Proc. Natl. Acad. Sci. U.S.A. *104*, 208–212.

Ramos-Vara, J.A. (2005). Technical aspects of immunohistochemistry. Vet Pathol *42*, 405–426.

Reichelt, M., Zerboni, L., and Arvin, A.M. (2008). Mechanisms of varicella-zoster virus neuropathogenesis in human dorsal root ganglia. J. Virol. *82*, 3971–3983.

Roizman, B., Knipe, D.M., and Whitley, R., eds. (2007). Herpes Simplex Viruses. In Fields' virology, Fields, B.N., Knipe, D.M., and Howley, P.M., eds. (Philadelphia, Wolters Kluwer Health/Lippincott Williams & Wilkins).

Sawyer, M.H., Wu, Y.N., Chamberlin, C.J., Burgos, C., Brodine, S.K., Bowler, W.A., LaRocco, A., Oldfield, E.C., 3rd, and Wallace, M.R. (1992). Detection of varicella-zoster virus DNA in the oropharynx and blood of patients with varicella. J. Infect. Dis. *166*, 885–888.

Sato, B., Ito, H., Hinchliffe, S., Sommer, M.H., Zerboni, L., and Arvin, A.M. (2003). Mutational analysis of open reading frames 62 and 71, encoding the varicella-zoster virus immediate-early transactivating protein, IE62, and effects on replication *in vitro* and in skin xenografts in the SCID-hu mouse *in vivo*. J. Virol. *77*, 5607–5620.

Schaap, A., Fortin, J.F., Sommer, M., Zerboni, L., Stamatis, S., Ku, C.C., Nolan, G.P., and Arvin, A.M. (2005). T-cell tropism and the role of ORF66 protein in pathogenesis of varicella-zoster virus infection. J. Virol. *79*, 12921–12933.

Schaap-Nutt, A., Sommer, M., Che, X., Zerboni, L., and Arvin, A.M. (2006). ORF66 protein kinase function is required for T-cell tropism of varicella-zoster virus *in vivo*. J. Virol. *80*, 11806–11816.

Sommer, M.H., Zagha, E., Serrano, O.K., Ku, C.C., Zerboni, L., Baiker, A., Santos, R., Spengler, M., Lynch, J., Grose, C., *et al.* (2001). Mutational analysis of the repeated open reading frames, ORFs 63 and 70 and ORFs 64 and 69, of varicella-zoster virus. J. Virol. *75*, 8224–8239.

Sulzer, D., Mosharov, E., Talloczy, Z., Zucca, F.A., Simon, J.D., and Zecca, L. (2008). Neuronal pigmented autophagic vacuoles: lipofuscin, neuromelanin, and ceroid as macroautophagic responses during aging and disease. J. Neurochem. *106*, 24–36.

Takahashi, M., Otsuka, T., Okuno, Y., Asano, Y., and Yazaki, T. (1974). Live vaccine used to prevent the spread of varicella in children in hospital. Lancet *2*, 1288–1290.

Theil, D., Derfuss, T., Paripovic, I., Herberger, S., Meinl, E., Schueler, O., Strupp, M., Arbusow, V., and Brandt, T. (2003). Latent herpesvirus infection in human trigeminal ganglia causes chronic immune response. Am J. Pathol. *163*, 2179–2184.

Vleck, S.E., Oliver, S.L., Reichelt, M., Rajamani, J., Zerboni, L., Jones, C., Zehnder, J., Grose, C., and Arvin, A.M. (2010). Anti-glycoprotein H antibody impairs the pathogenicity of varicella-zoster virus in skin xenografts in the SCID mouse model. J. Virol. *84*, 141–152.

Wang, K., Lau, T.Y., Morales, M., Mont, E.K., and Straus, S.E. (2005). Laser-capture microdissection: refining estimates of the quantity and distribution of latent herpes simplex virus 1 and varicella-zoster virus DNA in human trigeminal Ganglia at the single-cell level. J. Virol. *79*, 14079–14087.

Wang, L., Sommer, M., Rajamani, J., and Arvin, A.M. (2009). Regulation of the ORF61 promoter and ORF61 functions in varicella-zoster virus replication and pathogenesis. J. Virol. *83*, 7560–7572.

Yao, Z., Jackson, W., Forghani, B., and Grose, C. (1993). Varicella-zoster virus glycoprotein gpI/gpIV receptor: expression, complex formation, and antigenicity within the vaccinia virus-T7 RNA polymerase transfection system. J. Virol. *67*, 305–314.

Zerboni, L., Sommer, M., Ware, C.F., and Arvin, A.M. (2000). Varicella-zoster virus infection of a human CD4-positive T-cell line. Virology *270*, 278–285.

Zerboni, L., Hinchliffe, S., Sommer, M.H., Ito, H., Besser, J., Stamatis, S., Cheng, J., Distefano, D., Kraiouchkine, N., Shaw, A., *et al.* (2005a). Analysis of varicella zoster virus attenuation by evaluation of chimeric parent Oka/vaccine Oka recombinant viruses in skin xenografts in the SCIDhu mouse model. Virology *332*, 337–346.

Zerboni, L., Ku, C.C., Jones, C.D., Zehnder, J.L., and Arvin, A.M. (2005b). Varicella-zoster virus infection of human dorsal root ganglia *in vivo*. Proc. Natl. Acad. Sci. U.S.A. *102*, 6490–6495.

Zerboni, L., Reichelt, M., Jones, C.D., Zehnder, J.L., Ito, H., and Arvin, A.M. (2007). From the Cover: Aberrant infection and persistence of varicella-zoster virus in human dorsal root ganglia *in vivo* in the absence of glycoprotein I. Proc. Natl. Acad. Sci. U.S.A. *104*, 14086–14091.

Zerboni, L., and Arvin, A. (2008). The pathogenesis of varicella-zoster virus neurotropism and infection. In Neurotropic Viral Infections, Reiss, C., ed. (Cambridge, Cambridge University Press), pp. 225–251.

Zerboni, L., Sobel, R.A., Ramachandran, V., Rajamani, J., Ruyechan, W., Abendroth, A., and Arvin, A. (2010). The expression of varicella-zoster virus immediate early regulatory protein IE63 in neurons of latently infected human sensory ganglaia. Virol, J. *84*, 3421–3430.

HSV-1 Latency and the Roles of the LATs

17

David C. Bloom and Dacia L. Kwiatkowski

Abstract

Herpes simplex virus type 1 (HSV-1) latency is characterized by the persistence of viral genomes as episomes in the nuclei of sensory neurons. During this period only one region of the genome is abundantly transcribed: the region encoding the latency-associated transcripts (LATs). The LAT domain is transcriptionally complex, and while the predominant species that accumulates during latency is a 2.0 kb stable intron, other RNA species are transcribed from this region of the genome, including a number of lytic or acute-phase transcripts. In addition, a number of microRNA (miRNA) and non-miRNA small RNAs have recently been mapped to the LAT region of the genome. HSV-1 recombinant viruses with deletions of the LAT promoter exhibit reactivation deficits in a number of animal models, and there is evidence that other LAT deletion mutants also possess altered establishment and virulence properties. The phenotypic complexity associated with this region, as well as evidence that the LATs may play a role in suppressing latent gene expression, suggests that the LAT locus may function as a regulator to modulate the transcription of key lytic and latent genes. The goal of this review is to provide an overview of our current understanding of the role(s) of the HSV-1 latency associated transcripts (LATs) in the pathobiology of HSV-1 infections *in vivo*.

Introduction to HSV-1 latency

HSV-1 establishes a life-long latent infection within sensory ganglia. During latency, viral genomes are maintained in the nucleus of each cell as multiple episomal copies that do not integrate into the host genome. These circular episomes are associated with histones and are organized into nucleosomes (Deshmane and Fraser, 1989; Efstathiou *et al.*, 1986). During latency, transcription of viral productive-phase (lytic) RNAs is repressed, and abundant transcription is only seen from one region of the viral genome: the region encoding the latency-associated transcripts (LATs) (Spivack and Fraser, 1987; Stevens *et al.*, 1987).

HSV-1 latency is primarily studied in one of two well-established animal models, the rabbit and the mouse (for a review see Wagner and Bloom, 1997). Latent infection in the rabbit is typically established in the trigeminal ganglia (TG) following ocular infection, whereas in the mouse either the TG or dorsal root ganglia (DRG) are the major sites of latency following ocular or footpad infections, respectively. Following such infections, as many as 10–30% of the sensory neurons in a ganglion can be demonstrated to contain latent genomes. As in the case of human infections, latent HSV-1 genomes in some latently-infected neurons in mice and rabbits periodically reactivate, presumably in response to stress (for a review see Mitchell *et al.*, 2003). In experimental models of reactivation, the production of infectious virus can be induced by thermal stress and UV light in the mouse model, and adrenergic agents such as adrenaline in the rabbit model. An important point regarding experimental animal models of HSV-1 and -2 latency is that while the mouse does not shed virus at the periphery following reactivation, the rabbit and guinea pig models of HSV-1 and

HSV-2 latency respectively, do exhibit spontaneous shedding of virus at the original site of infection during latency (Berman and Hill, 1985; Stanberry *et al.*, 1982). The spontaneous shedding of those models more closely mirrors human HSV-1 and HSV-2 clinical shedding (Koelle and Corey, 2008).

Ever since their initial discovery, the fact that the LATs are the only gene products that can be abundantly and consistently detected during latency has made these RNAs a focus of intense study. Early on, it was shown that viruses with deletions of the LAT promoter were capable of establishing latent infections, but are severely restricted in their ability to be reactivated in models of induced reactivation (Hill *et al.*, 1990; Leib *et al.*, 1989). Interestingly, LAT promoter mutants can be induced to reactivate to near wild-type levels *ex vivo* following explant cultivation (Devi-Rao *et al.*, 1994). Therefore, it is important to highlight that while the LATs play a key role in induced reactivation (such as adrenaline-induced reactivation in the rabbit), not all models of reactivation (such as explant) show the same degree of LAT-dependence. In a similar vein, certain LAT mutants, particularly ones possessing large deletions as described below, exhibit 2- to 4-fold reductions in the number of latently-infected neurons (Perng *et al.*, 2000b; Thompson and Sawtell, 2001). This effect may be dependent on both the region of LAT that is deleted and the animal model (O'Neil *et al.*, 2004). The model-dependence of LAT phenotypes has made the definition of specific mechanisms for the LAT's role in these processes difficult to define. The following sections will attempt to summarize recent findings that have broadened our view of the range of phenotypes that LAT mutants possess. These findings suggest that the diverse functions attributed to the LATs may be due to regulatory functions contained within the LAT region that modulate the transcription of other genes during latency. These functions may influence the type of neuron in which latency is preferentially established and/or direct repressive epigenetic modifications to lytic genes on the latent genome in a manner that can be reversed to facilitate reactivation.

The LATs and their expression during HSV-1 latency

The LATs, or what are sometimes referred to singularly as the 'LAT,' are produced in neurons during latency as a primary 8.3–8.5 kb polyadenylated transcript (Fig. 17.1). This primary transcript is spliced to yield a stable 2.0 kb intron (also referred to as the major LAT), which is further spliced to yield a 1.5 kb intron in a subset of the neurons (Farrell *et al.*, 1991). The 2.0 kb intron accumulates abundantly in the nuclei of latently-infected neurons (Fig. 17.2), due largely to the fact that it is an extremely stable intron with a half-life on the order of 24 hours (Thomas *et al.*, 2002). This impressive stability results from inefficient debranching during splicing, resulting in the intron persisting as a lariat (Rodahl and Haarr, 1997; Wu *et al.*, 1996). This inefficient debranching has been proven to be the result of a non-consensus branch point, which when mutated to a consensus sequence, results in destabilization of the stable intron (Mukerjee *et al.*, 2004). Interestingly, this same study revealed that the virus containing the consensus branch point had no detectable differences from wild-type virus in replication or *in vivo* pathobiology.

Limited LAT transcription can be detected during an acute infection of sensory ganglia. When mice are infected with recombinant HSV-1 viruses containing a reporter gene under control of the LAT promoter, the LAT promoter is expressed almost exclusively in cells that do not express acute antigen (Margolis *et al.*, 1992b). These are likely cells where the virus is entering latency and strongly suggests that LAT expression and lytic gene expression are not compatible. While the major LAT accumulates abundantly in the nuclei of many latently-infected neurons, it is now clear that not all neurons that harbour HSV-1 DNA during latency produce detectable levels of LAT. Carefully controlled *in situ*/PCR hybridization analyses have revealed that only 1/3 of HSV DNA-containing neurons contain detectable levels of LAT (Gressens and Martin, 1994; Mehta *et al.*, 1995). This suggests that there are at least two different populations of latently-infected neurons as described by their transcriptional profiles: those expressing the LAT

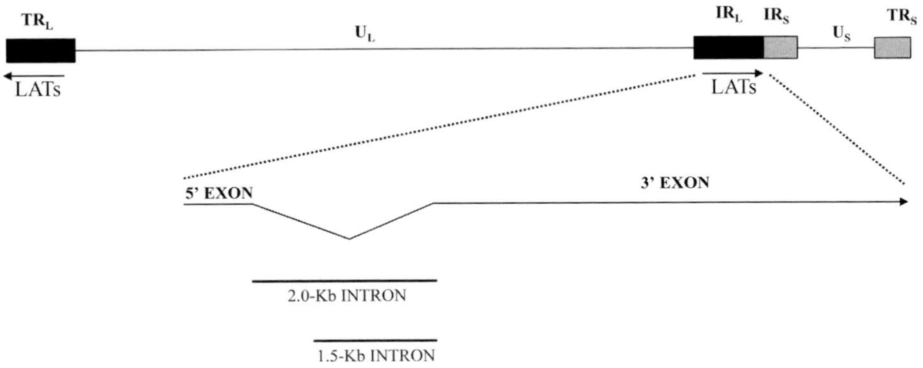

Figure 17.1 Diagram of the region encoding the HSV-1 LATs. Depiction of the HSV-1 genome showing the unique long (U_L) and short (U_S) regions that are flanked by the terminal repeat long (TR_L), internal repeat long (IR_L) and the internal repeat short (IR_S) and terminal repeat short (TR_S) respectively. The location of the LATs within the R_L is shown. The primary LAT RNA is an 8.5 kb transcript that is spliced to yield a stable intron of 2.0 kb, as well as a further spliced species of 1.5 kb that is seen in a subset of neurons.

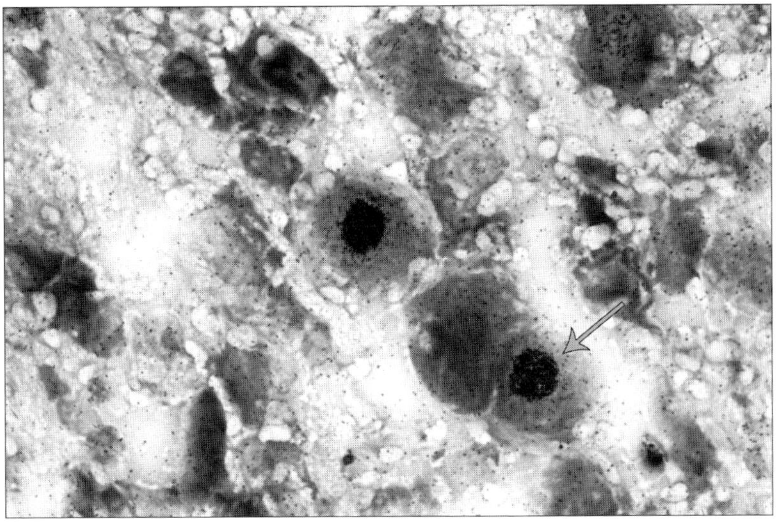

Figure 17.2 *In situ* hybridization of a latently infected rabbit TG showing the accumulation of the LAT 2.0 kb intron in the nuclei of two large sensory neurons. 5 µm section of a rabbit TG latently infected with HSV-1 strain 17*syn*+ probed with a tritium-labelled probe for the LAT 2.0 kb intron, and subsequently counterstained with Giemsa. The arrow is pointing to the nucleus of one of the two LAT positive neurons illustrating the abundance of silver grains indicating positive hybridization for LAT.

and those that are not. How the expression of the LATs alters transcription of lytic genes has not yet been mechanistically established, however recent findings and potential models are discussed later in this chapter.

The LAT seems to be expressed predominantly in certain phenotypic classes of sensory neurons, and not in others (Margolis *et al.*, 1992a; Yang *et al.*, 2000). The HSV-1 LATs are expressed in a sensory neuronal subtype classified by the marker A5, which differs from HSV-2 LATs which are predominately expressed in KH10 subtypes (Yang *et al.*, 2000). It was found that swapping the LAT regions between HSV-1 and HSV-2 to create chimeric HSV viruses was sufficient to cause switching of the preferential neuronal subtype, suggesting that the LATs influence the sub-type of neurons in which HSV establishes a latent infection (Margolis *et al.*, 2007). The region of the LAT responsible for this phenotype was further mapped to between the LAT TATA box and the splice site of the intron, in particular the sequences immediately after the LAT TATA region (Bertke *et al.*, 2009).

Regulation of LAT transcription during latency

The LAT promoter is relatively complex compared to most other HSV promoters and contains a number of known cellular transcription factor binding sites (Fig. 17.3). While the core promoter is contained within an approximately 200 bp fragment, additional *cis*-elements which confer neuronal-specific transcription extend some 650 bp further upstream (Fig. 17.3) (Dobson *et al.*, 1995). In addition, an enhancer element that has been shown to be critical for the expression of the LATs during latency resides several hundred bp downstream of the transcriptional start site (Fig. 17.4A) (Berthomme *et al.*, 2000; Lokensgard *et al.*, 1997).

Despite its relative complexity, the LAT promoter itself is not sufficient to maintain long-term latent expression of the LATs or that of heterologous transgenes during HSV-1 latency (Margolis *et al.*, 1993). Expression during latency has been shown to be due, at least in part, to enhancer activity located in the region encoding the 5′ exon of LAT. The Feldman lab demonstrated that this 5′ exon region was able to enhance the activities of the LAT and heterologous promoters (Berthomme *et al.*, 2000). In addition, this element plays a role in maintaining LAT promoter activity during latency and for this reason was named the long-term expression (LTE) element (Lokensgard *et al.*, 1994, 1997; Margolis *et al.*, 1993) (Fig. 17.4).

In order to study LAT processing and expression in the absence of *cis* and *trans* viral factors, a transgenic mouse was created that contains the

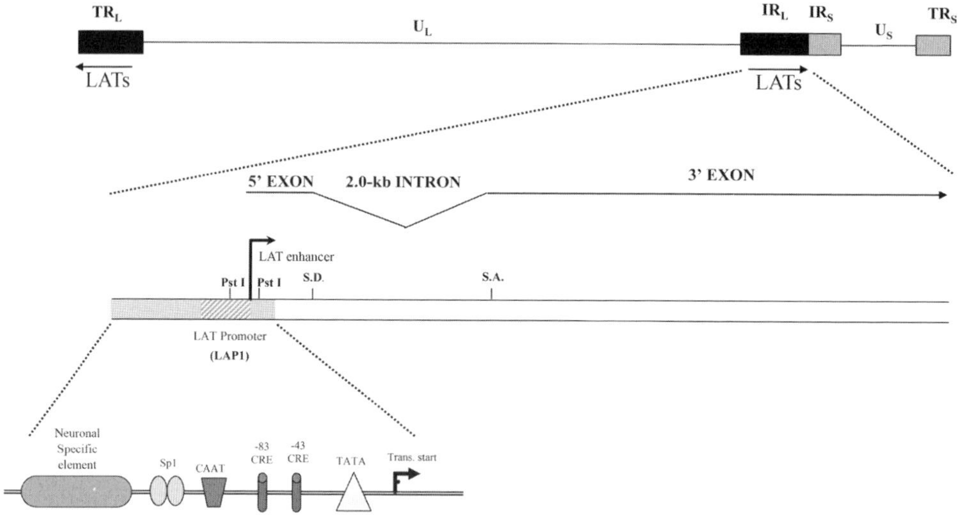

Figure 17.3 Diagram of the HSV-1 LAT promoter. Expanded view of the region encoding the 8.5 kb primary LAT transcript showing the LAT promoter (grey shaded region representing approximately 875 bp), location of the transcriptional start site (arrow) and the location of the splice donor (S.D.) and splice acceptor (S.A.) sites for the 2.0 kb intron. The core LAT promoter, containing elements necessary for basal transcription, is indicated by the cross-hatching. The expanded view of the promoter shows some of the transcription factor binding sites that have been identified, including a TATA box (Zwaagstra *et al.*, 1991), and two CRE sites (Kenny *et al.*, 1994; Leib *et al.*, 1991), as well as an upstream region shown to possess elements required for high-level expression of LAT in neurons (Dobson *et al.*, 1995; Zwaagstra *et al.*, 1990). The position of the LAT enhancer that contains elements required for sustained latent expression of the LAT is also indicated.

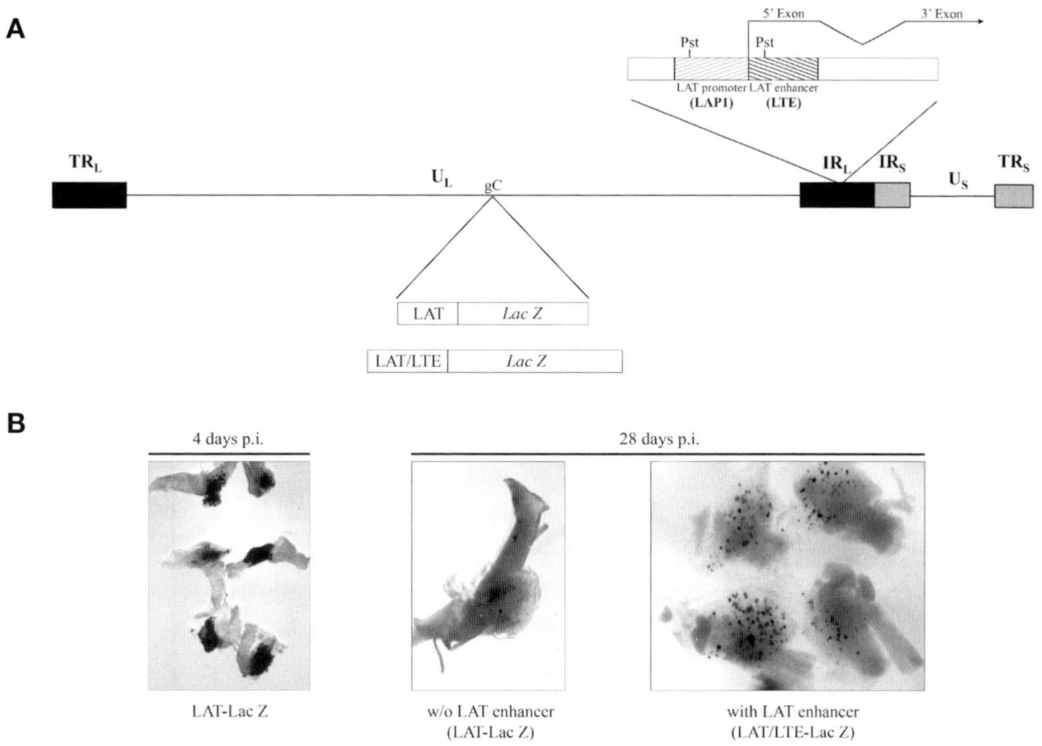

Figure 17.4 The LAT enhancer is required to maintain latent expression of a transgene from the LAT promoter. (A) Above the genome: An expanded view of the IR$_L$ shows the relative location of the LAT promoter (LAP1) and the LAT enhancer (LTE). Below the genome: the insertion of two *lac Z* reporter cassettes into the glycoprotein C (gC) gene. One cassette contains the ~875 bp LAT promoter without the LAT enhancer (LAT) and the other contains the LAT promoter and the enhancer (LAT/LTE). Mice were infected with these two recombinants by the footpad route and DRG (L4 + L5) were removed and stained with x-gal to visualize the reporter gene activity. (B) 4 days p.i. both recombinants exhibited strong x-gal staining (only the LAT-*Lac Z* is shown). In contrast at 28 days p.i., only a few x-gal stained neurons can be observed in the LAT-*LacZ* recombinant, whereas the LAT enhancer recombinant demonstrates sustained expression.

LAT promoter through the 2.0 kb intron. Though LAT expression could be detected in a number of tissues by RT-PCR, it was only in the sensory ganglia and several discrete regions of the CNS that accumulation of the 2.0 kb intron was abundant, thus revealing that regulation of expression of the LAT is highly cell-type specific and does not rely on viral *cis* or *trans* factors. Additionally, the presence of the LAT *in trans* did not alter the course of HSV-1 infection as measured by lytic replication, the establishment of latency and reactivation (Gussow *et al.*, 2006). Similar data were obtained in a transgenic mouse expressing the HSV-2 LAT (Wang *et al.*, 2009). It is interesting to note that the intron in both the HSV-1 and HSV-2 LAT transgenic mice did not show preferential accumulation in the TG vs. the DRG or in A5 vs. KH10 neurons, suggesting that the influence of the LAT on directing cellular tropism is not solely a reflection of cellular regulation of LAT transcription.

The transcriptional complexity of the LAT locus

The entire LAT locus is arguably one of the most transcriptionally complex regions in the HSV genome. In addition to the LATs, there are a number of other transcripts that have been mapped to this region. One is co-linear with the 2.0 kb intron and is actually produced from a promoter, termed the LAP-2 promoter, which is located within the region corresponding to the 5′ exon of the 8.5 kb

LAT (Fig. 17.5) (Goins *et al.*, 1994; Maggioncalda *et al.*, 1994; Nicosia *et al.*, 1993). Similarly, the 'upstream of LAT' (UOL) transcript originates upstream of the LAT promoter and is transcribed in a sense orientation to the LAT (Naito *et al.*, 2005). In this same area, a 0.7 kb transcript starts upstream of the LAT and extends into the promoter of LAT (Zhu *et al.*, 1999). Likewise, there are transcripts anti-sense to LAT in the region. The 'antisense to LAT' (AL-LAT) transcript, as described by Perng and Wechsler (Perng *et al.*, 2002) has a promoter that overlaps a portion of the LAP-2 promoter and is transcribed antisense to the LAT promoter. Another two transcripts antisense to the LAT were also identified; AL2 in the first 1.5 kb of LAT and AL3, which originates in the LAT intron (Jaber *et al.*, 2009).

Recently, smaller sized RNAs were added to the above list. Two small RNAs (sRNAs) were found encoded in the first 1.5 kb of LAT termed LAT sRNA1 and sRNA2, having lengths of 62 nt and 36 nt, respectively (Peng *et al.*, 2008). In addition, the LAT region was found to encode 8 miRNAs that can be detected in mouse TG during latency (Cui *et al.*, 2006; Umbach *et al.*, 2008; Umbach *et al.*, 2009). The primary LAT transcript itself encodes 7 of the miRNAs, while the eighth is derived from a transcript upstream of the LAT promoter (unpublished data). In addition, there have been a number of less abundant RNAs originating from the LAT region identified by deep sequencing that have the potential to be miRNAs (Jurak *et al.*, 2010). A more detailed discussion of the HSV-encoded miRNAs is found in Chapter 22 of this book. It should be pointed out that while *in vitro* analyses of these miRNAs suggest that they play a role in regulating lytic gene expression, to date there is no data demonstrating that deleting these miRNAs from the viral genome affects viral replication *in vitro*, or alters the pathobiology of the virus *in vivo*.

If we also look along the full length of the 8.5 kb LAT, we see that there are some very important and well-characterized lytic transcripts transcribed from this region including ICP4, ICP34.5, and ICP0. Clearly, the opportunities for complex regulation within this region are abundant. It is important to note that the LATs

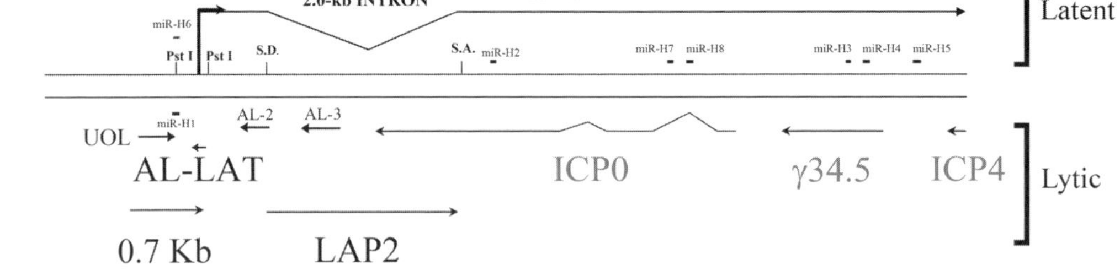

Figure 17.5 The transcriptional complexity of the HSV-1 LAT region. Expanded view of the LAT region shows transcripts that have been mapped to this region. These include transcripts that are transcribed primarily during latency: the 8.5 kb LAT, the 2.0 and 1.5 kb LAT intron species; the sRNAs; 6 LAT-encoded miRNAs and mir-H6, encoded by a transcript antisense to a portion of the LAT and to the LAT promoter. Also indicated are transcripts that are transcribed during the acute or lytic infection: AL-LAT, LAP2, ICP0, ICP34.5, ICP4, and miR-H1, and the names of transcripts that produce viral proteins are in grey. In addition the transcripts 'upstream of LAT' (UOL), and two anti-sense to LAT (AL) transcripts (AL2 and AL3) are present during lytic infection and at low levels n latently infected mouse TG.

(transcribed from the LAP1 promoter) are the only transcripts that are transcribed abundantly during latency, whereas these other transcripts templated in this region are transcribed primarily during the acute infection (Fig. 17.5).

Phenotypes associated with LAT deletion mutants

Given the transcriptional complexity of this region, it is probably not surprising that there have been many functions attributed to the LATs, including reactivation, establishment of latency, virulence, anti-apoptosis and apoptosis. A critical consideration is that it is very difficult to determine which of these phenotypes are directly encoded genetically within the LAT locus, and which of these functions might be secondary effects of regulatory properties of the LAT locus. For example, if there were direct effects of the LAT region on the establishment of latency, it would be difficult to separate the impact of these establishment effects on reactivation from direct effects of the LAT region on reactivation itself. Another example is that if there were basic LAT functions responsible for transcriptional regulation of some lytic genes (such as ICP0) during latency, it would be difficult to separate the loss of that function and the resulting indirect induction of apoptosis that might result from the downstream effects of leaky transcription from a direct role of the LAT in suppressing cellular apoptotic effectors (Bloom, 2004). Clearly, sorting out direct functions of the LATs has proven very difficult. Fig. 17.6 attempts to summarize the relative map positions of the phenotypic functions that have been mapped to the LAT locus. In constructing this functional genetic map, one has to consider that a range of LAT mutants that do not necessarily overlap in their deletions were used in the individual analyses and that these studies were performed in different viral backgrounds and characterized for their phenotypes in different animal models. With these caveats, the map is useful in pointing out a key point:

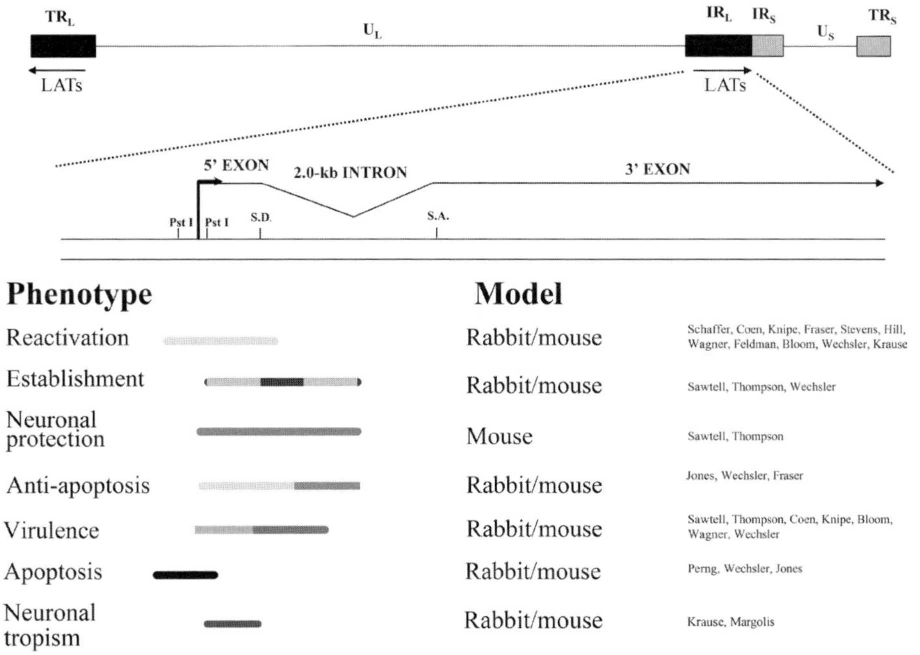

Figure 17.6 Phenotypes associated with the HSV-1 LAT region. Diagram illustrating the rough map positions for the various phenotypes that have been detected in LAT mutants. The model system (mouse or rabbit) in which these effects were noted is indicated, as well as the contributing laboratories. See the text for literature citations.

most of the phenotypes described for the LAT map to a region that extends from roughly the 5′ end of the LAT promoter through the region encoding the first 1 kb of the LAT intron, just 3′ of the ICP0 coding region. This is clearly a very important regulatory region for the virus. Below is a summary of the data supporting each of the phenotypes attributed to the LATs shown in Fig. 17.6.

Reactivation

The first phenotype attributed to HSV-1 recombinants containing deletions in the LAT region was decreased reactivation in both mouse and rabbit models (Bloom *et al.*, 1994; Hill *et al.*, 1990; Leib *et al.*, 1989; Perng *et al.*, 1994). This restriction is most evident in models of *in vivo* reactivation, such as hyperthermia induction in the mouse (Sawtell and Thompson, 1992) and adrenaline induction in the rabbit (Bloom *et al.*, 1994), and produces a less dramatic reduction in models of explant cultivation (Block *et al.*, 1993; Devi-Rao *et al.*, 1994). Reactivation of HSV-2 in the guinea pig vaginal model is also greatly reduced for HSV-1 mutants with LAT promoter (LAP1) deletions (Krause *et al.*, 1995). Deletions within the LAT region remain the only deletion/mutation in the viruses that specifically affect reactivation efficiencies while not influencing replicative yields of the virus either *in vivo* or *in vitro*.

The region of the LAT that is linked to decreased reactivation varies slightly between model systems and is largely based on the resolution afforded by the particular deletion mutants used to map the location. In all cases the reactivation phenotype resides within the region encompassing the LAT promoter and the first 1.5 kb of the primary LAT transcript (Bloom *et al.*, 1996a; Perng *et al.*, 1996). In the case of induced reactivation in the rabbit, the phenotype resides within the first 663 bp of the LAT transcript, or the region corresponding to roughly the 5′ exon (Jarman *et al.*, 2002). Significantly, the presence of the 2.0 kb intron, the most prominent LAT species, is not required for reactivation. However, deletions of ~350 nt or greater within the 5′ exon region reduce reactivation, whereas smaller deletions have no effect (Bloom *et al.*, 1996b). This region, termed the reactivation critical region (*rcr*), includes the LAT enhancer, and may suggest that the enhancer element, and not the LAT RNA itself, is important in reactivation.

Establishment

Initial analysis of LAT deletion mutants and their rescues suggested that they established equivalent latent infections, as assessed by total HSV-1 DNA present in ganglia during latent times by dot blot hybridization and semi-quantitative PCR analysis (Sedarati *et al.*, 1989; Steiner *et al.*, 1989). Recently, several lines of evidence have been put forth suggesting that certain LAT mutants exhibit decreased establishment. Using contextual analysis, where individual cells from latently-infected ganglia are assayed for the presence of latent HSV DNA by PCR, LAT deletion mutants were shown to be present in up to half as many cells in mouse TG as wild-type virus (Sawtell, 1997). This observation was extended in studies using HSV-1 recombinants engineered to express reporter genes, either in a wild-type or LAT-deletion viral backbone. Ganglia infected with the LAT deletion viruses exhibited roughly 1/3 to 1/2 as many reporter-positive cells as their wild-type (or rescued) counterparts (Perng *et al.*, 2000b; Thompson and Sawtell, 2001). Finally, in a well-controlled quantitative PCR (QT-PCR) analysis of a large number of murine TG demonstrated a 2- to 4-fold reduction in the amount of HSV-1 DNA present within the ganglia latently-infected with the LAT deletion viruses, compared to wild-type and rescue infected ganglia (Thompson and Sawtell, 2001). Collectively, these data support a role for the LATs in enhancing the establishment of latency, both in terms of total HSV DNA present, and in the total number of cells harbouring latent DNA. The mutants analysed in these studies contained relatively large deletions encompassing the LAT promoter and extending 3′ through a portion of the region encoding the LAT intron, leaving some debate whether the decreased establishment also would occur following infection with LAT mutants possessing smaller deletions. In fact, extensive analyses have been performed with recombinants possessing smaller LAT deletions and establishment defects have not been detected, though it must be noted that the sample sizes employed in these studies would not have allowed a statistically meaningful detection of just a 2- to 4-fold difference (O'Neil *et al.*, 2004). None-the-less, it seems clear that some genetic function that reduces establishment by 2- to 4-fold resides within at least the first 1.9 kb of the region encoding the LATs. Ultimately, the most important question is whether this degree of decreased establishment directly translates into decreased reactivation. The strongest argument against this comes from studies in the rabbit model. Following an ocular infection with wild-type HSV-1, there is approximately a 10-fold range in the amount of HSV-1 DNA present within the rabbit TG after latency has been established (O'Neil *et al.*, 2004). Even given this variation, 80–100% of the ganglia can be induced to reactivate the latent virus following adrenaline induction, and there is no observed correlation between amount of DNA and the ability to reactivate. When LAT promoter mutants are examined, there is no observed difference in the overall range of establishment by LAT promoter mutants, yet the LAT promoter mutants reactivate with 0–10% the efficiency of wild-type (O'Neil *et al.*, 2004). Therefore the evidence suggests that the subtle differences in the efficiency of establishment of some LAT mutants are not sufficient to explain the magnitude of decreased reactivation exhibited by these mutants. Similarly, it has been shown for HSV-2 that impairment of a LAT-deficient virus to reactivate from latency is not dependent on latent viral load (Hoshino *et al.*, 2009). While it seems likely that the LATs do influence establishment, there is strong evidence that reactivation and establishment functions of LAT are genetically separable.

Neuronal protection

A proposed mechanism for the LAT's role in establishment of latency is in protecting neurons from being killed during the resolution of the acute infection and the initial establishment of latency. This model proposes that LAT modulates replicative functions, perhaps by regulating ICP0 expression, and helps push HSV-1 towards a latent infection. Data supporting this model comes in part from experiments where fewer neurons containing latent genomes were detected in mice infected with LAT mutants, and from data in the mouse suggesting that infections with some LAT mutants actually result in the loss of twice as many neurons as is seen in a comparable infection with wild-type or rescue viruses (Thompson and Sawtell, 2001). While the data seem to strongly suggest that some function contained within the LAT region may help spare neurons and push the virus towards latency, the mutants still do establish a latent infection. Therefore, neuronal protection is clearly not an essential function for HSV-1 to establish latency, but instead may be important in protecting certain specific subtypes of sensory neurons. In addition, if the presence of the LATs biases the establishment of latency towards a specific population of neurons (which survive the lytic infection) this could have consequences if certain neurons respond to stress stimuli in a manner that is more conducive for reactivation.

Anti-apoptosis

The first indication that a function encoded within the LAT region promotes cell survival was put forth by Perng *et al.* (Perng *et al.*, 2000a) using a recombinant of HSV-1 McKrae (dLAT2903) which possesses a deletion spanning −161 to +1667 relative to the transcriptional start of the primary 8.5 kb LAT (see Fig. 17.2). This study demonstrated that large numbers of TUNEL-positive neurons could be observed in rabbit TG acutely infected (day 7 p.i) with dLAT2903, compared with only a few TUNEL-positive cells in TG from rabbits infected with the rescue of dLAT2903 or the parent strain, HSV-1 McKrae. TUNEL results were further confirmed by the demonstration that p85 cleavage products of poly-ADP-ribose polymerase (PARP) were detectable in >25% of TG neurons infected with the dLAT2903, whereas no PARP-positive cells were detected in TG infected with wild-type McKrae. This provided strong evidence that a function encoded within the LAT region ultimately resulted in the inhibition of caspase 3-induced PARP cleavage, initiated during an HSV-1 infection.

While questions were initially raised concerning the specificity of the anti-PARP p85 antisera, and the detection of cytoplasmic staining (Thompson and Sawtell, 2000), the observation that IMR-90 and CV-1 cells transiently transfected with a plasmid containing nt 301–2659 of the LAT primary transcript was able to protect >75% of the cells from ceramide or fumonisin B_1 induced apoptosis (relative to transfection with a control plasmid), strongly suggests that this region of the LAT protects against induction of apoptosis (Perng *et al.*, 2000a). The level of anti-apoptotic activity afforded by the LAT plasmid was comparable to a plasmid containing the baculovirus inhibitor of apoptosis (CpIAP). Finally, it was demonstrated the LAT region also protected Neuro-2a cells (mouse neuroblastoma) from etoposide-induced apoptosis. Since ceramide and fumonisin B_1 (FB_1) induce apoptosis through different pathways (protein kinase C and TNF/FAS respectively), transient transfection studies suggested that LAT's anti-apoptosis function acts through a common downstream effector.

Initial studies indicated that ocular infection of mice with the large LAT-deletion virus 17ΔAH (–1136 to + 828), resulted in depletion of over half the TG neurons relative to uninfected mice, mice infected with wild-type 17*syn*+, or the 17AH rescue (Thompson and Sawtell, 2001). Here the authors were unable to detect any increase in TUNEL- positive cells or evidence of the cleaved form of caspase 3, leading them to conclude that the loss of neurons was not due to an apoptotic mechanism. The possibility exists that the apoptosis observed with the LAT mutants might represent a species-specific effect, however two lines of evidence argue convincingly against this possibility. The first study employed a LAT deletion mutant (17N/H) whose deletion (–359 to +1491) overlaps with that of 17ΔAH, but extends further into the intron. 17N/H showed a significant increase in TUNEL-positive TG neurons compared to mice infected with 17*syn*+ or a smaller LAT deletion (17ΔSty) (Ahmed *et al.*, 2002). A second study using a transgenic mouse containing the LAT promoter and first 3 kb of the primary HSV-2 LAT transcript demonstrated diminished apoptosis, both in cell culture and *in vivo* (Wang *et al.*, 2002). In this study, hepatocytes and fibroblasts from transgenic or non-transgenic mice were treated with anti-TNF or anti-Fas antibody. Transgenic mice showed an approximate 50% reduction in numbers of cells exhibiting annexin V staining vs. non-transgenic controls. In addition, when transgenic mice were treated with anti-Fas antibody, they demonstrated 10-fold lower serum alanine transaminase (ALT) levels than the non-transgenic group. Only 37% of the transgenic mice died, compared to 100% of the non-transgenic mice. Collectively, these data suggest that the HSV LAT contains elements that are capable of blocking apoptosis both in rabbits and mice *in vivo*. One possible explanation for the authors' inability to detect evidence of apoptosis induced by 17AH is the smaller size of this recombinant's deletion.

It is unclear mechanistically how the LAT region acts to block apoptosis. Recent studies in mouse C1300 cells stably expressing portions of the LAT have suggested that the presence of the LAT, either directly or indirectly maintain both total and phosphorylated levels of AKT (protein kinase B) in order to promote cell survival (Li *et al.*, 2009). It will be of interest to determine whether the LAT maintains levels of AKT in neurons that survive the acute infection *in vivo* and go on to harbour latent genomes.

To date the body of evidence seems to support a role for the LAT in playing some role in protecting cells from induced apoptosis. Whether the *in vivo* effects are direct or indirect remains to be determined.

Virulence

Several studies have mapped regions within the LAT domain, that when deleted, alter the virulence of HSV-1. Most of these effects are relatively subtle and include phenotypes that differ in the mouse and rabbit models (Perng *et al.*, 2001; Perng *et al.*, 1999). In addition, some deletions have been shown to cause increases in virulence, as measured by viral replication in target tissues and death (Perng *et al.*, 1999), whereas other deletions have been shown to cause slight decreases in overall lethality and the production of infectious virus (Garber *et al.*, 1997). It is important to note that no LAT deletions to date have altered replication or viral yields in cell culture, so these effects seem to be true '*in vivo*' phenomena.

Apoptosis

The identification of another transcript termed 'antisense to LAT' or AL has been attributed to actually *inducing* apoptosis in the context of the viral infection (Perng *et al.*, 2002). This transcript is transcribed across the LAT promoter, and overlaps with just a small portion of the very 5′ end of the 8.5 kb LAT. The fact that this transcript might actually induce apoptosis initially seems paradoxical given the data suggesting that other regions of LAT inhibit apoptosis, however, the potential of this transcript to regulate other LAT functions in an antisense manner may suggest that the balance between inducing apoptosis and blocking apoptosis may be regulated. Since AL LAT is presumably expressed only during the lytic infection, its role in inducing apoptosis may be involved in viral spread.

Neuronal tropism

As mentioned previously, it has been shown in experimental ocular infections of mice that HSV-1 establishes latency preferentially within A5 neurons, a subtype of sensory neurons which express Galb1–4GlcNAc-R epitopes (Fenderson *et al.*, 1983; Yang *et al.*, 2000). In contrast, ocular infection with HSV-2 results in establishment of latency primarily in KH10 neurons, a completely different subpopulation of sensory neurons expressing Galα1–3Galβ1–4NAc-R epitopes (Fenderson *et al.*, 1983; Margolis *et al.*, 2007). Aside from suggesting that HSV-1 and HSV-2 have adapted to subtle differences between the sensory ganglia at the anatomic sites that they most often reside, it also suggests that the cell type provides a critical environment for the establishment of latency. Moreover, recent studies suggest that differences between the HSV-1 and HSV-2 LATs may play a role in directing the establishment of latency to these two distinct neuronal populations in the different ganglia (Imai *et al.*, 2009; Yoshikawa *et al.*, 1996), further implicating the LAT in regulatory pathways involved in establishment and maintenance of HSV latency.

The LAT's effect on lytic gene expression during latency

Given that the LAT is the most abundant transcript during latency and that LAT transcription seems to be incompatible with lytic transcription, it was hypothesized that the LAT might have a genome-wide effect on lytic gene transcription. By analysing the transcription profiles of LAT deletion mutants in KOS and 17*syn*+, it was clear that transcription of the LAT had either a repressing or activating effect (respectively for each strain) on lytic gene transcription during latency (Chen *et al.*, 1997; Garber *et al.*, 1997; Giordani *et al.*, 2008). In the absence of the LAT in KOS, more lytic transcripts were detected while in 17*syn*+ the opposite is true (Chen *et al.*, 1997; Giordani *et al.*, 2008). While these strains differ in the quantities of lytic transcripts that accumulate during latency, these studies clearly demonstrate that transcription of the LAT has an effect on lytic genes during latency. Also brought to light from these experiments is another important experimental factor, the effect of the host species. The studies using KOS were performed in the mouse TG, while those with 17*syn*+ were performed on rabbit TG. This suggests that in the mouse LAT acts as a repressor while in the rabbit, LAT acts as an activator. Whether or not species-specific neuronal factors play a role in transcriptional control of the latent virus is unknown. Given the epigenetic differences between KOS and 17*syn*+ (discussed below), it would seem that the effect of LAT is more dependent on the HSV-1 strain rather than the host animal but a definitive answer to this question is not yet available.

The LAT's influence on chromatin

Epigenetic regulation of latent HSV-1 genes relies on histone posttranslational modifications and not DNA methylation (Kubat *et al.*, 2004b). Chromatin immunoprecipitation (ChIP) has been utilized heavily to gain snapshot views of the chromatin residing on the HSV-1 genome (for reviews see Bloom *et al.*, 2010; Knipe and Cliffe, 2008). These analyses performed on murine DRG or TG latently-infected with HSV-1 have revealed domains of euchromatin and heterochromatin that correlate to the transcription of the LATs and transcriptional repression of the lytic genes, respectively (Giordani *et al.*, 2008; Kubat *et al.*, 2004a,b; Neumann *et al.*, 2007; Wang *et al.*, 2005) (Fig. 17.7). Specifically, the LATs are enriched in the following euchromatic markers: acetylation of lysines 9 and 14 (acetyl H3 K9, K14) of histone 3 as well as dimethylation of lysine 4 (diMe H3K4) of histone 3 (Giordani *et*

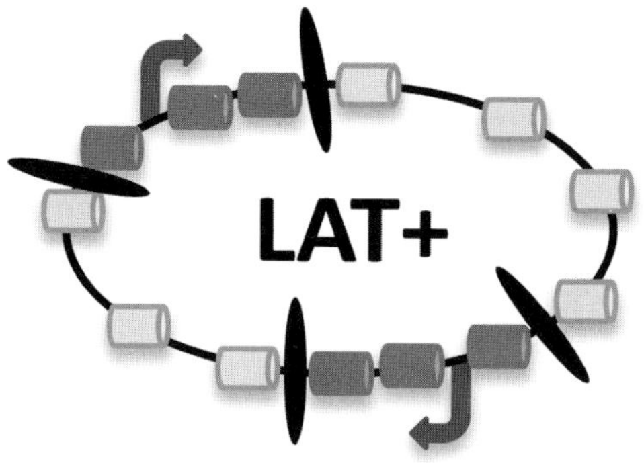

Figure 17.7 The HSV-1 latent genome is organized into transcriptionally permissive and non-permissive domains. ChIP analyses of the HSV-1 genome during a latent infection have shown that all lytic genes analysed are enriched in heterochromatic marks (i.e. TriMe H3K27) and under-enriched in euchromatic marks such as histone H3 acetyl (K9, 14) and histone H3 dimethyl K4, consistent with a transcriptionally repressed state (light coloured cylinders). In contrast, the LAT promoter and 5′ exon (LAT enhancer) region are enriched in acetylated and dimethylated histones (dark coloured cylinders), consistent with a transcriptionally active chromatin state. These regions are separately maintained by CTCF-binding domains (black ovals) present upstream and downstream of the LAT promoter. Taken together, this suggests that the HSV-1 genome is organized into distinct chromatin domains during latency that differ in their transcriptional permissivity. Note: regions of the genome depicted are not to scale.

al., 2008; Kubat *et al.*, 2004b). It was found that transcription of the LAT was not necessary to maintain the euchromatic region, since a LAT promoter deletion mutant displayed similar levels of acetyl H3 K9, K14 as wild type (strain KOS) (Kubat *et al.*, 2004a). This pattern was somewhat replicated with the same LAT mutant in strain 17*syn+* in that the enhancer retained enrichment of diMe H3K4 relative to lytic genes (Giordani *et al.*, 2008). However, the enrichment of diMe H3K4 on the LAT enhancer relative to wild type (17*syn+*) was decreased. This pattern was also observed on lytic genes ICP0 and ICP27, indicating that the genome in general may be in a somewhat less transcriptionally permissive environment in a non-LAT transcribing mutant relative to wild-type. This change in transcriptional permissiveness based on the lack of LAT transcription between strains KOS and 17syn+ has not yet been elucidated. Presumably, since euchromatic marks are still maintained, for the most part, in the absence of LAT in both strains, control of euchromatin in the LAT region for KOS and 17*syn+* is mostly dependent on *cis*-acting elements in the LAT enhancer that recruit proteins to establish and/or maintain transcriptional permissiveness.

Following the above mentioned studies, the Knipe lab performed a time-course experiment following an HSV-1 infection in the mouse from the lytic to latent stage examining the effect of LAT on deposition of a euchromatic marker, diMe H3K4, as well as a heterochromatic marker, dimethylation of H3 K9 (diMe H3K9). They concluded that the presence of the LAT increased the deposition of diMe H3K9 while lowering diMe H3K4 on lytic genes ICP4, tk, ICP27 and ICP8 (Wang *et al.*, 2005). This indicates that transcription of the LAT only affects chromatin deposition on lytic genes, an interesting observation considering that previous studies found the LAT unnecessary for euchromatin in the LAT region.

When the Knipe and Bloom labs published opposing data on this matter, the LAT's effect on chromatin became uncertain (Cliffe *et al.*, 2009; Kwiatkowski *et al.*, 2009). Though both labs are in agreement that facultative marks are present on the latent HSV-1 genome, it is the effect of deleting the LAT that results in remarkably different chromatin phenotypes. Knipe's group found that deletion of the LAT greatly reduced the amount of facultative heterochromatin (specifically trimethylation of lysine 27 on histone 3 [triMe H3K27]) suggesting that the LAT RNA is somehow responsible for recruitment of proteins that confer these modifications (Cliffe *et al.*, 2009). These data were in agreement with their previously published time-course data (Wang *et al.*, 2005). However, the Bloom lab performed a similar set of experiments and found that in the absence of LAT, an increased amount of facultative heterochromatin was detected on the genome (Kwiatkowski *et al.*, 2009). These contradictory outcomes of LAT and heterochromatin led the researchers to investigate what differences exist between the two experiments. Out of the possibilities that existed, that of strain-specific differences seemed the most plausible; one set of experiments employed the KOS wild type strain while the other used 17*syn*+ wild type strain. In addition to virulence phenotypes and transcription profile differences (previously mentioned), we may further add that chromatin profile differences exist as determined by LAT transcription. Direct comparison experiments between KOS and 17*syn*+ must be conducted to ultimately support this strain-specific chromatin effect. Additional factors such as anatomical site of latency may contribute to the LAT chromatin phenotype as well (KOS experiments performed in TG, 17*syn*+ experiments performed in DRG).

Regardless of the LAT's effect (increase or decrease in deposition of chromatin), it is clear that the LAT plays a role in the mechanism of heterochromatin assembly on the genome. Due to the variation in triMe H3K27 amounts as a result of LAT transcription, it is hypothesized that LAT is somehow interacting with a set of cellular proteins called Polycomb repressive complexes (PRCs). These complexes are responsible for the establishment and maintenance of the triMe H3K27 modification. To explain how the LAT is responsible for modifying the actions of the PRCs, we hypothesize that the LAT RNA is binding to these complexes to either recruit or saturate the available histone modifying proteins of the PRCs (Fig. 17.8). In the experiments performed with KOS, one could envisage a scenario where the LAT actively binds to and recruits the PRCs to trimethylate H3K27, and this action is abrogated in the absence of LAT transcription, resulting in lower amounts of triMe H3K27. However experiments performed with 17*syn*+ would support a hypothesis in which the LAT may bind to the PRCs and regulate how much activity it has on the genome, as lack of the LAT transcript results in greater than 2-fold enrichment of triMe H3K27.

In summary, if one examines a LAT-dependent reactivation model such as the rabbit, correlations can be drawn between the transcription/chromatin data and the biology of strain 17*syn*+ in this model. LAT mutants reactivate very inefficiently in the rabbit which can be hypothesized to be the result of deleting the LAT promoter, causing an *increase* in the amount of triMe H3K27 that is deposited on the lytic genes during latency and a *decrease* in the leakiness of lytic gene expression during latency (Giordani *et al.*, 2008; Kwiatkowski *et al.*, 2009). This suggests that the LAT acts to modulate the amount of triMe H3K27 deposition so as to make the repression easier to reverse, and therefore make induced reactivation more efficient.

How does LAT do all these things?

While over the past several years there have been a number of significant new developments regarding phenotypes associated with the LATs, how this region mediates these functions remains to be elucidated. Moreover, determining which of the observed phenotypes are dominant has been equally difficult. None-the-less, there have been a number of important recent findings that have contributed to eliminating some previous hypothetical mechanisms.

Antisense regulator

Ever since its initial discovery, the potential for the 2.0 kb LAT to regulate ICP0 through an antisense mechanism has been an obvious possibility (Stevens *et al.*, 1987). Work from the Steiner lab, and others, has demonstrated that expression of the 2.0 kb intron has the potential to reduce the expression of ICP0 and viral yields when one engineers LAT to be expressed constitutively during the acute infection; however, a direct demonstration of anti-sense regulation in the context of the natural infection has not been made (Mador *et al.*, 2003). In fact, the Glorioso lab performed a rather convincing analysis that failed to detect functional downregulation of ICP0 in the context of the viral genome (Burton *et al.*, 2003). It should be pointed out, however, that the door is

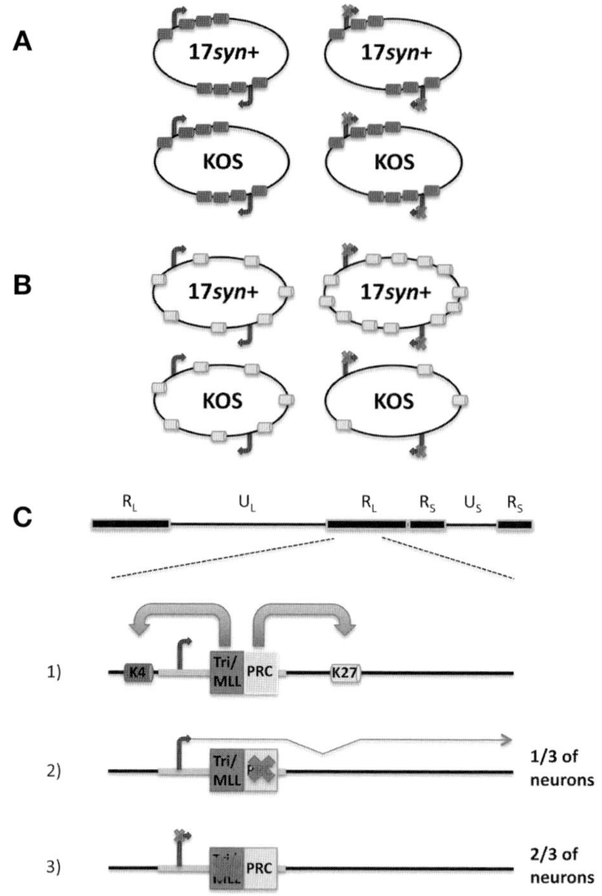

Figure 17.8 Model for the LAT's role in regulating heterochromatinization of the latent HSV-1 genome. The LAT enhancer acts as a regulatory region with the ability to bind groups of proteins responsible for establishing both *activating* (MLL/Trithorax) shown in dark grey) and *repressing* histone modifications (PRC shown in light grey). (A) ChIP analysis of HSV-1 strains 17*syn+* and KOS demonstrate the presence of euchromatic histone marks (acetyl H3 K9, K14 and dimethyl H3 K4) on the LAT promoter and enhancer regions of the genome during latency (left side of panel). When LAT promoter deletions are examined, strain 17*syn+* shows a reduction in the degree of euchromatin, whereas the KOS deletion mutant exhibits no change (right side of panel). (B) ChIP analysis of HSV-1 strains 17*syn+* and KOS demonstrate the presence of heterochromatin by the histone mark triMe H3K27 on the lytic gene regions of the genome during latency (left side of panel). When LAT promoter deletions are examined, strain 17ΔPst *shows* a *dramatic increase* in the degree of triMe H3K27 deposition, whereas the KOS deletion mutant exhibits *a decrease* (right side of panel). (C) Model to explain the LAT's involvement in modulating diH3K4 and triH3K27 modifications on the latent genome. (1) The LAT enhancer region is able to bind both MLL/trithorax complexes (direct dimeH3K4) and PRC complexes (direct triMeH3K27) to the HSV genome. (2) When the LAT is expressed, PRC binding and triMeH3K27 deposition to the LAT enhancer region is blocked and dimeH3K4 marks predominate. (3) When the LAT is not expressed, PRC binds and promotes deposition of triMe H3K27. In the case of strain 17*syn+* this results in an increase in triMeH3K27 on the entire genome. It should be noted that during latency only one-third of the neurons that contain HSV genomes express detectable LAT. This indicates that two transcriptional populations of genomes exist: those that express the LAT and those that do not. This could explain how PRC complexes can be detected on the LAT enhancer, a region associated with active transcription (Kwiatkowski *et al.*, 2009). It could be hypothesized that factors such as the number of viral genomes present within a given neuron or the phenotype of neuron could dictate the level of LAT expression. Consequently, the presence or absence of the LAT would then direct chromatinization of the genome, and ultimately the potential to reactivate.

not completely closed to the possibility that the LATs could mediate some (or all) of their effects by anti-sense regulation. One must not forget that the 8.5 kb LAT has the *potential* to regulate a number of very important lytic genes through anti-sense mechanisms including ICP0, ICP34.5, and even ICP4; however, the fact that expression of the LATs is very tightly regulated and is not observed in cells that express lytic transcripts suggests that it is unlikely that the LAT acts in an antisense manner.

LAT-encoded protein

Latently expressed proteins are well described among the gamma herpesviruses, but to date, while the LATs contain a number of potential ORFs that could be expressed, no LAT-encoded protein has been directly detected during a latent infection. It remains possible that future analyses will demonstrate that low levels of protein are made from some of these ORFs. However, it is important to note that if one analyzes the region of the LATs where the majority of the phenotypes reside (Fig. 17.6), potential ORFs are not well-conserved between different strains of HSV-1. This makes the hypothesis of a LAT-encoded protein less plausible.

LAT as a non-coding RNA

Given the current data, it seems most likely that the LATs act either as non-coding RNAs or that the LAT region acts as a DNA regulatory element. Given the recent identification of miRNAs encoded within the LAT region, a plausible hypothesis is that LAT-encoded miRNAs could act to regulate latency through reducing levels of either viral or cellular genes. Over-expression of several of these miRNAs *in vitro* suggest they have the potential to regulate ICP0, ICP4 and ICP34.5 (in the case of HSV-2) (Tang *et al.*, 2008; Umbach *et al.*, 2008). Another possibility is that some of these miRNAs target neuronal genes and facilitate the establishment or maintenance of the latent infection. Cellular miRNAs have been implicated in gene silencing through promoting heterochromatin formation (Zhang, 2009), so it is possible that the viral miRNAs could assist in repression of lytic genes during the establishment phase of latency through such a mechanism. To date there is no evidence for the effects of deleting these miRNAs in the context of the virus either *in vitro* or *in vivo*; however, these studies are under way.

A final hypothesis is that LAT could function as a non-coding structural RNA similar to the mammalian *XIST* or *Rox* of *Drosophila* that regulate expression of the sex chromosomes, through a mechanism that has yet to be worked out (Bloom *et al.*, 1996b). The *XIST* or *Rox* RNAs have been shown to associate *in cis* with the chromosome from which they are transcribed and 'paint it' (Clemson *et al.*, 1996; Meller *et al.*, 1997). This association has been postulated to help recruit specific histone modifying activities which ultimately alter the transcriptional status of the chromatin on that chromosome. There is no data, other than distant sequence similarities, to support this model, and association between the latent genome and LAT has been technically challenging due to difficulties detecting latent HSV-1 genomes by *in situ* hybridization. None-the-less, since non-coding XIST RNA made from the silent X-chromosome has been shown to facilitate PRC recruitment, it is possible that the LAT also interacts with the PRC complexes during the establishment of HSV latency to promote a reversible chromatin structure that represses lytic genes.

The LAT region as a *cis*-acting regulatory region

Recent data suggest that the 5′ exon-coding region of the LAT domain is a functionally important transcriptional regulator that regulates not only the expression of the LATs, but also has the potential to regulate or enhance the expression of surrounding genes (Berthomme *et al.*, 2000). This hypothesis is grounded in the fact that a number of phenotypes ascribed to the LATs, including reactivation, establishment and apoptosis, were mapped to the region encompassing the LAT promoter and the 5′ exon region (Fig. 17.6). Fine mapping of the reactivation critical region (*rcr*) has shown that deletions within the 5′ exon region, while still producing LAT, are sufficient to reduce reactivation (Bloom *et al.*, 1996b; Loutsch *et al.*, 1999). Due to the enhancer and chromatin control activities attributed to 5′ exon region, it is possible that the enhancer influences expression or repression of a number of other viral genes as part of a latency regulatory system (Kubat *et al.*, 2004a). This might explain why LAT deletions encompass a number of phenotypes.

A key component of this model would include a means of insulating neighbouring genes from the effects of the LAT enhancer during latency. It has been shown that the effects of the enhancer do not normally extend to the nearby ICP0 promoter. This promoter is associated with hypo-acetylated histone H3 during latency (Kubat *et al.*, 2004a), suggesting that a chromatin boundary with enhancer-blocking properties exists between the LAT enhancer and the ICP0 promoter. The recent identification of conserved binding sites for the cellular boundary protein CTCF and demonstration by ChIP that these CTCF sites are occupied during a latent infection suggests that these CTCF binding sites might form the core of a chromatin boundary that is functional during latency (Amelio *et al.*, 2006b). Indeed this element has been shown to be functional as an insulator when used in place of a *Drosophila* insulator (Chen *et al.*, 2007).

In this model, two chromatin boundaries flanking the LAT enhancer would establish a 'transcriptionally permissive zone,' as indicated by the region's enrichment in the specifically modified histone H3 acetyl (K9, 14). This hyperacetylated region encompasses the LAT promoter and the 5′ LAT exon, but does not extend to the ICP0 promoter (Fig. 17.7). A reactivation stimulus (such as heat stress or adrenaline induction) could cause a collapse of the boundary separating the LAT enhancer from ICP0, and allow the enhancer to facilitate transcription of ICP0. ICP0 has long been proposed to be a key player in the initial stages of HSV-1 reactivation (Cai *et al.*, 1993; Halford and Schaffer, 2001). Aside from the role of ICP0 as a promiscuous transactivator, work by the Everett and Knipe labs have implicated ICP0 in chromatin remodelling activities (Cliffe and Knipe, 2008; Everett *et al.*, 1999; Hancock *et al.*, 2010). Expression of ICP0 could therefore play an important role in altering the transcriptionally non-permissive histones that are associated with the other HSV-1 immediate early and early genes and potentially facilitate a productive reactivation event. In support of this model, it has recently been demonstrated that following explant-induced reactivation there is a transient decrease in LAT transcription followed by remodelling of the chromatin associated with the ICP0 promoter (Amelio *et al.*, 2006a). While the details of this model remain to be elucidated, it is easy to envisage how mutations in various components of such a multi-component LAT regulatory switch could ultimately modulate a large number of key viral functions and result in the multiple phenotypes that have been described for this region.

Is it possible that the LAT acts through a combination of the above mechanisms?

Given the complexity of the LAT locus, it would not be surprising if this region were to employ some or all of the above mechanisms to mediate its effects on latency and reactivation. In fact it is very likely that the multiple species of RNAs produced from this region exert multiple influences on the viral and cellular transcriptomes. Based on existing data, Fig. 17.8 illustrates one model that incorporates existing genetic and biochemical data that could explain how the LAT region acts to regulate the formation of reversible heterochromatin during latency, and perhaps gives insight into why this process is subject to viral differences. This model proposes that the LAT region functions through two of the above mechanisms: 1) *cis* elements in the LAT enhancer are able to bind both positive and negative regulatory histone modifying proteins, and 2) the LAT, as a large non-coding RNA, can also bind to these histone modifying proteins to modulate *the degree of heterochromatic repression on the lytic genes during latency.*

In the first part of this model, the LAT enhancer acts as a regulatory region with the ability to bind groups of proteins responsible for establishing both activating and repressing histone modifications (Fig. 17.8A). PRC proteins, which confer repressive histone modifications, are known to bind to the LAT enhancer (Kwiatkowski *et al.*, 2009). Additionally, there is evidence that activating complexes such as MLL, Trithorax and others have bound to this region by the presence of euchromatic histone modifications (in greatest enrichment on the LAT enhancer during latency). The fact that histone modifications representative of both MLL and PRC complexes can be detected in the absence of LAT expression supports the model that the LAT enhancer is a *cis*-acting regulatory region that can bind both of these classes of chromatin modellers. This part of the model explains why deletions within the LAT enhancer alter histone modifications on the HSV genome without altering LAT expression *per se*.

The second part of the model explains observations related to recombinants with LAT promoter deletions that eliminate almost all LAT expression. When LAT expression is abolished, lower amounts of diMe H3K4 and increased amounts of triMe H3K27 are detected on regions of the genome suggesting that the LAT RNA may play a role in *modulating* the recruitment/binding of these proteins regions of the viral genome. However, presence of the LAT RNA it is not necessary for binding to occur (Fig. 17.8A). As mentioned above, there is precedent for ncRNAs interacting with protein remodelling complexes to facilitate binding to the genome. This model also explains the observation that in rabbits latently infected with HSV-1 17*syn*+, that LAT promoter deletion results in a greater degree of suppression of lytic genes during latency (Giordani *et al.*, 2009). Said another way, it seems the LAT may prevent lytic genes from being suppressed too tightly, thereby allowing them to revert to a transcriptionally active state and facilitate efficient reactivation upon stimulation.

Conclusions and future directions

The LAT region and its encoded LATs have been implicated in a number of functions including establishment of latency, neuronal protection and anti-apoptosis, and reactivation from latency. The mechanism by which LAT modulates these processes is unclear, but focus has shifted to regulatory functions either at the non-coding RNA or DNA level. Given that the LAT locus is rich in transcription and gene expression control proteins, the idea of LAT acting as a regulatory component is an attractive model. In addition, the placement of an enhancer and an insulator within the LAT domain provides further potential to regulate expression from other viral genes. Expression of the LATs may play important roles as non-coding RNAs which may act to recruit histone modifying activities. Whatever the role of the LATs, if this region is shown to function as a dominant regulator of transcription, it could well explain why it has been so difficult to tease apart the phenotypes that have been attributed to the LAT region. Further characterization of the LAT enhancers and the boundary regions from a functional point of view should shed some additional light on how this region may function in modulating the balance between the lytic infection, latency and reactivation.

Acknowledgements

Thanks to Nicole Giordani for helpful comments on the preparation of the manuscript, and to Peterjon McAnany for help with the figures.

References

Ahmed, M., Lock, M., Miller, C.G., and Fraser, N.W. (2002). Regions of the herpes simplex virus type 1 latency-associated transcript that protect cells from apoptosis *in vitro* and protect neural cells *in vivo*. J. Virol. *76*, 717–729.

Amelio, A.L., Giordani, N.V., Kubat, N.J., O'Neil J., E., and Bloom, D.C. (2006a). Deacetylation of the herpes simplex virus type 1 latency-associated transcript (LAT) enhancer and a decrease in LAT abundance precede an increase in ICP0 transcriptional permissiveness at early times postexplant. J. Virol. *80*, 2063–2068.

Amelio, A.L., McAnany, P.K., and Bloom, D.C. (2006b). A chromatin insulator-like element in the herpes simplex virus type 1 latency-associated transcript region binds CCCTC-binding factor and displays enhancer-Blocking and silencing activities. J. Virol. *80*, 2358–2368.

Berman, E.J., and Hill, J.M. (1985). Spontaneous ocular shedding of HSV-1 in latently infected rabbits. Invest Opthalmol. Vis. Sci. *26*, 587–590.

Berthomme, H., Lokensgard, J., Yang, L., Margolis, T., and Feldman, L.T. (2000). Evidence for a bidirectional element located downstream from the herpes simplex virus type 1 latency-associated promoter that increases its activity during latency. J. Virol. *74*, 3613–3622.

Bertke, A.S., Patel, A., Imai, Y., Apakupakul, K., Margolis, T.P., and Krause, P.R. (2009). Latency-associated transcript (LAT) exon 1 controls herpes simplex virus species-specific phenotypes: reactivation in the guinea pig genital model and neuron subtype-specific latent expression of LAT. J. Virol. *83*, 10007–10015.

Block, T.M., Deshmane, S., Masonis, J., Maggioncalda, J., Valyi, N.T., and Fraser, N.W. (1993). An HSV LAT null mutant reactivates slowly from latent infection and makes small plaques on CV-1 monolayers. Virology *192*, 618–630.

Bloom, D.C. (2004). HSV LAT and Neuronal Survival. Int. Rev. Immunol. *23*, 187–198.

Bloom, D.C., Devi-Rao, G.B., Hill, J.M., Stevens, J.G., and Wagner, E.K. (1994). Molecular analysis of herpes simplex virus type 1 during epinephrine induced reactivation of latently infected rabbits *in vivo*. J. Virol. *68*, 1283–1292.

Bloom, D.C., Giordani, N.V., and Kwiatkowski, D.L. (2010). Epigenetic regulation of latent HSV-1 gene expression. Biochim Biophys Acta *1799*, 246–256.

Bloom, D.C., Hill, J.M., and Wagner, E. (1996a). Physical extent of the critical 5′ LAT element mediating efficient epinephrine-induced reactivation in the rabbit. in preparation.

Bloom, D.C., Hill, J.T., Wagner, E.K., Feldman, L.F., and Stevens, J.G. (1996b). A 348-bp region in the latency associated transcript facilitates herpes simplex virus type 1 reactivation. J. Virol. *70*, 2449–2459.

Burton, E.A., Hong, C.S., and Glorioso, J.C. (2003). The Stable 2.0-Kilobase Intron of the Herpes Simplex Virus Type 1 Latency-Associated Transcript Does Not Function as an Antisense Repressor of ICP0 in Nonneuronal Cells. J. Virol. *77*, 3516–3530.

Cai, W., Astor, T.L., Liptak, L.M., Cho, C., Coen, D.M., and Schaffer, P.A. (1993). The herpes simplex virus type 1 regulatory protein ICP0 enhances virus replication during acute infection and reactivation from latency. J. Virol. *67*, 7501–7512.

Chen, Q., Lin, L., Smith, S., Huang, J., Berger, S.L., and Zhou, J. (2007). CTCF-dependent chromatin boundary element between the latency-associated transcript and ICP0 promoters in the herpes simplex virus type 1 genome. J. Virol. *81*, 5192–5201.

Chen, S.H., Kramer, M.F., Schaffer, P.A., and Coen, D.M. (1997). A viral function represses accumulation of transcripts from productive-cycle genes in mouse ganglia latently infected with herpes simplex virus. J. Virol. *71*, 5878–5884.

Clemson, C.M., McNeil, J.A., Willard, H.F., and Lawrence, J.B. (1996). XIST RNA paints the inactive X chromosome at interphase: evidence for a novel RNA involved in nuclear/chromosomal structure. J. Cell Biol. *132*, 259–275.

Cliffe, A.R., Garber, D.A., and Knipe, D.M. (2009). Transcription of the herpes simplex virus latency-associated transcript promotes the formation of facultative heterochromatin on lytic promoters. J. Virol. *83*, 8182–8190.

Cliffe, A.R., and Knipe, D.M. (2008). Herpes simplex virus ICP0 promotes both histone removal and acetylation on viral DNA during lytic infection. J. Virol. *82*, 12030–12038.

Cui, C., Griffiths, A., Li, G., Silva, L.M., Kramer, M.F., Gaasterland, T., Wang, X.J., and Coen, D.M. (2006). Prediction and identification of herpes simplex virus 1-encoded microRNAs. J. Virol. *80*, 5499–5508.

Deshmane, S.L., and Fraser, N.W. (1989). During latency, herpes simplex virus type 1 DNA is associated with nucleosomes in a chromatin structure. J. Virol. *63*, 943–947.

Devi-Rao, G.B., Bloom, D.C., Stevens, J.G., and Wagner, E.K. (1994). Herpes simplex virus type 1 DNA replication and gene expression during explant induced reactivation of latently infected murine sensory ganglia. J. Virol. *68*, 1271–1282.

Dobson, A.T., Margolis, T.P., Gomes, W.A., and Feldman, L.T. (1995). *In vivo* deletion analysis of the Herpes simplex virus type 1 latency associated transcript promoter. J. Virol. *69*, 2264–2270.

Efstathiou, S., Minson, A.C., Field, H.J., Anderson, J.R., and Wildy, P. (1986). Detection of herpes simplex virus-specific DNA sequences in latently infected mice and in humans. J. Virol. *57*, 446–455.

Everett, R.D., Earnshaw, W.C., Findlay, J., and Lomonte, P. (1999). Specific destruction of kinetochore protein CENP-C and disruption of cell division by herpes simplex virus immediate-early protein Vmw110. EMBO J. *18*, 1526–1538.

Farrell, M.J., Dobson, A.T., and Feldman, L.T. (1991). Herpes simplex virus latency-associated transcript is a stable intron. Proc. Natl. Acad. Sci. U.S.A. *88*, 790–794.

Fenderson, B.A., Hahnel, A.C., and Eddy, E.M. (1983). Immunohistochemical localization of two monoclonal antibody-defined carbohydrate antigens during early murine embryogenesis. Dev. Biol. *100*, 318–327.

Garber, D.A., Schaffer, P.A., and Knipe, D.M. (1997). A LAT-associated function reduces productive-cycle gene expression during acute infection of murine sensory neurons with herpes simplex virus type 1. J. Virol. *71*, 5885–5893.

Giordani, N.V., Neumann, D.M., Kwiatkowski, D.L., Bhattacharjee, P.S., McAnany, P.K., Hill, J.M., and Bloom, D.C. (2008). During HSV-1 infection of rabbits, the ability to express the LAT increases latent-phase transcription of lytic genes. J. Virol. *82*, 6056–6060.

Goins, W.F., Sternberg, L.R., Croen, K.D., Krause, P.R., Hendricks, R.L., Fink, D.J., Straus, S.E., Levine, M., and Glorioso, J.C. (1994). A novel latency-active promoter is contained within the herpes simplex virus type 1 UL flanking repeats. J. Virol. *68*, 2239–2252.

Gressens, P., and Martin, J.R. (1994). In situ polymerase chain reaction: localization of HSV-2 DNA sequences in infections of the nervous system. J. Virol. Methods *46*, 61–83.

Gussow, A.M., Giordani, N.V., Tran, R.K., Imai, Y., Kwiatkowski, D.L., Rall, G.F., Margolis, T.P., and Bloom, D.C. (2006). Tissue-specific splicing of the herpes simplex virus type 1 latency-associated transcript (LAT) intron in LAT transgenic mice. J. Virol. *80*, 9414–9423.

Halford, W.P., and Schaffer, P.A. (2001). ICP0 is required for efficient reactivation of herpes simplex virus type 1 from neuronal latency. J. Virol. *75*, 3240–3249.

Hancock, M.H., Cliffe, A.R., Knipe, D.M., and Smiley, J.R. (2010). Herpes simplex virus VP16, but not ICP0, is required to reduce histone occupancy and enhance

histone acetylation on viral genomes in U2OS osteosarcoma cells. J. Virol. *84*, 1366–1375.
Hill, J.M., Sedarati, F., Javier, R.T., Wagner, E.K., and Stevens, J.G. (1990). Herpes simplex virus latent phase transcription facilitates *in vivo* reactivation. Virology *174*, 117–125.
Hoshino, Y., Pesnicak, L., Straus, S.E., and Cohen, J.I. (2009). Impairment in reactivation of a latency associated transcript (LAT)-deficient HSV-2 is not solely dependent on the latent viral load or the number of CD8(+) T cells infiltrating the ganglia. Virology *387*, 193–199.
Imai, Y., Apakupakul, K., Krause, P.R., Halford, W.P., and Margolis, T.P. (2009). Investigation of the mechanism by which herpes simplex virus type 1 LAT sequences modulate preferential establishment of latent infection in mouse trigeminal ganglia. J. Virol. *83*, 7873–7882.
Jaber, T., Henderson, G., Li, S., Perng, G.C., Carpenter, D., Wechsler, S.L., and Jones, C. (2009). Identification of a novel herpes simplex virus type 1 transcript and protein (AL3) expressed during latency. J. Gen. Virol. *90*, 2342–2352.
Jarman, R.G., Loutsch, J.M., Devi-Rao, G.B., Marquart, M.E., Banaszak, M.P., Zheng, X., Hill, J.M., Wagner, E.K., and Bloom, D.C. (2002). The region of the HSV-1 latency-associated transcript required for epinephrine-induced reactivation in the rabbit does not include the 2.0 kb intron. Virology *292*, 59–69.
Jurak, I., Kramer, M.F., Mellor, J.C., van Lint, A.L., Roth, F.P., Knipe, D.M., and Coen, D.M. (2010). Numerous Conserved and Divergent MicroRNAs Expressed by Herpes Simplex Viruses 1 and 2. J. Virol.
Kenny, J.J., Krebs, F.C., Hartle, H.T., Gartner, A.E., Chatton, B., Leiden, J.M., Hoeffler, J.P., Weber, P.C., and Wigdahl, B. (1994). Identification of a second ATF/CREB-like element in the Herpes Simplex Virus Type 1 (HSV-1) latency-associated transcript (LAT) promoter. Virology *200*, 220–235.
Knipe, D.M., and Cliffe, A. (2008). Chromatin control of herpes simplex virus lytic and latent infection. Nat. Rev. Microbiol. *6*, 211–221.
Koelle, D.M., and Corey, L. (2008). Herpes simplex: insights on pathogenesis and possible vaccines. Annu Rev. Med. *59*, 381–395.
Krause, P.R., Stanberry, L.R., Bourne, N., Connelly, B., Kurawadwala, J.F., Patel, A., and Straus, S.E. (1995). Expression of the herpes simplex virus type 2 latency-associated transcript enhances spontaneous reactivation of genital herpes in latently infected guinea pigs. J. Exp. Med. *181*, 297–306.
Kubat, N.J., Amelio, A.L., Giordani, N.V., and Bloom, D.C. (2004a). The Herpes Simplex Virus Type 1 Latency-Associated Transcript (LAT) Enhancer/rcr Is Hyperacetylated during Latency Independently of LAT Transcription. J. Virol. *78*, 12508–12518.
Kubat, N.J., Tran, R.K., McAnany, P., and Bloom, D.C. (2004b). Specific histone tail modification and not DNA methylation is a determinant of herpes simplex virus type 1 latent gene expression. J. Virol. *78*, 1139–1149.
Kwiatkowski, D.L., Thompson, H.W., and Bloom, D.C. (2009). The polycomb group protein Bmi1 binds to the herpes simplex virus 1 latent genome and maintains repressive histone marks during latency. J. Virol. *83*, 8173–8181.
Leib, D.A., Bogard, C.L., Kosz, V.M., Hicks, K.A., Coen, D.M., Knipe, D.M., and Schaffer, P.A. (1989). A deletion mutant of the latency-associated transcript of herpes simplex virus type 1 reactivates from the latent state with reduced frequency. J. Virol. *63*, 2893–2900.
Leib, D.A., Nadeau, K.C., Rundle, S.A., and Schaffer, P.A. (1991). The promoter of the latency-associated transcripts of herpes simplex virus type 1 contains a functional cAMP-reponse element: Role of the latency-associated transcripts and cAMP in reactivation and viral latency. Proc. Natl. Acad. Sci. U.S.A. *88*, 48–52.
Li, S., Carpenter, D., Hsiang, C., Wechsler, S.L., and Jones, C. (2009). The herpes simplex virus type 1 latency-associated transcript (LAT) inhibits apoptosis and promotes neurite sprouting in neuroblastoma cells following serum starvation by maintaining protein kinase B (AKT) levels. J. Gen Virol.
Lokensgard, J.R., Berthomme, H., and Feldman, L.T. (1997). The latency-associated promoter of herpes simplex virus type 1 requires. a region downstream of the transcription start site for long-term expression during latency. J. Virol. *71*, 6714–6719.
Lokensgard, J.R., Bloom, D.C., Dobson, A.T., and Feldman, L.T. (1994). Long-term promoter activity during herpes simplex virus latency. J. Virol. *68*, 7148–7158.
Loutsch, J.M., Perng, G.C., Hill, J.M., Zheng, X., Marquart, M.E., Block, T.M., Nesburn, A.B., and Wechsler, S.L. (1999). Identical 371-base-pair deletion mutations in the LAT genes of herpes simplex virus type 1 McKrae and 17syn+ results in different *in vivo* reactivation phenotypes. J. Virol. *73*, 767–771.
Mador, N., Braun, E., Haim, H., Ariel, I., Panet, A., and Steiner, I. (2003). Transgenic mouse with the herpes simplex virus type 1 latency-associated gene: expression and function of the transgene. J. Virol. *77*, 12421–12429.
Maggioncalda, J., Mehta, A., Fraser, N.W., and Block, T.M. (1994). Analysis of a Herpes Simplex Virus type 1 LAT mutant with a deletion between the putative promoter and the 5′ end of the 2.0 kilobase transcript. J. Virol. *68*, 7816–7824.
Margolis, T.P., Bloom, D.C., Dobson, A.T., Feldman, L.T., and Stevens, J.G. (1993). Decreased reporter gene expression during latent infection with HSV LAT promoter constructs. Virology *197*, 585–592.
Margolis, T.P., Dawson, C.R., and LaVail, J.H. (1992a). Herpes simplex viral infection of the mouse trigeminal ganglion. Immunohistochemical analysis of cell populations. Invest Ophthalmol. Vis. Sci. *33*, 259–267.
Margolis, T.P., Imai, Y., Yang, L., Vallas, V., and Krause, P.R. (2007). Herpes simplex virus type 2 (HSV-2) establishes latent infection in a different population of ganglionic neurons than HSV-1: role of latency-associated transcripts. J. Virol. *81*, 1872–1878.

Margolis, T.P., Sedarati, F., Dobson, A.T., Feldman, L.T., and Stevens, J.G. (1992b). Pathways of viral gene expression during acute neuronal infection with HSV-1. Virology *189*, 150–160.

Mehta, A., Maggioncalda, J., Bagasra, O., Thikkavarapu, S., Saikumari, P., Valyi-Nagy, T., Fraser, N.W., and Block, T.M. (1995). In situ DNA PCR and RNA hybridization of herpes simplex virus sequences in trigeminal ganglia of latently infected mice. Virology *206*, 633–640.

Meller, V.H., Wu, K.H., Roman, G., Kuroda, M.I., and Davis, R.L. (1997). roX1 RNA paints the X chromosome of male Drosophila and is regulated by the dosage compensation system. Cell *88*, 445–457.

Mitchell, B.M., Bloom, D.C., Cohrs, R.J., Gilden, D.H., and Kennedy, P.G. (2003). Herpes simplex virus-1 and varicella-zoster virus latency in ganglia. J. NeuroVirol. *9*, 194–204.

Mukerjee, R., Kang, W., Suri, V., and Fraser, N.W. (2004). A non-consensus branch point plays an important role in determining the stability of the 2-kb LAT intron during acute and latent infections of herpes simplex virus type-1. Virology *324*, 340–349.

Naito, J., Mukerjee, R., Mott, K.R., Kang, W., Osorio, N., Fraser, N.W., and Perng, G.C. (2005). Identification of a protein encoded in the herpes simplex virus type 1 latency associated transcript promoter region. Virus Res. *108*, 101–110.

Neumann, D.M., Bhattacharjee, P.S., Giordani, N.V., Bloom, D.C., and Hill, J.M. (2007). *In vivo* changes in the patterns of chromatin structure associated with the latent herpes simplex virus type 1 genome in mouse trigeminal ganglia can be detected at early times after butyrate treatment. J. Virol. *81*, 13248–13253.

Nicosia, M., Deshmane, S.L., Zabolotny, J.M., Valyi-Nagy, T., and Fraser, N.W. (1993). Herpes simplex virus type 1 latency-associated transcript (LAT) promoter deletion mutants can express a 2-kilobase transcript mapping to the LAT region. J. Virol. *67*, 7276–7283.

O'Neil, J.E., Loutsch, J.M., Aguilar, J.S., Hill, J.M., Wagner, E.K., and Bloom, D.C. (2004). Wide Variations in Herpes Simplex Virus Type 1 Inoculum Dose and Latency-Associated Transcript Expression Phenotype Do Not Alter the Establishment of Latency in the Rabbit Eye Model. J. Virol. *78*, 5038–5044.

Peng, W., Vitvitskaia, O., Carpenter, D., Wechsler, S.L., and Jones, C. (2008). Identification of two small RNAs within the first 1.5-kb of the herpes simplex virus type 1-encoded latency-associated transcript. J. NeuroVirol. *14*, 41–52.

Perng, G.-C., Dunkel, E.C., Geary, P.A., Slanina, S.M., Ghiasi, H., Kaiwar, R., Nesburn, A.B., and Wechsler, S.L. (1994). The latency-associated transcript gene of herpes simplex virus type 1 (HSV-1) is required for efficient *in vivo* spontaneous reactivation of HSV-1 from latency. J. Virol. *68*, 8045–8055.

Perng, G.-C., Ghiasi, H., Slanina, S.M., Nesburn, A.B., and Wechsler, S.L. (1996). The spontaneous reactivation function of the herpes simplex virus type 1 LAT gene resides completely within the first 1.5 kilobases of the 8.3-kilobase primary transcript. J. Virol. *70*, 976–984.

Perng, G.C., Esmaili, D., Slanina, S.M., Yukht, A., Ghiasi, H., Osorio, N., Mott, K.R., Maguen, B., Jin, L., Nesburn, A.B., *et al.* (2001). Three herpes simplex virus type 1 latency-associated transcript mutants with distinct and asymmetric effects on virulence in mice compared with rabbits. J. Virol. *75*, 9018–9028.

Perng, G.C., Jones, C., Ciacci-Zanella, J., Stone, M., Henderson, G., Yukht, A., Slanina, S.M., Hofman, F.M., Ghiasi, H., Nesburn, A.B., *et al.* (2000a). Virus-induced neuronal apoptosis blocked by the herpes simplex virus latency-associated transcript. Science *287*, 1500–1503.

Perng, G.C., Maguen, B., Jin, L., Mott, K.R., Kurylo, J., BenMohamed, L., Yukht, A., Osorio, N., Nesburn, A.B., Henderson, G., *et al.* (2002). A novel herpes simplex virus type 1 transcript (AL-RNA) antisense to thje 5′ end of the latency associated transcript produces a protein in infected rabbits. J. Virol. *76*, 8003–8010.

Perng, G.C., Slanina, S.M., Yukht, A., Drolet, B.S., Keleher, W., Jr., Ghiasi, H., Nesburn, A.B., and Wechsler, S.L. (1999). A herpes simplex virus type 1 latency-associated transcript mutant with increased virulence and reduced spontaneous reactivation. J. Virol. *73*, 920–929.

Perng, G.C., Slanina, S.M., Yukht, A., Ghiasi, H., Nesburn, A.B., and Wechsler, S.L. (2000b). The latency-associated transcript gene enhances establishment of herpes simplex virus type 1 latency in rabbits. J. Virol. *74*, 1885–1891.

Rodahl, E., and Haarr, L. (1997). Analysis of the 2-kilobase latency-associated transcript expressed in PC12 cells productively infected with herpes simplex virus type 1: evidence for a stable, nonlinear structure. J. Virol. *71*, 1703–1707.

Sawtell, N.M. (1997). Comprehensive quantification of herpes simplex virus latency at the single-cell level. J. Virol. *71*, 5423–5431.

Sawtell, N.M., and Thompson, R.L. (1992). Herpes simplex virus type 1 latency associated transcription unit promotes anatomical site-dependent establishment and reactivation from latency. J. Virol. *66*, 2157–2169.

Sedarati, F., Izumi, K.M., Wagner, E.K., and Stevens, J.G. (1989). Herpes simplex virus type 1 latency-associated transcription plays no role in establishment or maintenance of a latent infection in murine sensory neurons. J. Virol. *63*, 4455–4458.

Spivack, J.G., and Fraser, N.W. (1987). Detection of herpes simplex virus type 1 transcripts during latent infection in mice. J. Virol. *61*, 3841–3847.

Stanberry, L.R., Kern, E.R., Richards, J.T., Abbott, T.M., and Overall Jr., J.C. (1982). Genital herpes in guinea pigs: pathogenesis of primary infection and description of recurrent disease. J. Infect. Dis. *146*, 397–404.

Steiner, I., Spivack, J.G., Lirette, R.P., Brown, S.M., MacLean, A.R., Subak, S.J.H., and Fraser, N.W. (1989). Herpes simplex virus type 1 latency-associated transcripts are evidently not essential for latent infection. EMBO J. *8*, 505–511.

Stevens, J.G., Wagner, E.K., Devi, R.G.B., Cook, M.L., and Feldman, L.T. (1987). RNA complementary to a herpesvirus alpha gene mRNA is prominent in latently infected neurons. Science *235*, 1056–1059.

Tang, S., Bertke, A.S., Patel, A., Wang, K., Cohen, J.I., and Krause, P.R. (2008). An acutely and latently expressed herpes simplex virus 2 viral microRNA inhibits expression of ICP34.5, a viral neurovirulence factor. Proc. Natl. Acad. Sci. U.S.A. *105*, 10931–10936.

Thomas, D.L., Lock, M., Zabolotny, J.M., Mohan, B.R., and Fraser, N.W. (2002). The 2-kilobase intron of the herpes simplex virus type 1 latency-associated transcript has a half-life of approximately 24 hours in SY5Y and COS-1 cells. J. Virol. *76*, 532–540.

Thompson, R.L., and Sawtell, N.M. (2000). HSV latency-associated transcript and neuronal apoptosis. Science *289*, 1651.

Thompson, R.L., and Sawtell, N.M. (2001). Herpes simplex virus type 1 latency-associated transcript gene promotes neuronal survival. J. Virol. *75*, 6660–6675.

Umbach, J.L., Kramer, M.F., Jurak, I., Karnowski, H.W., Coen, D.M., and Cullen, B.R. (2008). MicroRNAs expressed by herpes simplex virus 1 during latent infection regulate viral mRNAs. Nature *454*, 780–783.

Umbach, J.L., Nagel, M.A., Cohrs, R.J., Gilden, D.H., and Cullen, B.R. (2009). Analysis of human alphaherpesvirus microRNA expression in latently infected human trigeminal ganglia. J. Virol. *83*, 10677–10683.

Wagner, E.K., and Bloom, D.C. (1997). Experimental investigation of herpes simplex virus latency. Clin. Micro. Reviews *10*, 419–443.

Wang, K., Mahalingam, G., Imai, Y., Pesnicak, L., Margolis, T.P., Straus, S.E., and Cohen, J.I. (2009). Cell type specific accumulation of the major latency-associated transcript (LAT) of herpes simplex virus type 2 in LAT transgenic mice. Virology *386*, 79–87.

Wang, K., Prikhodko, E., Pesnicak, L., Cohen, J.I., and Straus, S.E. (2002). Mice transgenic for the HSV-2 LAT resist Fas-mediated apoptosis and hepatic necrosis. Paper presented at: 27th International Herpesvirus Workshop (Cairns, Australia).

Wang, Q.Y., Zhou, C., Johnson, K.E., Colgrove, R.C., Coen, D.M., and Knipe, D.M. (2005). Herpesviral latency-associated transcript gene promotes assembly of heterochromatin on viral lytic-gene promoters in latent infection. Proc. Natl. Acad. Sci. U.S.A. *102*, 16055–16059.

Wu, T.-T., Su, Y.-H., Block, T.M., and Taylor, J.M. (1996). Evidence that two latency associated transcripts of herpes simplex type 1 are non-linear. J. Virol. *70*, 5962–5967.

Yang, L., Voytek, C.C., and Margolis, T.P. (2000). Immunohistochemical analysis of primary sensory neurons latently infected with herpes simplex virus type 1. J. Virol. *74*, 209–217.

Yoshikawa, T., Hill, J.M., Stanberry, L.R., Bourne, N., Kurawadwala, J.F., and Krause, P.R. (1996). The characteristic site-specific reactivation phenotypes of HSV-1 and HSV-2 depend upon the latency-associated transcript region. J. Exp. Med. *184*, 659–664.

Zhang, C. (2009). Novel functions for small RNA molecules. Curr. Opin. Mol. Ther. *11*, 641–651.

Zhu, J., Kang, W., Marquart, M.E., Hill, J.M., Zheng, X., Block, T.M., and Fraser, N.W. (1999). Identification of a novel 0.7 kb polyadenylated transcript in the LAT promoter region of HSV-1 that is strain specific and may contribute to virulence. Virology *265*, 296–307.

Zwaagstra, J.C., Ghiasi, H., Nesburn, A.B., and Wechsler, S.L. (1991). Identification of a major regulatory sequence in the latency associated transcript (LAT) promoter of herpes simplex virus type 1 (HSV-1). Virology *182*, 287–297.

Zwaagstra, J.C., Ghiasi, J.C., Slanina, S.M., Nesburn, A.B., Wheatley, S.C., Lillycrop, K., Wood, J., Latchman, D.S., Patel, K., and Wechsler, S.L. (1990). Activity of herpes simplex virus type 1 latency-associated transcript (LAT) promoter in neuron-derived cells: Evidence for neuron specificity and for a large LAT transcript. J. Virol. *64*, 5019–5028.

Vaccines and New Antiviral Strategies Against Herpes Simplex Virus

18

Timothy E. Dudek and David M. Knipe

Abstract

Vaccines have been among the most effective public health approaches for protecting individuals against viral disease, with two of the world's most successful vaccines being against smallpox virus and poliovirus. Herpes simplex virus 1 (HSV-1) is a nearly ubiquitous pathogen, and the worldwide prevalence of herpes simplex virus 2 (HSV-2) continues to increase. These two pathogens cause significant morbidity and mortality among the general population, but in particular in neonates and immunocompromised individuals. Perhaps most significantly, there is a 3–4 fold increased risk of HIV acquisition in HSV-2 infected individuals. To date, attempts at producing a vaccine against HSV have not been successful, but each attempt has brought insights into what may be required for an effective vaccine. Furthermore, intense studies into the immunology of HSV infection and the resources that have been put into vaccine design and development have recently yielded knowledge that will be necessary to achieve the goal of a highly effective vaccine against HSV. Here we will discuss several vaccine constructs that have contributed to the advancement of the HSV vaccine field and some of the new antiviral strategies to counter HSV.

Introduction

Since Edward Jenner showed that inoculation of humans with cowpox led to immunity against smallpox virus, vaccination has been and continues to be one of the most efficient means of protecting individuals and entire populations against viral infections. HSV-1 is a nearly ubiquitous pathogen that can cause serious ocular and central nervous disease, and HSV-2 can cause life-threatening infections in neonates and immunocompromised individuals. HSV-2 infection can also lead to life-threatening disease, and infection with HSV-2 has also been shown to increase the risk of human immunodeficiency virus 1 (HIV-1) acquisition by 3–4 fold (Freeman *et al.*, 2006; Roizman, Knipe, and Whitley, 2007; Wald and Link, 2002). Considering that regions of the world with the highest HIV-1 prevalence are also the regions of the world with the highest HSV-2 seroprevalence, an effective therapy against HSV infections could have a significant effect on HIV acquisition rates (Smith and Robinson, 2002; Weiss, 2004; UNAIDS/WHO). Although treatment of HSV-2 seropositive individuals with herpes suppressive drug therapy did not reduce their rate of HIV-1 acquisition (Celum *et al.*, 2008; Watson-Jones *et al.*, 2008), prevention of initial HSV-2 infection should have a protective effect against HIV-1 acquisition. Therefore a vaccine against HSV that effectively reduces the prevalence of HSV-2 should not only reduce the disease burden caused by HSV infection but also reduce HIV-1 acquisition rates. Much can be learned about the potential of a vaccine construct through the use of animal models, and we will discuss the advantages and disadvantages of the two most common animal models for HSV infection, mice and guinea pigs. We will then discuss previous vaccine attempts, vaccines presently being tested, and approaches for the future. We will conclude with discussion of some new anti-HSV therapeutic strategies.

HSV pathogenesis and clinical infections

HSV-1 undergoes a primary productive infection in the oral mucosa or oral cavity (gingivostomatitis) and then spreads to and establishes a latent infection in sensory neurons in the trigeminal ganglia. Reactivation can lead to recurrent infections of HSV-1 commonly known as cold sores or fever blisters. HSV-1 can also infect the cornea (herpes keratitis) as a result of primary infection or reactivation from latent infection. The host immune response to recurrent infections of the cornea can lead to progressive scarring of the cornea, clouding and eventual blindness. Furthermore, HSV-1 can travel along neuronal pathways into the central nervous system during primary infection or reactivation, resulting in very serious encephalitis. HSV-2 is acquired generally as a sexually transmitted disease. HSV-2 infects the genital mucosa, spreads via axonal transport to sacral ganglia where it establishes a latent infection in sensory neurons. Reactivation of this virus causes the recurrent lesions associated with genital herpes. In addition to genital infection, HSV-2 can spread systemically to cause meningitis. More serious, however, is the intrauterine or peripartum transmission of HSV-2 from a productively infected mother to her child, which can result in encephalitis and/or disseminated herpes infection in the newborn. Mortality of infected newborns is high despite the availability of antivirals to limit infection, and survivors frequently experience life-long sequelae. Thus, a herpes vaccine and new antiviral strategies are needed. The goals of a herpes vaccine would be to induce protective immunity that accomplishes any or all of the following: (1) Reduce viral infection at the primary site of infection, (2) reduce primary disease, 3) reduce spread of virus into sensory ganglia and reduce latent infection, (4) reduce spread of virus into the central nervous system and (5) reduce latent infection therapeutically and/or reactivation.

Animal models

Authentic animal infection models are key to the development of an effective vaccine against viral pathogens because the determination of which vaccine candidates are most efficacious and which immune responses are important for protection can only be done in an *in vivo* situation. In an ideal animal model, infection would be initiated via a route similar to that in which humans are commonly infected, i.e. inoculation of the mucosal epithelium for HSV. The host animal would then develop disease symptoms similar in both pathology and severity to those seen in humans. If the immune responses that the animal develops after infection protect the animal from disease, the host immune response can be dissected to determine which aspects are important for protection, including innate, humoral, and/or cellular immune responses. If the host cannot spontaneously control the pathogen, this provides a situation for the investigation of potential protective immune responses through immunization. Immune responses could then be elicited through immunization and those responses would be analysed, as mentioned above, before inoculation of the pathogen. If immunized animals are then protected from infection and/or disease, whereas unimmunized animals are not, these immune responses can be further dissected to determine the important aspects for protection. Although there is no perfect animal model for HSV infections, both guinea pigs and mice are commonly used and each model has its own strengths and weaknesses.

The guinea-pig has been considered by many to be the gold standard for genital HSV infection because HSV-2 infection of guinea pigs shares many features with that of humans. Intravaginal inoculation of HSV-2 into guinea pigs leads to acute replication and disease at the site of infection, establishment of a latent viral reservoir within the enervating sensory neurons, and periodic reactivation leading to viral shedding and even recurring disease as detected by visible lesions (Bourne *et al.*, 1994; Krause *et al.*, 1995). The development of both acute and recurrent disease provides the opportunity to test both the prophylactic and the therapeutic effects of potential vaccines, whereas the mouse can only be used to measure acute infection. Infected animals develop immune responses that consist of both humoral and cellular immunity. Due to the paucity of immunological reagents for the guinea pig, the full range of immune effector mechanisms is not known, but humoral immunity seems to be sufficient for protection in the guinea pig genital herpes model because passive immunization (Bourne *et al.*, 2002) and subunit vaccines are protective (Bourne *et al.*, 2003; Stanberry *et al.*, 1987). As discussed below, this may differ from the human situation.

Mice often provide good infection models as they are readily available in outbred and inbred lines, are relatively inexpensive, and have many immunological reagents available. Mice can be infected with HSV via several routes of infection, including footpad, flank, ocular, and intravaginal routes, each of which has its own unique properties. HSV infection at any of these sites leads to localized replication and shedding of virus, the development of lesions, establishment of latent infection within sensory neurons and persistence of latent viral DNA within local sensory ganglia and in some cases, central nervous disease. Reactivation of HSV from latent infection in mice occurs rarely *in vivo* but is readily observed when ganglia are explanted in culture (Stevens and Cook, 1971). Reactivation can be induced *in vivo* to a limited extent by ultraviolet irradiation (Laycock *et al.*, 1991) or elevated temperature (Sawtell and Thompson, 1992).

Furthermore, mice develop a wide range of innate, humoral and cellular immune responses against HSV that are readily measured with the many immunological reagents available for mice. Evidence suggests that both humoral and cellular immune responses can protect mice from HSV infection. In most cases, passive transfer of HSV-specific antibody can prevent encephalitis but not reduce mucosal replication (Eis-Hubinger, Schmidt, and Schneweis, 1993; McKendall, Klassen, and Baringer, 1979; Morrison and Knipe, 1997). Both $CD4^+$ and $CD8^+$ T-cell responses appear to play a role in protection from infection or reduction in levels of viral replication, latent viral loads and neuropathy, although $CD4^+$ T-cells appear to be the most important in protecting mice against corneal infection (Morrison and Knipe, 1997) and intravaginal infection (Morrison, 2008).

The mouse genital infection model now commonly used is derived from that of Parr *et al.* (1994) in which female mice are treated with medroxyprogesterone to thin the genital epithelium and make the mice uniformly susceptible to HSV-2 vaginal infection. These mouse models have been useful for vaccine studies in that one can quantify levels of acute viral replication and levels of latent viral loads found within the ganglia of sensory neurons that enervate the site of infection. Through these analyses, it has been shown that immunization can reduce viral shedding, latent viral loads and encephalitis and that protection is dose-dependent (Da Costa, Morrison, and Knipe, 2001). While HSV does not reactivate in mice, the potential effect of vaccination on recurrences can be inferred from the levels of latent viral DNA within neurons (Hoshino *et al.*, 2008), as these measurement have been shown to correlate with the number of *ex vivo* recurrence rates in mice (Hoshino *et al.*, 2007; Sawtell, 1997).

Measures of vaccine efficacy

A variety of parameters can be measured in animal models to determine the efficacy of a vaccine candidate.

Reduced encephalitis and lethality

Protection against lethal infection (survival) in an animal model was used as a measure of vaccine efficacy in the past. However, current animal care guidelines require that animals with paralytic disease be euthanized; therefore, protection against encephalitis or paralytic disease is now measured in most cases. A quantitative measure of the efficacy of a vaccine is the dose of vaccine required for 50% reduction in paralytic disease (protective dose 50 or PD_{50}; (Dudek, Mathews, and Knipe, 2008). As discussed above, humoral immunity can protect against encephalitis so this may be a measure of total immunity induced or just a measure of antibody induction.

Reduced local replication

Reduction of local replication and shedding by HSV in mice often requires cellular immunity and shows a relationship with reduction of latent infection in mice (Morrison and Knipe, 1994); therefore, this parameter can be a good quantitative measure of the efficacy of a vaccine in the mouse genital model (Da Costa *et al.*, 1997). Virus replication at the local sites can be measured as infectious virus or viral DNA shed at mucosal sites and captured on a swab. Measurement of the ability of a vaccine to prevent infection could determine the dose of challenge virus needed to infect immunized animals. The vaccine could be considered to be elevating the threshold of susceptibility of the animal or individual (Stanberry, 2004).

Reduced primary disease

Vaccines have traditionally prevented disease (Graham and Crowe, 2007); therefore prevention of primary herpetic lesions could be measured to determine the efficacy of a herpes vaccine in an animal model or even in a clinical trial. Quantitative differences in reduction of disease have been observed with different viral vaccine strains in the mouse genital model (Da Costa *et al.*, 2001).

Reduced latent infection

Latent infection can be measured as latent viral DNA content in sensory ganglia (Bourne *et al.*, 1996a), numbers of latency-associated transcript-(LAT) positive cells (Morrison and Knipe, 1996), or the ability of explanted ganglia to reactivate (Morrison and Knipe, 1994). All of these have been used as a measure of the ability of a vaccine to reduce latent infection. The references cited are examples of the use of these parameters to measure vaccine effects on latent infection.

Reduced reactivation

Recurrent lesions have been quantified in the guinea pig model as a measure of viral reactivation from latent infection. Reduction of recurrent lesions by prophylactic immunization measures effects of both the prevention of establishment and reactivation while the effects of therapeutic immunization are more directly the result of effects of immunization on reactivation.

Herpes vaccines and vaccine candidates

Traditionally, viral vaccines were either live, attenuated viruses or inactivated viruses. Newer forms of inactivated viral vaccines consist of viral protein subunits. There are several examples of successful vaccines of each type. Live, attenuated vaccines have included the smallpox vaccine, the yellow fever vaccine, the Sabin polio vaccine, and the measles vaccine. Killed viral vaccines have included the Salk polio vaccine. Viral protein vaccines have included hepatitis B vaccine and human papilloma virus vaccine. We will review the herpes vaccine approaches that have been tested in animal models and human clinical trials. None have been licensed by the US FDA, although a few are commercially available elsewhere.

Whole killed herpes virions

Initial attempts at production of a vaccine against HSV focused on the use of whole killed virus using methods to inactivate the virus including ultraviolet irradiation, heat inactivation and chemical inactivation. These vaccines have shown significant efficacy in animal models but have not shown efficacy in placebo-controlled, double-blinded studies (Whitley, 1993). Despite the lack of proven efficacy, three whole killed vaccines are commercially available against HSV infection, the Lupidon H and G vaccines and the Skinner vaccine.

Lupidon H and G

The Lupidon H (HSV-1) and G (HSV-2) vaccines contain heat-inactivated whole virus grown in the chorioallantoic membrane of chicken eggs. While these vaccines have claimed both prophylactic and therapeutic efficacy in small clinical trails, they have not been tested in double-blinded, placebo-controlled vaccine trials and have not been approved by the US Food and Drug Administration, but are commercially available in Germany.

Skinner vaccine

The NFU.Ac.HSV-1(S^-)MRC vaccine, known more commonly as the Skinner vaccine, is produced by infecting cells with the HSV-1 strain Triosbel, treating the cells with detergent to lyse cells and low-speed centrifugation to remove the cell nuclei, followed by formaldehyde inactivation of virus and ultracentrifugation to concentrate viral proteins to then be used as the vaccine. A double-blinded, placebo-controlled vaccine for the Skinner vaccine was performed on 303 individuals who had a history of genital HSV of more than a year with at least 6 recurrences in the year previous to the initiation of the trial. After six months, vaccinated women ($n = 84$; 0.5 recurrences/month) did show modest reductions in the median number of recurrences per month as compared to the placebo immunized women ($n = 86$; 0.7 recurrences/month), but no difference was seen in men (Skinner *et al.*, 1997). The Skinner vaccine is available commercially in the United Kingdom but has not been approved for use in the U.S. by the U.S. Food and Drug Administration.

Live-attenuated viruses as vaccines

There have been several attempts to produce an attenuated HSV strain that would be attenuated sufficiently for safety, while still being highly immunogenic. This has proven to be a difficult compromise to date, because highly attenuated viruses have been shown to lack strong immunogenicity.

R7017 and R7020

Perhaps the most studied live HSV vaccine strains to date are the R7017 and R7020 strains developed by Roizman and colleagues. This vaccine consists of the HSV-1 F strain with the modifications being a deletion of the thymidine kinase (*TK*) gene and a deletion of a 14.5-kbp fragment encompassing the junction of the U_L and U_s gene segments, into which genes encoding the HSV-2 glycoproteins D, G, and I were inserted. R7020 contains an additional insertion of the HSV-1 *TK* gene next to the HSV-2 genes. This was added as an additional biosafety measure by making the virus susceptible to the affects of the antiviral drug acyclovir and its derivatives (Meignier, Longnecker *et al.*, 1988). R7017 and R7020 were highly attenuated in rodent models but elicited protective immune responses (Meignier, Longnecker, and Roizman, 1988). The safety profile was confirmed in owl monkeys, with doses up to 10^5 plaque-forming units causing little or no noticeable disease (Meignier *et al.*, 1990). R7020 was tested in humans and was poorly immunogenic, this, along with difficulties with production of the virus in sufficient levels, led to the discontinuation of product development (Whitley, 1993).

Other live-attenuated vaccine candidates

The ICP10ΔPK strain of HSV-2 contains a deletion of the *ICP10* gene, formally known as the U_L39 gene encoding the large subunit of the viral ribonuclease reductase. ICP10 is believed to be involved in upregulation of Th2 cytokines such as IL-6 and IL-10 because deletion of this gene helps to shift the immune response towards a Th1 response (Gyotoku *et al.*, 2002). Evidence exists that ICP10 is required for efficient virus replication and latency establishment (Aurelian and Smith, 2000; Smith *et al.*, 1998), and ICP10ΔPK is attenuated for replication, shows reduced establishment of latency and lowered levels of viral DNA in latently infected ganglia (Aurelian *et al.*, 1999; Wachsman *et al.*, 2001). ICP10ΔPK has shown efficacy as a vaccine in mouse and guinea pig models of HSV-2 infection (Aurelian *et al.*, 1999; Gyotoku, Ono, and Aurelian, 2002; Wachsman *et al.*, 2001). This strain was reported to reduce recurrences in a phase I therapeutic trial in Mexico (Casanova *et al.*, 2002). The safety profile of this replication-competent HSV-2 strain has not been described, but it is capable of establishing latent infection, albeit at a reduced level (Wachsman *et al.*, 2001).

Biovex Inc. has developed a replication-competent HSV-2 virus as a genital herpes vaccine, and while the genotype and properties of this virus have not been reported in the scientific literature, the company states that four 'genes that collectively interfere with the functions of immune system have been deleted' (http://www.biovex.com/immunovex.html). A phase I clinical trial of the ImmunoVEX^{HSV-2} vaccine began in February 2010 in the United Kingdom.

Another replication-competent virus, the NS-gEnull strain of HSV-1 has a large deletion in the *gE* gene (U_S8), codons 124 to 512, and fails to express a functional gE protein (Nagashunmugam *et al.*, 1998). This strain has been shown to be defective for cell-to-cell spread, retrograde transport from the epithelium to the dorsal root ganglia and anterograde transport back out to the epithelium in mice (Dingwell *et al.*, 1994; Wang *et al.*, 2005a). This virus has shown protection in mice against wild-type HSV-1 and HSV-2 challenges (Brittle *et al.*, 2008).

Finally, an HSV-1 ICP0 mutant virus inoculated by the ocular route can establish latent infection in the trigeminal ganglia and render the mice resistant to challenge by WT HSV-1 via the ocular route (Halford *et al.*, 2006). The safety profile of a virus that establishes latent infection would have to be studied thoroughly.

Protein subunit vaccines

Protein subunit vaccines consist of proteins purified from infected cells, virions, or from genetically modified cells. Examples of successful subunit vaccines include the hepatitis B vaccine consisting of the hepatitis B surface antigen (HBsAg) and the virus-like particles of the human papillomavirus coat proteins used in the HPV vaccine. The most extensively studied subunit vaccines against HSV have been produced using HSV-2 glycoproteins B (gB2) and D (gD2). These viral glycoproteins have been produced as soluble, secreted proteins in mammalian cells and purified for use as a vaccine product.

gD2/gB2-MF59

Chiron developed a vaccine candidate consisting of the extracellular portions of gB2 and gD2. gD2 and gB2 with or without complete Freund's adjuvant showed protection in a guinea pig model, although the immunogenicity and protection were higher with the adjuvant (Stanberry *et al.*, 1989). Initial clinical studies focused on gD2 with alum as an adjuvant. This formulation elicited anti-HSV antibodies with neutralizing activity and lymphoproliferative responses in recipients (Straus *et al.*, 1993). Later studies used a mixture of both gD2 and gB2 in the squalene-based adjuvant MF59. Phase III studies of this vaccine showed that it elicited strong Th2 cellular immune responses and levels of neutralizing antibody titres that were comparable to those elicited in natural infection, but no efficacy was observed in prophylactic (Corey *et al.*, 1999) or therapeutic (Straus *et al.*, 1994, 1997) trials. These trials showed that neutralizing antibodies are not sufficient for long-term control of HSV infection. One hypothesis for the lack of protection was that MF59 had been shown to elicit Th2 type cellular immune responses (Singh *et al.*, 1998; Valensi *et al.*, 1994), and it was theorized that Type 1 cellular immune responses would be more important for protection against the viral infection in humans.

gD2-MPL

GlaxoSmithKline developed a vaccine candidate consisting of gD2 in a different adjuvant, 3-O-deacylated monophosphoryl lipid A (MPL). This adjuvant can induce Th1 type cellular immune responses (Thoelen *et al.*, 2001), the type of response implicated in protection against infection in animal models and in the clearance of lesions in humans. Phase III studies showed that this vaccine gave a 74% reduction in symptomatic HSV-2 infection in HSV-1-seronegative women, but did not show any efficacy in men or in HSV-1-seropositive women (Stanberry *et al.*, 2002). Although the protective effect was not seen among the study participants in general, this was the first HSV-2 vaccine candidate to show any efficacy within a double-blinded placebo controlled study, raising optimism that a herpes vaccine was feasible. An attractive hypothesis as to why this formulation had shown some efficacy was that the Th1 type cellular immune response elicited with the help of the MPL adjuvant. On the other hand, the limited breadth of the protection raised the possibility of the need for $CD8^{+}$ T cell responses for effective protection against HSV-2, as had been argued from studies of lymphocyte populations in HSV-2 lesions (Koelle *et al.*, 1998; Posavad *et al.*, 1996). An additional phase III study supported by NIH is currently ongoing in HSV-1 seronegative women to further determine the vaccine's efficacy in this population.

Other approaches for herpes vaccines

Because no vaccine approach has provided broad protection against HSV-2 infection, novel vaccine strategies including DNA vaccines, peptide vaccines, replication-defective mutant viral strains and single-cycle mutant viral strains have been tested as potential HSV-2 vaccines.

DNA vaccines

DNA vaccines targeting other pathogens have been effective in many animal model systems and in some cases in human clinical trials (Liu, 2010).

In the case of HSV, immunization with plasmids expressing gD2 reduces clinical disease and viral replication in guinea pig and rabbit models (Bourne *et al.*, 1996a; Bourne *et al.*, 1996b). Immunization with plasmids encoding full-length or secreted gD2 were protective in mice and guinea pigs, but a cytosolic form of gD2 was less effective (Strasser *et al.*, 2000). Immunization with plasmids encoding gB2 and gC2 were less effective as a prophylactic vaccine but reduced recurrences when used as a therapeutic vaccine (Mester *et al.*, 1999). Immunization of mice with a set of cosmids containing HSV-2 DNA induced protective immunity against an intravaginal challenge with the homologous HSV-2 strain MS virus (Braun *et al.*, 2008). Formulation of a plasmid gD2 vaccine with bupivacaine has enhanced the immunogenicity of the DNA vaccine (Bernstein *et al.*, 1999). A phase I clinical trial with a plasmid vaccine encoding full-length gD2 formulated with bupivacaine showed that the vaccine was safe but induced cellular responses in only one of four HSV-seronegative participants who received the highest dose (3000 μg) of vaccine (Cattamanchi *et al.*, 2008). Therefore, high doses of this DNA vaccine seem to be needed to induce cellular responses in humans.

Replication-impaired viruses as vaccines

Replication-impaired viruses are viruses that contain mutations in genes that are essential for virus replication and are grown on cell lines that express the deleted gene products, which complements the phenotypic defect and allows the virus to replicate in these genetically manipulated cells. Because these viruses are incapable of replication in normal cells and spread in the host, they are very safe vaccines (Dudek and Knipe, 2006). Two general types of replication-impaired viruses have been tested as vaccines, replication-defective viruses, which have defects at replication stages that prevent viral progeny virus assembly and single-cycle mutant viruses that assemble progeny virus particles that are non-infectious due to defective structural components.

Replication-defective viruses

Replication-defective mutant viruses are mutated in a viral gene needed for replication in normal cells, and their growth can be complemented in cells stably transformed with the missing gene product (DeLuca *et al.*, 1985). We showed that HSV-1 replication-defective mutants could protect against WT HSV-1 virus challenge (Nguyen *et al.*, 1992), arguing that infection of one round of cells can allow expression of sufficient immunogen for protection in this animal model. This was one of the first demonstrations that replication-defective mutant viruses could serve as a vaccine. We further showed that an HSV-2 replication-defective mutant could induce protection against genital HSV-2 infection in guinea pigs (Da Costa *et al.*, 1997). For a potential vaccine candidate, we constructed a double deletion mutant virus, *dl*5–29, which has two genes deleted, U_L5, which encodes a component of the helicase–primase complex, and U_L29, which encodes the single-stranded DNA binding protein, ICP8, each of which is essential for viral DNA replication. Therefore, *dl*5–29 can infect a normal cell and express a vast majority of viral immediate-early, early and late proteins, but it is unable to replicate its genome and produce infectious progeny (Da Costa *et al.*, 2000). The *dl*5–29 virus is highly effective for protection against acute and latent infection in mouse infection models (Da Costa *et al.*, 1999; Jones *et al.*, 2000) and for prophylactic and therapeutic immunization in guinea pigs (Da Costa *et al.* 1999; Hoshino *et al.*, 2005). Immunization with *dl*5–29 shows prophylactic immunization equivalent to HSV-2 gD in complete Freund's adjuvant (CFA) and stronger therapeutic efficacy than gD2 in CFA (Hoshino *et al.*, 2005). The *dl*5–29 virus is safe even by intracerebral inoculation in newborn mice or peripheral inoculation in immunodeficient mice (Hoshino *et al.*, 2008) and is also effective in HSV-1 pre-immune animals (Hoshino *et al.*, 2009). The HSV-2 *dl*5–29 replication-defective mutant is currently being developed by Sanofi Pasteur as a vaccine against HSV-2.

Single-cycle mutant viruses

A single-cycle mutant virus with a deletion in the glycoprotein H (*gH*, U_L24) gene has been tested as a candidate herpes vaccine. HSV-1 gH mutant viruses showed that gH is essential for formation of infectious virus particles (Forrester *et al.*, 1992). The gH deleted virus can infect a normal cell and express all HSV proteins except gH, leading the production of non-infectious virions, thereby stopping the cycle of viral infection. Immunization with the gH mutant virus pseudotyped with wild-type gH in its envelope gives immune protection against HSV-1 challenge in the ear pinna (Farrell *et al.*, 1994) and genital HSV-2 challenge in mice and guinea pigs (McLean *et al.*, 1994). These viruses may not be completely replication-defective *in vivo* because latent infection by an HSV-1 gH mutant virus, as detected by LAT expression in sensory neurons, was observed (Loudon *et al.*, 2001). These viruses were named defective infectious single cycle (DISC) viruses by Cantab Pharmaceuticals. An HSV-2 gH mutant virus was tested in clinical trials by Cantab and was safe but showed no clinical or virological benefit in a therapeutic phase II trial against genital HSV-2 disease (de Bruyn *et al.*, 2006). The DISC vaccines have been further developed by Xenova, which acquired Cantab, as oncolytic viruses, but there is no information on further development of the DISC viruses as herpes vaccines.

Future directions in HSV vaccines

Several areas provide potential areas for improvements in existing vaccine candidates or for new vaccine designs. Inactivation of viral genes that encode immune evasion functions could lead to improvements. For example, inactivation of the U_L41 gene encoding the virion host shutoff function leads to improvement in immunogenicity and protection in some situations (Dudek, Mathews, and Knipe, 2008; Geiss *et al.*, 2000; Walker and Leib, 1998) but not all situations (Hoshino *et al.*, 2008; Reszka, Dudek, and Knipe, 2010). Further study of these types of viral mutations could improve the vaccine candidates.

There is evidence that there may be 'protective' and 'non-protective' epitopes for HSV-specific $CD4^+$ and $CD8^+$ T cells (Chentoufi *et al.*, 2008; Hosken *et al.*, 2006). With the identification of the critical T cell responses that can control HSV acute infection and/or reactivation, it may be possible to construct vaccine vectors that express these key epitopes and induce immune responses that are more likely to be protective, while avoiding the elicitation of responses that are not protective.

Thus far, herpes vaccine design has targeted HSV-2 antigens and genetic backgrounds of North American and European isolates. Although there is very limited information on the genetic and antigenic diversity of HSV-2 viruses from around the world, there is evidence that the genetic diversity of HSV-2 is greater in HSV-2 strains from Africa (Norberg *et al.*, 2007). Initial results indicate that *dl*5–29 can induce protection against South African viruses but not as effectively as against US viruses (Dudek and Knipe, manuscript in preparation). This raises the possibility that antigenic diversity may also exist and that modified vaccine candidates will be required for optimal efficacy against HSV-2 in Africa and other regions of the world. Further studies are needed to determine the optimal immunogens for immunization against HSV-2 in epidemic areas such as sub-Saharan Africa.

New antiviral strategies

The broad area of antiviral strategies is outside the scope of this article, and this is reviewed elsewhere (Wilson *et al.*, 2009). Therefore, we will summarize three very interesting new strategies for countering HSV: RNA interference (RNAi), anti-HSV hammerhead ribozymes, and control of chromatin on the viral genome.

RNA interference

RNA interference (RNAi) has been shown to reduce specifically targeted gene expression. Small interfering RNAs (siRNA) bind to complementary sequences on mRNA and inhibit their translation and/or target the mRNA for degradation. This effect has been studied extensively *in vitro* and *in vivo* (Dykxhoorn *et al.*, 2006). RNAi has been used to reduce HSV gene expression and replication in cell culture (Bhuyan *et al.*, 2004). Furthermore, RNAi technology has been used to interfere successfully with HSV-2 infection *in vivo*. Initial studies showed that lipid complexed siRNA targeting the HSV-2 *U_L27*, or *U_L29* mRNAs can be applied as a pre-exposure prophylaxis against intravaginal HSV-2 infection in a mouse model and that a combination of the two siRNAs can be used as a post-exposure prophylaxis (Palliser *et al.*, 2006). A further study showed that the durability of the prophylaxis can be extended to up to one week post application by including cholesterol-conjugated siRNA targeting the HSV receptor nectin-1 along with the siRNA targeting *U_L29* (Wu *et al.*, 2009). The latter study did not measure any virological parameters so the extent of the protection against virus replication cannot be ascertained. While these studies used challenges with lower levels of WT HSV-2 than used in many vaccine studies, these studies do show the promise of siRNA for use as anti-HSV therapy, in particular for post-exposure therapy.

Ribozymes

Ribozymes are RNA enzymes involved in cleaving or splicing of RNA molecules. A RNase P ribozyme variant that targets the HSV ICP4 mRNA reduces viral replication in cell culture by more than 4000 fold (Trang *et al.*, 2000; Trang *et al.*, 2001). Hammerhead ribozymes, so called due to their folding into a structure that resembles the head of a hammer, specifically bind complementary sequences on mRNA and then cleave these mRNAs reducing the levels of protein produced within a cell (Fedor and Williamson, 2005; Long and Uhlenbeck, 1993). Hammerhead ribozyme technology was used to target four different HSV genes, *ICP4*, *ICP27*, *U_L20* and *U_L30*, with the *U_L20* ribozyme showing the greatest inhibition of viral replication in rabbit skin cell culture (Liu *et al.*, 2008). An adenoviral delivery system was then used to determine the ability of the *U_L20* ribozyme to protect mice from HSV-1 in a footpad infection model. After delivery of the Ad-*U_L20* or Ad-GFP or PBS controls into the mouse footpad, HSV-1 was inoculated onto the footpads following scarification. Ad-*U_L20*-treated mice showed (1) a reduction in the average levels of viral DNA found within the spinal cords as compared to control animals and (2) an increase in survival rates (Liu *et al.*, 2008). Although it remains to test this approach against HSV in a more biological context, these studies show the promise of hammerhead ribozymes as anti-HSV therapy *in vivo*.

Control of chromatin on the viral genome

Recent studies have defined the role of the chromatin in the decision of HSV to undergo a lytic infection versus a latent infection (Knipe and Cliffe, 2008). HSV-encoded proteins including the VP16 virion protein (Herrera and Triezenberg, 2004) and the immediate-early ICP0 protein (Cliffe and Knipe, 2008; Gu and Roizman, 2007) promote euchromatic modifications on chromatin associated with viral lytic gene promoters in epithelial cells to drive the productive infection cascade of viral gene expression. In contrast, in neurons, LAT (Cliffe *et al.*, 2009; Wang *et al.*, 2005b) and other factors (Kwiatkowski, 2009) promote the assembly of facultative heterochromatin assembly on viral lytic genes and the silencing of these genes. It has been proposed that latency might be interrupted or stabilized by therapeutic treatments that alter the modifications of chromatin (Knipe and Cliffe, 2008). VP16 and the cellular transcriptional coactivator host cell factor-1 recruit the lysine-specific demethylase-1 (LSD1) to viral IE promoters where LSD1 removes repressive methylation marks from the chromatin (Liang *et al.*, 2009). Monoamine oxidase inhibitors (MAOIs) inhibit LSD1 activity, and treatment of cells with MAOIs decreases HSV gene expression (Liang *et al.*, 2009). Furthermore, incubation of latently infected ganglia with an MAOI blocks reactivation of HSV (Liang *et al.*, 2009). It remains to be proven whether the inhibitors are blocking emergence from latent infection or a step common with lytic infection, but these results raise the potential of using MAOIs to prevent HSV reactivation.

Perspectives

Herpes simplex virus remains a major human pathogen, and a herpes vaccine is direly needed, not only to prevent herpetic disease but also to reduce the risk of HIV infection. Ongoing efforts are providing new vaccine candidates that need to be supported for clinical trials. New approaches for preventing HSV infection as drugs and microbicides are also being developed. In addition, approaches for keeping HSV silent in a latent infection have been discovered and promise a new class of anti-herpes drugs. Continued and even expanded support from pharmaceutical companies and the NIH need to focus on the development of these approaches, not only as potential therapies against herpes infections and disease, but also as therapeutics to reduce the HIV/AIDS pandemic.

Acknowledgements

Studies in the Knipe laboratory on HSV-2 vaccines are supported by NIH grant AI057552. DMK is a co-inventor on a patent on HSV replication-defective mutant vaccine technology that has been licensed by Harvard University to Sanofi-Pasteur Biologicals. DMK is a consultant to Sanofi-Pasteur Biologicals.

References

Aurelian, L., Kokuba, H., and Smith, C.C. (1999). Vaccine potential of a herpes simplex virus type 2 mutant deleted in the PK domain of the large subunit of ribonucleotide reductase (ICP10). Vaccine *17*, 1951–1963.

Aurelian, L., and Smith, C.C. (2000). Herpes simplex virus type 2 growth and latency reactivation by cocultivation are inhibited with antisense oligonucleotides complementary to the translation initiation site of the large subunit of ribonucleotide reductase (RR1). Antisense Nucleic Acid Drug Dev. *10*, 77–85.

Bernstein, D.I., Tepe, E.R., Mester, J.C., Arnold, R.L., Stanberry, L.R., and Higgins, T. (1999). Effects of DNA immunization formulated with bupivacaine in murine and guinea pig models of genital herpes simplex virus infection. Vaccine *17*, 1964–1969.

Bhuyan, P.K., Kariko, K., Capodici, J., Lubinski, J., Hook, L.M., Friedman, H.M., and Weissman, D. (2004). Short interfering RNA-mediated inhibition of herpes simplex virus type 1 gene expression and function during infection of human keratinocytes. J. Virol. *78*, 10276–10281.

Bourne, N., Bravo, F.J., Francotte, M., Bernstein, D.I., Myers, M.G., Slaoui, M., and Stanberry, L.R. (2003). Herpes simplex virus (HSV) type 2 glycoprotein D subunit vaccines and protection against genital HSV-1 or HSV-2 disease in guinea pigs. J. Infect. Dis. *187*, 542–549.

Bourne, N., Milligan, G.N., Scleiss, M.R., Bernstein, D.I., and Stanberry, L.R. (1996a). DNA immunization confers protective immunity on mice challenged intravaginally with herpes simplex virus type 2. Vaccine *14*, 1230–1234.

Bourne, N., Pyles, R.B., Bernstein, D.I., and Stanberry, L.R. (2002). Modification of primary and recurrent genital herpes in guinea pigs by passive immunization. J. Gen. Virol. *83*, 2797–2801.

Bourne, N., Stanberry, L.R., Bernstein, D.I., and Lew, D. (1996b). DNA immunization against experimental genital herpes simplex virus infection. J. Infect. Dis. *173*, 800–807.

Bourne, N., Stanberry, L.R., Connelly, B.L., Kurawadwala, J., Straus, S.E., and Krause, P.R. (1994). Quantity of latency-associated transcript produced by herpes simplex virus is not predictive of the frequency of experimental recurrent genital herpes. J. Infect. Dis. *169*, 1084–1087.

Braun, R.P., Dong, L., Jerome, S., Herber, R., Roberts, L.K., and Payne, L.G. (2008). Multi-antigenic DNA immunization using herpes simplex virus type 2 genomic fragments. *Hum Vaccin 4*, 36–43.

Brittle, E.E., Wang, F., Lubinski, J.M., Bunte, R.M., and Friedman, H.M. (2008). A replication-competent, neuronal spread-defective, live attenuated herpes simplex virus type 1 vaccine. J. Virol. *82*, 8431–8441.

Casanova, G., Cancela, R., Alonzo, L., Benuto, R., Magana Mdel, C., Hurley, D.R., Fishbein, E., Lara, C., Gonzalez, T., Ponce, R., Burnett, J.W., and Calton, G.J. (2002). A double-blind study of the efficacy and safety of the ICP10deltaPK vaccine against recurrent genital HSV-2 infections. Cutis *70*, 235–239.

Cattamanchi, A., Posavad, C.M., Wald, A., Baine, Y., Moses, J., Higgins, T.J., Ginsberg, R., Ciccarelli, R., Corey, L., and Koelle, D.M. (2008). Phase I study of a herpes simplex virus type 2 (HSV-2) DNA vaccine administered to healthy, HSV-2-seronegative adults by a needle-free injection system. Clin. Vaccine Immunol. 15, 1638–1643.

Celum, C., Wald, A., Hughes, J., Sanchez, J., Reid, S., Delany-Moretlwe, S., Cowan, F., Casapia, M., Ortiz, A., Fuchs, J., Buchbinder, S., Koblin, B., Zwerski, S., Rose, S., Wang, J., and Corey, L. (2008). Effect of aciclovir on HIV-1 acquisition in herpes simplex virus 2 seropositive women and men who have sex with men: a randomised, double-blind, placebo-controlled trial. Lancet *371*, 2109–2119.

Chentoufi, A.A., Binder, N.R., Berka, N., Durand, G., Nguyen, A., Bettahi, I., Maillere, B., and Benmohamed, L. (2008). Asymptomatic human $CD4^+$ cytotoxic T-cell epitopes identified from herpes simplex virus glycoprotein B. J. Virol. *8*, 11792–11802.

Cliffe, A.R., Garber, D.A., and Knipe, D.M. (2009). Transcription of the herpes simplex virus latency-associated transcript promotes the formation of facultative heterochromatin on lytic promoters. J. Virol. *83*, 8182–8190.

Cliffe, A.R., and Knipe, D.M. (2008). Herpes simplex virus ICP0 promotes both histone removal and acetylation on viral DNA during lytic infection J. Virol. *82*, 12030–12038.

Corey, L., Langenberg, A.G., Ashley, R., Sekulovich, R.E., Izu, A.E., Douglas, J.M., Jr., Handsfield, H.H., Warren, T., Marr, L., Tyring, S., DiCarlo, R., Adimora, A.A.,

Leone, P., Dekker, C.L., Burke, R.L., Leong, W.P., and Straus, S.E. (1999). Recombinant glycoprotein vaccine for the prevention of genital HSV-2 infection: two randomized controlled trials. Chiron HSV vaccine Study Group. JAMA *28*, 331–340.

Da Costa, X.J., Bourne, N., Stanberry, L.R., and Knipe, D.M. (1997). Construction and characterization of a replication-defective herpes simplex virus 2 ICP8 mutant strain and its use in immunization studies in a guinea pig model of genital herpes. Virology *232*, 1–12.

Da Costa, X.J., Jones, C.A., and Knipe, D.M. (1999). Immunization against genital herpes with a vaccine virus that has defects in productive and latent infection. Proc. Natl. Acad. Sci. U.S.A. *96*, 6994–6998.

Da Costa, X.J., Kramer, M.F., Zhu, J., Brockman, M.A., and Knipe, D.M. (2000). Construction, phenotypic analysis, and immunogenicity of a UL5/UL29 double deletion mutant of herpes simplex virus 2. J. Virol. *74*, 7963–7971.

Da Costa, X.J., Morrison, L.A., and Knipe, D.M. (2001). Comparison of different forms of herpes simplex replication-defective mutant viruses as vaccines in a mouse model of HSV-2 genital infection. Virology *288*, 256–263.

de Bruyn, G., Vargas-Cortez, M., Warren, T., Tyring, S.K., Fife, K.H., Lalezari, J., Brady, R.C., Shahmanesh, M., Kinghorn, G., Beutner, K.R., Patel, R., Drehobl, M.A., Horner, P., Kurtz, T.O., McDermott, S., Wald, A., and Corey, L. (2006). A randomized controlled trial of a replication defective (gH deletion) herpes simplex virus vaccine for the treatment of recurrent genital herpes among immunocompetent subjects. Vaccine *24*, 914–920.

DeLuca, N.A., McCarthy, A.M., and Schaffer, P.A. (1985). Isolation and characterization of deletion mutants of herpes simplex virus type 1 in the gene encoding immediate-early regulatory protein ICP4. J. Virol. *56*, 558–570.

Dingwell, K.S., Brunetti, C.R., Hendricks, R.L., Tang, Q., Tang, M., Rainbow, A.J., and Johnson, D.C. (1994). Herpes simplex virus glycoproteins E and I facilitate cell-to-cell spread *in vivo* and across junctions of cultured cells. J. Virol. *68*, 834–845.

Dudek, T., and Knipe, D.M. (2006). Replication-defective viruses as vaccines and vaccine vectors. Virology *344*, 230–239.

Dudek, T., Mathews, L.C., and Knipe, D.M. (2008). Disruption of the U(L)41 gene in the herpes simplex virus 2 dl5–29 mutant increases its immunogenicity and protective capacity in a murine model of genital herpes. Virology *372*, 165–175.

Dykxhoorn, D.M., Palliser, D., and Lieberman, J. (2006). The silent treatment: siRNAs as small molecule drugs. Gene Ther. *13*, 541–552.

Eis-Hubinger, A.M., Schmidt, D.S., and Schneweis, K.E. (1993). Anti-glycoprotein B monoclonal antibody protects T cell-depleted mice against herpes simplex virus infection by inhibition of virus replication at the inoculated mucous membranes. J. Gen. Virol. *74*, 379–385.

Farrell, H.E., McLean, C.S., Efstathlou, S., Inglls, S., and Minson, A.C. (1994). Vaccine potential of a herpes simplex virus type 1 mutant with an essential glycoprotein deleted. J. Virol. *68*, 927–932.

Fedor, M.J., and Williamson, J.R. (2005). The catalytic diversity of RNAs. Nat. Rev. Microbiol. 6, 399–412.

Forrester, A., Farrell, H., Wilkinson, G., Kaye, J., Davis-Poynter, N., and Minson, T. (1992). Construction and properties of a mutant of herpes simplex virus type 1 with glycoprotein H coding sequences deleted. J. Virol. *66*, 341–348.

Freeman, E.E., Weiss, H.A., Glynn, J.R., Cross, P.L., Whitworth, J.A., and Hayes, R.J. (2006). Herpes simplex virus 2 infection increases HIV acquisition in men and women: systematic review and meta-analysis of longitudinal studies. AIDS *20*, 73–83.

Geiss, B.J., Smith, T.J., Leib, D.A., and Morrison, L.A. (2000). Disruption of virion host shutoff activity improves the immunogenicity and protective capacity of a replication-incompetent herpes simplex virus type 1 vaccine strain. J. Virol. *74*, 11137–11144.

Graham, B.S., and Crowe, J.E., Jr. (2007). Immunization against viral diseases. 5th ed. *In* 'Fields Virology' (Knipe, D. M., and Howley, P. M., Eds.), Vol. 1, pp. 487–538. 2 vols. Lippincott, Williams and Wilkins, Philadelphia.

Gu, H., and Roizman, B. (2007). Herpes simplex virus-infected cell protein 0 blocks the silencing of viral DNA by dissociating histone deacetylases from the CoREST–REST complex. Proc. Natl. Acad. Sci. U.S.A. *104*, 17134–17139.

Gyotoku, T., Ono, F., and Aurelian, L. (2002). Development of HSV-specific $CD4^+$ Th1 responses and $CD8^+$ cytotoxic T lymphocytes with antiviral activity by vaccination with the HSV-2 mutant ICP10DeltaPK. Vaccine *20*, 2796–2807.

Halford, W.P., Weisend, C., Grace, J., Soboleski, M., Carr, D.J., Balliet, J.W., Imai, Y., Margolis, T.P., and Gebhardt, B.M. (2006). ICP0 antagonizes Stat 1-dependent repression of herpes simplex virus: implications for the regulation of viral latency. J. Virol. *3*, 44.

Herrera, F.J., and Triezenberg, S.J. (2004). VP16-dependent association of chromatin-modifying coactivators and underrepresentation of histones at immediate-early gene promoters during herpes simplex virus infection. J. Virol. *78*, 9689–9696.

Hoshino, Y., Dalai, S.K., Wang, K., Pesnicak, L., Lau, T.Y., Knipe, D.M., Cohen, J.I., and Straus, S.E. (2005). Comparative efficacy and immunogenicity of replication-defective, recombinant glycoprotein, and DNA vaccines for herpes simplex virus 2 infections in mice and guinea pigs. J. Virol. *79*, 410–418.

Hoshino, Y., Pesnicak, L., Cohen, J.I., and Straus, S.E. (2007). Rates of reactivation of latent herpes simplex virus from mouse trigeminal ganglia *ex vivo* correlate directly with viral load and inversely with number of infiltrating $CD8^+$ T cells. J. Virol. *81*, 8157–8164.

Hoshino, Y., Pesnicak, L., Dowdell, K.C., Burbelo, P.D., Knipe, D.M., Straus, S.E., and Cohen, J.I. (2009). Protection from herpes simplex virus (HSV)–2 infection with replication-defective HSV-2 or glycoprotein D2 vaccines in HSV-1-seropositive and HSV-1-seronegative guinea pigs. J. Infect. Dis. *200*, 1088–1095.

Hoshino, Y., Qin, J., Follmann, D., Cohen, J.I., and Straus, S.E. (2008). The number of herpes simplex virus-infected neurons and the number of viral genome copies per neuron correlate with the latent viral load in ganglia. Virology *372*, 56–63.

Hosken, N., McGowan, P., Meier, A., Koelle, D.M., Sleath, P., Wagener, F., Elliott, M., Grabstein, K., Posavad, C., and Corey, L. (2006). Diversity of the $CD8^+$ T-cell response to herpes simplex virus type 2 proteins among persons with genital herpes. J. Virol. *80*, 5509–5515.

Jones, C.A., Taylor, T.J., and Knipe, D.M. (2000). Biological properties of herpes simplex virus 2 replication-defective mutant strains in a murine nasal infection model. Virology *278*, 137–150.

Knipe, D.M., and Cliffe, A. (2008). Chromatin control of herpes simplex virus lytic and latent infection. Nat. Rev. Microbiol. *6*, 211–221.

Koelle, D.M., Posavad, C.M., Barnum, G.R., Johnson, M.L., Frank, J.M., and Corey, L. (1998). Clearance of HSV-2 from recurrent genital lesions correlates with infiltration of HSV-specific cytotoxic T lymphocytes. J. Clin. Invest. *101*, 1500–1508.

Krause, P.R., Stanberry, L.R., Bourne, N., Connelly, B., Kurawadwala, J.F., Patel, A., and Straus, S.E. (1995). Expression of the herpes simplex virus type 2 latency-associated transcript enhances spontaneous reactivation of genital herpes in latently infected guinea pigs. J. Exp. Med. 181, 297–306.

Kwiatkowski, D.L., Thompson, H.W., and Bloom, D.C. (2009). The polycomb group protein Bmi1 binds to the herpes simplex virus 1 latent genome and maintains repressive histone marks during latency. J. Virol. *83*, 8173–8181.

Laycock, K.A., Lee, S.F., Brady, R.H., and Pepose, J.S. (1991). Characterization of a murine model of recurrent herpes simplex viral keratitis induced by ultraviolet B radiation. Invest. Opthamol. Vis. Sci. *32*, 2741–2746.

Liang, Y., Vogel, J.L., Narayanan, A., Peng, H., and Kristie, T.M. (2009). Inhibition of the histone demethylase LSD1 blocks alpha-herpesvirus lytic replication and reactivation from latency. Nat. Med. *15*, 1312–1317.

Liu, J., Lewin, A.S., Tuli, S.S., Ghivizzani, S.C., Schultz, G.S., and Bloom, D.C. (2008). Reduction in severity of a herpes simplex virus type 1 murine infection by treatment with a ribozyme targeting the UL20 gene RNA. J. Virol. *82*, 7467–7474.

Liu, M.A. (2010). Gene-based vaccines: Recent developments. Curr. Opin. Mol. Ther. *12*, 86–93.

Long, D.M., and Uhlenbeck, O.C. (1993). Self-cleaving catalytic RNA. FASEB J. *7*, 25–30.

Loudon, P.T., Blakeley, D.M., Boursnell, M.E., Day, D.A., Duncan, I.A., Lowden, R.C., McLean, C.S., Martin, G., Miller, J.C., and Shaw, M.L. (2001). Preclinical safety testing of DISC-hGMCSF to support phase I clinical trials in cancer patients. J. Gene Med. *3*, 458–467.

McKendall, R.R., Klassen, T., and Baringer, J.R. (1979). Host defenses in herpes simplex infections of the nervous system: effect of antibody on disease and viral spread. Infect. Immun. *23*, 305–311.

McLean, C.S., Erturk, M., Jennings, R., Challanain, D.N., Minson, A.C., Duncan, I., Boursnell, M.E., and Inglis, S.C. (1994). Protective vaccination against primary and recurrent disease caused by herpes simplex virus (HSV) type 2 using a geneitcally disabled HSV-1. J. Infect. Dis. *170*, 1100–1109.

Meignier, B., Longnecker, R., and Roizman, B. (1988). *In vivo* behavior of genetically engineered herpes simplex viruses R7017 and R7020: construction and evaluation in rodents. J. Infect. Dis. *158*, 602–614.

Meignier, B., Martin, B., Whitley, R., and Roizman, B. (1990). *In vivo* behavior of genetically engineered herpes simplex viruses R7017 and R7020. II. Studies in immunocompetent and immunosuppressed owl monkeys (Aotus trivirgatus). J. Infect. Dis. *162*, 313–321.

Mester, J.C., Twomey, T.A., Tepe, E.T., and Bernstein, D.I. (1999). Immunity induced by DNA immunization with herpes simplex virus type 2 glycoproteins B and C. Vaccine *18*, 875–883.

Morrison, L.A. (2008). Replication-defective virus vaccine-induced protection of mice from genital herpes simplex virus 2 requires. CD4 T cells. Virology *376*, 205–210.

Morrison, L.A., and Knipe, D.M. (1994). Immunization with replication-defective mutants of herpes simplex virus type 1: Sites of immune intervention in pathogenesis of challenge virus infection. J. Virol. *68*, 689–696.

Morrison, L.A., and Knipe, D.M. (1996). Mechanisms of immunization with a replication-defective mutant of herpes simplex virus 1. Virology *220*, 402–413.

Morrison, L.A., and Knipe, D.M. (1997). Contributions of antibody and T cell subsets to protection elicited by immunization with a replication-defective mutant of herpes simplex virus type 1. Virology *239*, 315–326.

Nagashunmugam, T., Lubinski, J., Wang, L., Goldstein, L.T., Weeks, B.S., Sundaresan, P., Kang, E.H., Dubin, G., and Friedman, H.M. (1998). *In vivo* immune evasion mediated by the herpes simplex virus type 1 immunoglobulin G Fc receptor. J. Virol. *72*, 5351–5359.

Nguyen, L.H., Knipe, D.M., and Finberg, R.W. (1992). Replication-defective mutants of herpes simplex virus (HSV) induce cellular immunity and protect against lethal HSV infection. J. Virol. *66*, 7067–7072.

Norberg, P., Kasubi, M.J., Haarr, L., Bergstrom, T., and Liljeqvist, J.A. (2007). Divergence and recombination of clinical herpes simplex virus type 2 isolates. J. Virol. *81*, 13158–13167.

Palliser, D., Chowdhury, D., Wang, Q.Y., Lee, S.J., Bronson, R.T., Knipe, D.M., and Lieberman, J. (2006). An siRNA-based microbicide protects mice from lethal herpes simplex virus 2 infection. Nature *439*, 89–94.

Parr, M.B., Kepple, L., McDermott, M.R., Drew, M.D., Bozzola, J.J., and Parr, E.L. (1994). A mouse model for studies of mucosal immunity to vaginal infection by herpes simplex virus type 2. Lab. Invest. *70*, 369–380.

Posavad, C.M., Koelle, D.M., and Corey, L. (1996). High frequency of $CD8^+$ cytotoxic T-lymphocyte precursors specific for herpes simplex viruses in persons with genital herpes. J. Virol. *70*, 8165–8168.

Reszka, N.J., Dudek, T., and Knipe, D.M. (2010). Construction and properties of a herpes simplex virus 2 dl5-29 vaccine candidate strain encoding an HSV-1 virion host shutoff protein. Vaccine *20*, 2754–2762.

Roizman, B., Knipe, D.M., and Whitley, R.J. (2007). Herpes Simplex Virus. 5th ed. *In* 'Fields Virology' (Knipe, D. M., and Howley, P. M., Eds.), pp. 2501–2602. Lippincott, Williams and Wilkins, Philadelphia.

Sawtell, N.M. (1997). Comprehensive quantification of herpes simplex virus latency at the single-cell level. J. Virol. *71*, 5423–5431.

Sawtell, N.M., and Thompson, R.L. (1992). Rapid *in vivo* reactivation of herpes simplex virus in latently infected murine ganglionic neurons after transient hyperthermia. J. Virol. *66*, 2150–2156.

Singh, M., Carlson, J.R., Briones, M., Ugozzoli, M., Kazzaz, J., Barackman, J., Ott, G., and O'Hagan, D. (1998). A comparison of biodegradable microparticles and MF59 as systemic adjuvants for recombinant gD from HSV-2. Vaccine *16*, 1822–1827.

Skinner, G.R., Turyk, M.E., Benson, C.A., Wilbanks, G.D., Heseltine, P., Galpin, J., Kaufman, R., Goldberg, L., Hartley, C.E., and Buchan, A. (1997). The efficacy and safety of Skinner herpes simplex vaccine towards modulation of herpes genitalis; report of a prospective double-blind placebo-controlled trial. Med. Microbiol. Immunol. *186*, 31–36.

Smith, C.C., Peng, T., Kulka, M., and Aurelian, L. (1998). The PK domain of the large subunit of herpes simplex virus type 2 ribonucleotide reductase (ICP10) is required for immediate-early gene expression and virus growth. J. Virol. 72, 9131–9141.

Smith, J.S., and Robinson, N.J. (2002). Age-specific prevalence of infection with herpes simplex virus types 2 and 1: a global review. J. Infect. Dis. 186 Suppl *1*, S3–28.

Stanberry, L.R. (2004). Clinical trials of prophylactic and therapeutic herpes simplex virus Vaccines. *Herpes* 11 Suppl 3, 161A-169A.

Stanberry, L.R., Bernstein, D.I., Burke, R.L., Pachl, C., and Myers, M.G. (1987). Vaccination with recombinant herpes simplex virus glycoproteins: protection against initial and recurrent genital herpes. J. Infect. Dis. *155*, 914–920.

Stanberry, L.R., Harrison, C.J., Bernstein, D.I., Burke, R.L., Ott, G., and Myers, M.G. (1989). Herpes simplex virus glycoprotein immunotherapy of recurrent genital herpes: Factors influencing efficacy. Antiviral Res. *11*, 203–214.

Stanberry, L.R., Spruance, S.L., Bernstein, D.I., Mindel, A., Sacks, S., Tyring, S., Aoki, F.Y., Slaoui, M., Denis, M., Vandepapeliere, P., Dubin, G., and GlaxoSmithKline Herpes Vaccine Efficacy Study Group (2002). Glycoprotein-D-adjuvant vaccine to prevent genital herpes. N. Engl. J. Med. *347*, 1652–1661.

Stevens, J.G., and Cook, M.L. (1971). Latent herpes simplex virus in spinal ganglia of mice. Science *173*, 843–845.

Strasser, J.E., Arnold, R.L., Pachuk, C., Higgins, T.J., and Bernstein, D.I. (2000). Herpes simplex virus DNA vaccine efficacy: effect of glycoprotein D plasmid constructs. J. Infect. Dis. *182*, 1304–1310.

Straus, S.E., Corey, L., Burke, R.L., Savarese, B., Barnum, G., Krause, P.R., Kost, R.G., Meier, J.L., Sekulovich, R., Adair, S.F., *et al.* (1994). Placebo-controlled trial of vaccination with recombinant glycoprotein D of herpes simplex virus type 2 for immunotherapy of genital herpes. Lancet *343*, 1460–1463.

Straus, S.E., Savarese, B., Tigges, M., Freifeld, A.G., Krause, P.R., Margolis, D.M., Meier, J.L., Paar, D.P., Adair, S.F., Dina, D., *et al.* (1993). Induction and enhancement of immune responses to herpes simplex virus type 2 in humans by use of a recombinant glycoprotein D Vaccine. J. Infect. Dis. *167*, 1045–1052.

Straus, S.E., Wald, A., Kost, R.G., McKenzie, R., Langenberg, A.G., Hohman, P., Lekstrom, J., Cox, E., Nakamura, M., Sekulovich, R., Izu, A., Dekker, C., and Corey, L. (1997). Immunotherapy of recurrent genital herpes with recombinant herpes simplex virus type 2 glycoproteins D and B: results of a placebo-controlled vaccine trial. J. Infect. Dis. *176*, 1129–1134.

Thoelen, S., De Clercq, N., and Tornieporth, N. (2001). A prophylactic hepatitis B vaccine with a novel adjuvant system. Vaccine *19*, 2400–2403.

Trang, P., Kilani, A., Kim, J., and Liu, F. (2000). A ribozyme derived from the catalytic subunit of RNase P from Escherichia coli is highly effective in inhibiting replication of herpes simplex virus 1. J. Mol. Biol. (301), 817–826.

Trang, P., Lee, J., Kilani, A.F., Kim, J., and Liu, F. (2001). Effective inhibition of herpes simplex virus 1 gene expression and growth by engineered RNase P ribozyme. Nucleic Acids Res. *29*, 5071–5078.

Valensi, J.P., Carlson, J.R., and Van Nest, G.A. (1994). Systemic cytokine profiles in BALB/c mice immunized with trivalent influenza vaccine containing MF59 oil emulsion and other. advanced adjuvants. J. Immunol. *153*, 4029–4039.

Wachsman, M., Kulka, M., Smith, C.C., and Aurelian, L. (2001). A growth and latency compromised herpes simplex virus type 2 mutant (ICP10DeltaPK) has prophylactic and therapeutic protective activity in guinea pigs. Vaccine *19*, 1879–1890.

Wald, A., and Link, K. (2002). Risk of human immunodeficiency virus infection in herpes simplex virus type 2-seropositive persons: a meta-analysis. J. Infect. Dis. *185*, 45–52.

Walker, J., and Leib, D.A. (1998). Protection from primary infection and establishment of latency by vaccination with a herpes simplex virus type 1 recombinant deficient in the virion host shutoff (vhs) function. Vaccine *16*, 1–5.

Wang, F., Tang, W., McGraw, H.M., Bennett, J., Enquist, L.W., and Friedman, H.M. (2005a). Herpes simplex virus type 1 glycoprotein E is required for axonal localization of capsid, tegument, and membrane glycoproteins. J. Virol. *79*, 13362–13372.

Wang, Q.-Y., Zhou, C., Johnson, K.E., Colgrove, R.C., Coen, D.M., and Knipe, D.M. (2005b). Herpesviral latency-associated transcript gene promotes assembly of heterochromatin on viral lytic-gene promoters in latent infection. Proc. Natl Acad. of Sci. U.S.A. *102*, 16055–16059.

Watson-Jones, D., Weiss, H.A., Rusizoka, M., Changalucha, J., Baisley, K., Mugeye, K., Tanton, C., Ross, D., Everett, D., Clayton, T., Balira, R., Knight, L., Hambleton, I., Le Goff, J., Belec, L., and Hayes, R. (2008). Effect of herpes simplex suppression on incidence of HIV among women in Tanzania. N. Engl. J. Med. *358*, 1560–1571.

Weiss, H. (2004). Epidemiology of herpes simplex virus type 2 infection in the developing world. *Herpes* 11 Suppl *1*, 24A-35A.

Whitley, R.J. (1993). Prospects for vaccination against herpes simplex virus. Pediatric Annals *22*, 726–732.

Wilson, S.S., Fakioglu, E., and Herold, B.C. (2009). Novel approaches in fighting herpes simplex virus infections. Expert Rev. Anti Infect Ther. *7*, 559–568.

Wu, Y., Navarro, F., Lal, A., Basar, E., Pandey, R.K., Manoharan, M., Feng, Y., Lee, S.J., Lieberman, J., and Palliser, D. (2009). Durable protection from Herpes Simplex Virus-2 transmission following intravaginal application of siRNAs targeting both a viral and host gene. Cell Host Microbe *5*, 84–94.

Immunity to Herpes Simplex Virus

19

Keith R. Jerome

Abstract

HSV presents unique challenges to the human immune system. Most of these result from the ability of the virus to establish latency in neurons of the dorsal root ganglia. The first line of defence against the initial establishment of latent infection is the innate immune response. The innate response relies on a variety of cell types recognizing HSV infection via pattern recognition receptors, including toll-like receptors. After exposure, the adaptive immune response is triggered. However, the adaptive response must deal with reactivation of HSV from the latently infected neuron, which in turn seeds mucosal sites with virus. T cells are especially important in this, and likely control both the extent of reactivation from latently infected neurons as well as the extent of viral replication at mucosal sites. Not surprising, HSV has evolved a wide variety of immune evasion mechanisms to tip this balance in its favour and facilitate transmission to new hosts. The study of HSV and its interaction with the host immune system has provided insights into the function of both, and may ultimately facilitate the development of an effective HSV vaccine.

Clinical HSV disease

To understand the immunological control of HSV infections, it is helpful first to review the typical and atypical clinical courses of these infections.

Primary infection

Most HSV-1 infections are acquired early in childhood as subclinical or unrecognized infections (Spruance, 1992). In young children classic primary HSV-1 infection is associated with gingivostomatitis, fever, and marked submandibular lymphadenopathy. Oral lesions progress to ulceration, and heal without scarring over two to three weeks. Adolescents may present with pharyngitis and mononucleosis.

Primary HSV-2 infection classically presents as *herpes genitalis*, with extensive, bilateral vesicles, fever, inguinal lymphadenopathy, and dysuria (Corey *et al.*, 1983). Secondary lesions may develop in the second to third week. The lesions ulcerate and heal without scarring within three weeks. As with HSV-1, subclinical or unrecognized primary infection with HSV-2 is common.

The classic anatomical distributions of HSV-1 and HSV-2 are changing. The proportion of primary genital infections due to HSV-1 is increasing (Brugha *et al.*, 1997), from about 10% in 1983 to 32% in 1995 (Corey *et al.*, 1983; Wald *et al.*, 1995). HSV-1 may now represent the cause of the majority of first-episode of anogenital infections in certain populations (Roberts, 2005; Ryder *et al.*, 2009). This trend is thought to result from changing sexual practices, including increased oral-genital exposure (Roberts, 2005). The clinical presentation of primary genital HSV-1 cannot be distinguished from that of HSV-2. However, recurrent disease (see below) is less common with genital HSV-1 than HSV-2 (Corey *et al.*, 1983; Engelberg *et al.*, 2003).

Establishment of latency and recurrent disease

After primary infection with HSV-1 or HSV-2, the virus travels retrograde along nerve axons, and establishes latency in the dorsal root ganglia. Periodically the virus reactivates and travels anterograde via the nerve axon to oral or genital sites, resulting in release of infectious virus and potentially lesion formation. Recurrent disease has milder symptoms and a shorter time to lesion healing than primary episodes (Corey *et al.*, 1983; Spruance, 1992).

Asymptomatic or subclinical infection

Both HSV-1 and HSV-2 frequently infect individuals without causing recognized symptoms. Approximately 70–90% of individuals with HSV-2 antibodies have not been diagnosed with genital herpes (Fleming *et al.*, 1997; Xu *et al.*, 2006), although over half such persons will recognize symptoms after education regarding manifestations of HSV disease (Frenkel *et al.*, 1993; Langenberg *et al.*, 1989). Most patients with genital herpes have episodes of subclinical virus excretion from anogenital sites; these constitute an important source of transmission. In general, oral shedding of HSV-1 occurs somewhat less frequently than genital shedding of HSV-2 (Leone *et al.*, 2007). Oral shedding of HSV-2 or genital shedding of HSV-1 are comparatively rare, particularly outside the setting of newly acquired disease (Lafferty *et al.*, 1987; Leone *et al.*, 2007; Wald *et al.*, 2004).

Spectrum of disease severity

The frequency of HSV-2 genital recurrences varies widely among individuals, ranging from none to 12 or more per year (Benedetti *et al.*, 1994) with a mean rate of 0.33 recurrences/month. Orolabial HSV-1 recurrence rates are lower with a mean of 0.12 episodes/month while genital HSV-1 infection occurs even less frequently (mean 0.02 episodes/month) (Lafferty *et al.*, 1987). Orolabial HSV-2 recurrences are extremely infrequent (Lafferty *et al.*, 1987).

Herpes in the immunocompromised host

Immunosuppressed individuals with defective cell-mediated immunity frequently develop symptomatic HSV disease. HSV infections in such individuals can be severe, with extensive mucocutaneous necrosis and involvement of contiguous tissues leading to oesophagitis or proctitis. Disseminated HSV, which can lead to meningoencephalitis, pneumonitis, hepatitis, and coagulopathy, may be more common in immunocompromised and hospitalized patients than is generally recognized (Berrington *et al.*, 2009). Defects in NK function are associated with severe HSV disease (Biron *et al.*, 1989). Interestingly, individuals with defective humoral immunity generally control HSV infection quite effectively (Morison, 1975). These observations may provide some insight into the importance of these immune mechanisms.

Innate immunity and HSV

Upon exposure to HSV, whether from an external source or reactivation of latent disease, the earliest events involve the triggering of the innate immune response. A successful innate immune response may abort the initial infection and prevent the subsequent establishment of latency. During reactivation of latent HSV, a successful innate response may prevent the development of lesions and clinically relevant disease. Even *in situ*ations in which prevention of infection or symptoms is not possible, the innate response is essential for the subsequent establishment of an effective adaptive immune response. Thus, early recognition by the innate system of the presence of virus is central to the clinical control of HSV disease.

Detection of HSV by the innate immune system

Lacking the antigen-specific receptors of the adaptive immune system, the innate immune system relies upon the recognition of pathogen-associated molecular patterns (PAMPs) to identify the presence of viral infection. A number of pattern recognition receptors (PRRs) exist (reviewed in Baccala *et al.*, 2009), including the toll-like receptors (TLRs), RNA helicases such as RIG-I (retinoic acid-inducible gene 1 protein) and MDA5, NOD (nucleotide oligomerization domain) proteins, and certain double stranded RNA binding domain-containing proteins such as PKR (protein kinase R) and OAS (2′–5′-oligoadenylate synthetase. The TLRs appear to play an especially important role in the innate detection of HSV infections (reviewed in Gill *et al.*, 2008). Ten TLRs have been identified in humans, and exist as homo- or hetero-dimeric type I transmembrane proteins, allowing them to specifically recognize a variety of PAMPs. The TLRs can be expressed at the cell surface (TLR1, 2, 4–6, and 10) or on the membranes of endosomes (TLR3 and 7–9). While six of the TLRs have been shown to facilitate recognition of various viruses, only three, TLRs 2, 3, and 9, have been shown to be involved in recognition of HSV. TLR2 recognizes multiple molecular patterns found in various pathogens, including HSV (Kurt-Jones *et al.*, 2004, 2005). TLR3 recognizes double-stranded RNA, a characteristic signature of many viral infections (Botos *et al.*, 2009). TLR9 recognizes CpG motifs present in HSV DNA (Krug *et al.*, 2004; Lund *et al.*, 2003). TLRs 2 and 9 are present on epithelial cells at oral, ocular, and genital sites (Andersen *et al.*, 2006; Fazeli *et al.*, 2005), and mRNA for TLRs 1–6 have been detected in various sites throughout the female genital tract (Fazeli *et al.*, 2005; Herbst-Kralovetz *et al.*, 2008). Stimulation of TLRs on these cells leads to cytokine production (Herbst-Kralovetz *et al.*, 2008). Studies in mice suggest that TLRs 2 and 9 may act synergistically to control HSV infection (Sorensen *et al.*, 2008). TLR3 is present on monocytes and may be activated on such cells during HSV infection (Peri *et al.*, 2008). In addition, TLRs are present on human neuronal cells. Neuronal TLRs have been reported to be induced during HSV infection, resulting in production of IFN-alpha. Activation of neuronal TLRs prior to HSV exposure reduces the susceptibility of these cells to HSV infection (Zhou *et al.*, 2009).

In addition to the TLRs, cells possess molecules that act as both dsRNA sensors and antiviral effectors, some of which are triggered during HSV infection. Protein kinase R (PKR) is thought to become activated after dimerization due to RNA binding by its dsRNA binding domains (Heinicke *et al.*, 2009; Lemaire *et al.*, 2008). PKR can also be activated in an RNA-independent manner by the cellular protein PACT (Peters *et al.*, 2002). Activated PKR functions to limit protein synthesis within the cell by phosphorylating eukaryotic initiation factor 2 alpha (eIF-2α). Similarly, 2′–5′ oligoadenylate synthetase (OAS) recognizes dsRNA, via a positively charged groove in the molecule (Hartmann *et al.*, 2003). OAS functions by activating RNase L, which enhances the decay of mRNA (including viral mRNA) within infected cells (Floyd-Smith *et al.*, 1981; Wreschner *et al.*, 1981). As discussed later in this chapter, the cellular effects of PKR and OAS in HSV infection are antagonized by highly effective countermeasures encoded by the virus.

Response to innate detection One of the most important effects of PRR stimulation is the synthesis of cytokines and chemokines. The molecules synthesized depend on the cell type, the precise TLRs or PRRs activated, and potentially the sequence in which TLRs are stimulated (Sato *et al.*, 2006). Stimulation of TLR3 on uterine epithelial cells and epidermal keratinocytes using poly(I:C) has been shown to induce a broad spectrum of cytokines and chemokines, including IFN-alpha, IFN, beta, TNF-alpha, IL-6, GM-CSF, G-CSF, as well as chemokines including CXCL8/IL-8, CCL2/MCP-1, and CCL4/MIP-1beta (Schaefer *et al.*, 2005; Tohyama *et al.*, 2005). TLRs are also present on various immune cells, including dendritic cells, where they have been reported to cooperate in an orchestrated sequence to facilitate DC activation and cytokine secretion (Sato *et al.*, 2006). As discussed in detail below, the secretion of cytokines and chemokines is critical for early control of HSV infection, and for providing the proper setting for the subsequent generation of the adaptive immune response.

TLR agonists as therapeutics

The importance of TLRs in the initial detection of viral infection and the subsequent response has led to interest in TLR agonists as potential prophylactic or therapeutic agents. Poly I:C is a synthetic double-stranded RNA analogue that has been shown to activate TLR3 (Alexopoulou *et al.*, 2001), and to induce chemokine expression from uterine epithelial cells and epidermal keratinocytes (Schaefer *et al.*, 2005; Tohyama *et al.*, 2005). Poly I:C has been evaluated in various murine models of HSV infection, and has been shown to enhance resistance to HSV-2 challenge (Ashkar *et al.*, 2004; Herbst-Kralovetz and Pyles, 2006). Similarly, vaginal application of the bacterial-derived TLR2/6 agonist, FSL-1, increases resistance to HSV-2 infection by 25 fold (Rose *et al.*, 2009). The TLR agonists imiquimod and resiquimod have received much attention as potential anti-HSV agents (Miller *et al.*, 2002). These compounds are members of the imidazoquinoline family, and are recognized by TLRs 7 and 8. Imiquimod is approved by the US Food and Drug Administration (FDA) for the treatment of superficial basal cell carcinoma, and is also used in the topical treatment of anogenital warts and molluscum contagiosum (Gupta *et al.*, 2004). In preclinical studies, resiquimod suppressed HSV recurrences in the guinea pig model of HSV latency (Bernstein *et al.*, 2001). An initial small clinical trial showed similar efficacy in human beings, with resiquimod-treated individuals having an increased interval between clinical recurrences (Spruance *et al.*, 2001). However, follow-up studies have had mixed results (Fife *et al.*, 2008; Mark *et al.*, 2007), and thus the clinical utility, if any, of resiquimod remains to be established.

The TLR9 agonist CpG has also attracted much interest as a potential anti-herpes agent. CpG motifs have been shown to have direct immunostimulatory effects, and intravaginal administration provides significant protection from subsequent HSV challenge in mice (McCluskie *et al.*, 2006; Pyles *et al.*, 2002). The protective effect extends from 1 day prior to challenge to 4 days post challenge, suggesting that this compound might be useful as a preventative microbicide (Herbst-Kralovetz and Pyles, 2006; Pyles *et al.*, 2002). CpG motifs have also been evaluated as adjuvants in recent HSV-2 vaccine trials in mice (McCluskie *et al.*, 2006). These motifs activate dendritic cells (Lund *et al.*, 2003), thus facilitating the activation of HSV-specific T cells. Addition of CpG appears to favour the generation of protective Th1-type cellular immune responses (Tengvall *et al.*, 2006), and has been shown to significantly improve the efficacy of immunization with gB for protection of mice from vaginal infection (Gallichan *et al.*, 2001; Kwant and Rosenthal, 2004). While the optimal combination of TLR agonists for clinical use has not been established, recent work suggests that TLR 3 and TLR9 may induce similar responses, in that both provide protection against HSV infection that correlates best with induction of IFN-beta rather than other cytokines (Gill *et al.*, 2006).

Taken together, the preclinical experience suggests that TLR agonists may prove to be useful additions to the anti-HSV armamentarium, with the usual caveat that the murine experience is not always directly transferable to human beings. It should be noted that CpG has been suggested to promote the generation of autoreactive T cells (Azulay-Debby *et al.*, 2007). Some authors have therefore suggested that the comparable efficacy in preclinical models favours the use of TLR3 agonists such as poly I:C (Gill *et al.*, 2008), which in theory might be safer. Clinical trials of these compounds in humans will need to be performed carefully, paying careful attention to potential adverse effects.

Cells of the innate immune response

The initial recognition of HSV occurs in infected epithelial cells. However, a number of immune cells also play a role in the innate immune response to HSV. Perhaps the most important of these are NK cells (Nandakumar *et al.*, 2008). This critical role is highlighted by the severe HSV infections observed in persons with genetic defects in the NK cell response (Biron *et al.*, 1989). NK cells express a large number of activating and inhibitory receptors on their surface (Biassoni, 2009), and the balance of activating and inhibitory signals received by a given cell determines its activation status. In the case of HSV infection, the downregulation of MHC class I by ICP47 (Fruh *et al.*, 1995; Hill *et al.*, 1995; Jennings *et al.*, 1985) presumably contributes to the activation of NK cells by tipping the balance towards activating ligands. In addition, NK cells express TLRs, including TLR2 and TLR9 (Sivori *et al.*, 2004), and these are likely to contribute to NK cell activation during HSV infection. The protective effect of NK cells in HSV disease appears to depend upon their ability to localize to sites of infection. CCR5-deficient mice have impaired mobilization of NK cells and correspondingly increased susceptibility to HSV (Thapa *et al.*, 2007). Finally, NKT cells, which share lineage markers with NK cells but also express a limited repertoire of T cell receptors, also appear to play a role in resistance to HSV (Balato *et al.*, 2009; Grubor-Bauk *et al.*, 2008).

Other cells of the innate immune system play important roles in the innate response to HSV. Neutrophils express chemokine, cytokine, and pattern recognition receptors, and depletion of neutrophils leads to markedly enhanced susceptibility to vaginal HSV infection (Milligan, 1999; Milligan *et al.*, 2001), and to increased spread of virus after anterior chamber inoculation (Zheng *et al.*, 2008). Macrophages also express pattern recognition receptors, which presumably contribute to their detection of HSV. Finally, as outlined in detail later in this section, dendritic cells play a critical role in the innate response to HSV, by functionally bridging the innate and adaptive immune responses.

Complement and other molecules

Complement likely plays an underappreciated role in the initial response to HSV infection. The importance of complement in anti-HSV immunity is emphasized by the fact that HSV has evolved an antagonist of complement function. HSV glycoprotein C (gC) binds C3b, a critical component of the complement cascade, thereby preventing downstream events such as C5 activation and neutralization of virus (Hook *et al.*, 2006; Lubinski *et al.*, 2002). In C3-depleted animals, but not their normal counterparts, gC-minus virus is pathogenic (Lubinski *et al.*, 2002), supporting the conclusion that this represents a important means of immune evasion. In further support of this interpretation, mice unable to make C3 show increased susceptibility to HSV infection, which is corrected by engraftment of C3-producing bone marrow (Verschoor *et al.*, 2001). Finally, complement and complement receptor likely play an important role in the coordination of innate and adaptive immunity, since mice lacking complement receptors are unable to establish immunologic memory or to mount recall responses to replication-defective HSV (Brockman *et al.*, 2006). The relationship of complement to the establishment of an effective humoral immune response is the subject of a recent comprehensive review (Brockman and Knipe, 2008).

Besides complement, certain other humoral factors have been shown to contribute to anti-HSV immunity. Natural antibody can antagonize viral infection, possibly via direct neutralization of virus, activation of complement, or enhancement of subsequent adaptive responses (Ochsenbein and Zinkernagel, 2000). Certain peptides with biologic activity, known as defensins, also possess anti-HSV activity (Yasin *et al.*, 2000). The effects of defensins appear to be exerted at multiple pre- and post-entry steps (Hazrati *et al.*, 2006). In addition, both lactoferrin and human apolipoprotein E (apoE) have been reported to inhibit early steps of HSV infection (Dobson *et al.*, 2006; Valimaa *et al.*, 2009).

Other intracellular responses to HSV infection

Intracellular processes play an important role in the innate response to viral infection. For example, HSV infection of cells can lead to the induction of apoptosis. At its simplest, such triggering of apoptosis presumably renders the cell unable to efficiently support viral replication. In addition, in multicellular organisms apoptosis is intimately linked to the function and regulation of the immune system. Although cellular components from apoptotic cells can be presented by classical and cross-presentation pathways, in general apoptosis is thought to be tolerogenic (Peng *et al.*, 2007). Many viruses, including HSV, have evolved the ability to modulate apoptosis. This is likely a double-edged sword; inhibition of apoptosis may allow short-term viral replication, but the subsequent cell necrosis favours a vigorous immune response (Aubert and Jerome, 2003). For HSV, infection of epithelial cells is a pro-apoptotic stimulus, although cell death is effectively prevented by a variety of anti-apoptosis proteins (Nguyen and Blaho, 2007). Interestingly, though, in immune cells the balance favours the pro-apoptotic effects, and these cells typically undergo apoptosis upon HSV infection. Infection of dendritic cells induces apoptosis (Bosnjak *et al.*, 2005), and HSV induces apoptosis in a subset of infected T cells (Han *et al.*, 2007). Several factors appear to account for the different outcomes in immune vs. non-immune cells. Apoptosis of infected dendritic cells is due at least in part to HSV-induced degradation of the cellular anti-apoptotic protein c-FLIP (Kather *et al.*). In addition, specific HSV proteins can have different effects in different cell types. For example, ICP10 is anti-apoptotic in epithelial cells, but is markedly pro-apoptotic in Jurkat T cells (Han *et al.*, 2009).

Viral infection of cells can also lead to the induction of autophagy (Orvedahl and Levine, 2009; Schmid and Munz, 2007), a regulated catabolic process involving the degradation of a cell's own components through the lysosomal machinery. For at least two viruses, tobacco mosaic virus and sindbis virus, autophagy can successfully limit viral replication (Liang *et al.*, 1998; Liu *et al.*, 2005). Autophagy also facilitates the delivery of viral components to the endosome, where they are accessible to TLR3 and 7–9 (Lee *et al.*, 2007). It appears that HSV infection triggers autophagy pathways, but these pathways are antagonized by HSV-encoded proteins. The HSV early protein ICP34.5 inhibits autophagy via an interaction with beclin-1 (Orvedahl *et al.*, 2007), a protein that is required for the formation of autophagosomes and initiation of autophagy (Kihara *et al.*, 2001). However, the predominant role of ICP34.5 in cultured cells is likely the ability to prevent translational arrest via the PKR pathway as discussed earlier, since elimination of the autophagic pathway does not significantly alter replication of wild-type or ICP34.5-mutant virus (Alexander *et al.*, 2007). On the other hand, autophagy may play a much more important role *in vivo*. Mutant HSV with an engineered form of ICP34.5 that cannot bind beclin-1 is impaired in its ability to cause encephalitis in mice (Orvedahl *et al.*, 2007). Interestingly, neurovirulence with this mutant is restored in pkr ($^{-/-}$) mice, implying substantial interdependence of intracellular innate mechanisms.

Autophagy also likely plays a role in priming the adaptive immune response, by delivering viral antigens to late endosomal compartments, where they are degraded and ultimately loaded onto MHC class II (Schmid and Munz, 2007; Zwart *et al.*, 2005). It appears that HSV antagonizes this function, since a virus lacking the beclin 1-binding domain of ICP34.5 induces significantly stronger CD4 T-cell production of gamma interferon and interleukin-2 compared to wild type virus (Leib *et al.*, 2009). Autophagy may also play a role in the presentation of peptides via MHC class I. HSV-infected macrophages have been reported to present viral antigens on MHC class I molecules via a non-classical, vacuolar pathway of MHC class I loading that appears to be dependent upon a unique form of autophagy (English *et al.*, 2009). Further studies should reveal whether HSV has evolved countermeasures against this aspect of autophagy as well.

Linkage of innate and adaptive immunity

As noted previously, the innate immune response plays a critical role in initiating and orchestrating the adaptive immune response. The activation of innate immunity in a variety of cell types via pattern recognition receptors and the resulting cytokine and chemokine production provides the proper setting for the subsequent adaptive response. The most critical cell type for bridging the innate and adaptive immune responses is the dendritic cell. Dendritic cells recognize HSV infection via TLRs (Lund *et al.*, 2003; Sato *et al.*, 2006), and respond by activation, increased uptake of viral antigen, and ultimately migration to sites of interaction with T cells and presentation of antigen. The importance of the innate/adaptive linkage *in vivo* is highlighted by a recent study demonstrating that mice lacking TLR3 have a greatly reduced CD8 T cell response after flank infection by HSV (Davey *et al.*, 2010).

Adaptive immunity and HSV

While the initial response after exposure to HSV is mediated by the innate immune response, the ultimate clearance of primary or recurrent infection is dependent upon adaptive immunity. Unlike innate immunity, adaptive immunity is characterized by antigen-specific receptors, and by the concept of immune memory. Thus, most attempts at prevention or therapy of HSV disease via the immune system have focused on the adaptive response.

Humoral immunity

While the role of humoral immunity in HSV disease has been relatively understudied compared to that of the cellular immune response, it is clear that antibodies are intricately involved in the HSV disease process. Infection with HSV induces a broad antibody response, primarily directed towards envelope glycoproteins and the proteins of the viral tegument and capsid (Ashley *et al.*, 1985, 1994). The majority of antibody at genital mucosal sites is IgG, although secretory IgA is present as well (Russell and Mestecky, 2002). The detection of HSV antibodies in serum remains the mainstay of clinical diagnosis of HSV disease. In particular, newer approaches allowing the differentiation of antibodies to HSV-1 versus HSV-2, based on the type-specificity of the response to gG, facilitates proper counselling and management of patients (Ashley, 2001; Wald and Ashley-Morrow, 2002).

It is less clear, however, the degree to which the antibody response to HSV can prevent or ameliorate clinical disease. The ability of certain antibodies to neutralize HSV has been appreciated for many years (Vestergaard, 1980). Antibodies can also work with a number of cell types, including neutrophils and monocytes, to facilitate antibody-dependent cell-mediated cytotoxicity (ADCC) (Kohl, 1991).

Several lines of evidence suggest that antibodies can play at least a limited role in prevention of HSV disease. The most compelling evidence comes from the study of transmission of HSV-2 to neonates during the birth process. In general, the per-delivery risk of transmission of HSV-2 to neonates is quite low in the setting of established maternal infection (Corey and Wald, 2009). On the other hand, in the setting of primary maternal infection, the risk of transmission is 25- to 50-fold higher (Corey and Wald, 2009). The difference is attributed to the presence of maternal IgG antibodies, which are able to cross the placenta, providing protection to the infant (Ashley *et al.*, 1992; Brown *et al.*, 2005). Interestingly, however, maternal HSV-1 antibody appears to provide little if any protection against transmission of HSV-2 to the neonate (Brown *et al.*, 2007).

The evidence for a protective role for antibody in adults is less compelling. Although a common belief is that patients infected with a given strain of HSV generally do not become infected with a second strain, this may not be the case. One longitudinal study of patients with genital HSV-1 showed evidence of reinfection in 2 of 13 patients (Roest *et al.*, 2004). Another study of genital HSV-2 suggested that reinfection was less common, being detected in 2 of 65 patients (Roest *et al.*, 2004; Sakaoka *et al.*, 1995). On the other hand, prior infection with HSV-2 has been reported to provide protection against HSV-1 infection (Langenberg *et al.*, 1999), although the ability of prior HSV-1 infection to protect against HSV-2 infection remains controversial (Koelle and Corey, 2003). These observations have been attributed to antibody-mediated mechanisms, based largely on mouse models demonstrating that protection of mucosal surfaces from reinfection is mediated predominantly by antibody (Kuklin *et al.*, 1998; Parr and Parr, 2003). The protective effect of antibody in mice is mediated by both Fcγ receptor-dependent and -independent mechanisms (Chu *et al.*, 2008). The recent demonstration that humanized mice can mount a protective antibody response to immunization or infection (Kwant-Mitchell *et al.*, 2009) should allow greater insights into the applicability of these findings to human disease. In humans, one prospective case-control study found that low levels of total IgG1 and IgG3 antibodies correlated with increased frequency of recurrences, although levels of anti-HSV antibodies did not (Seppanen *et al.*, 2006).

On the other hand, several lines of evidence suggest that the protective role of antibody against HSV infection is limited. Isolated antibody deficiency generally does not lead to severe primary or recurrent HSV disease (Koelle and Corey, 2008). Persons with low levels of antibodies to HSV are more susceptible to experimental reactivation of virus by UV irradiation (Spruance *et al.*, 1995). However, low levels of HSV antibodies are not associated with frequent natural recurrences of HSV; in fact, symptomatic individuals appear to have higher levels of antibody than seropositive but asymptomatic persons, a finding which has been attributed to antibody levels being driven by antigen load (Spruance *et al.*, 1995). Finally, certain vaccine formulations have been demonstrated to induce neutralizing antibodies at levels similar to that of natural infection, yet have failed to prevent infection in large clinical trials (Corey *et al.*, 1999).

The biological basis for the failure of antibodies to provide clear-cut protection against

HSV remains unclear. In recurrent disease, the virus originates from latently infected neurons of the dorsal root ganglia, and the vast majority of its travel occurs intracellularly within axons (Cunningham *et al.*, 2006). Since the axons of the innervating neurons are intimately apposed to the basal cell layer of the mucosal or epithelial sites of reactivation (Zhu *et al.*, 2007, 2009), it is likely that the duration of virus exposure to antibody would be limited. The problem is compounded by the ability of HSV-2 to spread directly from cell to cell, both between epithelial cells (Dingwell *et al.*, 1994) and from infected cells to infiltrating immune cells (Aubert *et al.*, 2009). On the other hand, in genital lesions and secretions, HSV is thought to exist predominantly as free virus (Koelle and Corey, 2008), and in this setting neutralizing antibodies may play a role in resistance to infection. The most recent attempts to develop an anti-HSV vaccine have been designed to elicit both humoral and cellular immunity.

Importance of cellular immunity

It is generally assumed that T cells play a critical role in the control of HSV infections in humans, and this assumption seems well supported by the literature. Patients with severe immunosuppression are susceptible to severe HSV disease (Levin *et al.*, 2004; Tan and Goh, 2006), and given the lack of evidence of a critical role for B cells as outlined above, T cells would seem to be the most likely candidates. Additional evidence comes from cross-sectional studies of HIV/HSV co-infected men, which demonstrated that low levels of HSV-2-specific CD8 T cells (but not CD4 T cells) correlated with more severe HSV disease (Posavad *et al.*, 1997). Upon administration of highly active antiretroviral therapy (HAART), with improvement of T cell function, the severity of lesions decreased (Posavad *et al.*, 2004). However, HAART had no effect on the frequency of HSV-2 shedding, suggesting that the immune mechanisms controlling viral reactivation and shedding may be distinct from those limiting lesion formation.

The T cell response during human HSV infection

T cells are thought to play a critical role in the containment of both primary and recurrent HSV infection. The extended duration of primary compared to secondary lesions presumably provides sufficient time for the establishment of a primary T cell response. However, for practical reasons, the majority of studies regarding T-cell control of lesions in humans have been performed in the setting of recurrent disease. Both migratory and lymph node-resident dendritic cells play a role in the initial priming of the T cell response (Donaghy *et al.*, 2009; Lee *et al.*, 2009; Yoneyama *et al.*, 2005). This appears to occur via a mechanism that involves transfer of viral antigen from migratory dendritic cells to lymph-node resident dendritic cells (Allan *et al.*, 2006). Biopsy studies have demonstrated that the first T cells to infiltrate HSV lesions (12–24 hours after the onset of symptoms) consist primarily of CD4 positive lymphocytes (Cunningham *et al.*, 1985). Shortly thereafter, these early infiltrators are followed by infiltration of CD8 positive T cells, the appearance of which correlates with the elimination of infectious virus from the lesion (Koelle *et al.*, 1998). The generation of fully functional CD8 T cells appears to be dependent on CD4 T cell help (Rajasagi *et al.*, 2009). Mouse studies suggest that perforin- and Fas- mediated cytolytic mechanisms are involved in the clearance of HSV from genital epithelium (Dobbs *et al.*, 2005). While the infiltrating T cells are likely comprised of a variety of specificities, it is clear that a proportion are HSV specific. T cells from patients have been shown to recognize a variety of viral antigens (Koelle and Corey, 2003), although T cells with markers of skin homing show marked evidence of immunodominance and a focused response (Koelle *et al.*, 2003). While the number of infiltrating T cells decreases rapidly upon lesion healing, the density of the T cells remains significantly elevated for at least several months (Zhu *et al.*, 2007). A large proportion of these persistent normal T cells are HSV-specific, and as detailed below may play an important role in the control of subsequent episodes of viral shedding from the innervating axons.

Central versus peripheral T-cell control of HSV

There is an ongoing controversy regarding whether the clinically relevant control of HSV by T cells occurs at the dorsal root ganglion or instead at the peripheral mucosal sites of viral replication. There is clear evidence that T cells are present at both sites. HSV-infected ganglia show an increased infiltration of T cells, and these T cells appear to be activated via antigen presentation by parenchymal cells surrounding the infected neurons (van Lint *et al.*, 2005). In mice, the number of infiltrating CD8 T cells in the ganglia inversely correlates with the rate of HSV reactivation, implying a functional role for these cells (Hoshino *et al.*, 2007). A significant proportion of T cells within latently infected ganglia have been shown to be specific for HSV antigens (Khanna *et al.*, 2003; Sheridan *et al.*, 2009; Verjans *et al.*, 2007). CD8 T cells within the ganglia have been suggested to facilitate the maintenance of HSV latency, and their depletion leads to viral reactivation (Liu *et al.*, 2000; Sheridan *et al.*, 2009). Several mechanisms have been proposed for the ability of ganglion-infiltrating T cells to control HSV. There is no evidence of neuronal loss in HSV-infected ganglia, and it appears that cytotoxic mechanisms do not play an important role (Johnson *et al.*, 2008). Instead, ganglion-infiltrating T cells appear to rely upon non-cytotoxic mechanisms to control HSV reactivation. Several effector molecules have been implicated in this process, including granzyme A (Pereira *et al.*, 2000), gamma interferon (Decman *et al.*, 2005), and a non-lytic function mediated by granzyme B (Knickelbein *et al.*, 2008). Taken together, the invariant presence of T cells in infected ganglia, along with the depletion studies in mice, suggest that T cells exert a suppressive effect on HSV reactivation from neurons.

On the other hand, a large body of evidence also points to T cell activity at peripheral mucosal sites as being critical for T cell control of HSV. As noted above, the infiltration of T cells into HSV lesions is rapid and dramatic, and the infiltration of CD8 positive T cells correlates strongly with clearance of virus from lesions (Koelle *et al.*, 1998). Anecdotal experience suggests that HSV lesions recur at almost, but not exactly the same anatomic site, an observation that may be explained by recent findings demonstrating long-term persistence of T cells at the site of healed lesions, but not in the directly adjacent mucosa (Zhu *et al.*, 2009; Zhu *et al.*, 2007). Perhaps more impressively, these persisting T cells are localized at the junction between the innervating axons and the basal layer of the mucosa. At this site they appear optimally positioned to control bursts of viral release from the neuron, possibly before replication in epithelial cells can be completed.

How do we reconcile these two views? While experimental systems allowing a direct comparison of the importance of central versus peripheral control are not available, some new insights have become available from mathematical modellers. Using a stochastic model, Schiffer and colleagues have suggested that T cell control of HSV replication in the initially infected basal cells after release of virus from the neuron is a critical determinant of the ultimate outcome of HSV shedding (Schiffer *et al.*, 2009). Thus, if resident T cell density at the axon/basal cell junction is sufficient, viral replication is rapidly interrupted, and the ultimate outcome is a subclinical reactivation. On the other hand, if this control is inadequate, virus is rapidly and dramatically amplified, resulting in successive rounds of infection of surrounding mucosal cells with the development of clinically evident lesions. The model suggests that even a modest delay in T cell recognition of infected epithelial cells, or decrease in their efficacy, can tip the balance towards clinical disease (Schiffer *et al.*, 2009). Interestingly, the model predicts a low but relatively constant production of virus from the neuron, and the total burden of viral release from the infected neuron is predicted to influence the frequency of detectable episodes of shedding. Thus, these models may point towards a unified theory in which both central and peripheral control are important. This interpretation is supported by recent work in mice describing a functional role for long-lived HSV-specific memory T cells in both the skin and ganglia (Gebhardt *et al.*, 2009).

Immunopathology

Many of the immune mechanisms discussed in this chapter result in the death of virus-infected cells. In most circumstances and anatomical sites, the intensity and duration of the immune response is tightly regulated, such that there is not excessive collateral damage to surrounding cells. Even for extensive lesions at mucosal sites, which can result from a combination of virus and immune-mediated cell death, the inherent regenerative capacity of the epithelial cells limits long-term sequelae.

However, in certain circumstances HSV-induced immune pathology can be of major clinical importance. The most dramatic example of HSV-driven immunopathology is HSV stromal keratitis, one of the leading infectious causes of corneal blindness (Liesegang, 2001). Ocular HSV disease can manifest in a variety of clinical syndromes, of which HSV stromal keratitis is the most severe. HSV stromal keratitis is rare as an initial ocular presentation, but represents a significant proportion of recurrent disease (Knickelbein *et al.*, 2009). HSV stromal keratitis presents in both necrotizing and non-necrotizing forms. The necrotizing form, which can quickly lead to corneal perforation, is associated with active viral replication (Holbach *et al.*, 1991). On the other hand, the non-necrotizing form appears to be less dependent upon viral replication (Knickelbein *et al.*, 2009), and instead seems to be driven predominantly by immune mechanisms. Non-necrotizing HSV stromal keratitis is associated with severe stromal inflammation, with a predominance of T cells. HSV infection of corneal cells induces the expression of a number of pro-inflammatory cytokines (Li *et al.*, 2006), apparently via TLR signalling (Sarangi *et al.*, 2007), which likely contributes to the initiation of disease. Based largely on studies in mouse models, it appears that CD4 T cells play a critical role in the development of HSV stromal keratitis (Streilein *et al.*, 1997). CD4 T cells specific for HSV can be recovered from corneal tissue long after primary infection, implicating HSV in the disease process (Verjans *et al.*, 1998). Since the inflammation and tissue damage continue long after immune control of virus has been achieved, it has been suggested that autoinflammatory processes may contribute to the maintenance of the inflammatory state (Deshpande *et al.*, 2004). Balancing their role in the initiation of HSV stromal keratitis, however, CD4 T cells can also ameliorate the severity of the inflammation, via regulatory T cell activity and the production of IL-10 (Sarangi *et al.*, 2008; Sehrawat *et al.*, 2008).

Immune evasion by HSV

As detailed in the preceding sections, a broad variety of immune mechanisms play role in the control of HSV infections. Conversely, HSV itself deploys a broad spectrum of immune evasion mechanisms in its bid to co-exist with the infected host. In fact, viral evasion strategies have been described for nearly all of the immune mechanisms that threaten viral replication. This is perhaps not surprising for a large DNA virus, and likely reflects the long history of co-evolution between HSV and the human host.

As noted above, inmate immune mechanisms play a critical role in the initial response to HSV infection, and several HSV proteins antagonize innate immunity. Perhaps the earliest acting are those viral proteins that antagonize the interferon response, which have been recently comprehensively reviewed (Paladino and Mossman, 2009). Among these are ICP34.5, which inhibits the phosphorylation of EIF2α (He *et al.*, 1997); US 11, which antagonizes 2′–5′-OAS and PKR (Cassady *et al.*, 1998; Peters *et al.*, 2002; Sanchez and Mohr, 2007); and ICP0, ICP 27, US3, and vhs, which use a variety of mechanisms to antagonize the interferon response (Paladino and Mossman, 2009).

Complement is another aspect of the innate immune response that is targeted for disruption by HSV. The HSV glycoprotein C (gC) binds to C3b (Friedman *et al.*, 1984), inhibiting complement-mediated neutralization and cell lysis (Harris *et al.*, 1990; McNearney *et al.*, 1987). The ability to antagonize complement clearly plays a role in HSV pathogenesis, as viruses expressing a form of gC unable to bind C3b show reduced pathogenicity in a guinea pig model of infection (Lubinski *et al.*, 1998). Immunization with gC has been shown to block inhibition of complement, and thus may enhance vaccine efficacy (Awasthi *et al.*, 2009).

As noted earlier, intracellular processes, including apoptosis and autophagy, play an important role in the host response to HSV infection. An extensive body of literature exists demonstrating that HSV-2 is capable of inhibiting apoptosis (reviewed in Nguyen and Blaho, 2007), and the modulation of apoptosis is likely to have broad effects on the ability of the immune system to control the infection (Aubert and Jerome, 2003). Similarly, autophagy appears to play an important role in the response to infection, and again, viral antagonists exist, including ICP34.5 (Orvedahl *et al.*, 2007).

Dendritic cells play a critical role at the innate/adaptive immune interface, and therefore it is not surprising that HSV targets these cells for modulation. Dendritic cells can become infected by HSV (Mikloska *et al.*, 2001), and their infection has a number of physiological effects. First, after HSV infection a significant proportion of infected dendritic cells rapidly die via apoptosis (Jones *et al.*, 2003; Mikloska *et al.*, 2001). However, the induction of dendritic cell apoptosis may also have disadvantages for the virus, since apoptotic HSV-infected dendritic cells can be phagocytosed by uninfected bystander dendritic cells, which can then present HSV antigens to CD8 T cells (Bosnjak *et al.*, 2005). Even if HSV infection of dendritic cells does not result in their death, infection interferes with the function of these cells, by inhibiting their maturation and preventing efficient antigen presentation (Jones *et al.*, 2003). Furthermore, HSV infection of dendritic cells can lower expression of costimulatory molecules such as CD80 and CD86, again hindering their ability to activate the T cell response (Mikloska *et al.*, 2001). Similarly, HSV suppresses CD1d recycling, impairing the activation of natural killer T cells (Yuan *et al.*, 2006).

Finally, HSV employs a number of mechanisms to blunt adaptive immunity. The humoral response is antagonized by the gE/gI heterodimer, which forms a high affinity Fc receptor (Frank and Friedman, 1989). gE/gI binding to HSV-specific antibody blocks complement activation and antibody-dependent cellular cytotoxicity (ADCC) (Frank and Friedman, 1989), and mouse studies suggest that this effect is important *in vivo* (Lubinski *et al.*, 2002).

The broad variety of immune evasion mechanisms directed at T cells is evidence for the importance of these cells in the control of HSV infection. One of the major anti-T cell mechanisms is indirect, and occurs via the inhibition of peptide loading into MHC class I by ICP47 (Fruh *et al.*, 1995; Hill *et al.*, 1995). A similar effect against MHC class II is mediated via HSV gB, which diverts MHC class II molecules to the exosome pathway (Temme *et al.*, 2010). By inhibiting antigen presentation, HSV can render the infected cells invisible to T cells, allowing these cells to avoid recognition and destruction, or preventing the generation of efficient T cell help. However, the effects of ICP47 are quickly overcome by gamma interferon in lesions (Tigges *et al.*, 1996), and thus the virus has evolved several other anti-T cell mechanisms. As noted earlier, HSV has a number of anti-apoptosis proteins, and the expression of these proteins in infected cells renders those cells resistant to the induction of apoptosis by T cells (Jerome *et al.*, 1998). In addition to these indirect effects, HSV also has direct effects on T cells themselves. Although they are not fully permissive for viral replication, T cells can by infected by HSV, most efficiently via cell-to-cell spread during T cell contact with infected cells (Aubert *et al.*, 2009). As noted previously for dendritic cells, HSV infection of T cells is a pro-apoptotic stimulus, and a proportion of infected T cells undergo apoptosis (Han *et al.*, 2007). For those T cells escaping the induction of apoptosis, HSV has direct modulatory effects on T cell receptor signalling pathways, inhibiting most anti-viral effector mechanisms (Sloan *et al.*, 2006) and skewing T cells towards production of the immunosuppressive cytokine IL-10 (Sloan and Jerome, 2007).

Genetic factors predisposing to HSV disease

The marked variability in clinical symptoms between individuals infected with HSV has led to intense interest in host factors that might influence disease severity, and many of the prime candidates are components of the immune system. One of the most dramatic, but rare, manifestations of HSV infection is HSV encephalitis, which has been the subject of many such studies. Initial work in mice suggested that components of the interferon response pathway are major determinants of disease in mouse models of HSV encephalitis (reviewed in Sancho-Shimizu *et al.*, 2007).

Subsequent work in humans has confirmed the importance of this pathway as determinant of disease severity, with HSV encephalitis being associated with mutations or polymorphisms in STAT-1 (Dupuis *et al.*, 2003), NEMO (Niehues *et al.*, 2004), UNC-93B (Casrouge *et al.*, 2006), and TLR3 (Zhang *et al.*, 2007). Similarly, a polymorphism in lactoferrin has been reported to be associated with the development of HSV keratitis (Keijser *et al.*, 2008).

Although HSV encephalitis and keratitis are dramatic and devastating complications of HSV disease, a more broadly applicable issue is the determinants of susceptibility to HSV infection and the subsequent severity of orolabial or genital disease. An increased risk of HSV infection has been associated with a polymorphism in the 3′ untranslated region of the transcription factor T-bet (Svensson *et al.*, 2008), and also three polymorphisms in the human beta-defensin-1 molecule (Tesse *et al.*, 2009). After the establishment of latent infection, increased viral shedding and/or lesion severity have been associated with polymorphisms in TLR2 (Bochud *et al.*, 2007), mannose-binding lectin 2 (Seppanen *et al.*, 2009), and haptoglobin (Gennero *et al.*, 2010), and also with HLA-B27 and -Cw2 (Lekstrom-Himes *et al.*, 1999) or the absence of KIR2DL2 and KIR2DS2 (Estefania *et al.*, 2007). Currently it is not possible to reconcile these reports in a satisfying way, or to prioritize the importance of the various genetic risk factors. However, with the recent advances in inexpensive whole genome sequencing, the availability of large, well-defined patient cohorts may allow this issue to be evaluated comprehensively.

Vaccination against HSV

An effective vaccine against HSV would be a major advance in the control of this infection. Unfortunately, the development of such a vaccine has proven extremely difficult, and despite a large amount of effort over many years, most approaches have met with failure (Ramachandran and Kinchington, 2007). Perhaps most promising among the various vaccine candidates have been two glycoprotein subunit vaccines, both of which have completed large phase III clinical trials for efficacy. The first, comprising of the HSV glycoproteins gB and gD combined with the adjuvant MF59, induced high titres of neutralizing antibodies, but did not protect from HSV-2 acquisition or the development of genital ulcers (Corey *et al.*, 1999). The second vaccine was similar, comprised of a truncated form of gD combined with the adjuvant aluminium hydroxide linked to MPL. This vaccine showed moderate efficacy, but only among HSV-2 seronegative women (Stanberry *et al.*, 2002). Since the antigens in the two vaccines were similar, if the observed difference in efficacy has a biological basis, it may be result from the differences in adjuvants. The latter vaccine is currently being evaluated in another large phase III clinical trial, and if this study confirms efficacy, the vaccine could be licensed for clinical use in the near future. Nevertheless, given that the vaccine's efficacy is modest, and limited to only a subset of susceptible individuals, it is clear that more efficacious and broadly applicable vaccines are desperately needed. One would hope that our increasing understanding of the immune response to HSV would aid in this quest.

Acknowledgements

I thank Richard Pyles and Veijo Hukkanen for helpful insights regarding the innate response to HSV, and Martine Aubert for her careful review of the manuscript.

References

Alexander, D.E., Ward, S.L., Mizushima, N., Levine, B., and Leib, D.A. (2007). Analysis of the role of autophagy in replication of herpes simplex virus in cell culture. J. Virol. *81*, 12128–12134.

Alexopoulou, L., Holt, A.C., Medzhitov, R., and Flavell, R.A. (2001). Recognition of double-stranded RNA and activation of NF-kappaB by Toll-like receptor 3. Nature *413*, 732–738.

Allan, R.S., Waithman, J., Bedoui, S., Jones, C.M., Villadangos, J.A., Zhan, Y., Lew, A.M., Shortman, K., Heath, W.R., and Carbone, F.R. (2006). Migratory dendritic cells transfer antigen to a lymph node-resident dendritic cell population for efficient CTL priming. Immunity *25*, 153–162.

Andersen, J.M., Al-Khairy, D., and Ingalls, R.R. (2006). Innate immunity at the mucosal surface: role of toll-like receptor 3 and toll-like receptor 9 in cervical epithelial cell responses to microbial pathogens. Biol. Rep. *74*, 824–831.

Ashkar, A.A., Yao, X.D., Gill, N., Sajic, D., Patrick, A.J., and Rosenthal, K.L. (2004). Toll-like receptor (TLR)-3, but not TLR4, agonist protects against genital herpes infection in the absence of inflammation seen with CpG DNA. J. Infect. Dis. *190*, 1841–1849.

Ashley, R., Benedetti, J., and Corey, L. (1985). Humoral immune response to HSV-1 and HSV-2 viral proteins in patients with primary genital herpes. J. Med. Virol. *17*, 153–166.

Ashley, R.L. (2001). Sorting out the new HSV type specific antibody tests. Sex. Transm. Infect. *77*, 232–237.

Ashley, R.L., Corey, L., Dalessio, J., Wilson, P., Remington, M., Barnum, G., and Trethewey, P. (1994). Protein-specific cervical antibody responses to primary genital herpes simplex virus type 2 infections. J. Infect. Dis. *170*, 20–26.

Ashley, R.L., Dalessio, J., Burchett, S., Brown, Z., Berry, S., Mohan, K., and Corey, L. (1992). Herpes simplex virus-2 (HSV-2) type-specific antibody correlates of protection in infants exposed to HSV-2 at birth. J. Clin. Invest. *90*, 511–514.

Aubert, M., and Jerome, K.R. (2003). Apoptosis prevention as a mechanism of immune evasion. Int. Rev. Immunol. *22*, 361–371.

Aubert, M., Yoon, M., Sloan, D.D., Spear, P.G., and Jerome, K.R. (2009). The virological synapse facilitates herpes simplex virus entry into T cells. J. Virol. *83*, 6171–6183.

Awasthi, S., Lubinski, J.M., and Friedman, H.M. (2009). Immunization with HSV-1 glycoprotein C prevents immune evasion from complement and enhances the efficacy of an HSV-1 glycoprotein D subunit vaccine. Vaccine *27*, 6845–6853.

Azulay-Debby, H., Edry, E., and Melamed, D. (2007). CpG DNA stimulates autoreactive immature B cells in the bone marrow. Eur. J. Immunol. *37*, 1463–1475.

Baccala, R., Gonzalez-Quintial, R., Lawson, B.R., Stern, M.E., Kono, D.H., Beutler, B., and Theofilopoulos, A.N. (2009). Sensors of the innate immune system: their mode of action. Nat. Rev. Rheumatol 5, 448–456.

Balato, A., Unutmaz, D., and Gaspari, A.A. (2009). Natural killer T cells: an unconventional T-cell subset with diverse effector and regulatory functions. J. Invest Dermatol. *129*, 1628–1642.

Benedetti, J., Corey, L., and Ashley, R. (1994). Recurr. ence rates in genital herpes after symptomatic first-episode infection. Ann. Intern. Med. *121*, 847–854.

Bernstein, D.I., Harrison, C.J., Tomai, M.A., and Miller, R.L. (2001). Daily or weekly therapy with resiquimod (R-848) reduces genital recurr. ences in herpes simplex virus-infected guinea pigs during and after treatment. J. Infect. Dis. *183*, 844–849.

Berrington, W.R., Jerome, K.R., Cook, L., Wald, A., Corey, L., and Casper, C. (2009). Clinical correlates of herpes simplex virus viremia among hospitalized adults. Clin. Infect. Dis. *49*, 1295–1301.

Biassoni, R. (2009). Human natural killer receptors, co-receptors, and their ligands. In Current Protocols. J.E. Coligan, ed. (Hoboken, NJ, Wiley).

Biron, C.A., Byron, K.S., and Sullivan, J.L. (1989). Severe herpesvirus infections in an adolescent without natural killer cells. N. Engl. J. Med. 320, 1731–1735.

Bochud, P.Y., Magaret, A.S., Koelle, D.M., Aderem, A., and Wald, A. (2007). Polymorphisms in TLR2 are associated with increased viral shedding and lesional rate in patients with genital herpes simplex virus Type 2 infection. J. Infect. Dis. *196*, 505–509.

Bosnjak, L., Miranda-Saksena, M., Koelle, D.M., Boadle, R.A., Jones, C.A., and Cunningham, A.L. (2005). Herpes simplex virus infection of human dendritic cells induces apoptosis and allows cross-presentation via uninfected dendritic cells. J. Immunol. *174*, 2220–2227.

Botos, I., Liu, L., Wang, Y., Segal, D.M., and Davies, D.R. (2009). The toll-like receptor 3:dsRNA signaling complex. Biochim Biophys. Acta *1789*, 667–674.

Brockman, M.A., and Knipe, D.M. (2008). Herpes simplex virus as a tool to define the role of complement in the immune response to peripheral infection. Vaccine *26* Suppl *8*, I94–99.

Brockman, M.A., Verschoor, A., Zhu, J., Carroll, M.C., and Knipe, D.M. (2006). Optimal long-term humoral responses to replication-defective herpes simplex virus require CD21/CD35 complement receptor expression on stromal cells. J. Virol. *80*, 7111–7117.

Brown, E.L., Gardella, C., Malm, G., Prober, C.G., Forsgren, M., Krantz, E.M., Arvin, A.M., Yasukawa, L.L., Mohan, K., Brown, Z., *et al.* (2007). Effect of maternal herpes simplex virus (HSV) serostatus and HSV type on risk of neonatal herpes. Acta Obstet. Gynecol. Scand. *86*, 523–529.

Brown, Z.A., Gardella, C., Wald, A., Morrow, R.A., and Corey, L. (2005). Genital herpes complicating pregnancy. Obstet. Gynecol. *106*, 845–856.

Brugha, R., Keersmaekers, K., Renton, A., and Meheus, A. (1997). Genital herpes infection: a review. Int. J. Epidemiol. *26*, 698–709.

Casrouge, A., Zhang, S.Y., Eidenschenk, C., Jouanguy, E., Puel, A., Yang, K., Alcais, A., Picard, C., Mahfoufi, N., Nicolas, N., *et al.* (2006). Herpes simplex virus encephalitis in human UNC-93B deficiency. Science *314*, 308–312.

Cassady, K.A., Gross, M., and Roizman, B. (1998). The herpes simplex virus US11 protein effectively compensates for the gamma1(34.5) gene if present before activation of protein kinase R by precluding its phosphorylation and that of the alpha subunit of eukaryotic translation initiation factor 2. J. Virol. 72, 8620–8626.

Chu, C.F., Meador, M.G., Young, C.G., Strasser, J.E., Bourne, N., and Milligan, G.N. (2008). Antibody-mediated protection against genital herpes simplex virus type 2 disease in mice by Fc gamma receptor-dependent and -independent mechanisms. J. Reproduct. Immunol. 78, 58–67.

Corey, L., Adams, H.G., Brown, Z.A., and Holmes, K.K. (1983). Genital herpes simplex virus infections: clinical manifestations, course, and complications. Ann. Intern. Med. 98, 958–972.

Corey, L., Langenberg, A.G., Ashley, R., Sekulovich, R.E., Izu, A.E., Douglas, J.M., Jr., Handsfield, H.H., Warren, T., Marr, L., Tyring, S., *et al.* (1999). Recombinant glycoprotein vaccine for the prevention of genital HSV-2 infection: two randomized controlled trials. Chiron HSV vaccine Study Group. JAMA *282*, 331–340.

Corey, L., and Wald, A. (2009). Maternal and neonatal herpes simplex virus infections. N. Engl. J. Med. 361, 1376–1385.

Cunningham, A.L., Diefenbach, R.J., Miranda-Saksena, M., Bosnjak, L., Kim, M., Jones, C., and Douglas, M.W. (2006). The cycle of human herpes simplex virus infection: virus transport and immune control. J. Infect. Dis. *194* Suppl *1*, S11–18.

Cunningham, A.L., Turner, R.R., Miller, A.C., Para, M.F., and Merigan, T.C. (1985). Evolution of recurrent herpes simplex lesions. An immunohistologic study. J. Clin. Invest. *75*, 226–233.

Davey, G.M., Wojtasiak, M., Proietto, A.I., Carbone, F.R., Heath, W.R., and Bedoui, S. (2010). Cutting edge: priming of CD8 T cell immunity to herpes simplex virus type 1 requires. cognate TLR3 expression *in vivo*. J. Immunol. *184*, 2243–2246.

Decman, V., Kinchington, P.R., Harvey, S.A., and Hendricks, R.L. (2005). Gamma interferon can block herpes simplex virus type 1 reactivation from latency, even in the presence of late gene expression. J. Virol. *79*, 10339–10347.

Deshpande, S., Banerjee, K., Biswas, P.S., and Rouse, B.T. (2004). Herpetic eye disease: immunopathogenesis and therapeutic measures. Expert Rev. Mol. Med. *6*, 1–14.

Dingwell, K.S., Brunetti, C.R., Hendricks, R.L., Tang, Q., Tang, M., Rainbow, A.J., and Johnson, D.C. (1994). Herpes simplex virus glycoproteins E and I facilitate cell-to-cell spread *in vivo* and across junctions of cultured cells. J. Virol. *68*, 834–845.

Dobbs, M.E., Strasser, J.E., Chu, C.F., Chalk, C., and Milligan, G.N. (2005). Clearance of herpes simplex virus type 2 by $CD8^+$ T cells requires. gamma interferon and either. perforin- or Fas-mediated cytolytic mechanisms. J. Virol. *79*, 14546–14554.

Dobson, C.B., Sales, S.D., Hoggard, P., Wozniak, M.A., and Crutcher, K.A. (2006). The receptor-binding region of human apolipoprotein E has direct anti-infective activity. J. Infect. Dis. *193*, 442–450.

Donaghy, H., Bosnjak, L., Harman, A.N., Marsden, V., Tyring, S.K., Meng, T.C., and Cunningham, A.L. (2009). Role for plasmacytoid dendritic cells in the immune control of recurrent human herpes simplex virus infection. J. Virol. *83*, 1952–1961.

Dupuis, S., Jouanguy, E., Al-Hajjar, S., Fieschi, C., Al-Mohsen, I.Z., Al-Jumaah, S., Yang, K., Chapgier, A., Eidenschenk, C., Eid, P., *et al.* (2003). Impaired response to interferon-alpha/beta and lethal viral disease in human STAT1 deficiency. Nat. Genet. *33*, 388–391.

Engelberg, R., Carrell, D., Krantz, E., Corey, L., and Wald, A. (2003). Natural history of genital herpes simplex virus type 1 infection. Sex. Transm. Dis. *30*, 174–177.

English, L., Chemali, M., Duron, J., Rondeau, C., Laplante, A., Gingras, D., Alexander, D., Leib, D., Norbury, C., Lippe, R., *et al.* (2009). Autophagy enhances the presentation of endogenous viral antigens on MHC class I molecules during HSV-1 infection. Nat. Immunol. *10*, 480–487.

Estefania, E., Gomez-Lozano, N., Portero, F., de Pablo, R., Solis, R., Sepulveda, S., Vaquero, M., Gonzalez, M.A., Suarez, E., Roustan, G., *et al.* (2007). Influence of KIR gene diversity on the course of HSV-1 infection: resistance to the disease is associated with the absence of KIR2DL2 and KIR2DS2. Tissue Antigens *70*, 34–41.

Fazeli, A., Bruce, C., and Anumba, D.O. (2005). Characterization of Toll-like receptors in the female reproductive tract in humans. Hum. Reprod. *20*, 1372–1378.

Fife, K.H., Meng, T.C., Ferris, D.G., and Liu, P. (2008). Effect of resiquimod 0.01% gel on lesion healing and viral shedding when applied to genital herpes lesions. Antimicrob. Agents Chemother. *52*, 477–482.

Fleming, D.T., McQuillan, G.M., Johnson, R.E., Nahmias, A.J., Aral, S.O., Lee, F.K., and St. Louis, M.E. (1997). Herpes simplex virus type 2 in the United States, 1976 to 1994. N. Engl. J. Med. *337*, 1105–1111.

Floyd-Smith, G., Slattery, E., and Lengyel, P. (1981). Interferon action: RNA cleavage pattern of a (2′–5′) oligoadenylate–dependent endonuclease. Science *212*, 1030–1032.

Frank, I., and Friedman, H.M. (1989). A novel function of the herpes simplex virus type 1 Fc receptor: participation in bipolar bridging of antiviral immunoglobulin. J. Gen. Virol. *63*, 4479–4488.

Frenkel, L.M., Garratty, E.M., Shen, J.P., Wheeler, N., Clark, O., and Bryson, Y.J. (1993). Clinical reactivation of herpes simplex virus type 2 infection in seropositive pregnant women with no history of genital herpes. Ann. Int. Med. *118*, 414–418.

Friedman, H.M., Cohen, G.H., Eisenberg, R.J., Seidel, C.A., and Cines, D.B. (1984). Glycoprotein C of herpes simplex virus 1 acts as a receptor for the C3b complement component on infected cells. Nature *309*, 633–635.

Fruh, K., Ahn, K., Djaballah, H., Semp'e, P., van Endert, P.M., Tamp'e, R., Peterson, P.A., and Yang, Y. (1995). A viral inhibitor of peptide transporters for antigen presentation. Nature *375*, 415–418.

Gallichan, W.S., Woolstencroft, R.N., Guarasci, T., McCluskie, M.J., Davis, H.L., and Rosenthal, K.L. (2001). Intranasal immunization with CpG oligodeoxynucleotides as an adjuvant dramatically increases IgA and protection against herpes simplex virus-2 in the genital tract. J. Immunol. *166*, 3451–3457.

Gebhardt, T., Wakim, L.M., Eidsmo, L., Reading, P.C., Heath, W.R., and Carbone, F.R. (2009). Memory T cells in nonlymphoid tissue that provide enhanced local immunity during infection with herpes simplex virus. Nat. Immunol. *10*, 524–530.

Gennero, L., Roos, M.A., D'Amelio, P., Denysenko, T., Morra, E., Sperber, K., Ceroni, V., Panzone, M., Lesca, F., De Vivo, E., *et al.* (2010). Iron metabolism markers and haptoglobin phenotypes in susceptibility to HSV-1 or/and HSV-2 lesion relapses. Cell Biochem. Funct. *28*, 142–148.

Gill, N., Davies, E.J., and Ashkar, A.A. (2008). The role of toll-like receptor ligands/agonists in protection against genital HSV-2 infection. Am. J. Reprod Immunol. *59*, 35–43.

Gill, N., Deacon, P.M., Lichty, B., Mossman, K.L., and Ashkar, A.A. (2006). Induction of innate immunity against herpes simplex virus type 2 infection via local delivery of Toll-like receptor ligands correlates with beta interferon production. J. Virol. *80*, 9943–9950.

Grubor-Bauk, B., Arthur, J.L., and Mayrhofer, G. (2008). Importance of NKT cells in resistance to herpes simplex virus, fate of virus-infected neurons, and level of latency in mice. J. Virol. *82*, 11073–11083.

Gupta, A.K., Cherman, A.M., and Tyring, S.K. (2004). Viral and nonviral uses of imiquimod: a review. J. of cutaneous medicine and surg. *8*, 338–352.

Han, J.Y., Miller, S.A., Wolfe, T.M., Pourhassan, H., and Jerome, K.R. (2009). Cell type-specific induction and inhibition of apoptosis by Herpes Simplex virus type 2 ICP10. J. Virol. *83*, 2765–2769.

Han, J.Y., Sloan, D.D., Aubert, M., Miller, S.A., Dang, C.H., and Jerome, K.R. (2007). Apoptosis and antigen receptor function in T and B cells following exposure to herpes simplex virus. Virology *359*, 253–263.

Harris, S.L., Frank, I., Yee, A., Cohen, G.H., Eisenberg, R.J., and Friedman, H.M. (1990). Glycoprotein C of herpes simplex virus type 1 prevents complement-mediated cell lysis and virus neutralization. J. Infect. Dis. *162*, 331–337.

Hartmann, R., Justesen, J., Sarkar, S.N., Sen, G.C., and Yee, V.C. (2003). Crystal structure of the 2′-specific and double-stranded RNA-activated interferon-induced antiviral protein 2′–5′-oligoadenylate synthetase. Molecular cell *12*, 1173–1185.

Hazrati, E., Galen, B., Lu, W., Wang, W., Ouyang, Y., Keller, M.J., Lehrer, R.I., and Herold, B.C. (2006). Human alpha- and beta-defensins block multiple steps in herpes simplex virus infection. J. Immunol. *177*, 8658–8666.

He, B., Gross, M., and Roizman, B. (1997). The gamma (1) 34.5 protein of herpes simplex virus complexes with protein phosphatase 1 alpha to dephosphorylate the alpha subunit of the eukaryotic translation initiation factor 2 and preclude the shutoff of protein synthesis by double stranded RNA -activated protein kinase. Proc. Natl. Acad. Sci. U.S.A. *94*, 843–848.

Heinicke, L.A., Wong, C.J., Lary, J., Nallagatla, S.R., Diegelman-Parente, A., Zheng, X., Cole, J.L., and Bevilacqua, P.C. (2009). RNA dimerization promotes PKR dimerization and activation. J. Mol. Biol. *390*, 319–338.

Herbst-Kralovetz, M.M., and Pyles, R.B. (2006). Quantification of poly(I:C)-mediated protection against genital herpes simplex virus type 2 infection. J. Virol. *80*, 9988–9997.

Herbst-Kralovetz, M.M., Quayle, A.J., Ficarra, M., Greene, S., Rose, W.A., 2nd, Chesson, R., Spagnuolo, R.A., and Pyles, R.B. (2008). Quantification and comparison of toll-like receptor expression and responsiveness in primary and immortalized human female lower genital tract epithelia. Am J. Reprod Immunol. *59*, 212–224.

Hill, A., Jugovic, P., York, I., Russ, G., Bennink, J., Yewdell, J., Ploegh, H., and Johnson, D. (1995). Herpes simplex virus turns off the TAP to evade host immunity. Nature *375*, 411–415.

Holbach, L.M., Font, R.L., Baehr, W., and Pittler, S.J. (1991). HSV antigens and HSV DNA in avascular and vascularized lesions of human herpes simplex keratitis. Curr. Eye Res. *10 Suppl*, 63–68.

Hook, L.M., Lubinski, J.M., Jiang, M., Pangburn, M.K., and Friedman, H.M. (2006). Herpes simplex virus type 1 and 2 glycoprotein C prevents complement-mediated neutralization induced by natural immunoglobulin M antibody. J. Virol. *80*, 4038–4046.

Hoshino, Y., Pesnicak, L., Cohen, J.I., and Straus, S.E. (2007). Rates of reactivation of latent herpes simplex virus from mouse trigeminal ganglia *ex vivo* correlate directly with viral load and inversely with number of infiltrating $CD8^+$ T cells. J. Virol. *81*, 8157–8164.

Jennings, S.R., Rice, P.L., Kloszewski, E.D., Anderson, R.W., Thompson, D.L., and Tevethia, S.S. (1985). Effect of herpes simplex virus types 1 and 2 on surface expression of class I major histocompatibility complex antigens on infected cells. J. Virol. *56*, 757–766.

Jerome, K.R., Tait, J.F., Koelle, D.M., and Corey, L. (1998). Herpes Simplex Virus Type 1 Renders Infected Cells Resistant to Cytotoxic T-Lymphocyte-Induced Apoptosis. J. Virol. 72, 436–441.

Johnson, A.J., Chu, C.F., and Milligan, G.N. (2008). Effector $CD4^+$ T-cell involvement in clearance of infectious herpes simplex virus type 1 from sensory ganglia and spinal cords. J. Virol. *82*, 9678–9688.

Jones, C.A., Fernandez, M., Herc, K., Bosnjak, L., Miranda-Saksena, M., Boadle, R.A., and Cunningham, A. (2003). Herpes simplex virus type 2 induces rapid cell death and functional impairment of murine dendritic cells *in vitro*. J. Virol. *77*, 11139–11149.

Kather, A., Raftery, M.J., Devi-Rao, G., Lippmann, J., Giese, T., Sandri-Goldin, R.M., and Schonrich, Herpes, G. simplex virus type 1 (HSV-1)-induced apoptosis in human dendritic cells as a result of downregulation of cellular FLICE-inhibitory protein and reduced expression of HSV-1 antiapoptotic latency-associated transcript sequences. J. Virol. *84*, 1034–1046.

Keijser, S., Jager, M.J., Dogterom-Ballering, H.C., Schoonderwoerd, D.T., de Keizer, R.J., Krose, C.J., Houwing-Duistermaat, J.J., van der Plas, M.J., van Dissel, J.T., and Nibbering, P.H. (2008). Lactoferrin Glu561Asp polymorphism is associated with susceptibility to herpes simplex keratitis. Exp. Eye Res. *86*, 105–109.

Khanna, K.M., Bonneau, R.H., Kinchington, P.R., and Hendricks, R.L. (2003). Herpes simplex virus-specific memory $CD8^+$ T cells are selectively activated and retained in latently infected sensory ganglia. Immunity *18*, 593–603.

Kihara, A., Kabeya, Y., Ohsumi, Y., and Yoshimori, T. (2001). Beclin–phosphatidylinositol 3-kinase complex functions at the trans-Golgi network. EMBO reports *2*, 330–335.

Knickelbein, J.E., Hendricks, R.L., and Charukamnoetkanok, P. (2009). Management of herpes simplex virus stromal keratitis: an evidence-based review. Survey of ophthalmology *54*, 226–234.

Knickelbein, J.E., Khanna, K.M., Yee, M.B., Baty, C.J., Kinchington, P.R., and Hendricks, R.L. (2008). Noncytotoxic lytic granule-mediated CD8+ T cell inhibition of HSV-1 reactivation from neuronal latency. Science *322*, 268–271.

Koelle, D.M., and Corey, L. (2003). Recent progress in herpes simplex virus immunobiology and vaccine research. Clinical microbiology reviews *16*, 96–113.

Koelle, D.M., and Corey, L. (2008). Herpes simplex: insights on pathogenesis and possible vaccines. Annual review of medicine *59*, 381–395.

Koelle, D.M., Liu, Z., McClurkan, C.L., Cevallos, R.C., Vieira, J., Hosken, N.A., Meseda, C.A., Snow, D.C., Wald, A., and Corey, L. (2003). Immunodominance among herpes simplex virus-specific CD8 T cells expressing a tissue-specific homing receptor. Proc. Natl. Acad. Sci. U.S.A. *100*, 12899–12904.

Koelle, D.M., Posavad, C.M., Barnum, G.R., Johnson, M.L., Frank, J.M., and Corey, L. (1998). Clearance of HSV-2 from recurrent genital lesions correlates with infiltration of HSV-specific cytotoxic T lymphocytes. J. Clin. Invest. *101*, 1500–1508.

Kohl, S. (1991). Role of antibody-dependent cellular cytotoxicity in neonatal infection with herpes simplex virus. Rev. Infect. Dis. *13 Suppl 11*, S950–952.

Krug, A., Luker, G.D., Barchet, W., Leib, D.A., Akira, S., and Colonna, M. (2004). Herpes simplex virus type 1 activates murine natural interferon-producing cells through toll-like receptor 9. Blood *103*, 1433–1437.

Kuklin, N.A., Daheshia, M., Chun, S., and Rouse, B.T. (1998). Role of mucosal immunity in herpes simplex virus infection. J. Immunol. *160*, 5998–6003.

Kurt-Jones, E.A., Belko, J., Yu, C., Newburger, P.E., Wang, J., Chan, M., Knipe, D.M., and Finberg, R.W. (2005). The role of toll-like receptors in herpes simplex infection in neonates. J. Infect. Dis. *191*, 746–748.

Kurt-Jones, E.A., Chan, M., Zhou, S., Wang, J., Reed, G., Bronson, R., Arnold, M.M., Knipe, D.M., and Finberg, R.W. (2004). Herpes simplex virus 1 interaction with Toll-like receptor 2 contributes to lethal encephalitis. Proc. Natl. Acad. Sci. U.S.A. *101*, 1315–1320.

Kwant, A., and Rosenthal, K.L. (2004). Intravaginal immunization with viral subunit protein plus CpG oligodeoxynucleotides induces protective immunity against HSV-2. Vaccine *22*, 3098–3104.

Kwant-Mitchell, A., Ashkar, A.A., and Rosenthal, K.L. (2009). Mucosal innate and adaptive immune responses against herpes simplex virus type 2 in a humanized mouse model. J. Virol. *83*, 10664–10676.

Lafferty, W.E., Coombs, R.W., Benedetti, J., Critchlow, C., and Corey, L. (1987). Recurr. ences after oral and genital herpes simplex virus infection: influence of anatomic site and viral type. New Engl J. Med. *316*, 1444–1449.

Langenberg, A., Benedetti, J., Jenkins, J., Ashley, R., Winter, C., and Corey, L. (1989). Development of clinically recognizable genital lesions among women previously identified as having 'asymptomatic' herpes simplex virus type 2 infection. Ann. Int. Med. *110*, 882–887.

Langenberg, A.G., Corey, L., Ashley, R.L., Leong, W.P., and Straus, S.E. (1999). A prospective study of new infections with herpes simplex virus type 1 and type 2. Chiron HSV Vaccine Study Group. N. Engl. J. Med. 341, 1432–1438.

Lee, H.K., Lund, J.M., Ramanathan, B., Mizushima, N., and Iwasaki, A. (2007). Autophagy-dependent viral recognition by plasmacytoid dendritic cells. Science *315*, 1398–1401.

Lee, H.K., Zamora, M., Linehan, M.M., Iijima, N., Gonzalez, D., Haberman, A., and Iwasaki, A. (2009). Differential roles of migratory and resident DCs in T cell priming after mucosal or skin HSV-1 infection. The J. Exp. Med. 206, 359–370.

Leib, D.A., Alexander, D.E., Cox, D., Yin, J., and Ferguson, T.A. (2009). Interaction of ICP34.5 with Beclin. 1 modulates herpes simplex virus type 1 pathogenesis through control of CD4+ T-cell responses. J. Virol. *83*, 12164–12171.

Lekstrom-Himes, J.A., Hohman, P., Warren, T., Wald, A., Nam, J.M., Simonis, T., Corey, L., and Straus, S.E. (1999). Association of major histocompatibility complex determinants with the development of symptomatic and asymptomatic genital herpes simplex virus type 2 infections. J. Infect. Dis. *179*, 1077–1085.

Lemaire, P.A., Anderson, E., Lary, J., and Cole, J.L. (2008). Mechanism of PKR Activation by dsRNA. J. Mol. Biol. *381*, 351–360.

Leone, P., Warren, T., Hamed, K., Fife, K., and Wald, A. (2007). Famciclovir reduces viral mucosal shedding in HSV-seropositive persons. Sex. Transm. Dis. *34*, 900–907.

Levin, M.J., Bacon, T.H., and Leary, J.J. (2004). Resistance of herpes simplex virus infections to nucleoside analogues in HIV-infected patients. Clin. Infect. Dis. *39 Suppl 5*, S248–257.

Li, H., Zhang, J., Kumar, A., Zheng, M., Atherton, S.S., and Yu, F.S. (2006). Herpes simplex virus 1 infection induces the expression of proinflammatory cytokines, interferons and TLR7 in human corneal epithelial cells. Immunology *117*, 167–176.

Liang, X.H., Kleeman, L.K., Jiang, H.H., Gordon, G., Goldman, J.E., Berry, G., Herman, B., and Levine, B. (1998). Protection against fatal Sindbis virus encephalitis by beclin, a novel Bcl-2-interacting protein. J. Virol. *72*, 8586–8596.

Liesegang, T.J. (2001). Herpes simplex virus epidemiology and ocular importance. Cornea *20*, 1–13.

Liu, T., Khanna, K.M., Chen, X., Fink, D.J., and Hendricks, R.L. (2000). CD8+ T cells can block herpes simplex virus type 1 (HSV-1) reactivation from latency in sensory neurons. J. Exp. Med. *191*, 1459–1466.

Liu, Y., Schiff, M., Czymmek, K., Talloczy, Z., Levine, B., and Dinesh-Kumar, S.P. (2005). Autophagy regulates programmed. cell death during the plant innate immune response. Cell *121*, 567–577.

Lubinski, J.M., Jiang, M., Hook, L., Chang, Y., Sarver, C., Mastellos, D., Lambris, J.D., Cohen, G.H., Eisenberg, R.J., and Friedman, H.M. (2002). Herpes simplex virus type 1 evades the effects of antibody and complement *in vivo*. J. Virol. *76*, 9232–9241.

Lubinski, J.M., Wang, L., Soulika, A.M., Burger, R., Wetsel, R.A., Colten, H., Cohen, G.H., Eisenberg, R.J.,

Lambris, J.D., and Friedman, H.M. (1998). Herpes simplex virus type 1 glycoprotein gC mediates immune evasion *in vivo*. J. Virol. 72, 8257–8263.

Lund, J., Sato, A., Akira, S., Medzhitov, R., and Iwasaki, A. (2003). Toll-like receptor 9-mediated recognition of Herpes simplex virus-2 by plasmacytoid dendritic cells. The J. Exp. Med. 198, 513–520.

Mark, K.E., Corey, L., Meng, T.C., Magaret, A.S., Huang, M.L., Selke, S., Slade, H.B., Tyring, S.K., Warren, T., Sacks, S.L., *et al.* (2007). Topical resiquimod 0.01% gel decreases herpes simplex virus type 2 genital shedding: a randomized, controlled trial. J. Infect. Dis. *195*, 1324–1331.

McCluskie, M.J., Cartier, J.L., Patrick, A.J., Sajic, D., Weeratna, R.D., Rosenthal, K.L., and Davis, H.L. (2006). Treatment of intravaginal HSV-2 infection in mice: a comparison of CpG oligodeoxynucleotides and resiquimod (R-848). Antiviral research *69*, 77–85.

McNearney, T.A., Odell, C., Holers, V.M., Spear, P.G., and Atkinson, J.P. (1987). Herpes simplex virus glycoproteins gC-1 and gC-2 bind to the third component of complement and provide protection against complement-mediated neutralization of viral infectivity. The J. Exp. Med. 166, 1525–1535.

Mikloska, Z., Bosnjak, L., and Cunningham, A.L. (2001). Immature monocyte-derived dendritic cells are productively infected with herpes simplex virus type 1. J. Virol. *75*, 5958–5964.

Miller, R.L., Tomai, M.A., Harrison, C.J., and Bernstein, D.I. (2002). Immunomodulation as a treatment strategy for genital herpes: review of the evidence. Int. Immunopharmacol *2*, 443–451.

Milligan, G.N. (1999). Neutrophils aid in protection of the vaginal mucosae of immune mice against challenge with herpes simplex virus type 2. J. Virol. *73*, 6380–6386.

Milligan, G.N., Bourne, N., and Dudley, K.L. (2001). Role of polymorphonuclear leukocytes in resolution of HSV-2 infection of the mouse vagina. Journal of reproductive immunology *49*, 49–65.

Morison, W.L. (1975). Viral warts, herpes simplex and herpes zoster in patients with secondary immune deficiencies and neoplasms. Br J. Dermatol *92*, 625–630.

Nandakumar, S., Woolard, S.N., Yuan, D., Rouse, B.T., and Kumaraguru, U. (2008). Natural killer cells as novel helpers in anti-herpes simplex virus immune response. J. Virol. *82*, 10820–10831.

Nguyen, M.L., and Blaho, J.A. (2007). Apoptosis during herpes simplex virus infection. Advances in Virus Res. *69*, 67–97.

Niehues, T., Reichenbach, J., Neubert, J., Gudowius, S., Puel, A., Horneff, G., Lainka, E., Dirksen, U., Schroten, H., Doffinger, R., *et al.* (2004). Nuclear factor kappaB essential modulator-deficient child with immunodeficiency yet without anhidrotic ectodermal dysplasia. J. Allergy Clin. Immunol. *114*, 1456–1462.

Ochsenbein, A.F., and Zinkernagel, R.M. (2000). Natural antibodies and complement link innate and acquired immunity. Immunology today *21*, 624–630.

Orvedahl, A., Alexander, D., Talloczy, Z., Sun, Q., Wei, Y., Zhang, W., Burns, D., Leib, D.A., and Levine, B. (2007). HSV-1 ICP34.5 confers neurovirulence by targeting the Beclin. 1 autophagy protein. Cell host & microbe *1*, 23–35.

Orvedahl, A., and Levine, B. (2009). Eating the enemy within: autophagy in infectious diseases. Cell death and differentiation *16*, 57–69.

Paladino, P., and Mossman, K.L. (2009). Mechanisms employed by herpes simplex virus 1 to inhibit the interferon response. J. Interferon Cytokine Res. *29*, 599–607.

Parr, M.B., and Parr, E.L. (2003). Vaginal immunity in the HSV-2 mouse model. Int. Rev. Immunol. *22*, 43–63.

Peng, Y., Martin, D.A., Kenkel, J., Zhang, K., Ogden, C.A., and Elkon, K.B. (2007). Innate and adaptive immune response to apoptotic cells. Journal of autoimmunity *29*, 303–309.

Pereira, R.A., Simon, M.M., and A., S. (2000). Granzyme A, a noncytolytic component of $CD8^+$ cell granules, restricts the spread of herpes simplex virus in the peripheral nervous system of experimentally infected mice. J. Virol. *74*, 1029–1032.

Peri, P., Mattila, R.K., Kantola, H., Broberg, E., Karttunen, H.S., Waris, M., Vuorinen, T., and Hukkanen, V. (2008). Herpes simplex virus type 1 Us3 gene deletion influences toll-like receptor responses in cultured monocytic cells. Virology journal *5*, 140.

Peters, G.A., Khoo, D., Mohr, I., and Sen, G.C. (2002). Inhibition of PACT-mediated activation of PKR by the herpes simplex virus type 1 Us11 protein. J. Virol. *76*, 11054–11064.

Posavad, C.M., Koelle, D.M., Shaughnessy, M.F., and Corey, L. (1997). Severe genital herpes infections in HIV-infected individuals with impaired herpes simplex virus-specific $CD8^+$ cytotoxic T lymphocyte responses. Proc. Natl. Acad. Sci. U.S.A. *94*, 10289–10294.

Posavad, C.M., Wald, A., Kuntz, S., Huang, M.L., Selke, S., Krantz, E., and Corey, L. (2004). Frequent reactivation of herpes simplex virus among HIV-1-infected patients treated with highly active antiretroviral therapy. J. Infect. Dis. *190*, 693–696.

Pyles, R.B., Higgins, D., Chalk, C., Zalar, A., Eiden, J., Brown, C., Van Nest, G., and Stanberry, L.R. (2002). Use of immunostimulatory sequence-containing oligonucleotides as topical therapy for genital herpes simplex virus type 2 infection. J. Virol. *76*, 11387–11396.

Rajasagi, N.K., Kassim, S.H., Kollias, C.M., Zhao, X., Chervenak, R., and Jennings, S.R. (2009). $CD4^+$ T cells are required for the priming of $CD8^+$ T cells following infection with herpes simplex virus type 1. J. Virol. *83*, 5256–5268.

Ramachandran, S., and Kinchington, P.R. (2007). Potential prophylactic and therapeutic vaccines for HSV infections. Curr. Pharm Des *13*, 1965–1973.

Roberts, C. (2005). Genital herpes in young adults: changing sexual behaviours, epidemiology and management. Herpes *12*, 10–14.

Roest, R.W., Carman, W.F., Maertzdorf, J., Scoular, A., Harvey, J., Kant, M., Van Der Meijden, W.I., Verjans, G.M., and Osterhaus, A.D. (2004). Genotypic analysis of sequential genital herpes simplex virus type 1

(HSV-1) isolates of patients with recurrent HSV-1 associated genital herpes. J. Med. Virol. *73*, 601–604.
Rose, W.A., 2nd, McGowin, C.L., and Pyles, R.B. (2009). FSL-1, a bacterial-derived toll-like receptor 2/6 agonist, enhances resistance to experimental HSV-2 infection. Virology journal *6*, 195.
Russell, M.W., and Mestecky, J. (2002). Humoral immune responses to microbial infections in the genital tract. Microbes and infection/Institut Pasteur *4*, 667–677.
Ryder, N., Jin, F., McNulty, A.M., Grulich, A.E., and Donovan, B. (2009). Increasing role of herpes simplex virus type 1 in first-episode anogenital herpes in heterosexual women and younger men who have sex with men, 1992–2006. Sexually transmitted infections *85*, 416–419.
Sakaoka, H., Aomori, T., Gouro, T., and Kumamoto, Y. (1995). Demonstration of either. endogenous recurr. ence or exogenous reinfection by restriction endonuclease cleavage analysis of herpes simplex virus from patients with recrudescent genital herpes. J. Med. Virol. *46*, 387–396.
Sanchez, R., and Mohr, I. (2007). Inhibition of cellular 2–5′ oligoadenylate synthetase by the herpes simplex virus type 1 Us11 protein. J. Virol. *81*, 3455–3464.
Sancho-Shimizu, V., Zhang, S.Y., Abel, L., Tardieu, M., Rozenberg, F., Jouanguy, E., and Casanova, J.L. (2007). Genetic susceptibility to herpes simplex virus 1 encephalitis in mice and humans. Curr. Opin. Allergy Clin. Immunol. 7, 495–505.
Sarangi, P.P., Kim, B., Kurt-Jones, E., and Rouse, B.T. (2007). Innate recognition network driving herpes simplex virus-induced corneal immunopathology: role of the toll pathway in early inflammatory events in stromal keratitis. J. Virol. *81*, 11128–11138.
Sarangi, P.P., Sehrawat, S., Suvas, S., and Rouse, B.T. (2008). IL-10 and natural regulatory T cells: two independent anti-inflammatory mechanisms in herpes simplex virus-induced ocular immunopathology. J. Immunol. *180*, 6297–6306.
Sato, A., Linehan, M.M., and Iwasaki, A. (2006). Dual recognition of herpes simplex viruses by TLR2 and TLR9 in dendritic cells. Proc. Natl. Acad. Sci. U.S.A. *103*, 17343–17348.
Schaefer, T.M., Fahey, J.V., Wright, J.A., and Wira, C.R. (2005). Innate immunity in the human female reproductive tract: antiviral response of uterine epithelial cells to the TLR3 agonist poly(I:C). J. Immunol. *174*, 992–1002.
Schiffer, J.T., Abu-Raddad, L., Mark, K.E., Zhu, J., Selke, S., Magaret, A., Wald, A., and Corey, L. (2009). Frequent Release of Low Amounts of Herpes Simplex Virus From Neurons: Results of a Mathematical Model. Sci. Transl Med. *1*, 7ra16.
Schmid, D., and Munz, C. (2007). Innate and adaptive immunity through autophagy. Immunity *27*, 11–21.
Sehrawat, S., Suvas, S., Sarangi, P.P., Suryawanshi, A., and Rouse, B.T. (2008). *In vitro*-generated antigen-specific $CD4^+$ $CD25^+$ $Foxp3^+$ regulatory T cells control the severity of herpes simplex virus-induced ocular immunoinflammatory lesions. J. Virol. *82*, 6838–6851.
Seppanen, M., Lokki, M.L., Lappalainen, M., Hiltunen-Back, E., Rovio, A.T., Kares, S., Hurme, M., and Aittoniemi, J. (2009). Mannose-binding lectin 2 gene polymorphism in recurrent herpes simplex virus 2 infection. Hum Immunol. *70*, 218–221.
Seppanen, M., Meri, S., Notkola, I.L., Seppala, I.J., Hiltunen-Back, E., Sarvas, H., Lappalainen, M., Valimaa, H., Palikhe, A., Valtonen, V.V., *et al.* (2006). Subtly impaired humoral immunity predisposes to frequently recurr. ing genital herpes simplex virus type 2 infection and herpetic neuralgia. J. Infect. Dis. *194*, 571–578.
Sheridan, B.S., Cherpes, T.L., Urban, J., Kalinski, P., and Hendricks, R.L. (2009). Reevaluating the CD8 T-cell response to herpes simplex virus type 1: involvement of CD8 T cells reactive to subdominant epitopes. J. Virol. *83*, 2237–2245.
Sivori, S., Falco, M., Della Chiesa, M., Carlomagno, S., Vitale, M., Moretta, L., and Moretta, A. (2004). CpG and double-stranded RNA trigger human NK cells by Toll-like receptors: induction of cytokine release and cytotoxicity against tumors and dendritic cells. Proc. Natl. Acad. Sci. U.S.A. *101*, 10116–10121.
Sloan, D.D., Han, J.Y., Sandifer, T.K., Stewart, M., Hinz, A.J., Yoon, M., Johnson, D.C., Spear, P.G., and Jerome, K.R. (2006). Inhibition of TCR Signaling by Herpes Simplex Virus. J. Immunol. *176*, 1825–1833.
Sloan, D.D., and Jerome, K.R. (2007). Herpes Simplex Virus Remodels T-Cell Receptor Signaling, Resulting in p38-Dependent Selective Synthesis of Interleukin-10. J. Virol. *81*, 12504–12514.
Sorensen, L.N., Reinert, L.S., Malmgaard, L., Bartholdy, C., Thomsen, A.R., and Paludan, S.R. (2008). TLR2 and TLR9 synergistically control herpes simplex virus infection in the brain. J. Immunol. *181*, 8604–8612.
Spruance, S.L. (1992). The natural history of recurrent oral-facial herpes simplex virus infection. Semin Dermatol *11*, 200–206.
Spruance, S.L., Evans, T.G., McKeough, M.B., Thai, L., Araneo, B.A., Daynes, R.A., Mishkin, E.M., and Abramovitz, A.S. (1995). Th1/Th2-like immunity and resistance to herpes simplex labialis. Antiviral research *28*, 39–55.
Spruance, S.L., Tyring, S.K., Smith, M.H., and Meng, T.C. (2001). Application of a topical immune response modifier, resiquimod gel, to modify the recurr. ence rate of recurrent genital herpes: a pilot study. J. Infect. Dis. *184*, 196–200.
Stanberry, L.R., Spruance, S.L., Cunningham, A.L., Bernstein, D.I., Mindel, A., Sacks, S., Tyring, S., Aoki, F.Y., Slaoui, M., Denis, M., *et al.* (2002). Glycoprotein-D-adjuvant vaccine to prevent genital herpes. N. Engl. J. Med. 347, 1652–1661.
Streilein, J.W., Dana, M.R., and Ksander, B.R. (1997). Immunity causing blindness: five different paths to herpes stromal keratitis. Immunology today *18*, 443–449.
Svensson, A., Bergin, A.M., Lowhagen, G.B., Tunback, P., Bellner, L., Padyukov, L., and Eriksson, K. (2008). A 3′-untranslated region polymorphism in the TBX21 gene encoding T-bet is a risk factor for genital herpes

simplex virus type 2 infection in humans. The J. Gen. Virol. *89*, 2262–2268.

Tan, H.H., and Goh, C.L. (2006). Viral infections affecting the skin in organ transplant recipients: epidemiology and curr. ent management strategies. Am J. Clin. Dermatol 7, 13–29.

Temme, S., Eis-Hubinger, A.M., McLellan, A.D., and Koch, N. (2010). The Herpes Simplex Virus-1 encoded glycoprotein B diverts HLA-DR into the exosome pathway. J. Immunol. *184*, 236–243.

Tengvall, S., Lundqvist, A., Eisenberg, R.J., Cohen, G.H., and Harandi, A.M. (2006). Mucosal administration of CpG oligodeoxynucleotide elicits strong CC and CXC chemokine responses in the vagina and serves as a potent Th1-tilting adjuvant for recombinant gD2 protein vaccination against genital herpes. J. Virol. *80*, 5283–5291.

Tesse, R., Santoro, N., Giordano, P., Cardinale, F., De Mattia, D., and Armenio, L. (2009). Association between DEFB1 gene haplotype and herpes viruses seroprevalence in children with acute lymphoblastic leukemia. Pediatr Hematol Oncol. *26*, 573–582.

Thapa, M., Kuziel, W.A., and Carr, D.J. (2007). Susceptibility of CCR5-deficient mice to genital herpes simplex virus type 2 is linked to NK cell mobilization. J. Virol. *81*, 3704–3713.

Tigges, M.A., Leng, S., Johnson, D.C., and Burke, R.L. (1996). Human herpes simplex virus (HSV)-specific $CD8^+$ CTL clones recognize HSV-2-infected fibroblasts after treatment with IFN-gamma or when virion host shutoff functions are disabled. J. Immunol. *156*, 3901–3910.

Tohyama, M., Dai, X., Sayama, K., Yamasaki, K., Shirakata, Y., Hanakawa, Y., Tokumaru, S., Yahata, Y., Yang, L., Nagai, H., *et al.* (2005). dsRNA-mediated innate immunity of epidermal keratinocytes. Biochemical and biophysical research communications *335*, 505–511.

Valimaa, H., Tenovuo, J., Waris, M., and Hukkanen, V. (2009). Human lactoferrin but not lysozyme neutralizes HSV-1 and inhibits HSV-1 replication and cell-to-cell spread. Virology journal *6*, 53.

van Lint, A.L., Kleinert, L., Clarke, S.R., Stock, A., Heath, W.R., and Carbone, F.R. (2005). Latent infection with herpes simplex virus is associated with ongoing $CD8^+$ T-cell stimulation by parenchymal cells within sensory ganglia. J. Virol. *79*, 14843–14851.

Verjans, G.M., Hintzen, R.Q., van Dun, J.M., Poot, A., Milikan, J.C., Laman, J.D., Langerak, A.W., Kinchington, P.R., and Osterhaus, A.D. (2007). Selective retention of herpes simplex virus-specific T cells in latently infected human trigeminal ganglia. Proc. Natl. Acad. Sci. U.S.A. *104*, 3496–3501.

Verjans, G.M., Remeijer, L., van Binnendijk, R.S., Cornelissen, J.G., Volker-Dieben, H.J., Baarsma, S.G., and Osterhaus, A.D. (1998). Identification and characterization of herpes simplex virus-specific $CD4^+$ T cells in corneas of herpetic stromal keratitis patients. J. Infect. Dis. *177*, 484–488.

Verschoor, A., Brockman, M.A., Knipe, D.M., and Carroll, M.C. (2001). Cutting edge: myeloid complement C3 enhances the humoral response to peripheral viral infection. J. Immunol. *167*, 2446–2451.

Vestergaard, B.F. (1980). Herpes simplex virus antigens and antibodies: a survey of studies based on quantitative immunoelectrophoresis. Rev. Infect. Dis. *2*, 899–913.

Wald, A., and Ashley-Morrow, R. (2002). Serological testing for herpes simplex virus (HSV)–1 and HSV-2 infection. Clin. Infect. Dis. *35*, S173–182.

Wald, A., Ericsson, M., Krantz, E., Selke, S., and Corey, L. (2004). Oral shedding of herpes simplex virus type 2. Sex. Transm. Infect. *80*, 272–276.

Wald, A., Zeh, J., Selke, S., Ashley, R.L., and Corey, L. (1995). Virologic characteristics of subclinical and symptomatic genital herpes infections. N. Engl. J. Med. 333, 770–775.

Wreschner, D.H., McCauley, J.W., Skehel, J.J., and Kerr, I.M. (1981). Interferon action—sequence specificity of the ppp(A2′ p)nA-dependent ribonuclease. Nature *289*, 414–417.

Xu, F., Sternberg, M.R., Kottiri, B.J., McQuillan, G.M., Lee, F.K., Nahmias, A.J., Berman, S.M., and Markowitz, L.E. (2006). Trends in herpes simplex virus type 1 and type 2 seroprevalence in the United States. Jama *296*, 964–973.

Yasin, B., Pang, M., Turner, J.S., Cho, Y., Dinh, N.N., Waring, A.J., Lehrer, R.I., and Wagar, E.A. (2000). Evaluation of the inactivation of infectious Herpes simplex virus by host-defense peptides. Eur J. Clin. Microbiol. Infect. Dis. *19*, 187–194.

Yoneyama, H., Matsuno, K., Toda, E., Nishiwaki, T., Matsuo, N., Nakano, A., Narumi, S., Lu, B., Gerard, C., Ishikawa, S., *et al.* (2005). Plasmacytoid DCs help lymph node DCs to induce anti-HSV CTLs. The J. Exp. Med. 202, 425–435.

Yuan, W., Dasgupta, A., and Cresswell, P. (2006). Herpes simplex virus evades natural killer T cell recognition by suppressing CD1d recycling. Nat. Immunol. *7*, 835–842.

Zhang, S.Y., Jouanguy, E., Ugolini, S., Smahi, A., Elain, G., Romero, P., Segal, D., Sancho-Shimizu, V., Lorenzo, L., Puel, A., *et al.* (2007). TLR3 deficiency in patients with herpes simplex encephalitis. Science *317*, 1522–1527.

Zheng, M., Fields, M.A., Liu, Y., Cathcart, H., Richter, E., and Atherton, S.S. (2008). Neutrophils protect the retina of the injected eye from infection after anterior chamber inoculation of HSV-1 in BALB/c mice. Invest. Opthamol. Vis. Sci. *49*, 4018–4025.

Zhou, Y., Ye, L., Wan, Q., Zhou, L., Wang, X., Li, J., Hu, S., Zhou, D., and Ho, W. (2009). Activation of Toll-like receptors inhibits herpes simplex virus-1 infection of human neuronal cells. Journal of neuroscience research *87*, 2916–2925.

Zhu, J., Hladik, F., Woodward, A., Klock, A., Peng, T., Johnston, C., Remington, M., Magaret, A., Koelle, D.M., Wald, A., *et al.* (2009). Persistence of HIV-1 receptor-positive cells after HSV-2 reactivation is a potential mechanism for increased HIV-1 acquisition. Nat. Med. *15*, 886–892.

Zhu, J., Koelle, D.M., Cao, J., Vazquez, J., Huang, M.L., Hladik, F., Wald, A., and Corey, L. (2007). Virus-specific CD8$^+$ T cells accumulate near sensory nerve endings in genital skin during subclinical HSV-2 reactivation. The J. Exp. Med. 204, 595–603.

Zwart, W., Griekspoor, A., Kuijl, C., Marsman, M., van Rheenen, J., Janssen, H., Calafat, J., van Ham, M., Janssen, L., van Lith, M., *et al.* (2005). Spatial separation of HLA–DM/HLA-DR interactions within MIIC and phagosome-induced immune escape. Immunity 22, 221–233.

Immunity and Immune Evasion Strategies Induced by Varicella Zoster Virus

20

Paul R. Kinchington and Allison Abendroth

Abstract

Of three human alphaherpesviruses, only varicella zoster virus (VZV) induces a lifelong immunity that fully protects against clinical signs of exogenous re-infection and, for most of the population, from any sign of reactivation from the latent state. The importance of VZV specific immunity is exemplified by its absence: severity and morbidity of the primary infection (varicella) and incidence of reactivated disease (zoster) are greatly increased in those with immune compromise, particularly those impaired in the cell mediated immune responses. The protection afforded by VZV specific immunity underlies successful live attenuated vaccines that have greatly impacted the incidence of varicella, and reduce the incidence, severity and complications of zoster. Consequently, the components of VZV induced immunity and their contribution to the protective state has been well studied and is outlined in this chapter. Less is known of the strategies exploited by VZV to evade the innate and adaptive arms, but their activities are presumed to be critical to extend the life of the infected cell and to enhance viral production and dissemination. Evasion appears to include distinct strategies from those used by herpes simplex viruses and includes expression of novel immune evasion proteins. Our current understanding of these strategies and the underlying mechanisms is outlined, along with areas in which there is need for further study.

Introduction: the changing pattern of VZV pathogenesis

Both innate and adaptive components of the immune response are critical to effectively control VZV pathogenesis in humans, and compared to that against many viruses, they do it remarkably well. In an unvaccinated population, VZV primary disease is experienced only once by almost all immune competent individuals, and for some 80% of the population, no further clinical signs of virus reactivation from the latent state (i.e. zoster) ever become apparent. However, genetic defects in NK cell function (Etzioni *et al.*, 2005) or genetic, iatrogenic or disease-induced compromise in cell mediated immunity (CMI) results in increased morbidity, severity, chronicity and mortality of varicella (Buda *et al.*, 1996; Gershon *et al.*, 1997). Failure to acquire VZV antigen specific T lymphocytes correlates with risk of persistent viraemia and life-threatening virus dissemination (Arvin *et al.*, 1986; Gershon and Steinberg, 1979). Furthermore, CMI decline, whether caused by immune senescence from increasing age, underlying disease or iatrogenic cause, correlates with higher incidence of zoster (reviewed by Oxman (Oxman, 2009). Examples include markedly increased zoster after haematopoietic transplantation, in which CMI is highly depressed but antibody levels are maintained with intravenous gamma globulin; the finding that zoster is a common initial presentation in HIV infected patients with declining $CD4^+$ cell numbers: and that zoster is a major complication in patients with hematologic malignant neoplasms and in patients undergoing immunosuppressive therapy.

Immune-mediated protection underlies the basis for successful VZV vaccination. We are now in an age in which VZV remains the only human herpesvirus to which successful live attenuated vaccines are available for both primary and reactivated disease. The varicella vaccine was licensed in 1995 and is currently given in a two dose regimen prior to entering school in most US states. Widespread use has changed the epidemiology of varicella (Marin *et al.*, 2008; Seward *et al.*, 2008) and has dramatically reduced the school based epidemics seen prior to vaccination. The protection efficacy induced by the vaccine is variable (68–95%), but there is now an effective herd immunity that prevents most childhood epidemics (Civen *et al.*, 2009). Varicella now shows a more sporadic incidence, although some epidemics have been reported in highly vaccinated populations (Chaves *et al.*, 2008; Gould *et al.*, 2009). The recently licensed zoster vaccine (Oxman *et al.*, 2005) is based on the same attenuated virus as that for varicella prevention, but has 14X higher potency and is administered to VZV seropositive individuals in late adulthood. It boosts VZV specific immunity, reducing zoster incidence by greater than 50% and disease severity and the complications of zoster (measured as 'burden of illness') by two thirds. This includes reducing the problematic pain (postherpetic neuralgia: PHN) that often complicates zoster (Oxman *et al.*, 2005). The success of the varicella and zoster vaccines has cemented the importance of active VZV immunity in the prevention of human VZV disease.

The study of VZV pathogenesis has been complicated by the high degree of host specificity of the virus and its inability to replicate or cause disease in immune competent small animal models. Only the guinea pig has proven to be a usable immune competent small animal model, but it shows little or no disease and requires adapted virus. However, conceptions of varicella pathogenesis have undergone much of a reformulation as a result from the elegant and impressive studies examining VZV growth in human tissue (skin, neural tissue or a combination of thymus and liver to mimic T cells) harboured in severe combined immunodeficiency (SCID)-hu mice (Arvin, 2006; Ku *et al.*, 2004; Reichelt *et al.*, 2008; Zerboni *et al.*, 2005, 2007). While VZV pathogenesis is considered in more detail in Chapter 16, it is important to reiterate specific aspects learned from these and other studies because they have bearing on immunity and immune evasion. The current models of human disease are that primary varicella initiates following inhalation and the establishment of viral replication sites in respiratory lymphoid organs, such as the tonsils. A direct and systemic VZV dissemination ensues, mediated by circulatory haematopoietic cells that infiltrate such tissues, including T cells and antigen presenting cells (APCs). Recent work has particularly implicated a role for memory $CD4^+$ T cells with skin homing markers in dissemination (Ku *et al.*, 2002, 2005), although dendritic cells (DC), monocytes and macrophages are productively infected by VZV and may pass virus in lymph nodes to T cells for viremic delivery (Abendroth *et al.*, 2001b; Huch *et al.*, 2010; Morrow *et al.*, 2003a). When such cells infiltrate through the microcapillary vasculature in the deep dermal layers of skin at many sites over the body, a 'smouldering,' partially contained, cutaneous infection is established that eventually erupts after 10–21 days to yield the characteristic skin lesions typical of varicella. It has become apparent that components of innate immunity are primarily the activities that limit spread, since humoral and cellular immunity are detected only after appearance of skin lesions.

The VZV latent state establishes in sensory neurons during the primary infection at dorsal ganglia (Wang *et al.*, 2005). The widespread nature of varicella results in latency in multiple dorsal root and cranial ganglia. In contrast to the predominantly antigen-repressed state seen in herpes simplex virus (HSV)-1 latent infection, VZV latency is associated with expression of several viral lytic RNAs and proteins in VZV latently infected tissue obtained from animal models of latency or human autopsy ganglia (Cohrs *et al.*, 1996; Lungu *et al.*, 1995; Pevenstein *et al.*, 1999; Wang *et al.*, 2005). Up to 8 VZV genes have been reported to be transcribed in latently infected neurons by one group or another, including ORFs 4, 21, 61, 10, 29, 62, 63 and 66 (Cohen *et al.*, 2005a,b, 2007; Cohrs *et al.*, 1995, 1996; Hoover *et al.*, 2006; Lungu *et al.*, 1998; Mahalingam *et al.*, 1996; Sadzot-Delvaux *et al.*, 1998; Xia *et al.*, 2003; Zerboni *et al.*, 2010). The most commonly detected and most abundant transcripts are from ORF63 (Cohrs and Gilden, 2007), encoding a regulatory protein that has activities that modulate innate intracellular responses induced by type I IFNs (Ambagala *et al.*, 2009; Ambagala and Cohen, 2007; Di Valentin *et al.*, 2005). Regulatory proteins encoded by ORF4 and ORF62 have also been found to be expressed (Cohrs and Gilden, 2007; Meier *et al.*, 1993). Globally, it is not yet clear why persistent VZV encoded regulatory protein expression does not induce the rest of the viral lytic cycle in neurons. One possibility is that such proteins show abnormal cellular distribution during latency and do not gain access to the neuronal nucleus (Lungu *et al.*, 1998; Stallings *et al.*, 2006; Stallings and Silverstein, 2006; Walters *et al.*, 2008). A more important question is how neurons expressing VZV protein antigens avoid being recognized and cleared by surveying immunity components. Indeed, this indicates viral immune evasion may be contributing to maintenance of the latent state.

VZV reactivation is clinically seen as zoster, a localized skin disease in which lesions are pathologically similar to that of varicella, but are characteristically much larger and restricted to a skin area innervated by one ganglion (dermatome). Reactivation of VZV is quite different from HSV-1. The large lesions reflect peripheral delivery by multiple neurons originating from the host ganglia harbouring the reactivation, unlike the single or few neurons that deliver HSV-1 to the periphery. The involvement of multiple neurons reflects an extensive replication of VZV in the ganglionic tissue and intraneuronal spread prior to peripheral delivery, something that is rarely seen with HSV-1. Furthermore, and in contrast to the multiple recurrences seen with HSV, recurrence of zoster is rare, and if it occurs, reflects reactivation at different ganglia. The ganglionic replication not only boosts CMI and the protection it affords, but also may cause sufficient damage to pre-empt a particular ganglion for further reactivation.

The triggering of VZV reactivation is by means that are not yet clear: unlike HSV-1, VZV does not reactivate from explanted human autopsy ganglia or from latently infected tissue obtained from animal models, which makes it very difficult to study. The strong correlation of zoster with immune decline/senescence of CMI suggests the primary factor in controlling zoster is immune surveillance. We conjecture that VZV may be monitored by the cellular immune response in ways that are similar to the $CD8^+$ T cell regulation of reactivation seen in HSV-1, with some key differences as detailed below. While stress is not as obvious a factor in VZV reactivation as seen with HSV-1, a contribution of stress to VZV reactivation has been supported by the observation of asymptomatic VZV shedding in saliva of astronauts that show no clinical signs of reactivation (Cohrs *et al.*, 2008). Thus appearance of zoster probably requires stress induced changes in the neuronal chromatin configuration of the latent VZV genome (Gary *et al.*, 2006), which may be much more stable than that of HSV, in conjunction with a complete failure of CMI surveillance of the reactivating neuron to limit intraneuronal spread. Stress may also contribute to the failure of $CD8^+$ T cell surveillance, as found for latent HSV-1 infections in the murine model of latency (Freeman *et al.*, 2007).

Any consideration of VZV pathogenesis must address the association of pain with zoster. While reactivation of HSV-1 is rarely associated with severe or chronic pain or more than slight sensory loss, much of zoster is complicated by pain before, during, or long after the clinical manifestation. Over 30% of zoster cases are associated with pain after resolution of visible skin disease (PHN) that, at worst, may last months to years. PHN appears to have multiple causes, degrees of severity and a highly variable response to treatments, and over half of patients showing little response to any treatment strategy. It is likely that PHN is a result of a combination of ganglionic viral replication, inflammation, immune infiltration, VZV induced loss of neurons in the corresponding dorsal horn of the spinal cord, and altered physiology of surviving sensory neurons. The fact that boosting of CMI by the zoster vaccine may not prevent zoster from occurring but does reduce the severity of disease (and PHN), indicates higher CMI may limit the ganglionic spread and subsequent damage to neurons after VZV reactivation. However, ganglionic inflammatory immune responses may also contribute to pain, since some PHN cases (although far from all) can be alleviated with steroid treatment (Roxas, 2006). The contribution of host cell immune responses in human ganglia during and following zoster is difficult to study, largely as a result of the inherent difficulty in obtaining suitable ganglionic material. A new model of VZV induced pain in the rat (Garry *et al.*, 2005; Hasnie *et al.*, 2007) may yield a better picture of contributions of immunity to VZV induced pain.

VZV immunity

Innate immunity

In a naturally acquired primary VZV infection, only the innate responses are capable of containing viral spread and VZV replication until adaptive immunity develops. Innate immunity likely plays additional roles in the restriction of VZV infection at the mucosal epithelia and during the viremic phase, as well as aid the development of the VZV specific adaptive arm of the immune response. Innate responses include surveillance by activated natural killer (NK) cells and the secretion and activity of both type I interferons (IFNs: IFNα and IFNβ) and type II (IFNγ). NK cells have been shown to lyse VZV-infected fibroblasts *in vitro* (Bowden *et al.*, 1985; Ito *et al.*, 1996), express IFNγ and synthesize the cytolytic protein granulysin, which enhances the death of target cells *in vitro* (Hata *et al.*, 2001). Recent reports have linked fatal varicella to deficiencies in NK cell activity, supporting the importance of NK cells in containment of primary disease *in vivo* (Etzioni *et al.*, 2005; Notarangelo and Mazzolari, 2006). Similarly, Levy *et al.* reported NK-T cell deficiencies may predispose to severe and fatal varicella (Levy *et al.*, 2003). A study examining severe and fatal cases of varicella in 5 otherwise healthy patients suggested an absence of serum NK cells and primed $CD8^+$ T cells in these cases, and it was concluded that VZV could act to seclude these cells during primary infection from the circulatory pool in some patients (Vossen *et al.*, 2005). This remains an interesting immune deviation that needs to be explored further.

Type I IFN (IFNα and IFNβ) functions to inhibit VZV replication in both infected and uninfected cells by stimulating the production of a variety of interferon stimulated genes that induce an 'anti-viral state'. The importance of type I IFN *in vivo* is highlighted by the observation that treatment of immunocompromised individuals with IFNα can reduce the severity of varicella (Arvin *et al.*, 1982). Among the many IFN stimulated genes are those that encode protein kinase R (PKR), 2–5 oligoadenylate synthetase (2–5 OAS) and the Mx proteins that induce an anti-viral response by interfering with viral transcription, translation and possibly other viral processes (Sadler and Williams, 2008). While VZV is inhibited by both type I and type II IFN *in vitro* (Balachandra *et al.*, 1994; Desloges *et al.*, 2005a) the role of IFN secretion in VZV containment *in vivo* has not been fully elucidated. Studies in the SCID-hu skin mouse have shown that IFNα in SCID-hu VZV infected skin is expressed at high levels by uninfected cells surrounding VZV antigen positive tissue, but interestingly, not in the VZV infected epidermal cells themselves. In addition, depletion of IFNα by passive antibody treatment of SCID-hu mice resulted in more extensive viral growth in the human skin (Ku *et al.*, 2004, 2005). This supports the notion that IFNα production in uninfected bystander cells may contain VZV infection and spread. This is discussed in greater detail in the following section on immune evasion of IFN responses.

The humoral response

VZV specific antibodies can be detected in serum from 1–3 days after lesion appearance and have been shown to be directed to multiple viral proteins, including glycoproteins and structural proteins (Weigel and Grose, 1996). However, the magnitude of the humoral response appears to be less important than CMI in control of varicella, zoster and the maintenance of the protective state. Clinical evidence comes from the anecdotal reports that children with B cell immune deficiencies show similar clinical spectra of varicella to that seen in immune competent children. The incidence and morbidity of zoster correlates with CMI levels and not antibody levels. Indeed, higher antibody levels have been seen in more severe zoster patients, suggesting antibodies had little protective capability (Weinberg *et al.*, 2009). However, humoral immunity should not be dismissed. Passively acquired antibodies at the time of viral exposure can reduce the capacity of the virus to infect the host, and administration of exogenous zoster immune globulin to naive newborns and immune compromised individuals within 72 hr of VZV exposure results in a significant reduction in severity of disease (Miller *et al.*, 1989). It is presumed that antibodies neutralize VZV to block entry and/or uncoating, although they may block viral growth and cell fusion as well as viral release from infected cells, as recently shown for anti-gH antibodies *in vitro* and in the SCID-hu mouse model of VZV infection (Vleck *et al.*, 2010).

Antibodies to VZV remain throughout life and while levels may vary, their detection is used to indicate prior exposure to VZV. Such exposures are important in evaluating the need for vaccination in higher risk patients, since many varicella cases are mild, misdiagnosed or are not recalled by the patient. Sensitive techniques have been developed to detect and titre VZV specific antibody levels, and most often are variations of a solid phase enzyme immunoassay (EIA) or immunofluorescence assay, in which binding of antibody to immobilized VZV antigens are detected with antibody enzyme conjugates. These have been automated for detection of all three classes of antibodies (Forghani, 2001).

Development and maintenance of cell mediated immunity (CMI)

VZV specific CMI developing following primary infection is delayed and not seen in the incubation period, but its appears just after lesion appearance and correlates with viral clearance, resolution of disease and the maintenance of a long-lasting protective state to further clinical signs of VZV exposure. Both VZV specific $CD8^+$ and $CD4^+$ T cells are detected by 24–72 h after appearance of skin lesions, and high levels are found in most patients at 3 days post rash appearance (Arvin *et al.*, 1986). If acquisition of VZV specific T cells is delayed for more than 72 h, VZV may show persistent viraemia, more cutaneous lesions, and in extreme cases, may cause potential fatal virus dissemination (Abendroth and Arvin, 2001). The role of cellular immunity is particularly evident in children with immune compromise driven by diseases such as underlying HIV infection or congenital T cell deficiencies, in which varicella shows high morbidity and mortality. Both $CD4^+$ and $CD8^+$ T cells are evident following varicella, but studies have reported finding VZV specific T cells circulating in healthy VZV seropositive donors are largely of the $CD4^+$ T cell subset and only low levels of $CD8^+$ T cells are detected (Asanuma *et al.*, 2000; Jones *et al.*, 2006; Malavige *et al.*, 2008). During natural varicella infection, VZV specific $CD4^+$ T cells are predominantly of the Th1 subset, producing IL-2 and IFN-γ (Jenkins *et al.*, 1998; Bergen *et al.*, 1990; Zhang *et al.*, 1995). The spectrum of viral antigens targeted by both $CD4^+$ and $CD8^+$ T cell subsets is broad: that to the major regulatory transcription and virion tegument protein IE62 and the major VZV glycoprotein gE are recognized efficiently and equally by both $CD4^+$ and $CD8^+$ T cells (Arvin *et al.*, 1991; Sharp *et al.*, 1992). Additional viral target antigens include representatives of all transcriptional regulatory classes as well as different functional groups, including glycoproteins (gC, gB and gI), DNA replication proteins (ORF29) and viral tegument and transcriptional regulatory proteins (ORFs 4, 63 and 10) (Arvin *et al.*, 2002; Jones *et al.*, 2006, 2007; Sadzot-Delvaux *et al.*, 1997). Characterization of the specificity of T cell clones isolated from human patient corneas also revealed a similarly broad spectrum of multiple VZV antigens recognized (Milikan *et al.*, 2007). While it is widely accepted that $CD8^+$ T cells provide the ability to clear infected cell targets, $CD4^+$ T cells act to restrict viral replication through the secretion of cytokines and by directly targeting and killing of VZV infected cells. $CD4^+$ T cell clones isolated from patients with VZV induced ocular uveitis were shown to control VZV replication *in vitro* in major histocompatibility complex class II (MHC-II) expressing retinal pigment epithelial (RPE) cells that was partly mediated by cytotoxic mechanisms (Milikan *et al.*, 2009).

Memory immunity, protection and latency

Resolution of varicella is followed by persisting IgG levels and the establishment of a predominant $CD4^+$ T cell memory response with diverse antigen specificity. The VZV specific memory T cell load has been estimated, based on proliferation capacity and the ability to produce inflammatory cytokines following VZV antigen exposure, to be 1: 20,000 to 40,000 PBMCs (Hayward *et al.*, 1986). A recent study indicated that VZV responder cell frequency peaks in healthy individuals in early adulthood (Weinberg *et al.*, 2010), but thereafter showed a progressive decline in VZV specific CMI, as measured by tests including enumeration of responder cell frequencies and skin delayed hypersensitivity tests (Levin *et al.*, 2008). Studies demonstrating an association between the magnitude of the CMI response and the likelihood of zoster (Hata *et al.*, 2002; Levin *et al.*, 2008; Patterson-Bartlett *et al.*, 2007), in conjunction with the lack of zoster correlation with changing VZV antibody levels, establish CMI as the most important component of protection.

The correlation of zoster with increasing age and declining immunity was initially recognized in insightful studies in the 1960s by Hope Simpson (Hope-Simpson, 1965). He carefully recorded zoster cases in his community and recognized that epidemiology of zoster (incidence rising with age, occurring rarely as a second time, not correlating with time of varicella onset) correlated with waning immunity. He postulated that full virus reactivation and zoster appearance occurred if such immunity reduced to a critical stage. He also postulated that sporadic exposure to varicella throughout life (by 'contained reversions', reflecting internal subclinical reactivation or exogenous re-exposure) rapidly mobilized immune responses and increased protection against zoster to a higher level (see Fig. 20.1). Given that this model was developed long before many of the basics of immunology and its role in protection was well understood it is remarkable that the model has remained viable and well validated. The sporadic boosting of immunity following exogenous exposure has been substantiated by many studies, including the administration of the zoster vaccine and its ability to increase protection against zoster (Oxman *et al.*, 2005). CMI and VZV antibody levels are known to show significant fluctuations in specific populations followed longitudinally, and show a boosting following exogenous re-exposure to VZV. For example, Vossen *et al.* found exposure of immune individuals to external VZV induced the expansion of VZV antigen specific $CD4^+$ T cells in peripheral blood, with concomitant changes in $CD8^+$ T cells and NK cells and their induced secretion of IFN, TNF and IL-2 (Vossen *et al.*, 2004). This was despite no signs of disease in any of the 16 subjects, and no obvious detectable evidence of VZV DNA in the blood. Thus, external re-exposure to VZV (such as caring for a child with varicella) results in a subclinical infection and expression of viral antigens that boost the memory immune response. A more controversial issue is whether CMI is boosted by internal but contained subclinical reactivation events, and until recently, it was not known whether VZV reactivation occurred sub clinically. Evidence for subclinical reactivation has been mounting. $CD4^+$ T cells from healthy disease free donors have a phenotype indicative of recent or persistent antigen exposure (Malavige *et al.*, 2008). Human $CD4^+$ T cells that recognize IE63, the VZV latency associated protein, show a mixed central and effector memory differentiation phenotype indicative of persistent low level expression or possibly asymptomatic recurrence (Jones *et al.*, 2007). A recent 1 year longitudinal study of adult patients with little chance of exposure to exogenous varicella found unexplained peaks that suggested subclinical reactivation events (Klein *et al.*, 2006). Kraus and Klinman (Krause and Klinman, 2000) argued that variations in immunity were indicative of asymptomatic reactivation. The finding of subclinical viral shedding in saliva in high stress individuals (astronauts) has established the potential for subclinical reactivation and Hope Simpson's proposed 'internal boosting by contained reversions' (Mehta *et al.*, 2004, 2008; Cohrs *et al.*, 2008). This issue remains controversial and is in need of further assessment.

The means by which CMI controls reactivation of VZV from the latent state is far from clear. Adaptive immunity is not required for the establishment of latency, since latency is efficiently established in human ganglionic tissue implants in SCID hu mice lacking adaptive immunity (Zerboni *et al.*, 2005). It could be speculated that the persistent antigen expression seen in VZV latently infected ganglia enables efficient CMI recognition and surveillance by non cytolytic means. As stated previously, the model in which HSV-1 reactivation is controlled by ganglionic-resident $CD8^+$ T cells may apply to VZV. In this

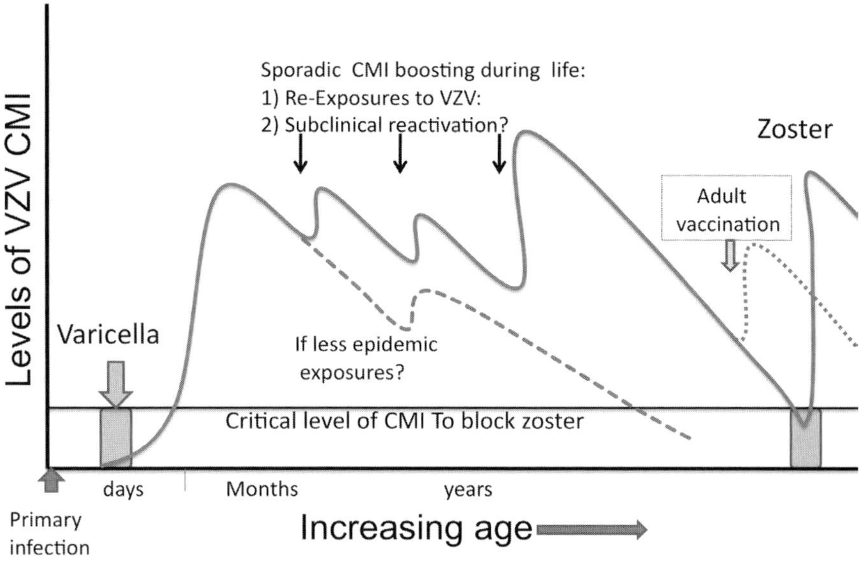

Figure 20.1 Models adapted from that of Hope Simpson, 1965, showing the level of VZV immunity as it rises and falls with increasing age. The solid line models the level of CMI and protection from disease, which appears after clinical varicella and reaches high levels. Sporadic boosting of immunity throughout life by subclinical exogenous VZV reinfection or by possible contained subclinical reactivation events delays the loss on immunity. For some, the declining level reaches a critical stage at which reactivation cannot be contained and zoster results. Vaccination may act to boost the level of CMI to delay and/or prevent zoster (dotted line). The dashed line represents a concern that fewer varicella epidemics may result in reduced sporadic immune boosting throughout life, with an earlier reaching of low CMI levels to the critical level needed to allow zoster to reactivate.

model, $CD8^+$ T cells remain adjacent to latently infected neurons such that reactivation events are closely monitored. A reactivation event resulting from changes in chromatin configuration leads to gene expression, rapid antigen presentation on the reactivating neuron in conjunction with MHC class I and rapid recognition by proximal $CD8^+$ T cells (Freeman *et al.*, 2007; Knickelbein *et al.*, 2009; Sheridan *et al.*, 2007). Their activation allows them to quickly exert a blockade on multiple stages of the reactivation process without killing off the latently infected neuron. The correlation of the CMI to the incidence of zoster is certainly indicative of control in this manner, and antigen expression seen in VZV latent infections may allow for even more efficient surveillance by CMI. However, while T cells have been observed in latently infected human ganglia, in contrast to expectations, one study found that latently infected neurons do not appear to include VZV specific $CD8^+$ T cells. Verjans *et al.* (2007) reported that human ganglia harbouring both latent HSV and VZV genomes were predominantly infiltrated with $CD8^+$ T cells that surrounded HSV latently infected neurons but not those latently infected with VZV and expressing viral antigens. Clonal expansion of the infiltrating T cells suggested they were predominantly HSV-1 antigen specific.

These interesting findings raise an unresolved conundrum, since they conflict with the clear role of CMI in protection from reactivation and the immune control seen in the HSV mouse model and of HSV latently infected human tissues. Maybe persistent VZV antigen expression during latency causes exhaustion of any resident VZV specific immune cells that infiltrate the ganglia. It is also possible that VZV antigen positive neurons do not efficiently present them on the neuronal surface due to immune evasion strategies used by VZV. Perhaps VZV latency is a highly stable state in which most viral genes outside of the specific VZV genes associated with VZV latency are very efficiently repressed. Those that are expressed under the umbrella of immune evasion strategies do not encourage the ganglionic retention and activation of T cells. Rare reactivation events would lead to viral infection of adjacent cells or satellite cells which, if an adequate CMI exists, would respond to contain the infection. If the CMI falls, ganglionic spread would not be contained and zoster would result. Is there, perhaps, an unrecognized range of subclinical zoster in which CMI blocks most of the disease, but is sufficiently boosted in its protective state? The puzzle regarding the immune control of latency requires much further study, and the work by Verjans *et al.* will likely stimulate similar studies to address human ganglionic immune infiltration.

Immunity following vaccination

The VZV vaccines are live and based on an isolate, Oka, obtained from a healthy Japanese child that was attenuated by passage in semipermissive guinea pig embryonic cell tissue culture. The vaccine against varicella (Takahashi *et al.*, 1974) was licensed by the Food and Drug Administration (FDA) for use in the United States in 1995, and is now also used widely in several developed countries (Gershon *et al.*, 2008c). The vaccine, commercially available under the name Varivax, was originally given in a single subcutaneous dose to children aged 12 months to 13 years, and as two doses in children 14 years and older, because one dose was not as effective in greatly increasing CMI and preventing varicella in adults as in children (Gershon *et al.*, 2008a). Based on increasing reports of varicella outbreaks and the decline of vaccine induced immunity over time (Chaves *et al.*, 2007; Gershon *et al.*, 2007; Lopez *et al.*, 2006; Michalik *et al.*, 2008), the Advisory Committee on Immunization Practices in the USA recommended a second dose of the varicella vaccine be administered routinely for all vaccinated children and adults (Marin *et al.*, 2007).

The effects of vaccination have been dramatic. Routine varicella immunization has led to a major decline in varicella incidence, varicella-associated hospitalizations and deaths. In the USA, where the vaccination rates are high, the incidence of varicella has decreased by ~80% (Seward *et al.*, 2002), varicella-associated hospitalizations have decreased by 88% (Zhou *et al.*, 2005) and mortality from varicella in immune competent individuals has decreased from ~100 deaths annually to 10 or fewer (Seward *et al.*, 2008). The varicella vaccine has proven to be very safe in healthy individuals and the main adverse effect is a usually mild break through rash in the first six weeks following immunization (Marin *et al.*, 2007). Such breakthroughs appear to be caused by clonal variants in the mixed vaccine population that may retain some ability to replicate at the skin (Quinlivan *et al.*, 2007)

Both cell mediated and humoral immunity have been demonstrated to last for many years after vaccine administration (Watson *et al.*, 1994; Watson and Starr, 1994; Zerboni *et al.*, 1998). ELISA assays show seroconversion occurs 4–6 weeks after single dose vaccination in ~97% of children aged 1–12 years, although adolescents (aged 13–17 years) have a lower seroconversion rate (79%) (White *et al.*, 1991). The VZV-specific antibody response to the vaccine at 6 weeks correlates with immune protection against varicella (Li *et al.*, 2002; White *et al.*, 1992). VZV-specific antibodies have been shown to persist in children and adults after a single varicella vaccine dose (Zerboni *et al.*, 1998). As expected, a second varicella dose in children produced an improved host immune response, higher antibody levels, a lower incidence of breakthrough varicella and improved vaccine efficacy (Kuter *et al.*, 2004). Using a VZV specific T cell proliferation assay, 98–100% of healthy children given a single dose of varicella vaccine have detectable T cell recognition of VZV antigens by 2–6 weeks post vaccination (Nader *et al.*, 1995). Furthermore, in a study following patients after 2 vaccine doses administered 4–6 years apart, VZV specific T cell proliferative responses were significantly higher at 6 weeks and 3 months after the second dose in comparison to the same time points following one dose (Watson *et al.*, 1995). A recent study indicated that responder cell frequencies induced by the varicella vaccine in normal or HIV infected children were as efficient as that induced by wild type infections (Weinberg *et al.*, 2010). The protective CMI generated following vaccination has been shown to be ~94% effective in preventing varicella over a 5 year period (Zerboni *et al.*, 1998). Studies of vaccinated individuals have revealed that breakthrough varicella disease is usually mild, even in vaccinees who had no detectable VZV antibodies, indicating the importance of CMI in providing immunological protection (Ampofo *et al.*, 2002).

The zoster vaccine was developed on models including that of Hope Simpson in that an induced vaccination late in life would increase CMI and delay its reduction to that 'critical level' that allowed clinical zoster appearance (see Fig. 20.1, dotted line) The zoster vaccine is 14 times higher dose than the Varivax vaccine but is based on the same virus, and was trialed for the prevention of zoster and PHN in 38,000 elderly people in the US (Oxman *et al.*, 2005). This study revealed a 51.3% decrease in the incidence of zoster and a 66.6% decrease in incidence of PHN in vaccinees. The burden of illness due to zoster was decreased by over 61% and like the varicella vaccine, has proven to be very safe (Oxman *et al.*, 2005). The vaccine boosts VZV specific CMI, and the magnitude of the CMI response correlated with the incidence of zoster (Hata *et al.*, 2002; Levin *et al.*, 2008; Patterson-Bartlett *et al.*, 2007). Recently, Weinberg *et al.* reported that memory VZV CMI between individuals following an episode of zoster and zoster vaccine recipients were similar, suggesting that the protection afforded by the zoster vaccine was as good as a natural VZV reactivation (Weinberg *et al.*, 2009). Currently in the USA, the CDC recommends a single dose of the zoster vaccine in healthy people over the age of 60 years. However, the best age to administer the zoster vaccine, the duration of protection conferred by this vaccine and whether booster doses may be needed to increase vaccine effectiveness still remain to be established.

Despite their success, both vaccines have several limitations that have, to some extent, tempered their widespread acceptance. In addition to breakthrough infection in healthy individuals and the capacity of the vaccine virus to cause zoster and be transmitted to others, the vaccine is largely unsuitable for administration to immune compromised individuals, as it may cause varicella or other serious disease (Marin *et al.*, 2007). Such individuals include cancer patients undergoing radiotherapy or chemotherapy, those with HIV/AIDS and others being treated with drugs that weaken the immune system. It is also not recommended for use in pregnant women or those planning pregnancy shortly after vaccination (Marin *et al.*, 2007). The vaccine virus (vOka) still retains the ability to access sensory ganglia, establish a life-long latent infection, and even reactivate, although such events are rare (Takayama *et al.*, 2000; Civen *et al.*, 2009; Tseng *et al.*, 2009). In addition, the varicella and zoster vaccines are changing epidemiology of VZV infections in ways that remain to be fully answered (Reynolds *et al.*, 2008). A concern is that universal varicella vaccination may remove the CMI boosting by exogenous re-exposure to virus (see Fig. 20.1 dashed line), with a secondary effect of accelerating the loss of VZV specific cellular immunity that normally occurs with increasing age to the critical levels that enable zoster development. Zoster may thus occur earlier in life for those harbouring a natural latent wild type infection (Brisson *et al.*, 2002). This alarm might last 30–50 years until the adult population is predominantly composed of vaccinated individuals with no varicella disease history. Studies to address this likelihood have been conflicting (for a comprehensive review see Reynolds *et al.*, 2008) and may be complicated by many unidentified factors (Rimland and Moanna, 2010).

The current live attenuated virus used for vaccination against varicella and zoster should be regarded as a 'first generation' vaccine. A 'second generation' vaccine should consist of specifically targeted modifications and defined mutations that limit replication at the inoculation site without causing a lesion; and one that prevents infection of T cells and neurons, yet induces immunity as effectively as natural infection. The development of such a vaccine, which is likely to be administered to a wider range of individuals, awaits a better understanding of the viral genes that function in cell tropism and evasion of host immune responses.

Immune evasion strategies of VZV

While the innate responses and VZV specific cell mediated immunity are critical to resolve varicella and induce the protective state, it has become clear that the virus has evolved strategies to evade components of both innate and adaptive responses. These presumably contribute to a more extended viral replication, enhanced virus access to skin sites of replication and more efficient transmission to others, and they may be deployed during the maintenance of the latent state. An analysis of the VZV genome reveals the absence of an orthologue of HSV-1 ICP47, which reduces MHC-I restricted surface antigen presentation required for $CD8^+$ T cell targeting of HSV infected cells. HSV-1 ICP47 blocks peptide antigen transport into the ER lumen for MHC-I assembly (Hill *et al.*, 1995; Tomazin *et al.*, 1996). VZV (and apparently all varicellovirus members) also lack equivalents of the HSV-1 γ34.5, which has functions that counteract the double stranded RNA sensor PKR, which if not counteracted, phosphorylates eukaryotic initiation factor (EIF2α) and halts translation (Chou and Roizman, 1994; He *et al.*, 1997). VZV also lacks orthologues of several HSV-1 short region proteins, including US5 and gD, which contribute to a block in apoptosis in HSV-1 infected cells. How VZV blocks these innate and adaptive responses is thus of considerable interest, as such events must still be counteracted during productive VZV infections.

Interference with interferons

While both type I and II IFNs can block VZV replication *in vitro* (Abendroth and Arvin, 2001; Arvin *et al.*, 1986), VZV appears to control expression of IFN and the downstream signalling events in order to provide a survival advantage to the virus during replication. Using the human skin xenograft-SCID mouse model, Ku *et al.* (2004) found IFNα immunohistochemically in epidermal skin cells in uninfected skin xenografts at low constitutive levels (Ku *et al.*, 2004), complete downregulation in VZV infected cells in xenografts, and strong upregulation in bystander uninfected epidermal cells. These studies were supported by demonstrating phosphorylation of the IFNα downstream intermediate Stat-1 and its translocation to the nuclei in uninfected cells surrounding VZV skin replication sites, but not in VZV-infected cells. These observations clearly indicate VZV infection of human skin cells impairs Stat1 activation and type IFN I production, but does not prevent signalling to bystander uninfected cells that up regulate an antiviral state. Administration of an IFNα/β receptor neutralizing antibody to SCID-hu skin mice resulted in enhanced VZV replication and lesion formation compared with mice that received no antibody (Ku *et al.*, 2004). Taken together, these suggest IFNα acts to modulate cutaneous VZV replication *in vivo*, but that VZV actively modulates the cutaneous innate immune response.

The identity of VZV proteins regulating IFN responses may have been identified. Using a viable recombinant VZV deleted for the IE63 gene, Ambagala and Cohen (2007) demonstrated that the virus was hypersensitive to the antiviral effects of IFNα but not IFNγ, and was inhibited by levels that were much lower than that required for blockade of VZV infection. IFNα appeared to block viral gene expression at multiple levels in the deletion virus, including a post transcriptional block. The underlying mechanisms remain to be fully determined. Very recently, IE62 was also shown to be involved in IFN evasion.

Interestingly, IE63 also acts to block intracellular signalling by the double stranded RNA sensor PKR, suggesting it acts as a partial functional equivalent to the γ34.5 gene of HSV-1. PKR is one of the many genes up regulated as a result of IFN signalling and if not counteracted, will lead to a block in translation. Ambagala and Cohen found that in cells infected with VZV lacking the ORF63 protein, EIF2α showed increased levels of phosphorylation (and thus inactivation) compared to cells infected with VZV expressing IE63, and that IE63 expressed transiently from transfected plasmids contained lower basal levels of the phosphorylated EIF2α (Ambagala *et al.*, 2009). While Desloges reported no significant alteration of PKR by VZV infection in human melanoma cells (Desloges *et al.*, 2005a,b), leading to some uncertainty on this issue, it is tempting to speculate that such activities of the IE63 latency-associated protein may contribute to a successful maintenance of the VZV latent state. IE63 is required for onset of latency in neurons in an animal model (Ambagala and Cohen, 2007; Cohen *et al.*, 2004, 2005a).

VZV has also been shown to block IFNγ signalling in T cells through the action of the ORF66 protein kinase. The ORF66 protein gene (orthologous to the US3 protein kinase in HSV-1 and PRV) encodes a basophilic serine-threonine specific protein kinase that has the unprecedented activity of modulating the nuclear import of the major transcriptional regulatory protein, IE62, to redirect it to the cytoplasm for subsequent virion incorporation (Eisfeld *et al.*, 2006). Schapp-Nutt and coworkers (Schaap-Nutt *et al.*, 2006; Schaap *et al.*, 2005) studied the formation of phospho-Stat in VZV infected T cells following IFNγ binding to its receptor, and found it significantly diminished when T cells were infected with VZV expressing ORF66 as compared to VZV infections lacking functional ORF66. It is not yet resolved as to how ORF66 blocks this activity, but it is notable that Roizman and colleagues have found that the HSV-1 US3 kinase phosphorylates the IFNγ receptor to prevent signalling (Liang and Roizman, 2008). Thus, we speculate VZV may target the receptor in a similar kinase dependent manner.

Blockade of the NFκB pathway

The NFκB transcription factor is maintained in most cells in an inactive state, sequestered in the cytoplasm by its association with the inhibitory protein IκBα. Degradation of IκBα results in translocation of NFκB proteins to the nucleus and stimulation of expression of a wide range of genes, including those involved in the immune response (Ghosh and Hayden, 2008; Hayden and Ghosh, 2008). Following microarray studies, in which many NFκB responsive genes were down regulated following VZV infection (Jones and Arvin, 2005), these authors reported that VZV infection inhibited degradation of IκBα and sequestered NFκB proteins (p50 and p65) in the cytoplasm of VZV infected cells (Jones and Arvin, 2006). As UV-inactivated virus did not inhibit the nuclear translocation of NFκB subunits, it was concluded that VZV *de novo* expression is required to block activation. The block of NFκB activation was also observed in skin xenografts of SCID-hu mice infected with VZV, where p50 and p65 remained in the cytoplasm of VZV infected cells, yet showed nuclear accumulation in bystander uninfected epidermal cells (Jones and Arvin, 2006). This finding is consistent with the lack of IFNα expression and Stat 1 activation in VZV infected but not in uninfected bystander cells (Ku *et al.*, 2004a), and also with the downregulation of the NFκB responsive ICAM-I (see below). Given that NFκB is a major inducer of IFNα transcription, its modulation by VZV is a major evasion strategy that may be central in VZV pathogenesis. The VZV gene product(s) responsible for the modulation of the NFκB pathway in HFs or any other cell-type has yet to be elucidated.

Blockade of apoptosis

Apoptosis is an innate response to block many viruses and its counteraction is often required for viral survival. VZV infected cells do show signs of apoptosis, but this is dependent on the host cell type analysed. VZV infected cultured cells were initially reported to have DNA fragmentation resulting from infection (Sadzot-Delvaux *et al.*, 1995). However, recent studies of cultured human neurons strongly suggest that VZV induces an activity that protects them from apoptosis (Hood *et al.*, 2006; Hood *et al.*, 2003). Deletion of ORF63 resulted in increased levels of neuronal apoptosis, suggesting that IE63 has anti-apoptotic functions in neurons. As a latency expressed protein, IE63 may contribute to establishment, maintenance and reactivation of latent infections by promoting survival of these post-mitotic cells, much in a manner that has been attributed to an anti-apoptotic role of the HSV-1 LAT RNA.

Studies in the SCID-hu model have suggested that ORF66 protein kinase is also an anti-apoptotic factor in T cells. Deletion of ORF66 results in a higher level of apoptosis of T cells infected with VZV (Schaap-Nutt *et al.*, 2006). VZV lacking ORF66, or expressing kinase inactive mutants (G102A) grew to levels 2 logs lower than parental virus in cultured human T cells but not in MeWo cells, and the G102A mutant showed increased levels of active caspase 3, the executioner protease in apoptosis. These findings imply that ORF66 has an important function in extending the survival of infected T cells until they are able to home in on the skin (Moffat *et al.*, 1998; Schaap-Nutt *et al.*, 2006; Schaap *et al.*, 2005). Interestingly, ORF66 kinase dependent regulation of apoptosis induced by VZV is not seen in culture in MeWos, skin fibroblasts or in primary corneal fibroblast cells which require the presence of ORF66 for

VZV growth (Erazo *et al.*, 2008). Recent work has shown that apoptosis in HSV-1 infection is more important in highly replicating or transformed cells than in primary cell lines (Nguyen *et al.*, 2007). VZV has also been reported to modulate the PI3K/Akt pathway, which can affect regulation of apoptosis (Rahaus *et al.*, 2007).

Modulation of surface MHC-I antigen presentation

MHC-I is required for correct presentation of unfamiliar antigens on the target cell surface for recognition by antigen specific $CD8^+$ T cells. Antigenic peptides are presented in a complex of peptide, β_2-microglobulin (β_2-M) and an MHC-I heavy chain (Hc), which may be of many hyper variable types. The MHC-I assemblon is targeted by many viruses, and VZV is no exception. VZV was initially shown to down-regulate surface MHC-I levels by Cohen (Cohen, 1998) in cultured cells and by Abendroth *et al.*, (2001a) on both cultured cells and on T cells obtained from SCID-Hu mice with Thy-liv implants. It was apparent that VZV did not down regulate levels of MHC-I Hc and β_2-M, but rather blocked the assembly of the MHC–I complex.

The MHC-I assemblon initiates with the generation of peptides by the host 26S proteasome and their active transport to the ER lumen by the transporter of antigen presentation (TAP), a heterodimer. TAP is a common point for virus directed blockade of surface MHC-I presentation, because its inactivation prevents the assembly and maturation of the main antigen presenting MHC-I complex types (MHC-I A and B), but does not affect the MHC-I types C and E that are primarily involved in the signalling to NK cells that the host cell is not foreign. While VZV has no orthologue of ICP47, some members of the varicellovirus group act on TAP through orthologues of the VZV ORF9A or 9.5 protein. The most well characterized is the bovine herpesvirus UL49.5 (Koppers-Lalic *et al.*, 2008), which encodes a non-glycosylated membrane protein that, in HSV-1, is linked to glycoprotein gM. Activity of VZV ORF9.5 protein on MHC-I surface levels was not seen when it was transiently over expressed in transfected cells, either alone or with expression of its predicted gM interacting protein encoded by ORF50 (Eisfeld *et al.*, 2007). As these studies have not yet been extended to the context of VZV infection, it still remains possible that VZV ORF 9a may have activity on the TAP transporter.

In the ER lumen, MHC-I Hc is associated with stabilizing chaperones, including tapasin, ER57 and calreticulin. A preassembly complex forms with peptide and Hc that can then stably couple with β_2-M. The peptide undergoes a trimming modification which increases its affinity in the complex, to the point where MHC-I becomes stable in the absence of chaperones. Loss of one of the tripartite components results in rapid turnover of Hc, and human cytomegalovirus encodes MHC-I modulators that dissociate Hc from the complex. No such turnover is apparent in VZV infected cells (Eisfeld *et al.*, 2007). The final stages of MHC-I assembly involves the maturing of the complex through the secretory pathway via the Golgi to the cell surface. At the point of *Cis⁻ to medial-* Golgi transport, high mannose glycoproteins undergo conversion to complex endoglycosidase-H (endo H)-resistant forms, with only properly assembled trimeric β_2-M-Hc-peptide components undergoing processing. Surface MHC-I antigens may undergo induced internal recycling and degradation by some viruses. Marginal reduced levels of the endo H resistant forms of MHC-I were observed by Cohen (1998) and Abendroth (Abendroth *et al.*, 2001a), suggesting MHC-I was not severely blocked at a stage prior to its Golgi maturation. Rather it was suggested that MHC-I failed to reach the surface because it was diverted to internal membranes.

Subsequent studies have supported the novel role of the ORF66 kinase in contributing to down regulated surface MHC-I levels. An initial screening (Abendroth *et al.*, 2001a) using transfected plasmids expressing VZV ORFs, found ORF66 expressing cells had lower surface MHC-I levels. A more extensive search for transiently expressed VZV proteins that down modulate MHC-I has revealed activity only for ORF66 kinase, out of 37 ORFs examined to date (Eisfeld *et al.*, 2007; Eisfeld and Kinchington, unpublished data). Using GFP tagged forms of ORF66 that was fully functional in regulating nuclear import of IE62

(Erazo *et al.*, 2008), the surface expression of MHC-I could be correlated with ORF66 expression kinetics using flow cytometry. This proved valuable in analyses of VZV infection, since cells infected with VZV recombinants expressing GFP tagged ORF66 could easily be identified and separated from the uninfected cells found in most VZV preparations.

Plasmid transfection-mediated expression or replication defective adenovirus-mediated transduction of the ORF66 kinase alone was shown to down regulate surface MHC-I, but similar expression of the kinase inactive proteins did not. This suggests that ORF66 expressed in the absence of other viral proteins is sufficient to reduce surface MHC-I, and that a host protein was phosphorylated by the kinase to drive this process. While the specific target has not yet been resolved, it is notable that several of the MHC-I assemblon components are phosphoproteins, including TAP, tapasin and MHC-I Hc itself. Recent work has also indicated that both the PRV and HSV US3 protein kinases have similar abilities when expressed independently of other viral proteins (Eisfeld and Kinchington, unpublished data). Thus, MHC-I downregulation may reflect a conserved activity of the alphaherpesvirus US3 kinases.

Using the GFP tagged 66 expressed from VZV recombinants, high efficiency VZV infections were established. Correlating surface MHC-I with GFP reporter activity revealed that VZV down regulated surface MHC-I in multiple VZV infected cell types much more efficiently if the protein kinase was expressed. However, if ORF66 was disrupted, recombinant VZV still showed somewhat lower surface MHC-I levels than that seen on uninfected cells, suggesting the kinase was only partly responsible. The ORF66 directed downregulation of surface MHC-I occurred at times that would be deemed late in VZV infection, although this remains difficult to assess, due to the asynchronous nature of VZV infections in culture. Nevertheless, it seems logical that VZV would benefit from encoding an efficient surface MHC-I down regulator at early stages of infection. The search for such activities continues and may be quite revealing.

How does the ORF66 kinase affect MHC-I surface expression? Classical pulse radiolabelling with 'chase' studies to follow the processing of MHC-I suggest the kinase exerts a block prior to MHC-I Golgi processing (Eisfeld *et al.*, 2007). Using the antibody W6/32 to identify assembled Hc/β_2-M heterodimers, it was demonstrated that MHC-I did not undergo rapid turnover in response to ORF66 expression, but rather was delayed in the timely acquisition of EndoH resistance, suggesting it was not processed efficiently by *cis* to *medial* Golgi transition. Consistent with this observation, immunofluorescent analyses revealed that MHC-I was associated with the ER or Golgi (Eisfeld *et al.*, 2007; Abendroth *et al.*, 2001a). This appears to represent a mechanistic block in MHC-I surface presentation that is not precedented, and is consequently under further scrutiny.

Inhibition of MHC class II expression

MHC-II expression mediates correct presentation of antigen to $CD4^+$ T cells and is largely restricted to professional antigen presenting cells. Unlike MHC-I, MHC-II surface presentation does not require the ER lumen-restricted assembly, but rather acquires peptides derived directly from the cytoplasm or derived from the extracellular milieu. Importantly, MHC-II is often up regulated in response to external signalling, including locally produced IFNγ from NK, NKT and adaptive immune cell infiltrates.

In vivo transport of VZV during varicella viraemia by the haematopoietic cell population may involve down modulation of surface MHC-II levels and modulation of IFNγ signalling, particularly given the importance of $CD4^+$ T cell responses in the resolution of varicella. Abendroth *et al.* (2000) demonstrated that IFNγ treatment up regulated a significantly lower level of cell-surface MHC class II in fibroblasts infected with VZV as compared to controls. Consistent with these studies, examination of naturally infected human skin biopsies for MHC class II and VZV RNA by *in situ* hybridization revealed that dermal and epidermal cells infected with VZV in varicella and zoster skin lesions, did not express detectable MHC class II transcripts (Abendroth *et al.*, 2000). In experimentally infected HFs,

VZV inhibited the expression of Stat1α and Jak2 signalling proteins, but had little effect on Jak1. VZV infection also inhibited transcription of the interferon regulatory factor (IRF-1) and class II transactivator (CIITA) (Abendroth *et al.*, 2000). Taken together, these data suggest VZV encodes an immunomodulatory function which directly interferes with IFNγ signal transduction via the Jak/Stat pathway that enables the virus to inhibit IFNγ induction of cell-surface MHC class II expression. This is reminiscent of the ORF66 directed downregulation of IFNγ signalling in T cells seen by Schaap Nutt *et al.* detailed earlier. Furthermore, the modulation of MHC class II expression was shown to impair the ability of VZV infected keratinocytes to stimulate $CD4^{+}$ T cells *in vitro* (Black *et al.*, 2009). Thus, the inability of naturally infected skin cells to express MHC class II during varicella and herpes zoster is likely to result from disruption of IFNγ mediated signal transduction with a consequence of decreased capacity of infected cells to stimulate a $CD4^{+}$ T cell response.

Downregulation of intercellular adhesion molecule (ICAM) expression

Skin has multiple cell-types with antigen presenting capacity, including epidermal Langerhan cells, keratinocytes and infiltrating monocytic cells that may have the capacity to act as APCs in a MHC class II restricted manner. In the presence of pro-inflammatory cytokines such as IFNγ and tumour necrosis factor (TNF), keratinocytes can be induced to express not only MHC class II molecules, but also surface intercellular adhesion molecule 1 (ICAM-1) to facilitate binding to leucocyte functional antigen (LFA-1) expressing T cells. Nikkels *et al.* (2004) performed an immunohistochemical analysis for a variety of different immune cell markers and cytokines in skin biopsies from zoster lesions (Nikkels *et al.*, 2004). While ICAM-1 is expressed at low levels in skin cells, its expression is up regulated in the periphery of VZV skin lesions. Despite increased expression of the cytokines IFNγ, TNFα and IL-6 in VZV infected skin there was a decrease in expression of both ICAM-1 and MHC class II in VZV infected keratinocytes within the centre of the zoster lesions, suggesting that VZV infection could modulate ICAM-1 expression in VZV-infected keratinocytes *in vivo.* Given that these cutaneous sites have high levels of viral replication, it is likely that diminishing immune surveillance may provide the virus with an advantage to replicate more efficiently. Black *et al.* (2009) also examined ICAM-1 expression on uninfected keratinocytes in comparison to VZV infected keratinocytes *in vitro* and observed that VZV inhibited the upregulation of ICAM-1 after IFNγ treatment (Black *et al.*, 2009). In combination, these observations demonstrate that VZV modulates the expression of ICAM-1 in keratinocytes both *in vitro* and *in vivo*.

Impact of VZV on DCs and cell-surface immune molecule expression

DCs are bone-marrow derived potent antigen presenting cells that are located at many sites, including the skin, blood, lymph and mucosal surfaces (Banchereau *et al.*, 2000; Banchereau and Steinman, 1998; Klagge and Schneider-Schaulies, 1999). DCs process antigen and present it to T-cell rich areas of the lymph nodes to activate naïve and resting antigen-specific T cells and drive effector T cell differentiation (Banchereau *et al.*, 2000). DCs may be the first cell type to encounter VZV during primary infection, and since human immature monocyte derived DCs are fully permissive to a productive VZV infection and can release infectious virus, they could mediate virus transmission to T cells for the viraemia following infection. The lack of altered cell surface levels of the immune molecules MHC class I, MHC class II, CD86, CD40 or CD1a, which are normally up regulated as they mature (Abendroth *et al.*, 2001b) suggested that VZV may modulate DC functions.

Normal mature DC up-regulate MHC I, MHC II, CD86, CD83 and CD80 (Banchereau and Steinman, 1998) as they present to T cells. VZV seeking to evade or delay the immune response would likely benefit from the disruption of immune function of such cells. Morrow and colleagues showed that VZV productively infects mature monocyte derived DC, and in doing so, impairs their ability to function properly. VZV infection selectively down regulated functionally important cell-surface immune molecules MHC class I, CD80, CD83 and CD86 (Morrow *et al.*, 2003), and significantly reduced their capacity to stimulate the proliferation of allogeneic T cells. The precise molecular mechanisms and the VZV gene(s) responsible for the disruption of immune molecules remain to be elucidated, but it is important to note the VZV ORF47 protein kinase gene has been shown to be required for virus replication in immature DC, but not in mature DC (Hu and Cohen, 2005). VZV ORF47 was, however, required for the full replicative cycle and for transmission of virus from these cells to other cells. This study also demonstrated that both immature and mature DCs infected with VZV had reduced surface Fas expression, which may protect infected cells from apoptosis, a finding consistent with lack of apoptosis in VZV infected immature DCs (Abendroth *et al.*, 2001b).

There have been additional studies examining the impact of VZV infection on different DC subsets *in vivo*. For example, the frequency of $CD1a^+$ DCs is reduced in VZV infected skin epidermis compared to uninfected skin, suggesting a VZV driven alteration in the distribution of DC in response to VZV infection of the skin (Nikkels *et al.*, 2004). In a separate study, plasmacytoid DC (PDC) were shown to infiltrate into the dermis of varicella lesions, suggesting that these cells contribute to the immune control of VZV infection (Gerlini *et al.*, 2006). More recent work on the DC subsets in skin biopsies from varicella and herpes zoster cases (Huch *et al.*, 2010) suggest that, in comparison to normal uninfected skin, the proportion of cells expressing DC-SIGN (a dermal DC marker) or DC-LAMP and CD83 (as mature DC markers) were not significantly altered, but that the frequency of Langerhans cells was significantly decreased in VZV infected skin, concomitant with a striking influx of PDC into VZV infected skin. The authors suggested that this loss was likely a consequence of migration of these cells to distal sites, such as lymph nodes. Within infected skin, the LCs and PDC were closely associated with VZV antigen positive cells, and a small proportion of both LC and PDC showed evidence of VZV infection. Despite only sporadic detection of VZV infected PDC and LC during natural cutaneous infection, these rare events may still be important, considering that other cell types that play crucial roles in the course of natural VZV infection are infected at very low levels. In this respect, the proportion of VZV infected lymphocytes in peripheral blood during natural VZV infection is very small, with estimates in the range of 1 in 100,000 PBMCs from healthy varicella patients. Yet, peripheral blood T cells are currently considered critical to transport virus to distal sites during VZV pathogenesis (Koropchak *et al.*, 1989, 1991).

In summary, VZV appears to induce changes in multiple DC subsets in the skin of individuals suffering from primary and recurrent VZV disease, and implicates DC as playing important roles in VZV pathogenesis. Furthermore, the capacity of VZV to infect and impair function of different DC subsets highlights VZV mediated immune control of these cells and the redirection of them to benefit viral infection and pathogenesis.

Concluding remarks

It is clear that innate and VZV specific immunity, particularly CMI, are efficient and capable of controlling the VZV primary infection, inducing the protective state, and blocking clinical signs of reactivation permanently in most of the population. For the unfortunate fraction that will suffer zoster, a drop in CMI is the most common predisposing factor, and this is clearly remediable for most through vaccination and boosting of CMI. Furthermore, immunity induced by the remarkably successful vaccination strategies prevent primary disease and greatly reduce incidence and severity of reactivated disease, zoster. It is thus likely that the VZV vaccines and the immunity they induce will eventually lead to an eventual dramatic reduction of most VZV wild types circulating in the population. There is already a significant fraction of the population in the USA who has never experienced varicella, and eventually this will supersede the fraction of adults harbouring wild type VZV. However, several issues are in need of further study regarding VZV immunity. These include innate immunity, zoster incidence in the era of vaccination, VZV immune evasion genes and immunity and improved vaccine candidates.

Immunity and reactivation

Exactly how do the varying arms of the immune response interact with the host neuron harbouring the VZV latent genome to control reactivation? The mechanisms and contributing important issues governing the VZV latent state, and how such events then interplay with the immune system, suffer from a paucity of experimental systems in which they can be experimentally assessed. Unlike HSV-1, VZV exhibits high host specificity, and there is a paucity of usable animal models for study of the interaction of the latent state with immunity. The fact that we cannot yet efficiently experimentally induce reactivation of VZV from human ganglionic tissues or from small immune competent animal models make VZV reactivation somewhat a mysterious black box. If such an animal model were to become available, it may open the gateway to assess its interaction with adaptive immunity. One possible model might be developed from the observation that enteric guinea pig neurons infected with VZV reactivate following adenovirus transduced expression of ORF61 or ICP0 (Gershon *et al.*, 2008b). Application of herpesviral modulators of gene expression to fresh viable human post mortem tissues might be interesting and revealing. A second model may be refinement of the SCID-human neuron models, which have so far allowed assessment of VZV latency in human neuronal tissues in the absence of immunity, to be combined with a partial reconstitution of human cellular immunity through bone marrow transplant implantation of human progenitor immune cells. There is also need for further study of human autopsy ganglionic tissues with T cells, since human VZV latently infected ganglia must be closely monitored by VZV specific CMI, but the identities of the CMI components have yet to be visualized. Are $CD4^+$ T cells the more important component and are they ganglion resident as seen with HSV-1 latent infections?

Zoster incidence in the era of vaccination

As outlined in the chapter, a critical question set by Hope Simpson's model is whether subclinical boosting of immunity is a contributing factor to zoster incidence. For those harbouring wild type VZV infections, the loss of circulating VZV epidemics resulting from the varicella vaccine has been predicted to lead to earlier incidence of zoster. While many studies have sought to address this issue, it is apparent that the answer is complicated. The recently raised possibility of subclinical VZV reactivation will likely stimulate a number of studies to determine circumstances in which subclinical reactivations occur. Any indications of such increased early zoster incidence in the absence of subclinical reactivation and external exposure might lead to the need for strategies for earlier vaccination for zoster.

VZV immune evasion genes

Modulation of immune function has emerged as a powerful strategy by which viruses evade or delay host defences during critical stages of infection, and VZV clearly employs several strategies to counteract both the innate and adaptive arms. However, few VZV immune evasion strategies have been tied to VZV genes and viral proteins. This may come to light in the very near future, as VZV genetics and derivation and testing of viral deletion mutants have become much more successful and feasible with the development of the VZV BAC and cosmid genetic systems. What may not be straightforward to resolve is where such strategies come to play in VZV pathogenesis. Once again, a good immunocompetent animal model is needed. The SCID hu mouse has been a powerful tool to identify roles for innate immune evasion strategies in increasing replication spread and dissemination of VZV, but where CMI and adaptive immune evasion strategies come to play in the course of VZV infection latency and disease remains to be defined, and will be a significant challenge.

Immunity and improved vaccine candidates

A significant outcome of elucidating mechanism and identifying viral genes that modulate host immune surveillance and infection will be for development of a better 'second generation' vaccine against VZV disease. The current vaccine is not ideal: it is a mixture, has a poorly understood basis for attenuation, and is not suitable for certain groups of high risk patients.. The requirement for multiple dosing to get the level of immunity to protective levels, and the fact that the vaccine has been shown to establish latency and can cause zoster, all point to a desire for an improved engineered vaccine with specifically targeted modifications. The 45 mutations accumulating in the vaccine strain (compared to its parental virus) affect multiple genes in ways that have not yet been determined. Once we have established how the mutations contribute to VZV growth and robustness, vaccines can be tailored to contain defined mutations or deletions so that attenuation can be correlated with maximum immunogenicity and maximum safety to form an improved vaccine. The identification of clonal breakthrough vaccine viruses may yield a partial picture of the important changes associated with attenuation. The ideal vaccine should be one that replicates at the inoculation site without causing a lesion, lacks viral genes that permit evasion of host defence mechanisms or infectivity for DCs, and yet induces immunity as effectively as natural infection.

The authors would like to highlight the very recent and important report showing that the main VZV regulatory protein, IE62, has been shown to modulate the phoshorylation and activity of the IRF3 factor, which lies upstream of the induction of IFN response gene expression. Thus one of the first genes made by VZV in infection contributes to the evasion of innate responses in an unprecedented manner (Sen *et al.*, 2010).

Acknowledgements

PRK would like to acknowledge support by NIH grants EY07897, NS064022, a CORE grant EY08098 and funds from the Eye & Ear foundation of Pittsburgh and the Research to Prevent Blindness Inc. AA would like to acknowledge the support from the Australian National Health and Medical Research Council.

References

Abendroth, A., and Arvin, A.M. (2001). Immune evasion as a pathogenic mechanism of varicella zoster virus. Semin. Immunol. *13*, 27–39.

Abendroth, A., Lin, I., Slobedman, B., Ploegh, H., and Arvin, A.M. (2001a). Varicella-zoster virus retains major histocompatibility complex class I proteins in the Golgi compartment of infected cells. J. Virol. *75*, 4878–4888.

Abendroth, A., Morrow, G., Cunningham, A.L., and Slobedman, B. (2001b). Varicella-zoster virus infection of human dendritic cells and transmission to T cells: implications for virus dissemination in the host. J. Virol. *75*, 6183–6192.

Abendroth, A., Slobedman, B., Lee, E., Mellins, E., Wallace, M., and Arvin, A.M. (2000). Modulation of major histocompatibility class II protein expression by varicella-zoster virus. J. Virol. *74*, 1900–1907.

Ambagala, A.P., Bosma, T., Ali, M.A., Poustovoitov, M., Chen, J.J., Gershon, M.D., Adams, P.D., and Cohen, J.I. (2009). Varicella-zoster virus immediate-early 63 protein interacts with human antisilencing function 1 protein and alters its ability to bind histones h3.1 and h3.3. J. Virol. *83*, 200–209.

Ambagala, A.P., and Cohen, J.I. (2007). Varicella-Zoster virus IE63, a major viral latency protein, is required to inhibit the alpha interferon-induced antiviral response. J. Virol. *81*, 7844–7851.

Ampofo, K., Saiman, L., LaRussa, P., Steinberg, S., Annunziato, P., and Gershon, A. (2002). Persistence of immunity to live attenuated varicella vaccine in healthy adults. Clin. Infect. Dis. *34*, 774–779.

Arvin, A.M. (2006). Investigations of the pathogenesis of Varicella zoster virus infection in the SCIDhu mouse model. Herpes *13*, 75–80.

Arvin, A.M., Koropchak, C.M., Williams, B.R., Grumet, F.C., and Foung, S.K. (1986). Early immune response in healthy and immunocompromised subjects with primary varicella-zoster virus infection. J. Infect. Dis. *154*, 422–429.

Arvin, A.M., Schmidt, N.J., Cantell, K., and Merigan, T.C. (1982). Alpha interferon administration to infants with congenital rubella. Antimicrob. Agents Chemother. *21*, 259–261.

Arvin, A.M., Sharp, M., Moir, M., Kinchington, P.R., Sadeghi-Zadeh, M., Ruyechan, W.T., and Hay, J. (2002). Memory cytotoxic T cell responses to viral tegument and regulatory proteins encoded by open reading frames 4, 10, 29, and 62 of varicella-zoster virus. Viral Immunol. *15*, 507–516.

Arvin, A.M., Sharp, M., Smith, S., Koropchak, C.M., Diaz, P.S., Kinchington, P., Ruyechan, W., and Hay, J. (1991). Equivalent recognition of a varicella-zoster virus immediate early protein (IE62) and glycoprotein I by cytotoxic T lymphocytes of either. $CD4^+$ or $CD8^+$ phenotype. J. Immunol. *146*, 257–264.

Asanuma, H., Sharp, M., Maecker, H.T., Maino, V.C., and Arvin, A.M. (2000). Frequencies of memory T cells specific for varicella-zoster virus, herpes simplex virus, and cytomegalovirus by intracellular detection of cytokine expression. J. Infect. Dis. *181*, 859–866.

Balachandra, K., Thawaranantha, D., Ayuthaya, P.I., Bhumisawasdi, J., Shiraki, K., and Yamanishi, K. (1994). Effects of human alpha, beta and gamma interferons on varicella zoster virus *in vitro*. Southeast Asian J. Trop. Med. Public Health *25*, 252–257.

Banchereau, J., Briere, F., Caux, C., Davoust, J., Lebecque, S., Liu, Y.J., Pulendran, B., and Palucka, K. (2000). Immunobiology of dendritic cells. Annu. Rev. Immunol. *18*, 767–811.

Banchereau, J., and Steinman, R.M. (1998). Dendritic cells and the control of immunity. Nature *392*, 245–252.

Bergen, R.E., Diaz, P.S., and Arvin, A.M. (1990). The immunogenicity of the Oka/Merck varicella vaccine in relation to infectious varicella-zoster virus and relative viral antigen content. J. Infect. Dis. *162*, 1049–1054.

Black, A.P., Jones, L., Malavige, G.N., and Ogg, G.S. (2009). Immune evasion during varicella zoster virus infection of keratinocytes. Clin. Exp. Dermatol. *34*, e941–944.

Bowden, R.A., Levin, M.J., Giller, R.H., Tubergen, D.G., and Hayward, A.R. (1985). Lysis of varicella zoster virus infected cells by lymphocytes from normal humans and immunosuppressed pediatric leukaemic patients. Clin. Exp. Immunol. *60*, 387–395.

Brisson, M., Gay, N.J., Edmunds, W.J., and Andrews, N.J. (2002). Exposure to varicella boosts immunity to herpes-zoster: implications for mass vaccination against chickenpox. Vaccine *20*, 2500–2507.

Buda, K., Tubergen, D.G., and Levin, M.J. (1996). The frequency and consequences of varicella exposure and varicella infection in children receiving maintenance therapy for acute lymphoblastic leukemia. J. Pediatr. Hematol. Oncol. *18*, 106–112.

Chaves, S.S., Gargiullo, P., Zhang, J.X., Civen, R., Guris, D., Mascola, L., and Seward, J.F. (2007). Loss of vaccine-induced immunity to varicella over time. N. Engl. J. Med. *356*, 1121–1129.

Chaves, S.S., Zhang, J., Civen, R., Watson, B.M., Carbajal, T., Perella, D., and Seward, J.F. (2008). Varicella disease among vaccinated persons: clinical and epidemiological characteristics, 1997–2005. J. Infect. Dis. *197* Suppl 2, S127–131.

Chou, J., and Roizman, B. (1994). Herpes simplex virus 1 gamma(1)34.5 gene function, which blocks the host response to infection, maps in the homologous domain of the genes expressed during growth arrest and DNA damage. Proc. Natl. Acad. Sci. U.S.A. *91*, 5247–5251.

Civen, R., Chaves, S.S., Jumaan, A., Wu, H., Mascola, L., Gargiullo, P., and Seward, J.F. (2009). The incidence and clinical characteristics of herpes zoster among children and adolescents after implementation of varicella vaccination. Pediatr. Infect. Dis. J. *28*, 954–959.

Cohen, J.I. (1998). Infection of cells with varicella-zoster virus down-regulates surface expression of class I major histocompatibility complex antigens. J. Infect. Dis. *177*, 1390–1393.

Cohen, J.I., Cox, E., Pesnicak, L., Srinivas, S., and Krogmann, T. (2004). The varicella-zoster virus open reading frame 63 latency-associated protein is critical for establishment of latency. J. Virol. *78*, 11833–11840.

Cohen, J.I., Krogmann, T., Bontems, S., Sadzot-Delvaux, C., and Pesnicak, L. (2005a). Regions of the varicella-zoster virus open reading frame 63 latency-associated protein important for replication *in vitro* are also critical for efficient establishment of latency. J. Virol. *79*, 5069–5077.

Cohen, J.I., Krogmann, T., Pesnicak, L., and Ali, M.A. (2007). Absence or overexpression of the Varicella-Zoster Virus (VZV) ORF29 latency-associated protein impairs late gene expression and reduces VZV latency in a rodent model. J. Virol. *81*, 1586–1591.

Cohen, J.I., Krogmann, T., Ross, J.P., Pesnicak, L., and Prikhod'ko, E.A. (2005b). Varicella-zoster virus ORF4 latency-associated protein is important for establishment of latency. J. Virol. *79*, 6969–6975.

Cohrs, R.J., Barbour, M., and Gilden, D.H. (1996). Varicella-zoster virus (VZV) transcription during latency in human ganglia: detection of transcripts mapping to genes 21, 29, 62, and 63 in a cDNA library enriched for VZV RNA. J. Virol. *70*, 2789–2796.

Cohrs, R.J., Barbour, M.B., Mahalingam, R., Wellish, M., and Gilden, D.H. (1995). Varicella-zoster virus (VZV) transcription during latency in human ganglia: prevalence of VZV gene 21 transcripts in latently infected human ganglia. J. Virol. *69*, 2674–2678.

Cohrs, R.J., and Gilden, D.H. (2007). Prevalence and abundance of latently transcribed varicella-zoster virus genes in human ganglia. J. Virol. *81*, 2950–2956.

Cohrs, R.J., Mehta, S.K., Schmid, D.S., Gilden, D.H., and Pierson, D.L. (2008). Asymptomatic reactivation and shed of infectious varicella zoster virus in astronauts. J. Med. Virol. *80*, 1116–1122.

Desloges, N., Rahaus, M., and Wolff, M.H. (2005a). Role of the protein kinase PKR in the inhibition of varicella-zoster virus replication by beta interferon and gamma interferon. J. Gen. Virol. *86*, 1–6.

Desloges, N., Rahaus, M., and Wolff, M.H. (2005b). Varicella-zoster virus does not significantly induce cell defence mechanism mediated by the 2–5A/RNase L pathway during its replication cycle. Med. Microbiol. Immunol. *194*, 25–31.

Di Valentin, E., Bontems, S., Habran, L., Jolois, O., Markine-Goriaynoff, N., Vanderplasschen, A., Sadzot-Delvaux, C., and Piette, J. (2005). Varicella-zoster virus IE63 protein represses the basal transcription machinery by disorganizing the pre-initiation complex. Biol. Chem. *386*, 255–267.

Eisfeld, A.J., Turse, S.E., Jackson, S.A., Lerner, E.C., and Kinchington, P.R. (2006). Phosphorylation of the varicella-zoster virus (VZV) major transcriptional regulatory protein IE62 by the VZV open reading frame 66 protein kinase. J. Virol. *80*, 1710–1723.

Eisfeld, A.J., Yee, M.B., Erazo, A., Abendroth, A., and Kinchington, P.R. (2007). Downregulation of class I major histocompatibility complex surface expression by varicella-zoster virus involves open reading frame 66 protein kinase-dependent and -independent mechanisms. J. Virol. *81*, 9034–9049.

Erazo, A., Yee, M.B., Osterrieder, N., and Kinchington, P.R. (2008). Varicella-zoster virus open reading frame 66 protein kinase is required for efficient viral growth in primary human corneal stromal fibroblast cells. J. Virol. *82*, 7653–7665.

Etzioni, A., Eidenschenk, C., Katz, R., Beck, R., Casanova, J.L., and Pollack, S. (2005). Fatal varicella associated with selective natural killer cell deficiency. J. Pediatr. *146*, 423–425.

Freeman, M.L., Sheridan, B.S., Bonneau, R.H., and Hendricks, R.L. (2007). Psychological stress compromises CD8+ T cell control of latent herpes simplex virus type 1 infections. J. Immunol. *179*, 322–328.

Garry, E.M., Delaney, A., Anderson, H.A., Sirinathsinghji, E.C., Clapp, R.H., Martin, W.J., Kinchington, P.R., Krah, D.L., Abbadie, C., and Fleetwood-Walker, S.M. (2005). Varicella zoster virus induces neuropathic changes in rat dorsal root ganglia and behavioral reflex sensitisation that is attenuated by gabapentin or sodium channel blocking drugs. Pain *118*, 97–111.

Gary, L., Gilden, D.H., and Cohrs, R.J. (2006). Epigenetic regulation of varicella-zoster virus open reading frames 62 and 63 in latently infected human trigeminal ganglia. J. Virol. *80*, 4921–4926.

Gerlini, G., Mariotti, G., Bianchi, B., and Pimpinelli, N. (2006). Massive recruitment of type I interferon producing plasmacytoid dendritic cells in varicella skin lesions. J. Invest. Dermatol. *126*, 507–509.

Gershon, A.A., Arvin, A.M., Levin, M.J., Seward, J.F., and Schmid, D.S. (2008a). Varicella vaccine in the United States: a decade of prevention and the way forward. J. Infect. Dis. *197* Suppl 2, S39–40.

Gershon, A.A., Arvin, A.M., and Shapiro, E. (2007). Varicella vaccine. N. Engl. J. Med. 356, 2648–2649; author reply 2648–2649.

Gershon, A.A., Chen, J., and Gershon, M.D. (2008b). A model of lytic, latent, and reactivating varicella-zoster virus infections in isolated enteric neurons. J. Infect. Dis. *197* Suppl 2, S61–65.

Gershon, A.A., M., T., and J., S. (2008c). Live attenuated varicella vaccine. In Vaccines, P. S, O. W, and O. P, eds. (Philadelphia, WB Saunders).

Gershon, A.A., Mervish, N., LaRussa, P., Steinberg, S., Lo, S.H., Hodes, D., Fikrig, S., Bonagura, V., and Bakshi, S. (1997). Varicella-zoster virus infection in children with underlying human immunodeficiency virus infection. J. Infect. Dis. *176*, 1496–1500.

Gershon, A.A., and Steinberg, S.P. (1979). Cellular and humoral immune responses to varicella-zoster virus in immunocompromised patients during and after varicella-zoster infections. Infect. Immun. *25*, 170–174.

Ghosh, S., and Hayden, M.S. (2008). New regulators of NF-kappaB in inflammation. Nat. Rev. Immunol. *8*, 837–848.

Gould, P.L., Leung, J., Scott, C., Schmid, D.S., Deng, H., Lopez, A., Chaves, S.S., Reynolds, M., Gladden, L., Harpaz, R., *et al.* (2009). An outbreak of varicella in elementary school children with two-dose varicella vaccine recipients–Arkansas, 2006. Pediatr. Infect. Dis. J. *28*, 678–681.

Hasnie, F.S., Breuer, J., Parker, S., Wallace, V., Blackbeard, J., Lever, I., Kinchington, P.R., Dickenson, A.H., Pheby, T., and Rice, A.S. (2007). Further. characterization of a rat model of varicella zoster virus-associated pain: Relationship between mechanical hypersensitivity and anxiety-related behavior, and the influence of analgesic drugs. Neuroscience *144*, 1495–1508.

Hata, A., Asanuma, H., Rinki, M., Sharp, M., Wong, R.M., Blume, K., and Arvin, A.M. (2002). Use of an inactivated varicella vaccine in recipients of hematopoietic-cell transplants. N. Engl. J. Med. *347*, 26–34.

Hata, A., Zerboni, L., Sommer, M., Kaspar, A.A., Clayberger, C., Krensky, A.M., and Arvin, A.M. (2001). Granulysin blocks replication of varicella-zoster virus and triggers apoptosis of infected cells. Viral Immunol. *14*, 125–133.

Hayden, M.S., and Ghosh, S. (2008). Shared principles in NF-kappaB signaling. Cell *132*, 344–362.

Hayward, A.R., Pontesilli, O., Herberger, M., Laszlo, M., and Levin, M. (1986). Specific lysis of varicella zoster virus-infected B lymphoblasts by human T cells. J. Virol. *58*, 179–184.

He, B., Gross, M., and Roizman, B. (1997). The gamma(1)34.5 protein of herpes simplex virus 1 complexes with protein phosphatase 1alpha to dephosphorylate the alpha subunit of the eukaryotic translation initiation factor 2 and preclude the shutoff of protein synthesis by double-stranded RNA-activated protein kinase. Proc. Natl. Acad. Sci. U.S.A. *94*, 843–848.

Hill, A., Jugovic, P., York, I., Russ, G., Bennink, J., Yewdell, J., Ploegh, H., and Johnson, D. (1995). Herpes simplex virus turns off the TAP to evade host immunity. Nature *375*, 411–415.

Hood, C., Cunningham, A.L., Slobedman, B., Arvin, A.M., Sommer, M.H., Kinchington, P.R., and Abendroth, A. (2006). Varicella-zoster virus ORF63 inhibits apoptosis of primary human neurons. J. Virol. *80*, 1025–1031.

Hood, C., Cunningham, A.L., Slobedman, B., Boadle, R.A., and Abendroth, A. (2003). Varicella-zoster virus-infected human sensory neurons are resistant to apoptosis, yet human foreskin fibroblasts are susceptible: evidence for a cell-type-specific apoptotic response. J. Virol. 77, 12852–12864.

Hoover, S.E., Cohrs, R.J., Rangel, Z.G., Gilden, D.H., Munson, P., and Cohen, J.I. (2006). Downregulation of varicella-zoster virus (VZV) immediate-early ORF62 transcription by VZV ORF63 correlates with virus replication *in vitro* and with latency. J. Virol. *80*, 3459–3468.

Hope-Simpson, R.E. (1965). The nature of herpes zoster: a long-term study and a new hypothesis. Proc. R. Soc. Med. *58*, 9–20.

Hu, H., and Cohen, J.I. (2005). Varicella-zoster virus open reading frame 47 (ORF47) protein is critical for virus replication in dendritic cells and for spread to other. cells. Virology 337, 304–311.

Huch, J., Cunningham, A., Arvin, A., Nasr, N., Santegoets, S., Slobedman, E., Slobedman, B., and Abendroth, A. (2010a). Impact of varicella zoster virus on dendritic cell subsets in human skin during natural infection. J. Virol. *84*, 4060–4072.

Ito, M., Watanabe, M., Kamiya, H., and Sakurai, M. (1996). Inhibition of natural killer (NK) cell activity against varicella-zoster virus (VZV)-infected fibroblasts and lymphocyte activation in response to VZV antigen by nitric oxide-releasing agents. Clin. Exp. Immunol. *106*, 40–44.

Jenkins, D.E., Redman, R.L., Lam, E.M., Liu, C., Lin, I., and Arvin, A.M. (1998). Interleukin (IL)–10, IL-12, and interferon-gamma production in primary and memory immune responses to varicella-zoster virus. J. Infect. Dis. *178*, 940–948.

Jones, J.O., and Arvin, A.M. (2005). Viral and cellular gene transcription in fibroblasts infected with small plaque mutants of varicella-zoster virus. Antiviral Res. *68*, 56–65.

Jones, J.O., and Arvin, A.M. (2006). Inhibition of the NF-kappaB pathway by varicella-zoster virus *in vitro* and in human epidermal cells *in vivo*. J. Virol. *80*, 5113–5124.

Jones, L., Black, A.P., Malavige, G.N., and Ogg, G.S. (2006). Persistent high frequencies of varicella-zoster virus ORF4 protein-specific $CD4^+$ T cells after primary infection. J. Virol. *80*, 9772–9778.

Jones, L., Black, A.P., Malavige, G.N., and Ogg, G.S. (2007). Phenotypic analysis of human $CD4^+$ T cells specific for immediate-early 63 protein of varicella-zoster virus. Eur. J. Immunol. *37*, 3393–3403.

Klagge, I.M., and Schneider-Schaulies, S. (1999). Virus interactions with dendritic cells. J. Gen. Virol. *80*, 823–833.

Klein, N.P., Holmes, T.H., Sharp, M.A., Heineman, T.C., Schleiss, M.R., Bernstein, D.I., Kemble, G., Arvin, A.M., and Dekker, C.L. (2006). Variability and gender differences in memory T cell immunity to varicella-zoster virus in healthy adults. Vaccine *24*, 5913–5918.

Knickelbein, J.E., Hendricks, R.L., and Charukamnoetkanok, P. (2009). Management of herpes simplex virus stromal keratitis: an evidence-based review. Surv. Ophthalmol. *54*, 226–234.

Koppers-Lalic, D., Verweij, M.C., Lipinska, A.D., Wang, Y., Quinten, E., Reits, E.A., Koch, J., Loch, S., Marcondes Rezende, M., Daus, F., *et al.* (2008). Varicellovirus UL 49.5 proteins differentially affect the function of the transporter associated with antigen processing, TAP. PLoS Pathog. *4*, e1000080.

Koropchak, C.M., Graham, G., Palmer, J., Winsberg, M., Ting, S.F., Wallace, M., Prober, C.G., and Arvin, A.M. (1991). Investigation of varicella-zoster virus infection by polymerase chain reaction in the immunocompetent host with acute varicella. J. Infect. Dis. *163*, 1016–1022.

Koropchak, C.M., Solem, S.M., Diaz, P.S., and Arvin, A.M. (1989). Investigation of varicella-zoster virus infection of lymphocytes by *in situ* hybridization. J. Virol. *63*, 2392–2395.

Krause, P.R., and Klinman, D.M. (2000). Varicella vaccination: evidence for frequent reactivation of the vaccine strain in healthy children. Nat. Med. *6*, 451–454.

Ku, C.C., Besser, J., Abendroth, A., Grose, C., and Arvin, A.M. (2005). Varicella-Zoster virus pathogenesis and immunobiology: new concepts emerging from investigations with the SCIDhu mouse model. J. Virol. *79*, 2651–2658.

Ku, C.C., Padilla, J.A., Grose, C., Butcher, E.C., and Arvin, A.M. (2002). Tropism of varicella-zoster virus for human tonsillar $CD4(^+)$ T lymphocytes that express activation, memory, and skin homing markers. J. Virol. *76*, 11425–11433.

Ku, C.C., Zerboni, L., Ito, H., Graham, B.S., Wallace, M., and Arvin, A.M. (2004a). Varicella-zoster virus transfer to skin by T Cells and modulation of viral replication by epidermal cell interferon-alpha. J. Exp. Med. *200*, 917–925.

Kuter, B., Matthews, H., Shinefield, H., Black, S., Dennehy, P., Watson, B., Reisinger, K., Kim, L.L., Lupinacci, L., Hartzel, J., *et al.* (2004). Ten year follow-up of healthy children who received one or two injections of varicella vaccine. Pediatr. Infect. Dis. J. *23*, 132–137.

Levin, M.J., Oxman, M.N., Zhang, J.H., Johnson, G.R., Stanley, H., Hayward, A.R., Caulfield, M.J., Irwin, M.R., Smith, J.G., Clair, J., *et al.* (2008). Varicella-zoster virus-specific immune responses in elderly recipients of a herpes zoster vaccine. J. Infect. Dis. *197*, 825–835.

Levy, O., Orange, J.S., Hibberd, P., Steinberg, S., LaRussa, P., Weinberg, A., Wilson, S.B., Shaulov, A., Fleisher, G., Geha, R.S., *et al.* (2003). Disseminated varicella infection due to the vaccine strain of varicella-zoster virus, in a patient with a novel deficiency in natural killer T cells. J. Infect. Dis. *188*, 948–953.

Li, S., Chan, I.S., Matthews, H., Heyse, J.F., Chan, C.Y., Kuter, B.J., Kaplan, K.M., Vessey, S.J., and Sadoff, J.C. (2002). Inverse relationship between six week postvaccination varicella antibody response to vaccine and likelihood of long term breakthrough infection. Pediatr. Infect. Dis. J. *21*, 337–342.

Liang, L., and Roizman, B. (2008). Expression of gamma interferon-dependent genes is blocked independently by virion host shutoff RNase and by US3 protein kinase. J. Virol. *82*, 4688–4696.

Lopez, A.S., Guris, D., Zimmerman, L., Gladden, L., Moore, T., Haselow, D.T., Loparev, V.N., Schmid, D.S., Jumaan, A.O., and Snow, S.L. (2006). One dose of varicella vaccine does not prevent school outbreaks: is it time for a second dose? Pediatrics *117*, e1070–1077.

Lungu, O., Annunziato, P.W., Gershon, A., Staugaitis, S.M., Josefson, D., LaRussa, P., and Silverstein, S.J. (1995). Reactivated and latent varicella-zoster virus in human dorsal root ganglia. Proc. Natl. Acad. Sci. U.S.A. *92*, 10980–10984.

Lungu, O., Panagiotidis, C.A., Annunziato, P.W., Gershon, A.A., and Silverstein, S.J. (1998). Aberrant intracellular localization of Varicella-Zoster virus regulatory proteins during latency. Proc. Natl. Acad. Sci. U.S.A. *95*, 7080–7085.

Mahalingam, R., Wellish, M., Cohrs, R., Debrus, S., Piette, J., Rentier, B., and Gilden, D.H. (1996). Expression of protein encoded by varicella-zoster virus open reading frame 63 in latently infected human ganglionic neurons. Proc. Natl. Acad. Sci. U.S.A. *93*, 2122–2124.

Malavige, G.N., Jones, L., Black, A.P., and Ogg, G.S. (2008). Varicella zoster virus glycoprotein E-specific $CD4^+$ T cells show evidence of recent activation and effector differentiation, consistent with frequent exposure to replicative cycle antigens in healthy immune donors. Clin. Exp. Immunol. *152*, 522–531.

Marin, M., Guris, D., Chaves, S.S., Schmid, S., and Seward, J.F. (2007). Prevention of varicella: recommendations of the Advisory Committee on Immunization Practices (ACIP). MMWR Recomm Rep *56*, 1–40.

Marin, M., Meissner, H.C., and Seward, J.F. (2008). Varicella prevention in the United States: a review of successes and challenges. Pediatrics *122*, e744–751.

Mehta, S.K., Cohrs, R.J., Forghani, B., Zerbe, G., Gilden, D.H., and Pierson, D.L. (2004). Stress-induced subclinical reactivation of varicella zoster virus in astronauts. J. Med. Virol. *72*, 174–179.

Mehta, S.K., Tyring, S.K., Gilden, D.H., Cohrs, R.J., Leal, M.J., Castro, V.A., Feiveson, A.H., Ott, C.M., and Pierson, D.L. (2008). Varicella-zoster virus in the saliva of patients with herpes zoster. J. Infect. Dis. *197*, 654–657.

Meier, J.L., Holman, R.P., Croen, K.D., Smialek, J.E., and Straus, S.E. (1993). Varicella-zoster virus transcription in human trigeminal ganglia. Virology *193*, 193–200.

Michalik, D.E., Steinberg, S.P., Larussa, P.S., Edwards, K.M., Wright, P.F., Arvin, A.M., Gans, H.A., and Gershon, A.A. (2008). Primary vaccine failure after 1 dose of varicella vaccine in healthy children. J. Infect. Dis. *197*, 944–949.

Milikan, J.C., Baarsma, G.S., Kuijpers, R.W., Osterhaus, A.D., and Verjans, G.M. (2009). Human ocular-derived virus-specific $CD4^+$ T cells control varicella zoster virus replication in human retinal pigment epithelial cells. Invest. Ophthalmol. Vis. Sci. *50*, 743–751.

Milikan, J.C., Kinchington, P.R., Baarsma, G.S., Kuijpers, R.W., Osterhaus, A.D., and Verjans, G.M. (2007). Identification of viral antigens recognized by ocular infiltrating T cells from patients with varicella zoster virus-induced uveitis. Invest. Ophthalmol. Vis. Sci. *48*, 3689–3697.

Miller, E., Cradock-Watson, J.E., and Ridehalgh, M.K. (1989). Outcome in newborn babies given anti-varicella-zoster immunoglobulin after perinatal maternal infection with varicella-zoster virus. Lancet *2*, 371–373.

Moffat, J.F., Zerboni, L., Sommer, M.H., Heineman, T.C., Cohen, J.I., Kaneshima, H., and Arvin, A.M. (1998). The ORF47 and ORF66 putative protein kinases of varicella-zoster virus determine tropism for human T cells and skin in the SCID-hu mouse. Proc. Natl. Acad. Sci. U.S.A. *95*, 11969–11974.

Morrow, G., Slobedman, B., Cunningham, A.L., and Abendroth, A. (2003a). Varicella-zoster virus productively infects mature dendritic cells and alters their immune function. J. Virol. *77*, 4950–4959.

Nader, S., Bergen, R., Sharp, M., and Arvin, A.M. (1995). Age-related differences in cell-mediated immunity to varicella-zoster virus among children and adults immunized with live attenuated varicella Vaccine. J. Infect. Dis. *171*, 13–17.

Nguyen, M.L., Kraft, R.M., and Blaho, J.A. (2007). Susceptibility of cancer cells to herpes simplex virus-dependent apoptosis. J. Gen. Virol. *88*, 1866–1875.

Nikkels, A.F., Sadzot-Delvaux, C., and Pierard, G.E. (2004). Absence of intercellular adhesion molecule 1 expression in varicella zoster virus-infected keratinocytes during herpes zoster: another. immune evasion strategy? Am. J. Dermatopathol. *26*, 27–32.

Notarangelo, L.D., and Mazzolari, E. (2006). Natural killer cell deficiencies and severe varicella infection. J. Pediatr. 148, 563–564; author reply 564.

Oxman, M.N. (2009). Herpes zoster pathogenesis and cell-mediated immunity and immunosenescence. J. Am. Osteopath. Assoc. *109*, S13–17.

Oxman, M.N., Levin, M.J., Johnson, G.R., Schmader, K.E., Straus, S.E., Gelb, L.D., Arbeit, R.D., Simberkoff, M.S., Gershon, A.A., Davis, L.E., *et al.* (2005). A vaccine to prevent herpes zoster and postherpetic neuralgia in older adults. N. Engl. J. Med. *352*, 2271–2284.

Patterson-Bartlett, J., Levin, M.J., Lang, N., Schodel, F.P., Vessey, R., and Weinberg, A. (2007). Phenotypic and functional characterization of *ex vivo* T cell responses to the live attenuated herpes zoster vaccine. Vaccine *25*, 7087–7093.

Pevenstein, S.R., Williams, R.K., McChesney, D., Mont, E.K., Smialek, J.E., and Straus, S.E. (1999). Quantitation of latent varicella-zoster virus and herpes simplex virus genomes in human trigeminal ganglia. J. Virol. 73, 10514–10518.

Quinlivan, M.L., Gershon, A.A., Al Bassam, M.M., Steinberg, S.P., LaRussa, P., Nichols, R.A., and Breuer, J. (2007). Natural selection for rash-forming genotypes of the varicella-zoster vaccine virus detected within immunized human hosts. Proc. Natl. Acad. Sci. U.S.A. *104*, 208–212.

Rahaus, M., Desloges, N., and Wolff, M.H. (2007). Varicella-zoster virus requires. a functional PI3K/Akt/GSK-3alpha/beta signaling cascade for efficient replication. Cell Signal *19*, 312–320.

Reichelt, M., Zerboni, L., and Arvin, A.M. (2008). Mechanisms of varicella-zoster virus neuropathogenesis in human dorsal root ganglia. J. Virol. *82*, 3971–3983.

Reynolds, M.A., Chaves, S.S., Harpaz, R., Lopez, A.S., and Seward, J.F. (2008). The impact of the varicella vaccination program on herpes zoster epidemiology in the United States: a review. J. Infect. Dis. *197* Suppl 2, S224–227.

Roxas, M. (2006). Herpes zoster and postherpetic neuralgia: diagnosis and therapeutic considerations. Altern. Med. Rev. *11*, 102–113.

Sadler, A.J., and Williams, B.R. (2008). Interferon-inducible antiviral effectors. Nat. Rev. Immunol. *8*, 559–568.

Sadzot-Delvaux, C., Arvin, A.M., and Rentier, B. (1998). Varicella-zoster virus IE63, a virion component expressed during latency and acute infection, elicits humoral and cellular immunity. J. Infect. Dis. *178* Suppl 1, S43–47.

Sadzot-Delvaux, C., Kinchington, P.R., Debrus, S., Rentier, B., and Arvin, A.M. (1997). Recognition of the latency-associated immediate early protein IE63 of varicella-zoster virus by human memory T lymphocytes. J. Immunol. *159*, 2802–2806.

Sadzot-Delvaux, C., Thonard, P., Schoonbroodt, S., Piette, J., and Rentier, B. (1995). Varicella-zoster virus induces apoptosis in cell culture. J. Gen. Virol. *76*, 2875–2879.

Schaap-Nutt, A., Sommer, M., Che, X., Zerboni, L., and Arvin, A.M. (2006). ORF66 protein kinase function is required for T-cell tropism of varicella-zoster virus *in vivo*. J. Virol. *80*, 11806–11816.

Schaap, A., Fortin, J.F., Sommer, M., Zerboni, L., Stamatis, S., Ku, C.C., Nolan, G.P., and Arvin, A.M. (2005). T-cell tropism and the role of ORF66 protein in pathogenesis of varicella-zoster virus infection. J. Virol. *79*, 12921–12933.

Sen, N., Sommer, M., Che, X., White, K., Ruyechan, W.T., and Arvin, A.M. (2010). Varicella-zoster virus immediate-early protein 62 blocks interferon regulatory factor 3 (IRF3) phosphorylation at key serine residues: a novel mechanism of IRF3 inhibition among herpesviruses. J. Virol. *84*, 9240–9253.

Seward, J.F., Marin, M., and Vazquez, M. (2008). Varicella vaccine effectiveness in the US vaccination program: a review. J. Infect. Dis. *197 Suppl 2*, S82–89.

Seward, J.F., Watson, B.M., Peterson, C.L., Mascola, L., Pelosi, J.W., Zhang, J.X., Maupin, T.J., Goldman, G.S., Tabony, L.J., Brodovicz, K.G., *et al.* (2002). Varicella disease after introduction of varicella vaccine in the United States, 1995–2000. JAMA *287*, 606–611.

Sharp, M., Terada, K., Wilson, A., Nader, S., Kinchington, P.E., Ruyechan, W.T., Hay, J., and Arvin, A.M. (1992). Kinetics and viral protein specificity of the cytotoxic T lymphocyte response in healthy adults immunized with live attenuated varicella vaccine. J. Infect. Dis. *165*, 852–858.

Sheridan, B.S., Knickelbein, J.E., and Hendricks, R.L. (2007). CD8 T cells and latent herpes simplex virus type 1: keeping the peace in sensory ganglia. Expert Opin. Biol. Ther. *7*, 1323–1331.

Stallings, C.L., Duigou, G.J., Gershon, A.A., Gershon, M.D., and Silverstein, S.J. (2006). The cellular localization pattern of Varicella-Zoster virus ORF29p is influenced by proteasome-mediated degradation. J. Virol. *80*, 1497–1512.

Stallings, C.L., and Silverstein, S.J. (2006). Posttranslational modification and cell type-specific degradation of varicella-zoster virus ORF29p. J. Virol. *80*, 10836–10846.

Takahashi, M., Otsuka, T., Okuno, Y., Asano, Y., and Yazaki, T. (1974). Live vaccine used to prevent the spread of varicella in children in hospital. Lancet *2*, 1288–1290.

Takayama, N., Takayama, M., and Takita, J. (2000). Herpes zoster in healthy children immunized with varicella vaccine. Pediatr. Infect. Dis. J. *19*, 169–170.

Tomazin, R., Hill, A.B., Jugovic, P., York, I., van Endert, P., Ploegh, H.L., Andrews, D.W., and Johnson, D.C. (1996). Stable binding of the herpes simplex virus ICP47 protein to the peptide binding site of TAP. EMBO J. *15*, 3256–3266.

Tseng, H.F., Smith, N., Marcy, S.M., Sy, L.S., and Jacobsen, S.J. (2009). Incidence of herpes zoster among children vaccinated with varicella vaccine in a prepaid health care plan in the United States, 2002–2008. Pediatr. Infect. Dis. J. *28*, 1069–1072.

Verjans, G.M., Hintzen, R.Q., van Dun, J.M., Poot, A., Milikan, J.C., Laman, J.D., Langerak, A.W., Kinchington, P.R., and Osterhaus, A.D. (2007). Selective retention of herpes simplex virus-specific T cells in latently infected human trigeminal ganglia. Proc. Natl. Acad. Sci. U.S.A. *104*, 3496–3501.

Vleck, S.E., Oliver, S.L., Reichelt, M., Rajamani, J., Zerboni, L., Jones, C., Zehnder, J., Grose, C., and Arvin, A.M. (2010). Anti-glycoprotein H antibody impairs the pathogenicity of varicella-zoster virus in skin xenografts in the SCID mouse model. J. Virol. *84*, 141–152.

Vossen, M.T., Biezeveld, M.H., de Jong, M.D., Gent, M.R., Baars, P.A., von Rosenstiel, I.A., van Lier, R.A., and Kuijpers, T.W. (2005). Absence of circulating natural killer and primed. $CD8^+$ cells in life-threatening varicella. J. Infect. Dis. *191*, 198–206.

Vossen, M.T., Gent, M.R., Weel, J.F., de Jong, M.D., van Lier, R.A., and Kuijpers, T.W. (2004). Development of virus-specific $CD4^+$ T cells on reexposure to Varicella-Zoster virus. J. Infect. Dis. *190*, 72–82.

Walters, M.S., Kyratsous, C.A., Wan, S., and Silverstein, S. (2008). Nuclear import of the varicella-zoster virus latency-associated protein ORF63 in primary neurons requires. expression of the lytic protein ORF61 and occurs in a proteasome-dependent manner. J. Virol. *82*, 8673–8686.

Wang, K., Lau, T.Y., Morales, M., Mont, E.K., and Straus, S.E. (2005). Laser-capture microdissection: refining estimates of the quantity and distribution of latent herpes simplex virus 1 and varicella-zoster virus DNA in human trigeminal Ganglia at the single-cell level. J. Virol. *79*, 14079–14087.

Watson, B., Boardman, C., Laufer, D., Piercy, S., Tustin, N., Olaleye, D., Cnaan, A., and Starr, S.E. (1995). Humoral and cell-mediated immune responses in healthy children after one or two doses of varicella vaccine. Clin. Infect. Dis. *20*, 316–319.

Watson, B., Gupta, R., Randall, T., and Starr, S. (1994). Persistence of cell-mediated and humoral immune responses in healthy children immunized with live attenuated varicella vaccine. J. Infect. Dis. *169*, 197–199.

Watson, B.A., and Starr, S.E. (1994). Varicella vaccine for healthy children. Lancet *343*, 928–929.

Weinberg, A., Lazar, A.A., Zerbe, G.O., Hayward, A.R., Chan, I.S., Vessey, R., Silber, J.L., Macgregor, R.R., Chan, K., Gershon, A.A., *et al.* (2010). Influence of age and nature of primary infection on varicella-zoster virus-specific cell-mediated immune responses. J. Infect. Dis. *201*, 1024–1030.

Weinberg, A., Zhang, J.H., Oxman, M.N., Johnson, G.R., Hayward, A.R., Caulfield, M.J., Irwin, M.R., Clair, J., Smith, J.G., Stanley, H., *et al.* (2009). Varicella-zoster virus-specific immune responses to herpes zoster in elderly participants in a trial of a clinically effective zoster vaccine. J. Infect. Dis. *200*, 1068–1077.

White, C.J., Kuter, B.J., Hildebrand, C.S., Isganitis, K.L., Matthews, H., Miller, W.J., Provost, P.J., Ellis, R.W., Gerety, R.J., and Calandra, G.B. (1991). Varicella vaccine (VARIVAX) in healthy children and adolescents: results from clinical trials, 1987 to 1989. Pediatrics *87*, 604–610.

White, C.J., Kuter, B.J., Ngai, A., Hildebrand, C.S., Isganitis, K.L., Patterson, C.M., Capra, A., Miller, W.J., Krah, D.L., Provost, P.J., *et al.* (1992). Modified cases of chickenpox after varicella vaccination: correlation of protection with antibody response. Pediatr. Infect. Dis. J. *11*, 19–23.

Xia, D., Srinivas, S., Sato, H., Pesnicak, L., Straus, S.E., and Cohen, J.I. (2003). Varicella-zoster virus open reading frame 21, which is expressed during latency, is essential for virus replication but dispensable for establishment of latency. J. Virol. *77*, 1211–1218.

Zerboni, L., Ku, C.C., Jones, C.D., Zehnder, J.L., and Arvin, A.M. (2005). Varicella-zoster virus infection of human dorsal root ganglia *in vivo*. Proc. Natl. Acad. Sci. U.S.A. *102*, 6490–6495.

Zerboni, L., Nader, S., Aoki, K., and Arvin, A.M. (1998). Analysis of the persistence of humoral and cellular immunity in children and adults immunized with varicella vaccine. J. Infect. Dis. *177*, 1701–1704.

Zerboni, L., Reichelt, M., Jones, C.D., Zehnder, J.L., Ito, H., and Arvin, A.M. (2007). Aberrant infection and persistence of varicella-zoster virus in human dorsal root ganglia *in vivo* in the absence of glycoprotein I. Proc. Natl. Acad. Sci. U.S.A. *104*, 14086–14091.

Zerboni, L., Sobel, R.A., Ramachandran, V., Rajamani, J., Ruyechan, W., Abendroth, A., and Arvin, A. (2010). The expression of varicella-zoster virus immediate early regulatory protein IE63 in neurons of latently infected human sensory ganglia. J. Virol. *84*, 3421–3430.

Zhang, Y., White, C.J., Levin, M., and Hayward, A. (1995). Cytokine production in varicella-zoster virus-stimulated lymphocyte cultures. Neurology *45*, S38–40.

Zhou, F., Harpaz, R., Jumaan, A.O., Winston, C.A., and Shefer, A. (2005). Impact of varicella vaccination on health care utilization. JAMA *294*, 797–802.

Herpesviruses and the Control of Autophagy

21

Philipe A.M. Gobeil and David A. Leib

Abstract

Autophagy is a rapidly growing area of biomedical research with broad relevance to fields including microbiology, cell biology, immunology, cancer biology, and neurodegeneration. In infection and immunity, autophagy is emerging as a pivotal pathway for mediating direct pathogen degradation as well as for the development of robust innate and adaptive immune responses. Successful pathogens have evolved to either evade or harness the autophagy pathway to further their replication and pathogenesis. In this chapter, the basic aspects of autophagy will be described, along with its role in cellular homeostasis, and the development of immunity. The primary focus is a survey of past and recent research defining the

Of design he was created thus, his own waste providing his own food, and all that he did or suffered taking place in and by himself. For the Creator conceived that a being which was self-sufficient would be far more excellent than one which lacked anything; and, as he had no need to take anything or defend himself against any one

Description of the Ouroboros from Plato's *Timaeus*

interplay of autophagy and the herpesviruses, with particular reference to immune evasion and pathogenesis.

Introduction and brief history

Autophagy, from the Greek for 'self eating', has become an intense area of research in biology and medicine. A purely numerical review of the literature shows that between the years 1990 and 1999 there were a total of 315 published papers with the keyword 'autophagy'. Between the years 2000 and 2009 there were 4051 papers published on autophagy, with a staggering 1376 papers published in 2009 alone. The journal *Autophagy* was launched in 2005 which has undoubtedly fuelled both interest and publication rates in the field. The exponential nature of this information explosion has been described as being analogous to that seen in apoptosis research through the 1980s and 1990s (Klionsky, 2007).

A major driving force for this expansion in autophagy research is the recognition of the ever-increasing areas of biological research upon which autophagy has a profound impact (Meijer and Codogno, 2006; Shintani and Klionsky, 2004; Todde *et al.*, 2009). Autophagy, also known as macroautophagy, is a cellular mechanism for the bulk degradation of cytosolic proteins, organelles and other macromolecules. It is a highly conserved pathway and is widely represented in yeast, plants, insects and mammals. By degrading these elements of the cell, metabolic precursors are made available for *de novo* synthesis of proteins and sub-cellular structures. In addition to its ubiquitous role in the maintenance of regular cellular turnover and homeostasis, autophagy is induced in response to stress stimuli. These include starvation and growth factor deprivation, accumulation of misfolded or insoluble protein aggregates, hypoxia, and infection. It is also involved in cellular remodelling including metamorphosis, differentiation, and development, ageing, atrophy and death signalling. Autophagy also plays a role in many human diseases, including liver disease, heart disease, cancer, neurodegeneration, various myopathies, and in the innate and adaptive immune responses to infection.

Autophagy is triggered in response to infection by all manner of microbes (see Orvedahl and Levine, 2009 for a review). Some microbes have evolved ways to utilize aspects of the autophagic response to promote their replication, while others are able to prevent or control the autophagy response. This chapter will explore the interaction of viruses and the autophagy pathway, with a focus on the herpesviruses, with particular reference to the alphaherpesviruses. Rather than an exhaustive and detailed review, this chapter is designed to be a gentle introduction to the field of autophagy, and to highlight some of the recent advances in our understanding of how the pathogenesis and cell biology of herpesvirus infections are impacted by the autophagy pathway. This new and exciting direction in the herpesvirus field will undoubtedly yield some important discoveries regarding host defence and viral counter-defence.

As described in an outstanding perspective of the growth and development autophagy field (Klionsky, 2007), an early pioneer of the autophagy field was Christian De Duve. He received the Nobel Prize in 1974 largely for his earlier work on the lysosomal system. Given that autophagy is a subset of the lysosomal system, it is not surprising that De Duve was among the first to describe the process, and he invented the term 'autophagy' in 1963. The process was described largely following electron microscopy studies of mammalian cells that showed double-membrane vesicles containing cytoplasm, often with organelles in varying degrees of degradation. This early work described the process as a natural component of normal cellular homeostasis, but also identified an enhancement under a wide variety of conditions including cellular differentiation, hormone exposure, metabolic stress, starvation, ageing, hypoxia, irradiation, and treatment with a variety of drugs (De Duve and Wattiaux, 1966). As a consequence of its apparent responses to these stimuli, induction of autophagy was originally associated with changes in cellular nutrition and the catabolic production of biosynthetic precursors. Subsequent work has investigated the plethora of stimuli that induce autophagy, the different forms of autophagy, and the many diverse consequences of its induction.

The multiple forms of autophagy

Autophagy involves the delivery of substrates to the lysosomal lumen and the breakdown of macromolecules therein. The process by which substrates for degradation enter this pathway differentiates the three major types of autophagy, chaperon-mediated autophagy, microautophagy, and macroautophagy (the major focus of this chapter). Chaperone-mediated autophagy (CMA) is a process specific to higher eukaryotes. It involves the digestion of targeted proteins that are specifically translocated across the lysosomal membrane (Chiang and Dice, 1988; Massey *et al.*, 2006). So while the macro- and microautophagy pathways digest whole cytosolic components, CMA is selective; indeed its protein targets contain a lysosomal targeting motif. This also limits CMA to soluble cytosolic proteins whereas the other forms of autophagy can target and degrade larger structures including whole organelles. In microautophagy, as in CMA, there is no phagosome; the lysosome itself sequesters cytosolic components for digestion (Uttenweiler and Mayer, 2008). Small portions of whole cytoplasm are engulfed by invagination and septation of the lysosomal membrane. As a result, both the cytoplasmic component as well as the budding lysosomal membrane are degraded. This has been best characterized in yeast, although microautophagy research is still in its relative infancy.

Macroautophagy, autophagy and xenophagy

Macroautophagy, which is often simply referred to as 'autophagy' refers to a process whereby endogenous cellular materials are transferred into vacuolar compartments for degradation via fusion with the lysosome. These engulfed and degraded endogenous materials can be cellular in origin,

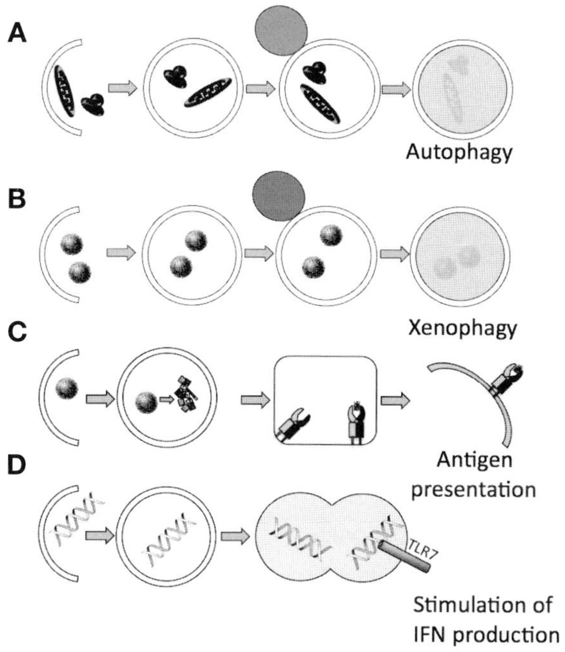

Figure 21.1 Autophagy and the contribution of xenophagy to host immunity. (A) Classical autophagy in response to stress or starvation involves formation of an isolation membrane which fuses to form an autophagosome. This autophagosome in turn fuses with the lysosome to form an autophagolysosome, with degradation of the engulfed cytoplasmic contents. (B) Xenophagy appears to involve a similar process to classical autophagy except that pathogens and components are engulfed and digested. (C and D) Xenophagic degradation of virus particles or nucleic acids can contribute to the immunological processes of antigen presentation and toll-like receptor mediated upregulation of interferon (IFN) production.

hence 'autophagy', or, may originate from invading intracellular pathogens, in which case the engulfment and degradation process is known as 'xenophagy'. Xenophagy, from the Greek 'Xeno' for foreign, generally refers to the degradation of intracellular bacteria or viruses. This chapter will most focus on the manner in which viruses, especially herpesviruses, evade or exploit the xenophagy response.

Autophagy is a conserved cellular pathway that eliminates defective proteins and organelles, prevents abnormal protein aggregate accumulation, and removes intracellular pathogens (Dunn, 1994; Klionsky and Emr, 2000; Mortimore and Poso, 1987; Virgin and Levine, 2009). This process begins with the formation and elongation of a double membrane that fuses to form an autophagosome. The cytoplasmic contents are non-specifically sequestered inside the autophagosome and are then degraded once the autophagosome fuses with the lysosome (Fig. 21.1A). As alluded to above, autophagy is not limited to degradation of self-proteins; it can also engulf and break down invading microorganisms in a process termed xenophagy (Levine, 2005). Xenophagy (Fig. 21.1B) can limit the replication of pathogens (Andrade *et al.*, 2006; Birmingham *et al.*, 2006; Ling *et al.*, 2006; Nakagawa *et al.*, 2004; Rich *et al.*, 2003), but some infectious agents can exploit autophagy to enhance their replication (Amer and Swanson, 2005; Jackson *et al.*, 2005; Prentice *et al.*, 2004).

Work from several groups have used the genetically tractable yeast system to identify the many genes involved in the induction and regulation of autophagy (Kabeya *et al.*, 2007; Tsukada and Ohsumi, 1993). Thirty-three autophagy-related genes (ATGs) have now been identified in yeast, and for most of them mammalian homologues have been identified (Kanki *et al.*, 2009). Autophagy is essential for mammalian development and homeostasis, and so it has not been possible to generate viable homozygous knockout mouse lines. Researchers have therefore utilized conditional knockout strategies to identify roles for autophagy in many important pathways, including the development of immunity (Hashimoto *et al.*, 2008; Stephenson *et al.*, 2009). Additionally, such approaches have shown specific roles for autophagy in the development of Crohn's Disease, cancer and neurodegeneration (Cadwell *et al.*, 2008; Hara *et al.*, 2006; Takahashi *et al.*, 2007).

Formation of the isolation membrane and elongation of the autophagosome is a complex process and has been the subject of many studies and reviews (Suzuki and Ohsumi; Yang and Klionsky, 2009). The induction phase requires a conjugate of Atg5-Atg12 as well as a second conjugation system in which LC3-I and LC3-II are produced from the cleavage of LC3 (Hanada *et al.*, 2007; Kuma *et al.*, 2002; Shintani *et al.*, 1999; Tanida *et al.*, 1999). Also required, is the protein Beclin 1 (Fig. 21.2). Beclin 1 is required for preautophagosome formation and is found in a complex with Vps34, a class III phosphoinositide 3-kinase (PI3K), at the *trans*-Golgi network (Furuya *et al.*, 2005). This complex may supply the phosphatidylinositol 3-phosphates to preautophagosome membranes facilitating localization of Atg proteins. As explored later in the chapter, Beclin 1-binding seems to be a common property of several viral proteins that inhibit xenophagy, suggesting that this may represent an 'Achilles heel' of the autophagy pathway that has been exploited by various pathogens.

Both PKR and phosphorylated eIF2α have been shown to promote autophagy, although the mechanism by which this occurs is unclear (Talloczy *et al.*, 2002). Gcn2, another eIF2α kinase, is required for starvation-induced autophagy in yeast, and PKR can functionally substitute for GCN2 in *GCN2*-disrupted yeast. Phosphorylation of eIF2α is necessary for effective translation of specific mRNAs including the transcriptional activator Gcn4 and several components of the autophagy pathway are targets of Gcn4 (Kubota *et al.*, 2003). Thus, phosphorylation of eIF2α resulting from starvation or viral infection may lead to the initiation of autophagy and other cellular programmes which promote cell survival following diverse stress stimuli. Several viral gene products that control PKR activity and eIF2α phosphorylation can thereby exert control on the autophagy pathway.

Several viral pathogens that actively perturb autophagy do so through interaction with Beclin 1 (Kyei *et al.*, 2009; Orvedahl *et al.*, 2007).

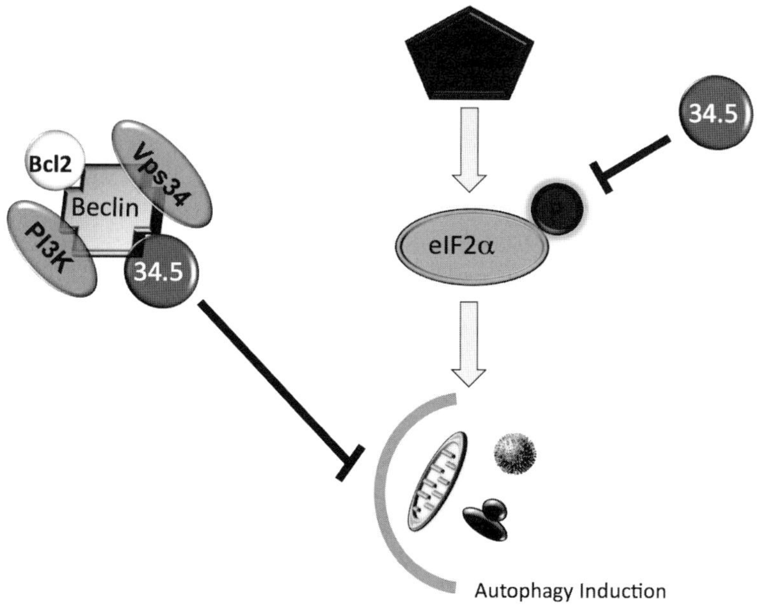

Figure 21.2 Inhibition of autophagy/xenophagy by HSV ICP34.5 occurs through two separable domains. A C-terminal domain, homologous to cellular GADD34, binds to and redirects a cellular phosphatase to dephosphorylate eIF2α, thereby reversing the effects of PKR and dampening translational arrest and the induction of autophagy. An N-terminal domain interferes with autophagy/xenophagy through its ability to bind the essential autophagy protein Beclin 1. The precise mechanism by which this occurs is unknown. The relative contributions of these domains to the inhibition of autophagy is not known, but the GADD34-hologous domain has a greater impact on neurovirulence.

Herpes simplex virus type 1 (HSV-1) ICP34.5 and γHV68 vBcl2 both interact with Beclin 1 to inhibit autophagy. It is also believed that influenza M2 protein, and HIV-1 Nef activate autophagy through interaction with Beclin 1. There are numerous other pathogens, bacteria, protozoa, viruses and fungi that perturb the autophagy pathway, although examples of these will be mentioned only in passing in this chapter (Orvedahl and Levine, 2009). It should be noted at this juncture that in addition to xenophagy playing a role in the direct control of pathogens through engulfment and degradation, xenophagy is also an important process for development of the innate and adaptive immune responses to infection. The relative contribution of xenophagy to direct engulfment and degradation of pathogens, versus its role in promoting the development of a cognate immune response is currently unknown, but will be a significant area of future research.

Xenophagy and immunity

How does autophagy/xenophagy mediate immunity to viruses?

Autophagy has been linked to many parts of the immune system, in both the innate and adaptive arms. In an excellent recent review, a dizzying array of immunological processes, including pathogen degradation, both activation and suppression of interferon production, lymphocyte development, tolerance, and antigen presentation were described as being impacted by the autophagy pathway (Virgin and Levine, 2009). With all of these aspects of immunity modulated by autophagy, the challenge is to determine exactly how autophagy and xenophagy control viral replication and pathogenesis. Three major possibilities have been considered in more detail, direct degradation of the pathogen (xenophagy), stimulation of interferon responses through toll-like receptors (TLRs) and PKR, and control of antigen presentation.

Direct xenophagy and degradation

Direct degradation of viral particles or viral components could act as a direct innate defence, although for some systems this remains unproven or controversial (Figs. 21.1B and 21.3). Evidence for this comes from a variety of work in plant and animal virus systems. Replication of tobacco mosaic virus (TMV) was enhanced in leaves of tobacco plants silenced for autophagy genes, supporting the idea that autophagy-related genes serve to limit virus replication (Liu *et al.*, 2005). Further support for direct degradation came from a study showing that ectopic expression of Beclin 1 in the nervous system by a Sindbis virus led to lower CNS titres and mortality (Liang *et al.*, 1998). In contrast, however, a recent paper, showed that CNS inactivation of Atg5, an essential autophagy protein, led to increased cell death, (Orvedahl *et al.*) but with no increase in viral replication. Similarly, a strain of HSV-1 which was engineered to be deficient in its ability to interfere with autophagy showed reduced replication and virulence *in vivo*, with increased numbers of virus particles observed inside xenophagosomes, suggesting that it was being directly degraded and destroyed (Orvedahl *et al.*, 2007). Subsequently, however, when this virus was examined *in vitro* in cells lacking Atg5, there was no change in virus replication (Alexander *et al.*, 2007a). While it remains possible, likely even, that xenophagic degradation of virus particles can serve to limit their replication *in vitro* and *in vivo*, there are surprisingly few conclusive virological examples of this mechanism. An intriguing example with HIV infection, however, has shown that while autophagy actually augments HIV replication, the Nef protein interacts with Beclin 1 and serves to inhibit maturation of the autophagosome (Kyei *et al.*, 2009). In the absence of Nef, HIV particles are degraded. It should be noted, however, that the xenophagic degradation of virus particles is not mutually exclusive from other autophagy-mediated mechanisms of limiting virus replication.

Xenophagy/autophagy, TLRs, and antiviral responses

The rapid antiviral responses observed following viral infection of a variety of cells frequently result from recognition of viral pathogen-associated molecular patterns (PAMPS) via TLRs. The TLRs have connections with the autophagy/xenophagy pathway, especially through the activities of TLR3, TLR7 and possibly TLR9. Delivery of PAMPS to endosomal TLRs via autophagy can strongly activate innate responses through type I interferon in cultured primary macrophages (Delgado *et al.*, 2008) (Fig. 21.1D). Double stranded RNA was shown to be an especially potent agonist of autophagy in a TLR-3, or a TLR7- and Myd88-dependent fashion. One study provided particularly compelling evidence for a role for autophagy in TLR-mediated virus recognition and response (Lee *et al.*, 2007). These authors showed that Atg5 is critical for cellular recognition of vesicular stomatitis infection and for the secretion of interferon α following infection of plasmacytoid dendritic cells.

The double stranded RNA-inducible protein kinase (PKR) is a pivotal sensor of virus infection, and a critical inducer of interferon production and the antiviral state (Balachandran and Barber, 2007). It has been shown that PKR, through its ability to phosphorylate eIF2α, is essential for the induction of autophagy, although the mechanism remains unknown (Talloczy *et al.*, 2002). Another related inducer of autophagy includes the unfolded protein response (UPR) which signals through both the PKR-like ER kinase, PERK, and PKR, both of which are activated and induced during virus infection to induce autophagy.

Autophagy/xenophagy and antigen presentation

Autophagy/xenophagy is also an important process for the adaptive immune response to infection through the delivery of antigens for major histocompatibility complex class I and II (MHC-I and II) presentation (Crotzer and Blum, 2005; Dengjel *et al.*, 2005; English *et al.*, 2009b; Lee *et al.*; Paludan *et al.*, 2005; Schmid *et al.*, 2007) (Fig. 21.1C). Endogenous antigens, or those synthesized by viruses (or other intracellular pathogens) are degraded by the proteasome and peptides are subsequently translocated to the ER, and loaded onto MHC-I molecules. Exogenous antigens are commonly endocytosed and then degraded by lysosomal proteases prior to binding of MHC-II molecules. Exception to this classical paradigm is the discovery that some antigen presenting cells can present extracellular antigens on MHC-II as well as MHC-I, a phenomenon known as cross-presentation (Bevan, 1976; Heath *et al.*, 1998). Cross-presentation allows a more effective adaptive response, allowing concomitant stimulation of $CD4^+$ and $CD8^+$ T cells. The inhibition of autophagy by pathogens therefore would serve as a common pathway by which to block CD4- and CD8-T cell responses and allow pathogens to remain under-recognized by the adaptive immune response (Crotzer and Blum, 2009).

Modulation of autophagy by herpesviruses

As described in the above sections, autophagy/xenophagy plays a role in antiviral immunity. In addition, there are some viruses that utilize autophagic pathway and its components to promote their replication (Sir and Ou). Autophagy is often strongly induced by virus infection, with co-localization of viral proteins and autophagy proteins, and this has sometimes led to assumptions and some controversy that autophagy is a requirement, or involved in virus replication (Prentice *et al.*, 2004; Zhao *et al.*, 2007). Such data should be interpreted with caution, and, given the core importance of autophagy to cellular homeostasis, it is not surprising that so many viruses perturb the pathway. The challenges have been to determine where this perturbation is truly critical to the lifecycle and pathogenesis of the virus. This issue notwithstanding, there is ample evidence that certain RNA viruses such as hepatitis C, dengue virus and poliovirus induce autophagosome formation, and that this induction initiates or enhances RNA replication (Dreux *et al.*, 2009; Jackson *et al.*, 2005; Lee *et al.*, 2008). In the following sections, the focus will be on the herpesviruses, especially the alphaherpesviruses, and their modulation of the autophagy pathway. The herpesviruses are perhaps the best-characterized viral examples of evasion of the autophagy pathway.

Modulation of autophagy by herpes simplex virus (HSV)

HSV was among the first herpesviruses to be associated with autophagy (Smith and de Harven, 1978). Much of the recent work on autophagy modulation by HSV, however, has focused almost exclusively on the role of ICP34.5. The HSV-encoded ICP34.5 is required for neurovirulence in mice (Bolovan *et al.*, 1994; Chou and Roizman, 1990). Viruses lacking the ability to code for ICP34.5 show up to a 10,000-fold increase in LD_{50} following intracranial injection of mice. Importantly, ICP34.5 is also required for neurovirulence in humans, since an ICP34.5-deleted virus has been used as an oncolytic treatment for gliomas (Harrow *et al.*, 2004; Kanai *et al.*, 2007). The major neurovirulence-enhancing domain of ICP34.5 is a GADD34 homology region (Chou and Roizman, 1994). This domain reverses PKR-mediated eIF2α phosphorylation by the recruitment of the cellular protein phosphatase-1α (He *et al.*, 1997). The activity of this phosphatase is redirected to the phosphorylated form of eIF2α, regenerating a pool of non-phosphorylated eIF2α available for translation of viral mRNA. ICP34.5 was thereby believed to function in neurovirulence solely by blocking eIF2α kinase signalling-dependent host translational arrest. This idea was bolstered by the observation that the replication and virulence defects of ICP34.5-null viruses are completely and specifically restored in both cells and mice deficient in PKR (Leib *et al.*, 2000).

An examination of other data, however, suggested that other functions of ICP34.5, in addition to inhibition of host-cell translational shutoff, might also mediate neurovirulence. For example, replacement of the ICP34.5 GADD34 domain with the murine homologue maintained the ability of HSV-1 to dephosphorylate eIF2α and normalize viral protein synthesis in cell culture, but resulted in a virus that remained attenuated in mice (Andreansky *et al.*, 1996; He *et al.*, 1996). Furthermore, an ICP34.5-null virus with a secondary mutation that leads to immediate-early expression of US11, a PKR agonist, exhibited wild-type ability to dephosphorylate eIF2α and prevention of host-cell shutoff, but exhibited persistently reduced neurovirulence (Mohr *et al.*, 2001; Mulvey *et al.*, 1999). A breakthrough in the search for other functions of ICP34.5 came in 2002 when it was shown that PKR activity and eIF2α phosphorylation were required for viral- and starvation- induced autophagy (Talloczy *et al.*, 2002). A logical extension of this observation in the same study showed that HSV-1 could antagonize autophagy through the activity of ICP34.5 (Fig. 21.2). A quantitative morphological analysis subsequently demonstrated PKR-dependent degradation of HSV virions and proteins in primary mouse embryo fibroblasts (Talloczy *et al.*, 2006). It was additionally shown that this degradation was antagonized by ICP34.5.

While the evidence that ICP34.5 could antagonize autophagy was building, a persistent challenge was to dissect the ability of ICP34.5 to inhibit translational arrest from its ability to inhibit autophagy. An important discovery in the Levine laboratory was that ICP34.5 had another binding partner (aside from PP1α), namely the essential autophagy protein, Beclin 1 (Orvedahl *et al.*, 2007). In this study, the interaction was demonstrated by yeast two-hybrid screen, and also by co-immune precipitation. The binding domain was localized to the N-terminus, a 20 amino acid domain between residues 68 and 87 being both necessary and sufficient for Beclin 1 binding, and for inhibition of starvation-induced autophagy. This domain thereby has become known as the Beclin-binding domain, or BBD. In the context of the viral genome, this domain was dispensable for dephosphorylation of eIF2α, and for blocking translational arrest. This implied that ICP34.5 is a modular protein, with the function of the BBD being separable from the GADD34 domain (Fig. 21.2). The key discoveries of this study was that the recombinant strain lacking BBD was significantly neuro-attenuated following intracerebral injection, and that this attenuation was restored in PKR-deficient mice. Alterations in the autophagy pathway could now be ascribed to ICP34.5's ability to bind Beclin 1, in addition to its mediation of eIF2α dephosphorylation (Orvedahl *et al.*, 2007). Moreover this BBD was significant in that it impacted HSV neuropathogenesis. A key area of future research will therefore be to unravel the precise mechanism by which binding of ICP34.5 to Beclin 1 interferes with autophagy/xenophagy. This finding has led some to propose that HSV may play a role in the aetiology of Alzheimer's disease through its ability to disrupt the autophagy pathway in the CNS (Itzhaki *et al.*, 2008).

Subsequent work has focused on the role of BBD in regulating viral replication, as well as its role in other disease models of HSV-1. The lead hypothesis from the original BBD study (Orvedahl *et al.*, 2007) was that BBD was preventing the formation of autophagosomes in infected cells, and thereby protecting the virus from degradation in the autophagolysosome. This remains an attractive idea, especially for protection of the central nervous system from infection with HSV and other intracellular pathogens. Xenophagy could represent a 'surgical strike' against such pathogens without calling in the destructive collateral damage associated with a full-blown inflammatory immune response, something that is not compatible with preserving the function of the CNS. Examination of other tissues, *in vitro* and *in vivo*, however, has cast some doubt whether this idea is applicable to all infected cells. For example, ICP34.5-deleted virus induced significantly more xenophagy in $Atg5^{+/+}$ fibroblasts than a wild-type virus, yet $Atg5^{-/-}$ fibroblasts produced comparable yields of HSV as $Atg5^{+/+}$ fibroblasts (Alexander *et al.*, 2007b). This suggested that xenophagy was not limiting ICP34.5-deficient virus production in these cultured cells. Further study revealed that mice infected by the ocular route with wild type HSV-1 and HSV-1 lacking the BBD showed no difference in the replication of these strains in several sites of infection at early times post infection (Leib *et al.*, 2009). This implied that once again there was no inherent difference in replication between wild-type and BBD-deleted viruses. At later times (>5 days), however, the BBD-deleted virus was cleared more rapidly from infected tissues than wild-type. The timing of this clearance suggested that the adaptive immune response may be involved in clearing the BBD-deleted virus. Consistent with this idea, compared to wild-type virus, the BBD-deleted virus stimulated a significantly stronger CD4 T-cell response and resulted in significantly more production of gamma interferon and interleukin-2 from HSV-specific CD4 T cells. These data suggested a role for the BBD of ICP34.5 in precluding autophagy-mediated class II antigen presentation, thereby enhancing HSV-1 virulence and pathogenesis.

Two other recent studies have demonstrated yet further complexity and intrigue in the interplay between xenophagy and ICP34.5. First, a study has shown that HSV-1 can induce a novel form of autophagy that leads to $CD8^+$ T cell activation (English *et al.*, 2009a; English *et al.*, 2009b). This study showed that in the absence of ICP34.5, as expected, there was significant accumulation of 'classical' autophagosomes in infected macrophages. In the presence of ICP34.5, autophagy was inhibited early after infection, but during the late phase infection a new from of autophagy was observed in which autophagosomes were formed by a coiling process of the nuclear membrane. This study also showed that autophagy was important for processing viral antigens for presentation on MHC class I molecules. An important future direction of research will be to determine the relative contributions of classical autophagy and nuclear- derived autophagy to antigen presentation and $CD8^+$ T cells activation. Second, a very recent study (Lee *et al.*) utilized a conditional deletion of Atg5 in dendritic cells using cre-lox recombination. They showed that dendritic cell Atg5-deleted mice exhibited impaired $CD4^+$ T cell priming and activation. These mice also showed increased disease following HSV-2 infection, and died more rapidly. Antigen processing and presentation of phagocytosed antigens for MHC class II was significantly impaired in these mice while cross-presentation of peptides on MHC I was unaffected. This study therefore underscores the importance of autophagy in dendritic cells for the processing of antigens for MHC II presentation.

Much emphasis has been placed on the role of ICP34.5 in the regulation of xenophagy, and much remains to be learned. Furthermore, it will be of interest to study other possible modulators of xenophagy/autophagy encoded by HSV. In particular, it would seem likely that US11 may contribute to the control of xenophagy through its regulation of PKR activity mediated by dsRNA or PACT (Cassady *et al.*, 1998; Khoo *et al.*, 2002; Peters *et al.*, 2002). Another possible candidate is ICP0, whose homologue in bovine herpesvirus, (bICP0, described below) has been shown to inhibit autophagy.

Modulation of autophagy by other alphaherpesviruses

Varicella zoster virus (VZV)

VZV has recently been shown to induce autophagy in both cultured cells, as well as in human skin (Takahashi *et al.*, 2009). This is an intriguing study because in contrast to the several studies on HSV infection which showed inhibition of autophagy, VZV appears to activate autophagy. The study used several criteria, including conversion of LC3 and morphological assays (for a review of autophagy quantification methods, see (Mizushima *et al.*)), to show that not only was xenophagy strongly induced, but that it appeared to cause depletion of cytoplasmic material. The observation of xenophagy in VZV infected skin from a biopsy was especially interesting, since this implies that this is not an artefact of cell culture, but rather a process that may have significant impact on pathogenesis. Some areas of future interest are readily apparent. First, the VZV gene product which is responsible for induction of the autophagy pathway was not identified. Second, the impact of inhibiting or activating the autophagy pathway on VZV replication was not studied. This may be an important area of study given the difficulty in growing high titre cell-free VZV in culture. Finally, the authors speculate whether the induction of autophagy, or defects in the autophagy pathway may be a potent determinant of pathogenesis in certain high-risk populations.

Bovine herpesvirus type 1 (BHV-1)

BHV-1, an alphaherpesvirus of cattle encodes an immediate-early gene, bICP0, that is analogous to ICP0 of HSV (Pereira *et al.*, 1977). BHV-1 is a powerful inducer of apoptosis, and overexpression of bICP0 alone can cause induction of caspase 3 and apoptotic cell death (Henderson *et al.*, 2005). Once again, analogous to HSV-1, a latency gene runs antisense to bICP0 that appears to be inhibitory of apoptosis(Ciacci-Zanella *et al.*, 1999). When the cytotoxicity of a bICP0 null virus was examined, it was notable that this recombinant virus induced high levels of apoptosis in rabbit skin cells, but not in bovine kidney cells, where non-apoptotic cell death was mostly observed (Geiser *et al.*, 2008). In these cells, there were increased numbers of cytoplasmic vesicles, suggestive of an induction of autophagy. Although further work will be required to definitively show that bICP0 is an inhibitor of autophagy, it is an intriguing observation that may have significant importance to the study of BHV-1 biology and pathogenesis.

Modulation of autophagy by betaherpesviruses

It is likely that all of the herpesviruses, once fully studied, will be found to modulate autophagy. In this section, autophagy research on the beta- and gammaherpesviruses will be briefly described. Although strictly speaking this area is beyond the scope of this chapter, is it both useful and important to note the breadth of research being conducted in this ever-growing field. A representative of the betaherpesviruses, human cytomegalovirus (HCMV) was probably the first herpesvirus to be associated with autophagy, along with HSV, in which it was noted that enveloped progeny viruses were located within cytoplasmic vacuoles containing lysosomal enzyme activity (Smith and de Harven, 1978). A much more recent study has demonstrated that HCMV can counteract autophagy in human fibroblasts (Chaumorcel *et al.*, 2008). A broad base of techniques were used to measure autophagy, and by all criteria autophagy was robustly reduced. Interestingly, HCMV inhibited autophagy induced through two independent pathways, and as early as 24 hours post infection. The authors propose that the viral gene responsible for the inhibition must be a structural or an early gene, and hypothesize that autophagy is an important host defence against HCMV. As with other pathogens, it is yet unknown whether xenophagy is functioning to directly degrade HCMV, to increase adaptive immunity, or a combination of both. Clearly much remains to be determined in the relationship between autophagy and HCMV infection.

Modulation of autophagy by gammaherpesviruses

Murine gammaherpesvirus 68

Murine gammaherpesvirus 68 has been an outstanding model for gammaherpesvirus biology and pathogenesis. A series of recent studies have placed this pathogen on the rapidly growing list of those that modulate autophagy to augment pathogenesis, and these have elucidated much of the molecular interactions that regulate this process (E *et al.*, 2009; Ku *et al.*, 2008; Sinha *et al.*, 2008). Most of the autophagy-modulating activities occur through the viral homologue of the cellular anti-apoptotic protein BCL-2, termed v-BCL-2, or M11. Moreover, given that all gammaherpesviruses express a homologue of BCL-2, this is likely to be a virus subfamily-common mechanism to perturb autophagy. M11 was shown to bind tightly to the BH3-like domain of Beclin 1, more tightly than cellular BCL-2 (Ku *et al.*, 2008; Sinha *et al.*, 2008). It was further shown that M11 potently competes with BCL-2 for binding to Beclin 1, and thereby inhibits autophagy. Another key observation in this series was to actually demonstrate that M11 was important for pathogenesis through its ability to inhibit autophagy (E *et al.*, 2009). Interestingly, several previous studies had shown that M11 was important for preventing apoptosis *in vitro*, and for promoting establishment of latency (de Lima *et al.*, 2005; Roy *et al.*, 2000; Wang *et al.*, 1999). The E *et al.* study, however, went a step further to actually show that it was the antagonism of host autophagy, rather than apoptosis, that was key for the maintenance of latency.

Epstein–Barr virus (EBV)

Although there have been only a small number of studies published on EBV and autophagy, it is significant that a study of endogenous MHC class II processing of Epstein–Barr nuclear antigen 1 (EBNA1) revealed that it accessed the pathway through autophagy (Paludan *et al.*, 2005). This was the first demonstration that antigens could enter the MHC class II pathway through autophagy. Subsequent studies have revealed that

EBV can regulate the unfolded protein response and autophagy in EBV-infected B cells through the activities of a viral oncogene, latent membrane protein 1 (LMP-1) (Lam *et al.*, 2004; Lee and Sugden, 2008a,b). The intriguing aspect of this series of studies is that LMP-1 can act as a kind of autophagy rheostat, it can induce autophagy is a highly dose-dependent fashion. LMP1 levels and the stage of autophagy control cell cycle progression in EBV-positive cell populations. Cells with low levels of LMP1 usually contain autophagosomes whereas cells with high levels of LMP-1 contain autolysosomes. Inhibition of autophagy with siRNA to Beclin 1 led to LMP-1 accumulation, revealing evidence for feedback and interplay between autophagy and LMP-1.

Kaposi's sarcoma-associated herpesvirus (KSHV) and herpesvirus saimiri (HVS)

As described above, gammaherpesvirus and cellular Bcl-2 molecules were shown to inhibit autophagy through Beclin 1 binding. A further study, however, has began to search in a broader fashion for viral factors that can perturb the autophagy pathway (Lee *et al.*, 2009). In this study, the authors used a screen based on redistribution of GFP-LC3 staining, a commonly used assay for autophagy (Mizushima *et al.*) The authors discovered that a KSHV gene termed K13 (also known as vFLIP) was able to suppress both starvation and rapamycin-induced autophagy. vFLIP homologues in both HVS and a poxvirus, Molluscum contagiosum virus, were also able to inhibit autophagy suggesting a common mechanism for these viruses. The authors also showed that FLIPs likely inhibit autophagy through preventing Atg3 from binding and processing LC3. Furthermore, a small peptide fragment of FLIP is sufficient to suppress Atg3–FLIP interaction, but does not affect interaction between Atg3 and LC3. The net result is that this peptide can facilitate Atg3–LC3 interaction, and leads to cell death with a high level of autophagy. Intriguingly, treatment of mice with this peptide in a primary effusion lymphoma model promoted regression of tumours. This provided a first glimpse of possible novel tumour therapies using modulation of the autophagy pathway.

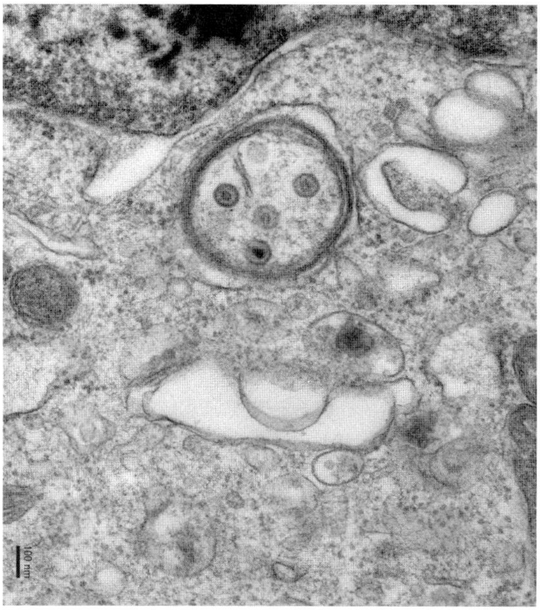

Figure 21.3 Electron micrograph of mouse embryo fibroblasts infected with an HSV-1 strain lacking ICP34.5. A double-membraned autophagosome, (the face of autophagy) is clearly visible containing five HSV particles. Scale bar =100nm. Reprinted from Alexander *et al.* (2007b) and used with permission.

Conclusions

Given the ever-burgeoning autophagy literature, our understanding of the interrelationship between viruses and the autophagy pathway is likely to grow exponentially in the coming years. The broad impact of xenophagy on a variety of viruses coupled with their diverse countermeasures leaves open many possibilities for the exploitation of autophagy modulation for therapeutic purposes. It is possible that a better understanding of these pathways could lead to a whole new class of antiviral therapies. These therapies could be designed to augment the xenophagic degradation of otherwise hard-to-treat intracellular pathogens, or to deny certain pathogens access to the autophagic machinery that serves to promote their replication and disease. In addition, work emerging from the gammaherpesviruses described above may also lead to novel therapies using autophagy alteration to treat cancer. Given the wide variety of human diseases for which autophagy is implicated (Eskelinen and Saftig, 2009), the discovery and development of safe modulators of autophagy may have broad therapeutic applications.

Acknowledgements

David Leib is supported by an NIH grant (RO1 EY09083), and by a Senior Scientific Investigator Award from Research to Prevent Blindness. We also acknowledge helpful comments on the chapter from Audra Charron, and Wandy Beatty for the electron micrograph in Fig. 21.3.

References

Alexander, D.E., Ward, S.L., Mizushima, N., Levine, B., and Leib, D.A. (2007a). Analysis of the role of autophagy in replication of herpes simplex virus in cell culture. J. Virol. *81*, 12128–12134.

Alexander, D.E., Ward, S.L., Mizushima, N., Levine, B., and Leib, D.A. (2007b). Analysis of the role of autophagy in replication of herpes simplex virus in cell culture. J. Virol. *81*, 12128–12134.

Amer, A.O., and Swanson, M.S. (2005). Autophagy is an immediate macrophage response to *Legionella* pneumophila. Cell. Microbiol. *7*, 765–778.

Andrade, R.M., Wessendarp, M., Gubbels, M.J., Striepen, B., and Subauste, C.S. (2006). CD40 induces macrophage anti-*Toxoplasma gondii* activity by triggering autophagy-dependent fusion of pathogen-containing vacuoles and lysosomes. J. Clin. Invest. *116*, 2366–2377.

Andreansky, S.S., He, B., Gillespie, G.Y., Soroceanu, L., Markert, J., Chou, J., Roizman, B., and Whitley, R.J. (1996). The application of genetically engineered herpes simplex viruses to the treatment of experimental brain tumors. Proc. Natl. Acad. Sci. U.S.A. *93*, 11313–11318.

Balachandran, S., and Barber, G.N. (2007). PKR in innate immunity, cancer, and viral oncolysis. Methods Mol. Biol. *383*, 277–301.

Bevan, M.J. (1976). Cross-priming for a secondary cytotoxic response to minor H antigens with H-2 congenic cells which do not cross-react in the cytotoxic assay. J. Exp. Med. *143*, 1283–1288.

Birmingham, C.L., Smith, A.C., Bakowski, M.A., Yoshimori, T., and Brumell, J.H. (2006). Autophagy controls *Salmonella* infection in response to damage to the *Salmonella*-containing vacuole. J. Biol. Chem. *281*, 11374–11383.

Bolovan, C.A., Sawtell, N.M., and Thompson, R.L. (1994). ICP34.5 mutants of herpes simplex virus type 1 strain 17syn+ are attenuated for neurovirulence in mice and for replication in confluent primary mouse embryo cell cultures. J. Virol. *68*, 48–55.

Cadwell, K., Liu, J.Y., Brown, S.L., Miyoshi, H., Loh, J., Lennerz, J.K., Kishi, C., Kc, W., Carrero, J.A., Hunt, S., *et al.* (2008). A key role for autophagy and the autophagy gene Atg16l1 in mouse and human intestinal Paneth cells. Nature *456*, 259–263.

Cassady, K.A., Gross, M., and Roizman, B. (1998). The second-site mutation in the herpes simplex virus recombinants lacking the gamma134.5 genes precludes shutoff of protein synthesis by blocking the phosphorylation of eIF-2alpha. J. Virol. *72*, 7005–7011.

Chaumorcel, M., Souquere, S., Pierron, G., Codogno, P., and Esclatine, A. (2008). Human cytomegalovirus controls a new autophagy-dependent cellular antiviral defense mechanism. Autophagy *4*, 46–53.

Chiang, H.L., and Dice, J.F. (1988). Peptide sequences that target proteins for enhanced degradation during serum withdrawal. J. Biol. Chem. *263*, 6797–6805.

Chou, J., and Roizman, B. (1990). The herpes simplex virus 1 gene for ICP34.5, which maps in inverted repeats, is conserved in several limited-passage isolates but not in strain 17syn+. J. Virol. *64*, 1014–1020.

Chou, J., and Roizman, B. (1994). Herpes simplex virus 1 gamma(1)34.5 gene function, which blocks the host response to infection, maps in the homologous domain of the genes expressed during growth arrest and DNA damage. Proc. Natl. Acad. Sci. U.S.A. *91*, 5247–5251.

Ciacci-Zanella, J., Stone, M., Henderson, G., and Jones, C. (1999). The latency-related gene of bovine herpesvirus 1 inhibits programmed. cell death. J. Virol. *73*, 9734–9740.

Crotzer, V.L., and Blum, J.S. (2005). Autophagy and intracellular surveillance: Modulating MHC class II antigen presentation with stress. Proc. Natl. Acad. Sci. U.S.A. *102*, 7779–7780.

Crotzer, V.L., and Blum, J.S. (2009). Autophagy and its role in MHC-mediated antigen presentation. J. Immunol. *182*, 3335–3341.

De Duve, C., and Wattiaux, R. (1966). Functions of lysosomes. Annu. Rev. Physiol *28*, 435–492.

de Lima, B.D., May, J.S., Marques, S., Simas, J.P., and Stevenson, P.G. (2005). Murine gammaherpesvirus 68 bcl-2 homologue contributes to latency establishment *in vivo*. J. Gen. Virol. *86*, 31–40.

Delgado, M.A., Elmaoued, R.A., Davis, A.S., Kyei, G., and Deretic, V. (2008). Toll-like receptors control autophagy. EMBO J. *27*, 1110–1121.

Dengjel, J., Schoor, O., Fischer, R., Reich, M., Kraus, M., Muller, M., Kreymborg, K., Altenberend, F., Brandenburg, J., Kalbacher, H., *et al.* (2005). Autophagy promotes MHC class II presentation of peptides from intracellular source proteins. Proc. Natl. Acad. Sci. U.S.A. *102*, 7922–7927.

Dreux, M., Gastaminza, P., Wieland, S.F., and Chisari, F.V. (2009). The autophagy machinery is required to initiate hepatitis C virus replication. Proc. Natl. Acad. Sci. U.S.A. *106*, 14046–14051.

Dunn, W.A., Jr. (1994). Autophagy and related mechanisms of lysosome-mediated protein degradation. Trends Cell Biol. *4*, 139–143.

E, X., Hwang, S., Oh, S., Lee, J.S., Jeong, J.H., Gwack, Y., Kowalik, T.F., Sun, R., Jung, J.U., and Liang, C. (2009). Viral Bcl-2-mediated evasion of autophagy aids chronic infection of gammaherpesvirus 68. PLoS Pathog. *5*, e1000609.

English, L., Chemali, M., and Desjardins, M. (2009a). Nuclear membrane-derived autophagy, a novel process that participates in the presentation of endogenous viral antigens during HSV-1 infection. Autophagy *5*, 1026–1029.

English, L., Chemali, M., Duron, J., Rondeau, C., Laplante, A., Gingras, D., Alexander, D., Leib, D., Norbury, C., Lippe, R., *et al.* (2009b). Autophagy enhances the presentation of endogenous viral antigens on MHC class I molecules during HSV-1 infection. Nat. Immunol. *10*, 480–487.

Eskelinen, E.L., and Saftig, P. (2009). Autophagy: a lysosomal degradation pathway with a central role in health and disease. Biochim. Biophys. Acta *1793*, 664–673.

Furuya, N., Yu, J., Byfield, M., Pattingre, S., and Levine, B. (2005). The evolutionarily conserved domain of Beclin. 1 is required for Vps34 binding, autophagy and tumor suppressor function. Autophagy *1*, 46–52.

Geiser, V., Rose, S., and Jones, C. (2008). Bovine herpesvirus type 1 induces cell death by a cell-type-dependent fashion. Microb Pathog *44*, 459–466.

Hanada, T., Noda, N.N., Satomi, Y., Ichimura, Y., Fujioka, Y., Takao, T., Inagaki, F., and Ohsumi, Y. (2007). The Atg12–Atg5 conjugate has a novel E3-like activity for protein lipidation in autophagy. J. Biol. Chem. *282*, 37298–37302.

Hara, T., Nakamura, K., Matsui, M., Yamamoto, A., Nakahara, Y., Suzuki-Migishima, R., Yokoyama, M., Mishima, K., Saito, I., Okano, H., *et al.* (2006). Suppression of basal autophagy in neural cells causes neurodegenerative disease in mice. Nature *441*, 885–889.

Harrow, S., Papanastassiou, V., Harland, J., Mabbs, R., Petty, R., Fraser, M., Hadley, D., Patterson, J., Brown, S.M., and Rampling, R. (2004). HSV1716 injection into the brain adjacent to tumour following surg. ical resection of high-grade glioma: safety data and long-term survival. Gene Ther. *11*, 1648–1658.

Hashimoto, D., Ohmuraya, M., Hirota, M., Yamamoto, A., Suyama, K., Ida, S., Okumura, Y., Takahashi, E., Kido, H., Araki, K., *et al.* (2008). Involvement of autophagy in trypsinogen activation within the pancreatic acinar cells. J. Cell Biol. *181*, 1065–1072.

He, B., Chou, J., Liebermann, D.A., Hoffman, B., and Roizman, B. (1996). The carboxyl terminus of the murine MyD116 gene substitutes for the corresponding domain of the gamma(1)34.5 gene of herpes simplex virus to preclude the premature shutoff of total protein synthesis in infected human cells. J. Virol. *70*, 84–90.

He, B., Gross, M., and Roizman, B. (1997). The gamma(1)34.5 protein of herpes simplex virus 1 complexes with protein phosphatase 1alpha to dephosphorylate the alpha subunit of the eukaryotic translation initiation factor 2 and preclude the shutoff of protein synthesis by double-stranded RNA-activated protein kinase. Proc. Natl. Acad. Sci. U.S.A. *94*, 843–848.

Heath, W.R., Kurts, C., Miller, J.F., and Carbone, F.R. (1998). Cross-tolerance: a pathway for inducing tolerance to peripheral tissue antigens. J. Exp. Med. *187*, 1549–1553.

Henderson, G., Zhang, Y., and Jones, C. (2005). The Bovine herpesvirus 1 gene encoding infected cell protein 0 (bICP0) can inhibit interferon-dependent transcription in the absence of other. viral genes. J. Gen. Virol. *86*, 2697–2702.

Itzhaki, R.F., Cosby, S.L., and Wozniak, M.A. (2008). Herpes simplex virus type 1 and Alzheimer's disease: the autophagy connection. J. NeuroVirol. *14*, 1–4.

Jackson, W.T., Giddings, T.H., Jr., Taylor, M.P., Mulinyawe, S., Rabinovitch, M., Kopito, R.R., and Kirkegaard, K. (2005). Subversion of cellular autophagosomal machinery by RNA viruses. PLoS Biol. *3*, e156.

Kabeya, Y., Kawamata, T., Suzuki, K., and Ohsumi, Y. (2007). Cis1/Atg31 is required for autophagosome formation in *Saccharomyces cerevisiae*. Biochim. Biophys. Res. Commun. *356*, 405–410.

Kanai, R., Tomita, H., Hirose, Y., Ohba, S., Goldman, S., Okano, H., Kawase, T., and Yazaki, T. (2007). Augmented therapeutic efficacy of an oncolytic herpes simplex virus type 1 mutant expressing ICP34.5 under the transcriptional control of musashi1 promoter in the treatment of malignant glioma. Hum. Gene Ther. *18*, 63–73.

Kanki, T., Wang, K., Baba, M., Bartholomew, C.R., Lynch-Day, M.A., Du, Z., Geng, J., Mao, K., Yang, Z., Yen, W.L., *et al.* (2009). A genomic screen for yeast mutants defective in selective mitochondria autophagy. Mol. Biol. Cell *20*, 4730–4738.

Khoo, D., Perez, C., and Mohr, I. (2002). Characterization of RNA determinants recognized by the arginine- and proline-rich region of Us11, a herpes simplex virus type 1-encoded double-stranded RNA binding protein that prevents PKR activation. J. Virol. *76*, 11971–11981.

Klionsky, D.J. (2007). Autophagy: from phenomenology to molecular understanding in less than a decade. Nat. Rev. Microbiol. 8, 931–937.

Klionsky, D.J., and Emr, S.D. (2000). Autophagy as a regulated pathway of cellular degradation. Science *290*, 1717–1721.

Ku, B., Woo, J.S., Liang, C., Lee, K.H., Hong, H.S., E., X., Kim, K.S., Jung, J.U., and Oh, B.H. (2008). Structural and biochemical bases for the inhibition of autophagy and apoptosis by viral BCL-2 of murine gamma-herpesvirus 68. PLoS Pathog. *4*, e25.

Kubota, H., Obata, T., Ota, K., Sasaki, T., and Ito, T. (2003). Rapamycin-induced translational derepression of GCN4 mRNA involves a novel mechanism for activation of the eIF2 alpha kinase GCN2. J. Biol. Chem. *278*, 20457–20460.

Kuma, A., Mizushima, N., Ishihara, N., and Ohsumi, Y. (2002). Formation of the approximately 350-kDa Apg12-Apg5.Apg16 multimeric complex, mediated by Apg16 oligomerization, is essential for autophagy in yeast. J. Biol. Chem. *277*, 18619–18625.

Kyei, G.B., Dinkins, C., Davis, A.S., Roberts, E., Singh, S.B., Dong, C., Wu, L., Kominami, E., Ueno, T., Yamamoto, A., *et al.* (2009). Autophagy pathway intersects with HIV-1 biosynthesis and regulates viral yields in macrophages. J. Cell Biol. *186*, 255–268.

Lam, N., Sandberg, M.L., and Sugden, B. (2004). High physiological levels of LMP1 result in phosphorylation of eIF2 alpha in Epstein–Barr virus-infected cells. J. Virol. *78*, 1657–1664.

Lee, D.Y., and Sugden, B. (2008a). The latent membrane protein 1 oncogene modifies B-cell physiology by regulating autophagy. Oncogene *27*, 2833–2842.

Lee, D.Y., and Sugden, B. (2008b). The LMP1 oncogene of EBV activates PERK and the unfolded protein response to drive its own synthesis. Blood *111*, 2280–2289.

Lee, H.K., Lund, J.M., Ramanathan, B., Mizushima, N., and Iwasaki, A. (2007). Autophagy-dependent viral recognition by plasmacytoid dendritic cells. Science *315*, 1398–1401.

Lee, H.K., Mattei, L.M., Steinberg, B.E., Alberts, P., Lee, Y.H., Chervonsky, A., Mizushima, N., Grinstein, S., and Iwasaki, A. (2010). *In vivo* requirement for atg5 in antigen presentation by dendritic cells immunity. Immunity *28*, 587–590.

Lee, J.S., Li, Q., Lee, J.Y., Lee, S.H., Jeong, J.H., Lee, H.R., Chang, H., Zhou, F.C., Gao, S.J., Liang, C., *et al.* (2009). FLIP-mediated autophagy regulation in cell death control. Nat. Cell Biol. *11*, 1355–1362.

Lee, Y.R., Lei, H.Y., Liu, M.T., Wang, J.R., Chen, S.H., Jiang-Shieh, Y.F., Lin, Y.S., Yeh, T.M., Liu, C.C., and Liu, H.S. (2008). Autophagic machinery activated by dengue virus enhances virus replication. Virology *374*, 240–248.

Leib, D.A., Alexander, D.E., Cox, D., Yin, J., and Ferguson, T.A. (2009). Interaction of ICP34.5 with Beclin. 1 modulates herpes simplex virus type 1 pathogenesis through control of CD4+ T-cell responses. J. Virol. *83*, 12164–12171.

Leib, D.A., Machalek, M.A., Williams, B.R., Silverman, R.H., and Virgin, H.W. (2000). Specific phenotypic restoration of an attenuated virus by knockout of a host resistance gene. Proc. Natl. Acad. Sci. U.S.A. *97*, 6097–6101.

Levine, B. (2005). Eating oneself and uninvited guests: autophagy-related pathways in cellular defense. Cell *120*, 159–162.

Liang, X.H., Kleeman, L.K., Jiang, H.H., Gordon, G., Goldman, J.E., Berry, G., Herman, B., and Levine, B. (1998). Protection against Fatal Sindbis Virus Encephalitis by Beclin, a Novel Bcl-2-Interacting Protein. J. Virol. *72*, 8586–8596.

Ling, Y.M., Shaw, M.H., Ayala, C., Coppens, I., Taylor, G.A., Ferguson, D.J., and Yap, G.S. (2006). Vacuolar and plasma membrane stripping and autophagic elimination of *Toxoplasma gondii* in primed. effector macrophages. J. Exp. Med. *203*, 2063–2071.

Liu, Y., Schiff, M., Czymmek, K., Talloczy, Z., Levine, B., and Dinesh-Kumar, S.P. (2005). Autophagy regulates programmed. cell death during the plant innate immune response. Cell *121*, 567–577.

Massey, A.C., Zhang, C., and Cuervo, A.M. (2006). Chaperone-mediated autophagy in aging and disease. Curr. Top Dev. Biol. *73*, 205–235.

Meijer, A.J., and Codogno, P. (2006). Signalling and autophagy regulation in health, aging and disease. Mol. Aspects Med. *27*, 411–425.

Mizushima, N., Yoshimori, T., and Levine, B. (2010). Methods in mammalian autophagy research. Cell *140*, 313–326.

Mohr, I., Sternberg, D., Ward, S., Leib, D., Mulvey, M., and Gluzman, Y. (2001). A herpes simplex virus type 1 gamma34.5 second-site suppressor mutant that exhibits enhanced growth in cultured glioblastoma cells is severely attenuated in animals. J. Virol. *75*, 5189–5196.

Mortimore, G.E., and Poso, A.R. (1987). Intracellular protein catabolism and its control during nutrient deprivation and supply. Annu. Rev. Nutr *7*, 539–564.

Mulvey, M., Poppers, J., Ladd, A., and Mohr, I. (1999). A herpesvirus ribosome-associated, RNA-binding protein confers a growth advantage upon mutants deficient in a GADD34-related function. J. Virol. *73*, 3375–3385.

Nakagawa, I., Amano, A., Mizushima, N., Yamamoto, A., Yamaguchi, H., Kamimoto, T., Nara, A., Funao, J., Nakata, M., Tsuda, K., *et al.* (2004). Autophagy defends cells against invading group A Streptococcus. Science *306*, 1037–1040.

Orvedahl, A., Alexander, D., Talloczy, Z., Sun, Q., Wei, Y., Zhang, W., Burns, D., Leib, D.A., and Levine, B. (2007). HSV-1 ICP34.5 confers neurovirulence by targeting the Beclin 1 autophagy protein. Cell Host Microbe *1*, 23–35.

Orvedahl, A., and Levine, B. (2009). Eating the enemy within: autophagy in infectious diseases. Cell Death Differ. *16*, 57–69.

Orvedahl, A., Macpherson, S., Sumpter, R., Jr., Talloczy, Z., Zou, Z., and Levine, B. (2010). Autophagy protects

against sindbis virus infection of the central nervous system. Cell Host Microbe 7, 115–127.
Paludan, C., Schmid, D., Landthaler, M., Vockerodt, M., Kube, D., Tuschl, T., and Munz, C. (2005). Endogenous MHC class II processing of a viral nuclear antigen after autophagy. Science *307*, 593–596.
Pereira, L., Wolff, M.H., Fenwick, M., and Roizman, B. (1977). Regulation of herpesvirus macromolecular synthesis. Properties, V. of alpha polypeptides made in H.S.V.-1 and H.S.V.-2 infected cells. Virology 77, 733–Peters, G.A., Khoo, D., Mohr, I., and Sen, G.C. (2002). Inhibition of PACT-mediated activation of PKR by the herpes simplex virus type 1 Us11 protein. J. Virol. *76*, 11054–11064.
Prentice, E., Jerome, W.G., Yoshimori, T., Mizushima, N., and Denison, M.R. (2004). Coronavirus replication complex formation utilizes components of cellular autophagy. J. Biol. Chem. *279*, 10136–10141.
Rich, K.A., Burkett, C., and Webster, P. (2003). Cytoplasmic bacteria can be targets for autophagy. Cell. Microbiol. *5*, 455–468.
Roy, D.J., Ebrahimi, B.C., Dutia, B.M., Nash, A.A., and Stewart, J.P. (2000). Murine gammaherpesvirus M11 gene product inhibits apoptosis and is expressed during virus persistence. Arch. Virol. *145*, 2411–2420.
Schmid, D., Pypaert, M., and Munz, C. (2007). Antigen-loading compartments for major histocompatibility complex class II molecules continuously receive input from autophagosomes. Immunity *26*, 79–92.
Shintani, T., and Klionsky, D.J. (2004). Autophagy in health and disease: a double-edged sword. Science *306*, 990–995.
Shintani, T., Mizushima, N., Ogawa, Y., Matsuura, A., Noda, T., and Ohsumi, Y. (1999). Apg10p, a novel protein-conjugating enzyme essential for autophagy in yeast. EMBO J. *18*, 5234–5241.
Sinha, S., Colbert, C.L., Becker, N., Wei, Y., and Levine, B. (2008). Molecular basis of the regulation of Beclin. 1-dependent autophagy by the gamma-herpesvirus 68 Bcl-2 homolog M11. Autophagy *4*, 989–997.
Sir, D., and Ou, J.H. (2010). Autophagy in viral replication and pathogenesis. Mol. Cells *29*, 1–7.
Smith, J.D., and de Harven, E. (1978). Herpes simplex virus and human cytomegalovirus replication in WI-38 cells. III. Cytochemical localization of lysosomal enzymes in infected cells. J. Virol. *26*, 102–109.
Stephenson, L.M., Miller, B.C., Ng, A., Eisenberg, J., Zhao, Z., Cadwell, K., Graham, D.B., Mizushima, N.N., Xavier, R., Virgin, H.W., *et al.* (2009). Identification of Atg5-dependent transcriptional changes and increases in mitochondrial mass in Atg5-deficient T lymphocytes. Autophagy *5*, 625–635.
Suzuki, K., and Ohsumi, Y. (2010). Current knowledge of the pre-autophagosomal structure (PAS). FEBS Lett. *584*, 1280–1286.
Takahashi, M.N., Jackson, W., Laird, D.T., Culp, T.D., Grose, C., Haynes, J.I., 2nd, and Benetti, L. (2009). Varicella-zoster virus infection induces autophagy in both cultured cells and human skin vesicles. J. Virol. *83*, 5466–5476.
Takahashi, Y., Coppola, D., Matsushita, N., Cualing, H.D., Sun, M., Sato, Y., Liang, C., Jung, J.U., Cheng, J.Q., Mule, J.J., *et al.* (2007). Bif-1 interacts with Beclin. 1 through UVRAG and regulates autophagy and tumorigenesis. Nat. Cell Biol. *9*, 1142–1151.
Talloczy, Z., Jiang, W., Virgin, H.W.t., Leib, D.A., Scheuner, D., Kaufman, R.J., Eskelinen, E.L., and Levine, B. (2002). Regulation of starvation- and virus-induced autophagy by the eIF2alpha kinase signaling pathway. Proc. Natl. Acad. Sci. U.S.A. *99*, 190–195.
Talloczy, Z., Virgin, H.W.I., and Levine, B. (2006). PKR-dependent autophagic degradation of herpes simplex virus type 1. Autophagy *2*, 24–29.
Tanida, I., Mizushima, N., Kiyooka, M., Ohsumi, M., Ueno, T., Ohsumi, Y., and Kominami, E. (1999). Apg7p/Cvt2p: A novel protein-activating enzyme essential for autophagy. Mol. Biol. Cell *10*, 1367–1379.
Todde, V., Veenhuis, M., and van der Klei, I.J. (2009). Autophagy: principles and significance in health and disease. Biochim. Biophys. Acta *1792*, 3–13.
Tsukada, M., and Ohsumi, Y. (1993). Isolation and characterization of autophagy-defective mutants of Saccharomyces cerevisiae. FEBS Lett. 333, 169–174.
Uttenweiler, A., and Mayer, A. (2008). Microautophagy in the yeast Saccharomyces cerevisiae. Methods Mol. Biol. *445*, 245–259.
Virgin, H.W., and Levine, B. (2009). Autophagy genes in immunity. Nat. Immunol. *10*, 461–470.
Wang, G.H., Garvey, T.L., and Cohen, J.I. (1999). The murine gammaherpesvirus-68 M11 protein inhibits Fas- and TNF-induced apoptosis. J. Gen. Virol. *80*, 2737–2740.
Yang, Z., and Klionsky, D.J. (2009). An overview of the molecular mechanism of autophagy. Curr. Top. Microbiol. Immunol. 335, 1–32.
Zhao, Z., Thackray, L.B., Miller, B.C., Lynn, T.M., Becker, M.M., Ward, E., Mizushima, N.N., Denison, M.R., and Virgin, H.W.t. (2007). Coronavirus replication does not require the autophagy gene ATG5. Autophagy 3, 581–585.

Human Alphaherpesvirus MicroRNAs: Expression and Functions

22

Jennifer L. Umbach and Bryan R. Cullen

Abstract
MicroRNAs (miRNAs) are an extensive class of ~22 nucleotide long regulatory RNAs expressed by all mammalian cells and also by several DNA viruses, including many members of the herpesvirus family. Using deep sequencing technology, it has now been demonstrated that Herpes Simplex Virus 1 (HSV-1) encodes at least eight viral miRNAs, seven of which are expressed in latently infected human neurons. Similarly, HSV-2 has also been shown to encode at least six miRNAs, four of which are evolutionarily conserved between HSV-2 and HSV-1. Perhaps surprisingly, varicella zoster virus does not appear to express any viral miRNAs in latently infected cells. This review discusses the potential functions of the currently known HSV-1 and HSV-2 miRNAs, focusing on a possible role in stabilizing viral latency in infected neurons.

Introduction

MicroRNAs (miRNAs) are a recently discovered family of ~22 nucleotide (nt) long regulatory RNAs that are expressed by all known multicellular eukaryotes (Bartel, 2009). Current evidence indicates that humans express over 700 different miRNAs. Almost all known miRNAs are initially expressed as part of a capped, polyadenylated transcript referred to as a primary miRNA (pri-miRNA)(Cai *et al.*, 2004; Kim, 2005). Within the pri-miRNA, the miRNA forms the upper ~22 nt of one arm of an ~32 base pair (bp) imperfect RNA stem bearing a large terminal loop (Zeng *et al.*, 2005). This stem–loop is recognized by a nuclear protein heterodimer, the microprocessor, which consists of the RNase III enzyme Drosha and an RNA binding co-factor called DGCR8. Cleavage of the pri-miRNA stem–loop results in the production of an ~60 nt hairpin RNA, called the pre-miRNA intermediate, and leaves an ~2 nt 3′ overhang (Kim, 2005).

The pre-miRNA is exported to the cytoplasm by Exportin 5 (Yi *et al.*, 2003) and there encounters a second RNase III enzyme, called Dicer. Dicer binds to the base of the pre-miRNA hairpin and then cleaves ~22 bp away, to remove the terminal loop and leave a second ~2 nt 3′ overhang (Kim, 2005). One strand of the resultant miRNA duplex intermediate, the miRNA strand, is then incorporated into the RNA induced silencing complex (RISC), where it acts as a guide RNA to direct RISC to complementary mRNA targets. The other strand, termed the passenger strand, is degraded during RISC loading. Of note, discrimination between the miRNA strand and the passenger strand is rarely absolute, so that low but detectable levels of the passenger strand may be loaded into RISC and recovered by conventional or deep sequencing techniques.

Once loaded into RISC, the miRNA can guide RISC to fully complementary mRNA targets, in which case mRNA cleavage usually results, or to partially complementary mRNA targets, resulting in translational inhibition and, in most cases, a modest but detectable drop in mRNA stability. Importantly, functionally relevant mRNA targets almost always show complete homology to at least the miRNA 'seed' sequence, extending from position 2 to 7 or 8 from the 5′ end of the miRNA (Bartel, 2009).

Because excision of the pre-miRNA intermediate by Drosha results in the nuclear cleavage of the pri-miRNA, a single transcript cannot function as both a pri-miRNA and as an mRNA. Therefore, almost all pri-miRNA hairpins are located either in non-coding RNAs or are located within introns present in mRNAs or non-coding RNAs (Cai *et al.*, 2004). A pri-miRNA may contain a single pri-miRNA stem–loop or may contain a cluster of multiple stem–loops. At least in the latter case, the full-length pri-miRNA is likely to be quite unstable and, hence, difficult to detect.

While miRNAs were first identified in nematodes, fruit flies and humans, it soon became apparent that some viruses also encode miRNAs. Viral miRNAs were first identified in the human gammaherpesvirus Epstein–Barr virus (EBV) (Pfeffer *et al.*, 2004) but subsequent work has identified miRNAs in a wide variety of herpesviruses and also in several polyomaviruses (Umbach and Cullen, 2009). However, miRNAs have not been detected in any RNA viruses, including retroviruses and flaviviruses. This likely reflects the fact that any encoded miRNA would, upon excision, lead to the degradation of the viral genomic RNA. Indeed, insertion of a pri-miRNA into a retroviral or lentiviral genome can result in a significant, but certainly not complete, loss of viral titre (E. Gottwein and B.R. Cullen, unpublished observations).

Identification of HSV-1 microRNAs

The identification of miRNAs encoded by EBV and several other beta- or gammaherpesvirus species suggested that human alphaherpesviruses might also encode viral miRNAs. The initial approach to this problem was primarily computational, based on the identification of potential miRNA hairpin structures. Using this approach, Pfeffer *et al.* (2005) hypothesized that herpes simplex virus 1 (HSV-1) would encode seven different viral miRNAs while HSV-2 might encode six. Interestingly, no viral miRNAs were computationally predicted for varicella zoster virus (VZV), the third, more distantly related, human alphaherpesvirus. A conceptually similar study (Cui *et al.*, 2006) also predicted a number of potential HSV-1 miRNAs using a computational approach, but this analysis also included Northern analyses of productively HSV-1-infected cells, resulting in the identification of the first HSV-1 miRNA, miR-H1 (Fig. 22.1). However, none of the other candidate HSV-1 miRNAs could

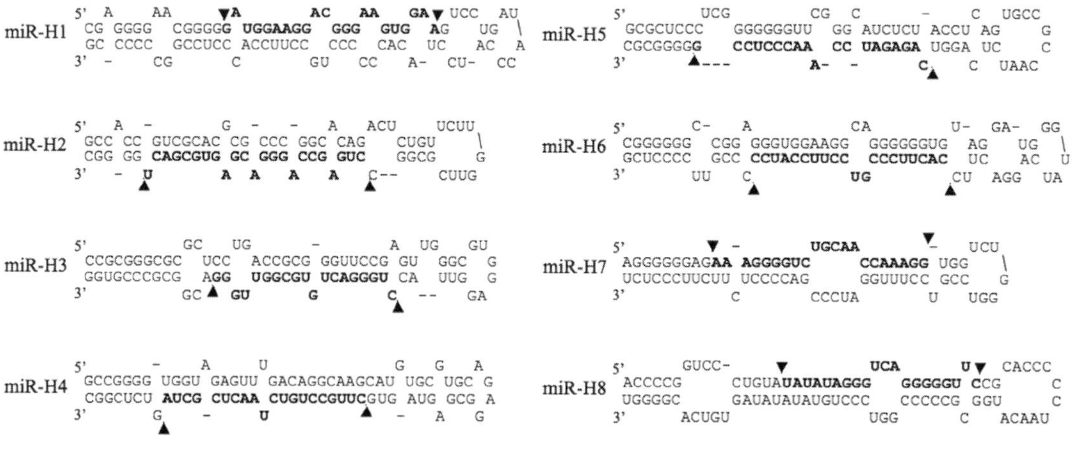

Figure 22.1 Predicted secondary structures of the HSV-1 primary microRNA stem–loops. These computationally predicted RNA secondary structures were generated using MFOLD software (Mathews *et al.*, 1999; Zuker, 2003). While this tends to insert small stems in the apical region of the pri-miRNA hairpins, experimental evidence suggests that the terminal loops are actually unstructured (Zeng *et al.*, 2005). The sequences of the major mature viral miRNAs are indicated in bold. Note that the pri-miRNA stem–loops encoding miR-H1 and miR-H6 lie perfectly antisense to one another.

be confirmed in this second study by analysis of productively infected cells.

An alternative approach to the identification of HSV-1-encoded miRNAs was reported by Umbach *et al.* (2008; 2009), who used deep sequencing technologies to identify HSV-1 miRNA expressed either in the latently HSV-1 infected trigeminal ganglia of experimentally infected mice (Umbach *et al.*, 2008) or expressed in naturally HSV-1-infected trigeminal ganglia derived post-mortem from humans (Umbach *et al.*, 2009). This effort resulted in the identification of seven additional HSV-1 miRNAs, miR-H2 though miR-H8 in mice and/or in humans (Fig. 22.1).

Analysis of the genomic location of these eight miRNAs revealed that all but two of the HSV-1 miRNAs derived from the second exon of the HSV-1 latency associated transcript LAT (Kang *et al.*, 2006; Umbach *et al.*, 2008; Umbach *et al.*, 2009). LAT is initially transcribed as an ~8 kb capped, polyadenylated RNA that is spliced to give rise to an ~2 kb intron and an ~6 kb exonic RNA (Kang *et al.*, 2006; Fig. 22.2). Remarkably, the ~6 kb exonic RNA is highly unstable while the ~2 kb intron accumulates to high levels in latently HSV-1-infected neurons (Farrell *et al.*, 1991; Kang *et al.*, 2006; Stevens *et al.*, 1987). Historically, LAT was thought to be the only viral transcript expressed at significant levels in latently HSV-1-infected neurons.

As shown in Fig. 22.2, six of the HSV-1 miRNAs map to the second exon of the unstable, exonic LAT while none map to the first exon or the stable, intronic segment. The remaining two HSV-1 miRNAs, miR-H1 and miR-H6, are encoded outside of the LAT. miR-H1 and miR-H6 in fact derive from 'mirror image' pri-miRNA stem–loops that are located exactly opposite to one another in the HSV-1 genome (Umbach *et al.*, 2008; Fig. 22.1). Interestingly, while miR-H6 is detected both during HSV-1 latency and productive infection, miR-H1 expression is low or undetectable during latency but high in productively HSV-1-infected cells (Cui *et al.*, 2006; Umbach *et al.*, 2008; Umbach *et al.*, 2009). Consistent with the high expression of LAT during latency, but not during productive HSV-1 replication (Kang *et al.*, 2006), five of the other HSV-1 miRNAs (miR-H2 through miR-H5 and miR-H7) are also expressed at high levels in latently infected neurons but at low, though detectable, levels during productive HSV-1 replication (Umbach *et al.*, 2008; Umbach *et al.*,

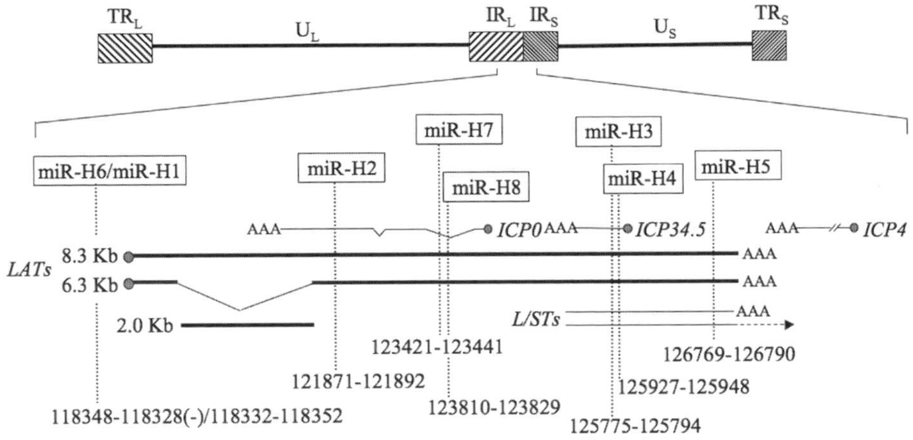

Figure 22.2 Genomic location of the HSV-1 miRNAs. Schematic of the HSV-1 linear DNA genome expanded to show details of the LAT locus. Relative sizes, locations and orientations of other viral transcripts in this region, including the viral L/ST transcripts, are also indicated. Sequence co-ordinates of virally-encoded miRNAs are given according to the HSV-1 strain 17 syn+ genome (NC_001806). All viral miRNAs are in the same transcriptional orientation as LAT, except for miR-H6. TR, terminal repeat; IR, internal repeat; U_L, unique long; U_S, unique short.

2009). HSV-1 miR-H8 is unusual in that it is nearly undetectable in latently infected neurons even through it is also derived from LAT.

Identification of HSV-2 microRNAs

In parallel to the identification of the HSV-1 described above, Tang *et al.* (Tang *et al.*, 2008; Tang *et al.*, 2009) used sequencing of small RNAs recovered from cells transfected with an HSV-2 LAT expression plasmid to identify three HSV-2 miRNAs. These workers went on to confirm expression of these three HSV-2 miRNAs in latently and/or lytically infected cells. While these three HSV-2 miRNAs were initially given the names miR-I, miR-II and miR-III, their sequence similarity to the HSV-1 miRNAs miR-H3, miR-H4 and miR-H2 respectively (Fig. 22.3), when viewed in combination with their shared locations within the genomes of these closely related herpesviruses (Figs. 22.2 and 22.4), led the miRBase database (Griffiths-Jones *et al.*, 2006) to rename them HSV-2 miR-H3, miR-H4 and miR-H2.

These initial studies have recently been extended by the use of deep sequencing to characterize small RNAs expressed in human sacral ganglia latently infected with HSV-2 (Umbach *et al.*, 2010). This work confirmed the existence of the previously reported HSV-2 miR-H3 and miR-H4 miRNAs and also identified three other HSV-2 miRNAs. One of these, HSV-2 miR-H7, is very similar in genomic location (Fig. 22.4) and sequence (Fig. 22.3) to HSV-1 miR-H7 (Fig. 22.2). However, HSV-2 miR-H9 and miR-H10 do not presently have any known HSV-1 equivalents. HSV-2 miR-H10 is of particular interest as it is currently unique among simplex miRNAs in not being encoded within or immediately adjacent to LAT (Fig. 22.4). Instead, miR-H10 is located in a short non-coding region located between the convergent viral genes encoding HSV-2 UL15 and UL18, in the unique long (U_L) region of the HSV-2 genome (Fig. 22.4). The identity of the pri-miRNA that gives rise to HSV-2 miR-H10 is currently unknown.

While HSV-2 has now been shown to encode at least six viral miRNAs, and HSV-1 at least eight, these numbers are unlikely to be complete. There is no report published so far that has used deep sequencing technology to examine HSV-1 or HSV-2 miRNA expression in productively infected cells. Indeed, we only know of one lytic phase specific HSV-1 miRNA, miR-H1, and this was not identified by sequencing (Cui *et al.*, 2006). Moreover, especially for HSV-2, the number of individual cDNA sequences recovered for each virally encoded miRNA *in vivo* was quite low, despite the use of deep sequencing technologies (Umbach *et al.*, 2010). While this result presumably reflects the small number of latently HSV-2 infected neurons in the three cadavers analysed, it clearly remains possible, indeed likely,

HSV-1 miR-H2-3p	CTGAGCCAGGGACGAGTGCGACT	70%
HSV-2 miR-H2-3p	CTGAGCCTGGGTCATGCGCGA--	
HSV-1 miR-H3-3p	-CTGGGACTGTGCGGTTGGG---	74%
HSV-2 miR-H3-3p	TTTGGGAGTCTGCGGTTGGGAGC	
HSV-1 miR-H4-5p	GGTAGAGTTTGACAGGCAAGCA---	52%
HSV-2 miR-H4-5p	---AGAGTTCACTCGGCACGCATGC	
HSV-1 miR-H7-5p	AAAGGGGTCTGCAACCAAAGG	91%
HSV-2 miR-H7-5p	AAAGGGGTCCGTAACCAAAGG	

Figure 22.3 Sequence alignment of the four conserved simplex virus miRNAs miR-H2, miR-H3, miR-H4 and miR-H7. Conserved bases are indicated by light shading and conserved miRNA seed sequences boxed.

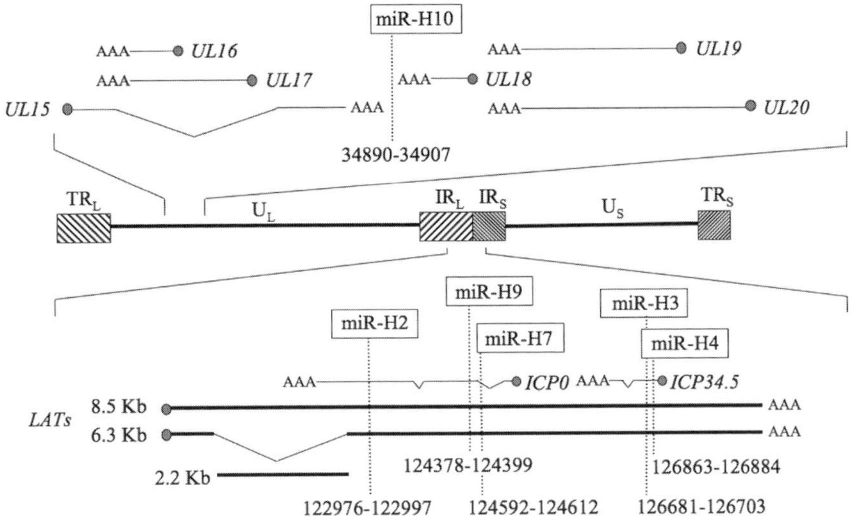

Figure 22.4 Genomic location of the HSV-2 miRNAs. This figure is closely similar in design to Fig. 22.2, except that the HSV-2 genome is shown. Both the LAT locus and the genomic region in U_L that flanks the miR-H10 miRNA are expanded. Sequence coordinates of virally encoded miRNAs are given according to the HSV-2 strain HG52 genome (NC_001798).

that HSV-2 expresses additional viral miRNAs in both latently and productively infected cells. In fact, as noted previously (Tang *et al.*, 2009), sequence analysis indicates that the HSV-1 miRNA miR-H5 is almost entirely conserved in HSV-2 and is also predicted to fold into a canonical miRNA stem–loop structure, yet this miRNA was not recovered from latently HSV-2 infected sacral ganglia (Umbach *et al.*, 2010).

VZV miRNAs: missing in action?

Deep sequencing analysis of small RNAs expressed within latently HSV-1-infected human trigeminal ganglia and HSV-2-infected human sacral ganglia failed to recover any miRNAs of VZV origin even though numerous HSV-1 or HSV-2 derived miRNAs could be identified (Umbach *et al.*, 2009; Umbach *et al.*, 2010). This negative result was obtained despite the fact that VZV genomic DNA, presumably present in latently VZV-infected neurons, could be readily detected by PCR. Moreover, analysis of trigeminal ganglia isolated from rhesus macaques experimentally infected with simian varicella virus, a close primate relative of VZV, also failed to identify any miRNAs of viral origin (Umbach *et al.*, 2009). Together, these data suggest that VZV, unlike HSV-1 and HSV-2, does not express viral miRNAs in latently infected neurons. Interestingly, while HSV-1 and HSV-2 both encode LAT transcripts, the more distantly related VZV lacks the genomic region homologous to the simplex LAT locus (Cohrs *et al.*, 2000; McGeoch, 1989). We also note that Pfeffer *et al.* (2005) computationally predicted several miRNAs in HSV-1 and HSV-2 (some, but not all of which, were then experimentally confirmed), yet did not predict any miRNA stem–loops in VZV. While it is therefore possible that VZV, currently uniquely among herpesviruses, does not encode any viral miRNAs, it is also possible that these do exist but are only expressed during productive viral replication.

Functions of HSV-1 and HSV-2 microRNAs

Inspection of the genomic locations of the various HSV-1 and HSV-2 miRNAs (Figs. 22.2 and 22.4) shows that HSV-1 and HSV-2 miR-H2 share a conserved genomic location antisense to the third exon of the viral mRNA encoding ICP0 while both HSV-1 and HSV-2 miR-H3 and miR-H4 share conserved genomic locations antisense to the viral ICP34.5 mRNA. ICP0 is an immediate early viral protein that functions as a *trans*-activator of viral early gene expression and that has been proposed to play a key role in both the initiation of lytic HSV-1 and HSV-2 replication and in reactivation from viral latency (Cai *et al.*, 1993; Everett, 2000; Halford and Schaffer, 2001; Loiacono *et al.*, 2003). On the other hand, ICP34.5 has no known role in reactivation from latency and, in fact, requires viral DNA synthesis for maximal expression (Randall and Roizman, 1997). ICP34.5 has been shown to inhibit the activation of the host innate antiviral factor PKR and to thereby prevent a virally induced shut-off of protein synthesis in infected cells (Chou and Roizman, 1994). ICP34.5 has also been shown to inhibit autophagy by targeting the host cell protein Beclin 1 (Orvedahl *et al.*, 2007). These actions of ICP34.5 appear essential for HSV-1 neurovirulence in experimentally infected mice.

Because miR-H2 lies antisense to the ICP0 mRNA, while miR-H3 and miR-H4 lie antisense to the ICP34.5 mRNA (Figs. 22.2 and 22.4), it might be expected that these viral miRNAs would be able to inhibit the expression of these viral gene products. Indeed, both HSV-1 miR-H2 and HSV-2 miR-H2 have been shown to inhibit ICP0 expression (Tang *et al.*, 2009; Umbach *et al.*, 2008), while HSV-2 miR-H3 and miR-H4 have both been reported to inhibit HSV-2 ICP34.5 expression (Tang *et al.*, 2008, 2009). In addition, Umbach *et al.* (2008) noted that HSV-1 miR-H6 has extended seed homology to the viral transcript encoding ICP4 and reported that miR-H6 expression can indeed strongly inhibit the expression of the ICP4 protein. Of note, ICP4 is an essential viral immediate early protein (Preston, 1979) that, like ICP0, can function to induce exit from latency (Halford and Schaffer, 2001). No miRNA equivalent to HSV-1 miR-H6 has been identified in HSV-2 to this point.

Previous work focusing on human cytomegalovirus (hCMV) had led to the suggestion that herpesvirus-encoded miRNAs might function to stabilize viral latency by inhibiting the expression of viral immediate early gene products (Murphy *et al.*, 2008) and the data reported above are certainly consistent with this hypothesis. In particular, the finding that miR-H2 can inhibit the expression of HSV-1 and HSV-2 ICP0, while miR-H6 inhibits the expression of the viral ICP4 protein, strongly suggests that these two viral miRNAs stabilize viral latency and inhibit entry into the productive virus replication cycle. It is less clear why both HSV-1 and HSV-2 would encode miRNAs that can inhibit the expression of the viral ICP34.5 protein, especially as ICP34.5 is a viral late protein that is not expressed either during latency or immediately after induction of lytic replication (Randall and Roizman, 1997). One possibility, however, is that inhibition of ICP34.5 expression, a key mediator of HSV-1 neurovirulence, may partially protect neurons from the cytotoxic effects of virus replication.

The observation that exonic LAT functions as a viral pri-miRNA can explain a number of previous reports concerning LAT function and also provides an explanation for the existence of exonic LAT, a highly unstable non-coding RNA whose function was previously obscure. (Note that these observations shed no light on the function of the stable, intronic LAT, as this ~2 kb RNA does not function as a pri-miRNA precursor). A number of groups have previously reported that LAT can inhibit ICP0 expression and promote viral latency (Garber *et al.*, 1997; Mador *et al.*, 1998; Maillet *et al.*, 2006; Thompson and Sawtell, 1997). Moreover, it has also been reported that LAT can protect neurons from apoptosis induced by herpes simplex virus infection, which could reflect the inhibition of IPC0 and/or ICP34.5 protein expression by LAT-derived miRNAs (Branco and Fraser, 2005; Perng *et al.*, 2000; Thompson and Sawtell, 2001). Finally, the identification of miR-H3 and miR-H4, and the demonstration that these viral miRNAs can inhibit ICP34.5 expression, explains the previous observation that ICP34.5 expression can be inhibited by the so-called L/ST transcripts of HSV-1. The HSV-1 L/ST RNAs are transcribed from an internal promoter, within the second exon of LAT (Fig. 22.2), but are only observed at significant levels in cells infected by HSV-1 mutants lacking a functional ICP4 protein, as ICP4 normally potently inhibits L/ST promoter activity. The L/ST transcripts lie antisense to the ICP34.5 mRNA and have previously been reported to inhibit ICP34.5 mRNA expression via an 'antisense mechanism' (Lee and Schaffer, 1998; Randall and Roizman, 1997). The observation that miR-H3 and miR-H4 are encoded within the L/ST RNAs (Fig. 22.2) suggests that it is, in fact, these two virally encoded miRNAs that exert this inhibitory effect on ICP34.5 expression.

While the above discussion identifies potential mRNA targets and functions for the HSV-1 and HSV-2 miRNAs miR-H2, miR-H3 and HSV-H4, as well as for the HSV-1 miR-H6 miRNA, it does not provide any insights into the functions of the other virally encoded miRNAs. With the exception of HSV-2 miR-H9, which lies antisense to exon 2 of the HSV-2 ICP0 mRNA (Fig. 22.4), and may therefore share the ability of miR-H2 to inhibit ICP0 gene expression (Umbach *et al.*, 2010), none of the other HSV-1 miRNAs (i.e. miR-H1, miR-H5, miR-H7 and miR-H8) or HSV-2 miRNAs (i.e. miR-H7 and miR-H10) lie antisense to a viral mRNA (Figs. 22.2 and 22.4) or show extensive and/or obvious homology to a viral transcript. It therefore seems likely that these other viral miRNAs function to repress neuronal mRNA transcripts that might, for example, encode proteins with antiviral activity. Moreover, the fact that miR-H2, miR-H3, miR-H4 and miR-H6 repress viral mRNAs does not preclude the possibility that they might also repress one or more cellular gene products. Indeed, a single miRNA encoded by hCMV, called miR-UL112–1, has been shown to repress both a viral mRNA and a cellular mRNA in infected cells (Grey *et al.*, 2007; Murphy *et al.*, 2008; Stern-Ginossar *et al.*, 2007).

Insight into whether viral miRNAs repress only viral transcripts, or instead also repress cellular mRNAs, can in principle be gained by a comparison of the sequences of the miRNAs conserved in HSV-1 and HSV-2. While these viruses are thought to have diverged ~8 million years ago, both HSV-1 and HSV-2 nevertheless encode a similar set of 74 protein-coding genes and show ~83% sequence conservation (Dolan *et al.*, 1998). Because miR-H2, miR-H3 and miR-H4 lie antisense to their known viral mRNA targets, i.e. the ICP0 and ICP34.5 mRNAs, any sequence change in the viral miRNA will, by definition, also occur in the viral mRNA and there should therefore be little pressure to maintain the sequence of the viral miRNA, other than to preserve the critical structural aspects of the viral pri-miRNA stem–loop (Fig. 22.1). Conversely, if a viral miRNA targets a cellular miRNA, which is likely to evolve very slowly, then one would predict that the critical seed region of the miRNA, i.e. position 2 to 7 or 8 from the 5′ end, would be evolutionarily conserved.

Alignment of the four miRNAs conserved from HSV-1 to HSV-2 reveals sequence conservation ranging from 52% (miR-H4) to 91% (miR-H7) (Fig. 22.3). The very high sequence conservation, including the miRNA seed sequence, seen with miR-H7 is striking as this miRNA is the only one of these four viral miRNAs that does not lie antisense to an exonic region of a target viral mRNA (Figs. 22.2 and 22.4) and miR-H7 is therefore the most likely of these four viral miRNAs to target a cellular mRNA(s). (miR-H7 does lie antisense to the ICP0 gene, but opposite an intronic region. As mammalian miRNAs function predominantly or exclusively in the cytoplasm of expressing cells, it is unlikely that miR-H7 inhibits ICP0 mRNA expression.)

In contrast to miR-H7, HSV-1 and HSV-2 miR-H3 and miR-H4 show less sequence conservation and also have different 5′ processing sites (Fig. 22.3). As a result, the seed regions of these two viral miRNAs are distinct and it is therefore very unlikely that these miRNAs could target the same cellular mRNAs, although their ability to target the antisense ICP34.5 mRNA would, as noted above, be unaffected. It is therefore likely that miR-H3 and miR-H4 only function to inhibit ICP34.5 expression. Finally, the miR-H2 miRNA expressed by both HSV-1 and HSV-2 does share a conserved miRNA seed sequence and it is therefore possible that the miR-H2 miRNA not only downregulates ICP0 expression in latently infected neurons but also inhibits the expression of one or more conserved cellular mRNA targets.

Conclusions

The recent identification of a series of miRNAs encoded by both HSV-1 and HSV-2 has the potential to revolutionize our understanding of the mechanisms underlying HSV-1 and HSV-2 latency. In particular, it is now clear that the previously mysterious exonic LAT–a capped, polyadenylated non-coding RNA that is highly unstable–actually functions as a pri-miRNA precursor that gives rise to multiple viral miRNAs in latently infected neurons. Evidence, summarized above, that several of these miRNAs target distinct viral mRNAs for inhibition, including the viral mRNAs encoding the key immediate early transactivators ICP0 and ICP4, as well as the viral pathogenicity factor ICP34.5, suggests that a subset of these viral miRNAs act to stabilize the latent state and may also serve to protect neurons from cytotoxicity after entry into the productive viral life cycle. The validation of the possibly key role of virally encoded miRNAs in the establishment of HSV-1 and HSV-2 latency, or of the alternative hypothesis that the viral miRNAs play only minor roles in the viral life cycles of HSV-1 and HSV-2, must await the full definition of the miRNA coding potential of these viruses and the functional analysis of viral mutants, lacking specific miRNAs, in experimentally infected animals.

References

Bartel, D.P. (2009). MicroRNAs: target recognition and regulatory functions. Cell *136*, 215–233.

Branco, F.J., and Fraser, N.W. (2005). Herpes simplex virus type 1 latency-associated transcript expression protects trigeminal ganglion neurons from apoptosis. J. Virol. *79*, 9019–9025.

Cai, W., Astor, T.L., Liptak, L.M., Cho, C., Coen, D.M., and Schaffer, P.A. (1993). The herpes simplex virus type 1 regulatory protein ICP0 enhances virus replication during acute infection and reactivation from latency. J. Virol. *67*, 7501–7512.

Cai, X., Hagedorn, C.H., and Cullen, B.R. (2004). Human microRNAs are processed from capped, polyadenylated transcripts that can also function as mRNAs. RNA *10*, 1957–1966.

Cohrs, R.J., Randall, J., Smith, J., Gilden, D.H., Dabrowski, C., van Der Keyl, H., and Tal-Singer, R. (2000). Analysis of individual human trigeminal ganglia for latent herpes simplex virus type 1 and varicella-zoster virus nucleic acids using real-time PCR. J. Virol. *74*, 11464–11471.

Cui, C., Griffiths, A., Li, G., Silva, L.M., Kramer, M.F., Gaasterland, T., Wang, X.J., and Coen, D.M. (2006). Prediction and identification of herpes simplex virus 1-encoded microRNAs. J. Virol. *80*, 5499–5508.

Dolan, A., Jamieson, F.E., Cunningham, C., Barnett, B.C., and McGeoch, D.J. (1998). The genome sequence of herpes simplex virus type 2. J. Virol. *72*, 2010–2021.

Everett, R.D. (2000). ICP0, a regulator of herpes simplex virus during lytic and latent infection. Bioessays *22*, 761–770.

Farrell, M.J., Dobson, A.T., and Feldman, L.T. (1991). Herpes simplex virus latency-associated transcript is a stable intron. Proc. Natl. Acad. Sci. U.S.A. *88*, 790–794.

Garber, D.A., Schaffer, P.A., and Knipe, D.M. (1997). A LAT-associated function reduces productive-cycle gene expression during acute infection of murine sensory neurons with herpes simplex virus type 1. J. Virol. *71*, 5885–5893.

Grey, F., Meyers, H., White, E.A., Spector, D.H., and Nelson, J. (2007). A human cytomegalovirus-encoded microRNA regulates expression of multiple viral genes involved in replication. PLoS Pathog. *3*, e163.

Griffiths-Jones, S., Grocock, R.J., van Dongen, S., Bateman, A., and Enright, A.J. (2006). miRBase: microRNA sequences, targets and gene nomenclature. Nucleic Acids Res. *34*, D140–144.

Halford, W.P., and Schaffer, P.A. (2001). ICP0 is required for efficient reactivation of herpes simplex virus type 1 from neuronal latency. J. Virol. *75*, 3240–3249.

Kang, W., Mukerjee, R., Gartner, J.J., Hatzigeorgiou, A.G., Sandri-Goldin, R.M., and Fraser, N.W. (2006). Characterization of a spliced exon product of herpes simplex type-1 latency-associated transcript in productively infected cells. Virology *356*, 106–114.

Kim, V.N. (2005). MicroRNA biogenesis: coordinated cropping and dicing. Nat. Rev. Microbiol. *6*, 376–385.

Lee, L.Y., and Schaffer, P.A. (1998). A virus with a mutation in the ICP4-binding site in the L/ST promoter of herpes simplex virus type 1, but not a virus with a mutation in open reading frame P, exhibits cell-type-specific expression of gamma(1)34.5 transcripts and latency-associated transcripts. J. Virol. *72*, 4250–4264.

Loiacono, C.M., Taus, N.S., and Mitchell, W.J. (2003). The herpes simplex virus type 1 ICP0 promoter is activated by viral reactivation stimuli in trigeminal ganglia neurons of transgenic mice. J. NeuroVirol. *9*, 336–345.

Mador, N., Goldenberg, D., Cohen, O., Panet, A., and Steiner, I. (1998). Herpes simplex virus type 1 latency-associated transcripts suppress viral replication and reduce immediate-early gene mRNA levels in a neuronal cell line. J. Virol. *72*, 5067–5075.

Maillet, S., Naas, T., Crepin, S., Roque-Afonso, A.M., Lafay, F., Efstathiou, S., and Labetoulle, M. (2006). Herpes simplex virus type 1 latently infected neurons differentially express latency-associated and ICP0 transcripts. J. Virol. *80*, 9310–9321.

Mathews, D.H., Sabina, J., Zuker, M., and Turner, D.H. (1999). Expanded sequence dependence of thermodynamic parameters improves prediction of RNA secondary structure. J. Mol. Biol. *288*, 911–940.

McGeoch, D.J. (1989). The genomes of the human herpesviruses: contents, relationships, and evolution. Annu. Rev. Microbiol. *43*, 235–265.

Murphy, E., Vanicek, J., Robins, H., Shenk, T., and Levine, A.J. (2008). Suppression of immediate-early viral gene expression by herpesvirus-coded microRNAs: implications for latency. Proc. Natl. Acad. Sci. U.S.A. *105*, 5453–5458.

Orvedahl, A., Alexander, D., Talloczy, Z., Sun, Q., Wei, Y., Zhang, W., Burns, D., Leib, D.A., and Levine, B.

(2007). HSV-1 ICP34.5 confers neurovirulence by targeting the Beclin. 1 autophagy protein. Cell Host Microbe *1*, 23–35.

Perng, G.C., Jones, C., Ciacci-Zanella, J., Stone, M., Henderson, G., Yukht, A., Slanina, S.M., Hofman, F.M., Ghiasi, H., Nesburn, A.B., *et al.* (2000). Virus-induced neuronal apoptosis blocked by the herpes simplex virus latency-associated transcript. Science *287*, 1500–1503.

Pfeffer, S., Sewer, A., Lagos-Quintana, M., Sheridan, R., Sander, C., Grasser, F.A., van Dyk, L.F., Ho, C.K., Shuman, S., Chien, M., *et al.* (2005). Identification of microRNAs of the herpesvirus family. Nat. Methods *2*, 269–276.

Pfeffer, S., Zavolan, M., Grasser, F.A., Chien, M., Russo, J.J., Ju, J., John, B., Enright, A.J., Marks, D., Sander, C., *et al.* (2004). Identification of virus-encoded microRNAs. Science *304*, 734–736.

Preston, C.M. (1979). Control of herpes simplex virus type 1 mRNA synthesis in cells infected with wild-type virus or the temperature-sensitive mutant tsK. J. Virol. *29*, 275–284.

Randall, G., and Roizman, B. (1997). Transcription of the derepressed open reading frame P of herpes simplex virus 1 precludes the expression of the antisense gamma(1)34.5 gene and may account for the attenuation of the mutant virus. J. Virol. *71*, 7750–7757.

Stern-Ginossar, N., Elefant, N., Zimmermann, A., Wolf, D.G., Saleh, N., Biton, M., Horwitz, E., Prokocimer, Z., Prichard, M., Hahn, G., *et al.* (2007). Host immune system gene targeting by a viral miRNA. Science *317*, 376–381.

Stevens, J.G., Wagner, E.K., Devi-Rao, G.B., Cook, M.L., and Feldman, L.T. (1987). RNA complementary to a herpesvirus alpha gene mRNA is prominent in latently infected neurons. Science *235*, 1056–1059.

Tang, S., Bertke, A.S., Patel, A., Wang, K., Cohen, J.I., and Krause, P.R. (2008). An acutely and latently expressed herpes simplex virus 2 viral microRNA inhibits expression of ICP34.5, a viral neurovirulence factor. Proc. Natl. Acad. Sci. U.S.A. *105*, 10931–10936.

Tang, S., Patel, A., and Krause, P.R. (2009). Novel less-abundant viral microRNAs encoded by herpes simplex virus 2 latency-associated transcript and their roles in regulating ICP34.5 and ICP0 mRNAs. J. Virol. *83*, 1433–1442.

Thompson, R.L., and Sawtell, N.M. (1997). The herpes simplex virus type 1 latency-associated transcript gene regulates the establishment of latency. J. Virol. *71*, 5432–5440.

Thompson, R.L., and Sawtell, N.M. (2001). Herpes simplex virus type 1 latency-associated transcript gene promotes neuronal survival. J. Virol. *75*, 6660–6675.

Umbach, J.L., and Cullen, B.R. (2009). The role of RNAi and microRNAs in animal virus replication and antiviral immunity. Genes Dev. *23*, 1151–1164.

Umbach, J.L., Kramer, M.F., Jurak, I., Karnowski, H.W., Coen, D.M., and Cullen, B.R. (2008). MicroRNAs expressed by herpes simplex virus 1 during latent infection regulate viral mRNAs. Nature *454*, 780–783.

Umbach, J.L., Nagel, M.A., Cohrs, R.J., Gilden, D.H., and Cullen, B.R. (2009). Analysis of human alphaherpesvirus microRNA expression in latently infected human trigeminal ganglia. J. Virol. *83*, 10677–10683.

Umbach, J.L., Wang, K., Tang, S., Krause, P.R., Mont, E.K., Cohen, J.I., and Cullen, B.R. (2010). Identification of viral microRNAs expressed in human sacral ganglia latently infected with herpes simplex virus 2. J. Virol. *84*, 1189–1192.

Yi, R., Qin, Y., Macara, I.G., and Cullen, B.R. (2003). Exportin-5 mediates the nuclear export of pre-microRNAs and short hairpin RNAs. Genes Dev. *17*, 3011–3016.

Zeng, Y., Yi, R., and Cullen, B.R. (2005). Recognition and cleavage of primary microRNA precursors by the nuclear processing enzyme Drosha. EMBO J. *24*, 138–148.

Zuker, M. (2003). Mfold web server for nucleic acid folding and hybridization prediction. Nucleic Acids Res. *31*, 3406–3415.

23

Oncolytic HSV Vectors for Cancer Therapy

Samuel Rabkin

Abstract

Oncolytic HSV (oHSV) virotherapy is a promising new strategy for cancer therapy, converting a human pathogen into a therapeutic agent. This takes advantage of the biology of HSV, by introducing genetic alterations that limit virus replication and cytotoxicity to transformed cancer cells while making the virus non-permissive in normal cells. HSV encodes a large number of genes that are non-essential for growth in tissue culture cells, but are nevertheless important for growth in post-mitotic cells and for interfering with intrinsic antiviral and innate immune responses. Many of the cellular pathways regulating growth and antiviral responses are disrupted in cancer cells, which means that viral gene products allowing replication in normal cells are not necessary in cancer cells. In considering the development of an infectious agent for human use, safety is a critical consideration. Therefore mutations targeting cancer cells must be combined with mutations in genes that play important roles *in vivo*; causing pathogenicity, spread through the nervous system and other organs, latency and reactivation, and adaptive immune responses. This review will focus more on the virological aspects of oHSV vectors and less on the cancer cell target, and describe the multiple strategies and genes involved in generating oHSV vectors. However, it is important to bear in mind that the effect of different HSV mutations will be highly dependent upon the physiology of the particular type of cancer cell and tumour, and that each oHSV vector will be more effective in some tumour types, so that it is unlikely that any one oHSV will be optimal for all types of cancer.

Introduction

Cancer plays a major role in human morbidity and mortality. Cancer cells develop through a series of accumulated genetic and epigenetic changes that contribute to their complexity and heterogeneity. A tumour consists not only of cancer cells but also its microenvironment, including fibroblasts, endothelial cells, immune cells, and extracellular matrix. This three dimensional 'organ' is shaped by its location in the body and interactions between the multiple cell types. Conventional therapies, chemotherapy and irradiation, have a narrow therapeutic index with significant adverse effects and are often limited due to the expansion of resistant cancer cells. Virotherapy, the use of viruses for cancer therapy, is an old concept with clinical studies going back over 50 years (Kelly and Russell, 2007). These early clinical studies used wild-type or passaged vaccine strain viruses, with the results being sporadic and limited or accompanied by adverse events due to virus pathogenicity, and interest in the strategy faded. Progress in virology and advances in genetic engineering provided the means to manipulate viruses so their pathogenicity could be attenuated and their replication targeted to cancer cells. The first description of a genetically engineered oncolytic virus was in 1991, with the use of a thymidine kinase (TK) deleted HSV for the treatment of glioma, due to the virus's ability to replicate in dividing but not non-dividing cells (Martuza *et al.*, 1991). Oncolytic viruses are able to selectively replicate in and kill cancer cells but are non-pathogenic in normal tissue. The transformed phenotype provides a permissive environment for some naturally-occurring viruses or viral mutants.

HSV has many features that make it attractive for cancer therapy: (1) it naturally undergoes a lytic infection that is cytotoxic; (2) it is highly infectious, so that 1 plaque forming unit (PFU) is sufficient to kill a cell and spread throughout a monolayer; (3) it infects most cell types in a broad range of species, including those used as experimental tumour models; (4) its genome is very stable; (5) it can exist in a latent state without detectably affecting cell physiology, a biological paradigm for oncolytic HSV (oHSV) infection of a non-permissive normal cell; (6) it does not integrate into the host chromosome, precluding potential insertional mutagenesis of normal cells (Mellerick and Fraser, 1987); (7) its large genome contains many non-essential genes that can be replaced (up to about 30 kb) with multiple therapeutic transgenes (Roizman, 1996); (8) many non-essential genes have been identified that contribute to pathogenicity (Mori and Nishiyama, 2006; Nishiyama, 1996); and (9) a number of effective anti-viral drugs are available to treat unfavourable virus replication and adverse events (Balfour, 1999; Cernik *et al.*, 2008).

oHSV

The first report of HSV being used to treat tumours in an experimental model was in 1990, where wild-type HSV was inoculated into implanted BHK-21 cells in hamsters (Cowan *et al.*, 1990). The same group earlier showed that HSV infection *in vitro* could reduce the tumourigenicity of BHK-21 cells (Skinner *et al.*, 1988). However, the associated pathogenicity of wild-type HSV clearly precluded its use as a therapeutic agent and necessitated the development of mutated HSVs that were 'safe' yet replication-competent in cancer cells. Once HSV has entered a cell, tumour selectivity can be achieved by exploiting the unique properties of the physiological milieu

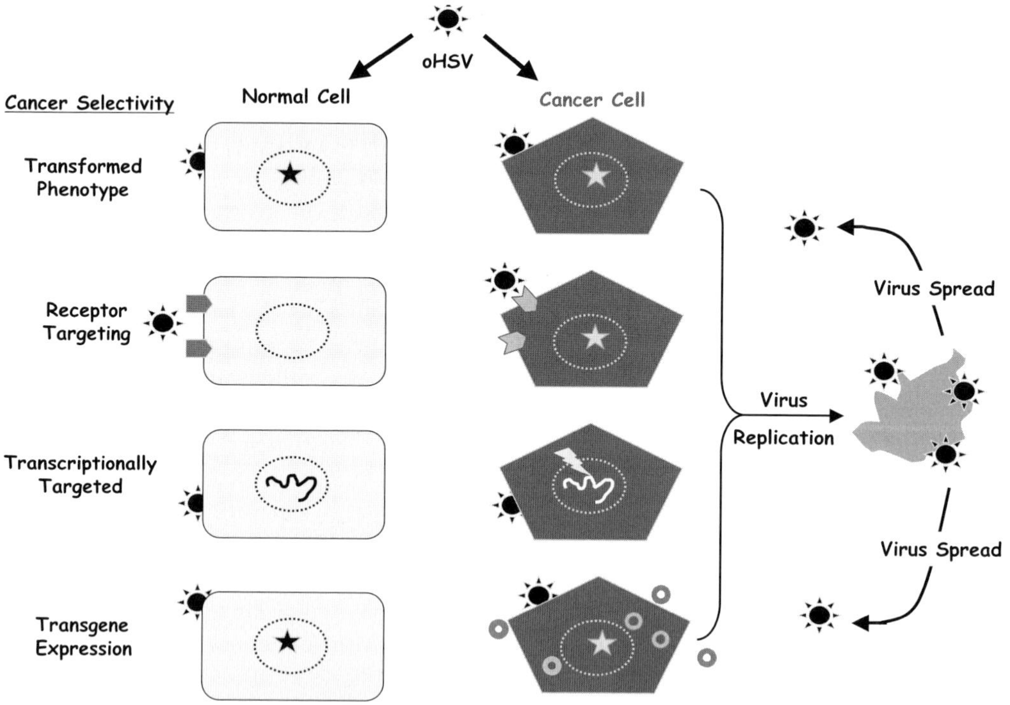

Figure 23.1 Strategies for oHSV targeting and therapy of tumours. Virus replication in permissive cancer cells leads to cell death, virus amplification and spread to new cancer cells where the process is repeated.

of the cancer cell (Fig. 23.1). Mutating viral genes necessary for replication in normal cells, but not in cancer cells can engender tumour selectivity. Examples of such genes that make the virus selectively permissive in cancer cells include those involved in: (i) nucleotide metabolism (TK), ribonucleotide reductase (RR) and uracil deglycosylase (UNG) (Varghese and Rabkin, 2002); (ii) blocking innate immune responses (γ34.5 and ICP0; Mohr, 2005; Chapters 7 and 13); and (iii) blocking apoptosis (Us3, ICP10PK) (Benetti and Roizman, 2004; Perkins *et al.*, 2003; Chapter 12).

Alternatively, vectors can be engineered that utilize the regulation of tumour-specific or tumour-elevated factors at the transcriptional level to drive virus replication in a tumour-specific fashion, transcriptional targeting (Fig. 23.1). Virus can also be targeted to cancer cells through interactions with cell surface receptors that are selectively or highly expressed on cancer cells (Campbell and Gromeier, 2005) (Fig. 23.1). This involves both introducing a retargeting ligand and detargeting HSV interactions with their natural receptors and normal cells, which could swamp out any specificity derived from novel virus–receptor interactions, by mutating viral glycoproteins (Menotti *et al.*, 2008; Zhou and Roizman, 2007). There are a number of recent reviews on oHSV that provide additional information on therapeutic strategies and cancer targets (Friedman *et al.*, 2009; Fukuhara and Todo, 2007; Grandi *et al.*, 2009; Jeyaretna and Kuroda, 2007; Jeyaretna *et al.*, 2007; Markert *et al.*, 2006).

Viral mutations that target replication to cancer cells

A range of different viral mutations make HSV non-permissive in 'normal' cells, yet maintain permissivity in cancer cells due to the 'transformed' phenotype (Fig. 23.2). As with many viruses that have evolved an exquisitely regulated replication cycle and interaction with their permissive host, genetically manipulating HSV for tumour selectivity often attenuates virus growth (a negative consequence) and pathogenesis (a positive consequence). Most oHSV are derived from laboratory strains of HSV-1, which have been adapted for tissue culture growth, and are often selected based on existing virus mutants. However, the phenotype of different laboratory strains varies (Ejercito *et al.*, 1968), including differences in neurovirulence (Dix *et al.*, 1983; Lopez, 1975; Sedarati and Stevens, 1987), and laboratory strains may not really be fully wild-type, for example strain F contains a temperature-sensitive mutation in ICP4 (Leopardi and Roizman, 1996). Based on the view that clinical isolates would be more virulent, an HSV-1 clinical isolate from a cold sore (JS1), which replicated better than laboratory strain 17⁺ in 5 human cancer cell lines, was used to construct oHSV OncoVEX (Liu *et al.*, 2003). We found that strain 17⁺ is generally less cytotoxic to human prostate cancer and glioma cells than other laboratory HSV-1 strains, such as Patton, which was more cytotoxic than 2 HSV-1 clinical isolates (unpublished results). The effect of genetic background on oncolytic activity *in vitro* and *in vivo*, and how this can be exploited has not been adequately considered. See Table 23.1 for a description of the oHSV vectors discussed.

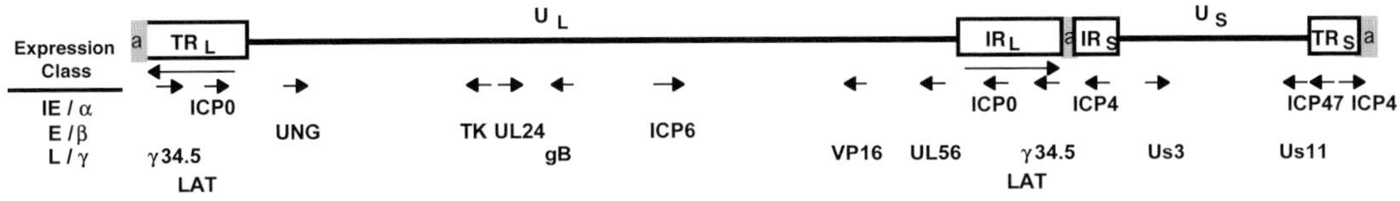

Figure 23.2 HSV genome. Schematic arrangement of viral genes mutated in oHSV. Arrows indicate the orientation of the transcripts. The general expression classes of the genes are indicated as immediate-early (IE/α), early (E/β), and late (L/γ). The boxes indicate the inverted repeat sequences, Terminal (T) and Internal (I), flanking the long (L) and short (S) unique regions. The a sequences contain direct repeats that function as packaging signals. LAT is a non-coding transcript that is expressed in latently infected cells.

Table 23.1 Oncolytic HSV-1 mutants

Virus	Gene modification	Parent	Features	References
1716	γ34.5Δ	17	Attenuated neurovirulence 759 bp deletion Currently in clinical trial	MacLean *et al.* (1991) Randazzo *et al.* (1995)
3616UB	γ34.5Δ, UNG⁻/ SV40 Epro-LacZ	F/R3616	Attenuated neurovirulence Track viral infection Hypersensitive to ACV/GCV	Pyles *et al.* (1997)
Baco-1	γ34.5Δ CMVpro-GFP	17/fHSV-Δ-pac	Insert of pac and GFP into BAC sequence Track viral infection	Fu *et al.* (2003)
C134	γ34.5Δ, CMVpro-IRS1	F/R3616 (C101)	HCMV epitope-tagged IRS1 inserted in UL3/4 intragenic region	Cassady (2005) Shah *et al.* (2007)
ΔPK (ICP10ΔPK)	ICP10PKΔ	HSV-2 G		Peng *et al.* (1996) Colunga *et al.*(2010)
*dl*8.36tk	TKΔ/TKpro-LacZ	*dl*sptk	Insensitive to ACV/GCV Track viral replication	Kaplitt *et al*, (1994)
*dl*sptk	TKΔ	KOS	Insensitive to ACV/GCV 360 bp deletion	Martuza *et al.* (1991)
DM33	γ34.5Δ, LAT⁻/ LATpro-GPF	McKrae (d34.5)	1.6 kb deletion of 2 kb LAT RNA	Samoto *et al.* (2001)
Fu-10	γ34.5Δ, ICP6⁻, Syn⁻	F	Attenuated neurovirulence Fusogenic (syncytial) Hypersensitive to ACV/GCV	Fu and Zhang (2002)
FusOn-H2	ICP10 PK-TMΔ, syn⁻ CMVpro-GFP/ICP10 RR	HSV-2 186	Track viral infection Unknown syn	Fu *et al.* (2006a)
G207	γ34.5Δ, ICP6⁻/ ICP6pro-LacZ	F/R3616	Attenuated neurovirulence ICP6/LacZ from hrR3 Track viral replication Hypersensitive to ACV/GCV	Mineta *et al.* (1995)

G47Δ	γ34.5Δ, ICP6⁻/ICP6pro-LacZ ICP47Δ, Us11proΔ	F/G207	Currently in clinical trial Normal MHC Class I expression α47pro-Us11 Entering clinical trial	Todo *et al.* (2001)
HF10	UL56Δ, LATΔ, gBsyn UL53-UL54 duplication	HF	Non-engineered Currently in clinical trial	Takakuwa *et al.* (2003)
hrR3	ICP6⁻/ICP6pro-LacZ	KOS	Hypersensitive to ACV/GCV Track viral replication	Goldstein and Weller (1988b)
HRV-34.5	γ34.5pro-HRV2-IRES-γ34.5	Patton/SPBg5e	Attenuated neurovirulence	Campbell *et al.* (2007)
KM100	$ICP0^{n212}$, $VP16^{in1814}$	KOS	VP16 is tegument transactivator	Mossman and Smiley (1999) Hummel *et al.* (2005)
KOS-SB	TKΔ	KOS	Insensitive to ACV/GCV	Jia *et al.*, (1994)
L1BR1	Us3 ICP8pro-LacZ	HSV-2 186	Track viral replication	Nishiyama *et al.* (1992) Kasuya *et al.* (2007)
MGH-1	γ34.5Δ, ICP6⁻/ ICP6pro-LacZ	F/R3616	Same construction as G207 Hypersensitive to ACV/GCV Track viral replication	Kramm *et al.* (1997)
mtHSV	γ34.5Δ SV40Epro-LacZ	F	Attenuated neurovirulence Track viral infection	Lan *et al.* (2003)
NV1020 (R7020)	UL24⁻, UL56⁻, IRΔ/HSV2 1 copy of α0, α4, γ34.5	F/HSV-2 G (I358)	HSV1/2 intertypic recombinant 14.5 kb deletion of IR HSV-2 HindIII L fragment insertion Currently in clinical trial	Meignier *et al.* (1988b) McAuliffe *et al.* (2000)
NV1066	IRΔ, UL56⁻/CMVpro-GFP (1 copy of α0, γ34.5, α4) TK⁻/BAC	F-BAC	Constructed in *E. coli* Contains BAC in TK Track viral infection	Wong *et al.* (2002)
Onc	IRΔ (UL55, UL56,	F/pYEbac102	Similar deletion as NV1020	Israyelyan *et al.* (2007)

Table 23.1 continued

Virus	Gene modification	Parent	Features	References
	1 copy of α0, γ34.5, α4)/		Contains BAC in UL3/UL4 intragenic region	
	EF-1αpro-RFP		Track viral infection	
OncoVEX	γ34.5$^-$, ICP47$^-$	JS1	Attenuated neurovirulence	Liu *et al.* (2003)
(JS1/34.5$^-$/47$^-$)			Normal MHC Class I expression	
OncSyn	gBsyn3	F/Onc	Syncytial mutant of Onc	Israyelyan *et al.* (2007)
	IRΔ/EF-1αpro-RFP		Track viral infection	
R3616	γ34.5Δ	F	Attenuated neurovirulence	Chou *et al.* (1990)
			1 kb deletion	Markert *et al.* (1993)
R3659	γ34.5Δ/α27-TK	F/Δ305	Attenuated neurovirulence	Lagunoff and Roizman (1995)
	TKΔ			Parker *et al.* (2000)
R7041	Us3Δ	F		Purves *et al.* (1987)
				Liu *et al.* (2007b)
RE6	γ34.5Δ	17/HG52	Attenuated neurovirulence	Thompson *et al.* (1983)
			HSV1/2 intertypic recombinant	Markert *et al.* (1993)
RH105	TKΔ/α4pro-LacZ	F	Insensitive to ACV/GCV	Ho and Mocarski (1988)
			Track viral infection	Boviatsis *et al.* (1994)
SPBg5e	γ34.5Δ/	Patton	Attenuated neurovirulence	Mohr and Gluzman (1996)
(Δ34.5)	γ34.5pro-Gluc		Track viral replication	Taneja *et al.* (2001)
SUP1	γ34.5Δ/γ34.5pro-Gluc,	Patton/SPBg5e	α47pro-Us11	Mohr and Gluzman (1996)
	ICP47Δ, Us11proΔ		Normal MHC class I expression	
Synco-2	γ34.5Δ	17/fHSV-Δ-pac	Insert of pac and GALV into BAC sequence	Fu *et al.* (2003)
	UL38pro-GALV.fus			

Transcriptionally targeted vectors				
bM24-TE	4F2 enh-TRE-CMVminpro-ICP4 ICP6pro-LacZ/ICP6$^-$, Us3$^-$	d120-BAC	Tandem synthetic TRE × 12 Constructed in *E. coli* Track viral replication	Kuroda *et al.* (2006b)
d12.CALP	4F2 enh-CALPpro-ICP4 TKpro-LacZ/TK$^-$, Us3$^-$	KOS d120	Selective for soft tissue and bone tumours Insensitive to ACV/GCV Track viral replication	Yamamura *et al.* (2001)
DF3, γ34.5	DF3/MUC1pro-γ34.5, CMV-GFP/TK-, γ34.5Δ	R3616	Selective for epithelial tumours Insensitive to ACV/GCV Track viral infection	Kasuya *et al.* (2004)
G92A	ALB enh-pro-ICP4, TKpro-LacZ/TK$^-$, Us3$^-$	KOS d120	Replication in albumin expressing cells Insensitive to ACV/GCV Track viral replication	Miyatake *et al.* (1997)
KeM34.5	γ34.5Δ musashi1pro-γ34.5/ICP6$^-$	G207	Selective for gliomas Hypersensitive to ACV/GCV	Kanai *et al.* (2007)
Myb34.5	B-myb pro-γ34.5/ICP6- γ34.5Δ	MGH1	Selective for dividing tumour cells Contains E2F binding site	Chung *et al.* (1999)
rQnestin34.5	nestin enh-Hsp68pro-γ34.5/ ICP6pro-EGFP, γ34.5Δ	MGH1-BAC	Constructed in *E. coli* Track viral replication	Kambara *et al.* (2005)
SS1	SRE-ICP4$^-$ IRES-DsRed/TK$^-$	KOS d120	Synthetic SRE × 5 and minimal TATA Selective for activated ELK	Pan *et al.* (2009)
Therapeutic transgene-expressing vectors				
G47Δ-IL18/B7	CMVpro-mIL-18-IRES-B7.1Ig, ICP6pro-LacZ/ICP6$^-$ ICP47Δ, γ34.5Δ	G47Δ-BAC	Constructed in *E. coli* Mouse IFNβ signal sequence-IL18 Track viral replication	Fukuhara *et al.* (2005a)
HSV1yCD	CMVpro-yCD CMVpro-GFP/ICP6$^-$	hrR3	5-FC-mediated cytotoxicity Track virus infection	Nakamura *et al.* (2001)
M002	Egr-1pro-IL-12/γ34.5Δ	R3659	Mouse IL-12 (p40-IRES-p35)	Parker *et al.* (2000)

Table 23.1 continued

Virus	Gene modification	Parent	Features	References
M010	Egr-1pro-CCL2/γ34.5Δ	R3659	Mouse CCL2 (MCP-1)	Parker *et al.* (2005)
M012	Egr-1pro-bCD/γ34.5Δ	R3659	5-FC-mediated cytotoxicity	Guffey *et al.* (2007)
M016 (HSV-PNP)	Egr-1pro-PNP/γ34.5Δ	R3659	MeP-dR mediated cytotoxicity	Bharara *et al.* (2005)
MGH2	HSVIE4/5pro-CYP2B1, CMVpro-shiCE, ICP6pro-GFP/ICP6⁻, γ34.5Δ	HsvQuik	Constructed in *E. coli*/MGH1-BAC CPA- and irinotecan-mediated cytotoxicity Track viral infection	Tyminski *et al.* (2005)
NV1034	α4pro-mGM-CSF/IRΔ, UL56-, α47pro-LacZ/ICP47Δ, Us11Δ HSV-2 gG	NV1020	Normal MHC class I expression Track viral infection	Wong *et al.* (2001b)
NV1042	α4pro-mIL-12/IRΔ, UL56-, α47pro-LacZ/ICP47Δ, Us11Δ HSV-2 gG	NV1020	Normal MHC class I expression Mouse IL-12 (p40–p35 fusion) Track viral infection	Wong *et al.* (2001b)
OncoVEX^{CD}	RSVpro-fused yCD-yUPRT/γ34.5Δ	OncoVEX	5-FC-mediated cytotoxicity	Simpson *et al.* (2006)
OncoVex^{GM-CSF}	CMVpro-GM-CSF/γ34.5Δ ICP47Δ	OncoVEX	In clinical trial Human GM-CSF	Liu *et al.* (2003)
R8306	Egr-1pro-IL-4/γ34.5Δ	R3659	Mouse IL-4	Andreansky *et al.* (1998)
rRp450	ICP6pro-CYP2B1/ICP6Δ	hrR3	CPA-mediated cytotoxicity	Chase *et al.* (1998)

Synco-2D	unknown Syn	Baco-1	Isolation of syn from mutagenized Baco-1	Nakamori *et al.* (2003)
	UL38pro-GALV.fus		GALV replaced EGFP in Baco-1syn-BAC	
			Constructed in *E. coli*	
vHSV-B7.1Ig	IE4/5pro-B7.1-Ig,	HsvQuik	Constructed in *E. coli*/MGH1-BAC	Ino *et al.* (2006)
	ICP6pro-EGFP/ICP6^{-}/34.5Δ		Mouse B7.1-human IgG Fc fusion	
			Track viral replication	
vP35	LATpro-IL-12 p35/LATΔ	DM33		Osorio and Ghiasi (2005)

4F2 enh, 4F2 heavy-chain gene 1st intron enhancer; ACV, acyclovir; ALB, mouse albumin enhancer (–10.5 to 85 kb) and promoter (–300 to +22); BAC, bacterial artificial chromosome; CALPpro, human calponin promoter (–260 to +73); CMV, cytomegalovirus; CD, cytosine deaminase; CPA, cyclophosphamide; CYP2B1, rat cytochrome P450; Egr-1, early-growth response-1 promoter; GCV, ganciclovir; GFP, green fluorescent protein; GALV.fus, gibbon ape leukaemia virus envelope fusogenic glycoprotein; Gluc, β-glucuronidase; HRV2, human rhinovirus type 2; IRΔ, internal repeat deletion; LacZ, *E. coli* LacZ; LAT, latency associated transcript; Myb, B-myb promoter; PNP, purine nucleoside phosphorylase; SRE, ELK DNA binding element; TRE, β-catenin/Tcf responsive element.

Nucleotide metabolism

In order for HSV to replicate in non-dividing or post-mitotic cells, the virus expresses enzymes involved in nucleotide metabolism so there are sufficient substrates for viral DNA replication. In dividing cells, these proteins are not required because cellular enzymes provide the same functions. Thus, mutations in these genes are complemented by their cellular counterparts in cancer cells. The following early HSV genes have been mutated in oHSV vectors: TK (UL23), a promiscuous kinase that phosphorylates deoxythymidine, deoxyuridine, deoxycytidine, thymidylate, and purine pentosides and nucleoside analogues, such as acyclovir (ACV) and ganciclovir (GCV) (Pilger *et al.*, 1999); RR, consisting of two subunits [the large subunit ICP6 (UL39) and the small subunit UL40] (Frame *et al.*, 1985), which reduces ribonucleotides to deoxyribonucleotides; and UNG (UL2), which repairs deaminated cytosine residues (Caradonna and Cheng, 1981).

TK

HSV mutants in TK replicate poorly, if at all, in non-dividing cells, and are attenuated for neurovirulence (Field and Wildy, 1978; Gordon *et al.*, 1983; Tenser, 1983). The effect of TK on HSV replication *in vivo* was demonstrated in a partial hepatectomy model, which induces proliferation of liver cells, where TK^- and TK^+ HSV titres were similar after liver injection, as opposed to greatly reduced in unoperated livers (Fang *et al.*, 1983). The first replication-competent HSV designed for oncolytic purposes was *dl*sptk, containing a deletion of the TK gene, so reversion could not occur (Martuza *et al.*, 1991). *In vitro* testing of this virus against human glioblastoma cells illustrated significant cytotoxic efficacy even at very low multiplicities of infection (MOI). This illustrated important features of oncolytic virotherapy; the large amplification of virus after infection with associated cell killing, and subsequent spread between cancer cells (Fig. 23.1). Moreover, a single inoculation of *dl*sptk into human gliomas growing in the brains of athymic mice extended their survival and even 'cured' some animals (Martuza *et al.*, 1991). Antitumour efficacy was also demonstrated with other TK deletion mutants (*dl*8.36tk, KOS-SB, and RH105) in rat brain tumours (Boviatsis *et al.*, 1994; Jia *et al.*, 1994; Kaplitt *et al.*, 1994). Despite substantial oncolytic efficacy *in vivo* in intracerebral tumour models, these vectors were neurotoxic at high doses. This overt toxicity is not treatable with commonly used anti-herpetic drugs like GCV or ACV, due to the lack of TK, which was seen as a major drawback to their clinical application. The presence of TK in oHSV vectors was thus viewed as an attractive insurance feature for clinical applications; and therefore, TK-deleted vectors have not been further pursued. However, antiherpetic drugs that do not depend upon TK activity are available, such as foscarnet and vidarabine (which target HSV-1 DNA polymerase), and this might permit re-consideration of TK mutants, with the goal of exploiting their potency while maintaining credible safety measures.

RR

Replication of a HSV-1 ICP6 deletion mutant is greatly diminished in serum-starved cells, where there was no effect on wild-type virus, demonstrating complementation by a cellular factor in dividing cells (Goldstein and Weller, 1988a). Cellular RR activity is cell cycle dependent, due to S phase specific expression of the M2 subunit (Cory and Sato, 1983; Engstrom *et al.*, 1985). However, it is not just cell proliferation that regulates RR, because $ICP6^-$ oHSV can replicate in quiescent MEFs with homozygous deletions of $p16^{INK4a}$, a tumour suppressor gene, as RR expression is maintained during G0 (Aghi *et al.*, 2008). Increased RR activity has been associated with malignant transformation (Fan *et al.*, 1996; Hurta and Wright, 1995). HSV RR is a key determinant of viral pathogenicity in the mouse and guinea pig after multiple routes of inoculation (Cameron *et al.*, 1988; Yamada *et al.*, 1991), likely due to decreased replication and inability to reactivate from latency (Jacobson *et al.*, 1989). Thus, in $p16^{-/-}$ knockout mice, where replication occurs the LD_{50} of $ICP6^-$ oHSV is two logs less than in wild-type mice (Aghi *et al.*, 2008). In contrast to HSV TK mutants, ICP6 mutants are hypersensitive to GCV and ACV, as well as non-nucleoside inhibitors like phosphonacetic acid (Coen *et al.*, 1989; Mineta *et al.*, 1994), an added safety feature for oHSV.

HrR3, a RR mutant where the *E. coli* Lac Z gene was inserted in frame into the ICP6 locus as a fusion protein, was originally constructed as a host-range mutant to test whether HSV RR was essential for virus replication (Goldstein and Weller, 1988b). The presence of LacZ provides a useful and sensitive marker to track viral infection. HrR3 was subsequently shown to have oncolytic activity (Mineta *et al.*, 1994). It preferentially replicated in metastatic tumour cells in the liver and not in normal liver cells despite the expression of low levels of RR in normal liver (Nomura *et al.*, 2009; Yoon *et al.*, 1998; Yoon *et al.*, 2000). Chemotherapeutic drugs that upregulate cellular RR synergize with ICP6$^-$ oHSV in killing cancer cells and inhibiting tumour growth (Aghi *et al.*, 2006; Nakano *et al.*, 2005b; Petrowsky *et al.*, 2001). This suggests that even in dividing cancer cells, RR is limiting for HSV. After intravenous delivery hrR3 not only replicated in liver tumours, but vascular endothelial cells (Nomura *et al.*, 2009), suggesting it may have anti-angiogenic properties

γ34.5

The key determinant of HSV-1 neuropathogenicity is the γ34.5 (RL1) gene, located in the long terminal and internal repeats overlapping the LAT region and therefore a diploid gene (Chou *et al.*, 1990; MacLean *et al.*, 1991; Thompson *et al.*, 1983). Neurovirulence (LD_{50} after intracerebral inoculation) of γ34.5 mutants is decreased by over 6 logs compared to wild-type HSV-1 (Chou *et al.*, 1990; MacLean *et al.*, 1991; Perng *et al.*, 1995; Valyi-Nagy *et al.*, 1994). Mutants in γ34.5 replicate in most cells in culture, but poorly, if at all, in sensory ganglia or the CNS, which may contribute to their poor ability to establish and reactivate from latency (Robertson *et al.*, 1992; Spivack *et al.*, 1995; Whitley *et al.*, 1993). The attenuation of neurovirulence was a positive attribute for oHSV, especially after the development of encephalitis in mice treated with oHSV TK mutants (Martuza *et al.*, 1991). HSV-1 deletion mutants of γ34.5, initially R3616 in strain F and 1716 in strain 17$^+$, were shown to be effective in a variety of brain tumour models (Andreansky *et al.*, 1997; Chambers *et al.*, 1995; Lasner *et al.*, 1996; Markert *et al.*, 1992; Randazzo *et al.*, 1995). R3616 and oHSV vectors derived from it also have a nonsense mutation in UL3, a non-essential protein of unknown function (Dambach *et al.*, 2006). It remains to be determined whether the UL3 mutation affects the oncolytic activity of these γ34.5Δ vectors. Even though it was initially developed for central nervous system (CNS) tumours, 1716 was the first oHSV shown to be efficacious in non-CNS tumours (Kucharczuk *et al.*, 1997; Randazzo *et al.*, 1997). At high doses, 1716 exhibited toxicity in a number of models; replication in and destruction of ependymal cells in BALB/c mice, morbidity of nude mice after ventricular injection (Kesari *et al.*, 1998; Lasner *et al.*, 1998), and weight loss and severe disease in AO rats and BALB/c mice after intracerebral injection (McMenamin *et al.*, 1998). This is likely why the maximum dose allowed in the first clinical trial of 1716 was only 10^5 PFU (Rampling *et al.*, 2000). A number of γ34.5 mutants have been constructed for use as oncolytic vectors, in different backgrounds and with different size deletions, including; RE6 (Brandt *et al.*, 1997; Markert *et al.*, 1993), Δ34.5 (Taneja *et al.*, 2001), DM33 (Samoto *et al.*, 2002), mtHSV (Lan *et al.*, 2003), and JS1/34.5$^-$ (Liu *et al.*, 2003).

In order to ensure the safety of γ34.5 mutants, screens for extragenic second site suppressors were conducted, based on serial passage in non-permissive SK-N-SH cells *in vitro* (Cassady *et al.*, 2002; Mohr and Gluzman, 1996). One such mutant, of unknown position, was found to have elevated neurovirulence compared to parental R3616, but still about 1000-fold less than wild-type HSV-1 (Cassady *et al.*, 2002). In a separate screen, it was found that passaging R3616 in human glioma D54-MG tumours *in vivo* selected for a mutant virus with slightly reduced neurovirulence, whereas passage *in vitro* selected for a mutant virus with about 10-fold increased neurovirulence (Shah *et al.*, 2006). This would suggest that *in vivo* growth of γ34.5 mutants in patients should not lead to more neurovirulent strains.

The γ34.5 protein (ICP34.5) has multiple activities. Among the most important for cancer cell selectivity is the inhibition of host cell protein synthesis shut-off that normally accompanies virus infection (Chou and Roizman, 1992). This is part of the virus's strategy to evade type

1 interferon (IFN) responses that thwart virus replication (Chapter 7). Normal cells upon sensing HSV infection activate interferon-inducible double-stranded RNA-dependent protein kinase R (PKR), which phosphorylates eIF2α to its inactive form, thereby inhibiting cellular, as well as viral, protein synthesis (Garcia *et al.*, 2007; Proud, 2005; Samuel *et al.*, 1997). ICP34.5 overcomes this by binding to and activating protein phosphatase 1α (PP1α), which de-phosphorylates eIF2α (He *et al.*, 1997). Mutation of 2 amino acids in the PP1 binding domain are sufficient to abrogate viral replication in the eye and neuroinvasion, behaving just like R3616 (Verpooten *et al.*, 2009). The C-terminal 64 amino acid domain of ICP34.5, which is homologous to the C-terminal domain of cellular growth arrest-DNA damage protein GADD34 (PPP1R15A or MyD116 in mice), mediates this activity (Chou and Roizman, 1994; Zhang *et al.*, 2008). Substitution of this region of ICP34.5 by the C-terminal domain of GADD34 complements the defect and restores the block to protein synthesis shutoff (He *et al.*, 1996). GADD34 also interacts with PP1 to dephosphorylate eIF2α as part of the cellular stress response (Novoa *et al.*, 2001). Treatment of cancer cells with chemotherapeutic agents that induce GADD34 expression can enhance the replication and cytotoxicity of γ34.5Δ oHSV (G207) (Aghi *et al.*, 2006; Bennett *et al.*, 2004). This combination strategy is very attractive for clinical translation because many of these chemotherapeutic agents are 'standards of care' that are typically given to patients, so any synergistic interaction with oHSV could lead to improved efficacy or reduced drug doses to minimize toxicity.

PKR activity is critical to the cell's innate response to HSV infection and in its resistance to γ34.5 mutants, so that γ34.5 mutants replicate in PKR$^{-/-}$ mouse embryo fibroblasts (MEFs) and are virulent in PKR$^{-/-}$, as well as IFN receptor$^{-/-}$ mice (Farassati *et al.*, 2001; Leib *et al.*, 1999; Leib *et al.*, 2000). IFNα treatment, which is upstream of PKR activation, inhibits γ34.5^{-} replication in wild-type MEFs, but not in PKR$^{-/-}$ MEFs and this activity maps to the PP1α binding domain of ICP34.5 (Cheng *et al.*, 2001). It is important to note that ICP34.5-PP1 dephosphorylation of eIF2α alone is not sufficient for full viral replication and other functions of ICP34.5 contribute (Cheng *et al.*, 2003). There are multiple eIF2 kinases in the cell, in addition to PKR: PKR-related endoplasmic reticulum kinase (PERK), stimulated by ER stress and unfolded proteins; general control non-derepressible 2 kinase (GCN2), activated by amino acid starvation; and haem-regulated inhibitor kinase (HRI), activated by haem deficiency (Proud, 2005). Furthermore, the level of eIF2B, which forms an initiation complex with eIF2α, can also impact the effect of eIF2α phosphorylation (Balachandran and Barber, 2004; Kim *et al.*, 2000). EIF2B is phosphorylated and inactivated by GSK3, which is inactivated by PI3K/Akt signalling, often elevated in cancer cells (Pavitt, 2005).

PKR activation is suppressed by MEK activation, which is often the case in cancer cells due to activating mutations in receptor tyrosine kinases, Ras or Raf. Thus cells with activated Ras/Raf/MEK signalling are permissive to replication of viruses that are sensitive to IFN (Battcock *et al.*, 2006), such as HSV-1 γ34.5 mutants (Farassati *et al.*, 2008; Farassati *et al.*, 2001). The susceptibility of cancer cells to R3616 was shown to correlate with the constitutive activity of MEK, but not the RAS genotype (Smith *et al.*, 2006). Among mouse malignant peripheral nerve sheath tumour (MPNST) cells, those with the highest levels of Ras signalling were most permissive to γ34.5Δ oHSV (G207) replication, whereas those with low Ras signalling had increased apoptosis and limited replication (Farassati *et al.*, 2008). Conversely, for human MPNST cells, there was no correlation between Ras activity and permissivity to γ34.5Δ oHSV (G207) (Mahller *et al.*, 2006). To further confound the role of Ras, it has been shown in pancreatic cancer cells that permissiveness to R3616 is correlated with PI3K signalling and not K-ras mutations or lack of PKR activation (Sarinella *et al.*, 2006). Despite the conflicting data, all of these aberrant signalling pathways are selectively upregulated in cancer and not normal cells, thus providing specificity to γ34.5 mutants.

Some of ICP34.5 other functions may also contribute to cancer cell permissivity of γ34.5Δ oHSV. ICP34.5 binds to proliferating cell nuclear antigen (PCNA), a cell cycle marker involved in DNA replication and repair that interacts with more than 100 proteins (Brown *et al.*, 1997; Harland *et al.*, 2003; Naryzhny, 2008). PCNA binding mapped to two locations in γ34.5, the C-terminal GADD34 homology domain and a region internal to the N-terminus (Harland *et al.*, 2003). PCNA is expressed in cancer cells, but whether this contributes to γ34.5Δ-oHSV permissivity or attenuated neuropathogenicity is unknown. ICP34.5 also suppresses cellular innate immunity by interacting with TANK-binding kinase 1 (TBK1) and disrupting IFN regulatory factor 3 (IRF3) activity early after infection and independent of eIF2α phosphorylation (Hiscott, 2007; Verpooten *et al.*, 2009). The TBK1 interaction requires an internal N-terminal domain of γ34.5 that partially overlaps a PCNA binding domain (Verpooten *et al.*, 2009).

Recently, it was shown that ICP34.5 binds to Beclin 1, to counteract the induction of cell autophagy (Talloczy *et al.*, 2002; see also Chapter 21). Beclin 1 recruits autophagy proteins to the preautophagosomal membrane (Sinha and Levine, 2008). Allelic loss of beclin 1 frequently occurs in cancer and defective autophagy promotes tumourigenesis (White and DiPaola, 2009). Autophagy, a process where cytosolic components and organelles are packaged in double-membrane vesicles for degradation, can be induced by stress and pathogens through PKR activation and eIF2α phosphorylation (He and Klionsky, 2009). However, the region of ICP34.5 that binds to beclin 1 is distinct from the domain that binds PP1, and maps to C-terminal amino acids 68–87 (Orvedahl *et al.*, 2007). Deletions of this region still block eIF2α phosphorylation, yet lose neurovirulence, except in $PKR^{-/-}$ mice, demonstrating its role in pathogenicity (Orvedahl *et al.*, 2007). The prevention of autophagy by ICP34.5 plays a limited role, if any, in virus replication, as opposed to its role in blocking of translational arrest (Alexander *et al.*, 2007). Therefore, deletion of the beclin 1 binding domain should be a useful mutation to introduce into oHSV vectors to attenuate neurovirulence. Finally, HSV-1 with deletions of the N-terminal domain of ICP34.5 grow poorly in mouse cells, although better than virus with complete deletions and independent of eIF2α phosphorylation, due to a defect in virion nuclear and cytoplasmic egress (Brown *et al.*, 1994; Jing *et al.*, 2004). It is not known whether this defect is specific for mouse cells, or also occurs in human cells. The N-terminal domain has also been reported to be involved in neurovirulence (Andreansky *et al.*, 1996).

Based on the functions of γ34.5, it was reasoned that vectors with deletions of the γ34.5 gene would not only attenuate neuropathogenicity of the virus, thereby mitigating the risk of encephalitis, but also limit the infection of normal non-dividing cells. Thus, all the oHSVs evaluated in patients with brain tumours have included deletions of both copies of γ34.5 (Parker *et al.*, 2009). However, a disadvantage of these mutants is that their replication even in permissive cells is decreased, probably due to premature shut off of host protein synthesis. Because of this, strategies to enhance the replication of γ34.5 deletion mutants without affecting neuropathogenicity have been explored (see Multi-mutated vectors and transcriptionally targeted vectors). Recently, it was found that γ34.5Δ oHSVs (R3616 and G207) are unable to replicate in glioma cancer stem cells isolated from patients (Wakimoto *et al.*, 2009). This is thought to be due to PKR signalling (Wakimoto *et al.*, 2009). As glioma stem cells are thought to represent a critical cell type in tumour progression and recurrence (Stiles and Rowitch, 2008), this is an important problem for γ34.5Δ vectors and suggests that more targeted mutations in γ34.5 might be useful. It is unknown if this attenuation applies to all cancer stem cells. Interestingly, human neural stem cells in culture are also resistant to γ34.5Δ oHSV replication, another demonstration of γ34.5's contribution to HSV neurovirulence.

ICP0

Targeting the defective IFN response of cancer cells with oHSV mutants unable to block IFN signalling led to the development of ICP0-null oHSV (Hummel *et al.*, 2005). ICP0 plays a critical role in blocking IFN-mediated inhibition of virus replication (Harle *et al.*, 2002), distinct from that of ICP34.5 (Mossman and Smiley, 2002). ICP0 is a RING finger E3 ubiquitin ligase that targets a number of cellular proteins for degradation, most notably PML, an IFN- and p53-inducible protein, leading to the disruption of nuclear domain (ND10) bodies (Everett, 2000; Maul *et al.*, 1993; see Chapter 4). Expression of PML is necessary for IFN-mediated inhibition of ICP0⁻ mutants (Chee *et al.*, 2003) and knockdown of PML expression increases ICP0⁻ replication (Everett *et al.*, 2006). PML protein is downregulated in a number of solid tumours, including adenocarcinomas and CNS tumours (Gurrieri *et al.*, 2004; Reineke and Kao, 2009). Interestingly, this is not due to decreases in mRNA or LOH, suggesting a post-transcriptional mechanism. PML acts as a tumour suppressor, likely due to its binding of p53, promotion of apoptosis through both intrinsic and extrinsic pathways, and induction of senescence (Bernardi *et al.*, 2008; Bourdeau *et al.*, 2009). Another mechanism by which ICP0 is thought to counteract type 1 IFN responses is to sequester activated IRF-3 and CBP in nuclear complexes, away from chromatin (Lin *et al.*, 2004; Melroe *et al.*, 2007). *In vivo*, ICP0⁻ virus replication and pathogenicity can be restored in $Stat1^{-/-}$ or IFN receptor knock-out mice (Halford *et al.*, 2006; Leib *et al.*, 1999).

ICP0⁻ mutants should be targeted to cancer cells due to selective downregulation of PML and defective IFN signalling. The oncolytic activity of ICP0⁻ mutants was first proposed and tested using a second-generation multi-mutated ICP0-null mutant, KM100 (Hummel *et al.*, 2005). Interestingly, it was recently shown that levels of PML somewhat correlate with ICP0⁻ oHSV sensitivity. Disrupting PML made ICP0⁻ oHSV-resistant prostate cancer cells sensitive (Sobol *et al.*, 2009) and PML overexpression abrogated this and also desensitized ICP0⁻ oHSV-sensitive prostate cancer cells (Sobol *et al.*, 2009). As PML overexpression in normal cells did not confer resistance to ICP0-null HSV (Lopez *et al.*, 2002), it is likely that the level of PML expression may have a more pronounced effect in cancer cells than normal IFN-responsive cells.

Us3

Defects in apoptotic pathways, a common feature of cancers, allow them to continue proliferating. For many viruses, including HSV, the ability to block apoptosis is important for replication in normal cells (see Chapter 10), and therefore mutations in viral genes that antagonize apoptosis can be used to potentially endow viruses with oncolytic activity (Liu and Kirn, 2005). The HSV Us3 gene presents such an opportunity. The non-essential Us3 gene product is a serine-threonine protein kinase that has the ability to inhibit apoptosis *in vitro* and *in vivo* (Asano *et al.*, 1999; Leopardi *et al.*, 1997; Ogg *et al.*, 2004). Us3 mutants in both HSV-1 (R7041) and HSV-2 (L1BR1) have been shown to have oncolytic activity. Infection of cancer cells with either L1BR1 or R7041 did not increase the proportion of apoptotic cells compared to wild-type infection (Kasuya *et al.*, 2007; Liu *et al.*, 2007b), whereas infection of normal cells did increase the apoptotic proportion (Hata *et al.*, 1999; Liu *et al.*, 2007b), as would be expected in cells with intact apoptosis pathways. Interestingly, apoptosis in L1BR1-infected cancer cells was significantly increased after treatment with chemotherapeutic agents, 5-FU and cisplatin, whereas no increase was observed with wild-type HSV-2, suggesting that the combination of Us3⁻ oHSV with chemotherapy would be a beneficial clinical strategy (Kasuya *et al.*, 2007).

Akt is phosphorylated/activated early after HSV infection, providing additional anti-apoptotic activity, which then decreases in a Us3-dependent fashion (Benetti and Roizman, 2006). This effect on activated Akt by Us3 has been taken advantage of to enhance the anti-cancer activity of an Us3Δ oHSV, R7041. Cancer cells infected with R7041 have elevated Akt activation compared to Us3-revertant R7306 or mock infected cells, making them more sensitive to PI3K/Akt inhibitors (Liu *et al.*, 2007b). Thus, the combination of R7041 with PI3K inhibitor LY294002 was synergistic in killing cancer cells, while the combination with wild-type was mostly antagonistic, and also had increased efficacy *in vivo* (Liu *et al.*, 2007b). From a clinical perspective, safety is a key concern. R7041 had an LD_{50} of 1.8×10^6 PFU after intracerebral inoculation of BALB/c mice (Meignier *et al.*, 1988a), although we have found increased neurovirulence in A/J mice (unpublished results). Intravenous injection

(10^7 PFU) in athymic mice or intraperitoneal injection (10^8 PFU) in BALB/c mice resulted in no clinical symptoms (Liu *et al.*, 2007b). L1BR1 was non-pathogenic when inoculated into the footpad or intraperitoneally (10^6 PFU) in ICR mice, but did have an LD_{50} of 12 after intracerebral inoculation (Nishiyama *et al.*, 1992). Both Us3 mutants showed selective replication in tumours after intratumoural or intravenous delivery (Kasuya *et al.*, 2007; Liu *et al.*, 2007b).

HSV-2 ICP10

The HSV-2 counterpart of HSV-1 ICP6 is ICP10, which contains a non-homologous N-terminal domain encoding a serine/threonine protein kinase (PK) (Smith, 2005). ICP10PK is a membrane-bound constitutively active kinase that signals through the Ras/MEK/ERK and PI3K/Akt pathways, consequently having mitogenic and anti-apoptotic activity (Smith *et al.*, 1998). The PK domain can be deleted without affecting its RR activity, but this mutant (ICP10ΔPK) still has significantly impaired replication in non-dividing cells (Smith *et al.*, 1998) and attenuated pathogenicity (Aurelian *et al.*, 1999). This is likely due to the lack of Ras signalling, as treatment of dividing cells with a MEK inhibitor (PD98059) similarly decreases wild-type HSV-2 replication (Smith *et al.*, 2000a). Based on these properties, ICP10ΔPK has been evaluated as a vaccine and shown to provide protection to lethal HSV-2 infection in mice (Aurelian *et al.*, 1999), and in phase I and II clinical trials significantly reduced recurrences (Aurelian, 2004). A cellular homologue of ICP10PK has been identified, H11 (Hsp22, HspB8), that is proposed to play a role in apoptosis, proliferation, and transformation (Aurelian *et al.*, 2001; Smith *et al.*, 2000b), which is cell type and expression level dependent (Anton *et al.*, 2008). Whether H11 has serine/threonine kinase activity remains controversial (Anton *et al.*, 2008). In a further twist linking H11 to antiviral responses, it induces phosphorylation of eIF2α and autophagy in a PERK-independent fashion (Carra *et al.*, 2009). It will be of interest to determine whether ICP10PK has similar activities and more importantly whether H11 plays any role in the oncolytic activity of ICP10PKΔ.

These properties of the ICP10PKΔ mutants prompted their examination as oncolytic vectors (Colunga *et al.*, 2010; Fu *et al.*, 2006a). The first such oHSV was FusOn-H2, where the PK and transmembrane (TM) domains were replaced by a CMV driven-GFP/RR domain fusion protein, while maintaining RR activity (Fu *et al.*, 2006a). During the construction of FusOn-H2, it was found to have a syncytial phenotype, in some, but not all cancer cells (Fu *et al.*, 2007a). It is unclear whether the syncytial phenotype is directly due to the ICP10 deletion or transgene insertion, but all isolated GFP recombinants had this phenotype (Fu *et al.*, 2006a). FusOn-H2 was similar or better than HSV-2 at killing a variety of human cancer cells, and did not replicate in primary mouse hepatocytes, while HSV-2 and oHSV Baco-1 did (Fu *et al.*, 2006a,b, 2007b). *In vivo*, FusOn-H2 was significantly better at inhibiting the growth of subcutaneous human breast tumours than the HSV-1 fusogenic oHSV Synco-2D (Fu *et al.*, 2006a). FusOn-H2 induced apoptosis in human cancer cells and in uninfected cells at low MOI, likely a bystander effect, as well as *in vivo* (Fu *et al.*, 2007a). From a safety standpoint, 10^7 PFU of FusOn-H2 caused no mortality in BALB/c mice with only a minimal transient increase in liver enzymes after intravenous injection, whereas 10^4 PFU of HSV-2 was uniformly fatal (Fu *et al.*, 2007b).

Based on the induction of programmed cell death by ICP10PK deleted HSV-2, ΔPK was tested as an oncolytic virus in melanoma (Colunga *et al.*, 2010). ΔPK virus yields were low in melanoma cells, but this was accompanied by a large induction of cell death, that was indicative of a bystander effect (Colunga *et al.*, 2010). This cell death was both calpain- and caspase-dependent, but not classical apoptosis. ΔPK infection also upregulated autophagy protein beclin1, H11, TNFα and macrophage infiltration (Colunga *et al.*, 2010). Whether any of these upregulated proteins contributed to anti-tumour activity remains to be determined. ΔPK induction of a variety of programmed cell death pathways provides another mechanism by which HSV mutants can engender oncolytic activity. The differences between ΔPK and FusOn-H2 may be due to the different size of the deletion, with ΔPK lacking only the ICP10 PK domain, or background strain.

Multi-mutated vectors

HSV-1 vectors with single gene mutations, such as those described above, could potentially revert, acquire new mutations, or recombine with wild type HSV to generate viruses with increased pathogenicity, thereby affecting safety in a clinical setting. *Second generation* vectors have been engineered with mutations in two viral genes, usually affecting pathogenicity, with the goal of improving safety. In addition, mutations have been incorporated into oHSV vectors to enhance their anti-tumour activity, including, suppressor mutations of 34.5 that enhance viral replication (Liu *et al.*, 2003; Taneja *et al.*, 2001; Todo *et al.*, 2001), ICP47 deletions that enhance immune recognition (Liu *et al.*, 2003; Todo *et al.*, 2001), and syncytial mutations that are fusogenic (Fu and Zhang, 2002).

The prototypic second-generation vector in the first class is G207, the first oHSV developed for clinical translation in glioma patients (Mineta *et al.*, 1995). Others include, MGH-1, 3616UB and DM33 (Kramm *et al.*, 1997; Pyles *et al.*, 1997; Samoto *et al.*, 2001). G207 was constructed from strain F, and contains deletions of both copies of γ34.5 (identical to R3616) and an *E. coli* LacZ gene insertion inactivating ICP6 (as in hrR3) (Mineta *et al.*, 1995). These widely spaced mutations make it unlikely that a neurovirulent isolate could arise through recombination. The presence of the LacZ reporter makes it easy to detect replicating virus (Huszthy *et al.*, 2008; Kooby *et al.*, 1999; Mineta *et al.*, 1995; Sundaresan *et al.*, 2000; Todo *et al.*, 1999; Yazaki *et al.*, 1995) and to uniquely identify G207, as opposed to clinical isolates, in patients undergoing treatment (Markert *et al.*, 2000).

The safety of G207, and the impact of ICP6 and γ34.5 on pathogenicity, was amply confirmed in two animal models, young BALB/c mice and New World owl monkeys (*Aotus nancymai*). *Aotus* monkeys are one of the most sensitive animals to HSV-1 infection and develop clinical disease that mimics that of human neonates (Hunt, 1993; Katzin *et al.*, 1967; Melendez *et al.*, 1969). Inoculation of G207 into the brain, liver, prostate, peripheral nerves, and by intravenous delivery ascertained safety without any complications in both species (Hunter *et al.*, 1999; Radbill *et al.*, 2007; Sundaresan *et al.*, 2000; Todo *et al.*, 2000a; Varghese *et al.*, 2001). Impressively, intracerebral inoculation of 1×10^9 PFU in *Aotus* monkeys was safe, whereas, 1×10^3 PFU of the wild-type parental strain F caused encephalitis and morbidity within 5 days (Hunter *et al.*, 1999). Re-challenge of monkeys with intracerebral or intraprostatic injection of G207 did not induce any pathology in the brain or virus shedding (Todo *et al.*, 2000a; Varghese *et al.*, 2001) and reactivation of latent virus in the brain by superinfection could not be detected in mice (Sundaresan *et al.*, 2000). Despite the lack of pathology, monkeys developed an anti-HSV antibody response (Hunter *et al.*, 1999; Todo *et al.*, 2000a). This safety has been further corroborated in human clinical trials (Markert *et al.*, 2000, 2009).

Efficacy of G207 has been demonstrated by a variety of investigators in a wide spectrum of tumour models (Todo *et al.*, 2002). Of 59 human cancer cell lines, only 5 were not susceptible to G207 (MDA-MB-231, EJ, K562, C18, SK-MEL3) (Todo *et al.*, 2002) and 4 of 10 human paediatric sarcomas were not susceptible (Bharatan *et al.*, 2002). *In vitro* susceptibility correlates well with *in vivo* efficacy (Toda *et al.*, 1998),which has been demonstrated in most solid tumours by a number of investigators (Todo *et al.*, 2002). During safety studies, it was noted that human bone marrow cells are resistant to G207 cytotoxicity, including the lack of any effect on the proportion or clonogenic capacity of $CD34^+$ cells (Wu *et al.*, 2001). This led to the testing of G207 as a biological purging agent in the setting of autologous haematopoietic stem cell transplantation. During tumour progression, cancer cells are constantly shed into the blood, so that contamination of bone marrow or peripheral blood is common and an impediment to successful autologous stem cell transplantation. G207, and OncoVEX, were very effective at purging human bone marrow of contaminating breast cancer cells (Hu *et al.*, 2006a; Wu *et al.*, 2001). 3616UB was derived from R3616, with an *E. coli* Lac Z gene insertion inactivating UNG rather than ICP6 (Pyles *et al.*, 1997). It was as efficacious as R3616 in inhibiting human tumour xenograft growth, but safer after intracerebral inoculation. Like the RR mutants, 3616UB is also hypersensitive to GCV (Pyles *et al.*, 1997).

Us11

HSV-1 that are mutated in γ34.5 have decreased replication (McKie *et al.*, 1996) and are unable to replicate in certain tumour cells (Mohr and Gluzman, 1996; Todo *et al.*, 2001). An extragenic, second-site suppressor of HSV-1 γ34.5Δ was isolated (SUP1) after serial passage in non-permissive human neuroblastoma SKNSH cells (Mohr and Gluzman, 1996). The suppressor mutations all contained deletions that placed the late γ_2Us11 gene under control of the immediate-early (IE) α47 promoter. Us11 is a RNA-binding protein that interacts directly with PKR and the dsRNA-independent protein activator of PKR (PACT) to block PKR activation and inhibition of protein synthesis (Cassady and Gross, 2002; Peters *et al.*, 2002; Poppers *et al.*, 2000). This is the same pathway that is blocked by γ34.5, except at a step prior to eIF2α phosphorylation, so Us11 would not counteract other eIF2α kinases. The C-terminal domain of Us11 binds to the N-terminal domain of PKR, and binding to PKR prevents PKR activation by dsRNA or PACT (Peters *et al.*, 2002). Us11, through its dsRNA binding domain, inhibits 2′–5′ oligoadenylate synthetase (OAS), which is activated by dsRNA and IFNα/β (Sanchez and Mohr, 2007). OAS products associate with and activate RNase L, which increases RNA degradation, a separate pathway in the host antiviral response (Sadler and Williams, 2008). Other activities of Us11 include; inhibiting apoptosis induced by heat shock or the broad kinase inhibitor staurosporine (Javouhey *et al.*, 2008), and interacting with homeodomain-interacting protein kinase 2 (HIPK2) (Giraud *et al.*, 2004). HIPK2 associates with ND10/PML bodies, can activate and phosphorylate p53 to induce apoptosis, and induces cell growth arrest, which can be altered by Us11 (Giraud *et al.*, 2004).

Early expression of Us11 is able to block activation of PKR and the associated shut down of protein synthesis (Cassady *et al.*, 1998). This suggested that IE Us11 should enhance the oncolytic activity of γ34.5Δ oHSV. SUP1 was then shown to have a greater than 100-fold increase in virus yield over the parental γ34.5Δ virus in human prostate cancer cells, resulting in significantly enhanced inhibition of tumour growth *in vivo*, even at low doses where Δ34.5 was not effective (Taneja *et al.*, 2001), with no increase in neurovirulence (Mohr *et al.*, 2001). G47Δ was constructed from G207 by deleting ICP47 (Todo *et al.*, 2001). G47Δ is more efficacious than G207 in all tumour models examined, both human and mouse (Fukuhara *et al.*, 2005b; Liu *et al.*, 2005b; Passer *et al.*, 2009; Todo *et al.*, 2001), while its safety after intracerebral injection remains similar to G207 (Todo *et al.*, 2001). It is important to note that Us11 is also needed late in infection to block protein shut-off (Mulvey *et al.*, 2003). This would indicate that the IE Us11 suppressor mutants only partially complement γ34.5 mutations and that further enhancement of these oHSV mutants (SUP1, G47Δ) should be possible in cancer cells.

Rather than using HSV-1 Us11 to complement the γ34.5Δ defects, it is possible to use proteins that block PKR-mediated protein shut-off from other viruses. An example of this approach is the expression of HCMV gene products, TRS1 or IRS1, in γ34.5Δ oHSV (Cassady, 2005). Both C130 (TRS1) and C134 (IRS1) block protein synthesis shut-off and phosphorylated eIF2α, replicate to almost wild-type titres in glioma cells, and inhibit tumour growth significantly better than parental γ34.5Δ C101 (Shah *et al.*, 2007). However, C130 was more neurovirulent than C134 or the parental γ34.5Δ C101, so C134 had a better therapeutic ratio (Shah *et al.*, 2007).

ICP47

HSV enlists multiple mechanisms to evade the host immune response, allowing the virus to co-exist in its host. One of the attributes of oHSV therapy is the induction of a tumour specific adaptive immune response, which might be negatively impacted by such evasion strategies. HSV IE ICP47 (Us12), is one such immune evasion gene that binds to the transporter associated with antigen presentation (TAP) and blocks the processing and presentation of antigens by MHC class I molecules on the surface of infected cells (Bauer and Tampe, 2002). The power of ICP47 to block MHC class I mediated immune responses has been demonstrated in a number of systems, for example in xenotransplantation and to extend transgene expression in gene therapy. Transduction of pig cells with ICP47 protected them from human CD8$^+$ T cell lysis, which did not occur with HCMV MHC class I blocking gene products (Us2, Us3, Us6, Us11) or adenovirus E3/19K (Crew and Phanavanh, 2003). Thus transduction of pigs with ICP47 could be beneficial for pig-to-human organ xenotransplants. A replication-deficient adenovirus vector expressing ICP47 was used to block adenovirus-specific CTL activity in rhesus monkeys after delivery to the lungs, however, this lead to increased NK activity (Scaria *et al.*, 2000), which would be expected with efficient MHC class I downregulation (Huard and Fruh, 2000).

Conversely, deletion of ICP47 in oHSV should enhance the immune response generated against infected tumour cells, compared to oHSV with intact ICP47, which is likely to be important for effective cancer therapy, as well as for HSV vaccines. This is in addition to the effect of ICP47 deletion on Us11 expression and enhancement of 34.5Δ oHSV replication. Unfortunately, ICP47 inhibition of TAP is species specific and the binding affinity for murine TAP is about 100-fold less than for human (Ahn *et al.*, 1996; Jugovic *et al.*, 1998; Song *et al.*, 1999), making it difficult to examine the immune effects in tumour-bearing mice. Therefore, the effect of oHSV infection of mouse tumours should be analogous to oHSV ICP47Δ infection in humans, which might lead to an inflated projection of immune-mediated responses of ICP47$^+$ oHSV in humans. Despite this species effect, ICP47$^-$ mutant HSV-1 is less neurovirulent in mice than wild-type, due to protective CD8$^+$ T cell responses and TAP1, likely due to low TAP1 and MHC class I expression levels in the brain (Burgos *et al.*, 2006; Goldsmith *et al.*, 1998). This increased safety should be more pronounced in humans.

G47Δ was the first oHSV developed to take advantage of deleting ICP47's TAP inhibitory activity, as well as its complementation of γ34.5Δ (Todo *et al.*, 2001). Infection of human fibroblasts and melanoma cells with G47Δ resulted in a smaller decrease in MHC class I presentation on the surface of infected cells compared to G207 or wild-type HSV-1 infected cells. G47Δ-infected melanoma cells were therefore better at stimulating matched tumour-infiltrating lymphocyte (TIL) lines than G207-infected cells (Todo *et al.*, 2001). JS1/ICP34.5/ICP47$^-$ (OncoVEX) is another oHSV containing an ICP47Δ. It has been combined with GM-CSF expression (OncoVEX^{GM-CSF}) to further enhance anti-tumour immunity (Liu *et al.*, 2003), which is being tested in humans.

Vaccine strains

Another approach for the development of oHSV vectors has been the use of attenuated vaccine strains, which have consequently undergone extensive safety studies. R7020 is a HSV-1/HSV-2 intertypic recombinant that was constructed as a vaccine candidate with a complex genome structure; deletion of the internal inverted repeat (including UL56 and 1 copy of ICP0, ICP4, and γ34.5) and insertion of a HSV-2 fragment (containing gG, gJ, gD, gI, Us2 and PK), deletion of TK and UL24, duplication of UL5 and UL6, and a HSV-1 α4promoter-TK transgene (Meignier *et al.*, 1988b). It was extensively tested for pathogenicity and safety in rodents (Meignier *et al.*, 1988b) and owl monkeys (Meignier *et al.*, 1990), but was unsuccessful as a vaccine in preliminary human trials (Bernstein and Stanberry, 1999; Whitley and Roizman, 2002). The attenuated pathogenicity may be due to deletion of: UL56, which contributes to intraperitoneal virulence in tree shrews and neuroinvasiveness in mice (Berkowitz *et al.*, 1994; Kehm *et al.*, 1996; Mori *et al.*, 2005; Rosen-Wolff *et al.*, 1991); UL24, which is important for virus replication in sensory ganglia (Jacobson *et al.*, 1998) and neurovirulence (Blakeney *et al.*, 2005); 1 copy of γ34.5 (Holman and MacLean, 2008; Perng *et al.*, 1996); or the internal repeat (Jenkins *et al.*, 1996). Building on this safety profile, R7020 was tested for oncolytic activity and found to be very effective, with virus titres recovered from tumours over 50-fold higher than input and still present 30 days after infection (Advani *et al.*, 1999). NV1020, a non-selected clonal derivative of R7020 that was isolated and characterized for clinical translation, is efficacious in a variety of non-CNS tumours (Currier *et al.*, 2005; Delman *et al.*, 2000; Ebright *et al.*, 2002;

Wong *et al.*, 2001a). This efficacy and safety data led to clinical trials of NV1020 for colorectal cancer metastatic to the liver (Kemeny *et al.*, 2006) (see Clinical Trial). NV1023 is a further modification of NV1020, where the TK-UL24 deletion has been repaired, the TK transgene at the L/S junction removed, and a LacZ reporter gene inserted downstream of the ICP47 promoter, deleting ICP47 and Us11 (Wong *et al.*, 2001b). It was constructed as the control vector for cytokine-expressing vectors NV1034 and NV1042 (see Therapeutic Transgene Expressing Vectors). NV1023 was effective at treating lymphatic metastases after travelling through the lymphatic system from the primary tumour (Wong *et al.*, 2002) and tumours established in the sciatic nerve where it caused no nerve damage (Gil *et al.*, 2007; Kelly *et al.*, 2008).

Syn mutants

To enhance the spread of oHSV in the tumour, syncytial mutants (syn) have been introduced into oHSV. In addition, syncytia formation itself is cytotoxic and can induce an immune response (Bateman *et al.*, 2002; Li *et al.*, 2007a). Syn mutants of HSV arise after virus passage in culture (Ruyechan *et al.*, 1979) and have been mapped to gB (Bzik *et al.*, 1984), gK (UL53) (Dolter *et al.*, 1994; Hutchinson *et al.*, 1992), UL20 (Melancon *et al.*, 2004), and UL24 (Tognon *et al.*, 1991). An unmapped syncytial mutant of G207, Fu-10, was isolated whose cell fusion was dependent on virus replication (Fu and Zhang, 2002). *In vitro*, Fu-10 was more cytotoxic than G207 in a number of human tumour cell lines. In a breast cancer lung metastasis model, intravenous delivery of Fu-10 was significantly better than G207 at reducing the number of lung metastases (Fu and Zhang, 2002). We also isolated syncytial mutants during the construction of G207 that were not mapped. However, in human glioma cells they were no more cytotoxic than G207 *in vitro* and were somewhat less effective at inhibiting tumour growth *in vivo* (unpublished results).

A separate approach has been to introduce defined syn mutants into oHSV. Onc was generated from a HSV1 strain F-BAC plasmid (Tanaka *et al.*, 2003) and contains a similar deletion of the internal repeat as NV1020 (from UL55 to ICP4), except instead of replacement with HSV-2 sequences, it contains a red fluorescent protein (Israyelyan *et al.*, 2007). The gBsyn3 mutant was recombined into Onc to produce OncSyn, which formed syncytial plaques and replicated to higher titres than Onc on human breast cancer cells (Israyelyan *et al.*, 2007). To further enhance cell fusion, a second syncytial mutation, gKsyn1, was introduced into OncSyn, using red-mediated recombination with pOncSyn plasmid in *E. coli*, to produce OncdSyn (Israyelyan *et al.*, 2008). Unfortunately, the second syn mutation decreased virus replication and did not increase anti-tumour efficacy *in vivo* (Israyelyan *et al.*, 2008). It should be borne in mind that both syn3 and syn1 mutations confer increased neurovirulence on HSV-1 (Goodman and Engel, 1991; Moyal *et al.*, 1992).

HF10 is a non-engineered clonal isolate of HF, a highly attenuated syncytial (likely in gB) HSV-1 strain passaged in embryonated eggs (Nishiyama *et al.*, 1991; Takakuwa *et al.*, 2003) that has a deletion of UL56 and the LAT promoter and duplication of UL53 and UL54 (Ushijima *et al.*, 2007). It is attenuated for neuroinvasiveness, but not neurovirulence (Mori *et al.*, 2005; Nawa *et al.*, 2008). *In vitro* replication in a murine macrophage cell line was restricted in a type I IFN fashion (Nawa *et al.*, 2008), suggesting that one of the mutations in HF10 is involved in inhibiting IFN-mediated innate immune responses. In a subcutaneous melanoma model in nude mice, where a clinical isolate caused severe dermatitis, HF10 replication was limited to the tumour through a mechanism involving NK cells (Nawa *et al.*, 2008).

Targeting virus entry to cancer selective receptors

HSV entry can be targeted away from normal cells to selected cancer cells by manipulating HSV glycoproteins so they loose interactions with their cognate cell surface receptors and introducing ligand sequences for cancer-selective cell surface molecules. Especially for systemic therapy, the broad tropism of HSV can be a drawback. Efforts to modify HSV attachment and entry to target tumour cells have been hampered by multiple HSV glycoproteins and cellular receptors and the close association between binding and viral envelope/

cell membrane fusion and entry. HSV attaches to the cell surface through interactions of viral glycoproteins gC and gB with cellular heparan sulfate proteoglycans (Akhtar and Shukla, 2009; Spear, 2004; see Chapter 8). Viral entry is then mediated by the binding of gD with cell surface receptors; HVEM (Hve A; TNF-receptor family), poliovirus receptor-related members of the immunoglobulin superfamily, nectin-1 (Hve C), and 3-O-sulphated heparin sulphate, and possibly gB with PILR-α (Akhtar and Shukla, 2009; Spear *et al.*, 2006). This is followed by fusion of the virus envelope and cell membrane, involving gB and the heterodimer gH-gL (Subramanian and Geraghty, 2007). It is worth noting that the level of nectin-1 expression on tumour cells can impact the efficacy of oHSV therapy (Huang *et al.*, 2007; Yu *et al.*, 2007). In human squamous cell carcinoma and thyroid cancer cells there was a statistically significant correlation between cytotoxicity and nectin-1 expression, but not HVEM expression or cell doubling time. A number of rodent tumour cell lines that are resistant to HSV infection, have been converted to susceptible cells by transduction with nectin-1 plasmids (Miller *et al.*, 2001; Wakimoto *et al.*, 2004).

Structure-function studies of gD have revealed independent HVEM and nectin-1 binding regions in the N-terminal domain and a profusogenic region in the C-terminus of the ectodomain (Spear *et al.*, 2006). This was exploited to generate an IL-13 receptor α2 retargeted virus by insertion of IL-13 into gD, disrupting the HVEM binding domain, deleting the heparan sulfate binding domain of gC, and subsequently mutating the gD nectin-1 binding domain and gB heparan sulfate binding domain (Zhou and Roizman, 2006). With the number of introduced mutations, it is not surprising that non-introduced mutations arose in gD during the generation of a similar virus expressing a urokinase plasminogen activator-gD chimera (Kamiyama *et al.*, 2006). These unplanned, but presumably selected mutations resulted in an independently expressed C-terminal domain of gD, thereby splitting the receptor-binding and profusogenic domains of gD into separate proteins (Zhou and Roizman, 2007). This should simplify the generation of viruses expressing a variety of receptor-ligands.

The Glorioso group has employed a different strategy, using a soluble gD-binding domain of nectin-1, to promote virus entry in the absence of gD receptor binding (Kwon *et al.*, 2006). Fusion of this soluble nectin-1 domain to a cell surface receptor single-chain antibody (scFv) can target entry to receptor-expressing cells (Nakano *et al.*, 2005a), allowing chimeric recombinant proteins to be used rather than genetic manipulation of the virus genome. The use of an scFv to target a cell surface receptor was employed in a strategy where the gD Ig-folded core (deleting the HVEM and nectin-1 binding domains) was replaced with the coding sequence for a scFv against HER-2 to target breast and ovarian cancers (Menotti *et al.*, 2008, 2009). This virus (R-LM249) was successfully retargeted to cancer cells expressing HER-2 both *in vitro* and *in vivo* after intratumoural injection (Menotti *et al.*, 2009), the first demonstration of *in vivo* retargeting of HSV. While R-LM249 did not display toxicity after intraperitoneal injection, indicating receptor-specificity, the encoded svFv is human HER2 specific, so mouse cells expressing HER2 should not be susceptible. The use of scFv allows one to potentially target any cell surface molecule.

Transcriptionally targeted vectors

An alternative strategy to target cancer cells is to take advantage of altered gene expression in transformed cells, in particular tumour-specific genes that are upregulated transcriptionally, and use these transcriptionally-regulated enhancer/promoter sequences to selectively drive viral oncolysis (Dorer and Nettelbeck, 2009; Hardcastle *et al.*, 2007). In such cases, cell specificity and level of expression is determined by the presence of appropriate transcription factors, the promoter/enhancer sequences, the position in the viral genome and aberrant regulation by adjacent viral elements, and potentially chromatin structure. Two strategies using a tumour-selective promoter/enhancer have been employed to transcriptionally regulate HSV replication: (i) driving expression of the essential IE gene ICP4 in a replication-deficient ICP4 deletion mutant (Kuroda *et al.*, 2006b; Miyatake *et al.*, 1997; Yamamura *et al.*, 2001), and (ii) driving γ34.5 to enhance the attenuated replication of γ34.5 mutants (Chung *et al.*, 1999; Kambara *et al.*, 2005; Kanai *et al.*, 2007).

Regulated ICP4

In the first example of this strategy, ICP4 was placed under control of the albumin enhancer/promoter and this cassette was inserted into the TK locus of ICP4 deletion mutant d120 to generate G92A (Miyatake *et al.*, 1997). Thus, the virus should only replicate only in cells expressing albumin, such as hepatoma and primary liver cells. Indeed, specificity of G92A replication *in vitro* was over a 1000-fold greater in hepatoma cells than in albumin non-expressing tumour cells (Miyatake *et al.*, 1997). *In vivo*, G92A was effective against subcutaneous and hepatic human liver tumours but not hrR3-sensitive prostate or colon cancer tumours (Chung *et al.*, 2006; Miyatake *et al.*, 1999). When G92A was injected into the spleen for hepatic delivery, X-gal staining was only apparent in liver tumour nodules, whereas hrR3 displayed some staining of normal liver (Chung *et al.*, 2006). Most importantly, no evidence of liver toxicity was observed when the virus was injected intrahepatically, despite high levels of albumin expression in normal liver (Miyatake *et al.*, 1999). This lack of toxicity is probably due to mutations in TK and Us3, which attenuate pathogenicity and replication in non-dividing cells.

A similar strategy has been applied to the generation of *d*l2.CALP, where the calponin promoter, active in soft tissue and bone tumours, drives the expression of ICP4 in the TK locus (Yamamura *et al.*, 2001). Again, *d*12.CALP was selective for human leiomyosarcoma, but not calponin-negative osteosarcoma xenografts (Yamamura *et al.*, 2001). In SS1, the ICP4 transgene is controlled by activated ELK, down-stream of Ras signalling, using a synthetic promoter containing 5 tandem repeats of the ELK DNA binding sequence (Pan *et al.*, 2009). SS1 replication in prostate cancer cells correlated with ELK activation, however, ELK was also activated in normal prostate epithelial cells (Pan *et al.*, 2009), so its safety is a question. The Wnt/β -catenin pathway is upregulated in most colorectal cancers and hepatoblastomas, as well as cancer stem cells, usually due to mutations in adenomatous polyposis coli (APC) or β-catenin genes (Gavert and Ben-Ze'ev, 2007; Polakis, 2007). This pathway was targeted with bM24-TE, which contains a synthetic β-catenin/Tcf promoter and the 4F2 heavy chain gene enhancer element driving ICP4 (Kuroda *et al.*, 2006b). In susceptible cancer cell lines, but not normal or other cancer cells, inclusion of the 4F2 enhancer increased ICP4 expression, virus yield, and cytotoxicity (Kuroda *et al.*, 2006b). As opposed to transgene insertion in the TK locus, which makes the vectors ACV resistant, the ICP4 transgene was inserted into UL39, disrupting ICP6. bM24-TE was constructed using the Flip-Flop HSV-BAC system, which permits rapid insertion of different promoter-ICP4 transgenes for recombinant virus generation (Kuroda *et al.*, 2006a). bM24-TE replication and inhibition of tumour growth was dependent upon β-catenin signalling (Kuroda *et al.*, 2006b).

This strategy does not always work. CEAICP4 contains the carcinoembryonic antigen (CEA) promoter driving ICP4, inserted into the ICP6 locus. Unfortunately, it replicated very poorly *in vitro* in a few cell lines and did not replicate at all in two CEA high-expressing tumour cell lines (Mullen *et al.*, 2002). Some promoters may not be strong enough to express sufficient levels of ICP4 to successfully drive virus replication. Even a CMV IE promoter–ICP4 construct (bM24-CMV) expresses very low levels of ICP4, with delayed replication kinetics and decreased virus yield in Vero cells compared to hrR3 ($ICP6^-$) (Kuroda *et al.*, 2006a). In HSPV-1, an HPV-16 upstream regulatory sequence (URR16) driving ICP4 was inserted between Us9 and Us10 in d120 for oral cancer cell selectivity (Griffith *et al.*, 2007). Unfortunately, ICP4 expression and virus replication was not regulated in a HPV fashion. It was further found that HSV infection, including with d120, induced expression from a URR16 reporter plasmid (Griffith *et al.*, 2007), suggesting that HSV IE genes other than ICP4 were activating this enhancer, likely inducing non-specific ICP4 expression in HSPV-1. It is important to recognize that transcriptional regulation of the HSV genome can affect/override the regulation of exogenous promoters/enhancers (Leary *et al.*, 1989; Panning and Smiley, 1989; Roemer *et al.*, 1991). For example, HSV–1 constructs containing glioma-selective regulatory sequences (TERT, surviving, VEGF-A, nestin) driving luciferase transgenes inserted between UL55 and UL56 were found to be relatively inactive in glioma cells, and behaved differently than the corresponding plasmids after transfection (Glass *et al.*, 2008).

Regulated γ34.5

A variation of this targeted approach is to transcriptionally regulate γ34.5 expression in a tumour specific fashion. Upregulating γ34.5 expression within tumour cells should enhance the oncolytic activity of γ34.5Δ vectors, in essence complementing the defect in selective cells. This strategy was initially examined with Myb34.5, where a γ34.5 transgene under control of the E2F responsive B-*myb* promoter was inserted into the ICP6 locus of MGH1 (Chung *et al.*, 1999). Accordingly, γ34.5 should be selectively expressed in cycling cells and those with deregulated E2F, i.e. tumour cells. Treatment of subcutaneous gliomas intratumourally or liver metastasis intravascularly with Myb34.5 was more effective than MGH1 (Chung *et al.*, 1999; Nakamura *et al.*, 2002). Replication of Myb34.5 was restricted to the tumour and Myb34.5 was less toxic than hrR3 after portal vein infusion or intracerebral inoculation, but slightly more neurovirulent than MGH1 (Chung *et al.*, 1999; Nakamura *et al.*, 2002).

A similar construct was generated with CEAγ34.5. While replication of the CEAγ34.5 virus was about 10-fold greater than MGH1, it did not vary between cells with high or low CEA expression, further illustrating problems with this promoter (Mullen *et al.*, 2002). DF3γ34.5, contains the DF3 promoter-γ34.5 transgene inserted into the TK locus of R3616 (Kasuya *et al.*, 2004). Mucin 1 (DF3 antigen) is aberrantly expressed on a variety of epithelial tumours (Abe and Kufe, 1993). In contrast to CEAγ34.5, DF3γ34.5 virus replication was the same as MGH1 in MUC1 non-expressing cells and 1 to 2 logs greater than MGH1 in MUC1-positive cells (Mullen *et al.*, 2002). This translated into selective inhibition of tumour growth *in vivo*, where DF3γ34.5 was more efficacious in MUC1-positive tumours than MGH1 and similar to MGH1 in MUC1-negative tumours (Kasuya *et al.*, 2004). A nestin-γ34.5 vector, rQnestin34.5, containing the nestin second intron enhancer and Hsp78 minimal promoter (Kambara *et al.*, 2005), was constructed using the HSVQuick method that involves a HSV-BAC plasmid and Flp-FRT recombination to insert the transgene into the ICP6 locus (Terada *et al.*, 2006). Nestin is expressed in neural progenitor cells, a variety of primary brain tumours, and glioma stem cells (Dell'Albani, 2008; Kambara *et al.*, 2005). rQnestin34.5 replicated better and was more cytotoxic than rHsvQ1 (parental virus) in glioma cell lines, but not in normal astrocytes, and was much more efficacious in prolonging the survival of mice with intracerebral gliomas (Kambara *et al.*, 2005). Another glioma-targeted vector, KeM34.5, has γ34.5 driven by the musashi1 promoter cloned into LacZ in ICP6 of G207 (Kanai *et al.*, 2007). It displayed increased replication in glioma cell lines compared to G207, but similar poor replication in non-glioma cancer cells, and no spread in primary mouse astrocytes (Kanai *et al.*, 2007).

Translational control is another way to regulate γ34.5 expression. The poliovirus IRES is partially responsible for its neurotropism, and when exchanged with the HRV2 IRES, poliovirus looses its neurovirulence but maintains replication and cytotoxicity in glioma cells (Gromeier *et al.*, 1996, 2000). To test this in HSV, HRV2 IRES was inserted in place of the γ34.5 UTR (HRV-34.5) (Campbell *et al.*, 2007). HRV-34.5 replicated well in human glioma cells, as opposed to the γ34.5Δ parent SPBg5e, and was attenuated for neurovirulence, but much less than SPBg5e (Campbell *et al.*, 2007). Use of the poliovirus IRES restored neurovirulence to almost that seen with wild-type Patton (Campbell *et al.*, 2007). This indicates that picornavirus IRES and UTRs can regulate gene expression in HSV.

miRNA

In place of promoter/enhancer DNA sequences regulating transcription, it is possible to repress gene expression and virus replication in normal cells using tissue specific microRNA (miRNA) target sequences (Kelly and Russell, 2009). miRNAs are small (~22 nucleotides) non-coding RNAs that pair with complementary target sequences, usually in the 3′ UTR of mRNA, to impair their translation or promote their degradation (Brown and Naldini, 2009). HSV encodes multiple miRNAs, especially in LAT, that regulate viral mRNAs (Tang *et al.*, 2008; Umbach *et al.*, 2008). This includes a number that are antisense to γ34.5 mRNA and reduce ICP34.5 expression (Tang *et al.*, 2008, 2009). Many miRNAs are expressed in a highly tissue specific fashion, with tumour suppressor miRNAs downregulated in cancer. To test the ability of miRNAs to repress HSV-1, target sequences to miR-143 or miR-145, which are expressed in normal tissues but downregulated in many solid tumours, were cloned into the 3′ UTR of ICP4 in an amplicon plasmid and propagated with an ICP4Δ helper virus (CgalΔ3) (Lee *et al.*, 2009). Virus replication was decreased when cancer cells were transfected with the corresponding pre-miRNA. *In vivo*, both amplicon/helper combinations markedly reduced LNCaP tumour volume, while CMV-ICP4-143T had much less spread to normal tissues (Lee *et al.*, 2009). It remains to be determined whether miRNA target sequences when inserted into the viral genome will be strong enough to block pathogenicity or sufficiently alter tropism.

Transgene expressing 'armed' oHSV

Suicide genes

Many chemotherapeutic drugs are converted into active agents by metabolism, often in the liver. One way to enhance the local concentration of toxic active metabolites in the tumour is to express pro-drug activating enzymes, which has been referred to a gene directed enzyme prodrug therapy (GDEPT) (Portsmouth *et al.*, 2007). The classic 'suicide gene' is HSV-TK, which converts nontoxic nucleoside analogues like GCV to GCV-monophosphate, which is further modified by cellular kinases to toxic GCV-triphosphate (Fyfe *et al.*, 1978), and is the basis for the antiviral activity of these compounds. While GCV has been tested in combination with TK^+ oHSV (Kasuya *et al.*, 1999; Kramm *et al.*, 1996; Luo *et al.*, 2007), it can be counterproductive because of its inhibition of virus replication (Todo *et al.*, 2000b; Yoon *et al.*, 1998). A key to the suicide gene strategy is the 'bystander effect', where the toxic metabolite diffuses to surrounding non-infected cells (Dachs *et al.*, 2009). Prodrug-activating enzymes that have been successfully utilized with oHSVs are: rat cytochrome P450 (*CYP2B1*) for the conversion of cyclophosphamide (CPA) to phosphoramide mustard and acrolein, in rRp450 (Chase *et al.*, 1998); yeast or bacterial cytosine deaminase (CD) for the conversion of 5-fluorocytosine (5-FC) to 5-FU, which is further modified by cellular enzymes to the toxic pyrimidine anti-metabolite 5-FdUTP, in HSV1yCD (Nakamura *et al.*, 2001) and M012 (Guffey *et al.*, 2007); secreted human intestinal carboxylesterase (CE) for the conversion of irinotecan (CPT-11) to SN-38, in MGH2 (Tyminski *et al.*, 2005); bacterial nitroreductase (NTR) for the conversion of CB1954 to a potent DNA-cross-linking 4-hydroxylamine, in HSV1790 (Braidwood *et al.*, 2009); and *E. coli* purine nucleoside phosphorylase (PNP) for the conversion of 6-methylpurine-2′-deoxyriboside to 6-methylpurine (6MeP), in HSV-PNP (Bharara *et al.*, 2005). Of these drugs, CPA (Fulci *et al.*, 2006; Li *et al.*, 2007c) and 5-FU (Eisenberg *et al.*, 2006; Nakano *et al.*, 2005b) have been examined in combination with oHSV in the absence of activating enzyme expression.

CPA was not inhibitory to the replication of rRp450 *in vitro* and enhanced its efficacy in suppressing the growth of subcutaneous glial tumours and diffuse liver tumours, due to elevated local metabolite concentrations (Chase *et al.*, 1998; Pawlik *et al.*, 2002). With safety demonstrated in preclinical studies (Currier *et al.*, 2008), rRp450 is currently moving towards clinical trial. To further safety in the brain and increase efficacy, both CYP2B1 and CE were inserted into MGH1 ($ICP6^-$, γ34.5Δ) using the HSVQuik system in *E. coli* to generate MGH2 (Terada *et al.*, 2006; Tyminski *et al.*, 2005). Neither CPA nor CPT-11 inhibited MGH2 replication in human glioma cells *in vitro*, and the combination significantly enhanced survival of mice bearing intracerebral gliomas (Tyminski *et al.*, 2005). For CD-expressing oHSVs, 5-FC inhibited replication *in vitro* to a similar extent as 5-FU with parental non-CD virus (Guffey *et al.*, 2007; Nakamura *et al.*, 2001). Despite this, administration of 5-FC significantly inhibited tumour growth *in vivo* (Guffey *et al.*, 2007; Nakamura *et al.*, 2001), and did not increase neurovirulence of M012 (Guffey *et al.*, 2007). In M012, CD is driven by the Egr-1 promoter, which is radiation-inducible (Kufe and Weichselbaum, 2003), and 5-FC radiosensitizes cancer cells expressing CD (Khil *et al.*, 1996). Therefore the combination of M012, 5-FC and radiotherapy, for example in glioma, could be clinically beneficial. To potentially enhance the conversion of 5-FC (although not directly tested), a fusion yeast CD-yeast UPRT transgene was inserted into OncoVEX (Simpson *et al.*, 2006). The fusion transgene was also inserted into fusogenic OncoVEX^{GALV}, but the combination was not any better at inhibiting tumour growth with 5-FC than with CD-UPRT alone (Simpson *et al.*, 2006).

Reporter genes for imaging

For many purposes non-invasive imaging of oHSV replication and spread is extremely beneficial for following activity *in vivo*. Bioluminescent imaging (BLI) of luciferase activity is useful for tracking virus spread in mouse models, including wild-type HSV for pathogenicity studies (Luker and Leib, 2005) and oHSV vectors (Mahller *et al.*, 2008; Terada *et al.*, 2006; Yamamoto *et al.*, 2006). Currently BLI technologies are not in clinical practice, however the following 2 technologies are applicable in patients. Cancer selective expression of GFP by oHSV, such as NV1066, is a potentially powerful way to image tumour cells *in situ* (Stiles *et al.*, 2003). In a number of studies, NV1066 infusion was able to localize lymph node metastases (Adusumilli *et al.*, 2006a; Eisenberg *et al.*, 2006), micrometastases (Stanziale *et al.*, 2004), neural invasion (Gil *et al.*, 2008), or to detect tumour margins (Adusumilli *et al.*, 2006b) using fluorescent endoscopy, which could be applied for minimally invasive surgery. While HSV-TK is not a transgene, its expression and therefore oHSV spread *in vivo* can be imaged non-invasively using positron emission tomography (PET) with [^{18}F]-FEAU or other analogues (Brader *et al.*, 2009). However, there are potential drawbacks to TK as a PET reporter; a weak signal due to limited expression from the endogenous gene (Bennett *et al.*, 2001; Jacobs *et al.*, 2001), and loss of signal due to oncolysis (Kuruppu *et al.*, 2007).

Fusogenic proteins

Cell fusion is an attractive strategy to enhance oHSV therapy, as illustrated with HSV syn mutants. Expression of exogenous fusogenic membrane glycoproteins can also be therapeutic on their own (Galanis *et al.*, 2001). The gibbon ape leukaemia virus envelope glycoprotein lacking the R-peptide (GALV.fus) is very fusogenic in human cells expressing the ubiquitous PiT-1 receptor (SLC20A1 transporter), but not in mouse cells (Overbaugh *et al.*, 2001), so that any pathology or fusion of normal cells associated with GALV.fus will not be apparent in mice. Synco-2 is a γ34.5Δ oHSV, derived from a HSV-BAC (fHSV-delta-pac), that expresses GALV.fus from the late UL38 promoter, so fusogenic activity will only occur after viral replication in tumour cells (Fu *et al.*, 2003). Synco-2 was as cytotoxic in Hep3B cells as Fu-10, and was more effective *in vivo* than Baco-1 (expressing GFP instead of GALV.fus) (Fu *et al.*, 2003). To further enhance the fusogenic activity, a syncytial mutant of Baco-1 was isolated by random mutagenesis and then the UL38pro-GALV.fus transgene was inserted into the HSV-BAC plasmid to generate an oHSV with 2 fusogenic activities (Synco-2D) (Nakamori *et al.*, 2003). Synco-2D was more cytotoxic in human prostate cells than Synco-2 or Baco-1, and more efficacious in inhibiting metastatic lung tumours after intravenous delivery (Nakamori *et al.*, 2004). In a syngeneic mouse breast cancer model, Synco-2D, compared to Synco-2 or FU-10, was the only oHSV able to significantly inhibit orthotopic tumour growth and metastasis, which was mediated by CD8^{+} T cells (Nakamori *et al.*, 2004). As GALV.fus (Synco-2) does not fuse mouse cells and Fu-10 doesn't induce an immune response, it is possible that GALV.fus is acting as a foreign antigen in the context of cell fusion to induce antitumour immunity. GALV.fus has also been inserted into OncoVEX under the CMV promoter (Simpson *et al.*, 2006). In 2 human tumour models, OncoVEX^{GALV} was only significantly better than OncoVEX at a low virus dose, but was better in the syngeneic rat 9L tumour model, even though the rat cells do not express the GALV receptor (Simpson *et al.*, 2006). Again, possibly indicating the immunogenic, rather than fusogenic, properties of GALV.

Immune-modulatory genes

One of the major effectors of oHSV efficacy in immune-competent mice and syngeneic tumour models is the generation of anti-tumour immunity, which involves tumour-specific adaptive immune responses, including CD8+ T-cells, Th1 cytokines, and memory that is effective against established tumours (Benencia *et al.*, 2005, 2008; Endo *et al.*, 2002; Li *et al.*, 2007b; Miller and Fraser, 2003; Toda *et al.*, 1999; Todo *et al.*, 1999). This likely is mediated by the induction of inflammation in the tumour and oncolytic release of tumour-associated antigens and 'danger signals' (Li *et al.*, 2008). Therefore the expression of cytokines or immunostimulatory factors that augment the 'in situ cancer vaccine' effects of oHSV are likely to be beneficial for therapy. A number of oHSV vectors have been generated that contain immune-modulatory transgenes, including; IL-4 (Andreansky *et al.*, 1998), IL-12 (Ino *et al.*, 2006; Parker *et al.*, 2000; Wong *et al.*, 2001b), IL-18 (Fukuhara *et al.*, 2005a; Ino *et al.*, 2006), GM-CSF (Liu *et al.*, 2003; Wong *et al.*, 2001b), soluble B7.1 (CD80) (Fukuhara *et al.*, 2005a; Ino *et al.*, 2006), CCL2 (Parker *et al.*, 2005), and TNFα (Han *et al.*, 2007). JS1/ICP34.5$^-$/ICP47$^-$/GM-CSF (OncoVex^{GM-CSF}) is the first cytokine-expressing oHSV to be taken into clinical trial (see Clinical Trials). In mouse tumour models it was only modestly better at high virus dose and for contralateral non-inoculated tumours than JS1/ICP34.5$^-$/ICP47$^-$ (Liu *et al.*, 2003).

In general, where compared, IL-12 has been found to be the most effective factor at enhancing oHSV therapy in syngeneic models (Ino *et al.*, 2006; Parker *et al.*, 2005; Varghese *et al.*, 2006a; Wong *et al.*, 2001b). For example, in a comparison of oHSVs expressing IL-12, IL-18, or B7.1Ig, at low virus dose (1×10^4 PFU), IL-12 was significantly better than the other factors, whereas at higher doses (2×10^6 PFU) all 3 factors were similarly effective at inhibiting tumour growth (Ino *et al.*, 2006). However, the combination of all 3 vectors was still significantly better than any of them alone. NV1042 expresses a protein fusion of the p40 and p35 subunits of mIL-12 and in all tumours tested was more efficacious than NV1034, expressing GM-CSF (Derubertis *et al.*, 2007; Varghese *et al.*, 2006a; Wong *et al.*, 2001b). M002, with a bicistronic mouse IL-12 transgene (p40-IRES-p35), had enhanced efficacy in syngeneic neuroblastoma and glioma models and IL-12 expression didn't alter the highly attenuated virulence of the γ34.5Δ background (Hellums *et al.*, 2005; Parker *et al.*, 2000). For clinical translation, M032 was constructed expressing human IL-12, due to the species specificity of IL-12 activity (Parker *et al.*, 2009). Tumour cells infected with either of M002 or NV1042 express measurable levels of IL-12 *in vitro* and *in vivo* (Varghese *et al.*, 2006b; Wong *et al.*, 2004) with enhanced infiltration of T cells within treated tumours (Hellums *et al.*, 2005; Jarnagin *et al.*, 2003; Parker *et al.*, 2000).

oHSV vectors should also be useful as HSV vaccines, as they are safe, maintain inflammatory features of HSV infection, often are defective in establishing latency or reactivating, and express a range of viral gene products. Expression of immune-modulatory transgenes should further enhance the immune response against the virus. This has been demonstrated in mouse models, with IL-12 expression (M002) being more effective than GM-CSF expression, or even wild-type (Parker *et al.*, 2006). In a separate study, it was found that expressing the p35 subunit of IL-12 was more efficacious than p40 alone or in combination, or IL4 in providing overall protection (Osorio and Ghiasi, 2005). It will be of interest to determine whether expressing the p35 subunit alone also enhances oncolytic activity.

Clinical trials

The ultimate goal of the development of oHSV vectors is to treat patients and the clinical testing of these vectors within a decade of the first description of an oHSV vector attests to the progress that has been made. Initiating a clinical trial with oHSV requires review and/or approval of a multitude of regulatory panels: the Recombinant DNA Advisory Committee (RAC) of NIH for review of human gene therapy clinical trials at NIH funded institutions; the local Institutional Review Board (IRB) for evaluation of research involving humans and the Informed Consent document; the local Institutional Biosafety Committee (IBC) for review of recombinant DNA research and potential biohazard risks; a Data and Safety Monitoring Board (DSMB) for monitoring patient safety during a clinical trial; and the Center for Biologics Evaluation and Research (CBER) at the FDA, for approval of an Investigational New Drug (IND) application to administer oHSV to humans in a clinical trial and ultimately to approve the agent for use. Information on cancer clinical trials in the US is available at NCI (http://www.cancer.gov/clinicaltrials/search/) or more broadly at http://clinicaltrials.gov/.

Successful safety and efficacy studies in pre-clinical animal models have led to phase I human clinical trials of G207, 1716, NV1020, OncoVEX^{GM-CSF}, and HF-10 in a variety of cancers that have either been completed or are ongoing (Liu *et al.*, 2007a; Nakao *et al.*, 2007). Other oHSV vectors currently entering, or moving towards clinical trial, include; G47Δ, rRp450, and M032. The first oHSV clinical trials were with G207 in the US (Markert *et al.*, 2000) and 1716 in the UK (Rampling *et al.*, 2000). Both studies enrolled patients with recurrent malignant glioma that was refractory to radiation and chemotherapy, and the virus was stereotactically injected into the tumour. The 1716 trial tested a dose range of 1×10^3 to 1×10^5 PFU, while the G207 trial tested a higher dose range (from 1×10^6 to 3×10^9 PFU) (Markert *et al.*, 2000; Rampling *et al.*, 2000). The low maximum dose of 1716 is reflective of it being slightly more neurovirulent than G207 (MacLean *et al.*, 1991; Sundaresan *et al.*, 2000) and not being tested in non-human primates. Each of the viruses was well tolerated by all patients with no adverse events ascribed to the viruses. Post-mortem analysis of patients who died showed no evidence of encephalitis, and a few patients from both trials survived for longer periods than might be expected from historical controls. In a follow-up study with 1716, 10^5 PFU of virus was injected into the tumour and then 4–9 days later, the tumour was resected to look for evidence of virus replication. Infectious virus was isolated from tumours at the injection sites from 2 of 12 patients; interestingly these were the only seronegative patients and both seroconverted (Papanastassiou *et al.*, 2002). In a second phase 1b trial of G207, virus was inoculated into the tumour through a catheter that was left in place and then 2–5 days later the tumour was resected en bloc and G207 injected into the brain surrounding the resection cavity (Markert *et al.*, 2009). Out of 6 patients, virus replication in tumours was detected in 3 patients (HSVpol mRNA RT-PCR), with anti-gC immunohistochemistry present in 2 patients and infectious virus isolated from 1 patient (Markert *et al.*, 2009). All 3 HSV-1 antibody seronegative patients seroconverted within 1–3 months post-treatment (Markert *et al.*, 2009). In a third 1716 glioma trial, virus was injected into the tumour bed after resection (Harrow *et al.*, 2004). A randomized phase III clinical trial jof 1716 for recurrent glioblastoma multiforme (GBM) is currently ongoing in the UK (GTAC-074). 1716 has also been tested in 20 patients with oral squamous cell carcinoma, administering 5×10^5 PFU intratumourally. Unfortunately there was no evidence of biological activity or virus replication detected, possibly suggesting too low a dose (Mace *et al.*, 2008).

NV1020 was the first oHSV to be administered intravascularly in patients. In a dose-escalating study, NV1020 (up to 10^8 PFU) was delivered via intra-hepatic artery infusion to patients with colorectal carcinoma metastatic to the liver and refractory to chemotherapy (Fong *et al.*, 2009; Kemeny *et al.*, 2006). Three of 5 HSV-2 seronegative patients seroconverted shortly after treatment and most patients displayed a significant decrease in CEA levels despite rising CEA levels prior to NV1020 (Fong *et al.*, 2009). No virus was detected in the hepatic vein at 10^7 PFU, indicating that all virus was cleared by the liver, whereas all patients at 10^8 PFU had NV1020-positive sera by PCR and one had infectious virus (Kemeny *et al.*, 2006). No maximum tolerated dose (MTD) was reached, but dose escalation was stopped at this point, due to; presumed saturation of liver/tumour uptake, the cost of clinical grade virus, the desire to proceed to a multiple dosing strategy that has been shown to be more effective in mouse models, and the lack of tolerance for serious adverse events in gene therapy trials. This trial has been followed by a multi-centre phase II clinical trial with dosing repeated weekly (NCT00149396) (Karrasch *et al.*, 2007).

HF10 has been tested in 2 phase I studies with 6 breast cancer patients and 3 head and neck squamous cell carcinoma patients, all HSV seropositive, and intumoural injection of $1–5 \times 10^5$ PFU (Nawa *et al.*, 2008). Because HF10 was not genetically-engineered it did not have to go through gene therapy regulatory review in Japan. HSV-infected cells were seen in excised tumour nodules about 2 weeks after treatment (Fujimoto *et al.*, 2006; Kimata *et al.*, 2006). A phase I clinical trial in the US for head and neck cancer sponsored by M's Science Corp has been approved (NCT0107185).

The first 'armed' transgene-expressing oHSV to enter clinical trial was OncoVEX^{GM-CSF} for accessible cutaneous or subcutaneous metastases (Hu *et al.*, 2006b). Local inflammation at the injected tumour lesion was dose limiting (10^7 PFU/ml) in the HSV-seronegative patients, so a multi-dosing schedule was tested to seroconvert the HSV-seronegative patients with a low dose (10^6 PFU/ml) followed by multiple high doses (10^8 PFU/ml), which was well tolerated (Hu *et al.*, 2006b). HSV antigen staining was mostly detected in necrotic tumour regions. This was followed by a phase II trial for unresectable metastatic melanoma, with intratumoural injection of accessible tumours (up to 10 tumours) with 4×10^6 PFU, followed by intratumoural 4×10^8 PFU every 2 weeks (Senzer *et al.*, 2009). All 13 HSV seronegative patients seroconverted, however, clinical responses did not correlate with HSV serostatus. Only 1 of 102 swabs from injection sites was weakly positive and all urine samples were negative (qPCR) (Senzer *et al.*, 2009), indicating that virus shedding is very rare, as has been the case with other oHSV trials that examined shedding. On the immune side, treatment induced local and systemic MART-1 T cells, and decreased Treg and T suppressor cells in injected lesions compared to non-treated melanoma patients (Kaufman *et al.*, 2009). The clinical responses were very promising, with durable responses seen at both injected and uninjected lesions, and a 1 year survival of 58% compared to 25.5% seen in a meta-analysis of phase II melanoma trials with a variety of agents (Senzer *et al.*, 2009). Based on this, BioVex has initiated a randomized phase III clinical trial in melanoma (NCT00769704), where OncoVEX^{GM-CSF} will be compared to recombinant GM-CSF. A second phase III clinical trial has also been approved for head and neck squamous cell carcinoma.

Overall, oHSV has been administered to over 200 cancer patients and so far there has been no evidence of virus-related serious adverse events. This demonstrates the safety of oHSV therapy, at least at the doses tested. While it is too earlier to know whether oHSV therapy will be sufficiently efficacious to be approved for clinical use, the early phase clinical trials provide anecdotal evidence of efficacy and optimism for success.

Conclusions

OHSV vectors have progressed rapidly from the initial concept of genetically-engineering HSV to target cancer cells to clinical translation. It is rather amazing that a highly pathogenic and occasionally lethal virus like HSV could be converted into a potentially beneficial therapeutic agent. The large genome size of HSV and wealth of knowledge about its biology and host interactions have provided for a variety of different virus manipulations engendering oncolytic activity. In addition to genetically modifying HSV, there are a number of strategies being pursued to enhance oHSV therapy: combinations with chemotherapy or radiation to increase virus replication and/or cytotoxicity (Advani *et al.*, 2006; Post *et al.*, 2004); injection of immune cells, such as DCs, to treated tumours to enhance anti-tumour immune responses (Farrell *et al.*, 2008); inhibition of acute innate immune responses to improve virus spread (Fulci *et al.*, 2006; Wakimoto *et al.*, 2003); extracellular matrix degradation (McKee *et al.*, 2006), angiostatic treatment (Aghi *et al.*, 2007; Kurozumi *et al.*, 2007), and alternative delivery mechanisms, such as blood–brain barrier disruption (Liu *et al.*, 2005a) or convection-enhanced delivery (CED) (Hadjipanayis *et al.*, 2008), to improve virus delivery within the tumour. Many hurdles remain before oHSV becomes a therapeutic reality, but the clinical results provide optimism.

References

Abe, M., and Kufe, D. (1993). Characterization of cis-acting elements regulating transcription of the human DF3 breast carcinoma-associated antigen (MUC1) gene. Proc. Natl. Acad. Sci. U.S.A. *90*, 282–286.

Adusumilli, P.S., Eisenberg, D.P., Stiles, B.M., Chung, S., Chan, M.K., Rusch, V.W., and Fong, Y. (2006a). Intraoperative localization of lymph node metastases with a replication-competent herpes simplex virus. J. Thorac. Cardiovasc Surg. 132, 1179–1188.

Adusumilli, P.S., Stiles, B.M., Chan, M.K., Eisenberg, D.P., Yu, Z., Stanziale, S.F., Huq, R., Wong, R.J., Rusch, V.W., and Fong, Y. (2006b). Real-time diagnostic imaging of tumors and metastases by use of a replication-competent herpes vector to facilitate minimally invasive oncological surgery. Faseb J. *20*, 726–728.

Advani, S.J., Chung, S.M., Yan, S.Y., Gillespie, G.Y., Markert, J.M., Whitley, R.J., Roizman, B., and Weichselbaum, R.R. (1999). Replication-competent, nonneuroinvasive genetically engineered herpes virus is highly effective in the treatment of therapy-resistant experimental human tumors. Cancer Res. *59*, 2055–2058.

Advani, S.J., Mezhir, J.J., Roizman, B., and Weichselbaum, R.R. (2006). ReVOLT: radiation-enhanced viral oncolytic therapy. Int. J. Radiat Oncol. Biol. Phys. *66*, 637–646.

Aghi, M., Rabkin, S., and Martuza, R.L. (2006). Effect of chemotherapy-induced DNA repair on oncolytic herpes simplex viral replication. J. Natl. Cancer Inst. *98*, 38–50.

Aghi, M., Rabkin, S.D., and Martuza, R.L. (2007). Angiogenic response caused by oncolytic herpes simplex virus-induced reduced thrombospondin expression can be prevented by specific viral mutations or by administering a thrombospondin-derived peptide. Cancer Res. *67*, 440–444.

Aghi, M., Visted, T., Depinho, R.A., and Chiocca, E.A. (2008). Oncolytic herpes virus with defective ICP6 specifically replicates in quiescent cells with homozygous genetic mutations in p16. Oncogene *27*, 4249–4254.

Ahn, K., Meyer, T.H., Uebel, S., Sempe, P., Djaballah, H., Yang, Y., Peterson, P.A., Fruh, K., and Tampe, R. (1996). Molecular mechanism and species specificity of TAP inhibition by herpes simplex virus ICP47. EMBO J. *15*, 3247–3255.

Akhtar, J., and Shukla, D. (2009). Viral entry mechanisms: cellular and viral mediators of herpes simplex virus entry. FEBS J. *276*, 7228–7236.

Alexander, D.E., Ward, S.L., Mizushima, N., Levine, B., and Leib, D.A. (2007). Analysis of the role of autophagy in replication of herpes simplex virus in cell culture. J. Virol. *81*, 12128–12134.

Andreansky, S., He, B., van Cott, J., McGhee, J., Markert, J.M., Gillespie, G.Y., Roizman, B., and Whitley, R.J. (1998). Treatment of intracranial gliomas in immunocompetent mice using herpes simplex viruses that express murine interleukins. Gene Ther. *5*, 121–130.

Andreansky, S., Soroceanu, L., Flotte, E.R., Chou, J., Markert, J.M., Gillespie, G.Y., Roizman, B., and Whitley, R.J. (1997). Evaluation of genetically engineered herpes simplex viruses as oncolytic agents for human malignant brain tumors. Cancer Res. *57*, 1502–1509.

Andreansky, S.S., He, B., Gillespie, G.Y., Soroceanu, L., Markert, J., Chou, J., Roizman, B., and Whitley, R.J. (1996). The application of genetically engineered herpes simplex viruses to the treatment of experimental brain tumors. Proc. Natl. Acad. Sci. U.S.A. *93*, 11313–11318.

Anton, A.S., Alim, S.S.-N., and Nikolai, B.G. (2008). Structure, properties, and functions of the human small heat-shock protein HSP22 (HspB8, H11, E2IG1): A critical review. J. Neurosci. Res. *86*, 264–269.

Asano, S., Honda, T., Goshima, F., Watanabe, D., Miyake, Y., Sugiura, Y., and Nishiyama, Y. (1999). US3 protein kinase of herpes simplex virus type 2 plays a role in protecting corneal epithelial cells from apoptosis in infected mice. J. Gen. Virol. *80*, 51–56.

Aurelian, L. (2004). Herpes simplex virus type 2 vaccines: new ground for optimism? Clin. Diagn. Lab. Immunol. *11*, 437–445.

Aurelian, L., Kokuba, H., and Smith, C.C. (1999). Vaccine potential of a herpes simplex virus type 2 mutant deleted in the PK domain of the large subunit of ribonucleotide reductase (ICP10). Vaccine *17*, 1951–1963.

Aurelian, L., Smith, C.C., Winchurch, R., Kulka, M., Gyotoku, T., Zaccaro, L., Chrest, F.J., and Burnett, J.W. (2001). A novel gene expressed in human keratinocytes with long-term *in vitro* growth potential is required for cell growth. J. Invest. Dermatol. *116*, 286–295.

Balachandran, S., and Barber, G.N. (2004). Defective translational control facilitates vesicular stomatitis virus oncolysis. Cancer Cell *5*, 51–65.

Balfour, H.H., Jr. (1999). Antiviral drugs. N. Engl. J. Med. *340*, 1255–1268.

Bateman, A.R., Harrington, K.J., Kottke, T., Ahmed, A., Melcher, A.A., Gough, M.J., Linardakis, E., Riddle, D., Dietz, A., Lohse, C.M., *et al.* (2002). Viral fusogenic membrane glycoproteins kill solid tumor cells by nonapoptotic mechanisms that promote cross presentation of tumor antigens by dendritic cells. Cancer Res. *62*, 6566–6578.

Battcock, S.M., Collier, T.W., Zu, D., and Hirasawa, K. (2006). Negative regulation of the alpha interferon-induced antiviral response by the Ras/Raf/MEK pathway. J. Virol. *80*, 4422–4430.

Bauer, D., and Tampe, R. (2002). Herpes viral proteins blocking the transporter associated with antigen processing TAP—from genes to function and structure. Curr. Top. Microbiol. Immunol. *269*, 87–99.

Benencia, F., Courreges, M.C., Conejo-Garcia, J.R., Mohamed-Hadley, A., Zhang, L., Buckanovich, R.J., Carroll, R., Fraser, N., and Coukos, G. (2005). HSV oncolytic therapy upregulates interferon-inducible chemokines and recruits immune effector cells in ovarian cancer. Mol. Ther. *12*, 789–802.

Benencia, F., Courreges, M.C., Fraser, N.W., and Coukos, G. (2008). Herpes virus oncolytic therapy reverses tumor immune dysfunction and facilitates tumor antigen presentation. Cancer Biol. Ther. *7*, 1194–1205.

Benetti, L., and Roizman, B. (2004). Herpes simplex virus protein kinase US3 activates and functionally overlaps protein kinase A to block apoptosis. Proc. Natl. Acad. Sci. U.S.A. *101*, 9411–9416.

Benetti, L., and Roizman, B. (2006). Protein kinase B/Akt is present in activated form throughout the entire replicative cycle of deltaU(S)3 mutant virus but only at early times after infection with wild-type herpes simplex virus 1. J. Virol. *80*, 3341–3348.

Bennett, J.J., Adusumilli, P., Petrowsky, H., Burt, B.M., Roberts, G., Delman, K.A., Zager, J.S., Chou, T.C., and Fong, Y. (2004). Up-regulation of GADD34 mediates the synergistic anticancer activity of mitomycin C and a gamma134.5 deleted oncolytic herpes virus (G207). Faseb J. *18*, 1001–1003.

Bennett, J.J., Tjuvajev, J., Johnson, P., Doubrovin, M., Akhurst, T., Malholtra, S., Hackman, T., Balatoni, J., Finn, R., Larson, S.M., *et al.* (2001). Positron emission tomography imaging for herpes virus infection: Implications for oncolytic viral treatments of cancer. Nat. Med. *7*, 859–863.

Berkowitz, C., Moyal, M., Rosen-Wolff, A., Darai, G., and Becker, Y. (1994). Herpes simplex virus type 1 (HSV-1) UL56 gene is involved in viral intraperitoneal pathogenicity to immunocompetent mice. Arch. Virol. *134*, 73–83.

Bernardi, R., Papa, A., and Pandolfi, P.P. (2008). Regulation of apoptosis by PML and the PML-NBs. Oncogene *27*, 6299–6312.

Bernstein, D.I., and Stanberry, L.R. (1999). Herpes simplex virus vaccines. Vaccine *17*, 1681–1689.

Bharara, S., Sorscher, E.J., Gillespie, G.Y., Lindsey, J.R., Hong, J.S., Curlee, K.V., Allan, P.W., Gadi, V.K., Alexander, S.A., Secrist, J.A., 3rd, *et al.* (2005). Antibiotic-mediated chemoprotection enhances adaptation of *E. coli* PNP for herpes simplex virus-based glioma therapy. Hum. Gene Ther. *16*, 339–347.

Bharatan, N.S., Currier, M.A., and Cripe, T.P. (2002). Differential susceptibility of pediatric sarcoma cells to oncolysis by conditionally replication-competent herpes simplex viruses. J. Pediatr. Hematol. Oncol. *24*, 447–453.

Blakeney, S., Kowalski, J., Tummolo, D., DeStefano, J., Cooper, D., Guo, M., Gangolli, S., Long, D., Zamb, T., Natuk, R.J., and Visalli, R.J. (2005). Herpes simplex virus type 2 UL24 gene is a virulence determinant in murine and guinea pig disease models. J. Virol. *79*, 10498–10506.

Bourdeau, V., Baudry, D., and Ferbeyre, G. (2009). PML links aberrant cytokine signaling and oncogenic stress to cellular senescence. Front Biosci. 14, 475–485.

Boviatsis, E.J., Scharf, J.M., Chase, M., Harrington, K., Kowall, N.W., Breakefield, X.O., and Chiocca, E.A. (1994). Antitumor activity and reporter gene transfer into rat brain neoplasms inoculated with herpes simplex virus vectors defective in thymidine kinase or ribonucleotide reductase. Gene Ther. *1*, 323–331.

Brader, P., Kelly, K., Gang, S., Shah, J.P., Wong, R.J., Hricak, H., Blasberg, R.G., Fong, Y., and Gil, Z. (2009). Imaging of lymph node micrometastases using an oncolytic herpes virus and [F]FEAU PET. PLoS One *4*, e4789.

Braidwood, L., Dunn, P.D., Hardy, S., Evans, T.R., and Brown, S.M. (2009). Antitumor activity of a selectively replication competent herpes simplex virus (HSV) with enzyme prodrug therapy. Anticancer Res. *29*, 2159–2166.

Brandt, C.R., Imesch, P.D., Robinson, N.L., Syed, N.A., Untawale, S., Darjatmoko, S.R., Chappell, R.J., Heinzelman, P., and Albert, D.M. (1997). Treatment of spontaneously arising retinoblastoma tumors in transgenic mice with an attenuated herpes simplex virus mutant. Virology *229*, 283–291.

Brown, B.D., and Naldini, L. (2009). Exploiting and antagonizing microRNA regulation for therapeutic and experimental applications. Nat. Rev. Genet. *10*, 578–585.

Brown, S.M., MacLean, A.R., Aitken, J.D., and Harland, J. (1994). ICP34.5 influences herpes simplex virus type 1 maturation and egress from infected cells *in vitro*. J. Gen. Virol. *75*, 3679–3686.

Brown, S.M., MacLean, A.R., McKie, E.A., and Harland, J. (1997). The herpes simplex virus virulence factor ICP34.5 and the cellular protein MyD116 complex with proliferating cell nuclear antigen through the 63-amino-acid domain conserved in ICP34.5, MyD116, and GADD34. J. Virol. *71*, 9442–9449.

Burgos, J.S., Serrano-Saiz, E., Sastre, I., and Valdivieso, F. (2006). ICP47 mediates viral neuroinvasiveness by induction of TAP protein following intravenous inoculation of herpes simplex virus type 1 in mice. J. NeuroVirol. *12*, 420–427.

Bzik, D.J., Fox, B.A., DeLuca, N.A., and Person, S. (1984). Nucleotide sequence of a region of the herpes simplex virus type 1 gB glycoprotein gene: mutations affecting rate of virus entry and cell fusion. Virology *137*, 185–190.

Cameron, J.M., McDougall, I., Marsden, H.S., Preston, V.G., Ryan, D.M., and Subak-Sharpe, J.H. (1988). Ribonucleotide reductase encoded by herpes simplex virus is a determinant of the pathogenicity of the virus in mice and a valid antiviral target. J. Gen. Virol. *69*, 2607–2612.

Campbell, S.A., and Gromeier, M. (2005). Oncolytic viruses for cancer therapy I. Cell-external factors: virus entry and receptor interaction. Onkologie *28*, 144–149.

Campbell, S.A., Mulvey, M., Mohr, I., and Gromeier, M. (2007). Attenuation of herpes simplex virus neurovirulence with picornavirus cis-acting genetic elements. J. Virol. *81*, 791–799.

Caradonna, S.J., and Cheng, Y.C. (1981). Induction of uracil-DNA glycosylase and dUTP nucleotidohydrolase activity in herpes simplex virus-infected human cells. J. Biol. Chem. *256*, 9834–9837.

Carra, S., Brunsting, J.F., Lambert, H., Landry, J., and Kampinga, H.H. (2009). HspB8 participates in protein quality control by a non-chaperone-like mechanism that requires. eIF2{alpha} phosphorylation. J. Biol. Chem. *284*, 5523–5532.

Cassady, K.A. (2005). Human cytomegalovirus TRS1 and IRS1 gene products block the double-stranded-RNA-activated host protein shutoff response induced by herpes simplex virus type 1 infection. J. Virol. *79*, 8707–8715.

Cassady, K.A., and Gross, M. (2002). The herpes simplex virus type 1 U(S)11 protein interacts with protein kinase R in infected cells and requires. a 30-amino-acid sequence adjacent to a kinase substrate domain. J. Virol. *76*, 2029–2035.

Cassady, K.A., Gross, M., Gillespie, G.Y., and Roizman, B. (2002). Second-site mutation outside of the U(S)10–12 domain of Deltagamma(1)34.5 herpes simplex virus 1 recombinant blocks the shutoff of protein synthesis induced by activated protein kinase R and partially restores. neurovirulence. J. Virol. *76*, 942–949.

Cassady, K.A., Gross, M., and Roizman, B. (1998). The herpes simplex virus US11 protein effectively compensates for the gamma1(34.5) gene if present before activation of protein kinase R by precluding its phosphorylation and that of the alpha subunit of eukaryotic translation initiation factor 2. J. Virol. *72*, 8620–8626.

Cernik, C., Gallina, K., and Brodell, R.T. (2008). The treatment of herpes simplex infections: an evidence-based review. Arch. Intern. Med. *168*, 1137–1144.

Chambers, R., Gillespie, G.Y., Soroceanu, L., Adreansky, S., Chatterjee, S., Chou, J., Roizman, B., and Whitley, R.J. (1995). Comparison of genetically engineered herpes simplex viruses for the treatment of brain tumors in a scid mouse model of human malignant glioma. Proc. Natl. Acad. Sci. U.S.A. *92*, 1411–1415.

Chase, M., Chung, R.Y., and Chiocca, E.A. (1998). An oncolytic viral mutant that delivers the CYP2B1 transgene and augments cyclophosphamide chemotherapy. Nat. Biotechnol. *16*, 444–448.

Chee, A.V., Lopez, P., Pandolfi, P.P., and Roizman, B. (2003). Promyelocytic leukemia protein mediates interferon-based anti-herpes simplex virus 1 effects. J. Virol. *77*, 7101–7105.

Cheng, G., Brett, M.E., and He, B. (2001). Val193 and Phe195 of the gamma 1, 34.5 protein of herpes simplex virus 1 are required for viral resistance to interferon-alpha/beta. Virology *290*, 115–120.

Cheng, G., Yang, K., and He, B. (2003). Dephosphorylation of eIF-2alpha mediated by the gamma(1)34.5 protein of herpes simplex virus type 1 is required for viral response to interferon but is not sufficient for efficient viral replication. J. Virol. *77*, 10154–10161.

Chou, J., Kern, E.R., Whitley, R.J., and Roizman, B. (1990). Mapping of herpes simplex virus-1 neurovirulence to gamma 34.5, a gene nonessential for growth in culture. Science *250*, 1262–1266.

Chou, J., and Roizman, B. (1992). The gamma 1(34.5) gene of herpes simplex virus 1 precludes neuroblastoma cells from triggering total shutoff of protein synthesis characteristic of programmed. cell death in neuronal cells. Proc. Natl. Acad. Sci. U.S.A. *89*, 3266–3270.

Chou, J., and Roizman, B. (1994). Herpes simplex virus 1 gamma(1)34.5 gene function, which blocks the host response to infection, maps in the homologous domain of the genes expressed during growth arrest and DNA damage. Proc. Natl. Acad. Sci. U.S.A. *91*, 5247–5251.

Chung, R.Y., Saeki, Y., and Chiocca, E.A. (1999). B-myb promoter retargeting of herpes simplex virus gamma34.5 gene-mediated virulence toward tumor and cycling cells. J. Virol. *73*, 7556–7564.

Chung, Y.S., Miyatake, S., Miyamoto, A., Miyamoto, Y., Dohi, T., and Tanigawa, N. (2006). Oncolytic recombinant herpes simplex virus for treatment of orthotopic liver tumors in nude mice. Int. J. Oncol. *28*, 793–798.

Coen, D.M., Goldstein, D.J., and Weller, S.K. (1989). Herpes simplex virus ribonucleotide reductase mutants are hypersensitive to acyclovir. Antimicrob. Agents Chemother. *33*, 1395–1399.

Colunga, A.G., Laing, J.M., and Aurelian, L. (2010). The HSV-2 mutant DeltaPK induces melanoma oncolysis through nonredundant death programs and associated with autophagy and pyroptosis proteins. Gene Ther. *17*, 315–327.

Cory, J.G., and Sato, A. (1983). Regulation of ribonucleotide reductase activity in mammalian cells. Mol. Cell Biochem. *53–54*, 257–266.

Cowan, M., Davies, J., Brookes, K., Billstrom, M., McLeish, P., Buchan, A., and Skinner, G.R. (1990). Inhibition of rate of tumour growth in rodent species by inoculation of herpesviruses and encephalomyocarditis virus. J. Med. Virol. *30*, 211–215.

Crew, M.D., and Phanavanh, B. (2003). Exploiting virus stealth technology for xenotransplantation: reduced human T cell responses to porcine cells expressing herpes simplex virus ICP47. Xenotransplantation *10*, 50–59.

Currier, M.A., Adams, L.C., Mahller, Y.Y., and Cripe, T.P. (2005). Widespread intratumoral virus distribution with fractionated injection enables local control of large human rhabdomyosarcoma xenografts by oncolytic herpes simplex viruses. Cancer Gene Ther. *12*, 407–416.

Currier, M.A., Gillespie, R.A., Sawtell, N.M., Mahller, Y.Y., Stroup, G., Collins, M.H., Kambara, H., Chiocca, E.A., and Cripe, T.P. (2008). Efficacy and safety of the oncolytic herpes simplex virus rRp450 alone and combined with cyclophosphamide. Mol. Ther. *16*, 879–885.

Dachs, G.U., Hunt, M.A., Syddall, S., Singleton, D.C., and Patterson, A.V. (2009). Bystander or no bystander for gene directed enzyme prodrug therapy. Molecules *14*, 4517–4545.

Dambach, M.J., Trecki, J., Martin, N., and Markovitz, N.S. (2006). Oncolytic viruses derived from the gamma34.5-deleted herpes simplex virus recombinant R3616 encode a truncated UL3 protein. Mol. Ther. *13*, 891–898.

Dell'Albani, P. (2008). Stem cell markers in gliomas. Neurochem. Res. *33*, 2407–2415.

Delman, K.A., Bennett, J.J., Zager, J.S., Burt, B.M., McAuliffe, P.F., Petrowsky, H., Kooby, D.A., Hawkins, W.G., Horsburgh, B.C., Johnson, P., and Fong, Y. (2000). Effects of preexisting immunity on the response to herpes simplex-based oncolytic viral therapy. Hum. Gene Ther. *11*, 2465–2472.

Derubertis, B.G., Stiles, B.M., Bhargava, A., Gusani, N.J., Hezel, M., D'Angelica, M., and Fong, Y. (2007). Cytokine-secreting herpes viral mutants effectively treat tumor in a murine metastatic colorectal liver model by oncolytic and T-cell-dependent mechanisms. Cancer Gene Ther. *14*, 590–597.

Dix, R.D., McKendall, R.R., and Baringer, J.R. (1983). Comparative neurovirulence of herpes simplex virus type 1 strains after peripheral or intracerebral inoculation of BALB/c mice. Infect. Immun. *40*, 103–112.

Dolter, K.E., Ramaswamy, R., and Holland, T.C. (1994). Syncytial mutations in the herpes simplex virus type 1 gK (UL53) gene occur in two distinct domains. J. Virol. *68*, 8277–8281.

Dorer, D.E., and Nettelbeck, D.M. (2009). Targeting cancer by transcriptional control in cancer Gene Ther. and viral oncolysis. Adv. Drug Deliv. Rev. *61*, 554–571.

Ebright, M.I., Zager, J.S., Malhotra, S., Delman, K.A., Weigel, T.L., Rusch, V.W., and Fong, Y. (2002). Replication-competent herpes virus NV1020 as direct treatment of pleural cancer in a rat model. J. Thorac Cardiovasc Surg. 124, 123–129.

Eisenberg, D.P., Adusumilli, P.S., Hendershott, K.J., Chung, S., Yu, Z., Chan, M.K., Hezel, M., Wong, R.J., and Fong, Y. (2006). Real-time intraoperative detection of breast cancer axillary lymph node metastases using a green fluorescent protein-expressing herpes virus. Ann. Surg. 243, 824–830; discussion 830–822.

Ejercito, P.M., Kieff, E.D., and Roizman, B. (1968). Characterization of herpes simplex virus strains differing in their effects on social behaviour of infected cells. J. Gen. Virol. *2*, 357–364.

Endo, T., Toda, M., Watanabe, M., Iizuka, Y., Kubota, T., Kitajima, M., and Kawakami, Y. (2002). In situ cancer vaccination with a replication-conditional HSV for the treatment of liver metastasis of colon cancer. Cancer Gene Ther. *9*, 142–148.

Engstrom, Y., Eriksson, S., Jildevik, I., Skog, S., Thelander, L., and Tribukait, B. (1985). Cell cycle-dependent expression of mammalian ribonucleotide reductase. Differential regulation of the two subunits. J. Biol. Chem. *260*, 9114–9116.

Everett, R.D. (2000). ICP0, a regulator of herpes simplex virus during lytic and latent infection. Bioessays *22*, 761–770.

Everett, R.D., Rechter, S., Papior, P., Tavalai, N., Stamminger, T., and Orr, A. (2006). PML contributes to a cellular mechanism of repression of herpes simplex virus type 1 infection that is inactivated by ICP0. J. Virol. *80*, 7995–8005.

Fan, H., Villegas, C., and Wright, J.A. (1996). Ribonucleotide reductase R2 component is a novel malignancy determinant that cooperates with activated oncogenes to determine transformation and malignant potential. Proc. Natl. Acad. Sci. U.S.A. *93*, 14036–14040.

Fang, Z.Y., Tenser, R.B., and Rapp, F. (1983). Hepatic infection by thymidine kinase-positive and thymidine kinase-negative herpes simplex virus after partial hepatectomy. Infect. Immun. *42*, 402–408.

Farassati, F., Pan, W., Yamoutpour, F., Henke, S., Piedra, M., Frahm, S., Al-Tawil, S., Mangrum, W.I., Parada, L.F., Rabkin, S.D., *et al.* (2008). Ras signaling influences permissiveness of malignant peripheral nerve sheath tumor cells to oncolytic herpes. Am. J. Pathol. *173*, 1861–1872.

Farassati, F., Yang, A.D., and Lee, P.W. (2001). Oncogenes in Ras signalling pathway dictate host-cell permissiveness to herpes simplex virus 1. Nat. Cell Biol. 3, 745–750.

Farrell, C.J., Zaupa, C., Barnard, Z., Maley, J., Martuza, R.L., Rabkin, S.D., and Curr. y, W.T., Jr. (2008). Combination immunotherapy for tumors via sequential intratumoral injections of oncolytic herpes simplex virus 1 and immature dendritic cells. Clin. Cancer Res. *14*, 7711–7716.

Field, H.J., and Wildy, P. (1978). The pathogenicity of thymidine kinase-deficient mutants of herpes simplex virus in mice. J. Hyg. (Lond) 81, 267–277.

Fong, Y., Kim, T., Bhargava, A., Schwartz, L., Brown, K., Brody, L., Covey, A., Karrasch, M., Getrajdman, G., Mescheder, A., *et al.* (2009). A herpes oncolytic virus can be delivered via the vasculature to produce biologic changes in human colorectal cancer. Mol. Ther. *17*, 389–394.

Frame, M.C., Marsden, H.S., and Dutia, B.M. (1985). The ribonucleotide reductase induced by herpes simplex virus type 1 involves minimally a complex of two polypeptides (136K and 38K). J. Gen. Virol. *66*, 1581–1587.

Friedman, G.K., Pressey, J.G., Reddy, A.T., Markert, J.M., and Gillespie, G.Y. (2009). Herpes simplex virus oncolytic therapy for pediatric malignancies. Mol. Ther. *17*, 1125–1135.

Fu, X., Tao, L., Cai, R., Prigge, J., and Zhang, X. (2006a). A mutant type 2 herpes simplex virus deleted for the protein kinase domain of the ICP10 gene is a potent oncolytic virus. Mol. Ther. *13*, 882–890.

Fu, X., Tao, L., Jin, A., Vile, R., Brenner, M.K., and Zhang, X. (2003). Expression of a fusogenic membrane glycoprotein by an oncolytic herpes simplex virus potentiates the viral antitumor effect. Mol. Ther. *7*, 748–754.

Fu, X., Tao, L., Li, M., Fisher, W.E., and Zhang, X. (2006b). Effective treatment of pancreatic cancer xenografts with a conditionally replicating virus derived from type 2 herpes simplex virus. Clin. Cancer Res. *12*, 3152–3157.

Fu, X., Tao, L., and Zhang, X. (2007a). An HSV-2-based oncolytic virus deleted in the PK domain of the ICP10 gene is a potent inducer of apoptotic death in tumor cells. Gene Ther. *14*, 1218–1225.

Fu, X., Tao, L., and Zhang, X. (2007b). An oncolytic virus derived from type 2 herpes simplex virus has potent therapeutic effect against metastatic ovarian cancer. Cancer Gene Ther. *14*, 480–487.

Fu, X., and Zhang, X. (2002). Potent systemic antitumor activity from an oncolytic herpes simplex virus of syncytial phenotype. Cancer Res. *62*, 2306–2312.

Fujimoto, Y., Mizuno, T., Sugiura, S., Goshima, F., Kohno, S., Nakashima, T., and Nishiyama, Y. (2006). Intratumoral injection of herpes simplex virus HF10 in recurrent head and neck squamous cell carcinoma. Acta Otolaryngol. *126*, 1115–1117.

Fukuhara, H., Ino, Y., Kuroda, T., Martuza, R.L., and Todo, T. (2005a). Triple gene-deleted oncolytic herpes simplex virus vector double-armed. with interleukin 18 and soluble B7–1 constructed by bacterial artificial chromosome-mediated system. Cancer Res. *65*, 10663–10668.

Fukuhara, H., Martuza, R.L., Rabkin, S.D., Ito, Y., and Todo, T. (2005b). Oncolytic herpes simplex virus vector g47delta in combination with androgen ablation for the treatment of human prostate adenocarcinoma. Clin. Cancer Res. *11*, 7886–7890.

Fukuhara, H., and Todo, T. (2007). Oncolytic herpes simplex virus type 1 and host immune responses. Curr. Cancer Drug Targets *7*, 149–155.

Fulci, G., Breymann, L., Gianni, D., Kurozomi, K., Rhee, S.S., Yu, J., Kaur, B., Louis, D.N., Weissleder, R., Caligiuri, M.A., and Chiocca, E.A. (2006). Cyclophosphamide enhances glioma virotherapy by inhibiting innate immune responses. Proc. Natl. Acad. Sci. U.S.A. *103*, 12873–12878.

Fyfe, J.A., Keller, P.M., Furman, P.A., Miller, R.I., and Elion, G.B. (1978). Thymidine kinase from herpes simplex virus phosphorylates the new antiviral compound, 9-(2-hydroxyethoxymethyl)guanine. J. Biol. Chem. *253*, 8721–8727.

Galanis, E., Bateman, A., Johnson, K., Diaz, R.M., James, C.D., Vile, R., and Russell, S.J. (2001). Use of viral fusogenic membrane glycoproteins as novel therapeutic transgenes in gliomas. Hum. Gene Ther. *12*, 811–821.

Garcia, M.A., Meurs, E.F., and Esteban, M. (2007). The dsRNA protein kinase PKR: virus and cell control. Biochimie *89*, 799–811.

Gavert, N., and Ben-Ze'ev, A. (2007). beta-Catenin signaling in biological control and cancer. J. Cell Biochem. *102*, 820–828.

Gil, Z., Kelly, K.J., Brader, P., Shah, J.P., Fong, Y., and Wong, R.J. (2008). Utility of a herpes oncolytic virus for the detection of neural invasion by cancer. Neoplasia *10*, 347–353.

Gil, Z., Rein, A., Brader, P., Li, S., Shah, J.P., Fong, Y., and Wong, R.J. (2007). Nerve-sparing therapy with oncolytic herpes virus for cancers with neural invasion. Clin. Cancer Res. *13*, 6479–6485.

Giraud, S., Diaz-Latoud, C., Hacot, S., Textoris, J., Bourette, R.P., and Diaz, J.J. (2004). US11 of herpes simplex virus type 1 interacts with HIPK2 and antagonizes HIPK2-induced cell growth arrest. J. Virol. *78*, 2984–2993.

Glass, M., Soling, A., and Messerle, M. (2008). Tumor-specific activity of cellular regulatory elements is down-regulated upon insertion into the herpes simplex virus genome. J. Neurovirol. 1–14.

Goldsmith, K., Chen, W., Johnson, D.C., and Hendricks, R.L. (1998). Infected cell protein (ICP)47 enhances herpes simplex virus neurovirulence by blocking the $CD8^+$ T cell response. J. Exp. Med. *187*, 341–348.

Goldstein, D.J., and Weller, S.K. (1988a). Factor(s) present in herpes simplex virus type 1-infected cells can compensate for the loss of the large subunit of the viral ribonucleotide reductase: Characterization of an ICP6 deletion mutant. Virology *166*, 41–51.

Goldstein, D.J., and Weller, S.K. (1988b). Herpes simplex virus type 1-induced ribonucleotide reductase activity is dispensable for virus growth and DNA synthesis: Isolation and characterization of an ICP6 lacZ insertion mutant. J. Virol. *62*, 196–205.

Goodman, J.L., and Engel, J.P. (1991). Altered pathogenesis in herpes simplex virus type 1 infection due to a syncytial mutation mapping to the carboxy terminus of glycoprotein B. J. Virol. *65*, 1770–1778.

Gordon, Y.J., Gilden, D.M., and Becker, Y. (1983). HSV-1 thymidine kinase promotes virulence and latency in the mouse. Invest. Ophthalmol. Vis. Sci. *24*, 599–602.

Grandi, P., Peruzzi, P., Reinhart, B., Cohen, J.B., Chiocca, E.A., and Glorioso, J.C. (2009). Design and application of oncolytic HSV vectors for glioblastoma therapy. Expert Rev. Neurother. *9*, 505–517.

Griffith, C., Noonan, S., Lou, E., and Shillitoe, E.J. (2007). An oncolytic mutant of herpes simplex virus type-1 in which replication is governed by a promoter/enhancer of human papillomavirus type-16. Cancer Gene Ther. *14*, 985–993.

Gromeier, M., Alexander, L., and Wimmer, E. (1996). Internal ribosomal entry site substitution eliminates neurovirulence in intergeneric poliovirus recombinants. Proc. Natl. Acad. Sci. U.S.A. *93*, 2370–2375.

Gromeier, M., Lachmann, S., Rosenfeld, M.R., Gutin, P.H., and Wimmer, E. (2000). Intergeneric poliovirus recombinants for the treatment of malignant glioma. Proc. Natl. Acad. Sci. U.S.A. *97*, 6803–6808.

Guffey, M.B., Parker, J.N., Luckett, W.S., Jr., Gillespie, G.Y., Meleth, S., Whitley, R.J., and Markert, J.M. (2007). Engineered herpes simplex virus expressing bacterial cytosine deaminase for experimental therapy of brain tumors. Cancer Gene Ther. *14*, 45–56.

Gurrieri, C., Capodieci, P., Bernardi, R., Scaglioni, P.P., Nafa, K., Rush, L.J., Verbel, D.A., Cordon-Cardo, C., and Pandolfi, P.P. (2004). Loss of the tumor suppressor PML in human cancers of multiple histologic origins. J. Natl. Cancer Inst. *96*, 269–279.

Hadjipanayis, C.G., Fellows-Mayle, W., and Deluca, N.A. (2008). Therapeutic efficacy of a herpes simplex virus with radiation or temozolomide for intracranial glioblastoma after convection-enhanced delivery. Mol. Ther. *16*, 1783–1788.

Halford, W.P., Weisend, C., Grace, J., Soboleski, M., Carr, D.J., Balliet, J.W., Imai, Y., Margolis, T.P., and Gebhardt, B.M. (2006). ICP0 antagonizes Stat 1-dependent repression of herpes simplex virus: implications for the regulation of viral latency. Virology J. *3*, 44.

Han, Z.Q., Assenberg, M., Liu, B.L., Wang, Y.B., Simpson, G., Thomas, S., and Coffin, R.S. (2007). Development of a second-generation oncolytic Herpes simplex virus expressing TNFalpha for cancer therapy. J. Gene Med. *9*, 99–106.

Hardcastle, J., Kurozumi, K., Chiocca, E.A., and Kaur, B. (2007). Oncolytic viruses driven by tumor-specific promoters. Curr. Cancer Drug Targets 7, 181–189.

Harland, J., Dunn, P., Cameron, E., Conner, J., and Brown, S.M. (2003). The herpes simplex virus (HSV) protein ICP34.5 is a virion component that forms a DNA-binding complex with proliferating cell nuclear antigen and HSV replication proteins. J. NeuroVirol. *9*, 477–488.

Harle, P., Sainz, B., Jr., Carr, D.J., and Halford, W.P. (2002). The immediate-early protein, ICP0, is essential for the resistance of herpes simplex virus to interferon-alpha/beta. Virology *293*, 295–304.

Harrow, S., Papanastassiou, V., Harland, J., Mabbs, R., Petty, R., Fraser, M., Hadley, D., Patterson, J., Brown, S.M., and Rampling, R. (2004). HSV1716 injection into the brain adjacent to tumour following surg. ical resection of high-grade glioma: safety data and long-term survival. Gene Ther. *11*, 1648–1658.

Hata, S., Koyama, A.H., Shiota, H., Adachi, A., Goshima, F., and Nishiyama, Y. (1999). Antiapoptotic activity of herpes simplex virus type 2: the role of US3 protein kinase gene. Microbes Infect. *1*, 601–607.

He, B., Chou, J., Liebermann, D.A., Hoffman, B., and Roizman, B. (1996). The carboxyl terminus of the murine MyD116 gene substitutes for the corresponding domain of the gamma(1)34.5 gene of herpes simplex virus to preclude the premature shutoff of total protein synthesis in infected human cells. J. Virol. *70*, 84–90.

He, B., Gross, M., and Roizman, B. (1997). The gamma(1)34.5 protein of herpes simplex virus 1 complexes with protein phosphatase 1alpha to dephosphorylate the alpha subunit of the eukaryotic translation initiation factor 2 and preclude the shutoff of protein synthesis by double-stranded RNA-activated protein kinase. Proc. Natl. Acad. Sci. U.S.A. *94*, 843–848.

He, C., and Klionsky, D.J. (2009). Regulation mechanisms and signaling pathways of autophagy. Annu. Rev. Genet. *43*, 67–93.

Hellums, E.K., Markert, J.M., Parker, J.N., He, B., Perbal, B., Roizman, B., Whitley, R.J., Langford, C.P., Bharara, S., and Gillespie, G.Y. (2005). Increased efficacy of an interleukin-12-secreting herpes simplex virus in a syngeneic intracranial murine glioma model. Neuro Oncol. *7*, 213–224.

Hiscott, J. (2007). Triggering the innate antiviral response through IRF-3 activation. J. Biol. Chem. *282*, 15325–15329.

Ho, D.Y., and Mocarski, E.S. (1988). Beta-galactosidase as a marker in the peripheral and neural tissues of the herpes simplex virus-infected mouse. Virology *167*, 279–283.

Holman, H.A., and MacLean, A.R. (2008). Neurovirulent factor ICP34.5 uniquely expressed in the herpes simplex virus type 1 Delta gamma 1, 34.5 mutant 1716. J. NeuroVirol. *14*, 28–40.

Hu, J.C., Booth, M.J., Tripuraneni, G., Davies, D., Zaidi, S.A., Tamburo de Bella, M., Slade, M.J., Marley, S.B., Gordon, M.Y., Coffin, R.S., *et al.* (2006a). A novel HSV-1 virus, JS1/34.5-/47-, purges contaminating breast cancer cells from bone marrow. Clin. Cancer Res. *12*, 6853–6862.

Hu, J.C., Coffin, R.S., Davis, C.J., Graham, N.J., Groves, N., Guest, P.J., Harrington, K.J., James, N.D., Love, C.A., McNeish, I., *et al.* (2006b). A phase I study of OncoVEXGM-CSF, a second-generation oncolytic herpes simplex virus expressing granulocyte macrophage colony-stimulating factor. Clin. Cancer Res. *12*, 6737–6747.

Huang, Y.Y., Yu, Z., Lin, S.F., Li, S., Fong, Y., and Wong, R.J. (2007). Nectin-1 is a marker of thyroid cancer sensitivity to herpes oncolytic therapy. J. Clin. Endocrinol Metab *92*, 1965–1970.

Huard, B., and Fruh, K. (2000). A role for MHC class I down-regulation in NK cell lysis of herpes virus-infected cells. Eur J. Immunol. *30*, 509–515.

Hummel, J.L., Safroneeva, E., and Mossman, K.L. (2005). The role of ICP0-Null HSV-1 and interferon signaling defects in the effective treatment of breast adenocarcinoma. Mol. Ther. *12*, 1101–1110.

Hunt, R.D. (1993). Herpesvirus simplex infection, In Nonhuman Primates I, Jones, T.C., Mohr, U., and Hunt, R.D., eds. (Berlin: Springer-Verlag), pp. 82–86.

Hunter, W.D., Martuza, R.L., Feigenbaum, F., Todo, T., Mineta, T., Yazaki, T., Toda, M., Newsome, J.T., Platenberg, R.C., Manz, H.J., and Rabkin, S.D. (1999). Attenuated, replication-competent herpes simplex virus type 1 mutant G207: safety evaluation of intracerebral injection in nonhuman primates. J. Virol. *73*, 6319–6326.

Hurta, R.A., and Wright, J.A. (1995). Malignant transformation by H-ras results in aberrant regulation of ribonucleotide reductase gene expression by transforming growth factor-beta 1. J. Cell Biochem. *57*, 543–556.

Huszthy, P.C., Goplen, D., Thorsen, F., Immervoll, H., Wang, J., Gutermann, A., Miletic, H., and Bjerkvig, R. (2008). Oncolytic herpes simplex virus type-1 therapy in a highly infiltrative animal model of human glioblastoma. Clin. Cancer Res. *14*, 1571–1580.

Hutchinson, L., Goldsmith, K., Snoddy, D., Ghosh, H., Graham, F.L., and Johnson, D.C. (1992). Identification and characterization of a novel herpes simplex virus glycoprotein, gK, involved in cell fusion. J. Virol. *66*, 5603–5609.

Ino, Y., Saeki, Y., Fukuhara, H., and Todo, T. (2006). Triple combination of oncolytic herpes simplex virus-1 vectors armed. with interleukin-12, interleukin-18, or soluble B7–1 results in enhanced antitumor efficacy. Clin. Cancer Res. *12*, 643–652.

Israyelyan, A., Chouljenko, V.N., Baghian, A., David, A.T., Kearney, M.T., and Kousoulas, K.G. (2008). Herpes simplex virus type-1(HSV-1) oncolytic and highly fusogenic mutants carrying the NV1020 genomic deletion effectively inhibit primary and metastatic tumors in mice. J. Virol. *5*, 68.

Israyelyan, A.H., Melancon, J.M., Lomax, L.G., Sehgal, I., Leuschner, C., Kearney, M.T., Chouljenko, V.N., Baghian, A., and Kousoulas, K.G. (2007). Effective treatment of human breast tumor in a mouse xenograft model with herpes simplex virus type 1 specifying the

NV1020 genomic deletion and the gBsyn3 syncytial mutation enabling high viral replication and spread in breast cancer cells. Hum. Gene Ther. *18*, 457–473.

Jacobs, A., Tjuvajev, J.G., Dubrovin, M., Akhurst, T., Balatoni, J., Beattie, B., Joshi, R., Finn, R., Larson, S.M., Herrlinger, U., *et al.* (2001). Positron emission tomography-based imaging of transgene expression mediated by replication-conditional, oncolytic herpes simplex virus type 1 mutant vectors *in vivo*. Cancer Res. *61*, 2983–2995.

Jacobson, J.G., Chen, S.H., Cook, W.J., Kramer, M.F., and Coen, D.M. (1998). Importance of the herpes simplex virus UL24 gene for productive ganglionic infection in mice. Virology *242*, 161–169.

Jacobson, J.G., Leib, D.A., Goldstein, D.J., Bogard, C.L., Schaffer, P.A., Weller, S.K., and Coen, D.M. (1989). A herpes simplex virus ribonucleotide reductase deletion mutant is defective for productive acute and reactivatable latent infections of mice and for replication in mouse cells. Virology *173*, 276–283.

Jarnagin, W.R., Zager, J.S., Klimstra, D., Delman, K.A., Malhotra, S., Ebright, M., Little, S., DeRubertis, B., Stanziale, S.F., Hezel, M., *et al.* (2003). Neoadjuvant treatment of hepatic malignancy: an oncolytic herpes simplex virus expressing IL-12 effectively treats the parent tumor and protects against recurr. ence-after resection. Cancer Gene Ther. *10*, 215–223.

Javouhey, E., Gibert, B., Arrigo, A.P., Diaz, J.J., and Diaz-Latoud, C. (2008). Protection against heat and staurosporine mediated apoptosis by the HSV-1 US11 protein. Virology *376*, 31–41.

Jenkins, F.J., Donoghue, A.M., and Martin, J.R. (1996). Deletion of the Herpes simplex 1 internal repeat sequences affects pathogenicity in the mouse. Front Biosci. 1, a59–68.

Jeyaretna, D.S., and Kuroda, T. (2007). Recent advances in the development of oncolytic HSV-1 vectors: 'arming' of HSV-1 vectors and application of bacterial artificial chromosome technology for their construction. Curr. Opin. Mol. Ther. *9*, 447–466.

Jeyaretna, D.S., Rabkin, S.D., and Martuza, R.L. (2007). Oncolytic herpes simplex virus therapy for peripheral nerve tumors. Neurosurg. Focus 22, E4.

Jia, W.W., McDermott, M., Goldie, J., Cyander, M., Tan, J., and Tufaro, F. (1994). Selective destruction of gliomas in immunocompetent rats by thymidine kinase-defective herpes simplex virus type 1. J. Natl. Cancer Inst. *86*, 1209–1215.

Jing, X., Cerveny, M., Yang, K., and He, B. (2004). Replication of herpes simplex virus 1 depends on the gamma 134.5 functions that facilitate virus response to interferon and egress in the different stages of productive infection. J. Virol. *78*, 7653–7666.

Jugovic, P., Hill, A.M., Tomazin, R., Ploegh, H., and Johnson, D.C. (1998). Inhibition of major histocompatibility complex class I antigen presentation in pig and primate cells by herpes simplex virus type 1 and 2 ICP47. J. Virol. *72*, 5076–5084.

Kambara, H., Okano, H., Chiocca, E.A., and Saeki, Y. (2005). An oncolytic HSV-1 mutant expressing ICP34.5 under control of a nestin promoter increases survival of animals even when symptomatic from a brain tumor. Cancer Res. *65*, 2832–2839.

Kamiyama, H., Zhou, G., and Roizman, B. (2006). Herpes simplex virus 1 recombinant virions exhibiting the amino terminal fragment of urokinase-type plasminogen activator can enter cells via the cognate receptor. Gene Ther. *13*, 621–629.

Kanai, R., Tomita, H., Hirose, Y., Ohba, S., Goldman, S., Okano, H., Kawase, T., and Yazaki, T. (2007). Augmented therapeutic efficacy of an oncolytic herpes simplex virus type 1 mutant expressing ICP34.5 under the transcriptional control of musashi1 promoter in the treatment of malignant glioma. Hum. Gene Ther. *18*, 63–73.

Kaplitt, M.G., Tjuvajev, J.G., Leib, D.A., Berk, J., Pettigrew, K.D., Posner, J.B., Pfaff, D.W., Rabkin, S.D., and Blasberg, R.G. (1994). Mutant herpes simplex virus induced regression of tumors growing in immunocompetent rats. J. Neurooncol. *19*. http://biomednet.com/db/medline/95054303, 137–147.

Karrasch, M., Sze, D.Y., Mescheder, A., and Reid, T. (2007). Repeated intrahepatic infusion of oncolytic herpes simplex virus NV1020 prior to second-line chemotherapy for patients suffering from colorectal cancer metastatic to the liver. Ann. Oncol. *18*, Suppl 9, 47P.

Kasuya, H., Nishiyama, Y., Nomoto, S., Goshima, F., Takeda, S., Watanabe, I., Nomura, N., Shikano, T., Fujii, T., Kanazumi, N., and Nakao, A. (2007). Suitability of a US3-inactivated HSV mutant (L1BR1) as an oncolytic virus for pancreatic cancer therapy. Cancer Gene Ther. *14*, 533–542.

Kasuya, H., Nishiyama, Y., Nomoto, S., Hosono, J., Takeda, S., and Nakao, A. (1999). Intraperitoneal delivery of hrR3 and ganciclovir prolongs survival in mice with disseminated pancreatic cancer. J. Surg. Oncol. *72*, 136–141.

Kasuya, H., Pawlik, T.M., Mullen, J.T., Donahue, J.M., Nakamura, H., Chandrasekhar, S., Kawasaki, H., Choi, E., and Tanabe, K.K. (2004). Selectivity of an oncolytic herpes simplex virus for cells expressing the DF3/MUC1 antigen. Cancer Res. *64*, 2561–2567.

Katzin, D.S., Connor, J.D., Wilson, L.A., and Sexton, R.S. (1967). Experimental herpes simplex infection in the Owl Monkey. Proc. Soc. Exp. Biol. Med. *125*, 391–398.

Kaufman, H.L., Kim, D.W., Deraffele, G., Mitcham, J., Coffin, R.S., and Kim-Schulze, S. (2009). Local and Distant Immunity Induced by Intralesional Vaccination with an Oncolytic Herpes Virus Encoding GM-CSF in Patients with Stage IIIc and IV Melanoma. Ann. Surg. Oncol.

Kehm, R., Rosen-Wolff, A., and Darai, G. (1996). Restitution of the UL56 gene expression of HSV-1 HFEM led to restoration of virulent phenotype; deletion of the amino acids 217 to 234 of the UL56 protein abrogates the virulent phenotype. Virus Res. *40*, 17–31.

Kelly, E., and Russell, S.J. (2007). History of oncolytic viruses: genesis to genetic engineering. Mol. Ther. *15*, 651–659.

Kelly, E.J., and Russell, S.J. (2009). MicroRNAs and the regulation of vector tropism. Mol. Ther. *17*, 409–416.

Kelly, K., Brader, P., Rein, A., Shah, J.P., Wong, R.J., Fong, Y., and Gil, Z. (2008). Attenuated multimutated herpes simplex virus-1 effectively treats prostate carcinomas with neural invasion while preserving nerve function. Faseb J. *22*, 1839–1848.

Kemeny, N., Brown, K., Covey, A., Kim, T., Bhargava, A., Brody, L., Guilfoyle, B., Haag, N.P., Karrasch, M., Glasschroeder, B., *et al.* (2006). Phase I, open-label, dose-escalating study of a genetically engineered herpes simplex virus, NV1020, in subjects with metastatic colorectal carcinoma to thc liver. Hum. Gene Ther. *17*, 1214–1224.

Kesari, S., Lasner, T.M., Balsara, K.R., Randazzo, B.P., Lee, V.M., Trojanowski, J.Q., and Fraser, N.W. (1998). A neuroattenuated ICP34.5-deficient herpes simplex virus type 1 replicates in ependymal cells of the murine central nervous system. J. Gen. Virol. *79*, 525–536.

Khil, M.S., Kim, J.H., Mullen, C.A., Kim, S.H., and Freytag, S.O. (1996). Radiosensitization by 5-fluorocytosine of human colorectal carcinoma cells in culture transduced with cytosine deaminase gene. Clin. Cancer Res. *2*, 53–57.

Kim, S.H., Forman, A.P., Mathews, M.B., and Gunnery, S. (2000). Human breast cancer cells contain elevated levels and activity of the protein kinase, PKR. Oncogene *19*, 3086–3094.

Kimata, H., Imai, T., Kikumori, T., Teshigahara, O., Nagasaka, T., Goshima, F., Nishiyama, Y., and Nakao, A. (2006). Pilot study of oncolytic viral therapy using mutant herpes simplex virus (HF10) against recurrent metastatic breast cancer. Ann. Surg. Oncol. *13*, 1078–1084.

Kooby, D.A., Carew, J.F., Halterman, M.W., Mack, J.E., Bertino, J.R., Blumgart, L.H., Federoff, H.J., and Fong, Y. (1999). Oncolytic viral therapy for human colorectal cancer and liver metastases using a multimutated herpes simplex virus type-1 (G207). Faseb J. *13*, 1325–1334.

Kramm, C.M., Chase, M., Herrlinger, U., Jacobs, A., Pechan, P.A., Rainov, N.G., Sena-Esteves, M., Aghi, M., Barnett, F.H., Chiocca, E.A., and Breakefield, X.O. (1997). Therapeutic efficiency and safety of a second-generation replication-conditional HSV1 vector for brain tumor Gene Ther. Hum. Gene Ther. *8*, 2057–2068.

Kramm, C.M., Rainov, N.G., Sena-Esteves, M., Barnett, F.H., Chase, M., Herrlinger, U., Pechan, P.A., Chiocca, E.A., and Breakefield, X.O. (1996). Long-term survival in a rodent model of disseminated brain tumors by combined intrathecal delivery of herpes vectors and ganciclovir treatment. Hum. Gene Ther. *7*, 1989–1994.

Kucharczuk, J.C., Randazzo, B., Chang, M.Y., Amin, K.M., Elshami, A.A., Sterman, D.H., Rizk, N.P., Molnar-Kimber, K.L., Brown, S.M., MacLean, A.R., *et al.* (1997). Use of a 'replication-restricted' herpes virus to treat experimental human malignant mesothelioma. Cancer Res. *57*, 466–471.

Kufe, D., and Weichselbaum, R. (2003). Radiation therapy: activation for gene transcription and the development of genetic radiotherapy-therapeutic strategies in oncology. Cancer Biol. Ther. *2*, 326–329.

Kuroda, T., Martuza, R.L., Todo, T., and Rabkin, S.D. (2006a). Flip-Flop HSV-BAC: bacterial artificial chromosome based system for rapid generation of recombinant herpes simplex virus vectors using two independent site-specific recombinases. BMC Biotechnol *6*, 40.

Kuroda, T., Rabkin, S.D., and Martuza, R.L. (2006b). Effective treatment of tumors with strong beta-catenin/T-cell factor activity by transcriptionally targeted oncolytic herpes simplex virus vector. Cancer Res. *66*, 10127–10135.

Kurozumi, K., Hardcastle, J., Thakur, R., Yang, M., Christoforidis, G., Fulci, G., Hochberg, F.H., Weissleder, R., Carson, W., Chiocca, E.A., and Kaur, B. (2007). Effect of tumor microenvironment modulation on the efficacy of oncolytic virus therapy. J. Natl. Cancer Inst. *99*, 1768–1781.

Kuruppu, D., Brownell, A.L., Zhu, A., Yu, M., Wang, X., Kulu, Y., Fuchs, B.C., Kawasaki, H., and Tanabe, K.K. (2007). Positron emission tomography of herpes simplex virus 1 oncolysis. Cancer Res. *67*, 3295–3300.

Kwon, H., Bai, Q., Baek, H.J., Felmet, K., Burton, E.A., Goins, W.F., Cohen, J.B., and Glorioso, J.C. (2006). Soluble V domain of Nectin-1/HveC enables entry of herpes simplex virus type 1 (HSV-1) into HSV-resistant cells by binding to viral glycoprotein D. J. Virol. *80*, 138–148.

Lagunoff, M., and Roizman, B. (1995). The regulation of synthesis and properties of the protein product of open reading frame P of the herpes simplex virus 1 genome. J. Virol. *69*, 3615–3623.

Lan, P., Dong, C., Qi, Y., Xiao, G., and Xue, F. (2003). Gene Ther. for mice sarcoma with oncolytic herpes simplex virus-1 lacking the apoptosis-inhibiting gene, icp34.5. J. Biochem. Mol. Biol. *36*, 379–386.

Lasner, T.M., Kesari, S., Brown, S.M., Lee, V.M.Y., Fraser, N.W., and Trojanowski, J.Q. (1996). Therapy of a murine model of pediatric brain tumors using a herpes simplex virus type-1 ICP34.5 mutant and demonstration of viral replication within the CNS. J. Neuropathol. Exp. Neurol. *55*, 1259–1269.

Lasner, T.M., Tal-Singer, R., Kesari, S., Lee, V.M., Trojanowski, J.Q., and Fraser, N.W. (1998). Toxicity and neuronal infection of a HSV-1 ICP34.5 mutant in nude mice. J. NeuroVirol. *4*, 100–105.

Leary, K., Yim, H.H., Zhou, L.B., Sekulovich, R.E., and Sandri-Goldin, R.M. (1989). The influence of the herpes simplex virus-1 DNA template environment on the regulation of gene expression. Virus Genes *3*, 57–68.

Lee, C.Y., Rennie, P.S., and Jia, W.W. (2009). MicroRNA regulation of oncolytic herpes simplex virus-1 for selective killing of prostate cancer cells. Clin. Cancer Res. *15*, 5126–5135.

Leib, D.A., Harrison, T.E., Laslo, K.M., Machalek, M.A., Moorman, N.J., and Virgin, H.W. (1999). Interferons regulate the phenotype of wild-type and

mutant herpes simplex viruses *in vivo*. J. Exp. Med. *189*, 663–672.
Leib, D.A., Machalek, M.A., Williams, B.R., Silverman, R.H., and Virgin, H.W. (2000). Specific phenotypic restoration of an attenuated virus by knockout of a host resistance gene [see comments]. Proc. Natl. Acad. Sci. U.S.A. *97*, 6097–6101.
Leopardi, R., and Roizman, B. (1996). The herpes simplex virus major regulatory protein ICP4 blocks apoptosis induced by the virus or by hyperthermia. Proc. Natl. Acad. Sci. U.S.A. *93*, 9583–9587.
Leopardi, R., Van Sant, C., and Roizman, B. (1997). The herpes simplex virus 1 protein kinase US3 is required for protection from apoptosis induced by the virus. Proc. Natl. Acad. Sci. U.S.A. *94*, 7891–7896.
Li, H., Dutuor, A., Fu, X., and Zhang, X. (2007a). Induction of strong antitumor immunity by an HSV-2-based oncolytic virus in a murine mammary tumor model. J. Gene Med. *9*, 161–169.
Li, H., Dutuor, A., Tao, L., Fu, X., and Zhang, X. (2007b). Virotherapy with a type 2 herpes simplex virus-derived oncolytic virus induces potent antitumor immunity against neuroblastoma. Clin. Cancer Res. *13*, 316–322.
Li, H., Zeng, Z., Fu, X., and Zhang, X. (2007c). Coadministration of a herpes simplex virus-2 based oncolytic virus and cyclophosphamide produces a synergistic antitumor effect and enhances tumor-specific immune responses. Cancer Res. *67*, 7850–7855.
Li, Q.X., Liu, G., and Wong-Staal, F. (2008). Oncolytic virotherapy as a personalized cancer vaccine. Int. J. Cancer *123*, 493–499.
Lin, R., Noyce, R.S., Collins, S.E., Everett, R.D., and Mossman, K.L. (2004). The herpes simplex virus ICP0 RING finger domain inhibits IRF3- and IRF7-mediated activation of interferon-stimulated genes. J. Virol. *78*, 1675–1684.
Liu, B.L., Robinson, M., Han, Z.Q., Branston, R.H., English, C., Reay, P., McGrath, Y., Thomas, S.K., Thornton, M., Bullock, P., *et al.* (2003). ICP34.5 deleted herpes simplex virus with enhanced oncolytic, immune stimulating, and anti-tumour properties. Gene Ther. *10*, 292–303.
Liu, R., Martuza, R.L., and Rabkin, S.D. (2005a). Intracarotid delivery of oncolytic HSV vector G47Delta to metastatic breast cancer in the brain. Gene Ther. *12*, 647–654.
Liu, R., Varghese, S., and Rabkin, S.D. (2005b). Oncolytic herpes simplex virus vector therapy of breast cancer in C3(1)/SV40 T-antigen transgenic mice. Cancer Res. *65*, 1532–1540.
Liu, T.C., Galanis, E., and Kirn, D. (2007a). Clinical trial results with oncolytic virotherapy: a century of promise, a decade of progress. Nat. Clin. Pract Oncol. *4*, 101–117.
Liu, T.C., and Kirn, D. (2005). Viruses with deletions in antiapoptotic genes as potential oncolytic agents. Oncogene *24*, 6069–6079.
Liu, T.C., Wakimoto, H., Martuza, R.L., and Rabkin, S.D. (2007b). Herpes simplex virus Us3(-) mutant as oncolytic strategy and synergizes with phosphatidylinositol 3-kinase-Akt targeting molecular therapeutics. Clin. Cancer Res. *13*, 5897–5902.
Lopez, C. (1975). Genetics of natural resistance to herpesvirus infection in mice. Nature *258*, 152–153.
Lopez, P., Jacob, R.J., and Roizman, B. (2002). Overexpression of promyelocytic leukemia protein precludes the dispersal of ND10 structures. and has no effect on accumulation of infectious herpes simplex virus 1 or its proteins. J. Virol. *76*, 9355–9367.
Luker, G.D., and Leib, D.A. (2005). Luciferase real-time bioluminescence imaging for the study of viral pathogenesis. Methods Mol. Biol. *292*, 285–296.
Luo, C., Mori, I., Goshima, F., Ushijima, Y., Nawa, A., Kimura, H., and Nishiyama, Y. (2007). Replication-competent, oncolytic herpes simplex virus type 1 mutants induce a bystander effect following ganciclovir treatment. J. Gene Med. *9*, 875–883.
Mace, A.T., Ganly, I., Soutar, D.S., and Brown, S.M. (2008). Potential for efficacy of the oncolytic Herpes simplex virus 1716 in patients with oral squamous cell carcinoma. Head Neck *30*, 1045–1051.
MacLean, A.R., ul-Fareed, M., Robertson, L., Harland, J., and Brown, S.M. (1991). Herpes simplex virus type 1 deletion variants 1714 and 1716 pinpoint neurovirulence-related sequences in Glasgow strain 17^+ between immediate early gene 1 and the 'a' sequence. J. Gen. Virol. *72*, 631–639.
Mahller, Y.Y., Rangwala, F., Ratner, N., and Cripe, T.P. (2006). Malignant peripheral nerve sheath tumors with high and low Ras-GTP are permissive for oncolytic herpes simplex virus mutants. Pediatr. Blood Cancer. *46*, 745–754.
Mahller, Y.Y., Vaikunth, S.S., Ripberger, M.C., Baird, W.H., Saeki, Y., Cancelas, J.A., Crombleholme, T.M., and Cripe, T.P. (2008). Tissue inhibitor of metalloproteinase-3 via oncolytic herpesvirus inhibits tumor growth and vascular progenitors. Cancer Res. *68*, 1170–1179.
Markert, J.M., Coen, D.M., Malick, A., Mineta, T., and Martuza, R.L. (1992). Expanded spectrum of viral therapy in the treatment of nervous system tumors. J. Neurosurg. 77, 590–594.
Markert, J.M., Liechty, P.G., Wang, W., Gaston, S., Braz, E., Karrasch, M., Nabors, L.B., Markiewicz, M., Lakeman, A.D., Palmer, C.A., *et al.* (2009). Phase Ib trial of mutant herpes simplex virus G207 inoculated pre-and post-tumor resection for recurrent GBM. Mol. Ther. *17*, 199–207.
Markert, J.M., Malick, A., Coen, D.M., and Martuza, R.L. (1993). Reduction and elimination of encephalitis in an experimental glioma therapy model with attenuated herpes simplex mutants that retain susceptibility to acyclovir. Neurosurgery *32*, 597–603.
Markert, J.M., Medlock, M.D., Rabkin, S.D., Gillespie, G.Y., Todo, T., Hunter, W.D., Palmer, C.A., Feigenbaum, F., Tornatore, C., Tufaro, F., and Martuza, R.L. (2000). Conditionally replicating herpes simplex virus mutant, G207 for the treatment of malignant glioma: results of a phase I trial. Gene Ther. *7*, 867–874.
Markert, J.M., Parker, J.N., Buchsbaum, D.J., Grizzle, W.E., Gillespie, G.Y., and Whitley, R.J. (2006). Oncolytic

HSV-1 for the treatment of brain tumours. Herpes *13*, 66–71.

Martuza, R.L., Malick, A., Markert, J.M., Ruffner, K.L., and Coen, D.M. (1991). Experimental therapy of human glioma by means of a genetically engineered virus mutant. Science *252*, 854–856.

Maul, G.G., Guldner, H.H., and Spivack, J.G. (1993). Modification of discrete nuclear domains induced by herpes simplex virus type 1 immediate early gene 1 product (ICP0). J. Gen. Virol. *74*, 2679–2690.

McAuliffe, P.F., Jarnagin, W.R., Johnson, P., Delman, K.A., Federoff, H., and Fong, Y. (2000). Effective treatment of pancreatic tumors with two multimutated herpes simplex oncolytic viruses. J. Gastrointest Surg. 4, 580–588.

McKee, T.D., Grandi, P., Mok, W., Alexandrakis, G., Insin, N., Zimmer, J.P., Bawendi, M.G., Boucher, Y., Breakefield, X.O., and Jain, R.K. (2006). Degradation of fibrillar collagen in a human melanoma xenograft improves the efficacy of an oncolytic herpes simplex virus vector. Cancer Res. *66*, 2509–2513.

McKie, E.A., MacLean, A.R., Lewis, A.D., Cruickshank, G., Rampling, R., Barnett, S.C., Kennedy, P.G., and Brown, S.M. (1996). Selective *in vitro* replication of herpes simplex virus type 1 (HSV-1) ICP34.5 null mutants in primary human CNS tumours—evaluation of a potentially effective clinical therapy. Br J. Cancer *74*, 745–752.

McMenamin, M.M., Byrnes, A.P., Charlton, H.M., Coffin, R.S., Latchman, D.S., and Wood, M.J. (1998). A gamma34.5 mutant of herpes simplex 1 causes severe inflammation in the brain. Neuroscience *83*, 1225–1237.

Meignier, B., Longnecker, R., Mavromara-Nazos, P., Sears, A.E., and Roizman, B. (1988a). Virulence of and establishment of latency by genetically engineered deletion mutants of herpes simplex virus. Virology *162*, 251–254.

Meignier, B., Longnecker, R., and Roizman, B. (1988b). *In vivo* behavior of genetically engineered herpes simplex viruses R7017 and R7020: construction and evaluation in rodents. J. Infect. Dis. *158*, 602–614.

Meignier, B., Martin, B., Whitley, R.J., and Roizman, B. (1990). *In vivo* behavior of genetically engineered herpes simplex viruses R7017 and R7020. II. Studies in immunocompetent and immunosuppressed owl monkeys (Aotus trivirgatus). J. Infect. Dis. *162*, 313–321.

Melancon, J.M., Foster, T.P., and Kousoulas, K.G. (2004). Genetic analysis of the herpes simplex virus type 1 UL20 protein domains involved in cytoplasmic virion envelopment and virus-induced cell fusion. J. Virol. *78*, 7329–7343.

Melendez, L.V., Espana, C., Hunt, R.D., Daniel, M.D., and Garcia, F.G. (1969). Natural herpes simplex infection in the owl monkey (aotus trivirgatus). Lab. Anim. Care *19*, 38–45.

Mellerick, D.M., and Fraser, N.W. (1987). Physical state of the latent herpes simplex virus genome in a mouse model system: evidence suggesting an episomal state. Virology *158*, 265–275.

Melroe, G.T., Silva, L., Schaffer, P.A., and Knipe, D.M. (2007). Recruitment of activated IRF-3 and CBP/p300 to herpes simplex virus ICP0 nuclear foci: Potential role in blocking IFN-beta induction. Virology *360*, 305–321.

Menotti, L., Cerretani, A., Hengel, H., and Campadelli-Fiume, G. (2008). Construction of a fully retargeted herpes simplex virus 1 recombinant capable of entering cells solely via human epidermal growth factor receptor 2. J. Virol. *82*, 10153–10161.

Menotti, L., Nicoletti, G., Gatta, V., Croci, S., Landuzzi, L., De Giovanni, C., Nanni, P., Lollini, P.L., and Campadelli-Fiume, G. (2009). Inhibition of human tumor growth in mice by an oncolytic herpes simplex virus designed to target solely HER-2-positive cells. Proc. Natl. Acad. Sci. U.S.A. *106*, 9039–9044.

Miller, C.G., and Fraser, N.W. (2003). Requirement of an integrated immune response for successful neuroattenuated HSV-1 therapy in an intracranial metastatic melanoma model. Mol. Ther. *7*, 741–747.

Miller, C.G., Krummenacher, C., Eisenberg, R.J., Cohen, G.H., and Fraser, N.W. (2001). Development of a syngenic murine B16 cell line-derived melanoma susceptible to destruction by neuroattenuated HSV-1. Mol. Ther. *3*, 160–168.

Mineta, T., Rabkin, S.D., and Martuza, R.L. (1994). Treatment of malignant gliomas using ganciclovir-hypersensitive, ribonucleotide reductase-deficient herpes simplex viral mutant. Cancer Res. *54*, 3963–3966.

Mineta, T., Rabkin, S.D., Yazaki, T., Hunter, W.D., and Martuza, R.L. (1995). Attenuated multi-mutated herpes simplex virus-1 for the treatment of malignant gliomas. Nat. Med. *1*, 938–943.

Miyatake, S., Iyer, A., Martuza, R.L., and Rabkin, S.D. (1997). Transcriptional targeting of herpes simplex virus for cell-specific replication. J. Virol. *71*, 5124–5132.

Miyatake, S.I., Tani, S., Feigenbaum, F., Sundaresan, P., Toda, H., Narumi, O., Kikuchi, H., Hashimoto, N., Hangai, M., Martuza, R.L., and Rabkin, S.D. (1999). Hepatoma-specific antitumor activity of an albumin enhancer/promoter regulated herpes simplex virus *in vivo*. Gene Ther. *6*, 564–572.

Mohr, I. (2005). To replicate or not to replicate: achieving selective oncolytic virus replication in cancer cells through translational control. Oncogene *24*, 7697–7709.

Mohr, I., and Gluzman, Y. (1996). A herpesvirus genetic element which affects translation in the absence of the viral GADD34 function. EMBO J. *15*, 4759–4766.

Mohr, I., Sternberg, D., Ward, S., Leib, D., Mulvey, M., and Gluzman, Y. (2001). A herpes simplex virus type 1 gamma34.5 second-site suppressor mutant that exhibits enhanced growth in cultured glioblastoma cells is severely attenuated in animals. J. Virol. *75*, 5189–5196.

Mori, I., Liu, B., Goshima, F., Ito, H., Koide, N., Yoshida, T., Yokochi, T., Kimura, Y., and Nishiyama, Y. (2005). HF10, an attenuated herpes simplex virus (HSV) type 1 clone, lacks neuroinvasiveness and protects mice against lethal challenge with HSV types 1 and 2. Microbes Infect. *7*, 1492–1500.

Mori, I., and Nishiyama, Y. (2006). Accessory genes define the relationship between the herpes simplex virus and its host. Microbes Infect. *8*, 2556–2562.

Mossman, K.L., and Smiley, J.R. (1999). Truncation of the C-terminal acidic transcriptional activation domain of herpes simplex virus VP16 renders expression of the immediate-early genes almost entirely dependent on ICP0. J. Virol. *73*, 9726–9733.

Mossman, K.L., and Smiley, J.R. (2002). Herpes simplex virus ICP0 and ICP34.5 counteract distinct interferon–induced barriers to virus replication. J. Virol. *76*, 1995–1998.

Moyal, M., Berkowitz, C., Rosen-Wolff, A., Darai, G., and Becker, Y. (1992). Mutations in the UL53 gene of HSV-1 abolish virus neurovirulence to mice by the intracerebral route of infection. Virus Res. *26*, 99–112.

Mullen, J.T., Kasuya, H., Yoon, S.S., Carroll, N.M., Pawlik, T.M., Chandrasekhar, S., Nakamura, H., Donahue, J.M., and Tanabe, K.K. (2002). Regulation of herpes simplex virus 1 replication using tumor-associated promoters. Ann. Surg. 236, 502–512; discussion 512–503.

Mulvey, M., Poppers, J., Sternberg, D., and Mohr, I. (2003). Regulation of eIF2alpha phosphorylation by different functions that act during discrete phases in the herpes simplex virus type 1 life cycle. J. Virol. *77*, 10917–10928.

Nakamori, M., Fu, X., Meng, F., Jin, A., Tao, L., Bast, R.C., Jr., and Zhang, X. (2003). Effective therapy of metastatic ovarian cancer with an oncolytic herpes simplex virus incorporating two membrane fusion mechanisms. Clin. Cancer Res. *9*, 2727–2733.

Nakamori, M., Fu, X., Pettaway, C.A., and Zhang, X. (2004). Potent antitumor activity after systemic delivery of a doubly fusogenic oncolytic herpes simplex virus against metastatic prostate cancer. Prostate *60*, 53–60.

Nakamura, H., Kasuya, H., Mullen, J.T., Yoon, S.S., Pawlik, T.M., Chandrasekhar, S., Donahue, J.M., Chiocca, E.A., Chung, R.Y., and Tanabe, K.K. (2002). Regulation of herpes simplex virus gamma(1)34.5 expression and oncolysis of diffuse liver metastases by Myb34.5. J. Clin. Invest. *109*, 871–882.

Nakamura, H., Mullen, J.T., Chandrasekhar, S., Pawlik, T.M., Yoon, S.S., and Tanabe, K.K. (2001). Multimodality therapy with a replication-conditional herpes simplex virus 1 mutant that expresses yeast cytosine deaminase for intratumoral conversion of 5-fluorocytosine to 5-fluorouracil. Cancer Res. *61*, 5447–5452.

Nakano, K., Asano, R., Tsumoto, K., Kwon, H., Goins, W.F., Kumagai, I., Cohen, J.B., and Glorioso, J.C. (2005a). Herpes simplex virus targeting to the EGF receptor by a gD-specific soluble bridging molecule. Mol. Ther. *11*, 617–626.

Nakano, K., Todo, T., Zhao, G., Yamaguchi, K., Kuroki, S., Cohen, J.B., Glorioso, J.C., and Tanaka, M. (2005b). Enhanced efficacy of conditionally replicating herpes simplex virus (G207) combined with 5-fluorouracil and surg. ical resection in peritoneal cancer dissemination models. J. Gene Med. 7, 638–648.

Nakao, A., Takeda, S., Shimoyama, S., Kasuya, H., Kimata, H., Teshigahara, O., Sawaki, M., Kikumori, T., Kodera, Y., Nagasaka, T., *et al.* (2007). Clinical experiment of mutant herpes simplex virus HF10 therapy for cancer. Curr. Cancer Drug Targets 7, 169–174.

Naryzhny, S.N. (2008). Proliferating cell nuclear antigen: a proteomics view. Cell. Mol. Life Sci. *65*, 3789–3808.

Nawa, A., Luo, C., Zhang, L., Ushjima, Y., Ishida, D., Kamakura, M., Fujimoto, Y., Goshima, F., Kikkawa, F., and Nishiyama, Y. (2008). Non-engineered, naturally oncolytic herpes simplex virus HSV1 HF-10: applications for cancer gene ther. Curr. Gene Ther. *8*, 208–221.

Nishiyama, Y. (1996). Herpesvirus genes: molecular basis of viral replication and pathogenicity. Nagoya J. Med. Sci. *59*, 107–119.

Nishiyama, Y., Kimura, H., and Daikoku, T. (1991). Complementary lethal invasion of the central nervous system by nonneuroinvasive herpes simplex virus types 1 and 2. J. Virol. *65*, 4520–4524.

Nishiyama, Y., Yamada, Y., Kurachi, R., and Daikoku, T. (1992). Construction of a US3 lacZ insertion mutant of herpes simplex virus type 2 and characterization of its phenotype *in vitro* and *in vivo*. Virology *190*, 256–268.

Nomura, N., Kasuya, H., Watanabe, I., Shikano, T., Shirota, T., Misawa, M., Sugimoto, H., Kanazumi, N., Nomoto, S., Takeda, S., and Nakao, A. (2009). Considerations for intravascular administration of oncolytic herpes virus for the treatment of multiple liver metastases. Cancer Chemother. Pharmacol. *63*, 321–330.

Novoa, I., Zeng, H., Harding, H.P., and Ron, D. (2001). Feedback inhibition of the unfolded protein response by GADD34-mediated dephosphorylation of eIF2alpha. J. Cell Biol. *153*, 1011–1022.

Ogg, P.D., McDonell, P.J., Ryckman, B.J., Knudson, C.M., and Roller, R.J. (2004). The HSV-1 Us3 protein kinase is sufficient to block apoptosis induced by overexpression of a variety of Bcl-2 family members. Virology *319*, 212–224.

Orvedahl, A., Alexander, D., Talloczy, Z., Sun, Q., Wei, Y., Zhang, W., Burns, D., Leib, D.A., and Levine, B. (2007). HSV-1 ICP34.5 confers neurovirulence by targeting the Beclin. 1 autophagy protein. Cell Host Microbe *1*, 23–35.

Osorio, Y., and Ghiasi, H. (2005). Recombinant herpes simplex virus type 1 (HSV-1) codelivering interleukin-12p35 as a molecular adjuvant enhances the protective immune response against ocular HSV-1 challenge. J. Virol. *79*, 3297–3308.

Overbaugh, J., Miller, A.D., and Eiden, M.V. (2001). Receptors and entry cofactors for retroviruses include single and multiple transmembrane-spanning proteins as well as newly described glycophosphatidylinositol-anchored and secreted proteins. Microbiol. Mol. Biol. Rev. 65, 371–389, table of contents.

Pan, W., Bodempudi, V., Esfandyari, T., and Farassati, F. (2009). Utilizing ras signaling pathway to direct selective replication of herpes simplex virus-1. PLoS One *4*, e6514.

Panning, B., and Smiley, J.R. (1989). Regulation of cellular genes transduced by herpes simplex virus. J. Virol. *63*, 1929–1937.

Papanastassiou, V., Rampling, R., Fraser, M., Petty, R., Hadley, D., Nicoll, J., Harland, J., Mabbs, R., and Brown, M. (2002). The potential for efficacy of the modified (ICP 34.5(-)) herpes simplex virus HSV1716 following intratumoural injection into human malignant glioma: a proof of principle study. Gene Ther. *9*, 398–406.

Parker, J.N., Bauer, D.F., Cody, J.J., and Markert, J.M. (2009). Oncolytic viral therapy of malignant glioma. Neurotherapeutics *6*, 558–569.

Parker, J.N., Gillespie, G.Y., Love, C.E., Randall, S., Whitley, R.J., and Markert, J.M. (2000). Engineered herpes simplex virus expressing IL-12 in the treatment of experimental murine brain tumors. Proc. Natl. Acad. Sci. U.S.A. *97*, 2208–2213.

Parker, J.N., Meleth, S., Hughes, K.B., Gillespie, G.Y., Whitley, R.J., and Markert, J.M. (2005). Enhanced inhibition of syngeneic murine tumors by combinatorial therapy with genetically engineered HSV-1 expressing CCL2 and IL-12. Cancer Gene Ther. *12*, 359–368.

Parker, J.N., Pfister, L.A., Quenelle, D., Gillespie, G.Y., Markert, J.M., Kern, E.R., and Whitley, R.J. (2006). Genetically engineered herpes simplex viruses that express IL-12 or GM-CSF as vaccine candidates. Vaccine *24*, 1644–1652.

Passer, B.J., Wu, C.L., Wu, S., Rabkin, S.D., and Martuza, R.L. (2009). Analysis of genetically engineered oncolytic herpes simplex viruses in human prostate cancer organotypic cultures. Gene Ther. *16*, 1477–1482.

Pavitt, G.D. (2005). eIF2B, a mediator of general and gene-specific translational control. Biochem. Soc. Trans 33, 1487–1492.

Pawlik, T.M., Nakamura, H., Mullen, J.T., Kasuya, H., Yoon, S.S., Chandrasekhar, S., Chiocca, E.A., and Tanabe, K.K. (2002). Prodrug bioactivation and oncolysis of diffuse liver metastases by a herpes simplex virus 1 mutant that expresses the CYP2B1 transgene. Cancer 95, 1171–1181.

Peng, T., Hunter, J.R., and Nelson, J.W. (1996). The novel protein kinase of the RR1 subunit of herpes simplex virus has autophosphorylation and transphosphorylation activity that differs in its ATP requirements for HSV-1 and HSV-2. Virology *216*, 184–196.

Perkins, D., Pereira, E.F., and Aurelian, L. (2003). The herpes simplex virus type 2 R1 protein kinase (ICP10 PK) functions as a dominant regulator of apoptosis in hippocampal neurons involving activation of the ERK survival pathway and upregulation of the antiapoptotic protein Bag-1. J. Virol. *77*, 1292–1305.

Perng, G.C., Chokephaibulkit, K., Thompson, R.L., Sawtell, N.M., Slanina, S.M., Ghiasi, H., Nesburn, A.B., and Wechsler, S.L. (1996). The region of the herpes simplex virus type 1 LAT gene that is colinear with the ICP34.5 gene is not involved in spontaneous reactivation. J. Virol. *70*, 282–291.

Perng, G.C., Thompson, R.L., Sawtell, N.M., Taylor, W.E., Slanina, S.M., Ghiasi, H., Kaiwar, R., Nesburn, A.B., and Wechsler, S.L. (1995). An avirulent ICP34.5 deletion mutant of herpes simplex virus type 1 is capable of *in vivo* spontaneous reactivation. J. Virol. *69*, 3033–3041.

Peters, G.A., Khoo, D., Mohr, I., and Sen, G.C. (2002). Inhibition of PACT-mediated activation of PKR by the herpes simplex virus type 1 Us11 protein. J. Virol. *76*, 11054–11064.

Petrowsky, H., Roberts, G.D., Kooby, D.A., Burt, B.M., Bennett, J.J., Delman, K.A., Stanziale, S.F., Delohery, T.M., Tong, W.P., Federoff, H.J., and Fong, Y. (2001). Functional interaction between fluorodeoxyuridine-induced cellular alterations and replication of a ribonucleotide reductase-negative herpes simplex virus. J. Virol. *75*, 7050–7058.

Pilger, B.D., Perozzo, R., Alber, F., Wurth, C., Folkers, G., and Scapozza, L. (1999). Substrate diversity of herpes simplex virus thymidine kinase. Impact Of the kinematics of the enzyme. J. Biol. Chem. *274*, 31967–31973.

Polakis, P. (2007). The many ways of Wnt in cancer. Curr. Opin. Genet Dev. *17*, 45–51.

Poppers, J., Mulvey, M., Khoo, D., and Mohr, I. (2000). Inhibition of PKR activation by the proline-rich RNA binding domain of the herpes simplex virus type 1 Us11 protein. J. Virol. *74*, 11215–11221.

Portsmouth, D., Hlavaty, J., and Renner, M. (2007). Suicide genes for cancer therapy. Mol. Aspects Med. *28*, 4–41.

Post, D.E., Fulci, G., Chiocca, E.A., and Van Meir, E.G. (2004). Replicative oncolytic herpes simplex viruses in combination cancer therapies. Curr. Gene Ther. *4*, 41–51.

Proud, C.G. (2005). eIF2 and the control of cell physiology. Semin Cell Dev. Biol. *16*, 3–12.

Purves, F.C., Longnecker, R.M., Leader, D.P., and Roizman, B. (1987). Herpes simplex virus 1 protein kinase is encoded by open reading frame US3 which is not essential for virus growth in cell culture. J. Virol. *61*, 2896–2901.

Pyles, R.B., Warnick, R.E., Chalk, C.L., Szanti, B.E., and Parysek, L.M. (1997). A novel multiply-mutated HSV-1 strain for the treatment of human brain tumors. Hum. Gene Ther. *8*, 533–544.

Radbill, A.E., Reddy, A.T., Markert, J.M., Wyss, J.M., Pike, M.M., Akella, N.S., Bharara, N., and Gillespie, G.Y. (2007). Effects of G207, a conditionally replication-competent oncolytic herpes simplex virus, on the developing mammalian brain. J. NeuroVirol. *13*, 118–129.

Rampling, R., Cruickshank, G., Papanastassiou, V., Nicoll, J., Hadley, D., Brennan, D., Petty, R., MacLean, A., Harland, J., McKie, E., *et al.* (2000). Toxicity evaluation of replication-competent herpes simplex virus (ICP 34.5 null mutant 1716) in patients with recurrent malignant glioma. Gene Ther. 7, 859–866.

Randazzo, B.P., Bhat, M.G., Kesari, S., Fraser, N.W., and Brown, S.M. (1997). Treatment of experimental subcutaneous human melanoma with a replication-restricted herpes simplex virus mutant. J. Invest. Dermatol. *108*, 933–937.

Randazzo, B.P., Kesari, S., Gesser, R.M., Alsop, D., Ford, J.C., Brown, S.M., MacLean, A., and Fraser, N.W. (1995). Treatment of experimental intracranial murine melanoma with a neuroattenuated herpes simplex virus 1 mutant. Virology *211*, 94–101.
Reineke, E.L., and Kao, H.Y. (2009). PML: An emerging tumor suppressor and a target with therapeutic potential. Cancer Ther. *7*, 219–226.
Robertson, L.M., MacLean, A.R., and Brown, S.M. (1992). Peripheral replication and latency reactivation kinetics of the non-neurovirulent herpes simplex virus type 1 variant 1716. J. Gen. Virol. *73*, 967–970.
Roemer, K., Johnson, P.A., and Friedmann, T. (1991). Activity of the simian virus 40 early promoter-enhancer in herpes simplex virus type 1 vectors is dependent on its position, the infected cell type, and the presence of Vmw175. J. Virol. *65*, 6900–6912.
Roizman, B. (1996). The function of herpes simplex virus genes: a primer for genetic engineering of novel vectors. Proc. Natl. Acad. Sci. U.S.A. *93*, 11307–11312.
Rosen-Wolff, A., Lamade, W., Berkowitz, C., Becker, Y., and Darai, G. (1991). Elimination of UL56 gene by insertion of LacZ cassette between nucleotide position 116030 to 121753 of the herpes simplex virus type 1 genome abrogates intraperitoneal pathogenicity in tree shrews and mice. Virus Res. *20*, 205–221.
Ruyechan, W.T., Morse, L.S., Knipe, D.M., and Roizman, B. (1979). Molecular genetics of herpes simplex virus. II. Mapping of the major viral glycoproteins and of the genetic loci specifying the social behavior of infected cells. J. Virol. *29*, 677–697.
Sadler, A.J., and Williams, B.R. (2008). Interferon-inducible antiviral effectors. Nat. Rev. Immunol. *8*, 559–568.
Samoto, K., Ehtesham, M., Perng, G.C., Hashizume, K., Wechsler, S.L., Nesburn, A.B., Black, K.L., and Yu, J.S. (2002). A herpes simplex virus type 1 mutant with gamma 34.5 and LAT deletions effectively oncolyses human U87 glioblastomas in nude mice. Neurosurgery *50*, 599–605; discussion 605–596.
Samoto, K., Perng, G.C., Ehtesham, M., Liu, Y., Wechsler, S.L., Nesburn, A.B., Black, K.L., and Yu, J.S. (2001). A herpes simplex virus type 1 mutant deleted for gamma34.5 and LAT kills glioma cells *in vitro* and is inhibited for *in vivo* reactivation. Cancer Gene Ther. *8*, 269–277.
Samuel, C.E., Kuhen, K.L., George, C.X., Ortega, L.G., Rende-Fournier, R., and Tanaka, H. (1997). The PKR protein kinase—an interferon-inducible regulator of cell growth and differentiation. Int. J. Hematol *65*, 227–237.
Sanchez, R., and Mohr, I. (2007). Inhibition of cellular 2′–5′ oligoadenylate synthetase by the herpes simplex virus type 1 Us11 protein. J. Virol. *81*, 3455–3464.
Sarinella, F., Calistri, A., Sette, P., Palu, G., and Parolin, C. (2006). Oncolysis of pancreatic tumour cells by a gamma34.5-deleted HSV-1 does not rely upon Ras-activation, but on the PI 3-kinase pathway. Gene Ther. *13*, 1080–1087.
Scaria, A., Sullivan, J.A., St George, J.A., Kaplan, J.M., Lukason, M.J., Morris, J.E., Plog, M., Nicolette, C., Gregory, R.J., and Wadsworth, S.C. (2000). Adenoviral vector expressing ICP47 inhibits adenovirus-specific cytotoxic T lymphocytes in nonhuman primates. Mol. Ther. *2*, 505–514.
Sedarati, F., and Stevens, J.G. (1987). Biological basis for virulence of three strains of herpes simplex virus type 1. J. Gen. Virol. *68*, 2389–2395.
Senzer, N.N., Kaufman, H.L., Amatruda, T., Nemunaitis, M., Reid, T., Daniels, G., Gonzalez, R., Glaspy, J., Whitman, E., Harrington, K., *et al.* (2009). Phase II Clinical Trial of a Granulocyte-Macrophage Colony-Stimulating Factor-Encoding, Second-Generation Oncolytic Herpesvirus in Patients With Unresectable Metastatic Melanoma. J. Clin. Oncol. *27*, 5763–5771.
Shah, A.C., Parker, J.N., Gillespie, G.Y., Lakeman, F.D., Meleth, S., Markert, J.M., and Cassady, K.A. (2007). Enhanced antiglioma activity of chimeric HCMV/HSV-1 oncolytic viruses. Gene Ther. *14*, 1045–1054.
Shah, A.C., Price, K.H., Parker, J.N., Samuel, S.L., Meleth, S., Cassady, K.A., Gillespie, G.Y., Whitley, R.J., and Markert, J.M. (2006). Serial passage through human glioma xenografts selects for a Deltagamma134.5 herpes simplex virus type 1 mutant that exhibits decreased neurotoxicity and prolongs survival of mice with experimental brain tumors. J. Virol. *80*, 7308–7315.
Simpson, G.R., Han, Z., Liu, B., Wang, Y., Campbell, G., and Coffin, R.S. (2006). Combination of a fusogenic glycoprotein, prodrug activation, and oncolytic herpes simplex virus for enhanced local tumor control. Cancer Res. *66*, 4835–4842.
Sinha, S., and Levine, B. (2008). The autophagy effector Beclin. 1: a novel BH3-only protein. Oncogene *27* Suppl 1, S137–148.
Skinner, G.R., Cowan, M., Davies, J., Brookes, K., Billstrom, M., and Buchan, A. (1988). Reduced tumorigenicity of rodent tumour cells and tumour explants following infection with wild type and mutant herpes simplex virus, bovine mammillitis virus and encephalomyocarditis virus. Br. J. Exp. Pathol. *69*, 495–504.
Smith, C.C. (2005). The herpes simplex virus type 2 protein ICP10PK: a master of versatility. Front. Biosci. 10, 2820–2831.
Smith, C.C., Nelson, J., Aurelian, L., Gober, M., and Goswami, B.B. (2000a). Ras-GAP binding and phosphorylation by herpes simplex virus type 2 RR1 PK (ICP10) and activation of the Ras/MEK/MAPK mitogenic pathway are required for timely onset of virus growth. J. Virol. *74*, 10417–10429.
Smith, C.C., Peng, T., Kulka, M., and Aurelian, L. (1998). The PK domain of the large subunit of herpes simplex virus type 2 ribonucleotide reductase (ICP10) is required for immediate-early gene expression and virus growth. J. Virol. *72*, 9131–9141.
Smith, C.C., Yu, Y.X., Kulka, M., and Aurelian, L. (2000b). A novel human gene similar to the protein kinase (PK) coding domain of the large subunit of herpes simplex virus type 2 ribonucleotide reductase (ICP10) codes for a serine-threonine PK and is expressed in melanoma cells. J. Biol. Chem. *275*, 25690–25699.

Smith, K.D., Mezhir, J.J., Bickenbach, K., Veerapong, J., Charron, J., Posner, M.C., Roizman, B., and Weichselbaum, R.R. (2006). Activated MEK suppresses activation of PKR and enables efficient replication and *in vivo* oncolysis by Deltagamma(1)34.5 mutants of herpes simplex virus 1. J. Virol. *80*, 1110–1120.

Sobol, P.T., Hummel, J.L., Rodrigues, R.M., and Mossman, K.L. (2009). PML has a predictive role in tumor cell permissiveness to interferon-sensitive oncolytic viruses. Gene Ther. *16*, 1077–1087.

Song, G.Y., DeJong, G., and Jia, W. (1999). Cell surface expression of MHC molecules in glioma cells infected with herpes simplex virus type-1. J. Neuroimmunol *93*, 1–7.

Spear, P.G. (2004). Herpes simplex virus: receptors and ligands for cell entry. Cell. Microbiol. *6*, 401–410.

Spear, P.G., Manoj, S., Yoon, M., Jogger, C.R., Zago, A., and Myscofski, D. (2006). Different receptors binding to distinct interfaces on herpes simplex virus gD can trigger events leading to cell fusion and viral entry. Virology *344*, 17–24.

Spivack, J.G., Fareed, M.U., Valyi-Nagy, T., Nash, T.C., O'Keefe, J.S., Gesser, R.M., McKie, E.A., MacLean, A.R., Fraser, N.W., and Brown, S.M. (1995). Replication, establishment of latent infection, expression of the latency-associated transcripts and explant reactivation of herpes simplex virus type 1 gamma 34.5 mutants in a mouse eye model. J. Gen. Virol. *76*, 321–332.

Stanziale, S.F., Stiles, B.M., Bhargava, A., Kerns, S.A., Kalakonda, N., and Fong, Y. (2004). Oncolytic herpes simplex virus-1 mutant expressing green fluorescent protein can detect and treat peritoneal cancer. Hum. Gene Ther. *15*, 609–618.

Stiles, B.M., Bhargava, A., Adusumilli, P.S., Stanziale, S.F., Kim, T.H., Rusch, V.W., and Fong, Y. (2003). The replication-competent oncolytic herpes simplex mutant virus NV1066 is effective in the treatment of esophageal cancer. Surgery *134*, 357–364.

Stiles, C.D., and Rowitch, D.H. (2008). Glioma stem cells: a midterm exam. Neuron *58*, 832–846.

Subramanian, R.P., and Geraghty, R.J. (2007). Herpes simplex virus type 1 mediates fusion through a hemifusion intermediate by sequential activity of glycoproteins D, H, L, and B. Proc. Natl. Acad. Sci. U.S.A. *104*, 2903–2908.

Sundaresan, P., Hunter, W.D., Martuza, R.L., and Rabkin, S.D. (2000). Attenuated, replication-competent herpes simplex virus type 1 mutant G207: safety evaluation in mice. J. Virol. *74*, 3832–3841.

Takakuwa, H., Goshima, F., Nozawa, N., Yoshikawa, T., Kimata, H., Nakao, A., Nawa, A., Kurata, T., Sata, T., and Nishiyama, Y. (2003). Oncolytic viral therapy using a spontaneously generated herpes simplex virus type 1 variant for disseminated peritoneal tumor in immunocompetent mice. Arch. Virol. *148*, 813–825.

Talloczy, Z., Jiang, W., Virgin, H.W. t., Leib, D.A., Scheuner, D., Kaufman, R.J., Eskelinen, E.L., and Levine, B. (2002). Regulation of starvation- and virus-induced autophagy by the eIF2alpha kinase signaling pathway. Proc. Natl. Acad. Sci. U.S.A. *99*, 190–195.

Tanaka, M., Kagawa, H., Yamanashi, Y., Sata, T., and Kawaguchi, Y. (2003). Construction of an excisable bacterial artificial chromosome containing a full-length infectious clone of herpes simplex virus type 1: viruses reconstituted from the clone exhibit wild-type properties *in vitro* and *in vivo*. J. Virol. *77*, 1382–1391.

Taneja, S., MacGregor, J., Markus, S., Ha, S., and Mohr, I. (2001). Enhanced antitumor efficacy of a herpes simplex virus mutant isolated by genetic selection in cancer cells. Proc. Natl. Acad. Sci. U.S.A. *98*, 8804–8808.

Tang, S., Bertke, A.S., Patel, A., Wang, K., Cohen, J.I., and Krause, P.R. (2008). An acutely and latently expressed herpes simplex virus 2 viral microRNA inhibits expression of ICP34.5, a viral neurovirulence factor. Proc. Natl. Acad. Sci. U.S.A. *105*, 10931–10936.

Tang, S., Patel, A., and Krause, P.R. (2009). Novel less-abundant viral microRNAs encoded by herpes simplex virus 2 latency-associated transcript and their roles in regulating ICP34.5 and ICP0 mRNAs. J. Virol. *83*, 1433–1442.

Tenser, R.B. (1983). Intracerebral inoculation of newborn and adult mice with thymidine kinase-deficient mutants of herpes simplex virus type 1. J. Infect. Dis. *147*, 956.

Terada, K., Wakimoto, H., Tyminski, E., Chiocca, E.A., and Saeki, Y. (2006). Development of a rapid method to generate multiple oncolytic HSV vectors and their *in vivo* evaluation using syngeneic mouse tumor models. Gene Ther.

Thompson, R.L., Wagner, E.K., and Stevens, J.G. (1983). Physical location of a herpes simplex virus type-1 gene function(s) specifically associated with a 10 million-fold increase in HSV neurovirulence. Virology *131*, 180–192.

Toda, M., Rabkin, S.D., Kojima, H., and Martuza, R.L. (1999). Herpes simplex virus as an *in situ* cancer vaccine for the induction of specific anti-tumor immunity. Hum. Gene Ther. *10*, 385–393.

Toda, M., Rabkin, S.D., and Martuza, R.L. (1998). Treatment of human breast cancer in a brain metastatic model by G207, a replication-competent multimutated herpes simplex virus 1. Hum. Gene Ther. *9*.

Todo, T., Ebright, M.I., Fong, Y., and Rabkin, S.D. (2002). Oncolytic herpes simplex virus (G207) therapy: from basic to clinical, In Tumor-suppressing viruses, genes, and drugs, Maruta, H., ed. (San Diego: Academic Press), pp. 45–75.

Todo, T., Feigenbaum, F., Rabkin, S.D., Lakeman, F., Newsome, J.T., Johnson, P.A., Mitchell, E., Belliveau, D., Ostrove, J.M., and Martuza, R.L. (2000a). Viral shedding and biodistribution of G207, a multimutated, conditionally replicating herpes simplex virus type 1, after intracerebral inoculation in aotus. Mol. Ther. *2*, 588–595.

Todo, T., Martuza, R.L., Rabkin, S.D., and Johnson, P.A. (2001). Oncolytic herpes simplex virus vector with enhanced MHC class I presentation and tumor cell killing. Proc. Natl. Acad. Sci. U.S.A. *98*, 6396–6401.

Todo, T., Rabkin, S.D., and Martuza, R.L. (2000b). Evaluation of ganciclovir-mediated enhancement of the antitumoral effect in oncolytic, multimutated herpes simplex virus type 1 (G207) therapy of brain tumors. Cancer Gene Ther. *7*, 939–946.

Todo, T., Rabkin, S.D., Sundaresan, P., Wu, A., Meehan, K.R., Herscowitz, H.B., and Martuza, R.L. (1999). Systemic antitumor immunity in experimental brain tumor therapy using a multimutated, replication-competent herpes simplex virus. Hum. Gene Ther. *10*, 2741–2755.

Tognon, M., Guandalini, R., Romanelli, M.G., Manservigi, R., and Trevisani, B. (1991). Phenotypic and genotypic characterization of locus Syn 5 in herpes simplex virus 1. Virus Res. *18*, 135–150.

Tyminski, E., Leroy, S., Terada, K., Finkelstein, D.M., Hyatt, J.L., Danks, M.K., Potter, P.M., Saeki, Y., and Chiocca, E.A. (2005). Brain tumor oncolysis with replication-conditional herpes simplex virus type 1 expressing the prodrug-activating genes, CYP2B1 and secreted human intestinal carboxylesterase, in combination with cyclophosphamide and irinotecan. Cancer Res. *65*, 6850–6857.

Umbach, J.L., Kramer, M.F., Jurak, I., Karnowski, H.W., Coen, D.M., and Cullen, B.R. (2008). MicroRNAs expressed by herpes simplex virus 1 during latent infection regulate viral mRNAs. Nature *454*, 780–783.

Ushijima, Y., Luo, C., Goshima, F., Yamauchi, Y., Kimura, H., and Nishiyama, Y. (2007). Determination and analysis of the DNA sequence of highly attenuated herpes simplex virus type 1 mutant HF10, a potential oncolytic virus. Microbes Infect. *9*, 142–149.

Valyi-Nagy, T., Fareed, M.U., O'Keefe, J.S., Gesser, R.M., MacLean, A.R., Brown, S.M., Spivack, J.G., and Fraser, N.W. (1994). The herpes simplex virus type 1 strain 17^+ gamma 34.5 deletion mutant 1716 is avirulent in SCID mice. J. Gen. Virol. *75*, 2059–2063.

Varghese, S., Newsome, J.T., Rabkin, S.D., McGeagh, K., Mahoney, D., Nielsen, P., Todo, T., and Martuza, R.L. (2001). Preclinical safety evaluation of G207, a replication-competent herpes simplex virus type 1, inoculated intraprostatically in mice and nonhuman primates. Hum. Gene Ther. *12*, 999–1010.

Varghese, S., and Rabkin, S.D. (2002). Oncolytic herpes simplex virus vectors for cancer virotherapy. Cancer Gene Ther. *9*, 967–978.

Varghese, S., Rabkin, S.D., Liu, R., Nielsen, P.G., Ipe, T., and Martuza, R.L. (2006a). Enhanced therapeutic efficacy of IL-12, but not GM-CSF, expressing oncolytic herpes simplex virus for transgenic mouse derived prostate cancers. Cancer Gene Ther. *13*, 253–265.

Varghese, S., Rabkin, S.D., Nielsen, P.G., Wang, W., and Martuza, R.L. (2006b). Systemic oncolytic herpes virus therapy of poorly immunogenic prostate cancer metastatic to lung. Clin. Cancer Res. *12*, 2919–2927.

Verpooten, D., Feng, Z., Valyi-Nagy, T., Ma, Y., Jin, H., Yan, Z., Zhang, C., Cao, Y., and He, B. (2009). Dephosphorylation of eIF2alpha mediated by the gamma134.5 protein of herpes simplex virus 1 facilitates viral neuroinvasion. J. Virol. *83*, 12626–12630.

Wakimoto, H., Fulci, G., Tyminski, E., and Chiocca, E.A. (2004). Altered expression of antiviral cytokine mRNAs associated with cyclophosphamide's enhancement of viral oncolysis. Gene Ther. *11*, 214–223.

Wakimoto, H., Johnson, P.R., Knipe, D.M., and Chiocca, E.A. (2003). Effects of innate immunity on herpes simplex virus and its ability to kill tumor cells. Gene Ther. *10*, 983–990.

Wakimoto, H., Kesari, S., Farrell, C.J., Curr. y, W.T., Jr., Zaupa, C., Aghi, M., Kuroda, T., Stemmer-Rachamimov, A., Shah, K., Liu, T.C., *et al.* (2009). Human glioblastoma-derived cancer stem cells: establishment of invasive glioma models and treatment with oncolytic herpes simplex virus vectors. Cancer Res. *69*, 3472–3481.

White, E., and DiPaola, R.S. (2009). The double-edged sword of autophagy modulation in cancer. Clin. Cancer Res. *15*, 5308–5316.

Whitley, R.J., Kern, E.R., Chatterjee, S., Chou, J., and Roizman, B. (1993). Replication, establishment of latency, and induced reactivation of herpes simplex virus gamma 1, 34.5 deletion mutants in rodent models. J. Clin. Invest. *91*, 2837–2843.

Whitley, R.J., and Roizman, B. (2002). Herpes simplex viruses: is a vaccine tenable? J. Clin. Invest. *110*, 145–151.

Wong, R.J., Chan, M.K., Yu, Z., Kim, T.H., Bhargava, A., Stiles, B.M., Horsburgh, B.C., Shah, J.P., Ghossein, R.A., Singh, B., and Fong, Y. (2004). Effective intravenous therapy of murine pulmonary metastases with an oncolytic herpes virus expressing interleukin 12. Clin. Cancer Res. *10*, 251–259.

Wong, R.J., Joe, J.K., Kim, S.H., Shah, J.P., Horsburgh, B., and Fong, Y. (2002). Oncolytic herpesvirus effectively treats murine squamous cell carcinoma and spreads by natural lymphatics to treat sites of lymphatic metastases. Hum. Gene Ther. *13*, 1213–1223.

Wong, R.J., Kim, S.H., Joe, J.K., Shah, J.P., Johnson, P.A., and Fong, Y. (2001a). Effective treatment of head and neck squamous cell carcinoma by an oncolytic herpes simplex virus. J. Am Coll Surg. 193, 12–21.

Wong, R.J., Patel, S.G., Kim, S., DeMatteo, R.P., Malhotra, S., Bennett, J.J., St-Louis, M., Shah, J.P., Johnson, P.A., and Fong, Y. (2001b). Cytokine gene transfer enhances herpes oncolytic therapy in murine squamous cell carcinoma. Hum. Gene Ther. *12*, 253–265.

Wu, A., Mazumder, A., Martuza, R.L., Liu, X., Thein, M., Meehan, K.R., and Rabkin, S.D. (2001). Biological purging of breast cancer cells using an attenuated replication-competent herpes simplex virus in human hematopoietic stem cell transplantation. Cancer Res. *61*, 3009–3015.

Yamada, Y., Kimura, H., Morishima, T., Daikoku, T., Maeno, K., and Nishiyama, Y. (1991). The pathogenicity of ribonucleotide reductase-null mutants of herpes simplex virus type1 in mice. J. Infect. Dis. *164*, 1091–1097.

Yamamoto, S., Deckter, L.A., Kasai, K., Chiocca, E.A., and Saeki, Y. (2006). Imaging immediate-early and strict-late promoter activity during oncolytic herpes simplex virus type 1 infection and replication in tumors. Gene Ther. *13*, 1731–1736.

Yamamura, H., Hashio, M., Noguchi, M., Sugenoya, Y., Osakada, M., Hirano, N., Sasaki, Y., Yoden, T., Awata, N., Araki, N., *et al.* (2001). Identification of the transcriptional regulatory sequences of human calponin promoter and their use in targeting a conditionally replicating herpes vector to malignant human soft tissue and bone tumors. Cancer Res. *61*, 3969–3977.

Yazaki, T., Manz, H.J., Rabkin, S.D., and Martuza, R.L. (1995). Treatment of human malignant meningiomas by G207, a replication-competent multimutated herpes simplex virus 1. Cancer Res. *55*, 4752–4756.

Yoon, S.S., Carroll, N.M., Chiocca, E.A., and Tanabe, K.K. (1998). Cancer Gene Ther. using a replication-competent herpes simplex virus type 1 vector [published erratum appears in Ann. Surg., 1998 Nov;228(5):following table of contents]. Ann. Surg. 228, 366–374.

Yoon, S.S., Nakamura, H., Carroll, N.M., Bode, B.P., Chiocca, E.A., and Tanabe, K.K. (2000). An oncolytic herpes simplex virus type 1 selectively destroys diffuse liver metastases from colon carcinoma. Faseb J. *14*, 301–311.

Yu, Z., Adusumilli, P.S., Eisenberg, D.P., Darr, E., Ghossein, R.A., Li, S., Liu, S., Singh, B., Shah, J.P., Fong, Y., and Wong, R.J. (2007). Nectin-1 expression by squamous cell carcinoma is a predictor of herpes oncolytic sensitivity. Mol. Ther. *15*, 103–113.

Zhang, C., Tang, J., Xie, J., Zhang, H., Li, Y., Zhang, J., Verpooten, D., He, B., and Cao, Y. (2008). A conserved domain of herpes simplex virus ICP34.5 regulates protein phosphatase complex in mammalian cells. FEBS Lett. *582*, 171–176.

Zhou, G., and Roizman, B. (2006). Construction and properties of a herpes simplex virus 1 designed to enter cells solely via the IL-13alpha2 receptor. Proc. Natl. Acad. Sci. U.S.A. *103*, 5508–5513.

Zhou, G., and Roizman, B. (2007). Separation of receptor-binding and profusogenic domains of glycoprotein D of herpes simplex virus 1 into distinct interacting proteins. Proc. Natl. Acad. Sci. U.S.A. *104*, 4142–4146.

Index

This index is not exhaustive. Page numbers listed correspond to the first appearance of each entry in each chapter.

E

F

G

H

I